About the Cover

The cover image illustrates a dodeca-spidroball formed by applying spidrons onto the surface of a dodecahedron. A spidron is a plane figure composed entirely of triangles, where, for every pair of triangles, each has a leg of the other as one of its legs, and neither has any point inside the interior of the other.

▶ Meet the Artist

The cover art was generated by Paul Nylander, a mechanical engineer with strong programming and mathematical skills that he uses to design complex engineering systems. Nylander says that he always enjoyed science and art as a hobby. When he was in high school, he had some aptitude for math, but programming was difficult for him. However, he became much more interested in programming when he began studying computer graphics. Most of Nylander's artwork is created in Mathematica, POV-Ray, and C++.

You can find a short bio and description of his work at:
http://virtualmathmuseum.org/mathart/ArtGalleryNylander/Nylanderindex.html.

Cover-Spidroball By Daniel Erdely and Paul Nylander (bugman123.com)

ALGEBRA 2

mheducation.com/prek-12

Copyright © 2018 McGraw-Hill Education

All rights reserved. No part of this publication may be reproduced or distributed in any form or by any means, or stored in a database or retrieval system, without the prior written consent of McGraw-Hill Education, including, but not limited to, network storage or transmission, or broadcast for distance learning.

Send all inquiries to:
McGraw-Hill Education
8787 Orion Place
Columbus, OH 43240

ISBN: 978-0-07-898519-5 *(Teacher Edition, Vol. 1)*
MHID: 0-07-898519-6 *(Teacher Edition, Vol. 1)*
ISBN: 978-0-07-898520-1 *(Teacher Edition, Vol. 2)*
MHID: 0-07-898520-X *(Teacher Edition, Vol. 2)*
ISBN: 978-0-07-903990-3 *(Student Edition)*
MHID: 0-07-903990-1 *(Student Edition)*

Printed in the United States of America.

4 5 6 7 8 9 QVS 21 20 19 18

Understanding by Design® is a registered trademark of the Association for Supervision and Curriculum Development ("ASCD").

McGraw-Hill is committed to providing instructional materials in Science, Technology, Engineering, and Mathematics (STEM) that give all students a solid foundation, one that prepares them for college and careers in the 21st century.

Contents in Brief

SUGGESTED PACING GUIDE

Each chapter includes multiple days for review and assessment.

Chapter 0	5 days
Chapter 1	15 days
Chapter 2	11 days
Chapter 3	12 days
Chapter 4	17 days
Chapter 5	11 days
Chapter 6	15 days
Chapter 7	12 days
Chapter 8	11 days
Chapter 9	11 days
Chapter 10	8 days
Total	**128 days**

Authors

Our lead authors ensure that the Macmillan/McGraw-Hill and Glencoe/McGraw-Hill mathematics programs are truly vertically aligned by beginning with the end in mind — success in Algebra 2 and beyond. By "backmapping" the content from the high school programs, all of our mathematics programs are well articulated in their scope and sequence.

LEAD AUTHORS

John A. Carter, Ph.D.

Mathematics Teacher
WINNETKA, ILLINOIS

Areas of Expertise:
Using technology and manipulatives to visualize concepts; mathematics achievement of English-language learners

Gilbert J. Cuevas, Ph.D.

Professor of Mathematics Education, Texas State University— San Marcos
SAN MARCOS, TEXAS

Areas of Expertise:
Applying concepts and skills in mathematically rich contexts; mathematical representations; use of technology in the development of geometric thinking

Roger Day, Ph.D., NBCT

Mathematics Department Chairperson, Pontiac Township High School
PONTIAC, ILLINOIS

Areas of Expertise:
Understanding and applying probability and statistics; mathematics teacher education

In Memoriam
Carol Malloy, Ph.D.

Dr. Carol Malloy was a fervent supporter of mathematics education. She was a Professor at the University of North Carolina, Chapel Hill, NCTM Board of Directors member, President of the Benjamin Banneker Association (BBA), and 2013 BBA Lifetime Achievement Award for Mathematics winner. She joined McGraw-Hill in 1996. Her influence significantly improved our programs' focus on real-world problem solving and equity. We will miss her inspiration and passion for education.

PROGRAM AUTHORS

Berchie Holliday, Ed.D.

National Mathematics Consultant
SILVER SPRING. MARYLAND

Areas of Expertise:
Using mathematics to model and understand real-world data; the effect of graphics on mathematical understanding

Ruth Casey

Regional Teacher Partner
UNIVERSITY OF KENTUCKY

Areas of Expertise:
Graphing technology and mathematics

CONTRIBUTING AUTHORS

Dinah Zike FOLDABLES

Educational Consultant
Dinah-Might Activities, Inc.
SAN ANTONIO. TEXAS

Jay McTighe

Educational Author and Consultant
COLUMBIA. MARYLAND

Consultants and Reviewers

These professionals were instrumental in providing valuable input and suggestions for improving the effectiveness of the mathematics instruction.

LEAD CONSULTANT

Viken Hovsepian

Professor of Mathematics
Rio Hondo College
WHITTIER, CALIFORNIA

CONSULTANTS

MATHEMATICAL CONTENT

Grant A. Fraser, Ph.D.
Professor of Mathematics
California State University, Los Angeles
LOS ANGELES, CALIFORNIA

Arthur K. Wayman, Ph.D.
Professor of Mathematics Emeritus
California State University, Long Beach
LONG BEACH, CALIFORNIA

GIFTED AND TALENTED

Shelbi K. Cole
Research Assistant
University of Connecticut
STORRS, CONNECTICUT

COLLEGE READINESS

Robert Lee Kimball, Jr.
Department Head, Math and Physics
Wake Technical Community College
RALEIGH, NORTH CAROLINA

DIFFERENTIATION FOR ENGLISH-LANGUAGE LEARNERS

Susana Davidenko
State University of New York
CORTLAND, NEW YORK

Alfredo Gómez
Mathematics/ESL Teacher
George W. Fowler High School
SYRACUSE, NEW YORK

GRAPHING CALCULATOR

Jerry Cummins
Former President
National Council of Supervisors of Mathematics
WESTERN SPRINGS, ILLINOIS

MATHEMATICAL FLUENCY

Robert M. Capraro
Associate Professor
Texas A&M University
COLLEGE STATION, TEXAS

PRE-AP

Dixie Ross
Lead Teacher for Advanced Placement Mathematics
Pflugerville High School
PFLUGERVILLE, TEXAS

READING AND WRITING

ReLeah Cossett Lent
Author and Educational Consultant
MORGANTON, GEORGIA

Lynn T. Havens
Director of Project CRISS
KALISPELL, MONTANA

REVIEWERS

Corey Andreasen
Mathematics Teacher
North High School
SHEBOYGAN, WISCONSIN

Mark B. Baetz
Mathematics Coordinating
  Teacher
Salem City Schools
SALEM, VIRGINIA

Kathryn Ballin
Mathematics Supervisor
Newark Public Schools
NEWARK, NEW JERSEY

Kevin C. Barhorst
Mathematics Department Chair
Independence High School
COLUMBUS, OHIO

Brenda S. Berg
Mathematics Teacher
Carbondale Community High
  School
CARBONDALE, ILLINOIS

Dawn Brown
Mathematics Department Chair
Kenmore West High School
BUFFALO, NEW YORK

Sheryl Pernell Clayton
Mathematics Teacher
Hume Fogg Magnet School
NASHVILLE, TENNESSEE

Bob Coleman
Mathematics Teacher
Cobb Middle School
TALLAHASSEE, FLORIDA

Jane E. Cotts
Mathematics Teacher
O'Fallon Township High School
O'FALLON, ILLINOIS

Michael D. Cuddy
Mathematics Instructor
Zypherhills High School
ZYPHERHILLS, FLORIDA

Melissa M. Dalton, NBCT
Mathematics Instructor
Rural Retreat High School
RURAL RETREAT, VIRGINIA

Tina S. Dohm
Mathematics Teacher
Naperville Central High School
NAPERVILLE, ILLINOIS

Laurie L.E. Ferrari
Mathematics Teacher
L'Anse Creuse High School–North
MACOMB, MICHIGAN

Steve Freshour
Mathematics Teacher
Parkersburg South High School
PARKERSBURG, WEST VIRGINIA

Shirley D. Glover
Mathematics Teacher
TC Roberson High School
ASHEVILLE, NORTH CAROLINA

Caroline W. Greenough
Mathematics Teacher
Cape Fear Academy
WILMINGTON, NORTH CAROLINA

Susan Hack, NBCT
Mathematics Teacher
Oldham County High School
BUCKNER, KENTUCKY

Michelle Hanneman
Mathematics Teacher
Moore High School
MOORE, OKLAHOMA

Theresalynn Haynes
Mathematics Teacher
Glenbard East High School
LOMBARD, ILLINOIS

Sandra Hester
Mathematics Teacher/AIG
  Specialist
North Henderson High School
HENDERSONVILLE, NORTH
CAROLINA

Jacob K. Holloway
Mathematics Teacher
Capitol Heights Junior High School
MONTGOMERY, ALABAMA

Robert Hopp
Mathematics Teacher
Harrison High School
HARRISON, MICHIGAN

Eileen Howanitz
Mathematics Teacher/Department
Chairperson
Valley View High School
ARCHBALD, PENNSYLVANIA

Charles R. Howard, NBCT
Mathematics Teacher
Tuscola High School
WAYNESVILLE, NORTH CAROLINA

Sue Hvizdos
Mathematics Department
Chairperson
Wheeling Park High School
WHEELING, WEST VIRGINIA

Elaine Keller
Mathematics Teacher
Mathematics Curriculum Director
K-12
Northwest Local Schools
CANAL FULTON, OHIO

Sheila A. Kotter
Mathematics Educator
River Ridge High School
NEW PORT RICHEY, FLORIDA

Frank Lear
Mathematics Department Chair
Cleveland High School
CLEVELAND, TENNESSEE

Jennifer Lewis
Mathematics Teacher
Triad High School
TROY, ILLINOIS

Catherine McCarthy
Mathematics Teacher
Glen Ridge High School
GLEN RIDGE, NEW JERSEY

Jacqueline Palmquist
Mathematics Department Chair
Waubonsie Valley High School
AURORA, ILLINOIS

Thom Schacher
Mathematics Teacher
Otsego High School
OTSEGO, MICHIGAN

Laurie Shappee
Teacher/Mathematics Coordinator
Larson Middle School
TROY, MICHIGAN

Jennifer J. Southers
Mathematics Teacher
Hillcrest High School
SIMPSONVILLE, SOUTH CAROLINA

Sue Steinbeck
Mathematics Department Chair
Parkersburg High School
PARKERSBURG, WEST VIRGINIA

Kathleen D. Van Sise
Mathematics Teacher
Mandarin High School
JACKSONVILLE, FLORIDA

Karen Wiedman
Mathematics Teacher
Taylorville High School
TAYLORVILLE, ILLINOIS

Solutions That Work

Glencoe High School Math Series is about connecting math content, rigor, and adaptive instruction for student success.

Glencoe High School Math Series provides you the flexibility to meet your classroom needs. As you build teacher-student relationships and a classroom environment that encourages learning, this program provides a flexible print and digital solution that allows you to:

- **Personalize learning** for all students with differentiated instruction and math content that meets rigorous standards.

- **Engage students** in a variety of ways with projects, activities, and resources that make math come to life and enable students to connect math to their world.

- **Drive success** on state assessments with dynamic digital tools that support and inform instructional decisions.

"Great...easy textbook to use, good online resources, and many examples, including great exit ticket ideas and warm up questions."

--*High School Teacher*

Rigor in Instruction

Glencoe High School Math Series is about connecting math content and rigor. Rigor is built-in throughout the entire program with a strong focus on promoting conceptual understanding and encouraging students to think critically.

Performance Tasks

A Performance Task is available for each chapter that enables students to practice persevering through a rich, multi-step task. The Performance Tasks are previewed at the beginning of the chapter. Students will learn concepts and skills as they work through each lesson that will help them to finish the task at the end of the chapter.

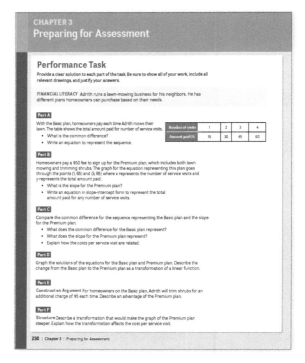

Multi-Step Questions

Multi-step questions allow students to practice reasoning, modeling, and looking for structure in different situations. These questions are incorporated into the Preparing for Assessment page at the end of each lesson and can also be found online in eAssessment.

68. MULTI-STEP A candle burns as shown in the graph.
MP 1, 2, 8

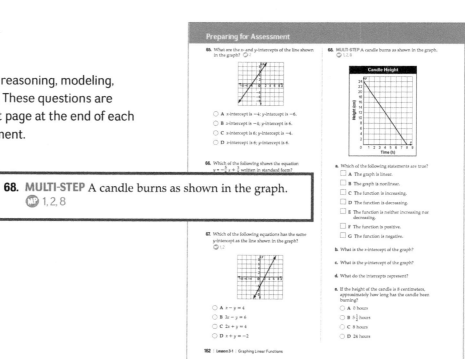

Standards for Mathematical Practice

The Standards for Mathematical Practice describe how students should approach mathematics. The goal of the practice standards is to instill in all students the abilities to be mathematically literate and to create a positive disposition for the importance of using math effectively.

You can use the following tools to incorporate the practices into your teaching.

> Mathematical Practice Study Tips located in the margins of the student edition.

> Look for the MP icon **MP** to see which practice is being taught.

> Mathematical Practice Strategies show you how to incorporate the practices in each chapter.

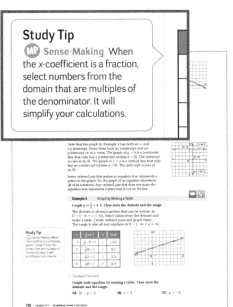

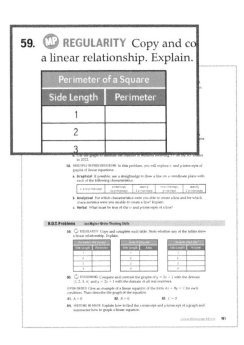

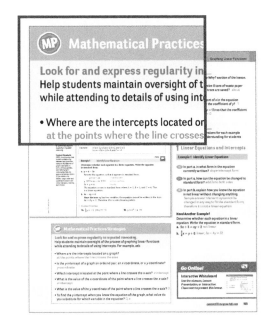

Interactive Student Guide

Working together, the Student Edition and *Interactive Student Guide* (ISG) promote a deep understanding of the content to ensure student success. Topics from the ISG correspond to the lessons within the Student Edition and point-of-use references are located in the margins of the Teacher Edition.

+

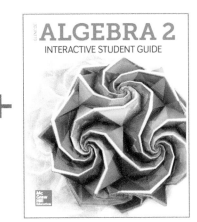

Understanding by Design

What should students know and be able to do? Understanding by Design can be used to help teachers identify learning goals, develop revealing assessments of student understanding, and plan effective and engaging learning activities.

Backward Design

Understanding by Design (UbD) is a framework that uses backward design to create a coherent curriculum by considering the desired results first and then planning instruction.

The backward design process guided the development of *Glencoe Algebra 1, Glencoe Geometry,* and *Glencoe Algebra 2.*

Identifying Desired Results

The first step in developing an effective curriculum using the UbD framework is to consider the goals. What should students know and be able to do?

Glencoe High School Math Series addresses the big ideas of algebra and focuses student attention on Essential Questions, which are located within each chapter of the Student and Teacher Editions.

An Essential Question is provided at the beginning of each chapter. These thought provoking questions can be used as:

> a discussion starter for your class; throughout the discussion identify what the students already know and what they would like to know about the topic. Revisit these notes throughout the chapter.

> a benchmark of understanding; post these questions in a prominent place and have students expand upon their initial response as their understanding of the subject material grows.

Follow-up Essential Questions can be found throughout each chapter. These questions challenge students to apply specific knowledge to a broader context, thus deepening their understanding.

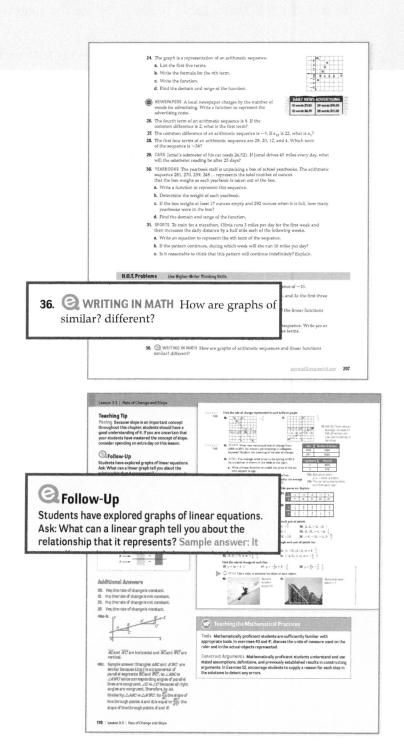

"In mathematics, essential questions are used to develop students' understanding of key concepts as well as core processes."

— **JAY MCTIGHE**, co-author of *Understanding by Design*

Determine Acceptable Evidence

A variety of assessment opportunities are available that enable students to show evidence of their understanding.

> Practice and Problem Solving and H.O.T. Problems allow students to explain, interpret, and apply mathematical concepts.

> Mid-Chapter Quizzes and Chapter Tests offer more traditional methods of assessment.

> McGraw-Hill eAssessment can also be used to customize and create assessments.

Plan Learning Experiences and Instruction

There are numerous performance activities available throughout the program, including:

> Algebra Labs that offer students hands-on learning experiences, and

> Graphing Technology Labs that use graphing calculators to aid student understanding.

You can also visit connectED.mcgraw-hill.com to choose from an extensive collection of resources to use when planning instruction, such as presentation tools, chapter projects, editable worksheets, digital animations, and eTools.

40. WRITING IN MATH Compare and contr[ast] functions with the graphs of linear fu[nctions] and minima.

Built-In Differentiated Instruction

Approximately 43% of teachers feel their classes are so mixed in terms of students' learning abilities that they can't teach them effectively.

Glencoe High School Math Series fully supports the 3-tier RtI model with print and digital resources to diagnose students, identify areas of need, and conduct short, frequent assessments for accurate data-driven decision making. Every lesson provides easy-to-use resources that consider the needs of all students.

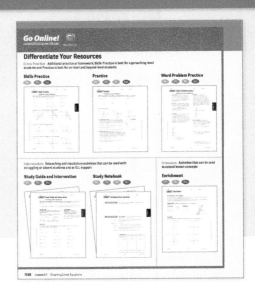

RtI: Response to Intervention

TIER 1: Daily Intervention

OL On grade level	**BL** Beyond grade level
Core instruction targets on-level students. Comprehensive instructional materials help you personalize instruction for every student.	At every step, resources and assignments are available for advanced learners.

On grade level

- Diagnostic Teaching
- Options for Differentiated Instruction
- Leveled Exercise Sets, Resources, and Technology
- Data-Driven Decision Making

Beyond grade level

- Higher-Order Thinking Questions
- Differentiated Homework Options
- Enrichment Masters
- Differentiated Instruction: Extension

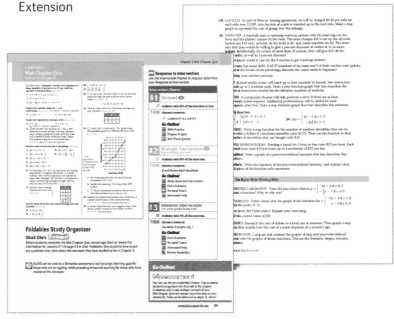

TIER 2: Strategic Intervention

AL Approaching grade level

You can choose from a myriad of intervention tips and ancillary materials to support struggling learners.

- Using Manipulatives
- Alternate Teaching Strategies
- Online resources, including animations, examples, and Personal Tutors

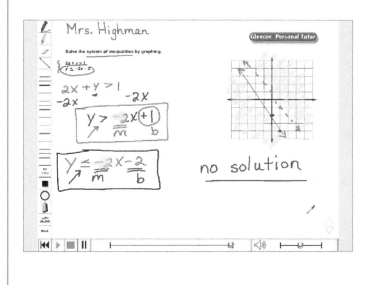

TIER 3: Intensive Intervention

AL Significantly below grade level

For students who are far below grade level, **Math Triumphs** provides step-by-step instruction, vocabulary support, and meaningful practice.

Assessment Tools that Inform Instruction

A comprehensive variety of print and online assessment options are built into **Glencoe High School Math Series**. Using traditional assessment combined with our online test generator and reporting system, you can use data to make on–the-spot instructional decisions for the entire class or individual students.

Chapter Assessment Resources

Prebuilt assessment resources are provided with every chapter to quickly print and deliver on demand. As an added feature, most are available in editable Word documents to allow for maximum flexibility and customization:

- Chapter Quizzes
- Mid-Chapter Tests
- Vocabulary Tests
- Extended Response Tests
- Standardized Tests

> The personalized study resources your students need today – to master state assessments tomorrow.

eAssessment

With a simple, intuitive interface that enables teachers to create customized assessments and homework, web-based eAssessment means easy, anytime access via a secure online test center. Manage your content, create and assign quizzes and tests that can be assigned online or in traditional format. The reporting system enables you to see performance at both a class and a student level which allows you to modify instruction or provide targeted support to improve student achievement.

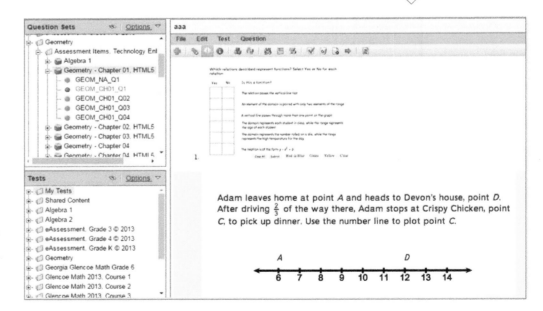

LEARNSMART®

Using this robust, online personalized study resource, students have a modernized study partner to practice for high-stakes testing. By following a personalized study plan, each student will see the course topics they have mastered or which they need to revisit. Learning resources are tagged to questions to ensure students have multiple opportunities to refresh their content knowledge. Advanced reporting enables you to maximize your instructional time leading up to end of course or end of year assessments.

ALEKS®

Differentiating instruction can be a challenge. It's **ALEKS®** to the rescue! This online resource uses adaptive questioning to provide each student with a personalized learning path.

Support each student's unique needs

- Knowledge checks identify what a student knows and where there are gaps
- Personalized instruction allows students to progress at their own pace
- Robust Reporting tools provide instructionally-actionable data
- Targeted instruction on Ready to Learn topics build learning momentum and confidence
- Open-response environment ensures students demonstrate understanding—no guessing!

Learn how **ALEKS** can enhance your current math curriculum. Contact your McGraw-Hill Education rep today.

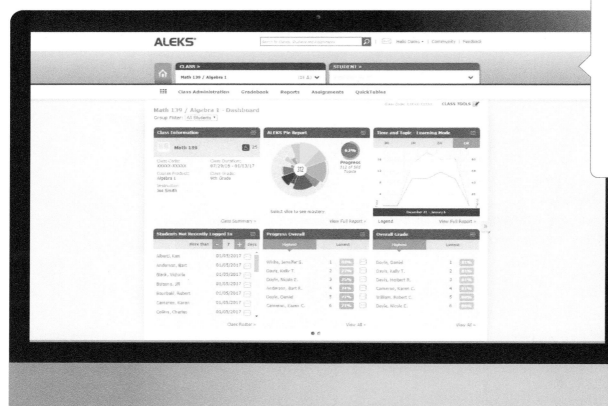

Use ConnectED as Your Portal

Nearly 84% of teachers say that they only spend half of their time teaching, as opposed to disciplining or doing administrative work.

ConnectED is a time-saving online portal that has all of your digital program resources in one place. Through this portal, you can access the complete Student Edition and Teacher Edition, digital animations and tutorials, editable worksheets, presentation tools, and assessment resources.

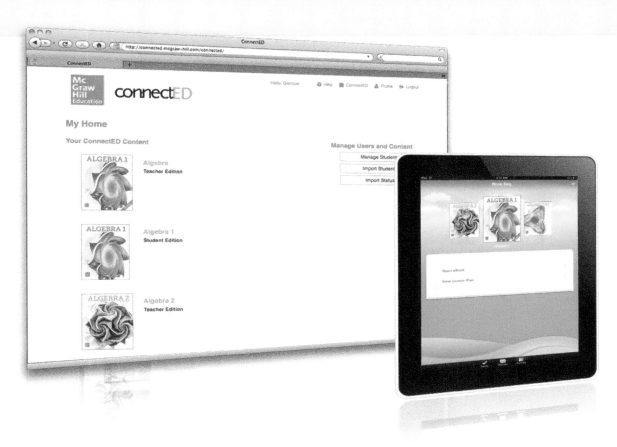

With ConnectED Mobile you can browse through your course content on the go. The app includes a powerful eBook engine where you can download, view, and interact with your books.

connectED allows you to:

- build lesson plans with easy-to-find print and digital resources
- search for activities to meet a variety of learning modalities
- teach with technology by providing virtual manipulatives, lesson animations, whole-class presentations, and more
- personalize instruction with print and digital resources
- provide students with anywhere, anytime access to student resources and tools, including eBooks, tutorials, animations, and the eGlossary
- access eAssessment, which allows you to assign online assessments, track student progress, generate reports, and differentiate instruction

Digital Tools to Enhance Learning Opportunities

Today's students have an unprecedented access to and appetite for technology and new media. They perceive technology as their friend and rely on it to study, work, play, relax, and communicate.

Your students are accustomed to the role that computers play in today's world. They may well be the first generation whose primary educational tool is a computer or a cell phone. The eStudentEdition allows your students to access their math curriculum anytime, anywhere. Along with your teacher edition, **Glencoe Algebra 2** provides a blended instructional solution for your next-generation students.

Investigate

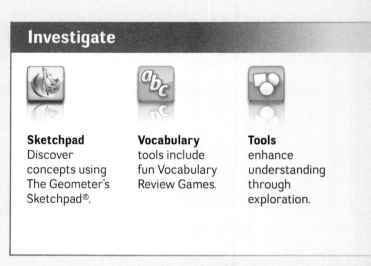

Sketchpad
Discover concepts using The Geometer's Sketchpad®.

Vocabulary
tools include fun Vocabulary Review Games.

Tools
enhance understanding through exploration.

Go Online!

Look for the *Go Online!* feature in your Teacher Edition to easily find digital tools to engage students and personalize instruction.

> Each chapter includes a curated list of differentiated resources as well as featured interactive whiteboard resources with suggested pacing.

> Each lesson includes a list of digital resources with implementation tips and suggested pacing.

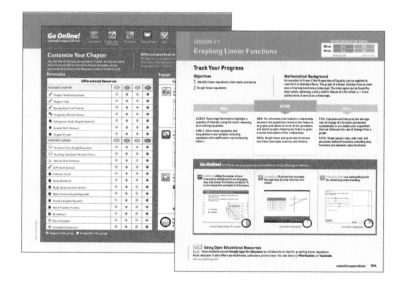

The Geometer's Sketchpad

The Geometer's Sketchpad is proven to increase engagement, understanding, and achievement. This premier digital learning tool makes math come alive by giving students a tangible, visual way to see the math in action through dynamic model manipulation of lines, shapes, and functions.

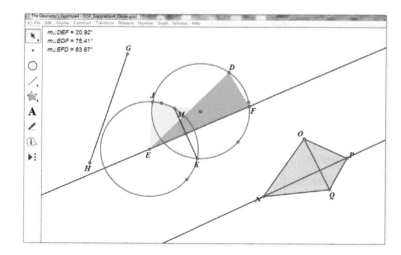

LearnSmart
Topic based
online
assessment

Animations
illustrate key
concepts
through step-
by-step tutorials
and videos.

Tutors
See and hear a
teacher explain
how to solve
problems.

**Calculator
Resources**
provides other
calculator
keystrokes for
each Graphing
Technology Lab.

**Self-Check
Practice**
allows students
to check their
understanding
and send results
to their teacher.

eBook
Interactive
learning
experience with
links directly to
assets

eToolkit

eToolkit enables students to use virtual manipulatives to extend
learning beyond the classroom by modifying concrete models in a
real-time, interactive format focused on problem-based learning.

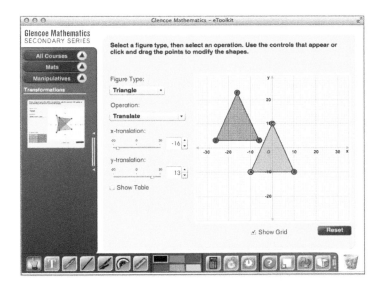

Interactive Student Guide

Interactive Student Guide is a dynamic digital resource to help
you and your students meet the increasing demands for rigor.

You can be empowered to teach confidently knowing every
lesson includes standards-aligned content and emphasizes the
Standards for Mathematical Practice. This guide works together
with the hard-case student edition to ensure that students can
reflect on comprehension and application, use mathematics in
real-world applications, and internalize concepts to develop
"second nature" recall.

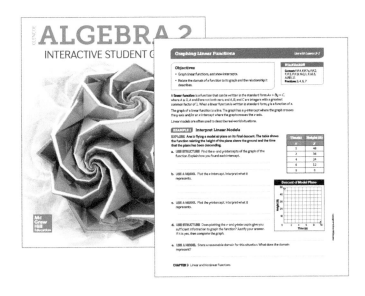

21st Century Skills

The U.S. Department of Labor estimates that 90% of 21st-century skilled workforce jobs will require post-secondary education. To be competitive in tomorrow's global workforce, American students must be prepared to succeed in college.

A strong high school curriculum is a good predictor of college readiness (Adelman, 2006). Students who take at least three years of college-preparatory mathematics using programs like **Glencoe Algebra 1**, **Glencoe Geometry**, and **Glencoe Algebra 2** are less likely to need remedial courses in college than students who do not (Abraham & Creech, 2002).

College Readiness

David Conley at the University of Oregon developed the following criteria for college readiness.

Key Content Knowledge

Glencoe High School Math Series is aligned to rigorous state and national standards, including the *NCTM Principles & Standards for School Mathematics,* the College Board Standards for College Success, and the American Diploma Project's Benchmarks. Correlations to these standards can be found at **connectED.mcgraw-hill.com.**

Habits of Mind

These include critical thinking skills such as analysis, interpretation, problem solving, and reasoning. Students can hone critical higher-order thinking skills through the use of **H.O.T. (Higher Order Thinking) Problems.**

Contextual Skills

These are practical skills like understanding the admissions process and financial aid, placement testing, and communicating with professors. Throughout each Glencoe mathematics program, students are required to write, explain, justify, prove, and analyze.

Academic Behaviors

These include general skills such as reading comprehension, time management, note-taking, and metacognition. Reading Math tips and Vocabulary Links help students with reading comprehension. Study Notebooks and Anticipation Guides help students build note-taking skills and aid with metacognition.

Developing STEM Careers

With **Glencoe High School Math Series**, you can unleash your students' curiosity about the world around them and prepare them for exciting **STEM** (**S**cience, **T**echnology, **E**ngineering, and **M**ath) careers.

Examples

Examples are relevant, connecting in-class experiences to the world beyond the classroom.

Real-World Careers

Real-World Careers are engaging, providing information on exciting careers.

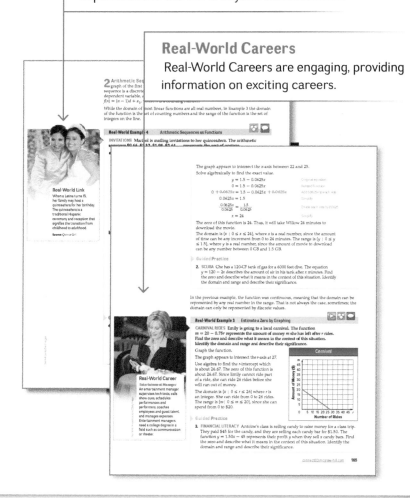

"The current and future health of America's 21st Century Economy depends directly on how broadly and deeply Americans reach a new level of literacy– '21st Century Literacy'–that includes strong academic skills, thinking, reasoning, teamwork skills, and proficiency in using technology."

–21st Century Workforce Commission National Alliance of Business

Developing 21st Century Skills

The Partnership for 21st Century Skills identifies the following key student elements of a 21st century education.

Core Subjects and 21st Century Themes

Glencoe High School Math Series has been aligned to rigorous state and national standards, *NCTM Principles & Standards for School Mathematics*, and the American Diploma Project's Benchmarks. Throughout each program, students solve problems that incorporate 21st century themes, such as financial literacy.

Learning and Innovation Skills

Students who are prepared for increasingly complex life and work environments are creative and innovative critical thinkers, problem solvers, effective communicators, and know how to work collaboratively. Throughout each **Glencoe High School Math Series**, students are required to write, explain, justify, prove, and analyze. Students use critical thinking skills through the use of H.O.T. (Higher Order Thinking) Problems and are encouraged to work collaboratively in labs.

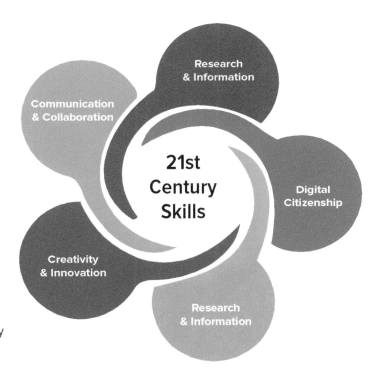

Information, Media, and Technology Skills

Students use technology, including graphing calculators and the Internet, to develop 21st century mathematics knowledge and skills throughout each program.

21st Century Assessments

Glencoe High School Math Series offers a variety of frequent and meaningful assessments built right into the curriculum structure and teacher support materials. These programs include both traditional and nontraditional methods of assessment, including quizzes and tests, performance tasks, and open-ended assessments. Digital assessment solutions offer additional options for creating, customizing, administering, and instantly grading assessments.

Professional Development

McGraw-Hill Education recognizes that learning is a lifelong endeavor. To ensure student and teacher success, we have built in resources to the Glencoe High School Math Series featuring best practices, implementation support, alternative teaching practices, and much more.

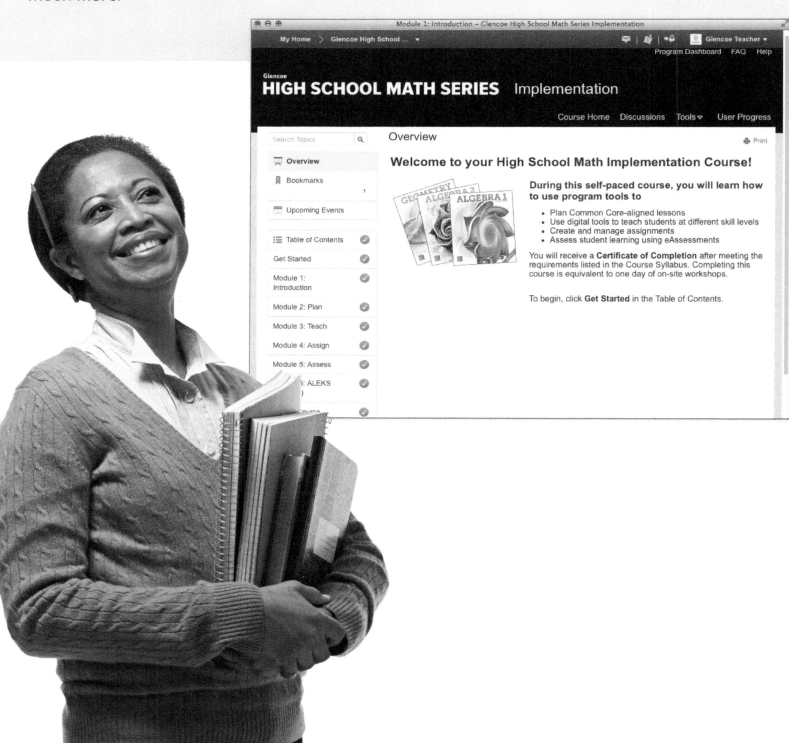

Correlations

Common Core State Standards, Traditional Algebra II Pathway, Correlated to *Glencoe Algebra 2*

Lessons in which the standard is the primary focus are indicated in **bold**.

Standards	Student Edition Lesson(s)	Student Edition Page(s)
Number and Quantity		
The Complex Number System N-CN		
Perform arithmetic operations with complex numbers. 1. Know there is a complex number i such that $i^2 = -1$, and every complex number has the form $a + bi$ with a and b real.	**3-3**	172–178
2. Use the relation $i^2 = -1$ and the commutative, associative, and distributive properties to add, subtract, and multiply complex numbers.	**3-3**	172–178
Use complex numbers in polynomial identities and equations. 7. Solve quadratic equations with real coefficients that have complex solutions.	**3-5**, Extend 3-5, **3-6**	**191–197**, 198, **199–207**
8. (+) Extend polynomial identities to the complex numbers.	**3-4, 3-5, 3-6**	**179–187, 191–197, 199–207**
9. (+) Know the Fundamental Theorem of Algebra; show that it is true for quadratic polynomials.	**4-9**	**293–300**
Algebra		
Seeing Structure in Expressions A-SSE		
Interpret the structure of expressions. 1. Interpret expressions that represent a quantity in terms of its context. ★ a. Interpret parts of an expression, such as terms, factors, and coefficients.	**3-1**	151–159
b. Interpret complicated expressions by viewing one or more of their parts as a single entity.	1-3, 2-2, **2-5**, 2-6, **3-1, 3-6**, 4-5, 9-4, 9-5, 9-6	21–27, 95–102, **118–124**, 125–132, **199–207**, 262–269, 635–642
2. Use the structure of an expression to identify ways to rewrite it.	3-4, **3-5**, 5-5, 6-2, **6-4**, 6-8, 6-9, 6-10	179–187, **191–197**, 345–350, 383–389, **397–404**, 430–436, 437–444, 445–451
Write expressions in equivalent forms to solve problems. 4. Derive the formula for the sum of a finite geometric series (when the common ratio is not 1), and use the formula to solve problems. ★	**6-3**	390–396

★ **Mathematical Modeling Standards**
(+) **Advanced Mathematics Standards**

Correlations

Standards	Student Edition Lesson(s)	Student Edition Page(s)
Arithmetic with Polynomials and Rational Expressions A-APR		
Perform arithmetic operations on polynomials.	4-1	229–235
1. Understand that polynomials form a system analogous to the integers, namely, they are closed under the operations of addition, subtraction, and multiplication; add, subtract, and multiply polynomials.		
Understand the relationship between zeros and factors of polynomials.	4-8	287–292
2. Know and apply the Remainder Theorem: For a polynomial $p(x)$ and a number a, the remainder on division by $x - a$ is $p(a)$, so $p(a) = 0$ if and only if $(x - a)$ is a factor of $p(x)$.		
3. Identify zeros of polynomials when suitable factorizations are available, and use the zeros to construct a rough graph of the function defined by the polynomial.	4-9	293–300
Use polynomial identities to solve problems.	4-7, Extend 4-9	282–286, 301
4. Prove polynomial identities and use them to describe numerical relationships.		
5. (+) Know and apply the Binomial Theorem for the expansion of $(x + y)^n$ in powers of x and y for a positive integer n, where x and y are any numbers, with coefficients determined for example by Pascal's Triangle.	4-2	237–241
Rewrite rational expressions.	4-3, Extend 4-3	242–249, 250
6. Rewrite simple rational expressions in different forms; write $\frac{a(x)}{b(x)}$ in the form $q(x) + \frac{r(x)}{b(x)}$, where $a(x)$, $b(x)$, $q(x)$, and $r(x)$ are polynomials with the degree of $r(x)$ less than the degree of $b(x)$, using inspection, long division, or, for the more complicated examples, a computer algebra system.		
7. (+) Understand that rational expressions form a system analogous to the rational numbers, closed under addition, subtraction, multiplication, and division by a nonzero rational expression; add, subtract, multiply, and divide rational expressions.	7-1, 7-2, 7-4, 7-6	467–475, 476–482, 491–498, 508–516

(+) **Advanced Mathematics Standards**

Standards	Student Edition Lesson(s)	Student Edition Page(s)
Creating Equations ★ A-CED		
Create equations that describe numbers or relationships. 1. Create equations and inequalities in one variable and use them to solve problems.	**1-1, 1-2, 3-2, 3-4, 3-5, 3-6, 3-7, 4-6, 4-8, 4-9, 6-2, 6-6, 6-7, 6-8, 6-9, 7-6**	**5–12, 13–19, 163–169, 179–187, 191–197, 199–207, 209–215, 274–281, 287–292, 293–300, 383–389, 416–422, 423–429, 430–436, 437–442, 509–516**
2. Create equations in two or more variables to represent relationships between quantities; graph equations on coordinate axes with labels and scales.	**1-3, Explore 1-6, 1-6, 2-4, 2-5, 2-7, 3-1, 3-2, Explore 4-4, 4-4, Explore 4-6, 5-4, 5-5, Extend 5-5, 6-1, 7-3, 7-4, Extend 7-4, 7-5, 9-6**	**21–27, 41,42–51, 111–116, 118–124, 133–137, 151–160, 163–170, 251–252, 253–261, 273, 338–344, 345–350, 351, 373–380, 483–489, 491–498, 499, 500–507, 635–642**
3. Represent constraints by equations or inequalities, and by systems of equations and/or inequalities, and interpret solutions as viable or non-viable options in a modeling context.	**1-2, 1-5, 1-6, 1-7, 1-8, 1-9, 2-5, 2-6, 3-7, 6-10, 7-6**	**13–19, 35–39, 42–51, 52–58, 60–66, 67–73, 118–124, 125–132, 209–215, 445–451, 508–516**
4. Rearrange formulas to highlight a quantity of interest, using the same reasoning as in solving equations.	**3-6**	**199–208**
Reasoning with Equations and Inequalities A-REI		
Understand solving equations as a process of reasoning and explain the reasoning. 2. Solve simple rational and radical equations in one variable, and give examples showing how extraneous solutions may arise.	**5-6, Extend 5-6, 7-6, Extend 7-6**	**352–358, 359–360, 508–516, 517–518**
Represent and solve equations and inequalities graphically. 11. Explain why the x-coordinates of the points where the graphs of the equations $y = f(x)$ and $y = g(x)$ intersect are the solutions of the equation $f(x) = g(x)$; find the solutions approximately, e.g., using technology to graph the functions, make tables of values, or find successive approximations. Include cases where $f(x)$ and/or $g(x)$ are linear, polynomial, rational, absolute value, exponential, and logarithmic functions.★	**1-6, Extend 3-2, 4-7, Extend 5-6, Explore 6-2, Extend 6-9, Extend 7-6**	**42–51, 171, 282–286, 359–260, 381–382, 443–444, 517–518**

★ **Mathematical Modeling Standards**

Correlations

Standards	Student Edition Lesson(s)	Student Edition Page(s)
Functions		
Interpreting Functions F-IF		
Interpret functions that arise in applications in terms of the context. 4. For a function that models a relationship between two quantities, interpret key features of graphs and tables in terms of the quantities, and sketch graphs showing key features given a verbal description of the relationship. ★	2-1, 2-2, 2-3, 2-4, 2-5, 2-6, 2-7, 3-1, **Extend 3-1, 3-2, Extend 3-6, Explore 4-4, 4-4, 4-5, Extend 4-5, Explore 4-6, 5-3, 5-4, 5-5, Extend 5-5, 6-1, 6-4, 7-3, 7-4, Extend 7-4, 7-5, 9-6**	**87–94,** 95–102, 103–109, 111–116, 118–124, 125–132, 133–137, 161–162, 163–170, 208, 251–252, 253–261, 262–269, 270–271, 273, 329–335, 338–344, 345–350, 351, 373–380, 397–404, 483–489, 491–498, 499, 500–507, 635–642
5. Relate the domain of a function to its graph and, where applicable, to the quantitative relationship it describes.	2-1, 2-5, 3-1, 4-4, 5-3, 5-4, 5-5, 6-1, 6-4, 7-3, 7-4, 9-6	87–94, 118–124, 151–159, 253–261, 329–335, 338–344, 345–350, 373–380, 635–642
6. Calculate and interpret the average rate of change of a function (presented symbolically or as a table) over a specified interval. Estimate the rate of change from a graph. ★	**1-3, Extend 3-6**	**21–27, 208**
Analyze functions using different representations. 7. Graph functions expressed symbolically and show key features of the graph, by hand in simple cases and using technology for more complicated cases. ★ b. Graph square root, cube root, and piecewise-defined functions, including step functions and absolute value functions.	**2-5, 5-4, 5-5, Extend 5-5**	118–124, **338–344, 345–350, 351**
c. Graph polynomial functions, identifying zeros when suitable factorizations are available, and showing end behavior.	**2-3, 4-4, 4-5, Extend 4-5, Explore 4-6, 4-8**	103–109, **253–261, 262–269, 270–271, 273,** 287–292
e. Graph exponential and logarithmic functions, showing intercepts and end behavior, and trigonometric functions, showing period, midline, and amplitude.	**6-1, 6-4, 9-6**	**373–380, 397–404, 635–642**
8. Write a function defined by an expression in different but equivalent forms to reveal and explain different properties of the function. a. Use the process of factoring and completing the square in a quadratic function to show zeros, extreme values, and symmetry of the graph, and interpret these in terms of a context.	**3-4, 3-5**	**179–187, 191–197**
b. Use the properties of exponents to interpret expressions for exponential functions.	**6-1, 6-10**	**373–380, 445–451**
9. Compare properties of two functions each represented in a different way (algebraically, graphically, numerically in tables, or by verbal descriptions).	**2-2, 2-6, 3-1, 4-4, 5-1, 5-4, 6-1, 7-4**	**95–102, 125–132,** 151–159, 253–261, 315–321, 338–344, 373–380, 491–498

★ **Mathematical Modeling Standards**

Standards	Student Edition Lesson(s)	Student Edition Page(s)
Building Functions F-BF		
1. Build a function that models a relationship between two quantities. b. Combine standard function types using arithmetic operations.	5-1, 5-2, Extend 6-10	315–321, 322–328, 452
Build new functions from existing functions. 3. Identify the effect on the graph of replacing $f(x)$ by $f(x) + k$, $k\,f(x)$, $f(kx)$, and $f(x + k)$ for specific values of k (both positive and negative); find the value of k given the graphs. Experiment with cases and illustrate an explanation of the effects on the graph using technology.	2-6, 5-4, 5-5, 6-1, 6-4, 7-3	125–132, 338–344, 345–350, 373–380, 397–404, 483–489
4. Find inverse functions. a. Solve an equation of the form $f(x) = c$ for a simple function f that has an inverse and write an expression for the inverse.	5-3	329–335
Linear, Quadratic, and Exponential Models F-LE		
Construct and compare linear and exponential models and solve problems.	6-2, 6-10	383–389, 445–451
4. For exponential models, express as a logarithm the solution to $ab^{ct} = d$ where a, c, and d are numbers and the base b is 2, 10, or e; evaluate the logarithm using technology.		
Trigonometric Functions F-TF		
Extend the domain of trigonometric functions using the unit circle. 1. Understand radian measure of an angle as the length of the arc on the unit circle subtended by the angle.	9-3, 9-5	612–618, 627–633
2. Explain how the unit circle in the coordinate plane enables the extension of trigonometric functions to all real numbers, interpreted as radian measures of angles traversed counterclockwise around the unit circle.	9-3, 9-5	612–618, 627–633
Model periodic phenomena with trigonometric functions. 5. Choose trigonometric functions to model periodic phenomena with specified amplitude, frequency, and midline. ★	9-6	635–642
Prove and apply trigonometric identities. 8. Prove the Pythagorean identity $\sin^2(\theta) + \cos^2(\theta) = 1$ and use it to calculate trigonometric ratios.	10-1, 10-2, 10-3, 10-4, 10-5	655–661, 662–667, 668–673, 675–681, 683–689

★ **Mathematical Modeling Standards**

Standards	Student Edition Lesson(s)	Student Edition Page(s)
Statistics and Probability		
Interpreting Categorical and Quantitative Data S-ID		
Summarize, represent, and interpret data on a single count or measurement variable.	8-6	566–572
4. Use the mean and standard deviation of a data set to fit it to a normal distribution and to estimate population percentages. Recognize that there are data sets for which such a procedure is not appropriate. Use calculators, spreadsheets, and tables to estimate areas under the normal curve.		
Making Inferences and Justifying Conclusions S-IC		
Understand and evaluate random processes underlying statistical experiments.	8-1	531–537
1. Understand statistics as a process for making inferences about population parameters based on a random sample from that population.		
2. Decide if a specified model is consistent with results from a given data-generating process, e.g., using simulation.	8-2	538–544
Make inferences and justify conclusions from sample surveys, experiments, and observational studies.	8-1	531–537
3. Recognize the purposes of and differences among sample surveys, experiments, and observational studies; explain how randomization relates to each.		
4. Use data from a sample survey to estimate a population mean or proportion; develop a margin of error through the use of simulation models for random sampling.	8-3, 8-6	545–550, 566–572
5. Use data from a randomized experiment to compare two treatments; use simulations to decide if differences between parameters are significant.	8-2	538–544
6. Evaluate reports based on data.	8-5	561–565
Using Probability to Make Decisions S-MD		
6. (+) Use probabilities to make fair decisions (e.g., drawing by lots, using a random number generator).	8-7	573–578
7. (+) Analyze decisions and strategies using probability concepts (e.g., product testing, medical testing, pulling a hockey goalie at the end of a game).	8-7	573–578

(+) **Advanced Mathematics Standards**

# Standards for Mathematical Practice

Glencoe Algebra 2 exhibits these practices throughout the entire program. All of the Standards for Mathematical Practice are covered in each chapter. The MP icon notes specific areas of coverage.

Mathematical Practices	Student Edition Lessons
1. Make sense of problems and persevere in solving them.	Throughout the text, for example: 1-7, 2-1, 2-3, 2-5, 3-1, 3-7, Extend 4-1, 4-4, 5-4, 5-5, 6-10, 7-5, 8-1, 8-2, 8-3, 8-4, 9-5, 10-2
2. Reason abstractly and quantitatively.	Throughout the text, for example: 1-4, 2-1, 2-3, 2-4, 3-4, 4-1, Extend 4-1, 5-4, 5-5, 6-1, 6-2, 7-5, 8-2, 9-2, 10-1
3. Construct viable arguments and critique the reasoning of others.	Throughout the text, for example: 1-1, 1-9, 2-2, 2-7, 3-2, 4-5, 5-2, 6-5, 7-2, 7-3, 8-2, 9-4, 10-3
4. Model with mathematics.	Throughout the text, for example: 1-2, 1-5, 1-8, 2-1, 2-3, 3-1, Extend 3-1, 4-2, 4-6, 5-1, 5-2, 5-4, 5-5, 5-6, Extend 5-6, 6-4, 6-5, 6-7, 6-8, 6-9, 7-5, 8-1, 8-2, 8-3, 8-5, 8-7, 9-4, 10-4, 10-5
5. Use appropriate tools strategically.	Throughout the text, for example: Extend 1-2, Extend 3-1, 2-3, Extend 3-5, Extend 4-5, Explore 4-6, Extend 4-9, Extend 5-5, Explore 6-2, Extend 7-6, 8-3, 8-7, Extend 9-1, Extend 9-5, Explore 10-5
6. Attend to precision.	Throughout the text, for example: 1-6, 2-6, 3-3, 4-3, 4-9, 5-6, 6-4, 6-5, 7-6, 8-5, 8-6, 9-1, 9-3, 10-3, 10-4, 10-5
7. Look for and make use of structure.	Throughout the text, for example: 1-3, 2-1, 2-2, 2-3, 2-4, 3-5, Extend 3-6, 4-7, 4-8, 5-1, 5-2, 5-3, 5-5, 6-4, 7-4, 8-4, 9-4, 10-1, 10-2
8. Look for and express regularity in repeated reasoning.	Throughout the text, for example: 1-1, 1-3, 1-8, 2-6, 3-6, 4-7, 5-3, 6-3, 6-6, 7-1, 8-6, Explore 9-1, 10-5

CHAPTER 0
Preparing for Advanced Algebra

Worksheets help to explain key concepts, let you practice your skills, and offer opportunities for extending the lessons. Find them in the Resources in ConnectED.

Go Online!
connectED.mcgraw-hill.com

jld3 Photography/Getty Images

CHAPTER 1
Linear Equations

Geometer's Sketchpad® allows you to interact with functions in a visual way. Investigate systems of equations and inequalities with sketches in ConnectED.

connectED.mcgraw-hill.com

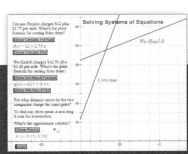

CHAPTER 2
Relations and Functions

Glow Images

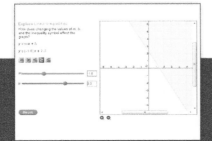

With the **Graphing Tools** in ConnectED, you can explore how changing parameters affects the graph of a function or an inequality.

Go Online!
connectED.mcgraw-hill.com

CHAPTER 3
Quadratic Functions

Jim Kolaczko/E+/Getty Images

Go Online!
connectED.mcgraw-hill.com

Review concepts with quick **Self-Check Quizzes** in ConnectED. Use them to check your own progress as you complete each lesson.

CHAPTER 4
Polynomials and Polynomial Functions

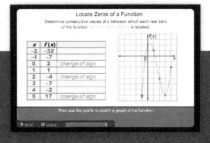

Animations demonstrate Key Concepts and topics from the chapter. Click to watch animations in ConnectED.

Go Online!
connectED.mcgraw-hill.com

Peathegee Inc/Blend Images/Getty Images

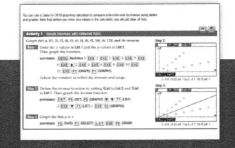

Erik Isakson/Blend Images LLC

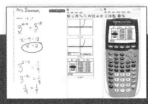

Personal Tutors that use graphing calculator technology show you every step to solving problems with this powerful tool. Find them in the Resources in ConnectED.

Go Online!
connectED.mcgraw-hill.com

CHAPTER 7
Rational Functions

Go Online!

Worksheets help to explain key concepts, let you practice your skills, and offer opportunities for extending the lessons. Find them in the Resources in ConnectED.

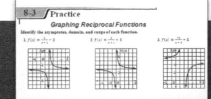

CHAPTER 8
Statistics and Probability

The number cube and coin toss tools can be helpful as you study this chapter. Find them in the **eToolkit** in ConnectED.

Go Online!

connectED.mcgraw-hill.com

CHAPTER 9
Trigonometric Functions

Vocabulary is important to learning the Key Concepts in this chapter. Find all the terms with animations, English pronunciations, and translations to 13 languages in the eGlossary in ConnectED.

connectED.mcgraw-hill.com

CHAPTER 10
Trigonometric Identities and Equations

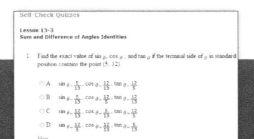

Use **Self-Check Quizzes** in ConnectED to check your understanding of trigonometric identities and equations.

Go Online!
connectED.mcgraw-hill.com

Student Handbook

Built-In Workbook

Reference

(MP) Standards for Mathematical Practice

Glencoe Algebra 2 exhibits these practices throughout the entire program. All of the Standards for Mathematical Practice will be covered in each chapter. The MP icon notes specific areas of coverage.

Mathematical Practices	What does it mean?
1. Make sense of problems and persevere in solving them.	Solving a mathematical problem takes time. Use a logical process to make sense of problems, understand that there may be more than one way to solve a problem, and alter the process if needed.
2. Reason abstractly and quantitatively.	You can start with a concrete or real-world context and then represent it with abstract numbers or symbols (decontextualize), find a solution, then refer back to the context to check that the solution makes sense (contextualize).
3. Construct viable arguments and critique the reasoning of others.	Sound mathematical arguments require a logical progression of statements and reasons. Mathematically proficient students can clearly communicate their thoughts and defend them.
4. Model with mathematics.	Modeling links classroom mathematics and statistics to everyday life, work, and decision-making. High school students at this level are expected to apply key takeaways from earlier grades to high-school level problems.
5. Use appropriate tools strategically.	Certain tools, including estimation and virtual tools are more appropriate than others. You should understand the benefits and limitations of each tool.
6. Attend to precision.	Precision in mathematics is more than accurate calculations. It is also the ability to communicate with the language of mathematics. In high school mathematics, precise language makes for effective communication and serves as a tool for understanding and solving problems.
7. Look for and make use of structure.	Mathematics is based on a well-defined structure. Mathematically proficient students look for that structure to find easier ways to solve problems.
8. Look for and express regularity in repeated reasoning.	Mathematics has been described as the study of patterns. Recognizing a pattern can lead to results more quickly and efficiently.

 **FOLDABLES®** by Dinah Zike

Folding Instructions

The following pages offer step-by-step instructions to make the Foldables® study guides.

Layered-Look Book

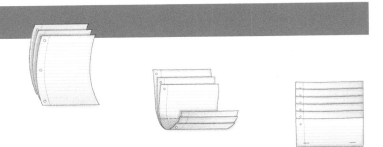

1. Collect three sheets of paper and layer them about 1 cm apart vertically. Keep the edges level.

2. Fold up the bottom edges of the paper to form 6 equal tabs.

3. Fold the papers and crease well to hold the tabs in place. Staple along the fold. Label each tab.

Shutter-Fold and Four-Door Books

1. Find the middle of a horizontal sheet of paper. Fold both edges to the middle and crease the folds. Stop here if making a shutter-fold book. For a four-door book, complete the steps below.

2. Fold the folded paper in half, from top to bottom.

3. Unfold and cut along the fold lines to make four tabs. Label each tab.

Concept-Map Book

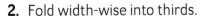
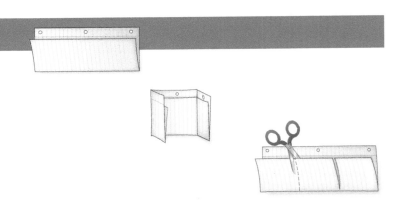

1. Fold a horizontal sheet of paper from top to bottom. Make the top edge about 2 cm shorter than the bottom edge.

2. Fold width-wise into thirds.

3. Unfold and cut only the top layer along both folds to make three tabs. Label the top and each tab.

Vocabulary Book

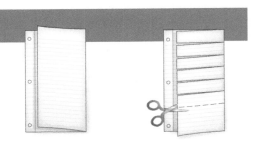

1. Fold a vertical sheet of notebook paper in half.

2. Cut along every third line of only the top layer to form tabs. Label each tab.

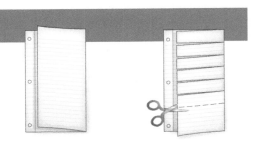

Pocket Book

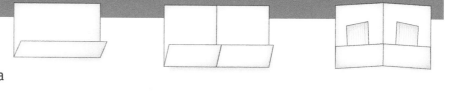

1. Fold the bottom of a horizontal sheet of paper up about 3 cm.

2. If making a two-pocket book, fold in half. If making a three-pocket book, fold in thirds.

3. Unfold once and dot with glue or staple to make pockets. Label each pocket.

Bound Book

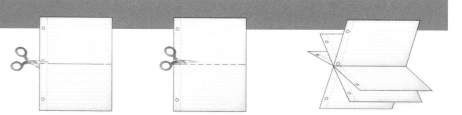

1. Fold several sheets of paper in half to find the middle. Hold all but one sheet together and make a 3-cm cut at the fold line on each side of the paper.

2. On the final page, cut along the fold line to within 3-cm of each edge.

3. Slip the first few sheets through the cut in the final sheet to make a multi-page book.

Top-Tab Book

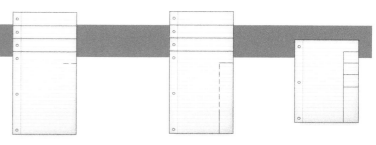

1. Layer multiple sheets of paper so that about 2–3 cm of each can be seen.

2. Make a 2–3-cm horizontal cut through all pages a short distance (3 cm) from the top edge of the top sheet.

3. Make a vertical cut up from the bottom to meet the horizontal cut.

4. Place the sheets on top of an uncut sheet and align the tops and sides of all sheets. Label each tab.

Accordion Book

1. Fold a sheet of paper in half. Fold in half and in half again to form eight sections.

2. Cut along the long fold line, stopping before you reach the last two sections.

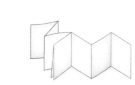

3. Refold the paper into an accordion book. You may want to glue the double pages together.

SUGGESTED PACING (DAYS)

90 min.	7	1	
45 min.	13		2
	Instruction		Review & Assess

Track Your Progress

This chapter focuses on content from the Creating Equations, Interpreting Functions, and Linear, Quadratic, and Exponential Models domains.

THEN	NOW	NEXT
F.BF.3 Identify the effect on the graph of replacing $f(x)$ by $f(x) + k$, $kf(x)$, $f(kx)$, and $f(x + K)$ for specific values of k (both positive and negative); find the value of k given the graphs. Experiment with cases and illustrate an explanation of the effects on the graph using technology. **F.IF.4** For a function that models a relationship between two quantities, interpret key features of graphs and tables in terms of the quantities, and sketch graphs showing key features given a verbal description of the relationship. **F.IF.7b** Graph square root, cube root, and piecewise-defined functions, including step functions and absolute value functions.	**A.CED.1** Create equations and inequalities in one variable and use them to solve problems. **F.IF.7e** Graph exponential and logarithmic functions, showing intercepts and end behavior, and trigonometric functions, showing period, midline, and amplitude. **F.IF.8b** Use the properties of exponents to interpret expressions for exponential functions. **F.LE.4** For exponential models, express as a logarithm the solution to $ab^{ct} = d$ where a, c, and d are numbers and the base b is 2, 10, or e; evaluate the logarithm using technology.	**A.CED.2** Create equations in two or more variables to represent relationships between quantities; graph equations on coordinate axes with labels and scales. **A.APR.7** Understand that rational expressions form a system analogous to the rational numbers, closed under addition, subtraction, multiplication, and division by a nonzero rational expression; add, subtract, multiply, and divide rational expressions. **A.REI.2** Solve simple rational and radical equations in one variable, and give examples showing how extraneous solutions may arise.

Standards for Mathematical Practice

All of the Standards for Mathematical Practice will be covered in this chapter. The MP icon notes specific areas of coverage.

 Teaching the Mathematical Practices
Help students develop the mathematical practices by asking questions like these.

Questioning Strategies As students approach problems in this chapter, help them develop mathematical practices by asking:

Sense-Making
· What is a geometric sequence and how is it used?
· What are natural bases and natural logarithms and how do you evaluate expressions containing them?

Reasoning
· Is it possible to find sums of geometric series? If so, how?
· How are logarithmic expressions evaluated?
· In what way can you simplify and evaluate expressions using the properties of logarithms?

Modeling
· How can real-world data be modeled by exponential growth and exponential decay functions?
· How do you find equations of best fit for data modeled by exponential and logarithmic functions?
· Given a data set, how do you choose the best model?

Precision
· How do you solve exponential equations and inequalities using common logarithms?
· How do you evaluate logarithmic expression using the Change of Base Formula?
· How do you solve logarithmic equations and inequalities?
· How do you use logarithms to solve problems involving logistic growth?

 Go Online!

 StudySync: SMP Modeling Videos

These demonstrate how to apply the Standards for Mathematical Practice to collaborate, discuss, and solve real-world math problems.

Go Online!
connectED.mcgraw-hill.com

 LearnSmart

 The Geometer's Sketchpad

 Vocabulary

 Personal Tutor

 Tools

 Calculator Resources

 Self-Check Practice

 Animations

Customize Your Chapter

Use the *Plan & Present*, *Assignment Tracker*, and *Assessment* tools in ConnectED to introduce lesson concepts, assign personalized practice, and diagnose areas of student need.

Differentiated Instruction

Throughout the program, look for the icons to find specialized content designed for your students.

- **AL** Approaching Level
- **OL** On Level
- **BL** Beyond Level
- **ELL** English Language Learners

Personalize

Differentiated Resources

FOR EVERY CHAPTER	AL	OL	BL	ELL
✓ Chapter Readiness Quizzes	●	●	◐	●
✓ Chapter Tests	●	●	●	●
✓ Standardized Test Practice	●	●	●	●
Vocabulary Review Games	●	●	◐	●
Anticipation Guide (English/Spanish)	●	●	◐	●
Student-Built Glossary	●	●	◐	●
Chapter Project	◐	●	●	●
FOR EVERY LESSON	**AL**	**OL**	**BL**	**ELL**
Personal Tutors (English/Spanish)	●	●	◐	●
Graphing Calculator Personal Tutors	●	●	●	●
▷ Step-by-Step Solutions	●	●	◐	●
✓ Self-Check Quizzes	●	●	●	●
5-Minute Check	●	●	●	●
Study Notebook	●	●	●	●
Study Guide and Intervention	●	●		●
Skills Practice	●	◐		●
Practice	◐	●	●	●
Word Problem Practice	◐	●	●	◐
Enrichment		●	●	●
✚ Extra Examples	●	◐		◐
Lesson Presentations	●	●	●	●

◐ Aligned to this group ● Designed for this group

Engage

Featured IWB Resources

 Geometer's Sketchpad provides students with a tangible, visual way to learn. *Use with Lessons 6-1 and 6-4.*

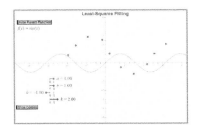

 eLessons engage students and help build conceptual understanding of big ideas. *Use with Lessons 6-1, 6-4, and 6-6.*

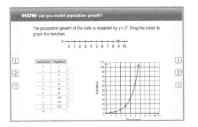

 Animations help students make important connections through motion. *Use with Lessons 6-4 and 6-7.*

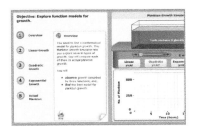

 Time Management How long will it take to use these resources? Look for the clock in each lesson interleaf.

Introduce the Chapter

Mathematical Background

An exponential equation is in the form $y = b^x$, where $b > 0$ and $b \neq 1$. The equation represents exponential growth when $b > 1$ and exponential decay when $0 < b < 1$. The inverse of an exponential function is the logarithmic function.

Essential Question

At the end of this chapter, students should be able to answer the Essential Question.

How can you make good decisions? Sample answer: Determine available options, compare advantages and disadvantages, and analyze consequences.

What factors can affect good decision making? Sample answers: amount of time available, process used, the environment, available options

Apply Math to the Real World

SCIENCE In this activity, students will use what they know about graphing data points to explore how the exponential growth of bacterial populations can be represented graphically. Have students complete this activity individually or in small groups. **MP** 1, 5, 7

CHAPTER 6
Exponential and Logarithmic Functions

THEN
You graphed functions and transformations of functions.

NOW
You will:
- Graph exponential and logarithmic functions.
- Solve exponential and logarithmic equations and inequalities.
- Solve problems involving exponential growth and decay.

MP WHY

SCIENCE Scientists often use aspects of mathematics to quantify their studies and to illustrate their observations.

Use the Mathematical Practices to complete the activity.

1. **Using Tools** Use the Internet to learn about bacterial growth.

2. **Applying Math** Use the Coordinate Plane tool to create a graph that reflects a bacterial growth curve, with the x-axis set as the time interval and the y-axis as the bacterial population. Use the line segment tool to connect the points.

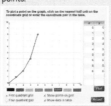

3. **Discuss** Is this the kind of growth typically seen in nature or do other factors impact the population? How might a scientist use mathematics to account for other factors that affect populations?

ALEKS®

Your Student Success Tool ALEKS is an adaptive, personalized learning environment that identifies precisely what each student knows and is ready to learn—ensuring student success at all levels.

- **Formative Assessment:** Dynamic, detailed reports monitor students' progress toward standards mastery.
- **Automatic Differentiation:** Strengthen prerequisite skills and target individual learning gaps.
- **Personalized Instruction:** Supplement in-class instruction with personalized assessment and learning opportunities.

Go Online!

Chapter Project
The Population Puzzle Students use what they have learned about exponential and logarithmic functions to complete a project.
This chapter project addresses environmental literacy, as well as several specific skills identified as being essential to student success by the Framework for 21st Century Learning. **MP** 1, 3, 4

Go Online to Guide Your Learning

Explore & Explain	Organize

The Geometer's Sketchpad

Use **The Geometer's Sketchpad** to illustrate how to graph exponential functions in Lesson 6-1 and to explore geometric sequences in Lesson 6-3.

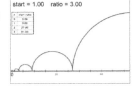

Graphing Tools: Logarithmic Functions

Use **Graphing Tools: Logarithmic Functions** to explore logarithmic functions and to enhance your understanding of logarithmic equations and inequalities in Lesson 6-9.

Foldables

Get organized! Create an **Exponential and Logarithmic Functions Foldable** before you start the chapter to arrange your notes.

Collaborate

Chapter Project

In the **The Population Puzzle** project, you will use what you have learned about exponential and logarithmic functions to complete a project that addresses environmental literacy.

eBook

Interactive Student Guide

Before starting the chapter, answer the **Chapter Focus** preview questions. Check your answers as you complete each lesson. At the end of the chapter, try the **Performance Task**.

Focus

 LEARNSMART

Need help studying? Complete the **Modeling with Functions** domain in LearnSmart to review for the chapter test.

ALEKS

You can use the **Exponential and LogarithmicFunctions** topic in ALEKS to explore what you know about relations and functions and what you are ready to learn.*

* Ask your teacher if this is part of your program.

Dinah Zike's **FOLDABLES**

Focus As students work through the lessons in this chapter, they write notes about exponential and logarithmic functions.

Teach Have students make and label their Foldables as illustrated. Have students label one page of their Foldables for every two lessons in the chapter and use the appropriate pages as they cover the material. Have students list the Key Concepts and the vocabulary terms and their definitions in their Foldables. Point out that the Foldables can also be used to record positive and negative experiences during learning.

When to Use It Encourage students to add to their Foldables as they work through the chapter and to use them to review for the chapter test.

Go Online!

Choosing Foldables

How do you know which Foldables strategy to use? In this video, you will learn best practices to use when choosing Foldables. **MP** 1, 5

Get Ready for the Chapter

RtI Response to Intervention

Use the Concept Check results and the Intervention Planner chart to help you determine your Response to Intervention.

Intervention Planner

TIER 1 On Level OL

IF students miss 25% of the exercises or less,

THEN choose a resource:

Go Online!

📄 Skills Practice, Chapter 5 and Chapter 6

📄 Chapter Project

TIER 2 Approaching Level AL

IF students miss 50% of the exercises,

THEN choose a resource:

Go Online!

📄 Study Guide and Intervention, Chapter 4 and Chapter 5

➕ Extra Examples

💬 Personal Tutors

📄 Homework Help

Quick Review Math Handbook

TIER 3 Intensive Intervention

IF students miss 75% of the exercises,

THEN use *Math Triumphs, Alg. 2*

Go Online!

➕ Extra Examples

💬 Personal Tutors

📄 Homework Help

🔤 Review Vocabulary

Additional Answers

1. add the exponent values together to simplify as a^{12}
2. Power of a Power Rule
3. Product of Powers Rule
4. Quotient of Powers Rule
5. no, not inverse functions

Get Ready for the Chapter

Go Online! for Vocabulary Review Games and key vocabulary in 13 languages.

Connecting Concepts	New Vocabulary		

Concept Check

Review the concepts used in this chapter by answering the questions below. 1–8. See margin.

1. Given $a^4a^3a^5$, where a does not equal zero, what would you do to simplify the expression?

2. Given $\dfrac{(a^3bc^2)^2}{a^4a^2b^2bc^5c^3}$, where no variable equals zero, what rule can be applied to simplify the numerator?

3. Given $\dfrac{(a^3bc^2)^2}{a^4a^2b^2bc^5c^3}$, where no variable equals zero, what rule can be applied to simplify the denominator?

4. Given $\dfrac{a^6b^2c^4}{a^6b^3c^8}$, where no variable equals zero, what rule can be applied to simplify this expression?

5. Are $f(x) = 2x + 5$ and $g(x) = 2x - 5$ inverse functions?

6. Are $f(x) = x - 6$ and $g(x) = x + 6$ inverse functions?

7. In the graph shown, one line is $f(x) = \dfrac{x-1}{2}$ and the other is the inverse of that function. Which line is the function and which is its inverse?

8. What are the steps to find the inverse of a function?

Performance Task Preview

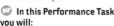

You can use the concepts and skills in the chapter to perform data analysis for a United Nations committee to determine the health of various nations' economies, populations, and food supplies. Understanding exponential and logarithmic functions will help you finish the Performance Task at the end of the chapter.

🔵 **In this Performance Task you will:**

• make sense of problems and persevere in solving them
• model with mathematics
• attend to precision

New Vocabulary

English		Español
exponential growth	p. 373	crecimiento exponencial
asymptote	p. 373	asíntota
growth factor	p. 375	factor de crecimiento
exponential decay	p. 375	desintegración exponencial
decay factor	p. 376	factor de desintegración
exponential equation	p. 383	ecuación exponencial
compound interest	p. 384	interés compuesto
exponential inequality	p. 385	desigualdad exponencial
geometric mean	p. 391	media geométrica
geometric series	p. 392	serie geométrica
logarithm	p. 397	logaritmo
logarithmic function	p. 398	función logarítmica
regression line	p. 408	reca de regresión
correlation coefficient	p. 409	coeficiente de correlación
common logarithm	p. 423	logaritmos communes
Change of Base Formula	p. 425	fórmula del cambio de base
natural base, e	p. 430	e base natural
natural logarithm	p. 430	logaritmo natural
logarithmic equation	p. 437	ecuación logarítmica
logarithmic inequality	p. 438	desigualdad logarítmica
rate of continuous growth	p. 445	el índice del crecimiento continuo
rate of continuous decay	p. 445	índice de desintegración continúa
logistic growth model	p. 448	modelo logístico del crecimiento

Review Vocabulary

domain dominio the set of all x-coordinates of the ordered pairs of a relation

function función a relation in which each element of the domain is paired with exactly one element in the range

range rango the set of all y-coordinates of the ordered pairs of a relation

{(−3, 1), (0, 2), (2, 4)}

Domain → Range

Key Vocabulary ELL

Introduce the key vocabulary in the chapter using the routine below.

Define In the function $x = b^y$, y is called the logarithm, base b, of x. Usually it is written as $y = \log_b x$ and is read "y equals log base b of x."

Example $3 = \log_2 8$

Ask How would you write the following in logarithmic form? $3^5 = 243$

$\log_3 243 = 5$

6. yes, inverse functions

7. line A is the function and line B is its inverse

8. replace $f(x)$ with y in the equation, interchange x and y, solve for y, and replace y with $f^{-1}(x)$

Graphing Exponential Functions

Track Your Progress

Objectives

1 Graph exponential growth functions.

2 Graph exponential decay functions.

Mathematical Background

The function $f(x) = b^x$, where b is a positive real number and $b \neq 1$, is an exponential function. When $b > 1$, the function has no x-intercepts and one y-intercept. It is an increasing function with a horizontal asymptote (the x-axis). When $0 < b < 1$, the function has no x-intercepts and one y-intercept. It is a decreasing function with a horizontal asymptote (the x-axis).

THEN	NOW	NEXT
A.REI.11 Explain why the x-coordinates of the points where the graphs of the equations $y = f(x)$ and $y = g(x)$ intersect are the solutions of the equation $f(x) = g(x)$; find the solutions approximately, e.g., using technology to graph the functions, make tables of values, or find successive approximations. Include cases where $f(x)$ and/or $g(x)$ are linear, polynomial, rational, absolute value, exponential, and logarithmic functions.*	**F.IF.4** For a function that models a relationship between two quantities, interpret key features of graphs and tables in terms of the quantities, and sketch graphs showing key features given a verbal description of the relationship.	**A.CED.1** Create equations and inequalities in one variable and use them to solve problems.

Go Online! All of these resources and more are available at connectED.mcgraw-hill.com

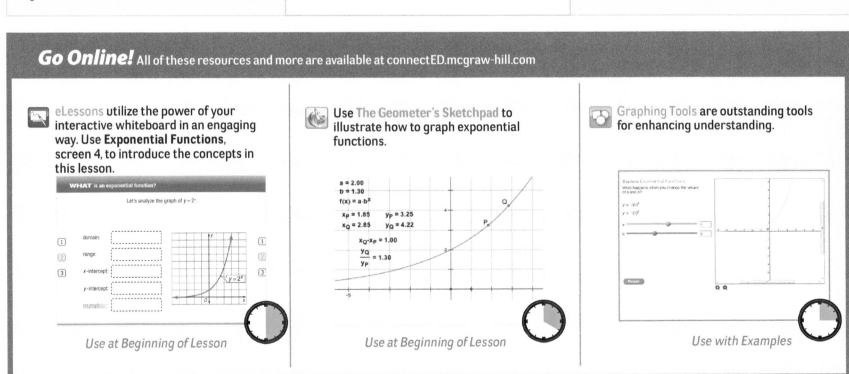

eLessons utilize the power of your interactive whiteboard in an engaging way. Use **Exponential Functions**, screen 4, to introduce the concepts in this lesson.

Use at Beginning of Lesson

Use **The Geometer's Sketchpad** to illustrate how to graph exponential functions.

Use at Beginning of Lesson

Graphing Tools are outstanding tools for enhancing understanding.

Use with Examples

OER **Using Open Educational Resources**

Games Look at the Growth and Decay Stations Game on **scribd.com** for cool new ways to introduce exponential growth and decay. Set this game up as five different stations around the room with your class divided into five small groups. *Use as classwork*

Worksheets

Differentiate Your Resources

Extra Practice Additional practice or homework; Skills Practice is best for approaching-level students and Practice is best for on-level and beyond-level students

Skills Practice

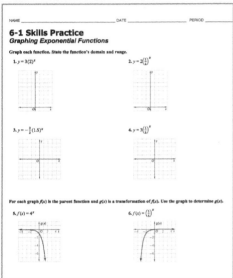

Practice

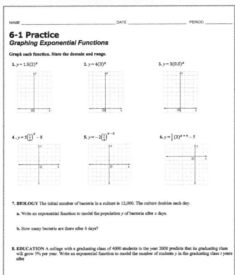

Word Problem Practice

Intervention Reteaching and vocabulary activities that can be used with struggling or absent students and as ELL support

Study Guide and Intervention

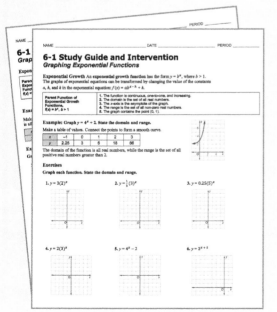

Study Notebook

Extension Activities that can be used to extend lesson concepts

Enrichment

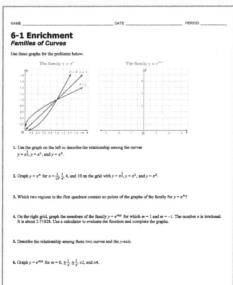

LESSON 1

Graphing Exponential Functions

:Then	:Now	:Why?
You graphed polynomial functions.	**1** Graph exponential growth functions. **2** Graph exponential decay functions.	If you see a funny video online and share it with 100 of your friends, and each of your friends shares the video with 100 of their friends, the number of people who see the video increases exponentially until you can say that the video has gone viral. The equation $y = 100^x$ can be used to represent this situation where x is the number of rounds that the video has been shared.

New Vocabulary
exponential function
exponential growth
asymptote
growth factor
exponential decay
decay factor

 Mathematical Practices
2 Reason abstractly and quantitatively.

Content Standards
A.CED.2 Create equations in two or more variables to represent relationships between quantities; graph equations on coordinate axes with labels and scales.
F.IF.4 For a function that models a relationship between two quantities, interpret key features of graphs and tables in terms of the quantities, and sketch graphs showing key features given a verbal description of the relationship.
F.IF.7e Graph exponential and logarithmic functions, showing intercepts and end behavior, and trigonometric functions, showing period, midline, and amplitude.
F.IF.8.b Use the properties of exponents to interpret expressions for exponential functions.
F.BF.3 Identify the effect on the graph of replacing $f(x)$ by $f(x) + k$, $k\,f(x)$, $f(kx)$, and $f(x + k)$ for specific values of k (both positive and negative); find the value of k given the graphs. Experiment with cases and illustrate an explanation of the effects on the graph using technology. Include recognizing even and odd functions from their graphs and algebraic expressions for them.

1 Exponential Growth A function like $y = 100^x$, where the base is a constant and the exponent is the independent variable, is an **exponential function**. An **exponential growth** function is a function of the form $f(x) = b^x$, where $b > 1$. The graph of an exponential function has an **asymptote**, which is a line that the graph approaches.

Key Concept Parent Function of Exponential Growth Functions

Parent Functions:	$f(x) = b^x$, $b > 1$
Type of graph:	continuous, one-to-one, and increasing
Domain:	$(-\infty, \infty)$, {all real numbers}, or $\{-\infty < x < \infty\}$
Range:	$(0, \infty)$, $\{f(x) \mid f(x) > 0\}$, or $\{0 < x < \infty\}$
Asymptote:	x-axis
Intercept:	$(0, 1)$
Symmetry:	none **Extrema:** none

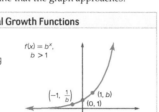

F.IF.7.e

Example 1 Graph Exponential Growth Functions

Graph $y = 3^x$. State the domain and range.

Make a table of values. Then plot the points and sketch the graph.

x	-3	-2	$-\frac{1}{2}$	0
$y = 3^x$	$3^{-3} = \frac{1}{27}$	$3^{-2} = \frac{1}{9}$	$3^{-\frac{1}{2}} = \frac{\sqrt{3}}{3}$	$3^0 = 1$

x	1	$\frac{3}{2}$	2
$y = 3^x$	$3^1 = 3$	$3^{\frac{3}{2}} = \sqrt{27}$	$3^2 = 9$

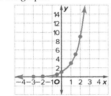

The domain is all real numbers, and the range is all positive real numbers.

▶ **Guided Practice**

1. Graph $y = 4^x$. State the domain and range. **See Ch. 6 Answer Appendix.**

 Mathematical Practices Strategies

Reason abstractly and quantitatively.
Help students predict the graphs of exponential functions by examining the function. For example, ask:

• How do you determine whether the function $y = b^x$ is increasing or decreasing? If the base is less than 1, the function is decreasing. If the base is greater than 1, the function is increasing.

• What is the y-intercept? the output when the input is 0

• How do you determine the equation for the horizontal asymptote? Add the vertical shift to 0 and set it equal to y.

Launch

Have students read the Why? section of the lesson. Ask:

• How many people will receive the video in the second round? 10,000

• How many people will receive the video in the fourth round? 100,000,000

• Consider a scenario similar to the one in the Why? section, only this video tells you to forward it on to 8 friends. Write an equation to represent this situation. $y = 8^x$

Teach

Ask the scaffolded questions for each example to build conceptual understanding for students at all levels.

1 Exponential Growth

Example 1 Graph Exponential Growth Functions

AL What is the y-intercept of $y = 3^x$? $(0, 1)$

OL What is the end behavior of $y = 3^x$? $y \to 0$ as $x \to -\infty$, $y \to \infty$ as $x \to \infty$

BL Explain why $y < 1$ for all $x < 0$ for the equation $y = 3^x$. If $x < 0$, then $y = \frac{1}{3^{|x|}}$ which is always less than 1.

Need Another Example?
Graph $y = 2^x$. State the domain and range.

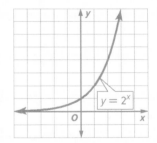

$D = (-\infty, +\infty)$ or {all real numbers}; $R = (0, +\infty)$ or $\{y \mid y > 0\}$

 Go Online!

Interactive Whiteboard
Use the *eLesson* or *Lesson Presentation* to present this lesson.

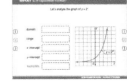

Example 2 Graph Transformations

AL Is $y = -b^x$ increasing or decreasing? decreasing

OL What is the equation that translates $y = 2^x$ to the right 3 units and down 5 units?
$y = 2^{(x-3)} - 5$

BL What is the range for any exponential function of the form $y = b^x + k$? (k, ∞)

Need Another Example?

Graph each function. State the domain and range.

a. $y = 3^x - 2$

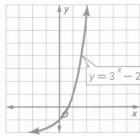

$D = (-\infty, +\infty)$, {all real numbers}, or $\{-\infty < x < +\infty\}$;
$R = (-2, +\infty)$, $\{y \mid y > -2\}$, or $\{-2 < y < +\infty\}$

b. $y = 2^{x-1}$

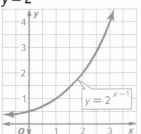

$D = (-\infty, +\infty)$, {all real numbers}, or $\{-\infty < x < +\infty\}$;
$R = (0, +\infty)$, $\{y \mid y > 0\}$, or $\{0 < y < +\infty\}$

The graph of $f(x) = b^x$ represents a parent graph of the exponential functions. The same techniques used to transform the graphs of other functions you have studied can be applied to the graphs of exponential functions.

Key Concept Transformations of Exponential Functions

$$f(x) = ab^{x-h} + k$$

h — Horizontal Translation	k — Vertical Translation
h units right if h is positive	k units up if k is positive
$\lvert h \rvert$ units left if h is negative	$\lvert k \rvert$ units down if k is negative

a — Orientation and Shape	
If $a < 0$, the graph is reflected in the x-axis.	If $\lvert a \rvert > 1$, the graph is stretched vertically. If $0 < \lvert a \rvert < 1$, the graph is compressed vertically.

F.IF.7.e

Example 2 Graph Transformations

Graph each function. State the domain and range.

a. $y = 2^x + 1$

The equation represents a translation of the graph of $y = 2^x$ one unit up.

x	$y = 2^x + 1$
−3	$2^{-3} + 1 = 1.125$
−2	$2^{-2} + 1 = 1.25$
−1	$2^{-1} + 1 = 1.5$
0	$2^0 + 1 = 2$
1	$2^1 + 1 = 3$
2	$2^2 + 1 = 5$
3	$2^3 + 1 = 9$

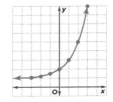

$D = (-\infty, \infty)$, {all real numbers}, or $\{-\infty < x < \infty\}$;
$R = (1, \infty)$, $\{f(x) \mid f(x) > 1\}$, or $\{1 < x < \infty\}$

b. $y = -\frac{1}{2} \cdot 5^{x-2}$

The equation represents a transformation of the graph of $y = 5^x$.

Graph $y = 5^x$ and transform the graph.

- $a = -\frac{1}{2}$: The graph is reflected in the x-axis and compressed vertically.
- $h = 2$: The graph is translated 2 units right.
- $k = 0$: The graph is not translated vertically.

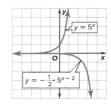

$D = (-\infty, \infty)$, {all real numbers}, or $\{-\infty < x < \infty\}$;
$R = (-\infty, 0)$ or $\{f(x) \mid f(x) < 0\}$, or $\{-\infty < x < 0\}$;

Guided Practice **2A, 2B.** See Ch. 6 Answer Appendix.

2A. $y = 2^{x+3} - 5$ **2B.** $y = 0.1(6)^x - 3$

Study Tip

MP Precision Remember that end behavior is the action of the graph as x approaches positive infinity or negative infinity. In Example **2a**, as x approaches infinity, y approaches infinity. In Example **2b**, as x approaches infinity, y approaches negative infinity.

Go Online!

Investigate how changing the values in the equation of an exponential function affects the graph by using the Graphing Tools in ConnectED.

MP **Teaching the Mathematical Practices**

Precision Mathematically proficient students try to communicate precisely, using clear definitions in discussion with others and in their own reasoning.

Go Online!

The most up-to-date resources available for your program can be found at connectED.mcgraw-hill.com.

Name_____

Graphing Exponential Functions

Date_____ Period_____

Sketch the graph of each function.

1) $y = 4 \cdot 2^x$

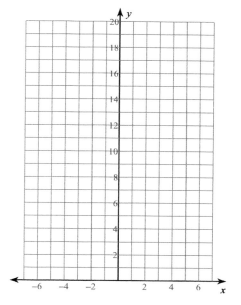

2) $y = 5 \cdot 2^x$

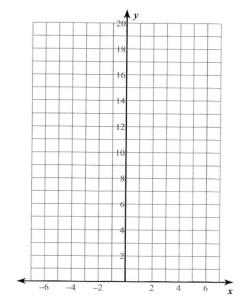

3) $y = 4 \cdot \left(\dfrac{1}{2}\right)^x$

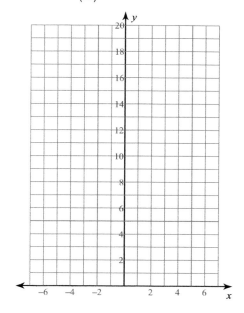

4) $y = 2 \cdot \left(\dfrac{1}{2}\right)^x$

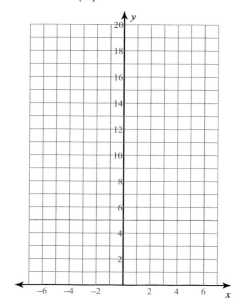

5) $y = 3 \cdot 2^{x-2} + 2$

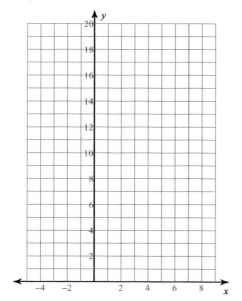

6) $y = 4 \cdot \left(\dfrac{1}{2}\right)^{x-1} - 2$

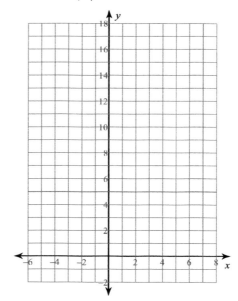

Write an equation for each graph.

7)

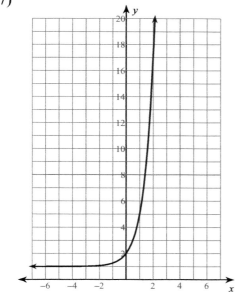

8)

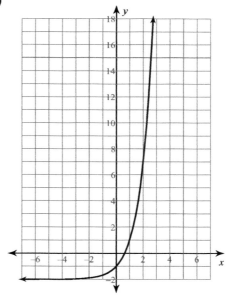

Name_____

Graphing Exponential Functions

Date_____ Period_____

Sketch the graph of each function.

1) $y = 4 \cdot 2^x$

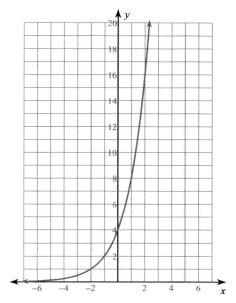

2) $y = 5 \cdot 2^x$

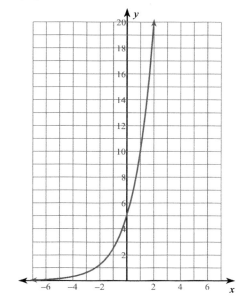

3) $y = 4 \cdot \left(\dfrac{1}{2}\right)^x$

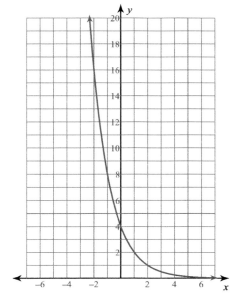

4) $y = 2 \cdot \left(\dfrac{1}{2}\right)^x$

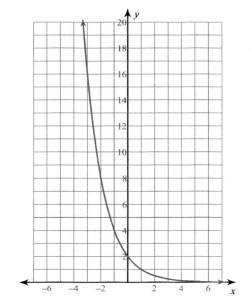

5) $y = 3 \cdot 2^{x-2} + 2$

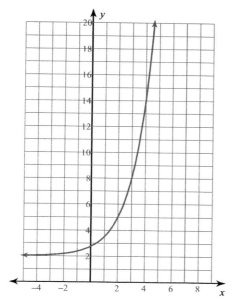

6) $y = 4 \cdot \left(\dfrac{1}{2}\right)^{x-1} - 2$

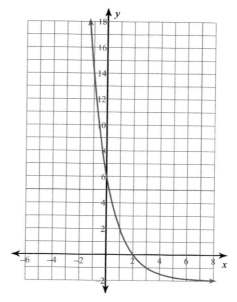

Write an equation for each graph.

7)

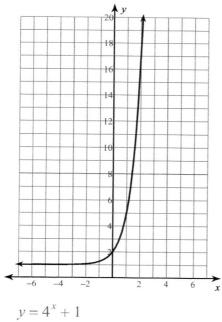

$y = 4^x + 1$

8)

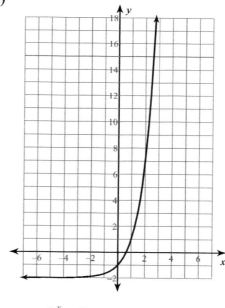

$y = 3^x - 2$

You can model exponential growth with a constant percent increase over specific time periods using the following function.

$$A(t) = a(1 + r)^t$$

The function can be used to find the amount $A(t)$ after t time periods, where a is the initial amount and r is the percent of increase per time period. Note that the base of the exponential expression, $1 + r$, is called the **growth factor**.

The exponential growth function is often used to model population growth.

F.IF.4, F.IF.7.e

Real-World Example 3 Graph Exponential Growth Functions

CENSUS The first U.S. census was conducted in 1790. At that time, the population was 3,929,214. Since then, the U.S. population has grown by approximately 2.03% annually. Draw a graph showing the population growth of the United States since 1790.

First, write an equation using $a = 3,929,214$, and $r = 0.0203$.

$y = 3,929,214(1.0203)^t$

Then graph the equation.

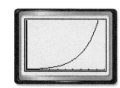

[0, 250] scl: 25 by [0, 400,000,000]
scl: 40,000,000

▶ Guided Practice 3. See Ch. 6 Answer Appendix.

3. FINANCIAL LITERACY Teen spending is expected to grow 3.5% annually from $208.7 billion in 2012. Draw a graph to show the spending growth.

2 Exponential Decay The second type of exponential function is **exponential decay**.

Key Concept Parent Function of Exponential Decay Functions

Parent Functions:	$f(x) = b^x, 0 < b < 1$
Type of graph:	continuous, one-to-one, and decreasing
Domain:	$(-\infty, \infty)$, {all real numbers}, or $\{-\infty < x < \infty\}$
Range:	$(0, \infty)$, $\{f(x) \mid f(x) > 0\}$, or $\{0 < x < \infty\}$
Asymptote:	x-axis
Intercept:	$(0, 1)$

Model
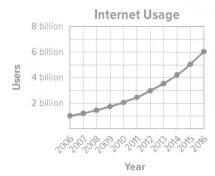

The graphs of exponential decay functions can be transformed in the same manner as those of exponential growth.

Real-World Link
The U.S. Census Bureau's American Community Survey is mailed to approximately 3.5 million randomly selected households.
Source: Census Bureau

Study Tip
Interest The formula for simple interest, $i = prt$, illustrates linear growth over time. However, the formula for compound interest, $A(t) = a(1 + r)^t$, illustrates exponential growth over time. This is why investments with compound interest make more money.

Example 3 Graph Exponential Growth Functions

AL **Use this equation to verify that the population in 1790 was 3,929,214.** Substitute 0 for t:
$3,929,214(1.0203)^0 = 3,929,214$.

OL **What does it mean in this context for the function to be increasing?** The population is increasing as time goes on.

BL **Use this function to write a new function that begins at the year 2013.**
$y = 625,996,637.7(1.0203)^t$

Need Another Example?

Internet In 2006, there were 1,020,000,000 people worldwide using the Internet. At that time, the number of users was growing by 19.5% annually. Draw a graph showing how the number of users would grow from 2006 to 2016 if that rate continued.

Internet Usage

Watch Out!

Common Misconceptions Be sure students do not confuse polynomial functions and exponential functions. While $y = x^2$ and $y = 2^x$ each have an exponent, $y = x^2$ is a polynomial function, and $y = 2^x$ is an exponential function.

Differentiated Instruction OL BL ELL

Verbal/Linguistic Learners Ask students where they have heard the term *exponential* before and what they think it might mean. Students may have heard terms like *exponential growth* on a television news program and they might think that exponential means "enormous." Use students' answers to introduce the concept of exponential functions.

2 Exponential Decay

Example 4 Graph Exponential Decay Functions

AL Why is $\left(\frac{1}{3}\right)^{-3} = 27$? $\left(\frac{1}{3}\right)^{-3} = \frac{1}{\left(\frac{1}{3}\right)^3} = \frac{1}{\frac{1}{27}} = 27$

OL What transformation is done to $y = 3^x$ to get $= \left(\frac{1}{3}\right)^x$? reflection in the y-axis

BL Write a rule about the end behavior of any equation of the form $y = b^x$ where $b > 0$.
If $b < 1$: $y \to \infty$ as $x \to -\infty$, $y \to 0$ as $x \to \infty$. If $b = 1$ the function is constant. If $b > 1$ $y \to 0$ as $x \to -\infty$, $y \to \infty$ as $x \to \infty$.

Need Another Example?

Graph each function. State the domain and range.

a. $y = \left(\frac{1}{5}\right)^x$

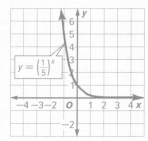

$D = (-\infty, \infty)$, {all real numbers}, or $\{-\infty < x < +\infty\}$;
$R = (0, +\infty)$, $\{y \mid y > 0\}$, or $\{0 < y < +\infty\}$

b. $y = -4\left(\frac{1}{2}\right)^{x-1} + 2$

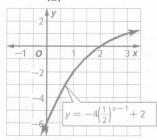

$D = (-\infty, +\infty)$, {all real numbers}, or $\{-\infty < x < +\infty\}$;
$R = (2, +\infty)$, $\{y \mid y < 2\}$, or $\{2 < y < +\infty\}$

Additional Answers (Guided Practice)

4A.

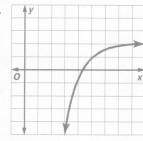

$D = (-\infty, +\infty)$, {all real numbers}, or $\{-\infty < x < +\infty\}$;
$R = (-\infty, 2)$, $\{y \mid y < 2\}$, or $\{-\infty < y < 2\}$

F.IF.7e, F.BF.3

Example 4   Graph Exponential Decay Functions

Graph each function. State the domain and range.

a. $y = \left(\frac{1}{3}\right)^x$

x	$y = \left(\frac{1}{3}\right)^x$
-3	$\left(\frac{1}{3}\right)^{-3} = 27$
-2	$\left(\frac{1}{3}\right)^{-2} = 9$
$-\frac{1}{2}$	$\left(\frac{1}{3}\right)^{-\frac{1}{2}} = \sqrt{3}$
0	$\left(\frac{1}{3}\right)^0 = 1$
1	$\left(\frac{1}{3}\right)^1 = \frac{1}{3}$
$\frac{3}{2}$	$\left(\frac{1}{3}\right)^{\frac{3}{2}} = \sqrt{\frac{1}{27}}$
2	$\left(\frac{1}{3}\right)^2 = \frac{1}{9}$

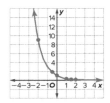

$D = (-\infty, \infty)$, {all real numbers}, or $\{-\infty < x < \infty\}$;
$R = (0, \infty)$, $\{f(x) \mid f(x) > 0\}$, or $\{0 < x < \infty\}$

b. $y = 2\left(\frac{1}{4}\right)^{x+2} - 3$

The equation represents a transformation of the graph of $y = \left(\frac{1}{4}\right)^x$.

Examine each parameter.

- $a = 2$: The graph is stretched vertically.
- $h = -2$: The graph is translated 2 units left.
- $k = -3$: The graph is translated 3 units down.

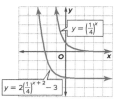

$D = (-\infty, \infty)$, {all real numbers}, or $\{-\infty < x < \infty\}$;
$R = (-3, \infty)$, $\{f(x) \mid f(x) > -3\}$, or $\{-3 < x < \infty\}$

Guided Practice   4A, 4B. See margin.

4A. $y = -3\left(\frac{2}{5}\right)^{x-4} + 2$ **4B.** $y = \frac{3}{8}\left(\frac{5}{6}\right)^{x-1} + 1$

Similar to exponential growth, you can model exponential decay with a constant percent of decrease over specific time periods using the following function.

$$A(t) = a(1 - r)^t$$

The base of the exponential expression, $1 - r$, is called the **decay factor**.

4B.

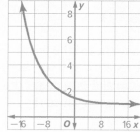

$D = (-\infty, +\infty)$, {all real numbers}, or $\{-\infty < x < +\infty\}$;
$R = (1, +\infty)$, $\{y \mid y > 1\}$, or $\{1 < y < +\infty\}$

5.

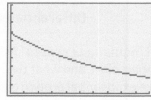

[0, 10] scl: 1 by [0, 100] scl: 10

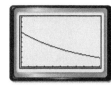

F.IF.4, F.IF.7e, F.BF.3

Real-World Example 5   Graph Exponential Decay Functions

TEA A cup of green tea contains 35 milligrams of caffeine. The average teen can eliminate approximately 12.5% of the caffeine from their system per hour.

a. Draw a graph to represent the amount of caffeine remaining after drinking a cup of green tea.

$$y = a(1 - r)^t$$
$$= 35(1 - 0.125)^t$$
$$= 35(0.875)^t$$

Graph the equation.

[0, 10] scl: 1 by [0, 50] scl: 5

b. Estimate the amount of caffeine in a teenager's body 3 hours after drinking a cup of green tea.

$$y = 35(0.875)^t$$ Equation from part a
$$= 35(0.875)^3$$ Replace t with 3
$$\approx 23.45$$ Use a calculator

The caffeine in a teenager will be about 23.45 milligrams after 3 hours.

Real-World Link

After water, tea is the most consumed beverage in the United States. It can be found in almost all American households. Just over half the American population drinks tea daily.

Source: The Tea Association of the USA

▶ **Guided Practice**

5. A cup of black tea contains about 68 milligrams of caffeine. Draw a graph to represent the amount of caffeine remaining in the body of an average teen after drinking a cup of black tea. Estimate the amount of caffeine in the body 2 hours after drinking a cup of black tea. **See margin for graph; 52.06 mg**

Check Your Understanding   ◯ = Step-by-Step Solutions begin on page R11.

✓ **Go Online!** for a Self-Check Quiz

Examples 1–2   **Graph each function. State the domain and range.** 1–6. See Ch. 6 Answer Appendix.
F.IF.7e
1. $f(x) = 2^x$
2. $f(x) = 5^x$
3. $f(x) = 3^{x-2} + 4$
4. $f(x) = 2^{x+1} + 3$
5. $f(x) = 0.25(4)^x - 6$
6. $f(x) = 3(2)^x + 8$

Example 3
F.IF.4,
F.IF.7e
7. ⓜⓓ **SENSE-MAKING** A virus spreads through a network of computers such that each minute, 25% more computers are infected. If the virus began at only one computer, graph the function for the first hour of the spread of the virus. See margin.

Example 4
F.IF.7e,
F.BF.3
Graph each function. State the domain and range. 8–11. See Ch. 6 Answer Appendix.
8. $f(x) = 2\left(\frac{2}{3}\right)^{x-3} - 4$
9. $f(x) = -\frac{1}{2}\left(\frac{3}{4}\right)^{x+1} + 5$
10. $f(x) = -\frac{1}{3}\left(\frac{4}{5}\right)^{x-4} + 3$
11. $f(x) = \frac{1}{8}\left(\frac{1}{4}\right)^{x+6} + 7$

Example 5
F.IF.4,
F.IF.7e,
F.BF.3
12. **FINANCIAL LITERACY** A new SUV depreciates in value each year by a factor of 15%. Draw a graph of the SUV's value for the first 20 years after the initial purchase. See margin.

All New Only $30,000

Additional Answers

7.

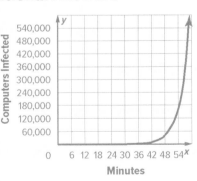

Minutes

12.

Years After Purchase

Extra Practice

See page R6 for extra exercises for students who are approaching level or for on-level students who need additional reinforcement.

Example 5 Graph Exponential Decay Functions

AL What does the *y*-intercept represent? the initial amount of caffeine in the system of the teenager

OL What inequality would you solve to find when the teenager has less than 20 milligrams of caffeine remaining in their system?
$35(0.875)^t < 20$

BL Why doesn't this model represent the situation perfectly? The model represents the amount of caffeine remaining on average. There will be a large variance among actual teenagers.

Need Another Example?

Air Pressure The pressure of the atmosphere is 14.7 lb/in² at Earth's surface. It decreases by about 20% for each mile of altitude up to about 50 miles.

a. Draw a graph to represent atmospheric pressure for altitude from 0 to 50 miles.

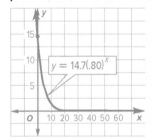

$y = 14.7(.80)^x$

b. Estimate the atmospheric pressure at an altitude of 10 miles. 1.6 lb/in²

Practice

Formative Assessment Use Exercises 1–12 to assess students' understanding of the concepts in this lesson.

The Practice and Problem Solving exercises assess the content taught in the lesson. The Preparing for Assessment page is meant to be used as preparation for end-of-course assessments.

Go Online!   eBook

Interactive Student Guide

Use the *Interactive Student Guide* to deepen conceptual understanding.
· Graphing Exponential Functions

ALGEBRA 2
INTERACTIVE STUDENT GUIDE

Levels of Complexity Chart

The levels of the exercises progress from 1 to 3, with Level 1 indicating the lowest level of complexity.

Exercises	13–25	26–29, 39–45	30–38
▷ Level 3			●
▷ Level 2		○	
Level 1	●		

Teaching the Mathematical Practices

Reasoning Mathematically proficient students make sense of quantities and their relationships in problem situations. Quantitative reasoning entails habits of creating a coherent representation of the problem at hand; considering the units involved; attending to the meaning of quantities, not just how to compute them; and knowing and having flexibility with using different properties of operations and objects.

Follow-Up

Students have explored exponential functions.
Ask:

- How can being financially literate help you to make good decisions? Sample answer: If you are financially literate, you understand the vocabulary of financial terms and know how to analyze data and trends. Successfully applying these skills when considering your available options can help you to make good decisions in many real-world situations such as opening a bank account, applying for college loans, and buying a house.

Additional Answer

26.

Practice and Problem Solving

Extra Practice is on page R6.

Examples 1–2
F.IF.7.e

Graph each function. State the domain and range. 13–18. See Ch. 6 Answer Appendix.

13. $f(x) = 2(3)^x$
14. $f(x) = -2(4)^x$
15. $f(x) = 4^{x+1} - 5$
16. $f(x) = 3^{2x} + 1$
17. $f(x) = -0.4(3)^{x+2} + 4$
18. $f(x) = 1.5(2)^x + 6$

Example 3
F.IF.4,
F.IF.7.e

19. **SCIENCE** The population of a colony of beetles grows 30% each week for 10 weeks. If the initial population is 65 beetles, graph the function that represents the situation. See Ch. 6 Answer Appendix.

Example 4
F.IF.7.e,
F.BF.3

Graph each function. State the domain and range. 20–25. See Ch. 6 Answer Appendix.

20. $f(x) = -4\left(\frac{3}{5}\right)^{x+4} + 3$
21. $f(x) = 3\left(\frac{2}{5}\right)^{x-3} - 6$
22. $f(x) = \frac{1}{2}\left(\frac{1}{5}\right)^{x+5} + 8$
23. $f(x) = \frac{3}{4}\left(\frac{2}{3}\right)^{x+4} - 2$
24. $f(x) = -\frac{1}{2}\left(\frac{3}{8}\right)^{x+2} + 9$
25. $f(x) = -\frac{5}{4}\left(\frac{4}{5}\right)^{x+4} + 2$

Example 5
F.IF.4,
F.IF.7.e,
F.BF.3

26. ▷ **ATTENDANCE** The attendance for a basketball team declined at a rate of 5% per game throughout a losing season. Graph the function modeling the attendance if 15 home games were played and 23,500 people were at the first game. See margin.

27. **SKATING** The function $P(x) = 3.9(0.9^x)$ can be used to model the number of inline skaters in millions x years since 2004. a–b. See margin.
   a. Classify the function as either exponential *growth* or *decay*, and identify the growth or decay factor. Then graph the function.
   b. Explain what the $P(x)$-intercept and the asymptote represent in this situation.

28. **HEALTH** Each day, 10% of a certain drug dissipates from the system.
   a. Classify the function representing this situation as either exponential *growth* or *decay*, and identify the growth or decay factor. Then graph the function. See margin.
   b. How much of the original amount remains in the system after 9 days? a little less than 40%
   c. If a second dose should not be taken if more than 50% of the original amount is in the system, when should the label say it is safe to redose? Design the label and explain your reasoning. Sample answer: The 7th day; see students' work.

29. ⓂⓅ **REASONING** A sequence of numbers follows a pattern in which the next number is 125% of the previous number. The first number in the pattern is 18.
   a. Write the function that represents the situation. $f(x) = 18(1.25)^{x-1}$
   b. Classify the function as either exponential *growth* or *decay*, and identify the growth or decay factor. Then graph the function for the first 10 numbers. See margin.
   c. What is the value of the tenth number? Round to the nearest whole number. 134

▷ **For each graph, $f(x)$ is the parent function and $g(x)$ is a transformation of $f(x)$. Use the graph to determine the equation of $g(x)$.** 31. $g(x) = 4(2)^{x-3}$ or $g(x) = \frac{1}{2}(2^x)$

30. $f(x) = 3^x$ $g(x) = 3^{x-4} + 5$
31. $f(x) = 2^x$
32. $f(x) = 4^x$ $g(x) = -2(4)^{x+1} + 3$

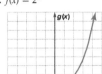 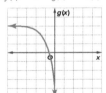

Differentiated Homework Options

Levels	🅐🅛 Basic	🅞🅛 Core	🅑🅛 Advanced
Exercises	13–25, 34–45	13–25 odd, 26–29, 31, 33–45	26–45
2-Day Option	13–25 odd, 39–45	13–25, 39–45	
	14–24 even, 34–38	26–38	

 You can use ALEKS to provide additional remediation support with personalized instruction and practice.

33 **MULTIPLE REPRESENTATIONS** In this problem, you will use the tables below for exponential functions $f(x)$, $g(x)$, and $h(x)$. **a, b. See Ch. 6 Answer Appendix.**

x	−1	0	1	2	3	4	5
f(x)	2.5	2	1	−1	−5	−13	−29

x	−1	0	1	2	3	4	5
g(x)	5	11	23	47	95	191	383

x	−1	0	1	2	3	4	5
h(x)	3	2.5	2.25	2.125	2.0625	2.0313	2.0156

a. **Graphical** Graph the functions for $-1 \leq x \leq 5$ on separate graphs.

b. **Verbal** List any function with a negative coefficient. Explain your reasoning.

c. **Analytical** List any function with a graph that is translated to the left. $g(x)$ and $h(x)$

d. **Analytical** Determine which functions are growth models and which are decay models. **See Ch. 6 Answer Appendix.**

A.CED.2, F.IF.4, F.IF.7e, F.IF.8b, F.BF.3

H.O.T. Problems Use Higher-Order Thinking Skills

34. **MP REASONING** Determine whether each statement is *sometimes*, *always*, or *never* true. Explain your reasoning.

a. An exponential function of the form $y = ab^{x-h} + k$ has a y-intercept.

b. An exponential function of the form $y = ab^{x-h} + k$ has an x-intercept.

c. The function $f(x) = |b|^x$ is an exponential growth function if b is an integer. Sometimes; Sample answer: The function is not exponential if $b = 1$ or -1.

35. **ERROR ANALYSIS** Vince and Grady were asked to graph the following functions. Vince thinks they are the same, but Grady disagrees. Who is correct? Explain your reasoning.

x	y
0	2
1	1
2	0.5
3	0.25
4	0.125
5	0.0625
6	0.03125

an exponential function with rate of decay of $\frac{1}{2}$ and an initial amount of 2

Vince; the graphs of the function would be the same.

36. **CHALLENGE** A substance decays 35% each day. After 8 days, there are 8 milligrams of the substance remaining. How many milligrams were there initially? about 251 mg

37. **OPEN-ENDED** Give an example of a value of b for which $f(x) = \left(\frac{8}{b}\right)^x$ represents exponential decay. Sample answer: 10

38. **WRITING IN MATH** Write the procedure for transforming the graph of $g(x) = b^x$ to the graph of $f(x) = ab^{x-h} + k$. Justify each step.

38. Sample answer: The parent function, $g(x) = b^x$, is stretched if a is greater than 1 or compressed if a is less than 1. The parent function is translated up k units if k is positive and down $|k|$ units if k is negative. The parent function is translated h units to the right if h is positive and $|h|$ units to the left if h is negative.

34a. Always; Sample answer: The domain of exponential functions is all real numbers, so $(0, y)$ always exists.

34b. Sometimes; sample answer: The graph of an exponential function crosses the x-axis when $k < 0$.

Additional Answers

27a. decay; 0.9

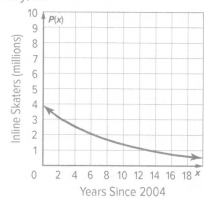

27b. The $P(x)$-intercept represents the number of inline skaters in 2004. The asymptote is the x-axis. The number of inline skaters can approach 0, but will never equal 0. This makes sense as there will probably always be some who continue to skate.

28a. decay; 0.9

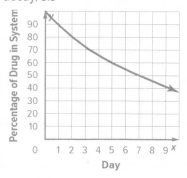

29b. growth; 1.25

MP Standards for Mathematical Practice

Emphasis On	Exercises
1 Make sense of problems and persevere in solving them.	13–18, 20–25, 40
2 Reason abstractly and quantitatively.	29–32, 34, 37, 39, 41, 42
3 Construct viable arguments and critique the reasoning of others.	35, 38
4 Model with mathematics.	26–28
5 Use appropriate tools strategically.	36
6 Attend to precision.	43–45

Go Online!

eSolutions Manual
Create worksheets, answer keys, and solutions handouts for your assignments.

Assess

Ticket Out the Door Make several copies each of five exponential functions. Give one function to each student. As the students leave the room, ask them to tell you whether their functions represent exponential growth or decay.

Preparing for Assessment

Exercises 39-45 require students to use the skills they will need on standardized assessments. Each exercise is dual-coded with content standards and mathematical practice standards.

Dual Coding		
Items	Content Standards	**MP** Mathematical Practices
39	F.IF.4	2
40	F.IF.8b	1
41	F.IF.4	2
42	F.LE.2	2
43	F.IF.4	6
44	F.LE.2	6
45	F.LE.2	6

Diagnose Student Errors

Survey student responses for each item. Class trends may indicate common errors and misconceptions.

39.

A	Translated the graph down 1 unit instead of up 1 unit
B	CORRECT
C	Misinterpreted the roles of the 3 and the 5 in the equation
D	Misinterpreted the roles of the 3 and the 5 in the equation, and translated down 1 unit instead of up 1 unit

Go Online!

Self-Check Quiz
Students can use *Self-Check Quizzes* to check their understanding of this lesson.

Preparing for Assessment

39. Which is the best model for the data shown?
MP 2 F.IF.4 **A**

x	−1	0	1	2	3
y	1.6	4	16	76	376

- ○ A $y = 3(5)^x + 1$
- ○ B $y = 3(5)^x - 1$
- ○ C $y = 5(3^x) + 1$
- ○ D $y = 5(3^x) - 1$

40. The value of Louise's comic book collection has been increasing by 11% every year. Her collection began with a value of $150.00. About how long did it take her collection to double in value? **MP** 1 F.IF.8b **B**

- ○ A between 5 and 6 years
- ○ B between 6 and 7 years
- ○ C between 7 and 8 years
- ○ D between 8 and 9 years
- ○ E between 9 and 10 years

41. Which is the best model for the data shown?
MP 2 F.IF.4 **B**

x	−3	−2	−1	0	1	2	3
y	64	32	16	8	4	2	1

- ○ A $y = 8(2)^x$
- ○ B $y = 8\left(\frac{1}{2}\right)^x$
- ○ C $y = 2(8^x)$
- ○ D $y = 2\left(\frac{1}{8}\right)^x$

42. A bacteria grows exponentially so that it doubles every 12 days. There are 500 bacteria initially. **MP** 2 F.LE.2

a. Write a function that models the number of bacteria after t days.

$f(t) = 500 \cdot 2^{\frac{t}{12}}$

b. Find the number of bacteria after 60 days.

16,000

43. **MULTI-STEP** The function $g(x)$ is transformed from the function $f(x)$.

$$g(x) = 1 - \left(\frac{1}{3}\right)^{x-2}$$ **MP** 6 F.IF.4

a. Describe the transformations from $f(x) = \left(\frac{1}{3}\right)^x$.

translated right 2 units, up 1 unit, and reflected in the x-axis

b. Is the function $g(x)$ increasing or decreasing?

increasing

c. Identify any asymptotes.

horizontal asymptote y = 1

d. What is the y-intercept?

−8

e. Sketch its graph. See margin.

44. **MULTI-STEP** Suppose you accidentally open a canister of plutonium in your living room and 200 units of radiation leak out. Every year, there is half as much radiation as there was the year before. **MP** 6 F.LE.2

a. Write a function that represents the amount of radiation in your living room after t years.

$f(t) = 200 \cdot \left(\frac{1}{2}\right)^t$

b. How many units of radiation will there be after 4 years?

12.5 units.

c. Will your living room ever be free of radiation? Explain.

No, as the horizontal asymptote is $y = 0$, there will never be zero units of radiation.

45. Your grandfather would like you to invest in his company. He guarantees that if you invest in the stock of his company, you will earn 15% on your money every year. If you invest $100, and your grandfather is right, how much money will you have after 20 years? Round your answer to the nearest cent. **MP** 6 F.LE.2

$1636.65

Differentiated Instruction **BL**

Extension Have students flip 50 pennies and count the number of heads. Then have students remove those pennies that landed on heads and repeat the activity. Students should record their results and make a plot of the trial number versus the number of heads counted in that trial. Have students graph their data and then explain why, in theory, their data should be modeled by the equation $y = \left(\frac{1}{2}\right)^x$.

Additional Answer

40.

A	Used a flawed graph to solve the problem or misread the graph
B	CORRECT
C	Used a flawed graph to solve the problem
D	Used a flawed graph to solve the problem

43e.

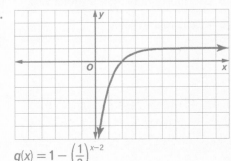

$$g(x) = 1 - \left(\frac{1}{3}\right)^{x-2}$$

Solving Exponential Equations and Inequalities

Track Your Progress

Objectives

1 Solve exponential equations.

2 Solve exponential inequalities.

Mathematical Background

Simple exponential equations can be solved by rewriting one or both sides of the equation so that the bases are the same. Once that has been achieved, the Property of Equality for Exponential Functions can be used to solve for the variable.

THEN	NOW	NEXT
F.IF.4 For a function that models a relationship between two quantities, interpret key features of graphs and tables in terms of the quantities, and sketch graphs showing key features given a verbal description of the relationship.	**A.CED.1** Create equations and inequalities in one variable and use them to solve problems.	**A.SSE.4** Derive the formula for the sum of a finite geometric series (when the common ratio is not 1), and use the formula to solve problems.

Go Online! All of these resources and more are available at connectED.mcgraw-hill.com

Personal Tutors (for every example) let students hear real teachers solve problems. Students can pause and repeat as many times as necessary.

eToolkit contains outstanding tools for enhancing understanding.

Use **Self-Check Quiz** to assess students' understanding of the concepts in this lesson.

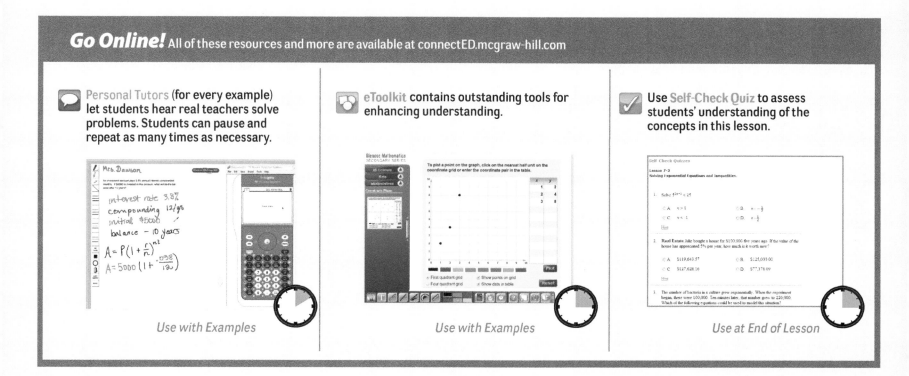

Use with Examples   *Use with Examples*   *Use at End of Lesson*

OER Using Open Educational Resources

Practice Have students practice solving exponential equations on **MathWarehouse.com**. They can view worked out step-by-step solutions to exponential equations. *Use as review or remediation*

Go Online!
connectED.mcgraw-hill.com

Worksheets

Differentiate Your Resources

Extra Practice Additional practice or homework; Skills Practice is best for approaching-level students and Practice is best for on-level and beyond-level students

Skills Practice

6-2 Skills Practice
Solving Exponential Equations and Inequalities

Practice

6-2 Practice
Solving Exponential Equations and Inequalities

Word Problem Practice

6-2 Word Problem Practice
Solving Exponential Equations and Inequalities

Intervention Reteaching and vocabulary activities that can be used with struggling or absent students and as ELL support

Study Guide and Intervention

6-2 Study Guide and Intervention
Solving Exponential Equations and Inequalities

Study Notebook

6-2 Solving Exponential Equations and Inequalities

Extension Activities that can be used to extend lesson concepts

Enrichment

6-2 Enrichment
Musical Relationships

EXPLORE 6-2

Graphing Technology Lab

Solving Exponential Equations and Inequalities

You can use a TI-83/84 Plus graphing calculator to solve exponential equations by graphing or by using the table feature. To do this, you will write the equations as systems of equations.

Mathematical Practices

5 Use appropriate tools strategically.

Content Standards

A.REI.11 Explain why the *x*-coordinates of the points where the graphs of the equations $y = f(x)$ and $y = g(x)$ intersect are the solutions of the equation $f(x) = g(x)$; find the solutions approximately, e.g., using technology to graph the functions, make tables of values, or find successive approximations. Include cases where $f(x)$ and/or $g(x)$ are linear, polynomial, rational, absolute value, exponential, and logarithmic functions.

Activity 1

Solve $3^{x-4} = \frac{1}{9}$.

Step 1 Graph each side of the equation as a separate function. Enter 3^{x-4} as **Y1**. Be sure to include parentheses around the exponent. Enter $\frac{1}{9}$ as **Y2**. Then graph the two equations.

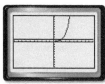

[−10, 10] scl: 1 by [−1, 1] scl: 0.1

Step 2 Use the intersect feature.

You can use the **intersect** feature on the **CALC** menu to approximate the ordered pair of the point at which the graphs cross.

The calculator screen shows that the *x*-coordinate of the point at which the curves cross is 2. Therefore, the solution of the equation is 2.

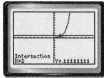

[−10, 10] scl: 1 by [−1, 1] scl: 0.1

Step 3 Use the **TABLE** feature.

You can also use the **TABLE** feature to locate the point at which the curves intersect.

The table displays *x*-values and corresponding *y*-values for each graph. Examine the table to find the *x*-value for which the *y*-values of the graphs are equal.

At $x = 2$, both functions have a *y*-value of $0.\overline{1}$ or $\frac{1}{9}$. Thus, the solution of the equation is 2.

CHECK   Substitute 2 for *x* in the original equation.

$$3^{x-4} \stackrel{?}{=} \frac{1}{9} \qquad \text{Original equation}$$

$$3^{2-4} \stackrel{?}{=} \frac{1}{9} \qquad \text{Substitute 2 for } x$$

$$3^{-2} \stackrel{?}{=} \frac{1}{9} \qquad \text{Simplify}$$

$$\frac{1}{9} = \frac{1}{9} \checkmark \qquad \text{The solution checks.}$$

A similar procedure can be used to solve exponential inequalities.

(continued on the next page)

Focus

Objective Use a graphing calculator to solve exponential equations by graphing or by using the table feature.

Materials

● TI-83/84 Plus or other graphing calculator

Teaching Tip

In Step 1 of Activity 1, remind students to include parentheses around the exponent.

Teach ELL

Working in Cooperative Groups Put students in groups of two or three, mixing abilities. Then have groups complete Activities 1 and 2 and Exercises 1 and 10.

Activity 1

● Before discussing Activity 1, use a simple equation such as $2x = 6$ to remind students how the equation can be solved by graphing. Graph the equations $y = 2x$ and $y = 6$ and then identify the point of intersection of the graphs.

● Ask students why it is necessary in Step 1 to enter the equation using parentheses around the exponent.

● Have students substitute the solution to Activity 1 into the original equation to verify that it is correct.

Activity 2

- In Activity 2, make sure students understand why the inequality needs to be rewritten as a system of inequalities.

Practice Have students complete Exercises 2–9.

Assess

Formative Assessment Use Exercises 4 and 9 to assess whether students comprehend how to use a graphing calculator to solve exponential equations and inequalities.

From Concrete to Abstract Have students explain how the solution set for Activity 2 would change if the inequality were $2^{x-2} \leq 0.5^{x-3}$.

Solving Exponential Equations and Inequalities *Continued*

Activity 2

Solve $2^{x-2} \geq 0.5^{x-3}$.

Step 1 Enter the related inequalities.

Rewrite the problem as a system of inequalities.

The first inequality is $2^{x-2} \geq y$ or $y \leq 2^{x-2}$. Because this inequality includes the *less than or equal to* symbol, shade below the curve.

First enter the boundary, and then use the arrow and ENTER keys to choose the shade below icon, ◣.

The second inequality is $y \geq 0.5^{x-3}$. Shade above the curve since this inequality contains *greater than or equal to*.

KEYSTROKES: Y= ◀ ◀ ENTER ENTER ENTER ▶ ▶ 2 △ (
X,T,θ,n − 2) ENTER ◀ ◀ ENTER ENTER ▶
▶ .5 △ (X,T,θ,n − 3)

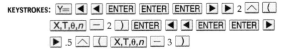

Step 2 Graph the system.

keystrokes: GRAPH

The *x*-values of the points in the region where the shadings overlap is the solution set of the original inequality. Using the **intersect** feature, you can conclude that the solution set is $\{x \mid x \geq 2.5\}$.

[−10, 10] scl: 1 by [−10, 10] scl: 1

Step 3 Use the **TABLE** feature.

Verify using the **TABLE** feature. Set up the table to show *x*-values in increments of 0.5.

KEYSTROKES: 2nd [TBLSET] 0 ENTER .5 ENTER 2nd [TABLE]

Notice that for *x*-values greater than $x = 2.5$, **Y1 > Y2**. This confirms that the solution of the inequality is $\{x \mid x \geq 2.5\}$.

10. Because the system consists of the expressions on both sides of the original equation or inequality, whatever values are solutions of the system will satisfy the original equation or inequality.

Exercises

Solve each equation or inequality.

1. $9^{x-1} = \frac{1}{81}$ −1

2. $4^{x+3} = 25x$ 2

3. $5^{x-1} = 2x$ 1.76

4. $3.5^{x+2} = 1.75^{x+3}$ −1.2

5. $-3^{x+4} = -0.5^{2x+3}$ −2.6

6. $6^{2-x} - 4 < -0.25^{x-2.5}$ $\{x \mid x > 1.8\}$

7. $16^{x-1} > 2^{2x+2}$ $\{x \mid x > 3\}$

8. $3^{x-4} \leq 5^{\frac{x}{2}}$ $\{x \mid x < 2\}$

9. $5^{x+3} \leq 2^{x+4}$ $\{x \mid x \leq -2.2\}$

10. **WRITING IN MATH** Explain why this technique of graphing a system of equations or inequalities works to solve exponential equations and inequalities.

LESSON 2

Solving Exponential Equations and Inequalities

::Then	::Now	::Why?
● You graphed exponential functions.	● **1** Solve exponential equations. **2** Solve exponential inequalities.	● Feral hogs are a highly adaptive species that causes significant ecological and economic damage. Suppose the population of feral hogs is growing exponentially and can be modeled by $y = 2.5(1.45)^x$, where x is the number of years since 2016 and y is the number of hogs in millions.

You can use $y = 2.5(1.45)^x$ to determine how many hogs there will be in a given year or to determine the year in which the hog population is at a certain level.

 New Vocabulary
exponential equation
compound interest
exponential inequality

MP **Mathematical Practices**
2 Reason abstractly and quantitatively.

Content Standards
A.CED.1 Create equations and inequalities in one variable and use them to solve problems.
F.LE.4 For exponential models, express as a logarithm the solution to $ab^{ct} = d$ where a, c, and d are numbers and the base b is 2, 10, or e; evaluate the logarithm using technology.

1 **Solve Exponential Equations** In an **exponential equation**, variables occur as exponents.

⚙ Key Concept Property of Equality for Exponential Functions

Words	Let $b > 0$ and $b \neq 1$. Then $b^x = b^y$ if and only if $x = y$.
Example	If $3^x = 3^5$, then $x = 5$. If $x = 5$, then $3^x = 3^5$.

The Property of Equality can be used to solve exponential equations.

F.LE.4

Example 1 Solve Exponential Equations

Solve each equation.

a. $2^x = 8^3$

$2^x = 8^3$	Original equation
$2^x = (2^3)^3$	Rewrite 8 as 2^3.
$2^x = 2^9$	Power of a Power
$x = 9$	Property of Equality for Exponential Functions

b. $9^{2x-1} = 3^{6x}$

$9^{2x-1} = 3^{6x}$	Original equation
$(3^2)^{2x-1} = 3^{6x}$	Rewrite 9 as 3^2.
$3^{4x-2} = 3^{6x}$	Power of a Power
$4x - 2 = 6x$	Property of Equality for Exponential Functions
$-2 = 2x$	Subtract 4x from each side.
$-1 = x$	Divide each side by 2.

▶ **Guided Practice**

1A. $4^{2n-1} = 64$ 2
1B. $5^{5x} = 125^{x+2}$ 3

MP **Mathematical Practices Strategies**

Make sense of problems and persevere in solving them. Help students solve equations and inequalities. For example, ask:

• How is solving an exponential equation similar to solving a linear equation? The solution processes are similar, however in an exponential equation, the independent variable is an exponent.

• What must be true in order to use the Property of Equality for Exponential Equations? The Property of Equality requires that each side of an exponential equation has the same base.

• If both sides of an exponential equation do not have the same base, can the equation be solved? Explain. Yes; the Properties of Exponents can be used to transform the equation so that the bases are the same.

Launch

Have students read the Why? section of the lesson. Ask:

● What value of x represents the year 2026? 10

● How many hogs are represented by $y = 25$? 25,000,000

● How many hogs will there be in 2024? about 48,900,000

Teach

Ask the scaffolded questions for each example to build conceptual understanding for students at all levels.

1 Solve Exponential Equations

Example 1 Solve Exponential Equations

AL How can you check your answers? Plug your solution in and make sure the equation holds true.

OL In part **a**, why must 8 be written as 2^3? So that both sides of the equation have the same base.

BL Why must the restriction of $b \neq 1$ be put on the Property of Equality for Exponential Functions? Because 1 raised to any exponent is 1, so the property does not hold true for $b = 1$.

Need Another Example?
Solve each equation.
a. $3^x = 9^4$ 8
b. $2^{5x} = 4^{2x-1}$ −2

Go Online!

Interactive Whiteboard

Use the *eLesson* or *Lesson Presentation* to present this lesson.

Example 2 Write an Exponential Function

AL Why is this equation only an approximation? The fourth root of 1.323 was approximated.

OL Why does an exponential function model the growth of bacteria? The more bacteria there are the faster the population grows.

BL How can you use your calculator to estimate when the number of bacteria will have doubled? Graph $y = 7500(1.323)^x$ and $y = 15,000$ and find the intersection.

Need Another Example?

Population In 2010, the population of Phoenix was 1,445,632. By 2015, it was estimated at 1,563,025.

a. Write an exponential function that could be used to model the population of Phoenix. Write x in terms of the number of years since 2010.
$y = 1,445,632(1.0157)^x$

b. Predict the population of Phoenix in 2025.
1,826,163

You can use information about growth or decay to write the equation of an exponential function.

A.CED.1, F.LE.4

Real-World Example 2   Write an Exponential Function

SCIENCE Kristin starts an experiment with 7500 bacteria cells. After 4 hours, there are 23,000 cells.

a. Write an exponential function that could be used to model the number of bacteria after x hours if the number of bacteria changes at the same rate.

At the beginning of the experiment, the time is 0 hours and there are 7500 bacteria cells. Thus, the y-intercept, and the value of a, is 7500.

When $x = 4$, the number of bacteria cells is 23,000. Substitute these values into an exponential function to determine the value of b.

$y = ab^x$	Exponential function
$23,000 = 7500 \cdot b^4$	Replace x with 4, y with 23000, and a with 7500.
$3.067 \approx b^4$	Divide each side by 7500.
$\sqrt[4]{3.067} \approx b$	Take the 4th root of each side.
$1.323 \approx b$	Use a calculator.

An equation that models the number of bacteria is $y \approx 7500(1.323)^x$.

b. How many bacteria cells can be expected in the sample after 12 hours?

$y \approx 7500(1.323)^x$	Modeling equation
$\approx 7500(1.323)^{12}$	Replace x with 12.
$\approx 215,665$	Use a calculator.

There will be approximately 215,665 bacteria cells after 12 hours.

Guided Practice

2. **RECYCLING** A manufacturer distributed 3.2 million cans in 2010 made from recycled aluminum.

   A. In 2015, the manufacturer distributed 420,000 cans made from recycled aluminum. Assuming that the recycling rate continues, write an equation to model the distribution each year of cans that are made from recycled aluminum. $y = 3.2(0.67)^x$

   B. How many cans made from recycled aluminum can be expected in the year 2055? none

Exponential functions are used in situations involving compound interest. **Compound interest** is interest paid on the principal of an investment and any previously earned interest.

Key Concept Compound Interest

You can calculate compound interest using the following formula.
$$A = P\left(1 + \frac{r}{n}\right)^{nt},$$
where A is the amount in the account after t years, P is the principal amount invested, r is the annual interest rate, and n is the number of compounding periods each year.

Real-World Link

In 2011, the U.S. recycling rate for metals of 34% prevented the release of approximately 20 million metric tons of carbon into the air—roughly the amount emitted annually by 4 million cars.

Source: Environmental Protection Agency

Go Online!

Watch the Personal Tutor describe how to solve a compound interest problem with a partner. Then try describing how to solve a problem for them. Have them ask questions to help your understanding. **ELL**

A.CED.1, F.LE.4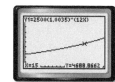

Example 3   Compound Interest

An investment account pays 4.2% annual interest compounded monthly. If $2500 is invested in this account, what will be the balance after 15 years?

Analyze   Find the total amount in the account after 15 years.

Formulate   Use the compound interest formula.
$$P = 2500, r = 0.042, n = 12, \text{ and } t = 15$$

Determine
$$A = P\left(1 + \frac{r}{n}\right)^{nt} \qquad \text{Compound Interest Formula}$$
$$= 2500\left(1 + \frac{0.042}{12}\right)^{12 \cdot 15} \qquad P = 2500, r = 0.042, n = 12, t = 15$$
$$\approx 4688.87 \qquad \text{Use a calculator.}$$

Justify   Graph the corresponding equation $y = 2500(1.0035)^{12t}$. Use **CALC: value** to find y when $x = 15$.

The y-value 4688.8662 is very close to 4688.87, so the answer is reasonable.

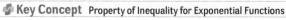

Evaluate   The compound interest formula is a model used to solve problems in the real world.

[0, 20] scl: 1 by [0, 10,000] scl: 1000

Watch Out!

Percents Remember to convert all percents to decimal form: 4.2% is 0.042.

▷ **Guided Practice**

3. Find the account balance after 20 years if $100 is placed in an account that pays 1.2% interest compounded twice a month. **$127.12**

2 **Solve Exponential Inequalities**   An **exponential inequality** is an inequality involving exponential functions.

🔊 **Key Concept**   Property of Inequality for Exponential Functions
Words   Let $b > 1$. Then $b^x > b^y$ if and only if $x > y$, and $b^x < b^y$ if and only if $x < y$.
Example   If $2^x > 2^6$, then $x > 6$. If $x > 6$, then $2^x > 2^6$.

This property also holds true for ≤ and ≥.

F.LE.4

Example 4   Solve Exponential Inequalities

Solve $16^{2x-3} < 8$.

$$16^{2x-3} < 8 \qquad \text{Original inequality}$$
$$(2^4)^{2x-3} < 2^3 \qquad \text{Rewrite 16 as } 2^4 \text{ and 8 as } 2^3.$$
$$2^{8x-12} < 2^3 \qquad \text{Power of a Power}$$
$$8x - 12 < 3 \qquad \text{Property of Inequality for Exponential Functions}$$
$$8x < 15 \qquad \text{Add 12 to each side.}$$
$$x < \frac{15}{8} \qquad \text{Divide each side by 8.}$$

▷ **Guided Practice**

Solve each inequality.

4A. $3^{2x-1} \ge \frac{1}{243}$ $\{x \mid x \ge -2\}$ **4B.** $2^{x+2} > \frac{1}{32}$ $\{x \mid x > -7\}$

Example 3   Compound Interest

AL   Why is $n = 12$? The interest is compounded monthly which means it will be compounded 12 times a year.

OL   Use your calculator to estimate when the account will have $3000. $t \approx 4.35$ years

BL   What is the analog of the principle in the function in example 2 modeling the growth of bacteria? the initial number of bacteria

Need Another Example?

An investment account pays 5.4% annual interest compounded quarterly. If $4000 is placed in this account, find the balance after 8 years. $6143.56

2 **Solve Exponential Inequalities**

Example 4   Solve Exponential Inequalities

AL   What is the solution to $16^{2x-3} > 8$? $x > \frac{15}{8}$

OL   How could you use your calculator to check your solution? Graph $y = 16^{2x-3}$ and $y = 8$ and the solution will be where the graph of $y = 16^{2x-3}$ is above the graph of $y = 8$.

BL   Write a rule similar to the Property of Inequality for Exponential Functions for $0 < b < 1$. $b^x > b^y$ if and only if $x < y$, and $b^x < b^y$ if and only if $x > y$.

Need Another Example?

Solve $5^{3-2x} > \frac{1}{625}$. $x < \frac{7}{2}$

Differentiated Instruction   **AL**   **OL**

IF   you want students to check to see if their answer is correct,

THEN   remind students to choose any value in the solution interval and see if it satisfies the original inequality.

Practice

Formative Assessment Use Exercises 1–8 to assess students' understanding of the concepts in this lesson.

The Practice and Problem Solving exercises assess the content taught in the lesson. The Preparing for Assessment page is meant to be used as preparation for end-of-course assessments.

Extra Practice

See page R6 for extra exercises for students who are approaching level or for on-level students who need additional reinforcement.

Levels of Complexity Chart

The levels of the exercises progress from 1 to 3, with Level 1 indicating the lowest level of complexity.

Exercises	9–29	30–38, 50–57	39–49
▶ Level 3			●
▶ Level 2		●	
Level 1	●		

 Teaching the Mathematical Practices

Modeling Mathematically proficient students can apply the mathematics they know to solve problems arising in everyday life, analyze relationships mathematically to draw conclusions, and interpret their mathematical results in the context of a situation.

Go Online!

eBook

Interactive Student Guide

Use the *Interactive Student Guide* to deepen conceptual understanding.
· Solving Exponential Equations and Inequalities

ALGEBRA 2
INTERACTIVE STUDENT GUIDE

Check Your Understanding   ◯ = Step-by-Step Solutions begin on page R11.

 **Go Online!** for a Self-Check Quiz

Example 1
F.LE.4

Solve each equation.

1. $3^{5x} = 27^{2x-4}$ 12

2. $16^{2y-3} = 4^{y+1}$ $\frac{7}{3}$

3. $2^{6x} = 32^{x-2}$ −10

4. $49^{x+5} = 7^{8x-6}$ $\frac{8}{3}$

Example 2
A.CED.1,
F.LE.4

5. **SCIENCE** Mitosis is a process in which one cell divides into two. The *Escherichia coli* is one of the fastest growing bacteria. It can reproduce itself in 15 minutes.

   a. Write an exponential function to represent the number of cells c after t minutes.

   b. If you begin with one *Escherichia coli* cell, how many cells will there be in one hour?

   5a. $c = 2^{\frac{t}{15}}$
   5b. 16 cells

Example 3
A.CED.1,
F.LE.4

6. A certificate of deposit (CD) pays 2.25% annual interest compounded biweekly. If you deposit $500 into this CD, what will the balance be after 6 years? $572.23

Example 4
F.LE.4

Solve each inequality.

7. $4^{2x+6} \le 64^{2x-4}$ $\{x \mid x \ge 4.5\}$

8. $25^{y-3} \le \left(\frac{1}{125}\right)^{y+2}$ $\{y \mid y \le 0\}$

Practice and Problem Solving   Extra Practice is on page R6.

Example 1
F.LE.4

Solve each equation.

9. $8^{4x+2} = 64$ 0

10. $5^{x-6} = 125$ 9

11. $81^{a+2} = 3^{3a+1}$ −7

12. $256^{b+2} = 4^{2-2b}$ −1

13. $9^{3c+1} = 27^{3c-1}$ $\frac{5}{3}$

14. $8^{2y+4} = 16^{y+1}$ −4

Example 2
A.CED.1,
F.LE.4

15. **MODELING** In 2015, My-Lien received $10,000 from her grandmother. Her parents invested all of the money, and by 2027, the amount will have grown to $16,960.

    a. Write an exponential function that could be used to model the money y. Write the function in terms of x, the number of years since 2015. $y = 10{,}000(1.045)^x$

    b. Assume that the amount of money continues to grow at the same rate. What would be the balance in the account in 2037? about $26,336.52

Write an exponential function for the graph that passes through the given points.

16. (0, 6.4) and (3, 100) $y = 6.4(2.5)^x$

17. (0, 256) and (4, 81) $y = 256(0.75)^x$

18. (0, 128) and (5, 371,293) $y = 128(4.926)^x$

19. (0, 144), and (4, 21,609) $y = 144(3.5)^x$

Example 3
A.CED.1,
F.LE.4

20. Find the balance of an account after 7 years if $700 is deposited into an account paying 4.3% interest compounded monthly. $945.34

21. Determine how much is in a retirement account after 20 years if $5000 was invested at 6.05% interest compounded weekly. $16,755.63

22. A savings account offers 1.1% interest compounded bimonthly. If $110 is deposited in this account, what will the balance be after 15 years? $129.71

23. A college savings account pays 8.2% annual interest compounded semiannually. What is the balance of an account after 12 years if $21,000 was initially deposited? $55,085.44

Example 4
F.LE.4

Solve each inequality.

24. $625 \ge 5^{a+8}$ $\{a \mid a \le -4\}$

25. $10^{5b+2} > 1000$ $\left\{b \mid b > \frac{1}{5}\right\}$

26. $\left(\frac{1}{64}\right)^{c-2} < 32^{2c}$ $\left\{c \mid c > \frac{3}{4}\right\}$

27. $\left(\frac{1}{27}\right)^{2d-2} \le 81^{d+4}$ $\{d \mid d \ge -1\}$

28. $\left(\frac{1}{9}\right)^{3t+5} \ge \left(\frac{1}{243}\right)^{t-6}$ $\{t \mid t \le -40\}$

29. $\left(\frac{1}{36}\right)^{w+2} < \left(\frac{1}{216}\right)^{4w}$ $\left\{w \mid w < \frac{2}{5}\right\}$

Differentiated Homework Options

Levels	**AL** Basic	**OL** Core	**BL** Advanced
Exercises	9–29, 42–57	9–29 odd, 30, 31–37 odd, 38–57	30–57
2-Day Option	9–29 odd, 50–57	9–29, 50–57	
	10–28 even, 42–49	30–49	

⊕ You can use ALEKS to provide additional remediation support with personalized instruction and practice.

30. MULTI-STEP George is leaving money in a college savings plan that earns 4% interest for his newborn granddaughter, Jenny. Currently, 4-year tuition at his alma mater costs $15,000 per year and has been increasing at an annual rate of 5%.

a. How much money would he need to deposit now in order to cover her tuition expenses at his alma mater? $76,804

b. Describe your solution process.

c. What assumptions did you make?
   b–c. See Ch. 6 Answer Appendix.

31 **ANIMALS** Studies show that an animal will defend a territory, with area in square yards, that is directly proportional to the 1.31 power of the animal's weight in pounds.

a. If a 45-pound beaver will defend 170 square yards, write an equation for the area a defended by a beaver weighing w pounds. $a = 1.16w^{1.31}$

b. Scientists believe that thousands of years ago, the beaver's ancestors were 11 feet long and weighed 430 pounds. Use your equation to determine the area defended by these animals. about 3268 yd²

Solve each equation.

32. $\left(\frac{1}{2}\right)^{4x+1} = 8^{2x+1}$ $-\frac{2}{5}$ **33.** $\left(\frac{1}{5}\right)^{x-5} = 25^{3x+2}$ $\frac{1}{7}$ **34.** $216 = \left(\frac{1}{6}\right)^{x+3}$ -6

35. $\left(\frac{1}{8}\right)^{3x+4} = \left(\frac{1}{4}\right)^{-2x+4}$ $-\frac{4}{13}$ **36.** $\left(\frac{2}{3}\right)^{5x+1} = \left(\frac{27}{8}\right)^{x-4}$ $\frac{11}{8}$ **37.** $\left(\frac{25}{81}\right)^{2x+1} = \left(\frac{729}{125}\right)^{-3x+1}$ 1

38. **MODELING** In 1950, the world population was about 2.556 billion. By 2000, it had increased to about 6.08 billion.

40a. $A = 5000$ $\left(\frac{4.065}{4}\right)^{4t}$;

$A = 5000$ $\left[\left(\frac{12.042}{12}\right)^{12t} + \left(\frac{52.023}{52}\right)^{52t}\right]$

a. Write an exponential function of the form $y = ab^x$ that could be used to model the world population y in billions for 1950 to 2000. Write the equation in terms of x, the number of years since 1950. (Round the value of b to the nearest ten-thousandth.) $y = 2.556(1.0175)^x$

b. Suppose the population continued to grow at that rate. Estimate the population in 2013. 7.625 billion

c. In 2013, the population of the world was about 7.164 billion. Compare your estimate to the actual population. The prediction was about 461 million greater than the actual.

d. Use the equation you wrote in part **a** to estimate the world population in the year 2040. How accurate do you think the estimate is? Explain your reasoning. See margin.

39. **TREES** The diameter of the base of a tree trunk in centimeters varies directly with the $\frac{3}{2}$ power of its height in meters.

a. A young sequoia tree is 6 meters tall, and the diameter of its base is 19.1 centimeters. Use this information to write an equation for the diameter d of the base of a sequoia tree if its height is h meters high. $d = 1.30h^{\frac{3}{2}}$

b. The General Sherman Tree in Sequoia National Park, California, is approximately 84 meters tall. Find the diameter of the General Sherman Tree at its base. about 1001 cm

40. **FINANCIAL LITERACY** Mrs. Jackson has two different retirement investment plans from which to choose.

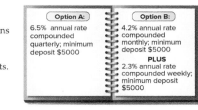

Option A:
6.5% annual rate compounded quarterly; minimum deposit $5000

Option B:
4.2% annual rate compounded monthly; minimum deposit $5000
PLUS
2.3% annual rate compounded weekly; minimum deposit $5000

a. Write equations for Option A and Option B given the minimum deposits.

b. Draw a graph to show the balances for each investment option after t years. See margin.

c. Explain whether Option A or Option B is the better investment choice. See margin.

Additional Answers

Additional Answers

38d. About 12.180 billion; because the prediction for 2013 was greater than the actual population, this prediction is probably even higher than the actual population will be at the time.

40b.

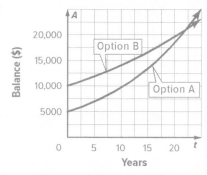

40c. Sample answer: During the first 22 years, Option B is the better choice because the total is greater than that of Option A. However, after about 22 years, the balance of Option A exceeds that of Option B, so Option A is the better choice.

Differentiated Instruction OL BL

Extension Allow students to develop their sense of consumerism by providing them with an initial deposit amount and having them shop around for the best interest rates. Students should record relevant information including bank name, account type, interest rate, how often interest is compounded, and restrictions on the account. Ask students to graph the growth of their initial deposit over time.

MP Teaching the Mathematical Practices

Construct Arguments Mathematically proficient students understand and use stated assumptions, definitions, and previously established results in constructing arguments. They make conjectures and build a logical progression of statements to explore the truth of their conjectures. And they are able to analyze situations by breaking them into cases, and can recognize and use counterexamples.

Assess

Name the Math Have students describe the set of values for b that are possible in an exponential function of the form $y = b^x$.

Watch Out!

Error Analysis In Exercise 43, students may fail to recognize the use of basic exponent rules when variables are involved. Remind them that $(a^m)^n = a^{mn}$.

---

41. MULTIPLE REPRESENTATIONS In this problem, you will explore the rapid increase of an exponential function. A large sheet of paper is cut in half, and one of the resulting pieces is placed on top of the other. Then the pieces in the stack are cut in half and placed on top of each other. Suppose this procedure is repeated several times.

   a. Concrete Perform this activity and count the number of sheets in the stack after the first cut. How many pieces will there be after the second cut? How many pieces after the third cut? How many pieces after the fourth cut? **2, 4, 8, 16**

   b. Tabular Record your results in a table. **See Ch. 6 Answer Appendix.**

   c. Symbolic Use the pattern in the table to write an equation for the number of pieces in the stack after x cuts. $y = 2^x$

   d. Analytical The thickness of ordinary paper is about 0.003 inch. Write an equation for the thickness of the stack of paper after x cuts. $y = 0.003(2)^x$

   e. Analytical How thick will the stack of paper be after 30 cuts? **about 3,221,225.47 in.**

A.CED.1, F.LE.4

H.O.T. Problems	Use Higher-Order Thinking Skills

46a. Always; 2^x will always be positive, and -8^{20x} will always be negative.

46b. Always; by definition the graph will always be increasing even if it is a small increase.

46c. Never; by definition the graph will always be decreasing even if it is a small decrease.

49. Sample answer: Divide the final amount by the initial amount. If n is the number of time intervals that pass, take the nth root of the answer.

42. WRITING IN MATH In a problem about compound interest, describe what happens as the compounding period becomes more frequent while the principal and overall time remain the same. **See Ch. 6 Answer Appendix.**

43. ERROR ANALYSIS Beth and Liz are solving $6^{x-3} > 36^{-x-1}$. Is either of them correct? Explain your reasoning.

Beth	Liz
$6^{x-3} > 36^{-x-1}$	$6^{x-3} > 36^{-x-1}$
$6^{x-3} > (6^2)^{-x-1}$	$6^{x-3} > (6^2)^{-x-1}$
$6^{x-3} > 6^{-2x-2}$	$6^{x-3} > 6^{-x+1}$
$x - 3 > -2x - 2$	$x - 3 > -x + 1$
$3x > 1$	$2x > 4$
$x > \frac{1}{3}$	$x > 2$

43. Sample answer: Beth; Liz added the exponents instead of multiplying them when taking the power of a power.

45. Reducing the term will be more beneficial. The multiplier is 1.3756 for the 4-year and 1.3828 for the 6.5%.

44. CHALLENGE Solve for x: $16^{18} + 16^{18} + 16^{18} + 16^{18} + 16^{18} = 4^x$. **37.1610**

45. OPEN-ENDED What would be a more beneficial change to a 5-year loan at 8% interest compounded monthly: reducing the term to 4 years or reducing the interest rate to 6.5%?

46. MP CONSTRUCT ARGUMENTS Determine whether the following statements are *sometimes*, *always*, or *never* true. Explain your reasoning.

   a. $2^x > -8^{20x}$ for all values of x.

   b. The graph of an exponential growth equation is increasing.

   c. The graph of an exponential decay equation is increasing.

47. OPEN-ENDED Write an exponential inequality with a solution of $x \le 2$. **Sample answer:** $4^x \le 4^2$

48. PROOF Show that $27^{2x} \cdot 81^{x+1} = 3^{2x+2} \cdot 9^{4x+1}$. **See Ch. 6 Answer Appendix.**

49. WRITING IN MATH If you were given the initial and final amounts of a radioactive substance and the amount of time that passes, how would you determine the rate at which the amount was increasing or decreasing in order to write an equation?

MP Standards for Mathematical Practice

Emphasis On	Exercises
1 Make sense of problems and persevere in solving them.	44, 52, 54, 56–57
2 Reason abstractly and quantitatively.	47, 49, 51, 53, 55
3 Construct viable arguments and critique the reasoning of others.	30, 42–43, 45–46, 48
4 Model with mathematics.	5, 15, 31, 38–41
6 Attend to precision.	50

Go Online! 🄴

eSolutions Manual

Create worksheets, answer keys, and solutions handouts for your assignments.

Preparing for Assessment

50. Carmen invested $3000 in an account that pays 3.6% annual interest compounded monthly. To the nearest whole dollar, what will be the balance in her account after 6 years? ⓂⓅ 6 A.CED.1 **D**

○ **A** $3223

○ **B** $3648

○ **C** $3709

○ **D** $3722

○ **E** $38,285

51. Let $p \# q = p^q - q^p$ for all nonzero values of p and q. What is the value of $-2 \# 4$? ⓂⓅ 2 A.CED.1 **C**

○ **A** $-16\frac{1}{16}$

○ **B** 0

○ **C** $15\frac{15}{16}$

○ **D** $16\frac{1}{16}$

○ **E** 32

52. What is the value of x in the equation $8^{12-4x} = 4^{x+18}$? ⓂⓅ 1 A.CED.1 **C**

○ **A** $-5\frac{1}{7}$

○ **B** -4

○ **C** 0

○ **D** $1\frac{1}{5}$

○ **E** $3\frac{3}{5}$

53. Let m^x be greater than 1 for all negative values of x. What are the possible values of m? ⓂⓅ 2 F.LE.4 **E**

○ **A** $-\infty < m < -1$

○ **B** $-\infty < m < 0$

○ **C** $-1 < m < 0$

○ **D** $-1 < m < 1$

○ **E** $0 < m < 1$

54. What is the value of x in $\left(\frac{1}{9}\right)^{3x-4} = 27^{2x+1}$? ⓂⓅ 1 A.CED.1 **C**

○ **A** -1

○ **B** -0.8

○ **C** 0.42

○ **D** 5

55. Solve the following equations for x. ⓂⓅ 2 A.CED.1

  a. $9^{2x} = \left(\frac{1}{27}\right)^{x+1}$

    $\boxed{-\frac{3}{7}}$

  b. $25(5^{3x}) = \left(\frac{1}{5}\right)^{-x}$

    $\boxed{-1}$

56. **MULTI-STEP** Suppose the amount of a radioactive substance $a(t)$, in grams, after t years is given by the function $a(t) = 900(2)^{\frac{t}{10}}$. ⓂⓅ 1 F.LE.4

  a. How much of the substance remains after 3 years? Round your answer to the nearest gram.

    $\boxed{731\ g}$

  The half-life of the substance is how many years it takes for half of the substance to decay.

  b. Write down an equation whose solution gives the half-life of the substance.

    $\boxed{\text{Sample answer: } 2^{-\frac{t}{10}} = \frac{1}{2}}$

  c. What is the half-life of the substance?

    $\boxed{10\ \text{years}}$

57. Solve the equation and inequalities: ⓂⓅ 1 A.CED.1

  a. $\frac{2}{5^x} = 0.08$ $\boxed{x = 2}$

  b. $4^{\frac{x}{2}} > 8$ $\boxed{x > 3}$

  c. $\frac{3}{2^x} \le 0.375$ $\boxed{x \ge 3}$

Differentiated Instruction ⓄⓁ ⒷⓁ

Extension Have students extend the solution to Example 3 for an increasing number of compounding periods. Try daily compounding ($n = 365$), and then explore what happens if n is varied up to tens of thousands of times per year. The final amount approaches an upper limit, which in this case is about $4694.03.

52.

A	Rewrote $36 - 12x = 2x + 36$ as $72 = -14x$
B	Rewrote the equation as $4^{12-4x} = 4^{2(x+18)}$
C	CORRECT
D	Equated $12 - 4x$ with $x + 18$

53.

A	Thought negative power results in a sign change
B	Misinterpreted the effect of negative powers
C	Did not consider odd powers of negatives
D	Did not consider odd powers of negatives
E	CORRECT

Preparing for Assessment

Exercises 50–57 require students to use the skills they will need on standardized assessments. Each exercise is dual-coded with content standards and mathematical practice standards.

Dual Coding		
Items	Content Standards	ⓂⓅ Mathematical Practices
50	A.CED.1	6
51	A.CED.1	2
52	A.CED.1	1
53	F.LE.4	2
54	A.CED.1	1
55	A.CED.1	2
56	F.LE.4	1
57	A.CED.1	1

Diagnose Student Errors

Survey student responses for each item. Class trends may indicate common errors and misconceptions.

50.

A	Evaluated $3000\left(1 + \frac{0.036}{6}\right)^{12}$
B	Evaluated $3000 + 6(3000(0.036))$
C	Evaluated $3000(1 + 0.036)^6$
D	CORRECT

51.

A	Evaluated -2^4 instead of $(-2)^4$
B	Believed $p^q = q^p$
C	CORRECT
D	Evaluated $(-q)^p$ instead of subtracting q^p
E	Thought negative power results in a sign change.

Go Online!

Self-Check Quiz
Students can use *Self-Check Quizzes* to check their understanding of this lesson. You can also give the *Chapter Quiz*, which covers the content in Lessons 6-1 and 6-2.

Geometric Sequences and Series

Track Your Progress

Objectives

1 Use geometric sequences.

2 Find sums of geometric series.

Mathematical Background

The formula for the sum of a geometric series $S_n = \dfrac{a_1 - a_n r}{1 - r}$ is used when the number of terms is not given. If n is known, the formula $S_n = \dfrac{a_1(1 - r^n)}{1 - r}$ can be used as it is not necessary to calculate a_n.

THEN	NOW	NEXT
A.CED.1 Create equations and inequalities in one variable and use them to solve problems.	**A.SSE.4** Derive the formula for the sum of a finite geometric series (when the common ratio is not 1), and use the formula to solve problems.	**F.IF.7e** Graph exponential and logarithmic functions, showing intercepts and end behavior, and trigonometric functions, showing period, midline, and amplitude.

Go Online! All of these resources and more are available at connectED.mcgraw-hill.com

Use **The Geometer's Sketchpad** to explore geometric sequences and series.

Personal Tutors (for every example) let students hear real teachers solve problems. Students can pause and repeat as many times as necessary.

Use **Self-Check Quiz** to assess students' understanding of the concepts in this lesson.

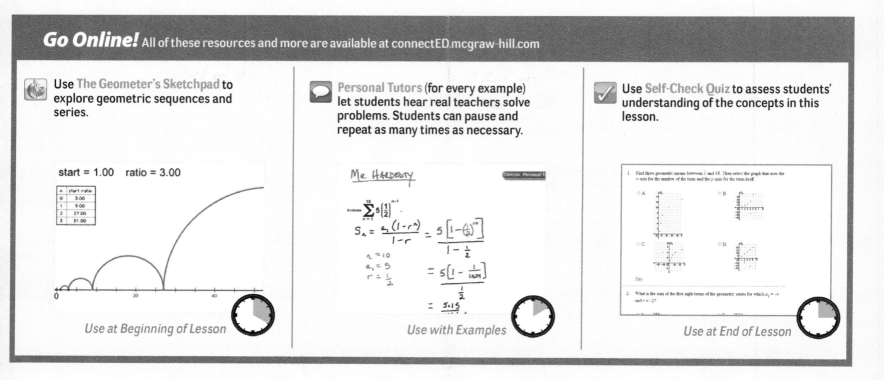

Use at Beginning of Lesson

Use with Examples

Use at End of Lesson

OER **Using Open Educational Resources**

Lesson Planning For new ways to introduce the concepts in this lesson, visit the geometric sequences and exponential functions page on **ck12.org**. This page contains videos, exploration activities, practice problems, and sample quizzes. *Use as planning*

Go Online!
connectED.mcgraw-hill.com

Worksheets

Differentiate Your Resources

Extra Practice Additional practice or homework; Skills Practice is best for approaching-level students and Practice is best for on-level and beyond-level students

Skills Practice

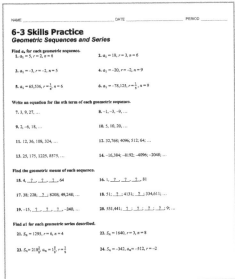

Practice

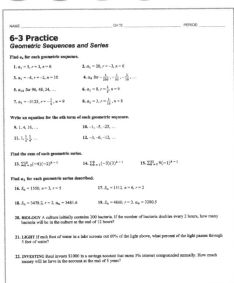

Word Problem Practice

Intervention Reteaching and vocabulary activities that can be used with struggling or absent students and as ELL support

Extension Activities that can be used to extend lesson concepts

Study Guide and Intervention

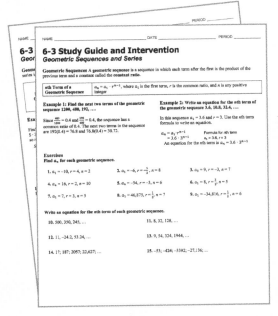

Study Notebook

Enrichment

Launch

Have students read the Why? section of the lesson. Ask:

- What are the first 3 terms of the pattern?
  5, 25, 125

- Why is this sequence not an arithmetic sequence? There is no common difference between the terms.

- What is the common ratio in the sequence? $r = 5$

- Compare a common difference with a common ratio. The first involves addition; the second involves multiplication.

Teach

Ask the scaffolded questions for each example to build conceptual understanding for students at all levels.

1 Geometric Sequences

Example 1 Find the nth Term

AL To what power are you taking 5? 7

OL Why is $a_1 = 5$? Five of Julian's friends reposted the link.

BL Write the general equation that would find a_n for this sequence. $a_n = 5 \cdot 5^{n-1}$

Need Another Example?

Find the sixth term of a geometric sequence for which $a_1 = -3$ and $r = -2$. $a_6 = 96$

LESSON 3
Geometric Sequences and Series

Then	Now	Why?
You determined whether a sequence was geometric.	**1** Use geometric sequences. **2** Find sums of geometric series.	Julian sees a band at a concert. He posts a link for the band's website on his social network page. Five of his friends repost the link, then five of each of their friends repost the link, and so on. If this pattern continues, how many people will post the link on the eighth round?

New Vocabulary
geometric means
geometric series

MP Mathematical Practices
8 Look for and express regularity in repeated reasoning.

Content Standards
A.SSE.4 Derive the formula for the sum of a finite geometric series (when the common ratio is not 1), and use the formula to solve problems.

1 Geometric Sequences As with arithmetic sequences, there is a formula for the nth term of a geometric sequence.

> **Key Concept** nth Term of a Geometric Sequence
>
> The nth term a_n of a geometric sequence in which the first term is a_1 and the common ratio is r is given by the following formula, where n is any natural number.
>
> $$a_n = a_1 r^{n-1}$$

You will prove this formula in Exercise 68.

A.SSE.4

Real-World Example 1   Find the nth Term

MUSIC If the pattern continues, how many people will post the link on the eighth round?

Analyze You need to determine how many people posted the link on the eighth round. Five people posted the link on the first round. Each of those people had five people repost the link on the second round.

Formulate This is a geometric sequence, and the common ratio is 5. Use the formula for the nth term of a geometric sequence.

Determine $a_n = a_1 r^{n-1}$ nth term of a geometric sequence

      $a_8 = 5(5)^{8-1}$ $a_1 = 5, r = 5,$ and $n = 8$

      $a_8 = 5(78,125)$ or 390,625 $5^7 = 78,125$

Justify There will be 390,625 posts on the 8th round. Write out the first eight terms by multiplying by the common ratio.

      5, 25, 125, 625, 3125, 15,625, 78,125, 390,625 ✓

Evaluate The formula for the nth term of a geometric sequence is more efficient than calculating each term using the common ratio.

> **Guided Practice**
>
> 1. **SOCIAL NETWORKING** Shira posts a joke on her social network page. Four of her friends repost the joke on their pages. Four of each of their friends also repost the joke, and so on. How many people will post the joke on the ninth round? 262,144

MP **Mathematical Practices Strategies**

Look for and express regularity in repeated reasoning.
Help students understand the similarities between arithmetic and geometric sequences and series. For example, ask:

- **How is a geometric sequence similar to an arithmetic sequence?** In a geometric sequence, each term is determined by a multiplying a nonzero constant, called a common ratio, by the previous term. In an arithmetic sequence, each term is determined by adding or subtracting a nonzero constant, called a common difference, by the previous term.

- **How is a geometric series similar to an arithmetic series?** A geometric series is the sum of the terms of a geometric sequence. An arithmetic series is the sum of the terms of an arithmetic sequence.

If you are given some of the terms of a geometric sequence, you can determine an equation for finding the nth term of the sequence.

A.SSE.4

Example 2   Write an Equation for the nth Term

Write an equation for the nth term of each geometric sequence.

a. 0.5, 2, 8, 32, ...

$r = 8 \div 2$ or 4; 0.5 is the first term.

$a_n = a_1 r^{n-1}$ nth term of a geometric sequence

$a_n = 0.5(4)^{n-1}$ $a_1 = 0.5$ and $r = 4$

b. $a_4 = 5$ and $r = 6$

Step 1 Find a_1.

$a_n = a_1 r^{n-1}$ nth term of a geometric sequence

$5 = a_1 (6^{4-1})$ $a_n = 5, r = 6$ and $n = 4$

$5 = a_1 (216)$ Evaluate the power.

$\frac{5}{216} = a_1$ Divide each side by 216.

Step 2 Write the equation.

$a_n = a_1 r^{n-1}$ nth term of a geometric sequence

$a_n = \frac{5}{216}(6)^{n-1}$ $a_1 = \frac{5}{216}$ and $r = 6$

▷ Guided Practice   **2A.** $a_n = -0.25(-8)^{n-1}$

Write an equation for the nth term of each geometric sequence.

2A. −0.25, 2, −16, 128, ...   **2B.** $a_3 = 16, r = 4$ $a_n = 1(4)^{n-1}$

Like arithmetic means, **geometric means** are the terms between two nonconsecutive terms of a geometric sequence. The common ratio r can be used to find the geometric means.

A.SSE.4

Example 3   Find Geometric Means

Find three geometric means between 2 and 1250.

Step 1 Because there are three terms between the first and last term, there are 3 + 2 or 5 total terms, so $n = 5$.

Step 2 Find r.

$a_n = a_1 r^{n-1}$ nth term of a geometric sequence

$1250 = 2r^{5-1}$ $a_n = 1250, a_1 = 2,$ and $n = 5$

$625 = r^4$ Divide each side by 2.

$\pm 5 = r$ Take the 4th root of each side.

Step 3 Use r to find the four arithmetic means.

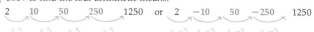

2   10   50   250   1250   or   2   −10   50   −250   1250

×5   ×5   ×5   ×5   ×−5 ×−5 ×−5 ×−5

The geometric means are 10, 50, and 250 or −10, 50, and −250.

▷ Guided Practice

3. Find four geometric means between 0.5 and 512.   2, 8, 32, 128

Reading Math

Geometric Means
A geometric mean can also be represented geometrically. In the figure below, h is the geometric mean between x and y.

Example 2 Write an Equation for the nth Term

AL What variable represents the common ratio? r

OL What would the fifth term of the sequence in part **b** be? 30

BL Explain why in part **a** you cannot simplify a_n to 2^{n-1}. In order of operations, exponents are applied before multiplication.

Need Another Example?

Write an equation for the nth term of each geometric sequence.

a. 5, 10, 20, 40, . . . $a_n = 5 \cdot 2^{n-1}$

b. $a_5 = 4$ and $r = 3$ $a_n = \frac{4}{81}(3^{n-1})$

Example 3 Find Geometric Means

AL What are a_n, a_1, and n? $a_n = 1250, a_1 = 2,$ and $n = 5$

OL Why do we get two solutions in this example? Because there are two solutions to the equation $r^4 = 625$, r could either be 5 or −5, so either of the sequences with those r values can be solutions.

BL Would it be possible to find 3 geometric means between 2 and −1250? no

Need Another Example?

Find three geometric means between 3.12 and 49.92. 6.24, 12.48, 24.96, or −6.24, 12.48, −24.96

Watch Out!

Preventing Errors Encourage students to begin a geometric sequence problem by writing the known values for each of the variables n, a, and r.

Differentiated Instruction ELL

Beginning Help students access text by working through finding the sum of a geometric series using an interactive whiteboard. Point out each variable in the formula and each step of the solution, explaining using short phrases.

Intermediate Provide students with a study guide to make it accessible to all students. Paraphrasing content helps students make connections more easily. The Study Guide and Intervention worksheets on ConnectED offer concise explanations with examples.

Teaching Tip

Sense-Making You may wish to have students calculate one or more sums with and without using the formula in order to verify the formula. This will also demonstrate the formula's efficiency.

> ### Watch Out!
> **Preventing Errors** Emphasize the importance of writing every step of calculations as an equation, so that each numeric value found during the process is clearly identified.

2 Geometric Series

Example 4 Find the Sum of a Geometric Series

(AL) **Why is $n = 8$?** We are solving for the number after 8 reposts.

(OL) **Write S_n in terms of r for this situation.**
$$S_n = \frac{5 - 5r^n}{1 - r}$$

(BL) **After how many rounds would there be 305,175,780 reposts?** 13

Need Another Example?

Music Use the information in Example 4. How many people reposted the link after the sixth round? 19,530

Example 5 Sum in Sigma Notation

(AL) **Write out the terms of the series you are adding in this example.** 16, 32, 64, 128, 256, 512, 1024, 2048

(OL) **How many terms will it take for the sum to be greater than 5000?** $n = 9$

(BL) **Write another sum in sigma notation that would have the same sum but would start at $k = 1$.** $\sum_{k=1}^{8} 4(2)^{k+1}$

Need Another Example?

Find $\sum_{n=1}^{12} 3 \cdot 2^{n-1}$. 12,285

---

2 Geometric Series A **geometric series** is the indicated sum of the terms of a geometric sequence. The sum of the first n terms of a series is denoted S_n. You can use either of the following formulas to find the partial sum S_n of the first n terms of a geometric series.

Key Concept Partial Sum of a Geometric Series

Given	The sum S_n of the first n terms is:
a_1 and n	$S_n = \dfrac{a_1 - a_1 r^n}{1 - r}, r \neq 1$
a_1 and a_n	$S_n = \dfrac{a_1 - a_n r}{1 - r}, r \neq 1$

A.SSE.4

Go Online!

Watch **Personal Tutor** videos to hear descriptions of problem solving. Try describing how to solve a problem for a partner. Have them ask you questions to help your understanding. **ELL**

Real-World Example 4 Find the Sum of a Geometric Series

MUSIC Refer to the beginning of the lesson. If the pattern continues, how many total people posted the link after the eighth round?

Five people reposted the link in the first round and there are 8 rounds of reposts. So, $a_1 = 5$, $r = 5$ and $n = 8$.

$$S_n = \frac{a_1 - a_1 r^n}{1 - r} \qquad \text{Sum formula}$$
$$S_8 = \frac{5 - 5 \cdot 5^8}{1 - 5} \qquad a_1 = 5, r = 5, \text{ and } n = 8$$
$$S_8 = \frac{-1,953,120}{-4} \qquad \text{Simplify the numerator and denominator.}$$
$$S_8 = 488,280 \qquad \text{Divide.}$$

There will be 488,280 total people who reposted the link after 8 rounds.

▸ **Guided Practice**

Find the sum of each geometric series.

4A. $a_1 = 2$, $n = 10$, $r = 3$ 59,048

4B. $a_1 = 2000$, $a_n = 125$, $r = \frac{1}{2}$ 3875

As with arithmetic series, sigma notation can also be used to represent geometric series.

A.SSE.4

Example 5 Sum in Sigma Notation

> **Watch Out!**
> **Sigma Notation** Notice in Example 5 that you are being asked to evaluate the sum from the 3rd term to the 10th term.

Find $\sum_{k=3}^{10} 4(2)^{k-1}$.

Find a_1, r, and n. In the first term, $k = 3$ and $a_1 = 4 \cdot 2^{3-1}$ or 16. The base of the exponential function is r, so $r = 2$. There are $10 - 3 + 1$ or 8 terms, so $n = 8$.

$$S_n = \frac{a_1 - a_1 r^n}{1 - r} \qquad \text{Sum formula}$$
$$= \frac{16 - 16(2)^8}{1 - 2} \qquad a_1 = 16, r = 2, \text{ and } n = 8$$
$$= 4080 \qquad \text{Use a calculator.}$$

▸ **Guided Practice**

Find each sum.

5A. $\sum_{k=4}^{12} \frac{1}{4} \cdot 3^{k-1}$ 66,426.75

5B. $\sum_{k=2}^{9} \frac{2}{3} \cdot 4^{k-1}$ 58,253.333

Differentiated Instruction **OL** **BL** **ELL**

IF you think students might be interested in learning how this lesson applies to real-world situations,

THEN have students research how biologists and ecologists use geometric series in their work to count and predict the population changes for various organisms.

You can use the formula for the sum of a geometric series to help find a particular term of the series.

Example 6 Find the First Term of a Series

A.SSE.4

Find a_1 in a geometric series for which $S_n = 13{,}116$, $n = 7$, and $r = 3$.

$$S_n = \frac{a_1 - a_1 r^n}{1 - r} \qquad \text{Sum formula}$$

$$13{,}116 = \frac{a_1 - a_1(3^7)}{1 - 3} \qquad S_n = 13{,}116, r = 3 \text{ and } n = 7$$

$$13{,}116 = \frac{a_1(1 - 3^7)}{1 - 3} \qquad \text{Distributive Property}$$

$$13{,}116 = \frac{-2186 a_1}{-2} \qquad \text{Subtract}$$

$$13{,}116 = 1093 a_1 \qquad \text{Simplify}$$

$$12 = a_1 \qquad \text{Divide each side by 1093}$$

▶ **Guided Practice**

6. Find a_1 in a geometric series for which $S_n = -26{,}240$, $n = 8$, and $r = -3$. **16**

Check Your Understanding ◯ = Step-by-Step Solutions begin on page R11.

Go Online! for a Self-Check Quiz

Example 1 **1.** 🔵 REGULARITY Dean is making a family tree for his grandfather. He was able to trace
A.SSE.4 many generations. If Dean could trace his family back 10 generations, how many
ancestors are in the 10th generation back? **1024**

Example 2 Write an equation for the nth term of each geometric sequence. **3.** $a_n = 18\left(\frac{1}{3}\right)^{n-1}$
A.SSE.4
2. 2, 4, 8, ... $a_n = 2 \cdot 2^{n-1}$ **3.** 18, 6, 2, ...   **4.** −4, 16, −64, ... $a_n = -4(-4)^{n-1}$

⑤ $a_2 = 4$, $r = 3$ $a_n = \frac{4}{3}(3)^{n-1}$ **6.** $a_6 = \frac{1}{8}$, $r = \frac{3}{4}$ **7.** $a_2 = -96$, $r = -8$ $a_n = 12(-8)^{n-1}$

6. $a_n = \frac{128}{243}\left(\frac{3}{4}\right)^{n-1}$

Example 3 Find the geometric means of each sequence.
A.SSE.4
8. 0.25, ?, ?, ?, 64 1, 4, 16 or −1, 4, −16   **9.** 0.20, ?, ?, ?, 125 1, 5, 25 or −1, 5, −25

Example 4 **10.** GAMES Miranda arranges some rows of dominoes so that after she knocks over the
A.SSE.4 first one, each domino knocks over two more dominoes when it falls. If there are ten
rows, how many dominoes does Miranda use? **1023**

Example 5 Find the sum of each geometric series.
A.SSE.4
11. $\displaystyle\sum_{k=1}^{6} 3(4)^{k-1}$ **4095**   **12.** $\displaystyle\sum_{k=1}^{8} 4\left(\frac{1}{2}\right)^{k-1}$ **7.96875**

Example 6 Find a_1 for each geometric series described.
A.SSE.4
13. $S_n = 85\frac{5}{16}$, $r = 4$, $n = 6$ $\frac{1}{16}$ **14.** $S_n = 91\frac{1}{12}$, $r = 3$, $n = 7$ $\frac{1}{12}$

15. $S_n = 1020$, $a_n = 4$, $r = \frac{1}{2}$ 512   **16.** $S_n = 121\frac{1}{3}$, $a_n = \frac{1}{3}$, $r = \frac{1}{3}$ 81

Example 6 Find the First Term of a Series

AL What sum is 13,116? S_7

OL Which values would we need to use this formula to solve for r? S_n, a_1, n

BL Give a problem statement where the answer would be $a_1 = 1$. Example: $S_n = 7$, $n = 3$, $r = 2$

Need Another Example?

Find a_1 in a geometric series for which $S_8 = 765$, $n = 8$, and $r = 2$. 3

Practice

Formative Assessment Use Exercises 1–16 to assess students' understanding of the concepts in this lesson.

The Practice and Problem Solving exercises assess the content taught in the lesson. The Preparing for Assessment page is meant to be used as preparation for assessment.

> **MP Teaching the Mathematical Practices**
>
> **Regularity** Mathematically proficient students apply math to problems arising in everyday life. As they work to solve Exercise 1, encourage students to relate other situations in their lives that can be modeled with sequences and series.

Teaching the Mathematical Practices

Perseverance Mathematically proficient students evaluate their own problem solving process. In Exercise 41, encourage students to talk about how they solved the problem and any things they learned that could help them solve a similar problem more efficiently.

Additional Answers

64. $S_n = \dfrac{a_1 - a_1 r^n}{1-r}$

$= \dfrac{a_1 - a_1 r^{n-1} \cdot r}{1-r}$

$= \dfrac{a_1 - a_n r}{1-r}$

65. $S_n = \dfrac{a_1 - a_n r}{1-r}$

(Alternate sum formula)

$a_n = a_1 \cdot r^{n-1}$

(Formula for nth term)

$\dfrac{a_n}{r^{n-1}} = a_1$

(Divide each side by r^{n-1}.)

$S_n = \dfrac{\dfrac{a_n}{r^{n-1}} - a_n r}{1-r}$

(Substitution.)

$= \dfrac{\dfrac{a_n}{r^{n-1}} - \dfrac{a_n r \cdot r^{n-1}}{r^{n-1}}}{1-r}$

$\left(\text{Multiply by } \dfrac{\frac{r^{n-1}}{r^{n-1}}}{1}. \right)$

$= \dfrac{\dfrac{a_n(1-r^n)}{r^{n-1}}}{1-r}$

(Simplify.)

$= \dfrac{a_n(1-r^n)}{r^{n-1}(1-r)}$

(Divide by $(1-r)$.)

$= \dfrac{a_n(1-r^n)}{r^{n-1} - r^n}$

(Simplify.)

67. Sample answer: $n-1$ needs to change to n, and the 10 needs to change to a 9. When this happens, the terms for both series will be identical (a_1 in the first series will equal a_0 in the second series, and so on), and the series will be equal to each other.

Practice and Problem Solving

Extra Practice is on page R6.

Example 1
A.SSE.4

17. WEATHER Heavy rain in Brieanne's town caused the Pecos River to rise. The river rose three inches the first day, and each day after rose twice as much as the previous day. How much did the river rise in five days? 93 in.

Find a_n for each geometric sequence.

18. $a_1 = 2400$, $r = \frac{1}{4}$, $n = 7$ $\frac{75}{128}$ or 0.5859375 **19.** $a_1 = 800$, $r = \frac{1}{2}$, $n = 6$ 25

20. $a_1 = \frac{2}{9}$, $r = 3$, $n = 7$ 162 **21.** $a_1 = -4$, $r = -2$, $n = 8$ 512

22. BIOLOGY A certain bacteria grows at a rate of 3 cells every 2 minutes. If there were 260 cells initially, how many are there after 21 minutes? 864,567

Example 2
A.SSE.4

Write an equation for the nth term of each geometric sequence.

23. $-3, 6, -12, \ldots$

24. $288, -96, 32, \ldots$

25. $-1, 1, -1, \ldots$

26. $\frac{1}{3}, \frac{2}{9}, \frac{4}{27}, \ldots$

27. $8, 2, \frac{1}{2}, \ldots$

28. $12, -16, \frac{64}{3}, \ldots$

29. $a_3 = 28$, $r = 2$

30. $a_4 = -8$, $r = 0.5$

31. $a_6 = 0.5$, $r = 6$

32. $a_3 = 8$, $r = \frac{1}{2}$

33. $a_4 = 24$, $r = \frac{1}{3}$

34. $a_4 = 80$, $r = 4$
$a_n = \frac{5}{4}(4)^{n-1}$

Example 3
A.SSE.4

Find the geometric means of each sequence.

35. $810, \underline{?}, \underline{?}, \underline{?}, 10$

36. $640, \underline{?}, \underline{?}, \underline{?}, 2.5$

37. $\frac{7}{2}, \underline{?}, \underline{?}, \underline{?}, \frac{56}{81}$

38. $\frac{729}{64}, \underline{?}, \underline{?}, \underline{?}, \frac{324}{9}$

39. Find two geometric means between 3 and 375. 15, 75

40. Find two geometric means between 16 and -2. $-8, 4$

Example 4
A.SSE.4

41. PERSEVERANCE A certain water filtration system can remove 70% of the contaminants each time a sample of water is passed through it. If the same water is passed through the system four times, what percent of the original contaminants will be removed from the water sample? 99.19%

Find the sum of each geometric series.

42. $a_1 = 36$, $r = \frac{1}{3}$, $n = 8$ 53.9918

43. $a_1 = 16$, $r = \frac{1}{2}$, $n = 9$ 31.9375

44. $a_1 = 240$, $r = \frac{3}{4}$, $n = 7$ 831.855

45. $a_1 = 360$, $r = \frac{4}{3}$, $n = 8$ 9707.82

46. VACUUMS A vacuum claims to pick up 80% of the dirt every time it is run over the carpet. Assuming this is true, what percent of the original amount of dirt is picked up after the seventh time the vacuum is run over the carpet? 99.99%

Example 5
A.SSE.4

Find the sum of each geometric series.

47. $\displaystyle\sum_{k=1}^{7} 4(-3)^{k-1}$ 2188

48. $\displaystyle\sum_{k=1}^{8} (-3)(-2)^{k-1}$ 255

49. $\displaystyle\sum_{k=1}^{9} (-1)(4)^{k-1}$ $-87,381$

50. $\displaystyle\sum_{k=1}^{10} 5(-1)^{k-1}$ 0

Example 6
A.SSE.4

Find a_1 for each geometric series described.

51. $S_n = -2912$, $r = 3$, $n = 6$ -8

52. $S_n = -10,922$, $r = 4$, $n = 7$ -2

53. $S_n = 1330$, $a_n = 486$, $r = \frac{3}{2}$ 64

54. $S_n = 4118$, $a_n = 128$, $r = \frac{2}{3}$ 1458

55. $a_n = 1024$, $r = 8$, $n = 5$ 0.25

56. $a_n = 1875$, $r = 5$, $n = 7$ $\frac{3}{25}$

Answers to margin (right column):

23. $a_n = (-3)(-2)^{n-1}$

24. $a_n = 288\left(-\frac{1}{3}\right)^{n-1}$

25. $a_n = (-1)(-1)^{n-1}$

26. $a_n = \frac{1}{3}\left(\frac{2}{3}\right)^{n-1}$

27. $a_n = 8 \cdot \left(\frac{1}{4}\right)^{n-1}$

28. $a_n = 12\left(-\frac{4}{3}\right)^{n-1}$

29. $a_n = 7(2)^{n-1}$

30. $a_n = -64(0.5)^{n-1}$

31. $a_n = \frac{1}{15,552}(6)^{n-1}$

32. $a_n = 32\left(\frac{1}{2}\right)^{n-1}$

33. $a_n = 648\left(\frac{1}{3}\right)^{n-1}$

35. 270, 90, 30 or $-270, 90, -30$

36. 160, 40, 10 or $-160, 40, -10$

37. $\frac{7}{3}, \frac{14}{9}, \frac{28}{27}$ or $-\frac{7}{3}, \frac{14}{9}, -\frac{28}{27}$

38. $\frac{243}{16}, \frac{81}{4}, 27$ or $-\frac{243}{16}, \frac{81}{4}, -27$

Differentiated Homework Options

Levels	AL Basic	OL Core	BL Advanced
Exercises	17–57, 64–80	17–57 odd, 58–80	58–80
2-Day Option	17–57 odd, 73–80	17–57, 73–80	
	18–56 even, 64–72	58–72	

You can use ALEKS to provide additional remediation support with personalized instruction and practice.

57 SCIENCE One minute after it is released, a gas-filled balloon has risen 100 feet. In each succeeding minute, the balloon rises only 50% as far as it rose in the previous minute. How far will it rise in 5 minutes? **193.75 ft**

58. CHEMISTRY Radon has a half-life of about 4 days. This means that about every 4 days, half of the mass of radon decays into another element. How many grams of radon remain from an initial 60 grams after 4 weeks? **about 0.46875 g**

59. **MP** REASONING A virus goes through a computer, infecting the files. If one file was infected initially and the total number of files infected doubles every minute, how many files will be infected in 20 minutes? **524,288**

60. GEOMETRY In the figure, the sides of each equilateral triangle are twice the size of its inscribed triangle. If the pattern continues, find the sum of the perimeters of the first eight triangles. **about 119.5 cm**

20 cm

61. PENDULUMS The first swing of a pendulum travels 30 centimeters. If each subsequent swing travels 95% as far as the previous swing, find the total distance traveled by the pendulum after the 30th swing. **about 471 cm**

62. PHONE CHAINS A school established a phone chain in which every staff member calls two other staff members to notify them when the school closes due to weather. The first round of calls begins with the superintendent calling both principals. If there are 94 total staff members and employees at the school, how many rounds of calls are there? **7**

63. TABLETS High Tech Electronics advertises a monthly installment plan for the purchase of a popular tablet. A buyer pays $40 at the end of the first month, $44 at the end of the second month, $48.40 at the end of the third month, and so on for one year.

63a. $53.24, $64.42, $94.32

a. What will the payments be at the end of the 4th, 6th, and 10th months?

b. Find the total cost of the tablet. **$855.37**

c. Why is the cost found in part **b** not entirely accurate?

63c. Each payment made is rounded to the nearest cent, so the sum of the payments may be off by several cents.

A.SSE.4

H.O.T. Problems   Use Higher-Order Thinking Skills

64. PROOF Derive the General Sum Formula using the Alternate Sum Formula. See margin.

65. PROOF Derive a sum formula that does not include a_1. See margin.

66. OPEN-ENDED Write a geometric series for which $r = \frac{3}{4}$ and $n = 6$.

66. Sample answer:
$256 + 192 + 144 + 108 + 81 + \frac{243}{4}$

67. **MP** REASONING Explain how $\sum_{k=1}^{10} 3(2)^{k-1}$ needs to be altered to refer to the same series if $k = 1$ changes to $k = 0$. Explain your reasoning. See margin.

68. PROOF Prove the formula for the nth term of a geometric sequence. See margin.

69. CHALLENGE The fifth term of a geometric sequence is $\frac{1}{27}$th of the eighth term. If the ninth term is 702, what is the eighth term? **234**

70. CHALLENGE Use the fact that h is the geometric mean between x and y in the figure at the right to find h^4 in terms of x and y. x^2y^2

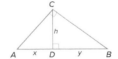

71. OPEN-ENDED Write a geometric series with 6 terms and a sum of 252.   Sample answer:
$4 + 8 + 16 + 32 + 64 + 128$

72. **e** WRITING IN MATH How can you classify a sequence? Explain your reasoning. See margin.

Levels of Complexity Chart

The levels of the exercises progress from 1 to 3, with Level 1 indicating the lowest level of complexity.

Exercises	18–57	58–61, 73–80	62–72
Level 3			●
Level 2		●	
Level 1	●		

Additional Answers

68. Sample answer:

Let $a_n =$ the nth term of the sequence and $r =$ the common ratio.

$a_2 = a_1 \cdot r$	Definition of the second term of a geometric sequence
$a_3 = a_2 \cdot r$	Definition of the third term of a geometric sequence
$a_3 = a_1 \cdot r \cdot r$	Substitution
$a_3 = a_1 \cdot r^2$	Associative Property of Multiplication
$a_3 = a_1 \cdot r^{3-1}$	$3 - 1 = 2$
$a_n = a_1 \cdot r^{n-1}$	$n = 3$

72. Sample answer: A series is arithmetic if every pair of consecutive terms shares a common difference. A series is geometric if every pair of consecutive terms shares a common ratio. If the series displays both qualities, then it is both arithmetic and geometric. If the series displays neither quality, then it is neither geometric nor arithmetic.

Assess

Crystal Ball Ask students to describe how they think their knowledge of sums of geometric series might help them learn about sums of infinite geometric series in the next lesson.

MP **Standards for Mathematical Practice**

Emphasis On	Exercises
1 Make sense of problems and persevere in solving them.	41, 68–69, 76
2 Reason abstractly and quantitatively.	59, 67, 70–72, 77–79
3 Construct viable arguments and critique reasoning of others.	64–66
4 Model with mathematics.	10, 17, 22, 46, 57–58, 61–63, 73
6 Attend to precision.	80
8 Look for and express regularity in repeated reasoning.	1, 60, 74–75

Go Online!

Self-Check Quiz

Students can use *Self-Check Quizzes* to check their understanding of this lesson.

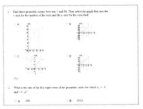

Preparing for Assessment

Exercises 73–80 require students to use the skills they will need on standardized tests. Each exercise is dual-coded with content standards and mathematical practice standards.

Dual Coding		
Items	Content Standards	ⓂⒶ Mathematical Practices
73	A.SSE.4	4
74	A.SSE.4	8
75	A.SSE.4	8
76	A.SSE.4	1
77	A.SSE.4	2
78	F.LE.2	2
79	F.LE.2	2
80	A.SSE.4	6

Diagnose Student Errors

Survey student responses for each item. Class trends may indicate common errors and misconceptions.

74.

A	Used 4 instead of 8 for n
B	Set $a_1 = 30$, the number of minutes
C	Raised r to the $n-1$ power instead of the nth power
D	CORRECT

76.

A	Used $a_1 - r^n$ as the numerator in the formula instead of $a_1 - a_1 r^n$
B	Used r^{n-1} instead of r^n in the formula
C	Divided by r instead of $1 - r$ in the formula
D	CORRECT
E	Mixed up a_1 and r in the formula

77.

A	Evaluated expression for $k = 2$
B	Evaluated expression for $k = 12$
C	CORRECT
D	Evaluated the expression starting with $k = 1$ instead of $k = 2$
E	Evaluated the expression $9k$ instead of $9k - 4$

78.

A	Chose incorrect function
B	Chose incorrect function
C	Chose incorrect function
D	Chose incorrect function
E	CORRECT

Preparing for Assessment

73. Janelle makes jewelry. She e-mails a link for her website to four potential customers. They each forward the link to three friends. If the link is forwarded again following the same pattern, how many people will receive the link on the tenth round of e-mails? ⓂⒶ 4 A.SSE.4 **78,732**

74. A certain bacteria doubles in number every 30 minutes. If there were 140 cells initially, how many cells are there after 4 hours? ⓂⒶ 8 A.SSE.4 **D**
- ○ A 1120
- ○ B 3840
- ○ C 17,920
- ○ D 35,840

75. Radon has a half-life of 4 days. This means that about every 4 days, half of the mass of radon decays into another element. How many grams of radon remain from an initial 120 grams after 28 days? ⓂⒶ 8 A.SSE.4 **0.9375**

76. Refer to question 73. After the tenth round, what is the total number of people who have received e-mails containing a link to Janelle's website? ⓂⒶ 1 A.SSE.4 **D**
- ○ A 29,522.5
- ○ B 39,364
- ○ C 78,731
- ○ D 118,096
- ○ E 1,048,576

77. Find $\sum_{k=3}^{7} \left(\frac{1}{2} \cdot 4^{k-1}\right)$. ⓂⒶ 2 A.SSE.4 **C**
- ○ A 170.5
- ○ B 680
- ○ C 2728
- ○ D 5456
- ○ E 43,688

78. The curve below could be part of the graph of which function? ⓂⒶ 2 F.LE.2 **E**

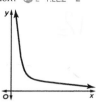

- ○ A $y = \sqrt{x}$
- ○ B $y = x^2 - 5x + 4$
- ○ C $y = -x + 20$
- ○ D $y = \log x$
- ○ E $xy = 4$

79. A bacteria grows exponentially so that it doubles every 12 days. If there are 500 bacteria initially, find the number of bacteria after: ⓂⒶ 2 F.LE.2
- **a.** 12 days
  > 1000
- **b.** 60 days
  > 16,000

80d. Sample answer: The fifth term is $\frac{5}{2^3}$ and $\left(\frac{5}{2^5}\right)^2 = \frac{5}{2} \cdot \frac{5}{2^9}$

80. MULTI-STEP Given the geometric sequence
$$a_n = \frac{5}{2^n}:$$ ⓂⒶ 6 A.SSE.4
- **a.** Find the first term.
  > $\frac{5}{2}$
- **b.** Find the common ratio.
  > $\frac{1}{2}$
- **c.** Find the ninth term.
  > $\frac{5}{2^9} = \frac{5}{512}$
- **d.** Show that the fifth term is the geometric mean between the first term and the ninth term.
  >
- **e.** Find the sum of the first nine terms.
  > $\frac{2555}{512}$

Differentiated Instruction ⓄⓁ ⒷⓁ

Extension Write the expression for the nth term of a geometric sequence, $a_1 r^{n-1}$, on the board. Ask students to write an expression for the next term, with term number $(n+1)$, by replacing n with $(n+1)$ in the expression $a_1 r^{n-1}$. Then have them show that $\dfrac{a_{n+1}}{a_n}$ is equal to the common ratio r. $\dfrac{a_1 r^{(n+1)-1}}{a_1 r^{n-1}} = \dfrac{r^n}{r^{n-1}} = r^{n-(n-1)} = r^1 = r$

Logarithms and Logarithmic Functions

SUGGESTED PACING (DAYS)

90 min.	0.75
45 min.	1.5

Instruction

Track Your Progress

Objectives

1 Evaluate logarithmic expressions.

2 Graph logarithmic functions.

Mathematical Background

The equation $y = \log_b x$ is read "y equals the logarithm to the base b of the number x." The base b is always positive and $b \neq 1$. Because the equation $y = \log_b x$ is equivalent to the exponential equation $x = b^y$, a logarithm is an exponent. It is the exponent that the base b requires in order to equal the number x.

THEN	NOW	NEXT
F.IF.7b Graph functions expressed symbolically and show key features of the graph, by hand in simple cases and using technology for more complicated cases. Graph square root, cube root, and piecewise-defined functions, including step functions and absolute value functions.	**A.SSE.2** Use the structure of an expression to identify ways to rewrite it. **F.IF.7e** Graph exponential and logarithmic functions, showing intercepts and end behavior, and trigonometric functions, showing period, midline, and amplitude.	**A.SSE.2** Use the structure of an expression to identify ways to rewrite it.

Go Online! All of these resources and more are available at connectED.mcgraw-hill.com

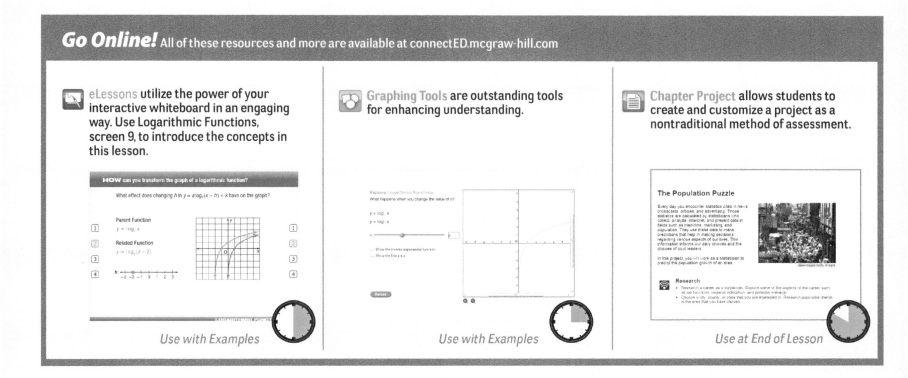

eLessons utilize the power of your interactive whiteboard in an engaging way. Use Logarithmic Functions, screen 9, to introduce the concepts in this lesson.

Use with Examples

Graphing Tools are outstanding tools for enhancing understanding.

Use with Examples

Chapter Project allows students to create and customize a project as a nontraditional method of assessment.

Use at End of Lesson

OER **Using Open Educational Resources**

Exploration Have students explore the Exponential and Logarithmic Functions tools on **Hotmath.com** to make connections between the two functions before beginning the lesson. *Use as homework*

Differentiate Your Resources

Extra Practice Additional practice or homework; Skills Practice is best for approaching-level students and Practice is best for on-level and beyond-level students

Skills Practice

Practice

Word Problem Practice

Intervention Reteaching and vocabulary activities that can be used with struggling or absent students and as ELL support

Extension Activities that can be used to extend lesson concepts

Study Guide and Intervention

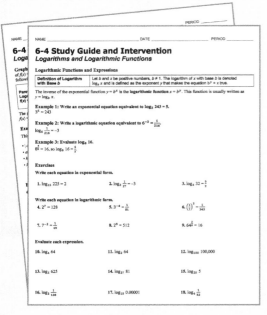

Study Notebook

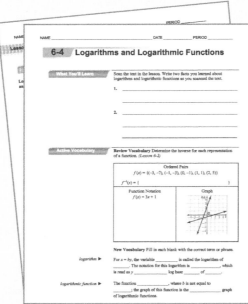

Enrichment

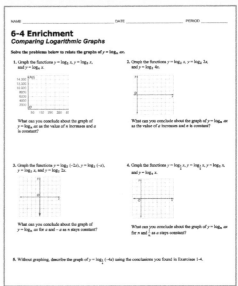

LESSON 4

Logarithms and Logarithmic Functions

Then	Now	Why?
• You found the inverse of a function.	**1** Evaluate logarithmic expressions. **2** Graph logarithmic functions.	• Many scientists believe the extinction of the dinosaurs was caused by an asteroid striking Earth. Astronomers use the Palermo scale to classify objects near Earth based on the likelihood of impact. To make comparing several objects easier, the scale was developed using *logarithms*. The Palermo scale value of any object can be found using the equation $PS = \log_{10} R$, where R is the relative risk posed by the object.

New Vocabulary
logarithm
logarithmic function

MP **Mathematical Practices**
4 Model with mathematics.
6 Attend to precision.
7 Look for and make use of structure.

Content Standards
A.SSE.2 Use the structure of an expression to identify ways to rewrite it.
F.IF.4 For a function that models a relationship between two quantities, interpret key features of graphs and tables in terms of the quantities, and sketch graphs showing key features given a verbal description of the relationship.
F.IF.7.e Graph exponential and logarithmic functions, showing intercepts and end behavior, and trigonometric functions, showing period, midline, and amplitude.
F.BF.3 Identify the effect on the graph of replacing $f(x)$ by $f(x) + k, k\, f(x), f(kx)$, and $f(x + k)$ for specific values of k (both positive and negative); find the value of k given the graphs. Experiment with cases and illustrate an explanation of the effects on the graph using technology. Include recognizing even and odd functions from their graphs and algebraic expressions for them.

1 **Logarithmic Functions and Expressions** Consider the exponential function $f(x) = 2^x$ and its inverse. Recall that you can graph an inverse function by interchanging the x- and y-values in the ordered pairs of the function. Notice that the domain and range of the functions are switched.

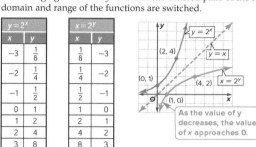

$y = 2^x$	
x	y
-3	$\frac{1}{8}$
-2	$\frac{1}{4}$
-1	$\frac{1}{2}$
0	1
1	2
2	4
3	8

$x = 2^y$	
x	y
$\frac{1}{8}$	-3
$\frac{1}{4}$	-2
$\frac{1}{2}$	-1
1	0
2	1
4	2
8	3

As the value of y decreases, the value of x approaches 0.

The inverse of $y = 2^x$ can be defined as $x = 2^y$. In general, the inverse of $y = b^x$ is $x = b^y$. In $x = b^y$, the variable y is called the **logarithm** of x. This is usually written as $y = \log_b x$, which is read *y equals log base b of x.*

🔷 Key Concept Logarithm with Base *b*

Words   Let b and x be positive numbers, $b \neq 1$. The *logarithm* of x with base b is denoted $\log_b x$ and is defined as the exponent y that makes the equation $b^y = x$ true.

Symbols   Suppose $b > 0$ and $b \neq 1$. For $x > 0$, there is a number y such that

$$\log_b x = y \text{ if and only if } b^y = x.$$

Example   If $\log_3 27 = y$, then $3^y = 27$.

The definition of logarithms can be used to express logarithms in exponential form.
A.SSE.2

Example 1   Logarithmic to Exponential Form

Write each equation in exponential form.
a. $\log_2 8 = 3$

$\log_2 8 = 3 \rightarrow 8 = 2^3$

b. $\log_4 \frac{1}{256} = -4$

$\log_4 \frac{1}{256} = -4 \rightarrow \frac{1}{256} = 4^{-4}$

▷ **Guided Practice**
1A. $\log_4 16 = 2$ $16 = 4^2$
1B. $\log_3 729 = 6$ $729 = 3^6$

Launch

Have students read the Why? section of the lesson. Ask:
- **What kinds of objects are found near Earth?**
  Sample answer: asteroids, meteors, comets

- **What is meant by "likelihood of impact"?**
  Sample answer: probability that an object hits Earth

Teach

Ask the scaffolded questions for each example to build conceptual understanding for students at all levels.

1 Logarithmic Functions and Expressions

Example 1 Logarithmic to Exponential Form

AL What is exponential form? $b^x = y$

OL Which part of the exponential form is the logarithm? the exponent

BL If $b > 1$, when will a logarithm be negative? when $x < 1$

Need Another Example?
Write each equation in exponential form.
a. $\log_3 9 = 2$ $9 = 3^2$
b. $\log \frac{1}{100} = -2$ $\frac{1}{100} = 10^{-2}$

MP **Teaching the Mathematical Practices**
Help students develop the mathematical practices by asking questions like these.

Attend to precision.
Help students recognize the relationships when graphing a logarithmic function. For example, ask:

- **What is the domain of a logarithmic function?** real numbers greater than 0

- **What is the range of a logarithmic function?** real numbers

- **What is the inverse of a logarithmic function?** exponential function

---

Go Online!

Interactive Whiteboard
Use the *eLesson* or *Lesson Presentation* to present this lesson.

Science Photo Library - ANDRZEJ WOJCICKI/Brand X Pictures/Getty Images

Example 2 Exponential to Logarithmic Form

AL What is logarithmic form? $y = \log_b x$

OL How does the base of an exponential equation compare to the base of the related logarithmic equation? They are the same.

BL Why can't $4^{\frac{1}{2}} = 2$ be put into radical form and then translated into logarithmic form? $4^{\frac{1}{2}} = 2$ in radical form would be $\sqrt{4} = 2$. The closest thing to putting this in logarithmic form would be $\sqrt{} = \log_4 2$. The radical sign by itself does not make sense.

Need Another Example?

Write each equation in logarithmic form.
a. $5^3 = 125$ $\log_5 125 = 3$
b. $27^{\frac{1}{3}} = 3$ $\log_{27} 3 = \frac{1}{3}$

Example 3 Evaluate Logarithmic Expressions

AL Why is 16 written as 4^2? So that both sides of the equation have the same base and we can apply the Property of Equality for Exponential Functions.

OL If $x = b^2$, to what is $\log_b x$ equal? 2

BL If x is the nth root of b, what is $\log_b x$ equal to? $\frac{1}{n}$

Need Another Example?

Evaluate $\log_3 243$. 5

The definition of logarithms can also be used to write exponential equations in logarithmic form.

A.SSE.2

Example 2   Exponential to Logarithmic Form

Write each equation in logarithmic form.

a. $15^3 = 3375$

$15^3 = 3375 \rightarrow \log_{15} 3375 = 3$

b. $4^{\frac{1}{2}} = 2$

$4^{\frac{1}{2}} = 2 \rightarrow \log_4 2 = \frac{1}{2}$

▸ **Guided Practice**

2A. $4^3 = 64$ $\log_4 64 = 3$

2B. $125^{\frac{1}{3}} = 5$ $\log_{125} 5 = \frac{1}{3}$

You can use the definition of a logarithm to evaluate a logarithmic expression.

A.SSE.2

Example 3   Evaluate Logarithmic Expressions

Evaluate $\log_{16} 4$.

$\log_{16} 4 = y$	Let the logarithm equal y.
$4 = 16^y$	Definition of logarithm
$4^1 = (4^2)^y$	$16 = 4^2$
$4^1 = 4^{2y}$	Power of a Power
$1 = 2y$	Property of Equality for Exponential Functions
$\frac{1}{2} = y$	Divide each side by 2.

Thus, $\log_{16} 4 = \frac{1}{2}$.

▸ **Guided Practice**

Evaluate each expression.

3A. $\log_3 81$ 4

3B. $\log_{\frac{1}{2}} 256$ −8

> **Watch Out!**
>
> **Logarithmic Base** It is easy to get confused about which number is the base and which is the exponent in logarithmic equations. Consider highlighting each number as you solve to help organize your calculations.

2 Graphing Logarithmic Functions $y = \log_b x$, where $b \neq 1$, is a **logarithmic function**. The graph of $f(x) = \log_b x$ represents a parent graph of the logarithmic functions.

Key Concept Parent Function of Logarithmic Functions

Parent function:	$f(x) = \log_b x$	Type of graph:	continuous, one-to-one
Domain:	$(0, \infty)$ or $\{x \mid x > 0\}$	Range:	{all real numbers}
Asymptote:	$f(x)$-axis	Intercept:	$(1, 0)$
Symmetry:	none	Extrema:	none

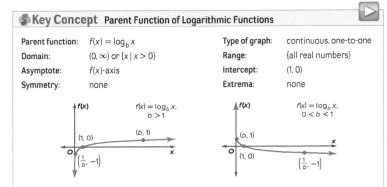

Differentiated Instruction **OL** **BL**

Logical Learners After discussing the definition of logarithm, write $y = 2x$ on the board and ask students to rewrite the equation with x in terms of y. $x = \frac{1}{2}y$ Repeat for $y = x^2$. $x = \pm\sqrt{y}$ Now write $y = 2^x$ on the board and ask students to rewrite this equation with x in terms of y. This will likely have students baffled. Explain that the rewritten equation is $x = \log_2 y$. Stress that a logarithm is defined as the inverse of an exponential function.

Go Online!

The most up-to-date resources available for your program can be found at <u>connectED.mcgraw-hill.com</u>.

Example 4   Graph Logarithmic Functions

F.IF.7.e

Graph each function.

a. $f(x) = \log_5 x$

> **Step 1** Identify the base.
> $b = 5$

> **Step 2** Determine points on the graph.
> Because $5 > 1$, use the points $\left(\frac{1}{b}, -1\right)$, $(1, 0)$, and $(b, 1)$.

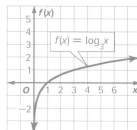

> **Step 3** Plot the points and sketch the graph.
> $\left(\frac{1}{b}, -1\right) \rightarrow \left(\frac{1}{5}, -1\right)$
> $(1, 0)$
> $(b, 1) \rightarrow (5, 1)$

b. $f(x) = \log_{\frac{1}{3}} x$

> **Step 1** $b = \frac{1}{3}$

> **Step 2** $0 < \frac{1}{3} < 1$, so use the points $\left(\frac{1}{3}, 1\right)$, $(1, 0)$ and $(3, -1)$.

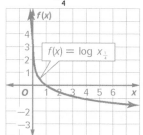

> **Step 3** Sketch the graph.

$\frac{1}{\frac{1}{3}} = \frac{1}{1} \cdot \frac{3}{1} = 3$

▶ **Guided Practice** **4A, 4B.** See margin.

4A. $f(x) = \log_2 x$ **4B.** $f(x) = \log_{\frac{1}{8}} x$

The same techniques used to transform the graphs of other functions you have studied can be applied to the graphs of logarithmic functions.

🔑 Key Concept Transformations of Logarithmic Functions

$$f(x) = a \log_b (x - h) + k$$

h – Horizontal Translation	k – Vertical Translation
h units right if h is positive $\lvert h \rvert$ units left if h is negative	k units up if k is positive $\lvert k \rvert$ units down if k is negative

a – Orientation and Shape	
If $a < 0$, the graph is reflected across the x-axis.	If $\lvert a \rvert > 1$, the graph is stretched vertically. If $0 < \lvert a \rvert < 1$, the graph is compressed vertically.

Differentiated Instruction **ELL**

Intermediate/Advanced The word *logarithm* derives from the Greek *logos* which means *reason or proportion* and *arithmos* which means *number*. Have students discuss how this can help them remember the definition of *logarithm*.

Additional Answers (Guided Practice)

4A.

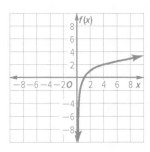

4B.

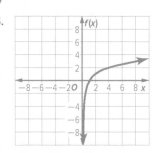

---

2 Graphing Logarithmic Functions

Example 4 Graph Logarithmic Functions

AL How could you use the graph of $y = 5^x$ to graph $f(x) = \log_5 x$? Take points from $y = 5^x$ and switch the x and y values to get points on $f(x) = \log_5 x$.

OL What is the end behavior of $y = \log_b x$?
If $b > 1$: As $x \to 0$, $y \to -\infty$ and as $x \to \infty$, $y \to \infty$. If $0 < b < 1$: As $x \to 0$, $y \to \infty$ and as $x \to \infty$, $y \to -\infty$.

BL If b could be equal to 1, what would the domain of $f(x) = \log_1 x$ be? Why? {1}, because if you raise 1 to any power you get 1

Need Another Example?

Graph each function.
a. $f(x) = \log_3 x$

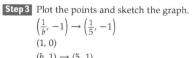

b. $f(x) = \log_{\frac{1}{4}} x$

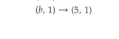

Teaching Tip

Pacing Students have not covered logarithmic functions before and are likely to find them confusing. Expect students to need extra time to absorb the material in this lesson before continuing with the rest of the chapter.

Example 5 Graph Logarithmic Functions

AL Of what function is $f(x) = 3\log_{10} x + 1$ a transformation? $f(x) = \log_{10} x$

OL Write the function that would translate $f(x) = \log_3 x$ up 2 and to the left 5. $f(x) = \log_3(x + 5) + 2$

BL If you reflect $f(x) = \log_b x$ in the x-axis, is your transformed function an increasing or decreasing function? If $b > 1$, decreasing. If $0 < b < 1$, then increasing.

Need Another Example?

Graph each function.

a. $f(x) = \frac{1}{3}\log_6 x - 1$

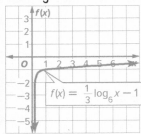

$f(x) = \frac{1}{3}\log_6 x - 1$

b. $f(x) = 4\log_{\frac{1}{3}}(x + 2)$

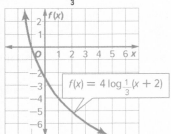

$f(x) = 4\log_{\frac{1}{3}}(x + 2)$

Example 6 Find Inverses of Exponential Functions

AL What value on the Richter scale would represent an intensity of 1? 1

OL What is the difference in intensity of an earthquake with a 3.2 rating compared to an earthquake with a rating of 4.5? 3003.79

BL Explain the meaning of the inputs and outputs of $y = \log_{10} x + 1$ in this context. The inputs are the intensity of the earthquake and the output is the Richter scale rating.

Need Another Example?

Air Pressure At Earth's surface, the air pressure is defined as 1 atmosphere. Pressure decreases by about 20% for each mile of altitude. Atmospheric pressure can be modeled by $P = 0.8^x$, where x measures altitude in miles.

a. Find the atmospheric pressure in atmospheres at an altitude of 8 miles. 0.168 atmospheres

b. Write an equation for the inverse of the function. $P = \log_{0.8} x$

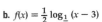

F.IF.7.e

Example 5   Graph Logarithmic Functions

Graph each function.

Study Tip
End Behavior In Example 5a, as x approaches infinity, $f(x)$ approaches infinity.

a. $f(x) = 3\log_{10} x + 1$

This represents a transformation of the graph of $f(x) = \log_{10} x$.

- $|a| = 3$: The graph stretches vertically
- $h = 0$: There is no horizontal shift.
- $k = 1$: The graph is translated 1 unit up.

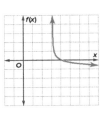

b. $f(x) = \frac{1}{2}\log_{\frac{1}{4}}(x - 3)$

This is a transformation of the graph of $f(x) = \log_{\frac{1}{4}} x$.

- $|a| = \frac{1}{2}$: The graph is compressed vertically.
- $h = 3$: The graph is translated 3 units to the right.
- $k = 0$: There is no vertical shift.

▸ **Guided Practice**

Graph each function. 5A, 5B. See margin.

5A. $f(x) = 2\log_3(x - 2)$

5B. $f(x) = \frac{1}{4}\log_{\frac{1}{2}}(x + 1) - 5$

A.SSE.2

Real-World Example 6   Find Inverses of Exponential Functions

EARTHQUAKES The Richter scale measures earthquake intensity. The increase in intensity between each number is 10 times. For example, an earthquake with a rating of 7 is 10 times more intense than one measuring 6. The intensity of an earthquake can be modeled by $y = 10^{x-1}$, where x is the Richter scale rating.

Real-World Link
The largest recorded earthquake in the United States was a magnitude 9.2 that struck Prince William Sound, Alaska, on Good Friday, March 28, 1964.

Source: United States Geological Survey

a. Use the information at the left to find the intensity of the strongest recorded earthquake in the United States.

$$\begin{aligned} y &= 10^{x-1} && \text{Original equation} \\ &= 10^{9.2-1} && \text{Substitute 9.2 for } x. \\ &= 10^{8.2} && \text{Simplify.} \\ &= 158{,}489{,}319.2 && \text{Use a calculator.} \end{aligned}$$

b. Write an equation of the form $y = \log_{10} x + c$ for the inverse of the function.

$$\begin{aligned} y &= 10^{x-1} && \text{Original equation} \\ x &= 10^{y-1} && \text{Replace } x \text{ with } y, \text{ replace } y \text{ with } x, \text{ and solve for } y. \\ y - 1 &= \log_{10} x && \text{Definition of logarithm} \\ y &= \log_{10} x + 1 && \text{Add 1 to each side.} \end{aligned}$$

▸ **Guided Practice**

6. Write an equation for the inverse of the function $y = 0.5^x$. $y = \log_{0.5} x$

Differentiated Instruction **ELL**

Write several exponential functions and their related logarithmic functions on the board. Ask students to read aloud each expression. For example, have students say *two to the third power is equal to eight* for $2^3 = 8$. Then have students say *the logarithm of eight with base two is equal to three* for $\log_2 8 = 3$.

Additional Answers (Guided Practice)

5A.

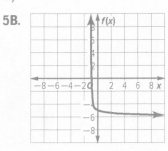

5B.

Check Your Understanding

 = Step-by-Step Solutions begin on page R11.

 **Go Online!** for a Self-Check Quiz

Example 1
A.SSE.2
Write each equation in exponential form.
1. $\log_8 512 = 3$ $8^3 = 512$
2. $\log_5 625 = 4$ $5^4 = 625$

Example 2
A.SSE.2
Write each equation in logarithmic form.
3. $11^3 = 1331$ $\log_{11} 1331 = 3$
4. $16^{\frac{3}{4}} = 8$ $\log_{16} 8 = \frac{3}{4}$

Example 3
A.SSE.2
Evaluate each expression.
5. $\log_{13} 169$ 2
6. $\log_2 \frac{1}{128}$ -7
7. $\log_6 1$ 0

Examples 4–5
F.IF.7e
Graph each function. 8–9. See margin.
8. $f(x) = \log_3 x$
9. $f(x) = \log_{\frac{1}{6}} x$
10. $f(x) = 4 \log_4 (x - 6)$
11. $f(x) = 2 \log_{\frac{1}{10}} x - 5$
10, 11. See margin on page 402.

Example 6
A.SSE.2
12. SCIENCE Use the information at the beginning of the lesson. The Palermo scale value of any object can be found using the equation $PS = \log_{10} R$, where R is the relative risk posed by the object. Write an equation in exponential form for the inverse of the function. $PS = 10^R$

Practice and Problem Solving

Extra Practice is on page R6.

Example 1
A.SSE.2
Write each equation in exponential form.
13. $\log_2 16 = 4$ $2^4 = 16$
14. $\log_7 343 = 3$ $7^3 = 343$
15. $\log_9 \frac{1}{81} = -2$ $9^{-2} = \frac{1}{81}$
16. $\log_3 \frac{1}{27} = -3$ $3^{-3} = \frac{1}{27}$
17. $\log_{12} 144 = 2$ $12^2 = 144$
18. $\log_9 1 = 0$ $9^0 = 1$

Example 2
A.SSE.2
Write each equation in logarithmic form.
19. $9^{-1} = \frac{1}{9}$ $\log_9 \frac{1}{9} = -1$
20. $6^{-3} = \frac{1}{216}$ $\log_6 \frac{1}{216} = -3$
21. $2^8 = 256$ $\log_2 256 = 8$
22. $4^6 = 4096$ $\log_4 4096 = 6$
23. $27^{\frac{2}{3}} = 9$ $\log_{27} 9 = \frac{2}{3}$
24. $25^{\frac{3}{2}} = 125$ $\log_{25} 125 = \frac{3}{2}$

Example 3
A.SSE.2
Evaluate each expression.
25. $\log_3 \frac{1}{9}$ -2
26. $\log_4 \frac{1}{64}$ -3
27. $\log_8 512$ 3
28. $\log_6 216$ 3
29. $\log_{27} 3$ $\frac{1}{3}$
30. $\log_{32} 2$ $\frac{1}{5}$
31. $\log_9 3$ $\frac{1}{2}$
32. $\log_{121} 11$ $\frac{1}{2}$
33. $\log_{\frac{1}{5}} 3125$ -5
34. $\log_{\frac{1}{8}} 512$ -3
35. $\log_{\frac{1}{3}} \frac{1}{81}$ 4
36. $\log_{\frac{1}{6}} \frac{1}{216}$ 3

Examples 4–5
F.IF.7e
PRECISION Graph each function. 37–48. See Ch. 6 Answer Appendix.
37. $f(x) = \log_6 x$
38. $f(x) = \log_{\frac{1}{5}} x$
39. $f(x) = 4 \log_2 x + 6$
40. $f(x) = \log_{\frac{1}{9}} x$
41. $f(x) = \log_{10} x$
42. $f(x) = -3 \log_{\frac{1}{12}} x + 2$
43. $f(x) = 6 \log_{\frac{1}{8}} (x + 2)$
44. $f(x) = -8 \log_3 (x - 4)$
45. $f(x) = \log_{\frac{1}{4}} (x + 1) - 9$
46. $f(x) = \log_5 (x - 4) - 5$
47. $f(x) = -\frac{1}{6} \log_8 (x - 3) + 4$
48. $f(x) = -\frac{1}{3} \log_{\frac{1}{6}} (x + 2) - 5$

Differentiated Homework Options

Levels	AL Basic	OL Core	BL Advanced
Exercises	13–56, 60–73	13–55 odd, 57–73	57–73
2-Day Option	13–55 odd, 67–73	13–56, 67–73	
	14–56 even, 60–66	57–66	

 You can use ALEKS to provide additional remediation support with personalized instruction and practice.

Practice

Formative Assessment Use Exercises 1–12 to assess students' understanding of the concepts in this lesson.

The Practice and Problem Solving exercises assess the content taught in the lesson. The Preparing for Assessment page is meant to be used as preparation for end-of-course assessments.

MP Teaching the Mathematical Practices

Precision Mathematically proficient students try to use clear definitions in their reasoning, calculate accurately and efficiently, and make explicit use of definitions.

Extra Practice

See page R6 for extra exercises for students who are approaching level or for on-level students who need additional reinforcement.

Levels of Complexity Chart

The levels of the exercises progress from 1 to 3, with Level 1 indicating the lowest level of complexity.

Exercises	13–56	57, 58, 67–73	59–66
Level 3			●
Level 2		●	
Level 1	●		

Additional Answers

8.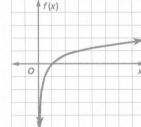

9.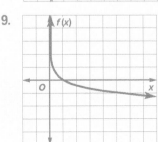

Additional Answers

10.

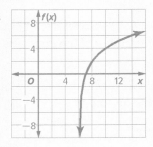

11.

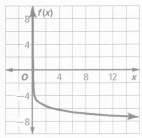

49b.

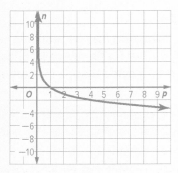

Example 6
A.SSE.2

49. PHOTOGRAPHY The formula $n = \log_2 \frac{1}{p}$ represents the change in the f-stop setting n to use in less light where p is the fraction of sunlight.

a. Benito's camera is set up to take pictures in direct sunlight, but it is a cloudy day. If the amount of sunlight on a cloudy day is $\frac{1}{4}$ as bright as direct sunlight, how many f-stop settings should he move to accommodate less light? 2

b. Graph the function. See margin.

c. Use the graph in part b to predict what fraction of daylight Benito is accommodating if he moves down 3 f-stop settings. Is he allowing more or less light into the camera? $\frac{1}{8}$; less light

50. EDUCATION To measure a student's retention of knowledge, the student is tested after a given amount of time. A student's score on an Algebra 2 test t months after the school year is over can be approximated by $y(t) = 85 - 6 \log_2 (t + 1)$, where $y(t)$ is the student's score as a percent.

a. What was the student's score at the time the school year ended ($t = 0$)? 85

b. What was the student's score after 3 months? 73

c. What was the student's score after 15 months? 61

Graph each function. 51–56. See Ch. 6 Answer Appendix.

51 $f(x) = 4 \log_2 (2x - 4) + 6$

52. $f(x) = -3 \log_{12} (4x + 3) + 2$

53. $f(x) = 15 \log_{14} (x + 1) - 9$

54. $f(x) = 10 \log_5 (x - 4) - 5$

55. $f(x) = -\frac{1}{6} \log_8 (x - 3) + 4$

56. $f(x) = -\frac{1}{3} \log_6 (6x + 2) - 5$

B **57.** **MODELING** In general, the more money a company spends on advertising, the higher the sales. The amount of money in sales for a company, in thousands, can be modeled by the equation $S(a) = 10 + 20 \log_4(a + 1)$, where a is the amount of money spent on advertising in thousands, when $a \geq 0$.

57b. If $3000 is spent on advertising, $30,000 is returned in sales. If $15,000 is spent on advertising, $50,000 is returned in sales. If $63,000 is spent on advertising, $70,000 is returned in sales.

a. The value of $S(0) = 10$, which means that if no money is spent on advertising, $10,000 is returned in sales. Find the values of $S(3)$, $S(15)$, and $S(63)$. $S(3) = 30$, $S(15) = 50$, $S(63) = 70$

b. Interpret the meaning of each function value in the context of the problem.

c. Graph the function. See Ch. 6 Answer Appendix.

d. Use the graph in part c and your answers from part a to explain why the money spent in advertising becomes less "efficient" as it is used in larger amounts.

57d. Because eventually the graph plateaus, and no matter how much money you spend you are still returning about the same in sales.

58. BIOLOGY The generation time for bacteria is the time that it takes for the population to double. The generation time G for a specific type of bacteria can be found using experimental data and the formula $G = \dfrac{t}{3.3 \log_b f}$, where t is the time period, b is the number of bacteria at the beginning of the experiment, and f is the number of bacteria at the end of the experiment.

a. The generation time for mycobacterium tuberculosis is 16 hours. How long will it take four of these bacteria to multiply into 1024 bacteria? 264 h or 11 days

b. An experiment involving rats that had been exposed to salmonella showed that the generation time for the salmonella was 5 hours. After how long would 20 of these bacteria multiply into 8000? 49.5 h or 2 days 1.5 h

c. *E. coli* are fast growing bacteria. If 6 *E. coli* can grow to 1296 in 4.4 hours, what is the generation time of *E. coli*? $\frac{1}{3}$ h or 20 min

---

Go Online!

eBook

Interactive Student Guide

Use the *Interactive Student Guide* to deepen conceptual understanding.

· Logarithms and Logarithmic Functions

ALGEBRA 2
INTERACTIVE STUDENT GUIDE

59 FINANCIAL LITERACY Jacy has charged $2000 on a credit card. The credit card company charges 24% interest, compounded monthly. The credit card company uses $\log_{\left(1 + \frac{0.24}{12}\right)} \frac{A}{2000} = 12t$ to determine how much time it will be until Jacy's debt reaches a certain amount, if A is the amount of debt after a period of time, and t is time in years.

a. Graph the function for Jacy's debt. See margin.
b. Approximately how long will it take Jacy's debt to double? ≈ 3 years
c. Approximately how long will it be until Jacy's debt triples? ≈ 4.5 years

60. GRAPHING CALCULATOR Graph $f(x) = \log_2 x$ and the transformation graph, See Ch. 6 Answer $g(x)$. Determine the effects on each of the key attributes of the graph of $f(x)$. Appendix.

a. $g(x) = 3f(x)$ **b.** $g(x) = -2f(x)$ **c.** $g(x) = f(x) + 4$
d. $g(x) = f(x) - 6$ **e.** $g(x) = f(x + 5)$ **f.** $g(x) = f(x - 8)$

A.SSE.2, F.IF.4, F.IF.7e, F.BF.3

H.O.T. Problems   Use Higher-Order Thinking Skills

61. CRITIQUE ARGUMENTS Consider $y = \log_b x$ in which b, x, and y are real numbers. Zero can be in the domain *sometimes*, *always* or *never*. Justify your answer.

61. Never; if zero were in the domain, the equation would be $y = \log_b 0$. Then $b^y = 0$. However, for any real number b, there is no real power that would let $b^y = 0$.

62. ERROR ANALYSIS Betsy says that the graphs of all logarithmic functions cross the y-axis at $(0, 1)$ because any number to the zero power equals 1. Tyrone disagrees. Is either of them correct? Explain your reasoning.
Tyrone; Sample answer: The graphs of logarithmic functions pass through $(1, 0)$, not $(0, 1)$.

63. REASONING Without using a calculator, compare $\log_7 51$, $\log_8 61$, and $\log_9 71$. Which of these is the greatest? Explain your reasoning.

64. OPEN-ENDED Write a logarithmic equation of the form $y = \log_b x$ for each of the following conditions. a–e. See Ch. 6 Answer Appendix.

a. y is equal to 25.   **b.** y is negative.
c. y is between 0 and 1.   **d.** x is 1.
e. x is 0.

63. $\log_7 51$; Sample answer: $\log_7 51$ equals a little more than 2. $\log_8 61$ equals a little less than 2. $\log_9 71$ equals a little less than 2. Therefore, $\log_7 51$ is the greatest.

65. ERROR ANALYSIS Elisa and Matthew are evaluating $\log_{\frac{1}{7}} 49$. Is either of them correct? Explain your reasoning. See Ch. 6 Answer Appendix.

Elisa	Matthew
$\log_{\frac{1}{7}} 49 = y$	$\log_{\frac{1}{7}} 49 = y$
$\frac{1}{7}^y = 49$	$49^y = \frac{1}{7}$
$(7^{-1})^y = 7^2$	$(7^2)^y = (7)^{-1}$
$(7)^{-y} = 7^2$	$7^{2y} = (7)^{-1}$
$y = 2$	$2y = -1$
	$y = -\frac{1}{2}$

66. WRITING IN MATH A transformation of $\log_{10} x$ is $g(x) = a \log_{10}(x - h) + k$. Explain the process of graphing this transformation. See Ch. 6 Answer Appendix.

Standards for Mathematical Practice

Emphasis On	Exercises
3 Construct viable arguments and critique reasoning of others.	61–63, 65
4 Model with mathematics.	8–12, 37–59, 67, 72
5 Use appropriate tools strategically.	60
6 Attend to precision.	5–7, 25–36, 66, 69, 72–73
7 Look for and make use of structure.	1–4, 13–24, 61–64, 68, 70–72

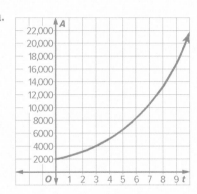

Preparing for Assessment

Exercises 67–73 require students to use the skills they will need on standardized assessments. Each exercise is dual-coded with content.

Dual Coding		
Items	Content Standards	Mathematical Practices
67	F.IF.7e, F.BF.3	4
68	F.BF.3	7
69	F.IF.7e	6
70	A.SSE.2	7
71	F.BF.3	7
72	F.IF.7e	4, 6, 7
73	A.SSE.2	6

Diagnose Student Errors

67.

A	No vertical shift
B	Shifted left 2 units instead of right
C	CORRECT
D	No horizontal shift

68.

A	Did not realize n must be negative and believed it needed to be less than p
B	Did not realize n must be negative and believed it needed to be less than 1
C	Did not realize n must be negative and believed it needed to be less than m
D	CORRECT

69.

A	Misinterpreted the impact of adding 1 in the equation
B	CORRECT
C	Believed the graph of $y = a \log b(x - h) + k$ has a vertical asymptote at $y = k$
D	Believed the graph of $y = a \log b(x - h) + k$ has a vertical asymptote at $y = b$

Go Online!

Self-Check Quiz
Students can use *Self-Check Quizzes* to check their understanding of this lesson.

Preparing for Assessment

67. Which graph is the function $f(x) = -\log(x - 2)$?
🔵 4 F.IF.7e, F.BF.3 **C**

○ A
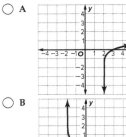

○ B
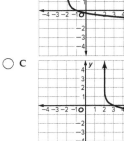

○ C
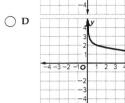

○ D

68. Let $n = \log_m p$. If $p > 1 > m > 0$, which statement must be true? 🔵 7 F.BF.3 **D**

○ A $p > n > 1$

○ B $1 > n > m$

○ C $m > n > 0$

○ D $0 > n$

69. Which of the following is an asymptote for the graph of $y = \log_4 x + 1$? 🔵 6 F.IF.7e **B**

○ A $x = -1$

○ B $x = 0$

○ C $y = 1$

○ D $y = 4$

70. If $-3 = \log_3 x$, what is the value of x?
🔵 7 A.SSE.2 **C**

○ A -27 ○ C $\frac{1}{27}$

○ B $-\frac{1}{27}$ ○ D 27

71. Select all of the transformations for the function $f(x) = -\log_5(x + 4) - 5$ based on the parent function $f(x) = \log_5 x$. 🔵 7 F.BF.3 **B, C, F**

☐ A translated up 5 units

☐ B translated down 5 units

☐ C translated left 4 units

☐ D translated right 4 units

☐ E reflected in the y-axis

☐ F reflected in the x-axis

72. MULTI-STEP The function $f(x) = 10 - \log(x + 10)$ represents the home attendance of a football team in ten thousands, when $f(x)$ is based on the number of home games x. 🔵 4, 6, 7 F.IF.7e

a. Sketch a graph of the function. a-c. See Ch. 6 Answer Appendix
b. What is the domain of the function?
c. What is the range of the function?
d. What is the attendance for the second home game?

89,208

e. What is the attendance for the eighth home game?

87,447

f. Explain the general trend in attendance based on the graph of the function.
The attendance is decreasing for each game.

73. A logarithmic function, $f(x) = \log_a(x - b)$ is increasing with a vertical asymptote of $x = 3$. The point $(5, 1)$ lies on its graph. Find the value of a and of b. 🔵 6 A.SSE.2

$a = 2, b = 3$

Differentiated Instruction OL BL

Extension Point out that the division problem $\frac{32}{4} = 8$ can be written as $\frac{2^5}{2^2} = 2^3$.

Ask students to identify the base 2 logarithm in the dividend, the divisor, and the resulting quotient. Then have them write an equation that relates the logarithms.
$\log_2 32 = 5$; $\log_2 4 = 2$; $\log_2 8 = 3$; $\log_2 32 - \log_2 4 = \log_2 8$

70.

A	Evaluated -3^3
B	Made a sign error when evaluating
C	CORRECT
D	Forgot to use the negative and evaluated 3^3

71.

A	Misinterpreted the vertical shift
B	CORRECT
C	CORRECT
D	Misinterpreted the horizontal shift
E	Misinterpreted the reflection
F	CORRECT

LESSON 6-5

Modeling Data

SUGGESTED PACING (DAYS)

90 min.	0.5
45 min.	1.0

Introduction

Track Your Progress

Objectives

1 Find equations of best fit for data modeled by exponential and logarithmic functions.

2 Choose the best model for a data set.

Mathematical Background

Modeling data has multiple real-world applications and can be used to predict future events. Choosing the best model for a data set ensures the accuracy and reliability of your predictions.

THEN

F.IF.1 Understand that a function from one set (called the domain) to another set (called the range) assigns to each element of the domain exactly one element of the range. If f is a function and x is an element of its domain, then $f(x)$ denotes the output of f corresponding to the input x. The graph of f is the graph of the equation $y = f(x)$.

NOW

A.CED.2 Create equations in two or more variables to represent relationships between quantities: graph equations on coordinate axes with labels and scales.

F.IF.4 For a function that models a relationship between two quantities, interpret key features of graphs and tables in terms of the quantities, and sketch graphs showing key features given a verbal description of the relationship.

F.LE.4 For exponential models, express as a logarithm the solution to $ab^{ct} = d$ when a, c, and d are numbers and the base b is 2, 10, or e; evaluate the logarithm using technology.

NEXT

A.REI.11 Explain why the x-coordinates of the points where the graphs of the equations $y = f(x)$ and $y = g(x)$ intersect are the solutions of the equation $f(x) = g(x)$; find the solutions approximately, e.g., using technology to graph the functions, make tables of values, or find successive approximations. Include cases where $f(x)$ and/or $g(x)$ are linear, polynomial, rational, absolute value, exponential, and logarithmic functions.

Go Online! All of these resources and more are available at connectED.mcgraw-hill.com

 Use the eGlossary to define correlation coefficient and other key vocabulary in the lesson.

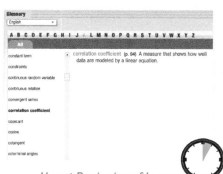

Use at Beginning of Lesson

Graphing Calculator Personal Tutors let students hear real teachers solve problems. Students can pause and repeat as many times as necessary.

Use with Examples

The Chapter Project allows students to create and customize a project as a nontraditional method of assessment.

Use at End of Lesson

OER Using Open Educational Resources

GAMES Have students work in pairs to play Correlation and Regression Jeopardy at **Super Teacher Tools** to learn and reinforce the concepts of using a regression (variables, coefficients, etc.) to model the fit of a data set. *Use as classwork*

Differentiate Your Resources

Extra Practice Additional practice or homework; Skills Practice is best for approaching-level students and Practice is best for on-level and beyond-level students

Skills Practice

6-5 Skills Practice
Modeling Data

Describe the scatterplot and an appropriate scale for each data set.

1.

x	0	−1.4	3	−2.8	5	6.2	8.3	12.1	−4.5
y	2	0.9	5.5	−0.4	7.5	8.3	10.9	14.5	−2

2.

x	0	4	0.5	0.8	−0.5	0.4	0.2	2
y	1.4	1.7	4	0.9	1	0.3	0.6	0.05	4.5

Sketch the graph and write if the data set is exponential or linear. If it is linear write a linear equation for the line of best fit.

3. (0, 1), (−1, −3), (10, 41), (2, 9), (7, 29), (−9, −35), (5, 21)

4. (0, 1), (6, 64), (2, 4), (4, 16), (5, 8), (5, 32), (1, 2)

Write each exponential function as a logarithmic function, then solve for x.

5. $8^x + 3 = 32{,}771$ 6. $17^x = 4{,}913$

7. $12^x = 639.55$ 8. $5^x − 15 = 15{,}607$

Each equation represents a data set. Choose the best model for each.

9. $y = 2x^2 + 3$ 10. $y = \frac{1}{2}x^3$

11. $y = -0.4x − 11$ 12. $y = 3.5^x$

Describe the correlation for each value of r.

13. $r = 0.965$ 14. $r = -1$ 15. $r = 0.72$ 16. $r = -0.59$

Practice

6-5 Practice
Modeling Data

Describe the scatterplot and an appropriate scale for each data set.

1.

x	0	4	3	0.5	3.5	5	2	3.5	1
y	1	16.5	10	1.2	7	25	5	12.5	2.2

2.

x	0	2	−5	3	1	−4	8	−6	−1
y	−3	1	−13	3	−1	−11	13	−15	−5

Sketch the graph and write if the data set is exponential or linear. If it is linear write a linear equation for the line of best fit.

3. (0, −4), (−1, −2), (2, −8), (4, −12), (−10, 16), (8, −20), (−5, 6)

4. (0, 1), (3, 0.125), (6, 0.0156), (4, 0.0625), (2, 0.25), (4, 0.5), (5, 0.0313)

Write each exponential function as a logarithmic function, then solve for x.

5. $6^x + 14 = 1{,}310$ 6. $11^x = 161{,}051$

Identify the type of regression for each of the following line of best fit equations.

7. $y = 2(4)^x$ 8. $y = 3x^5$

9. $y = 9x + 4$ 10. $y = x^3 + 2x + 8$

11. A manufacturing firm kept track of how the amount of money the company spent on advertising impacted its sales. The results are shown in the table. Graph the data in the table and pick the best model.

Advertising	Sales
$80	$1,000
$120	$1,150
$160	$1,530
$200	$1,750
$320	$2,780
$400	$3,150
$640	$4,760
$1200	$6,270

Word Problem Practice

6-5 Word Problem Practice
Modeling Data

1. **POPULATION** The population per square mile of a country has changed considerably over a period of years. The table shows the number of people per square mile for several years. Use a graphing calculator to graph the scatter plot that models the relationship between the variables.

Population Density

Year	People per mi²	Year	People per mi²
1996	2	2006	12
1997	5	2007	14
1998	6	2008	14.5
1999	4	2009	15
2000	5	2010	15.5
2001	7	2011	16
2002	7	2012	16
2003	8	2013	17
2004	10	2014	18
2005	10	2015	20

2. **SAVINGS** Brittany deposited $100 into an account, then forgot about it and made no further deposits or withdrawal. The table shows the account balance for a period of 10 years.

Time (years)	Account Balance ($)
0	100
2	107.60
4	116.10
6	125.04
8	139.17
10	155.80

a. Use the graphing calculator to draw a scatterplot of the data.

b. Use the regression feature of the graphing calculator to determine the best model. Round coefficients and constants to the nearest hundredth.

3. **TEST GRADES** A professor conducted a survey by asking 12 of his students how many hours they spent preparing for the final exam. He then matched the students' responses with their final exam grades.

Grade	Hours	Grade	Hours
96	8	93	7.5
84	6	86	6.5
89	5.5	77	5
55	1.5	60	2.25
52	1.25	77	5
60	2	83	6

a. Graph and analyze the data set. State if it is linear or exponential. If the scatterplot is linear, write an equation for the line of best fit.

b. Based on your graph, what is the minimum number of hours that you should spend studying if you want to get a passing grade of 60% or better on the final exam?

4. **BACTERIAL GROWTH** Dominique is calculating bacterial growth in a culture in his science class. He places 20 bacteria on a slide under a microscope and documented the population growth of the bacteria every hour.

Hours (x)	0	1	2	3	4
Population (y)	20	60	180	540	1,620

a. Graph the data.

b. What is the relationship between the variables?

c. What is the best model?

d. What is the regression line?

e. When will the bacteria exceed 40,000?

Intervention Reteaching and vocabulary activities that can be used with struggling or absent students and as ELL support

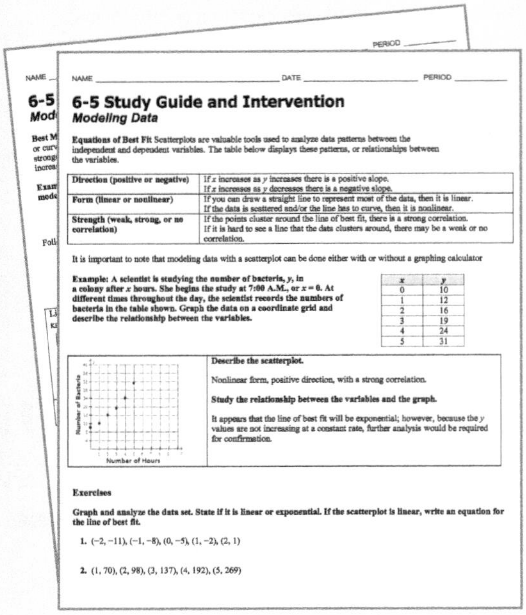

Study Guide and Intervention

6-5 Study Guide and Intervention
Modeling Data

Equations of Best Fit Scatterplots are valuable tools used to analyze data patterns between the independent and dependent variables. The table below displays these patterns, or relationship between the variables.

Direction (positive or negative)	If x increases as y increases there is a positive slope. If x increases as y decreases there is a negative slope.
Form (linear or nonlinear)	If you can draw a straight line to represent most of the data, then it is linear. If the data is scattered and/or the line has to curve, then it is nonlinear.
Strength (weak, strong, or no correlation)	If the points cluster around the line of best fit, there is a strong correlation. If it is hard to see a line that the data clusters around, there may be a weak or no correlation.

It is important to note that modeling data with a scatterplot can be done either with or without a graphing calculator.

Example: A scientist is studying the number of bacteria, y, in a colony after x hours. She begins the study at 7:00 A.M., or x = 0. At different times throughout the day, the scientist records the numbers of bacteria in the table shown. Graph the data on a coordinate grid and describe the relationship between the variables.

x	y
0	10
1	12
2	16
3	19
4	24
5	31

Describe the scatterplot.

Nonlinear form, positive direction, with a strong correlation.

Study the relationship between the variables and the graph.

It appears that the line of best fit will be exponential; however, because the y values are not increasing at a constant rate, further analysis would be required for confirmation.

Exercises

Graph and analyze the data set. State if it is linear or exponential. If the scatterplot is linear, write an equation for the line of best fit.

1. (−2, −11), (−1, −8), (0, −5), (1, −2), (2, 1)

2. (1, 70), (2, 98), (3, 137), (4, 192), (5, 269)

Study Notebook

6-5 Modeling Data

What You'll Learn Scan the text under the *Now* heading. List two things you will learn about in the lesson.

1.

2.

Active Vocabulary New Vocabulary Write the definition next to each term.

correlation coefficient ▶

regression line or curve ▶

Vocabulary Link Give an example of what might have a *strong* correlation in the context of the getting a good grade on a test. Give an example of what might have a *weak* correlation in the context of the getting a good grade on a test.

Extension Activities that can be used to extend lesson concepts

Enrichment

6-5 Enrichment
Median-Fit Lines

A median-fit line is a particular type of line of fit. Follow the steps below to find the equation of the median-fit line for the data.

Approximate Percentage of Violent Crimes Committed by Juveniles That Victims Reported to Law Enforcement

Year	1980	1982	1984	1986	1988	1990	1992	1994	1996
Offenders	36	36	33	32	31	30	29	29	30

Source: U.S. Bureau of Justice Statistics

1. Divide the data into three approximately equal groups. There should always be the same number of points in the first and third groups. In this case, there will be three data points in each group.

2. Find x_1, x_2, and x_3, the medians of the x values in groups 1, 2, and 3, respectively. Find y_1, y_2, and y_3, the medians of the y values in groups 1, 2, and 3, respectively.

3. Find an equation of the line through (x_1, y_1) and (x_3, y_3).

4. Find Y, the y-coordinate of the point on the line in Exercise 2 with an x-coordinate of x_2.

5. The median-fit line is parallel to the line in Exercise 2, but is one-third closer to (x_2, y_2). This means it passes through $\left(x_2, \frac{2}{3}Y + \frac{1}{3}y_2\right)$. Find this ordered pair.

6. Write an equation of the median-fit line.

7. Use the median-fit line to predict the percentage of juvenile violent crime offenders in 2010 and 2020.

LESSON 5

Modeling Data

:: Then	:: Now	:: Why?
● You learned to understand the relationship between an independent and dependent variable and how that relationship is represented in a function.	**1** Find equations of best fit for data modeled by exponential and logarithmic functions. **2** Choose the best model for a data set.	● Leila made a study card showing the steps needed to use her graphing calculator to make a scatter plot for life expectancy based on year of birth.

New Vocabulary
correlation coefficient
regression line
regression curve

Mathematical Practices

3 Construct viable arguments and critique the reasoning of others.

4 Model with mathematics.

6 Attend to precision.

Content Standards
A.CED.2 Create equations in two or more variables to represent relationships between quantities; graph equations on coordinate axes with labels and scales.
F.IF.4 For a function that models a relationship between two quantities, interpret key features of graphs and tables in terms of the quantities, and sketch graphs showing key features given a verbal description of the relationship.
F.LE.4 For exponential models, express as a logarithm the solution to $ab^{ct} = d$ when a, c, and d are numbers and the base b is 2, 10, or e; evaluate the logarithm using technology.

Make a scatter plot.
- Enter the years of birth in L1 and the ages in L2.

KEYSTROKES: [STAT] [ENTER] 1980 [ENTER] 1983 [ENTER] 1990 [ENTER] ...

- Set the viewing window to fit the data.

KEYSTROKES: [WINDOW] 1975 [ENTER] 2010 [ENTER] 5 [ENTER] 70 [ENTER] 90 [ENTER] 2

- Use STAT PLOT to graph the scatter plot.

KEYSTROKES: [2nd] [STAT PLOT] [ENTER] [ENTER] [GRAPH]

---

1 Equations of Best Fit An equation of best fit helps to analyze data patterns between two variables. A scatter plot is one of the tools used to display data and provides information on the relationship between the variables. This relationship is between the independent, or the predictor, variable on the x-axis and the dependent, or response, variable on the y-axis. When you describe a data set displayed on a scatterplot, the relationship between the variables is described by the direction, form, and strength.

Direction (positive or negative)
If y increases as x increases, there is a positive slope. If y decreases as x increases, there is a negative slope.

Form (linear or nonlinear)
If all the data lies on the same line, then it is linear. If the data is scattered and the line has to curve, then it is nonlinear.

Strength (weak, strong, no correlation)
If the points cluster around the line of best fit, there is a strong correlation. If it is hard to see a line that the data clusters around, there may be a weak or no correlation.

Example 1 Analyze Patterns using a Scatterplot

POPULATION The population per square mile of a country has changed dramatically over a period of years. The table shows the number of people per square mile for several years. Graph the data and describe its direction, form, and strength.

Use a graphing calculator to enter the data. Then draw a scatter plot that shows how the number of people per square mile is related to year.

Step 1 Enter the years in L1. These are the x-values in the domain. Enter the people per square mile in L2. These are the y-values in the range.

Step 2 Determine the scale for the graph. The years are from 1790 to 2000, so use a scale of $1790 \leq x \leq 2000$ in increments of 10. The population per square mile ranges from 4.5 to 80, so use a scale of $0 \leq y \leq 115$ in increments of 5.

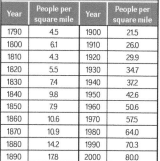

U.S. Population Density			
Year	People per square mile	Year	People per square mile
1790	4.5	1900	21.5
1800	6.1	1910	26.0
1810	4.3	1920	29.9
1820	5.5	1930	34.7
1830	7.4	1940	37.2
1840	9.8	1950	42.6
1850	7.9	1960	50.6
1860	10.6	1970	57.5
1870	10.9	1980	64.0
1880	14.2	1990	70.3
1890	17.8	2000	80.0

Source: Northeast-Midwest Institute

MP Mathematical Practices Strategies

Make Sense of Problems
Help students analyze relationships between the variables, the regression line, and the regression models, recognizing that every step helps to ensure accuracy and reliability of the data. Ask:

- Why is labeling the axes just as important as accurately inputting the data set into the calculator? Sample answer: Switching the x- and y-values in the input table will result in an incorrect graph.

- After a model is determined and an equation of best fit is written, how could you check your work? Sample answer: Choose a test point and substitute it into the equation to determine if it holds true.

- Is the line of best fit an accurate measure of the data set? Sample answer: Yes, if you have chosen the best model.

Launch

Have students read the Why? section of the lesson. Ask:

- **Describe Leila's viewing window based on the given keystrokes.** The x-axis is to go from 1975–2010 in increments of 5 and the y-axis is to be labeled 70–90 in increments of 2.

- **How many variables are being graphed and what are their titles?** 2, life expectancy and year of birth

- **How does Leila indicate a new data point?** by pressing enter after each entry

Teach

Ask the scaffolded questions for each example to build conceptual understanding for students at all levels.

1 Equations of Best Fit

Example 1 Analyze Patterns using a Scatterplot

AL If the data is scattered and random, then how do you determine the direction and correlation? There is not a direction and there is no correlation.

OL Define what a scatterplot is. A part of a coordinate grid with plotted points that visually display the relationship between the variables.

BL If x is the predictor variable and y is the response variable, what would happen if you switched the y and x variables while graphing? The graph would be reflected in the line $y = x$, which would make interpreting the data less intuitive.

Go Online!

Interactive Whiteboard

Use the *eLesson* or *Lesson Presentation* to present this lesson.

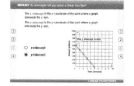

Need Another Example?

Graphic Art A graphic artist used a math equation to create his art. Describe the scatterplot positive, nonlinear, strong

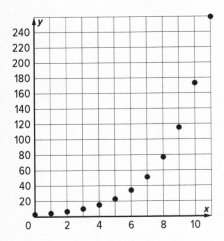

Teaching Tip

Structure Explain the different equations and how the variables change to represent different situations and events while the structure of the formula remains the same.

(MP) Teaching the Mathematical Practices

Precision Mathematically proficient students make generalizations about the meaning of symbols. They work carefully to increase accuracy. They are careful and they perform each step, process the results, and analyze, understanding the process and calculations being performed by the graphing calculator. Have students practice drawing graphs and using the graphing calculator to help them connect mathematical ideas.

(e) Essential Question

How do you make accurate predictions from several different types of data sets? To make accurate predictions for a data set, you must find and use the regression line.

Example 2 Graph and Analyze a Data Set

(AL) What is the formula for an exponential equation? $y = ab^x$.

(OL) How do you know the graph is not a parabola? Parabolas are formed by quadratic equations and have positive and negative values, so they do not have a starting or ending point.

Study Tip

Structure Organizing the domain and range values from least to greatest can help you determine the best scale.

Step 3 Graph the scatterplot.

Step 4 Describe the scatterplot.

The scatterplot is nonlinear, with a positive direction and a strong correlation.

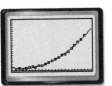

[1780, 2020] scl: 10 by [0, 115] scl: !

1. linear, positive, strong

> **Guided Practice**

1. The table shows the cost of a smartphone in different years. Graph the data and describe its direction, form, and strength.

Year	Cost
2012	325.50
2013	406.88
2014	508.60
2015	635.75
2016	794.69

Sometimes a graphing calculator is not available. Understanding how to model data without a calculator is an essential skill.

A.CED.2

Example 2 Graph and Analyze a Data Set

Sahara is selling cups of lemonade as a fundraiser. She records the grams of lemon powder she uses per cup to make six batches of lemonade and the time in hours it takes to sell each batch. Her data set is: (0, 0.05), (1, 0.1), (2, 0.2), (3, 0.4), (4, 0.8), (5, 1.6). Graph the data and describe its direction, form, and strength.

Create a table to organize the data.   Graph the data on a coordinate grid.

x	y
0	0.05
1	0.1
2	0.2
3	0.4
4	0.8
5	1.6

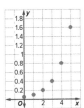

Study Tip

(MP) Reasoning The line of best fit does not have to intersect all or even most of the points. Most of the points just have to cluster around the line.

Describe the scatterplot.

Positive direction, nonlinear form, with a strong correlation

If this were a linear scatterplot, we could use a ruler to draw a line that represents the data set and then write a linear equation, but this data set curves.

Study the relationship between the variables and the graph.

The y values are increasing by a constant rate of 2. This data set appears to be exponential.

> **Guided Practice**

Use a graph to describe the direction, form, and strength of the data set. If the scatterplot is linear, then write the linear equation for the line of best fit.

2A. (−3, 0), (−2, 2), (−2.5, 1), (−1, 4) linear, positive, strong

2B. (0, 6), (1, 12), (2, 24), (3, 48) nonlinear, positive, strong

2C. (2, 150), (10, 141), (15, 132.54), (20, 124.56) nonlinear, negative, strong

(BL) If you drew a line to represent this data set, what would be the equation? Would this be an accurate model? $y = 0.29x − 0.19$; no

Need Another Example?

Graph the following data set. If it is linear, then write a linear equation for the line of best fit. (0, 4), (1, 6), (2, 12), (3, 22)
This is not a linear equation. ($y = 2x^2 + 4$)

When data is nonlinear, a linear equation, or line of best fit, will not accurately represent the data. Nonlinear data may be modeled by a curve of fit based on a quadratic, exponential, or logarithmic function. For data that fits an exponential curve, use the relationship between exponents and logarithms to find an equation for the curve of fit.

Exponential Function
$f(x) = a^x, a > 0$
When $a = 2$:

x	y
0	1
1	2
2	4
3	8

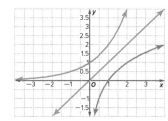

Logarithmic Function
$f(x) = \log_a x, a > 0 \text{ and } a \neq 1$
When $a = 2$:

x	y
1	0
2	1
4	2
8	3

The exponential data are the inverse of the logarithmic data. One graph is the reflection of the other in the line $y = x$. Using the idea of inverse operations, an exponential expression can be written as a logarithmic expression. For example, the exponential expression $4^3 = 64$ can be rewritten as $\log_4 (64) = 3$.

The following formulas are helpful when working with exponential and logarithmic functions.

Key Concepts Important Formulas

Exponential growth or decay:
$A = Pe^{rt}$

Natural exponential function:
$f(x) = e^x$

Natural logarithmic function:
$f(x) = \log_e(x) \text{ or } f(x) = \ln(x)$

Standard form of an exponential function:
$a^x = b$

Standard form of a logarithmic function:
$\log_a(b) = x$

To use a graphing calculator to find a curve of fit, you may need to change the base of a logarithm.

Change the base of a logarithm from b to a: $\log_b x = \dfrac{\log_a x}{\log_a b}$

F.LE.4

Example 3 Convert an Exponential Function to a Logarithmic Function

VIRUS In 1991, the Internet was introduced and the first computer virus threatened 500,000 computers. Computer engineers calculated that one person's computer would have infected 5 other computers every minute. They created an exponential function to calculate how long it would take for the virus to spread to all 500,000 computers.

How many minutes would it take for the computer virus to destroy 500,000 computers?

Convert the exponential function $500,000 = 5^x$ to a logarithmic function to solve for x.

Step 1 Rewrite the exponential function as a logarithmic function.

$500,000 = 5^x$ becomes $\log_5 500,000 = x$

Step 2 Study the logarithm. If you know the power of 5, that would equal 500,000 then solve for x. If not, then change the base to a base of 10 and use a calculator.

Step 3 Change the base 5 to base 10.

$$\log_5 500,000 = \frac{\log_{10} 500,000}{\log_{10} 5}$$

> 5 is the old base
> 10 is the new base

Step 4 Input the equation into the calculator and solve for x. Round to the nearest hundredth.

The 500,000 computers would be destroyed in 8.15 minutes.

> **Study Tip**
>
> Precision Take the time to study the formula and walk through the steps a few times to increase understanding and accuracy. Use simple exponential formulas that you can solve mentally, then use a calculator to change the base and check your work.

Guided Practice

Use a scientific or graphing calculator to solve for x.

3A. $4^x = 262,144$ $x = 9$

3B. $27^x = 19,683$ $x = 3$

3C. $7^x = 2401$ $x = 4$

2 Best Model for a Data Set A **regression line** or **regression curve** is an equation of a function that represents a line or curve of fit. Choosing the best model for a data set allows you to find the regression line or curve with the strongest correlation; an increased *correlation coefficient* means an increase in the precision of a prediction using the equation of the regression line or curve.

> **Study Tip**
>
> Modeling Knowing the general shape of the regression models will help you choose the best model for the data set.

Key Concepts Types of Regression Equations and Models

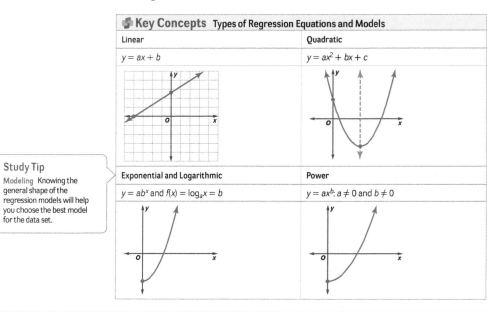

Linear	Quadratic
$y = ax + b$	$y = ax^2 + bx + c$

Exponential and Logarithmic	Power
$y = ab^x$ and $f(x) = \log_a x = b$	$y = ax^b$; $a \neq 0$ and $b \neq 0$

The graphing calculator is a good tool to find the line of best fit for nonlinear scatter plots, and will automatically calculate the **correlation coefficient** r. The correlation coefficient is a measure that shows how well a data set is modeled by a given equation. It is similar to analyzing the relationship between the variables, but it analyzes the relationship between the data set and the line of best fit, represented by an equation.

- Positive correlation: when x increases, y increases.
- Negative correlation: when x increases, y decreases.
- No correlation: the data are randomly dispersed (r is close to 0).
- Perfect correlation: the data lie exactly on the regression line ($r = \pm 1$).
- Strong correlation: $|r| \geq 0.8$
- Weak correlation: $|r| \leq 0.5$

A.CED.2

Example 4   Use Regression Equation and Models

Use the data set from Example 1. What is the equation for the line or curve of best fit?

Find a regression equation.

To find an equation that best fits the data, use the regression feature of the calculator. Examine various regressions to determine the best model.

Recall that the calculator returns the correlation coefficient r, which is used to indicate how well the model fits the data. The closer r is to 1 or −1, the better the fit.

Linear regression   Quadratic regression

KEYSTROKES: [STAT] [▶] 4 [ENTER]   KEYSTROKES: [STAT] [▶] 5 [ENTER]

$$r^2 = 0.9974003374$$
$$r = \sqrt{0.9974003374}$$
$$r \approx 0.9986993228$$

Exponential regression   Power regression

KEYSTROKES: [STAT] [▶] 0 [ENTER]   KEYSTROKES: [STAT] [▶] [ALPHA] [A] [ENTER]

Compare the r-values.

Linear: 0.945411996   Quadratic: 0.9986993228

Exponential: 0.991887235   Power: 0.9917543535

The r-value of the quadratic regression is closest to 1, so the equation of the curve of best fit is about $y = 0.002x^2 - 7.499x + 6798.7$. You can examine the equation visually by graphing the regression equation with the scatter plot.

KEYSTROKES: [STAT] [▶] 5 [ENTER] [Y=] [VARS] 5 [▶] [▶] 1 [GRAPH]

2 Best Model for a Data Set

Example 4 Use Regression Equations and Models

(AL) **What information on the graphing calculator helps you choose the best model?** the r-values

(OL) **Why is it important to find exact r-values?** If the values are rounded or estimated, the model of best fit will not be accurate.

(BL) **How does the information on the model regression screen of your calculator help you write the regression line?** It gives you the values of the variables, so you can create the equation.

Need Another Example?

World Population It is estimated that the world population reached 1 billion around 1804. Since then, it has been increasing at a dramatic pace. The table shows the estimated world population in billions since 1900.

Years Since 1900	Population (billions)
0	1.656
50	2.558
60	3.043
70	3.713
80	4.451
90	5.289
100	6.089
110	6.892

What is the equation for the line or curve of best fit?

about $y = 0.0005x^2 - 0.0041x + 1.6292$

Example 5 Make a Prediction

AL Why is it important to know how to do all of these steps without the use of a calculator? Understanding the math behind the calculator is important for cases where a calculator cannot be used.

OL How does the description of the scatterplot help you check the reasonability of your answer? If you know the direction and form of the scatterplot and your answer doesn't seem to fit within those parameters, your answer may be unreasonable.

BL What other predictions could you make from this data set? Check students' work.

Need Another Example?

Bri invested $285 dollars into stocks and watched the value of her investment grow. If this trend continues, how much money would she have in ten years? $347.41

Years	0	1	2	3	4
Investment	285	290.70	296.51	302.44	308.49

Teaching Tip

Sense-Making Students automatically assume that logarithmic base should be 10. Ask students to write examples of numbers with 10 as the exponent.

> **Guided Practice**

4. Use the data set from Example 2. What is the equation for the line or curve of best fit? $y = 0.5(2^x)$

The statistical value r^2 represents the *correlation of determination*. It can be used to increase the accuracy of the equation used to model data. It considers variation in the data (outliers) and represents the accuracy of the correlation coefficient to the data.

F.IF.4

Example 5 Making a Prediction

Use the data set from Example 1 and a graphing calculator to predict the population per square mile in 2020.

Enter the data and make a scatterplot.

[1780, 2020] scl: 10 by [0, 115] scl: !

To determine the population per square mile in 2020, find the value of y when $x = 20$.

KEYSTROKES: 2nd [CALC] ENTER 2020

If the trend continues, there will be approximately 94.9 people per square mile in 2020. To check your work, you can substitute the values into the equation.

U.S. Population Density			
Year	People per square mile	Year	People per square mile
1790	4.5	1900	21.5
1800	6.1	1910	26.0
1810	4.3	1920	29.9
1820	5.5	1930	34.7
1830	7.4	1940	37.2
1840	9.8	1950	42.6
1850	7.9	1960	50.6
1860	10.6	1970	57.5
1870	10.9	1980	64.0
1880	14.2	1990	70.3
1890	17.8	2000	80.0

Source: Northeast-Midwest Institute

> **Guided Practice**

Fredrick has 20 weeks to train for a 15K race. He tracks his progress by running a practice race every Wednesday and Thursday. The table shows the distance of the race he is able to run each week. At the end of 7 weeks he stops keeping track of his progress, but continues to train for the race.

Weeks	Distance (km)
1	1.5
2	1.75
3	1.3
4	1.8
5	2.1
6	2.45
7	2.5

5A. Will Fredrick be able to run the 15K in 20 weeks? Approximately how far will he be able to run? No; 12.3 km

5B. How many weeks will it take for him to confidently run the 15K? 23 weeks

5C. Will Fredrick be able to run a 20K race in 6 months? no

Differentiated Instruction **OL** **BL**

IF   students are successful with converting exponential functions into logarithmic functions,

THEN   have them work in groups to solve the following problem.
Create a game where you use exponential word problems, graphs of logarithmic functions, and functions that are written in exponential and logarithmic form.

Check Your Understanding

⚪ = Step-by-Step Solutions begin on page R11.

✓ **Go Online!** for a Self-Check Quiz

Example 1
F.IF.4

Describe the direction, form, and strength of each scatterplot.

1.
nonlinear, positive, weak

2.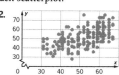
nonlinear, positive, weak

3.
linear, negative, strong

4.
nonlinear, positive, strong

6. The x-values are put in the L1 category and the y-values are put into the L2 category.

5. Why is the x variable sometimes referred to as the predictor variable?

6. What values are input into L1 and L2?

5. The x variable is sometimes referred to as the predictor variable because when you solve for x you are finding the solution to a prediction of unknown events.

Example 2
A.CED.2

Use a graph to describe the direction, form, and strength of the data set. If the scatterplot is linear, then write the linear equation for the line of best fit.

7. (0, 3), (−1, 2.5), (−2, 2), (1, 3.5), (2, 4), (3, 4.5) linear; $y = \frac{1}{2}x + 3$

8. (0,6), (1, 18), (2, 54), (3, 162) exponential

9. (3, 10), (7, 16), (4, 20), (8, 35) exponential

Example 3
F.LE.4

Solve for x.

10. $3^x = 27$ $x = 3$

11. $99^x = 1$ $x = 0$

12. $12^x = 288$ $x = 2.28$

13. $11^x = 121$ $x = 2$

14. $45^x = 91{,}125$ $x = 3$

15. $52^x = 7{,}311{,}616$ $x = 4$

Hours	Grade
10	100
8	83
5	92
2.5	57
4	74
9	95
6	75
6.5	85

Examples 1, 4, 5
F.IF.4, A.CED.2

A teacher asked eight random students how many hours they spent each week studying. She then matched the students' answers with their first quarter report card.

16. Make a graph of the data. What type of regression model works best with the data? linear regression model

17. What is the minimum number of hours that you should spend studying if you want to get a 90 in this class? 8 hours

Practice and Problem Solving

Extra Practice is found on page R6.

Example 1
F.IF.4

18. Describe the direction, form, and strength of the data set.

x	0	−2.1	−3	−2.1	−3	−3.25	5	9.4	11.5
y	1	4	7	7	7	10.6	0.3	0.2	0

18. nonlinear, negative, strong

19. Describe the direction, form, and strength of the data set.

x	0	5	5	3	0.5	16	4	7	4
y	7	7	12	12	8	8	6	14	9

19. nonlinear, no correlation

Practice

Formative Assessment Use Exercises 1–17 to assess students' understanding of the concepts in this lesson.

The Practice and Problem Solving exercises assess the content taught in the lesson.

Levels of Complexity Chart

The levels of the exercises progress from 1 to 3, with Level 1 indicating the lowest level of complexity.

Exercises	1–21	22–39, 51–56	40–50
◖ Level 3			●
◗ Level 2		●	
Level 1	●		

Extra Practice

See page R6 for extra exercises for students who are approaching level or for on-level students who need additional reinforcement.

Teaching Tip

Analyze Relationships Because graphing calculators are an important tool we use in mathematics, students should have a solid understanding on how to use them and how to ensure they are inputting the data accurately.

Additional Answers

41b.

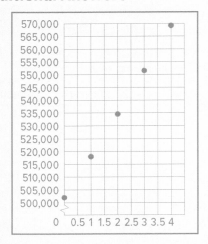

42a.

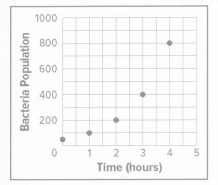

Example 2
A.CED.2 Use a graph to describe the direction, form, and strength of the data set. If the scatterplot is linear, then write the linear equation for the line of best fit.

20. (0, 1), (3, 27), (6, 729), (5, 243), (4, 81), (2, 9), (1, 3) exponential

21. (3, 8.5), (1, 5.5), (5, 11.5), (12, 22), (20, 34), (17, 29.5), (2, 7) linear; $y = 1.5x + 4$

22. (1, 6), (2, 12), (3, 24), (4, 48), (5, 96) exponential

Example 3
F.LE.4 **Solve for x.**

23. $9^x + 5 = 6566$ $x = 4$ **24.** $22^x = 484$ $x = 2$

25. $36^x + 2 = 14739$ $x = 2.68$ **26.** $5^x = 20$ $x = 1.86$

27. $e^x = 1$ $x = 0$ **28.** $10^x = 1778.28$ $x = 3.25$

Examples 1, 4, 5 **Each equation represents a data set. Choose the best model for each.**
F.IF.4, A.CED.2 **29.** $y = \frac{1}{4}x^3$ power   **30.** $y = 4x^2 + 2x - 3$ quadratic

31. $y = 4x + 7$ linear   **32.** $y = 24^x$ exponential

39. Sample answer: If the right side of the graph is higher than the left side, the direction is positive. If the right side of the graph is lower than the left side, the direction is negative.

Marcy studied a poster at her dentist's office. It showed the average number of grams of sugar a person consumes per day and the number of decayed teeth that person had.

Sugar (g)	140	25	100	85	95	10	40	60	75	130	135	110
Teeth	10	1	5	6	7	0	3	5	8	9	7	6

33. What is the regression model and line? linear, $y = 0.062756x + 0.32753$

34. Estimate approximately how many of Marcy's teeth will have some decay if she eats 55 grams of sugar a day. 3.779

Describe the correlation for each value of r.

35. $r = 0.5$ weak   **36.** $r = 0.82$ strong   **37.** $r = 1$ perfect   **38.** $r = 0$ no correlation

39. How do you visually determine the direction of data when describing a scatter plot?

40. Graph the data in the table.

The scatterplot is nonlinear, with a strong correlation, and a positive direction.

  a. Describe the scatterplot.

  b. Give an example of a regression model that does not accurately describe the set. Explain. Linear regression; the scatterplot is nonlinear. The regression line should curve.

x	y
0	0
1	1
2	4
3	9
4	16

41 REASONING In the mid-1990s, several natural disasters plagued a large city, so its inhabitants left. The surrounding cities witnessed an influx of population growth. Starting in 1994, one city had a population of 502,000 and began to grow at a constant rate of 3.2% every year.

  a. Let x represent years since 1994. Make a table for $x = 0, 1, 2, 3,$ and 4. Round to the nearest whole number. 502,000, 518,064, 534,642, 551,750, 569,406

  b. Graph the data. See margin.

  c. Find the equation that models the given situation. $y = 502,000(1 + 0.032)^t$

  d. What is the best model? exponential

  e. What will the population be in 2018? 1,069,095

  f. In how many years will the population be 1.5 million? 35 years

Differentiated Homework Options

Levels	AL Basic	OL Core	BL Advanced
Exercises	1–9, 37–40	10–17, 25–56	35–56
2-Day Option	10–15, 16–17	22–23, 31–36	43–44
	31–34	43–56	49–56

 You can use ALEKS to provide additional remediation support with personalized instruction and practice.

Go Online!   eBook

Interactive Student Guide

Use the *Interactive Student Guide* to deepen conceptual understanding.

· Modeling: Exponential and Logarithmic Functions

ALGEBRA 2
INTERACTIVE STUDENT GUIDE

43. Sample answer: It helps you eliminate possibilities and check your work. The linear regression model is very different than the quadratic regression model and by studying a scatterplot you can sometimes see those differences.

44e. No. Sample answer: The r-value of the exponential regression is 0.9936096203 and the r-value of the quadratic regression is 0.9980751795, which is closer to 1.

44f. The correlation coefficient establishes the strength of the relationship between the data and the regression line.

47. Sample answer: We need the line of best fit to solve for x. If we choose the wrong model, the regression model will not have the highest correlation rate and the prediction will be inaccurate.

49a. Sketching the graph could be more accurate if the data has a few outliers that are skewing the data.

49b–d. The sketched line of best fit would be linear, but with the graphing calculator the regression model would be a curve. So the graphing calculator would be more accurate in these cases.

42. PRECISION Francine is calculating bacterial growth in a culture in her science class. She put 50 bacteria on a slide under a microscope and documented the population growth of the bacteria every hour.

Hours (x)	0	1	2	3	4
Bacteria population (y)	50	100	200	400	800

a. Graph the data. See margin.

b. What is the relationship between the variables? The y-values are doubling.

c. What is the best model? exponential regression model

d. What is the regression line? $y = 50(2)^x$

e. When will the bacteria exceed a million? a little over 14 hours

43. How does describing the scatter plot help you determine the best regression model?

44. MULTI-STEP Jewel deposits $50 into a savings account and does not make any deposits or withdrawals. The table shows the balance in the account after 12 years.

Time	Balance
0	$50
2	$55.80
4	$64.80
6	$83.09
8	$101.40
10	$123.14
12	$162.67

a. Make a scatterplot of the data. 44a–b. Check students' work.

b. Calculate and graph a curve of fit using an exponential regression.

c. Write the equation of best fit. Sample answer: $y = 46.47(1.10^x)$

d. Based on the model, what will the account balance be after 25 years? $558.58

e. Is an exponential model the best fit for the data? Explain.

f. What does the variable r represent and why is it important?

45 Paleontologists think they discovered a new dinosaur that is half the size of a T-rex. They measured the length and diameter, in centimeters, of the arms of a miniature T-rex.

Diameter	Length
1.76	15.99
2.6	20.69
3.19	23.68
4.58	30.06
5.12	32.36
5.81	35.17
6.47	37.76
6.67	38.41
8.08	43.72
8.29	44.47

a. Graph the data. See margin.

b. What is the scale? Sample answer: $0 \le x \le 10$ by 1 and $10 \le y \le 50$ by 5

c. What is the best regression model? power regression

d. Is there a strong correlation? Explain.

e. Predict the diameter if the length is 30.57 centimeters. 4.7cm

45d. Yes, the correlation coefficient is close to 1

F.IF.4, A.CED.2

H.O.T. Problems Use Higher-Order Thinking Skills

46. CRITIQUE ARGUMENTS Malcolm changed the quadratic function $7^x = 343$ into the logarithmic function $\log_7 343 = x$ and entered the function into his calculator. The answer it gave was $x = 2.54$ (rounded to the nearest hundredth). Is he correct? Explain your answer. Sample answer: No. He forgot that most calculators have logs with base 10.

47. OPEN-ENDED Explain why we have to choose the best model before making a prediction.

48. CHALLENGE Write a word problem that would lead to a data set for one of the regression models. Graph the data set using the regression model you chose, and find the regression line and correlation coefficient. See margin.

49. PRECISION For each type of regression model, explain whether you would get a more accurate result by using a graphing calculator or by sketching the graph and using a linear equation for the line of best fit.

a. linear regression   **b.** quadratic regression
c. power regression   **d.** exponential regression

50. WRITING IN MATH Describe the difference between the process of how to make a prediction using a graphing calculator and using a sketch. See margin.

Standards for Mathematical Practice

Emphasis On	Exercises
1 Make sense of problems and persevere in solving them.	5, 40, 56
2 Reason abstractly and quantitatively.	1–4, 18–19, 39, 41, 48, 50–52, 54–55
3 Construct viable arguments and critique the reasoning of others.	46–47
4 Model with mathematics.	16–17, 33–34, 43–45
5 Use appropriate tools strategically.	22
6 Attend to precision.	42, 49, 53–55

---

e **Follow-Up**

Students have explored using a graphing calculator to make predictions. Ask: What real-world situations or jobs would require this skill? Sample answer: Investing, banking, scientific investigation, medical professions

Assess

Ticket Out the Door Make several copies of each model, exponential equations, and data sets and have the students determine the best model, convert the exponential functions into logarithmic functions, and practice using the graphing calculator.

Additional Answers

45a.

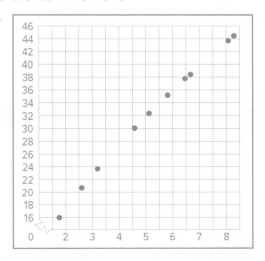

50. Sample answer: The difference would be the accuracy, the graphing calculator calculates all of the values to create a regression line and could interchange between the regression models to show which has the highest correlation coefficient.

Go Online!

eSolutions Manual

Create worksheets, answer keys, and solutions handouts for your assignments.

Preparing for Assessment

Exercises 51–56 require students to use the skills they will need on assessments. Each exercise is dual-coded with content.

Dual Coding		
Items	Content Standards	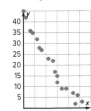 Mathematical Practices
51	F.IF.4	2
52	A.CED.2	2
53	F.LE.4	6
54	A.CED.2	2, 6
55	F.IF.4, A.CED.2	2, 6
56	A.CED.2	1

Diagnose Student Errors

51.

A	Needs to learn strategies to visualize data
B	CORRECT
C	Understands negative direction, but needs to review correlation
D	Needs to review direction

52.

A	Understands exponential equations should have an exponent, but not that the *x* is the exponent
B	Needs to review linear and exponential equations
C	CORRECT
D	Understands exponential equations should have an exponent, but not that the *x* is the exponent

Go Online!

Quizzes

Students can use *Self-Check Quizzes* to check their understanding of this lesson. You can also give the *Chapter Quiz*, which covers the content in Lessons 6-3, 6-4, and 6-5.

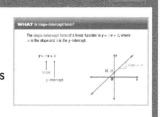

Preparing for Assessment

51. Describe the scatter plot. $\bigcirc$ 2 F.IF.4 **B**

- A nonlinear, positive direction, weak correlation
- B linear, negative direction, strong correlation
- C nonlinear, negative direction, weak correlation
- D linear, positive direction, strong correlation

52. Choose the data set that represents an exponential model. $\bigcirc$ 2 A.CED.2 **C**

- A $y = 3x^2 + x - 17$
- B $y = 0.5x - 9$
- C $y = 2(1.67)^x$
- D $y = 2x^2$

53. Convert the exponential function $4(20)^x = 3,200,000$ into a logarithmic function to solve for *x*. What is the value of *x*? Round to the nearest hundredth. $\bigcirc$ 6 F.LE.4 **A**

- A 4.54
- C 10.80
- B 5.00
- D 8.64

54. Choose the best model for the data set. $\bigcirc$ 2, 6 A.CED.2 **B**

linear: $r = .945421996$

quadratic: $r = .9984103974$

exponential: $r = .994897438$

power: $r = .9917643588$

- A linear
- C exponential
- B quadratic
- D power

55. MULTI-STEP Every year the local golf club sponsors a tournament. The tournament starts with 512 participants. During each round, one half of the players are eliminated. $\bigcirc$ 2, 6 F.IF.4, A.CED.2

a. What values go into L1?

> 0,1,2,3,4, 5, 6, 7, 8, 9

b. What values go into L2?

> 512, 256, 128, 64, 32, 16, 8, 4, 2, 1

c. Describe the scatter plot.

> nonlinear, negative direction, strong correlation

d. Choose the best model. Explain.

> Exponential, because it has the strongest correlation.

e. How many players remain after 5 rounds?

> 16

f. How many players remain after 8 rounds?

> 2

g. How many rounds are in the tournament?

> 9

56. The fish population in pond grows by 10% every year. At the beginning of the observation, there are 100 fish in this pond. $\bigcirc$ 1 A.CED.2

a. Write an exponential function to model the fish population in the pond.

> Sample answer: $P(x) = 100 \cdot 1.1^x$

b. How many fish will be in the pond after 2 years?

> 121

c. After how many years will the fish population in the pond reach 146? Round your answer to the nearest whole number.

> 4 years

53.

A	CORRECT
B	Understands structure, but needs to review how to divide the variable *a* on each side
C	Needs to review the meaning of the variables and the structure of the formula
D	Understands the need to divide a variable on each side, but isn't sure about the structure of the formulas

54.

A	Understands the need for the correlation to be greater than 0.8
B	CORRECT
C	Understands the need for the correlation to be close to 1
D	Understands the need for the correlation to be close to 1

CHAPTER 6
Mid-Chapter Quiz
Lessons 6-1 through 6-5

Graph each function. State the domain and range.
(Lesson 6-1)

1. $f(x) = 3(4)^x$ **1–4. See Ch. 6 Answer Appendix.**

2. $f(x) = -(2)^x + 5$

3. $f(x) = -0.5(3)^{x+2} + 4$

4. $f(x) = -3\left(\frac{2}{3}\right)^{x-1} + 8$

5. SCIENCE You are studying a bacteria population. The population originally started with 6000 bacteria cells. After 2 hours, there were 28,000 bacteria cells. (Lesson 6-1)

   a. Write an exponential function that could be used to model the number of bacteria after x hours if the number of bacteria changes at the same rate. $f(x) = 6000(2.16025)^x$

   b. How many bacteria cells can be expected after 4 hours? **about 130,667**

   c. What mathematical practice did you use to solve this problem? **See students' work.**

6. MULTIPLE CHOICE Which exponential function has a graph that passes through the points at (0, 125) and (3, 1000)? (Lesson 6-1) **D**

   A $f(x) = 125(3)^x$ **C** $f(x) = 125(1000)^x$

   B $f(x) = 1000(3)^x$ **D** $f(x) = 125(2)^x$

7. POPULATION In 2002, a certain city had a population of 45,000. It increased to 68,000 by 2014. (Lesson 6-2)

   a. What is an exponential function that could be used to model the population of this city x years after 2002? $f(x) = 45,000(1.0350)^x$

   b. Use your model to estimate the population in 2027. **106,346**

Solve each equation or inequality. Check your solution. (Lesson 6-2)

8. $4^{3x-1} = 16^x$ **1**

9. $\frac{1}{9} = 243^{2x+1}$ $-\frac{7}{10}$

10. $16^{2x+3} < 64$ $\left\{x \mid x < -\frac{3}{4}\right\}$

11. $\left(\frac{1}{32}\right)^{x+3} \geq 16^{3x}$ $\left\{x \mid x \leq -\frac{15}{17}\right\}$

Find the indicated term for each geometric sequence. (Lesson 6-3)

12. $a_2 = 8, r = 2, a_8 = ?$ **512**

13. $a_3 = 0.5, r = 8, a_{10} = ?$ **1,048,576**

14. Find the geometric means of the sequence below. (Lesson 6-3)

   $0.5,$ ___**4**___, ___**3**___, ___**256**___, 2048

Evaluate the sum of each geometric series. (Lesson 6-3)

15. $\sum_{k=1}^{8} 3 \cdot 2^{k-1}$ **765**

16. $\sum_{k=1}^{9} 4 \cdot (-1)^{k-1}$ **4**

17. INCOME Peter works for a house building company for 4 months per year. He starts out making $3000 per month. At the end of each month, his salary increases by 5%. How much money will he make in those 4 months? (Lesson 6-3) **$12,930.38**

18. MULTIPLE CHOICE Find the value of x for $\log_3 (x^2 + 2x) = \log_3 (x + 2)$. (Lesson 6-4) **C**

   A $x = -2, 1$ **C** $x = 1$

   B $x = -2$ **D** no solution

Graph each function. (Lesson 6-4) **19, 20. See Ch. 6 Answer Appendix.**

19. $f(x) = 3 \log_2 (x - 1)$

20. $f(x) = -4 \log_3 (x - 2) + 5$

Evaluate each expression. (Lesson 6-4)

21. $\log_4 32$ $\frac{5}{2}$

22. $\log_5 5^{12}$ **12**

23. $\log_{16} 4$ $\frac{1}{2}$

24. Write $\log_9 729 = 3$ in exponential form. (Lesson 6-4) $9^3 = 729$

25. Determine the type of function that would best model the data in the table. Then determine the regression equation. (Lesson 6-5) **exponential; $y = 1.4246(2.3685)^x$**

x	1	2	3	4	5
y	3	9	20	45	100

Foldables Study Organizer

Dinah Zike's

Before students complete the Mid-Chapter Quiz, encourage them to review the information for Lessons 6-1 through 6-5 in their Foldables. Students may benefit from sharing their Foldable with a partner and taking turns summarizing what they have learned about exponential and logarithmic functions, while the other partner listens carefully. They should seek clarification of any concepts, as needed.

ALEKS can be used as a formative assessment tool to target learning gaps for those who are struggling, while providing enhanced learning for those who have mastered the concepts.

RtI Response to Intervention
Use the Intervention Planner to help you determine your Response to Intervention.

Intervention Planner

TIER 1 **On Level** OL

IF students miss 25% of the exercises or less,

THEN choose a resource:

   SE Lessons 6-1 through 6-5

   Go Online!

   Skills Practice

   Chapter Project

   Self-Check Quizzes

TIER 2 **Strategic Intervention** AL
Approaching grade level

IF students miss 50% of the exercises,

THEN *Quick Review Math Handbook*

   Go Online!

   Study Guide and Intervention

   Extra Examples

   Personal Tutors

   Homework Help

TIER 3 **Intensive Intervention**
2 or more grades below level

IF students miss 75% of the exercises,

THEN choose a resource:

   Use *Math Triumphs, Alg. 2*

   Go Online!

   Extra Examples

   Personal Tutors

   Homework Help

   Review Vocabulary

Go Online!

eAssessment

You can use the premade Mid-Chapter Test to assess students' progress in the first half of the chapter. Customize and create multiple versions of your Mid-Chapter Quiz and answer keys that align to your standards. Tests can be delivered on paper or online.

Properties of Logarithms

SUGGESTED PACING (DAYS)

90 min.	0.5
45 min.	1.0

Instruction

Track Your Progress

Objectives

1 Simplify and evaluate expressions using the properties of logarithms.

2 Solve logarithmic equations using the properties of logarithms.

Mathematical Background

In addition to the Product, Quotient, and Power Properties of Logarithms, there are other properties such as the four listed below that can be helpful.

$\log_b 1 = 0$
$\log_b b = 1$
$\log_b b^x = x$
$b^{\log_b x} = x$

THEN	**NOW**	**NEXT**
A.SSE.2 Use the structure of an expression to identify ways to rewrite it.	**A.CED.1** Create equations and inequalities in one variable and use them to solve problems.	**A.CED.1** Create equations and inequalities in one variable and use them to solve problems.

Go Online! All of these resources and more are available at connectED.mcgraw-hill.com

eLessons utilize the power of your interactive whiteboard in an engaging way. Use **Logarithms**, screen 13, to introduce the concepts in this lesson.

Personal Tutors (for every example) let students hear real teachers solve problems. Students can pause and repeat as many times as necessary.

Use **Self-Check Quiz** to assess students' understanding of the concepts in this lesson.

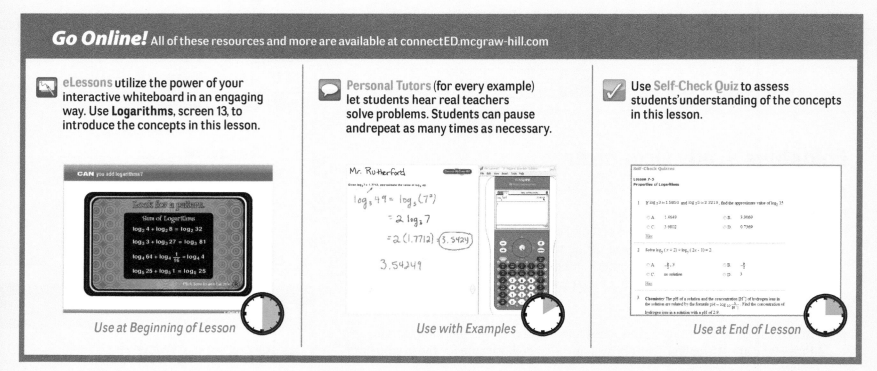

Use at Beginning of Lesson

Use with Examples

Use at End of Lesson

OER **Using Open Educational Resources**

Tutorials Have students review the tutorial on **purplemath** about properties of logarithms. They can also see worked-out examples and tips.

Go Online!
connectED.mcgraw-hill.com

Worksheets

Differentiate Your Resources

Extra Practice Additional practice or homework; Skills Practice is best for approaching-level students and Practice is best for on-level and beyond-level students

Skills Practice

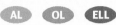

AL OL ELL

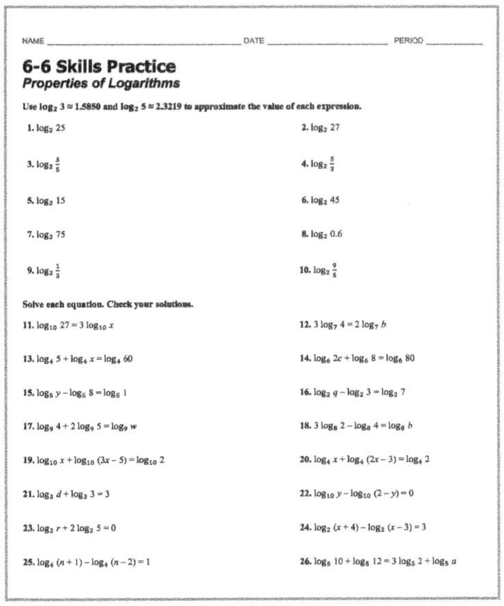
NAME ______ DATE ______ PERIOD ______

6-6 Skills Practice
Properties of Logarithms

Use $\log_2 3 \approx 1.5850$ and $\log_2 5 \approx 2.3219$ to approximate the value of each expression.

1. $\log_2 25$ 2. $\log_2 27$
3. $\log_2 \frac{3}{5}$ 4. $\log_2 \frac{5}{3}$
5. $\log_2 15$ 6. $\log_2 45$
7. $\log_2 75$ 8. $\log_2 0.6$
9. $\log_2 \frac{1}{3}$ 10. $\log_2 \frac{9}{5}$

Solve each equation. Check your solutions.

11. $\log_{10} 27 = 3 \log_{10} x$ 12. $3 \log_7 4 = 2 \log_7 b$
13. $\log_4 5 + \log_4 x = \log_4 60$ 14. $\log_6 2c + \log_6 8 = \log_6 80$
15. $\log_5 y - \log_5 8 = \log_5 1$ 16. $\log_2 q - \log_2 3 = \log_2 7$
17. $\log_9 4 + 2 \log_9 5 = \log_9 w$ 18. $3 \log_8 2 - \log_8 4 = \log_8 b$
19. $\log_{10} x + \log_{10} (3x - 5) = \log_{10} 2$ 20. $\log_4 x + \log_4 (2x - 3) = \log_4 2$
21. $\log_3 d + \log_3 3 = 3$ 22. $\log_{10} y - \log_{10} (2 - y) = 0$
23. $\log_2 r + 2 \log_2 5 = 0$ 24. $\log_2 (x + 4) - \log_2 (x - 3) = 3$
25. $\log_4 (n + 1) - \log_4 (n - 2) = 1$ 26. $\log_5 10 + \log_5 12 = 3 \log_5 2 + \log_5 a$

Practice

AL OL BL ELL

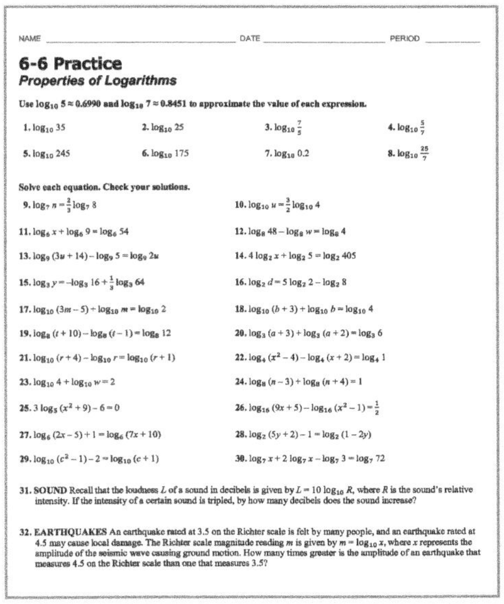
NAME ______ DATE ______ PERIOD ______

6-6 Practice
Properties of Logarithms

Use $\log_{10} 5 \approx 0.6990$ and $\log_{10} 7 \approx 0.8451$ to approximate the value of each expression.

1. $\log_{10} 35$ 2. $\log_{10} 25$ 3. $\log_{10} \frac{7}{5}$ 4. $\log_{10} \frac{5}{7}$
5. $\log_{10} 245$ 6. $\log_{10} 175$ 7. $\log_{10} 0.2$ 8. $\log_{10} \frac{25}{7}$

Solve each equation. Check your solutions.

9. $\log_7 n = \frac{2}{3} \log_7 8$ 10. $\log_{10} u = \frac{3}{2} \log_{10} 4$
11. $\log_6 x + \log_6 9 = \log_6 54$ 12. $\log_8 48 - \log_8 w = \log_8 4$
13. $\log_9 (3u + 14) - \log_9 5 = \log_9 2u$ 14. $4 \log_2 x + \log_2 5 = \log_2 405$
15. $\log_3 y = -\log_3 16 + \frac{1}{3} \log_3 64$ 16. $\log_2 d = 5 \log_2 2 - \log_2 8$
17. $\log_{10} (3m - 5) + \log_{10} m = \log_{10} 2$ 18. $\log_{10} (b + 3) + \log_{10} b = \log_{10} 4$
19. $\log_8 (t + 10) - \log_8 (t - 1) = \log_8 12$ 20. $\log_3 (a + 3) + \log_3 (a + 2) = \log_3 6$
21. $\log_{10} (r + 4) - \log_{10} r = \log_{10} (r + 1)$ 22. $\log_4 (x^2 - 4) - \log_4 (x + 2) = \log_4 1$
23. $\log_{10} 4 + \log_{10} w = 2$ 24. $\log_8 (n - 3) + \log_8 (n + 4) = 1$
25. $3 \log_5 (x^2 + 9) - 6 = 0$ 26. $\log_{16} (9x + 5) - \log_{16} (x^2 - 1) = \frac{1}{2}$
27. $\log_6 (2x - 5) + 1 = \log_6 (7x + 10)$ 28. $\log_2 (5y + 2) - 1 = \log_2 (1 - 2y)$
29. $\log_{10} (c^2 - 1) - 2 = \log_{10} (c + 1)$ 30. $\log_7 x + 2 \log_7 x - \log_7 3 = \log_7 72$

31. **SOUND** Recall that the loudness L of a sound in decibels is given by $L = 10 \log_{10} R$, where R is the sound's relative intensity. If the intensity of a certain sound is tripled, by how many decibels does the sound increase?

32. **EARTHQUAKES** An earthquake rated at 3.5 on the Richter scale is felt by many people, and an earthquake rated at 4.5 may cause local damage. The Richter scale magnitude reading m is given by $m = \log_{10} x$, where x represents the amplitude of the seismic wave causing ground motion. How many times greater is the amplitude of an earthquake that measures 4.5 on the Richter scale than one that measures 3.5?

Word Problem Practice

AL OL BL ELL

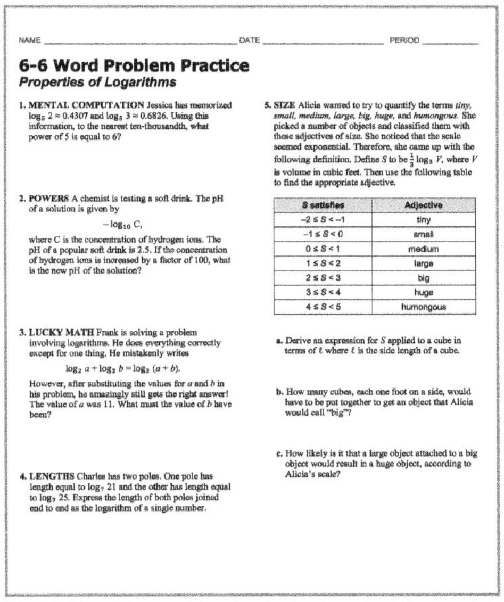
NAME ______ DATE ______ PERIOD ______

6-6 Word Problem Practice
Properties of Logarithms

1. **MENTAL COMPUTATION** Jessica has memorized $\log_5 2 \approx 0.4307$ and $\log_5 3 \approx 0.6826$. Using this information, to the nearest ten-thousandth, what power of 5 is equal to 6?

2. **POWERS** A chemist is testing a soft drink. The pH of a solution is given by
$-\log_{10} C$,
where C is the concentration of hydrogen ions. The pH of a popular soft drink is 2.5. If the concentration of hydrogen ions is increased by a factor of 100, what is the new pH of the solution?

3. **LUCKY MATH** Frank is solving a problem involving logarithms. He does everything correctly except for one thing. He mistakenly writes
$\log_b a + \log_b b = \log_b (a + b)$.
However, after substituting the values for a and b in his problem, he amazingly still gets the right answer! The value of a was 11. What must the value of b have been?

4. **LENGTHS** Charles has two poles. One pole has length equal to $\log_7 21$ and the other has length equal to $\log_7 25$. Express the length of both poles joined end to end as the logarithm of a single number.

5. **SIZE** Alicia wanted to try to quantify the terms *tiny, small, medium, large, big, huge,* and *humongous*. She picked a number of objects and classified them with those adjectives of size. She noticed that the scale seemed exponential. Therefore, she came up with the following definition. Define S to be $\frac{1}{3} \log_3 V$, where V is volume in cubic feet. Then use the following table to find the appropriate adjective.

S satisfies	Adjective
$-2 \le S < -1$	tiny
$-1 \le S < 0$	small
$0 \le S < 1$	medium
$1 \le S < 2$	large
$2 \le S < 3$	big
$3 \le S < 4$	huge
$4 \le S < 5$	humongous

a. Derive an expression for S applied to a cube in terms of ℓ where ℓ is the side length of a cube.

b. How many cubes, each one foot on a side, would have to be put together to get an object that Alicia would call "big"?

c. How likely is it that a large object attached to a big object would result in a huge object, according to Alicia's scale?

Intervention Reteaching and vocabulary activities that can be used with struggling or absent students and as ELL support

Study Guide and Intervention

AL OL ELL

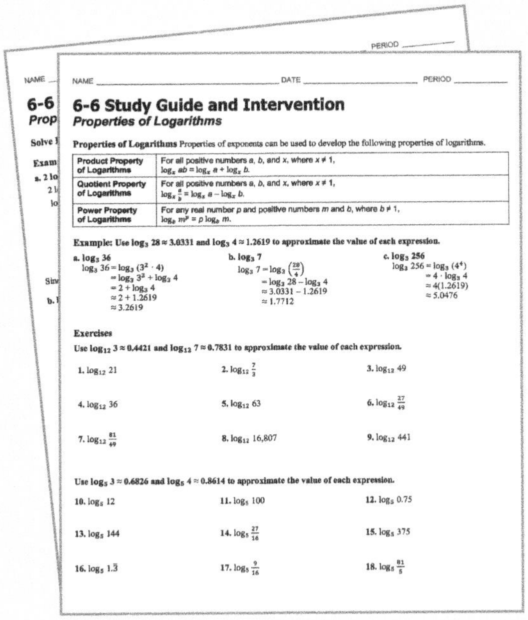
NAME ______ DATE ______ PERIOD ______

6-6 Study Guide and Intervention
Properties of Logarithms

Properties of Logarithms Properties of exponents can be used to develop the following properties of logarithms.

Product Property of Logarithms	For all positive numbers a, b, and x, where $x \ne 1$, $\log_x ab = \log_x a + \log_x b$.
Quotient Property of Logarithms	For all positive numbers a, b, and x, where $x \ne 1$, $\log_x \frac{a}{b} = \log_x a - \log_x b$.
Power Property of Logarithms	For any real number p and positive numbers m and b, where $b \ne 1$, $\log_b m^p = p \log_b m$.

Example: Use $\log_3 28 \approx 3.0331$ and $\log_3 4 \approx 1.2619$ to approximate the value of each expression.

a. $\log_3 36$
$\log_3 36 = \log_3 (3^2 \cdot 4)$
$= \log_3 3^2 + \log_3 4$
$= 2 + \log_3 4$
$\approx 2 + 1.2619$
≈ 3.2619

b. $\log_3 7$
$\log_3 7 = \log_3 \left(\frac{28}{4}\right)$
$= \log_3 28 - \log_3 4$
$\approx 3.0331 - 1.2619$
≈ 1.7712

c. $\log_3 256$
$\log_3 256 = \log_3 (4^4)$
$= 4 \cdot \log_3 4$
$\approx 4(1.2619)$
≈ 5.0476

Exercises

Use $\log_{12} 3 \approx 0.4421$ and $\log_{12} 7 \approx 0.7831$ to approximate the value of each expression.

1. $\log_{12} 21$ 2. $\log_{12} \frac{7}{3}$ 3. $\log_{12} 49$
4. $\log_{12} 36$ 5. $\log_{12} 63$ 6. $\log_{12} \frac{27}{49}$
7. $\log_{12} \frac{81}{49}$ 8. $\log_{12} 16{,}807$ 9. $\log_{12} 441$

Use $\log_5 3 \approx 0.6826$ and $\log_5 4 \approx 0.8614$ to approximate the value of each expression.

10. $\log_5 12$ 11. $\log_5 100$ 12. $\log_5 0.75$
13. $\log_5 144$ 14. $\log_5 \frac{27}{16}$ 15. $\log_5 375$
16. $\log_5 1.3$ 17. $\log_5 \frac{9}{16}$ 18. $\log_5 \frac{81}{5}$

Study Notebook

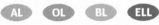

AL OL BL ELL

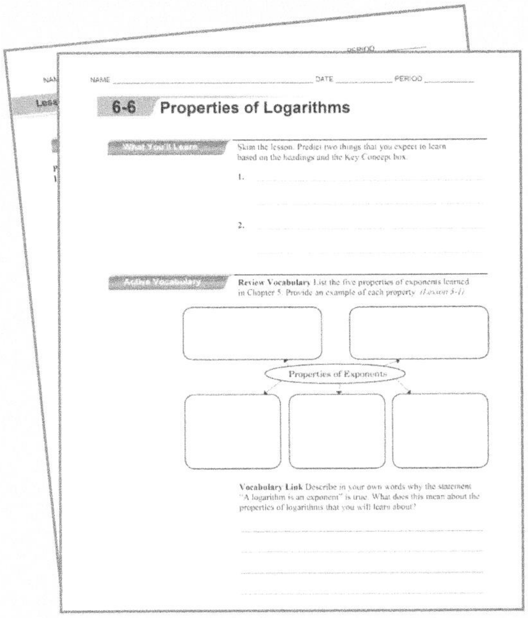
NAME ______ DATE ______ PERIOD ______

6-6 Properties of Logarithms

What You'll Learn — Skim the lesson. Predict two things that you expect to learn based on the headings and the Key Concepts box.

1.

2.

Active Vocabulary — Review Vocabulary List the five properties of exponents learned in Chapter 5. Provide an example of each property. (Lesson 5-1)

Properties of Exponents

Vocabulary Link Describe in your own words why the statement "A logarithm is an exponent" is true. What does this mean about the properties of logarithms that you will learn about?

Extension Activities that can be used to extend lesson concepts

Enrichment

OL BL ELL

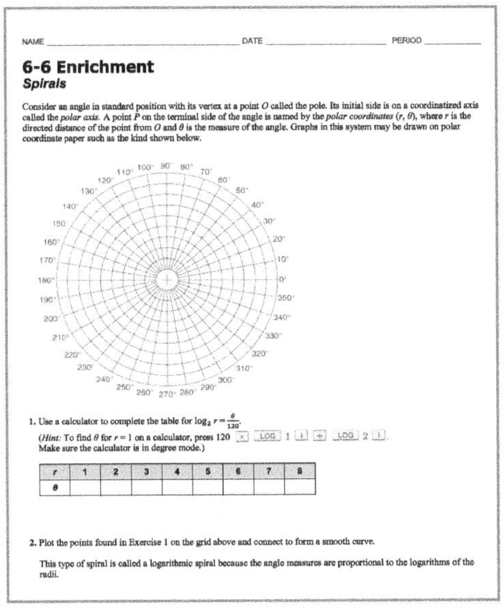
NAME ______ DATE ______ PERIOD ______

6-6 Enrichment
Spirals

Consider an angle in standard position with its vertex at a point O called the pole. Its initial side is on a coordinatized axis called the *polar axis*. A point P on the terminal side of the angle is named by the *polar coordinates* (r, θ), where r is the directed distance of the point from O and θ is the measure of the angle. Graphs in this system may be drawn on polar coordinate paper such as the kind shown below.

1. Use a calculator to complete the table for $\log_2 r = \frac{\theta}{120}$.
(*Hint:* To find θ for $r = 1$ on a calculator, press 120 × LOG 1 ÷ LOG 2 .)
Make sure the calculator is in degree mode.

r	1	2	3	4	5	6	7	8
θ								

2. Plot the points found in Exercise 1 on the grid above and connect to form a smooth curve.

This type of spiral is called a logarithmic spiral because the angle measures are proportional to the logarithms of the radii.

Launch

Have students read the Why? section of the lesson. Ask:

- Is lemon juice acidic or basic? acidic
- Is milk of magnesia acidic or basic? basic
- Milk of magnesia is how many times as basic as neutral water? $10^{10-7} = 10^3 = 1000$

Teach

Ask the scaffolded questions for each example to build conceptual understanding for students at all levels.

1 Properties of Logarithms

Example 1 Use the Product Property

AL What is the first step of this problem? Factor 192 into its prime factorization.

OL Why is $\log_4 3$ an irrational number? 3 cannot be written as 4 raised to a rational number.

BL For what values of x is this property useful in approximating $\log_b x$? when x has factors of b in its prime factorization.

Need Another Example?
Use $\log_5 2 \approx 0.4307$ to approximate the value of $\log_5 250$. 3.4307

LESSON 6
Properties of Logarithms

			Product	pH Level
			lemon juice	2.1
			hot sauce	3.5
			tomatoes	4.2
			black coffee	5.0
			milk	6.4
			pure water	7.0
			eggs	7.8
			milk of magnesia	10.0

::Then	::Now	::Why?
You evaluated logarithmic expressions and solved logarithmic equations.	**1** Simplify and evaluate expressions using the properties of logarithms. **2** Solve logarithmic equations using the properties of logarithms.	The level of acidity in food is important to some consumers with sensitive stomachs. Most of the foods that we consume are more acidic than basic. The pH scale measures acidity; a low pH indicates an acidic solution, and a high pH indicates a basic solution. It is another example of a logarithmic scale based on powers of ten. Black coffee has a pH of 5, while neutral water has a pH of 7. Black coffee is one hundred times as acidic as neutral water, because $10^{7-5} = 10^2$ or 100.

MP **Mathematical Practices**
8 Look for and express regularity in repeated reasoning.

Content Standards
A.CED.1 Create equations and inequalities in one variable and use them to solve problems.

1 **Properties of Logarithms** Because logarithms are exponents, the properties of logarithms can be derived from the properties of exponents. The Product Property of Logarithms can be derived from the Product of Powers Property of Exponents.

> **Key Concept** Product Property of Logarithms
>
> **Words** The logarithm of a product is the sum of the logarithms of its factors.
>
> **Symbols** For all positive numbers a, b, and x, where $x \neq 1$, $\log_x ab = \log_x a + \log_x b$.
>
> **Example** $\log_2 [(5)(6)] = \log_2 5 + \log_2 6$

To show that this property is true, let $b^x = a$ and $b^y = c$. Then, using the definition of logarithm, $x = \log_b a$ and $y = \log_b c$.

$b^x b^y = ac$	Substitution
$b^{x+y} = ac$	Product of Powers
$\log_b b^{x+y} = \log_b ac$	Property of Equality for Logarithmic Functions
$x + y = \log_b ac$	Inverse Property of Exponents and Logarithms
$\log_b a + \log_b c = \log_b ac$	Replace x with $\log_b a$ and y with $\log_b c$.

You can use the Product Property of Logarithms to approximate logarithmic expressions.

A.CED.1

Example 1   Use the Product Property

Use $\log_4 3 \approx 0.7925$ to approximate the value of $\log_4 192$.

$\log_4 192 = \log_4 (4^3 \cdot 3)$	Replace 192 with $64 \cdot 3$ or $4^3 \cdot 3$
$= \log_4 4^3 + \log_4 3$	Product Property
$= 3 + \log_4 3$	Inverse Property of Exponents and Logarithms
$\approx 3 + 0.7925$ or 3.7925	Replace $\log_4 3$ with 0.7925.

▷ **Guided Practice**

1. Use $\log_4 2 = 0.5$ to approximate the value of $\log_4 32$. 2.5

MP **Teaching the Mathematical Practices**

Look for and express regularity in repeated reasoning. Help students look for and express regularity in properties and equations of logarithms. For example, ask:

- What is the product property of logarithms? $\log_x (ab) = \log_x a + \log_x b$
- What is the quotient property of logarithms? $\log_x (a/b) = \log_x a - \log_x b$
- What is the power property of logarithms? $\log_x m^p = p \log_x m$
- When would a logarithmic equation have an extraneous solution? Sample answer: when the solution of the equation results in a calculation of a negative log value

---

Go Online!

Interactive Whiteboard
Use the *eLesson* or *Lesson Presentation* to present this lesson.

Recall that the quotient of powers is found by subtracting exponents. The property for the logarithm of a quotient is similar. Let $b^x = a$ and $b^y = c$. Then $\log_b a = x$ and $\log_b c = y$.

$$\frac{b^x}{b^y} = \frac{a}{c}$$

$$b^{x-y} = \frac{a}{c} \qquad \text{Quotient Property}$$

$$\log_b b^{x-y} = \log_b \frac{a}{c} \qquad \text{Property of Equality for Logarithmic Equations}$$

$$x - y = \log_b \frac{a}{c} \qquad \text{Inverse Property of Exponents and Logarithms}$$

$$\log_b a - \log_b c = \log_b \frac{a}{c} \qquad \text{Replace } x \text{ with } \log_b a \text{ and } y \text{ with } \log_b c$$

📕 Key Concept Quotient Property of Logarithms

Words	The logarithm of a quotient is the difference of the logarithms of the numerator and the denominator.
Symbols	For all positive numbers a, b, and x, where $x \neq 1$, $\log_x \frac{a}{b} = \log_x a - \log_x b$.
Example	$\log_2 \frac{5}{6} = \log_2 5 - \log_2 6$

A.CED.1

Real-World Example 2 Quotient Property

SCIENCE The pH of a substance is defined as the concentration of hydrogen ions $[H^+]$ in moles. It is given by the formula $\text{pH} = \log_{10} \frac{1}{H^+}$. Find the amount of hydrogen in a liter of acid rain that has a pH of 4.2.

Understand The formula for finding pH and the pH of the rain is given.

Plan Write the equation. Then, solve for $[H^+]$.

Solve

$$\text{pH} = \log_{10} \frac{1}{H^+} \qquad \text{Original equation}$$

$$4.2 = \log_{10} \frac{1}{H^+} \qquad \text{Substitute 4.2 for pH.}$$

$$4.2 = \log_{10} 1 - \log_{10} H^+ \qquad \text{Quotient Property}$$

$$4.2 = 0 - \log_{10} H^+ \qquad \log_{10} 1 = 0$$

$$4.2 = -\log_{10} H^+ \qquad \text{Simplify.}$$

$$-4.2 = \log_{10} H^+ \qquad \text{Multiply each side by } -1.$$

$$10^{-4.2} = H^+ \qquad \text{Definition of logarithm}$$

There are $10^{-4.2}$, or about 0.000063, mole of hydrogen in a liter of this rain.

Check

$$4.2 = \log_{10} \frac{1}{10^{-4.2}} \qquad \text{pH} = 4.2, H^+ = 10^{-4.2}$$

$$4.2 \stackrel{?}{=} \log_{10} 1 - \log_{10} 10^{-4.2} \qquad \text{Quotient Property}$$

$$4.2 \stackrel{?}{=} 0 - (-4.2) \qquad \text{Simplify.}$$

$$4.2 = 4.2 \checkmark$$

Using the Quotient Property, the equation is simplified to two terms that can easily be evaluated.

▶ **Guided Practice**

2. **SOUND** The loudness L of a sound, measured in decibels, is given by $L = 10 \log_{10} R$, where R is the sound's relative intensity. Suppose one person talks with a relative intensity of 10^6 or 60 decibels. How much louder would 100 people be, talking at the same intensity?   **20 decibels louder**

Real-World Link

Acid rain is more acidic than normal rain. Smoke and fumes from burning fossil fuels rise into the atmosphere and combine with the moisture in the air to form acid rain. Acid rain can be responsible for the erosion of statues, as in the photo above.

Example 2 Quotient Property

AL What is the domain of this function? $H^+ > 0$

OL A substance is acidic when pH is less than 7. For what values of H^+ is a substance acidic? $H^+ > 10^{-7}$

BL If the pH of a substance increases by 2, what is the change in the concentration of hydrogen ions? It is divided by 100.

Need Another Example?

Science Find the amount of hydrogen in a liter of acid rain that has a pH of 5.5. $10^{-5.5}$, or 0.0000032 moles

Teaching Tip

Problem Solving When discussing the Product Property of Logarithms, point out that the logarithms used in the example show that the property applies to all logarithms, not just those that can be simplified.

Go Online!

The most up-to-date resources available for your program can be found at <u>connectED.mcgraw-hill.com</u>.

Example 3 Power Property of Logarithms

AL Approximate $\log_2 125$. 6.9657

OL How can you use this property to approximate $\log_2 \sqrt{5}$? $\log_2 \sqrt{5} = \log_2 5^{\frac{1}{2}} = \frac{1}{2} \log_2 5 \approx \frac{1}{2}(2.3219)$ or 1.16095

BL Prove $\log_b m^p = p \log_b m$. Proof. $y = \log_b m^p$ if and only if $b^y = m^p$ if and only if $(b^y)^{\frac{1}{p}} = m$. $(b^y)^{\frac{1}{p}} = b^{\frac{y}{p}}$, so $m = b^{\frac{y}{p}}$. $m = b^{\frac{y}{p}}$ if and only if $\log_b m = \frac{y}{p}$ if and only if $y = p \log_b m$. Therefore, $\log_b m^p = p \log_b m$.

Need Another Example?
Given $\log_5 6 \approx 1.1133$, approximate the value of $\log_5 216$. 3.3399

2 Solve Logarithmic Equations

Example 4 Solve Equations Using Properties of Logarithms

AL What is the domain of $\log_6 x + \log_6 (x - 9)$? $x > 9$

OL How do you know that $x = -3$ cannot be a solution? It is not in the domain of $\log_6 x + \log_6 (x - 9)$.

BL How could you use your calculator to check your solution? Graph $y = \log_6 x + \log_6 (x - 9)$ and $y = 2$ and find the intersection point.

Need Another Example?
Solve the equation.
$4 \log_2 x - \log_2 5 = \log_2 125$ 5

Teaching Tip
Sense-Making Encourage students to check their answers by estimating. For instance, in Example 3, $2^4 = 16$ and $2^5 = 32$. So, it is reasonable for $\log_2 25$ to be between 4 and 5.

Recall that the power of a power is found by multiplying exponents. The property for the logarithm of a power is similar.

> ### 🔑 Key Concept Power Property of Logarithms
>
> **Words**   The logarithm of a power is the product of the logarithm and the exponent.
>
> **Symbols**   For any real number p, and positive numbers m and b, where $b \neq 1$, $\log_b m^p = p \log_b m$.
>
> **Example** $\log_2 6^5 = 5 \log_2 6$

A.CED.1

> **Study Tip**
> 🔧 **Tools** You can check this answer by evaluating $2^{4.6438}$ on a calculator. The calculator should give a result of about 25, because $\log_2 25 \approx 4.6438$ means $2^{4.6438} \approx 25$.

Example 3   Power Property of Logarithms

Given $\log_2 5 \approx 2.3219$, approximate the value of $\log_2 25$.

$$\log_2 25 = \log_2 5^2 \qquad \text{Replace 25 with } 5^2.$$
$$= 2 \log_2 5 \qquad \text{Power Property}$$
$$\approx 2(2.3219) \text{ or } 4.6438 \qquad \text{Replace } \log_2 5 \text{ with 2.3219.}$$

▶ **Guided Practice**

  3. Given $\log_3 7 \approx 1.7712$, approximate the value of $\log_3 49$. ≈ 3.5424

2 Solve Logarithmic Equations You can use the properties of logarithms to solve equations involving logarithms.

A.CED.1

Example 4   Solve Equations Using Properties of Logarithms

Solve $\log_6 x + \log_6 (x - 9) = 2$.

$$\log_6 x + \log_6 (x - 9) = 2 \qquad \text{Original equation}$$
$$\log_6 x (x - 9) = 2 \qquad \text{Product Property}$$
$$x(x - 9) = 6^2 \qquad \text{Definition of logarithm}$$
$$x^2 - 9x - 36 = 0 \qquad \text{Subtract 36 from each side.}$$
$$(x - 12)(x + 3) = 0 \qquad \text{Factor.}$$
$$x - 12 = 0 \quad \text{or} \quad x + 3 = 0 \qquad \text{Zero Product Property}$$
$$x = 12 \qquad\qquad x = -3 \qquad \text{Solve each equation.}$$

CHECK $\log_6 x + \log_6 (x - 9) = 2$
$\log_6 12 + \log_6 (12 - 9) \stackrel{?}{=} 2$
$\log_6 12 + \log_6 3 \stackrel{?}{=} 2$
$\log_6 (12 \cdot 3) \stackrel{?}{=} 2$
$\log_6 36 \stackrel{?}{=} 2$
$2 = 2$ ✓

$\log_6 x + \log_6 (x - 9) = 2$
$\log_6 (-3) + \log_6 (-3 - 9) \stackrel{?}{=} 2$
$\log_6 (-3) + \log_6 (-12) \stackrel{?}{=} 2$

Because $\log_6 (-3)$ and $\log_6 (-12)$ are undefined, -3 is an extraneous solution.

The solution is $x = 12$.

▶ **Guided Practice**

  4A. $2 \log_7 x = \log_7 27 + \log_7 3$ 9   **4B.** $\log_6 x + \log_6 (x + 5) = 2$ 4

> **Go Online!**
> You will want to reference the Properties of Logarithms often. Log into your eStudent Edition to bookmark this lesson.

Differentiated Instruction AL OL BL ELL

Interpersonal Learners Immediately after discussing Example 4, have pairs of students rework the Example together without looking at the solution in the text. Have the partners take turns explaining the solution steps to each other. Have them also discuss the reasonableness of their solutions.

Check Your Understanding

○ = Step-by-Step Solutions begin on page R11.

 **Go Online!** for a Self-Check Quiz

Example 1
A.CED.1

Use $\log_4 3 \approx 0.7925$ and $\log_4 5 \approx 1.1610$ to approximate the value of each expression.

1. $\log_4 18$ 2.085

2. $\log_4 15$ 1.9535   (3.5)

3. $\log_4 \frac{5}{3}$ 0.3685

4. $\log_4 \frac{3}{4}$ −0.2075

Example 2
A.CED.1

5. MOUNTAIN CLIMBING As elevation increases, the atmospheric air pressure decreases. The formula for pressure based on elevation is $a = 15{,}500(5 - \log_{10} P)$, where a is the altitude in meters and P is the pressure in pascals (1 psi $\approx$ 6900 pascals). What is the air pressure at the summit in pascals (Pa) for each mountain listed in the table at the right?   See margin.

Mountain	County	Height (m)
Guadalupe Peak	Culberson	2667
Emory Peak	Brewster	2385
Anthony's Nose	El Paso	2111
Panther Peak	Hudspeth	1936
Buck Mountain	Jeff Davis	1789

Example 3
A.CED.1

Given $\log_3 5 \approx 1.465$ and $\log_5 7 \approx 1.2091$, approximate the value of each expression.

6. $\log_3 25$ 2.93

7. $\log_5 49$ 2.4182

Example 4
A.CED.1

Solve each equation. Check your solutions.

8. $\log_4 48 - \log_4 n = \log_4 6$ 8

9. $\log_3 2x + \log_3 7 = \log_3 28$ 2

10. $3 \log_2 x = \log_2 8$ 2

11. $\log_{10} a + \log_{10} (a - 6) = 2$ 13.4403

Practice and Problem Solving

Extra Practice is on page R6.

Example 1
A.CED.1

Use $\log_4 2 = 0.5$, $\log_4 3 \approx 0.7925$, and $\log_4 5 \approx 1.1610$ to approximate the value of each expression.

12. $\log_4 30$ 2.4535

13. $\log_4 20$ 2.1610

14. $\log_4 \frac{2}{3}$ −0.2925

15. $\log_4 \frac{4}{3}$ 0.2075

16. $\log_4 9$ 1.5850

17. $\log_4 8$ 1.5

18a. $10^{2.7}$ or about 500 times more intense

Example 2
A.CED.1

18. SCIENCE The magnitude M of an earthquake is measured on the Richter scale using the formula $M = \log 10x$, where x is the intensity of the seismic wave causing the ground motion. In 2007, an earthquake near San Francisco registered approximately 5.6 on the Richter scale. The famous San Francisco earthquake of 1906 measured 8.3 in magnitude.

  a. How many times more intense, was the 1906 earthquake than the 2007 earthquake?

  b. Richter himself classified the 1906 earthquake as having a magnitude of 8.3. More recent research indicates it was most likely a 7.9. How many times greater in intensity was Richter's measure of the earthquake?
Richter thought the earthquake was $10^{0.4}$ or about $2\frac{1}{2}$ times greater than it actually was.

Example 3
A.CED.1

Given $\log_6 8 \approx 1.1606$ and $\log_7 9 \approx 1.1292$, approximate the value of each expression.

19. $\log_6 48$ 2.1606

20. $\log_7 81$ 2.2584

21. $\log_6 512$ 3.4818

22. $\log_7 729$ 3.3876

Example 4
A.CED.1

MP PERSEVERANCE Solve each equation. Check your solutions.

23. $\log_3 56 - \log_3 n = \log_3 7$ 8

24. $\log_2 (4x) + \log_2 5 = \log_2 40$ 2

25. $5 \log_2 x = \log_2 32$ 2

26. $\log_{10} a + \log_{10} (a + 21) = 2$ 4

Differentiated Homework Options

Levels	**AL** Basic	**OL** Core	**BL** Advanced
Exercises	12–26, 61–76	13–49 odd, 50, 51–57 odd, 59–76	27–76
2-Day Option	13–25 odd, 69–76	13–25, 69–76	
	12–26 even, 61–68	27–68	

 You can use ALEKS to provide additional remediation support with personalized instruction and practice.

Practice

Formative Assessment Use Exercises 1–11 to assess students' understanding of the concepts in this lesson.

The Practice and Problem Solving exercises assess the content taught in the lesson. The Preparing for Assessment page is meant to be used as preparation for end-of-course assessments.

Extra Practice

See page R6 for extra exercises for students who are approaching level or for on-level students who need additional reinforcement.

Levels of Complexity Chart

The levels of the exercises progress from 1 to 3, with Level 1 indicating the lowest level of complexity.

Exercises	12–26	27–59, 69–76	60–68
C Level 3			●
B Level 2		●	
Level 1	●		

Perseverance Mathematically proficient students start by explaining to themselves the meaning of a problem and looking for entry points to its solution. They analyze givens, constraints, relationships, and goals and make conjectures about the form and meaning of the solution and plan a solution pathway rather than simply jumping into a solution attempt.

Additional Answer

5. Guadalupe Peak: 67,287.7 Pa; Emory Peak: 70,166.37 Pa; Anthony's Nose: 73,081.33 Pa; Panther Peak; 75,006.13 Pa; Buck Mountain: 76,662.09 Pa

Go Online!   eBook

Interactive Student Guide

Use the *Interactive Student Guide* to deepen conceptual understanding.
· Properties of Logarithms

 ALGEBRA 2 INTERACTIVE STUDENT GUIDE

Teaching the Mathematical Practices

Reasoning Mathematically proficient students make sense of quantities and their relationships in problem situations. Quantitative reasoning entails habits of creating a coherent representation of the problem at hand; considering the units involved; attending to the meaning of quantities, not just how to compute them; and knowing and flexibly using different properties of operations and objects.

Additional Answer

50b. 40 yr; There will likely be environmental or political factors that affect the rate of change in the whale population but without those this answer is reasonable.

60c.

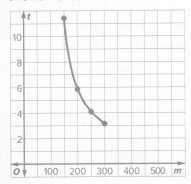

B **27. PROBABILITY** In the 1930s, Dr. Frank Benford demonstrated a way to determine whether a set of numbers has been randomly chosen or manually chosen. If the sets of numbers were not randomly chosen, then the Benford formula, $P = \log_{10}\left(1 + \frac{1}{d}\right)$, predicts the probability of a digit d being the first digit of the set. For example, there is a 4.6% probability that the first digit is 9.

a. Rewrite the formula to solve for the digit if given the probability. $d = \frac{1}{10^P - 1}$

b. Find the digit that has a 9.7% probability of being selected. 4

c. Find the probability that the first digit is 1 ($\log_{10} 2 \approx 0.30103$). 30.1%

Use $\log_5 3 \approx 0.6826$ and $\log_5 4 \approx 0.8614$ to approximate the value of each expression.

28. $\log_5 40$ 2.2921

29. $\log_5 30$ 2.1133

30. $\log_5 \frac{3}{4}$ −0.1788

31. $\log_5 \frac{4}{3}$ 0.1788

32. $\log_5 9$ 1.3652

33. $\log_5 16$ 1.7228

34. $\log_5 12$ 1.5440

35. $\log_5 27$ 2.0478

Solve each equation. Check your solutions.

36. $\log_3 6 + \log_3 x = \log_3 12$ 2

37. $\log_4 a + \log_4 8 = \log_4 24$ 3

38. $\log_{10} 18 - \log_{10} 3x = \log_{10} 2$ 3

39. $\log_7 100 - \log_7 (y + 5) = \log_7 10$ 5

40. $\log_2 n = \frac{1}{3} \log_2 27 + \log_2 36$ 108

41. $3 \log_{10} 8 - \frac{1}{2} \log_{10} 36 = \log_{10} x$ $85\frac{1}{3}$

Solve for n.

42. $\log_a 6n - 3 \log_a x = \log_a x$ $\frac{x^4}{6}$

43. $2 \log_b 16 + 6 \log_b n = \log_b (x - 2)$ $\left(\frac{x-2}{256}\right)^{\frac{1}{6}}$

Solve each equation. Check your solutions. **46.** no solution

44. $\log_{10} z + \log_{10} (z + 9) = 1$ 1

45. $\log_3 (a^2 + 3) + \log_3 3 = 3$ $\sqrt{6}, -\sqrt{6}$

46. $\log_2 (15b - 15) - \log_2 (-b^2 + 1) = 1$

47. $\log_4 (2y + 2) - \log_4 (y - 2) = 1$ 5

48. $\log_6 0.1 + 2 \log_6 x = \log_6 2 + \log_6 5$ 10

49. $\log_7 64 - \log_7 \frac{8}{3} + \log_7 2 = \log_7 4p$ 12

50. REASONING Suppose there are 5000 humpback whales in existence today, and the population decreases at a rate of 4% per year.

a. Write a logarithmic function for the time in years based upon population. $t = \log_{0.96}\left(\frac{P}{5000}\right)$

b. After how many years will the population drop below 1000? Is this reasonable? See margin.

State whether each identity is *true* or *false*.

51. $\log_8 (x - 3) = \log_8 x - \log_8 3$ false

52. $\log_5 22x = \log_5 22 + \log_5 x$ true

53. $\log_{10} 19k = 19 \log_{10} k$ false

54. $\log_2 y^5 = 5 \log_2 y$ true

55. $\log_7 \frac{x}{3} = \log_7 x - \log_7 3$ true

56. $\log_4 (z + 2) = \log_4 z + \log_4 2$ false

57. $\log_8 p^4 = (\log_8 p)^4$ false

58. $\log_9 \frac{x^2 y^3}{z^4} = 2 \log_9 x + 3 \log_9 y - 4 \log_9 z$ true

59. MULTI-STEP Teresa's retirement account balance is currently $320,000 and it is increasing about 15% per year from deposits and interest. When she retires, she plans to invest this money in a CD with 5% annual interest. Each year, she will only spend the interest and reinvest the principal in a new CD. She also expects to earn $1050 per month from Social Security when she retires. Her goal is to have enough saved in order to earn $50,000 per year before taxes.

a. In how many years will she be able to retire? Sample answer: about 6 yr

b. Describe and evaluate your solution process. **b, c.** See Ch. 6 Answer Appendix.

c. What assumptions did you make?

60. FINANCIAL LITERACY The average American carries a credit card debt of approximately $8600 with an annual percentage rate (APR) of 18.3%. The formula $m = \dfrac{b\left(\frac{r}{n}\right)}{1 - \left(1 + \frac{r}{n}\right)^{-nt}}$

can be used to compute the monthly payment m that is necessary to pay off a credit card balance b in a given number of years t, where r is the annual percentage rate and n is the number of payments per year.

a. What monthly payment should be made in order to pay off the debt in exactly three years? What is the total amount paid? **$312.21; $11,239.56**

b. The equation $t = \dfrac{\log\left(1 - \frac{br}{mn}\right)}{-n\log\left(1 + \frac{r}{n}\right)}$ can be used to calculate the number of years necessary for a given payment schedule. Copy and complete the table. **See margin.**

Payment (m)	Years (t)
$50	non-real
$100	non-real
$150	11.42
$200	5.87
$250	4.09
$300	3.16

c. Graph the information in the table from part **b.**

d. If you could only afford to pay $100 a month, will you be able to pay off the debt? If so, how long will it take? If not, why not? **No; the monthly interest is $131.15, so the payments do not even cover the interest.**

e. What is the minimum monthly payment that will work toward paying off the debt? **$131.16**

A.CED.1

H.O.T. Problems Use Higher-Order Thinking Skills

61. OPEN-ENDED Write a logarithmic expression for each condition. Then write the expanded expression.

a. a product and a quotient **Sample answer:** $\log_b \dfrac{xz}{5} = \log_b x + \log_b z - \log_b 5$

b. a product and a power **Sample answer:** $\log_b m^4 p^6 = 4\log_b m + 6\log_b p$

c. a product, a quotient, and a power

Sample answer: $\log_b \dfrac{j^8 k}{h^5} = 8\log_b j + \log_b k - 5\log_b h$

62. CONSTRUCT ARGUMENTS Use the properties of exponents to prove the Power Property of Logarithms. **See margin.**

63. WRITING IN MATH Explain why the following are true. **63b.** $\log_b b = 1$, because $b^1 = b$.

a. $\log_b 1 = 0$ **b.** $\log_b b = 1$ **c.** $\log_b b^x = x$
$\log_b 1 = 0$, because $b^0 = 1$. $\log_b b^x = x$, because $b^x = b^x$.

64. CHALLENGE Simplify $\log_{\sqrt{a}}\left(a^2\right)$ to find an exact numerical value. **See margin.**

65. WHICH ONE DOESN'T BELONG? Find the expression that does not belong. Explain.

$\log_b 24 = \log_b 2 + \log_b 12$	$\log_b 24 = \log_b 20 + \log_b 4$
$\log_b 24 = \log_b 8 + \log_b 3$	$\log_b 24 = \log_b 4 + \log_b 6$

$\log_b 24 \neq \log_b 20 + \log_b 4$; all other choices are equal to $\log_b 24$.

66. REASONING Use the properties of logarithms to prove that $\log_a \dfrac{1}{x} = -\log_a x$. **See Ch. 6 Answer Appendix.**

67. CHALLENGE Simplify $x^{3\log_x 2 - \log_x 5}$ to find an exact numerical value. **See Ch. 6 Answer Appendix.**

68. See Ch. 6 Answer Appendix.

68. WRITING IN MATH Explain how the properties of exponents and logarithms are related. Include examples like the one shown at the beginning of the lesson illustrating the Product Property, but with the Quotient Property and Power Property of Logarithms.

Standards for Mathematical Practice

Emphasis On	Exercises
1 Make sense of problems and persevere in solving them.	5, 18, 27, 50, 59, 60, 75, 76
4 Model with mathematics.	8–11, 23–26, 36–49, 51–58, 61–72, 74, 75
8 Look for and express regularity in repeated reasoning.	1–4, 6–7, 12–17, 19–22, 28–35, 73

Assess

Crystal Ball Tell students that in the next lesson they will learn to solve equations using common (base 10) logarithms. Ask them to write how they think what they learned today will connect with the next lesson they will study.

Additional Answers

62. For $b > 0$, $m > 0$, $m \neq 1$, and any real number p, $m = b^{\log_b m}$ and $m^p = b^{\log_b (m^p)}$.

Use the identity $m^p = m^p$ and substitute for m on the right and m^p on the left.

$b^{\log_b (m^p)} = \left(b^{\log_b m}\right)^p$

$b^{\log_b (m^p)} = \left(b^{\log_b m}\right)^p$

$\log_b (m^p) = (\log_b m) \cdot p$

$\log_b (m^p) = p\log_b m$

64. $\log_{\sqrt{a}}\left(a^2\right) = x$
$\left(\sqrt{a}\right)^x = a^2$
$\left(a^{\frac{1}{2}}\right)^x = a^2$
$a^{\frac{x}{2}} = a^2$
$\dfrac{x}{2} = 2$
$x = 4$

Go Online!

eSolutions Manual

Create worksheets, answer keys, and solutions handouts for your assignments.

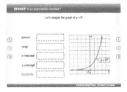

Preparing for Assessment

Exercises 69–76 require students to use the skills they will need on standardized assessments. Each exercise is dual-coded with content.

Dual Coding		
Items	Content Standards	Mathematical Practices
69	A.CED.1	8
70	A.CED.1	8
71	A.CED.1	8
72	A.CED.1	8
73	A.CED.1	1
74	A.CED.1	8
75	A.CED.1	8
76	A.CED.1	8

Diagnose Student Errors

69.

A	Did not eliminate extraneous solution
B	Used the extraneous solution
C	CORRECT
D	Rewrote the equation incorrectly

70.

A	Rewrote the equation as $(3x + 2) - 2 = 7^4$
B	Rewrote the equation as $(3x + 2) - 2^4 = 7^4$
C	Rewrote the equation as $(3x + 2) - 2 = 4^7$
D	CORRECT

71.

A	Used division instead of subtraction for the logarithm of a quotient
B	Used addition instead of multiplication for the logarithm of a power and division instead of subtraction for the logarithm of a quotient
C	CORRECT
D	Used addition instead of multiplication for the logarithm of a power

Preparing for Assessment

69. What is the solution set in $\log_4 x + \log_4 (x - 12) = 3$? 8 A.CED.1 **C**

- ○ **A** $\{-4, 16\}$
- ○ **B** $\{-4\}$
- ○ **C** $\{16\}$
- ○ **D** ∅

70. What is the value of x in $\log_4 (3x + 2) - 2 = 7$? 8 A.CED.1 **D**

- ○ **A** $800\frac{1}{3}$
- ○ **B** 805
- ○ **C** $5461\frac{1}{3}$
- ○ **D** $87,380\frac{2}{3}$

71. Let $\log_q m = p$ and $\log_q n = r$. What is the value of $\log_q \frac{m^3}{n}$? 8 A.CED.1 **C**

- ○ **A** $\frac{3p}{r}$
- ○ **B** $\frac{3p}{r}$
- ○ **C** $3p - r$
- ○ **D** $3 + p - r$

72. What is the solution set of $\log_3 (9x^2) + 1 = \log_3 (4)$? 8 A.CED.1 **A**

- ○ **A** $\left\{ -\frac{2\sqrt{3}}{9}, \frac{2\sqrt{3}}{9} \right\}$
- ○ **B** $\left\{ -\frac{1}{3}, \frac{1}{3} \right\}$
- ○ **C** $\left\{ \frac{2\sqrt{3}}{9} \right\}$
- ○ **D** ∅

73. If $\log_2 y = 0.451$, what is the value of $\log_2 16y^2$? 1 A.CED.1 **4.902**

74. What is the extraneous solution to $\log_2 x + \log_2 (x - 2) = 3$? 8 A.CED.1 **B**

- ○ **A** $\{4, -2\}$
- ○ **B** $\{-2\}$
- ○ **C** $\{4\}$
- ○ **D** ∅

75. MULTI-STEP Charles Richter defined the magnitude of an earthquake to be the equation

$$M = \log_{10} \left(\frac{I}{S} \right)$$

where I is the intensity of the earthquake (measured by the amplitude in centimeters) and S is the intensity of a standard earthquake (which is a value of 10^{-4} cm). 8 A.CED.1

- **a.** What is the replacement set for I?
  all real numbers greater than or equal to 0
- **b.** What are the possible values of M?
  all real numbers greater than or equal to 0
- **c.** If one earthquake is 20,000 times as intense as a standard earthquake, then the value of $\frac{I}{S} = 20,000$. What is the magnitude of the earthquake? **4.3**
- **d.** If one earthquake is 200,000 times as intense as a standard earthquake, then what is its magnitude? **5.3**
- **e.** If one earthquake is 2,000,000 times as intense as a standard earthquake, then what is its magnitude? **6.3**
- **f.** Explain the pattern from the responses in **c**, **d**, and **e**. Sample answer: The magnitude of each earthquake increases by 1 unit when the intensity increases by a factor of 10.

76. The pH of an aqueous solution is given by the formula, pH $= -\log [H^+]$, where H^+ is the hydrogen ion concentration. 8 A.CED.1

- **a.** What is the pH of an aqueous solution in which $H^+ = 2.7 \times 10^{-3}$ M?

  2.57

- **b.** If the pH of an aqueous solution is 6.52, what is the hydrogen ion concentration?

  3.02×10^{-7} M

Differentiated Instruction BL

Extension Show students the following:

Ask students to predict $\log_{10} 30{,}000$. 4.4771

$\log_{10} 3 \approx 0.4771$
$\log_{10} 30 \approx 1.4771$
$\log_{10} 300 \approx 2.4771$
$\log_{10} 3000 \approx 3.4771$

Have students use properties of logarithms to explain this pattern. Sample explanation: 3, 30, 300, and 3000 can be written as $3 \times 10^0, 3 \times 10^1, 3 \times 10^2$, and 3×10^3 respectively. Then the base 10 logarithms of each can be rewritten as a sum of two logarithms. For example, $\log_{10} 3000$ can be written as $\log_{10} (3 \cdot 10^3)$. Then it follows that $\log_{10} (3 \cdot 10^3) = \log_{10} 3 + \log_{10} 10^3 = \log_{10} (3) + 3 = 3.4771$.

72.

A	CORRECT
B	Rewrote the equation as $9x^2 + 3 = 4$
C	Believed the negative solution was an extraneous solution
D	Rewrote the equation as $9x^2 + 3 = 4$ and dismissed the solutions as extraneous

Common Logarithms

Track Your Progress

Objectives

1 Solve exponential equations and inequalities using common logarithms.

2 Evaluate logarithmic expressions using the Change of Base Formula.

Mathematical Background

Base 10 logarithms are called common logarithms. When the base of a logarithm is not shown, the base is assumed to be 10. When you have a logarithmic expression of any base, evaluate it using the Change of Base Formula to translate the expression into one that involves common logarithms.

THEN	NOW	NEXT
F.IF.4 For a function that models a relationship between two quantities, interpret key features of graphs and tables in terms of the quantities, and sketch graphs showing key features given a verbal description of the relationship.	**A.CED.1** Create equations and inequalities in one variable and use them to solve problems.	**F.LE.4** For exponential models, express as a logarithm the solution to $ab^{ct} = d$ where a, c, and d are numbers and the base b is 2, 10, or e; evaluate the logarithm using technology.

Go Online! All of these resources and more are available at connectED.mcgraw-hill.com

Chapter Project allows students to create and customize a project as a nontraditional method of assessment.

Personal Tutors (for every example) let students hear real teachers solve problems. Students can pause and repeat as many times as necessary.

Use **Self-Check Quiz** to assess students' understanding of the concepts in this lesson.

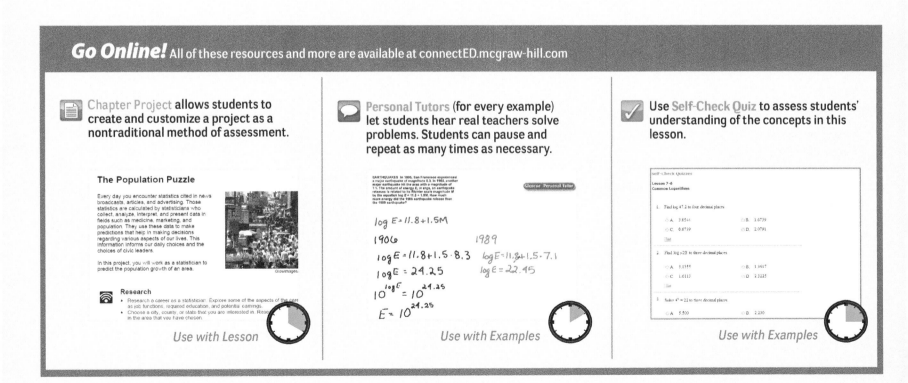

Use with Lesson   *Use with Examples*   *Use with Examples*

OER Using Open Educational Resources

Tutorials Have students watch the common logarithms video on virtualnerd.com in order to prepare for the lesson or to reinforce key concepts. Students can also watch background tutorials to refresh their knowledge of prerequisite skills. *Use as flipped learning or review*

Differentiate Your Resources

Extra Practice Additional practice or homework; Skills Practice is best for approaching-level students and Practice is best for on-level and beyond-level students

Skills Practice

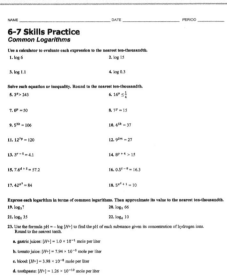

Practice

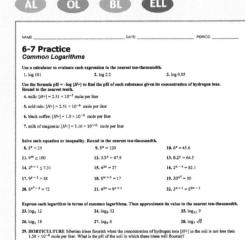

Word Problem Practice

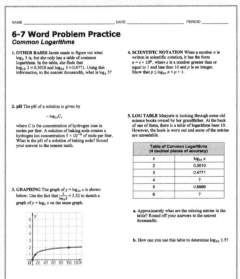

Intervention Reteaching and vocabulary activities that can be used with struggling or absent students and as ELL support

Extension Activities that can be used to extend lesson concepts

Study Guide and Intervention

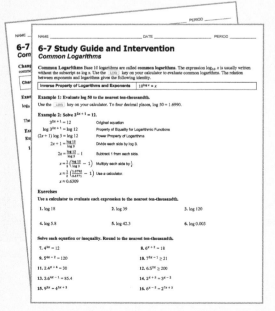

Study Notebook

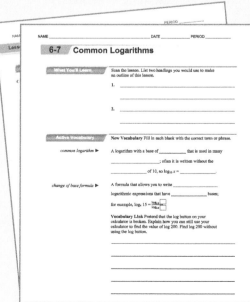

Enrichment

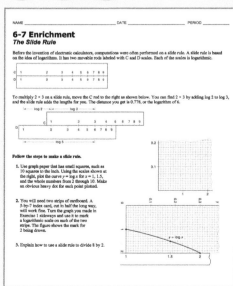

LESSON 7

Common Logarithms

::Then	::Now	::Why?

::Then

● You simplified expressions and solved equations using properties of logarithms.

::Now

1. Solve exponential equations and inequalities using common logarithms.

2. Evaluate logarithmic expressions using the Change of Base Formula.

::Why?

● Seismologists use the Richter scale to measure the strength or magnitude of earthquakes. The magnitude of an earthquake is determined using the logarithm of the amplitude of waves recorded by seismographs.

The logarithmic scale used by the Richter scale is based on the powers of 10. For example, a magnitude 6.4 earthquake can be represented by $6.4 = \log_{10} x$.

Richter Number	Intensity
1	10^1 micro
2	10^2 minor
3	10^3 minor
4	10^4 light
5	10^5 moderate
6	10^6 strong
7	10^7 major
8	10^8 great

 New Vocabulary

common logarithm
Change of Base Formula

 Mathematical Practices

4 Model with mathematics.

Content Standards
A.CED.1 Create equations and inequalities in one variable and use them to solve problems.

1 Common Logarithms You have seen that the base 10 logarithm function, $y = \log_{10} x$, is used in many applications. Base 10 logarithms are called **common logarithms**. Common logarithms are usually written without the subscript 10.

$$\log_{10} x = \log x, x > 0$$

Most scientific calculators have a LOG key for evaluating common logarithms. The graph of the common log is shown.

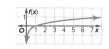

A.CED.1

Example 1 Find Common Logarithms

Use a calculator to evaluate each expression to the nearest ten-thousandth.

a. log 5
KEYSTROKES: LOG 5 ENTER
.6989700043 $\log 5 \approx 0.6990$

b. log 0.3
KEYSTROKES: LOG 0.3 ENTER
−.5228787453 $\log 0.3 \approx -0.5229$

▶ **Guided Practice**

1A. log 7 0.8451

1B. log 0.5 −0.3010

The common logarithms of numbers that differ by integral powers of ten are closely related. Remember that a logarithm is an exponent. For example, in the equation $y = \log x$, y is the power to which 10 is raised to obtain the value of x.

$\log x = y$	$\rightarrow$ means $\rightarrow$	$10^y = x$
$\log 1 = 0$	because	$10^0 = 1$
$\log 10 = 1$	because	$10^1 = 10$
$\log 10^m = m$	because	$10^m = 10^m$

(MP) Teaching the Mathematical Practices

Attend to precision Help students communicate precisely to others. For example, ask:

● What is the base of the log when it is not written in an expression? 10

● What is the change of base formula? $\log_b a = \dfrac{\log b}{\log a}$

● How can a logarithmic equation be rewritten? Sample answer: The equation can be rewritten in exponential form.

Launch

Have students read the Why? section of the lesson. Ask:

● How does the Richter number of a "Great" earthquake compare with the number for a "Light" earthquake? It is twice as large.

● How does the intensity of a "Great" earthquake compare to the intensity of a "Light" earthquake? It is 10,000 times as large.

● Where would an earthquake with an intensity half as large as 10^8 appear in the table? between "Major" and "Great"

Teach

Ask the scaffolded questions for each example to build conceptual understanding for students at all levels.

1 Common Logarithms

Example 1 Find Common Logarithms

AL Write log 5 with a subscript base. $\log_{10} 5$

OL Why is it useful to make $\log_{10} x = \log x$? $\log_{10} x$ is used often in applications, so writing it as log x allows us to save time writing the subscript.

BL What equations could you use your calculator to find the intersection point of that would also evaluate log 5? $y = 10^x$ and $y = 5$.

Need Another Example?

Use a calculator to evaluate each expression to the nearest ten-thousandth.

a. log 6 about 0.7782

b. log 0.35 about −0.4559

Go Online!

Interactive Whiteboard

Use the *eLesson* or *Lesson Presentation* to present this lesson.

Example 2 Solve Logarithmic Equations

AL What is the domain of the original function?
$I > 0$

OL Use the Power Property of Logarithms to rewrite $10 \log \frac{I}{m}$. $\log\left(\frac{I}{m}\right)^{10}$

BL How many more times intense is 70 decibels than 66.6 decibels? 2.19

Need Another Example?

Jet Engines Refer to Example 2. The sound of a jet engine can reach a loudness of 125 decibels. How many times the minimum intensity of audible sound is this? $10^{12.5}$, or about 3×10^{12}, which is 3 trillion

Example 3 Solve Exponential Equations Using Logarithms

AL How do you put $\frac{\log 19}{\log 4}$ in your calculator?
$\log (19) \div \log 4$.

OL Does this seem like a reasonable answer? Why? Yes, because $4^2 = 16$, so we know the answer should be just over 2.

BL Rewrite $4^x = 19$ in logarithmic form, then use the two different ways that we can express x to write a statement of equality. $\log_4 19 = x$; $\log_4 19 = \frac{\log 19}{\log 4}$

Need Another Example?

Solve $5^x = 62$. about 2.5643

Common logarithms are used in the measurement of sound. Soft recorded music is about 36 decibels (dB).

A.CED.1

Real-World Example 2 Solve Logarithmic Equations

ROCK CONCERT The loudness L, in decibels, of a sound is $L = 10 \log \frac{I}{m}$, where I is the intensity of the sound and m is the minimum intensity of sound detectable by the human ear. Residents living several miles from a concert venue can hear the music at an intensity of 66.6 decibels. How many times the minimum intensity of sound detectable by the human ear was this sound, if m is defined to be 1?

$L = 10 \log \frac{I}{m}$	Original equation
$66.6 = 10 \log \frac{I}{1}$	Replace L with 66.6 and m with 1.
$6.66 = \log I$	Divide each side by 10 and simplify.
$I = 10^{6.66}$	Exponential form
$I \approx 4{,}570{,}882$	Use a calculator.

The sound heard by the residents was approximately 4,570,000 times the minimum intensity of sound detectable by the human ear.

▶ **Guided Practice** about 2×10^{25} ergs

2. EARTHQUAKES The amount of energy E in ergs that an earthquake releases is related to its Richter scale magnitude M by the equation $\log E = 11.8 + 1.5M$. Use the equation to find the amount of energy released by the 2004 Sumatran earthquake, which measured 9.0 on the Richter scale and led to a tsunami.

If both sides of an exponential equation cannot easily be written as powers of the same base, you can solve by taking the logarithm of each side.

A.CED.1

Example 3 Solve Exponential Equations Using Logarithms

Solve $4^x = 19$. Round to the nearest ten-thousandth.

$4^x = 19$	Original equation
$\log 4^x = \log 19$	Property of Equality for Logarithmic Functions
$x \log 4 = \log 19$	Power Property of Logarithms
$x = \frac{\log 19}{\log 4}$	Divide each side by log 4
$x \approx 2.1240$	Use a calculator.

The solution is approximately 2.1240.

CHECK You can check this answer graphically by using a graphing calculator. Graph the line $y = 4^x$ and the line $y = 19$. Then use the **CALC** menu to find the intersection of the two graphs. The intersection is very close to the answer that was obtained algebraically. ✓

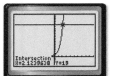

[−10, 10] scl: 1 by [−5, 25] scl: 1

▶ **Guided Practice**

3A. $3^x = 15$ ≈2.4650

3B. $6^x = 42$ ≈2.0860

The same strategies that are used to solve exponential equations can be used to solve exponential inequalities.

A.CED.1

Example 4   Solve Exponential Inequalities Using Logarithms

Solve $3^{5y} < 7^{y-2}$. Round to the nearest ten-thousandth.

$$3^{5y} < 7^{y-2}$$ Original inequality

$$\log 3^{5y} < \log 7^{y-2}$$ Property of Inequality for Logarithmic Functions

$$5y \log 3 < (y-2) \log 7$$ Power Property of Logarithms

$$5y \log 3 < y \log 7 - 2 \log 7$$ Distributive Property

$$5y \log 3 - y \log 7 < -2 \log 7$$ Subtract $y \log 7$ from each side.

$$y(5 \log 3 - \log 7) < -2 \log 7$$ Distributive Property

$$y < \frac{-2 \log 7}{5 \log 3 - \log 7}$$ Divide each side by $5 \log 3 - \log 7$.

$$\{y \mid y < -1.0972\}$$ Use a calculator.

CHECK Test $y = -2$.

$$3^{5y} < 7^{y-2}$$ Original inequality

$$3^{5(-2)} \stackrel{?}{<} 7^{(-2)-2}$$ Replace y with -2.

$$3^{-10} \stackrel{?}{<} 7^{-4}$$ Simplify.

$$\frac{1}{59,049} < \frac{1}{2401} \checkmark$$ Negative Exponent Property

▸ **Guided Practice**

Solve each inequality. Round to the nearest ten-thousandth.

4A. $3^{2x} \geq 6^{x+1}$ $\{x \mid x \geq 4.4190\}$ **4B.** $4^y < 5^{2y+1}$ $\{y \mid y > -0.8782\}$

2 Change of Base Formula The **Change of Base Formula** allows you to write equivalent logarithmic expressions that have different bases.

⬛ **Key Concept**   Change of Base Formula

Symbols   For all positive numbers a, b, and n, where $a \neq 1$ and $b \neq 1$,

$$\log_a n = \frac{\log_b n}{\log_b a}. \quad \begin{array}{l} \leftarrow \text{log base } b \text{ of original number} \\ \leftarrow \text{log base } b \text{ of old base} \end{array}$$

Example $\log_3 11 = \dfrac{\log_{10} 11}{\log_{10} 3}$

To prove this formula, let $\log_a n = x$.

$$a^x = n$$ Definition of logarithm

$$\log_b a^x = \log_b n$$ Property of Equality for Logarithmic Functions

$$x \log_b a = \log_b n$$ Power Property of Logarithms

$$x = \frac{\log_b n}{\log_b a}$$ Divide each side by $\log_b a$

$$\log_a n = \frac{\log_b n}{\log_b a}$$ Replace x with $\log_a n$

Differentiated Instruction **OL** **BL**

Logical Learners Ask students to recall that an equation like $4^x = 19$ from Example 3 could be written in logarithmic form as $\log_4 19 = x$. Although this logarithm cannot be directly evaluated, the Change of Base Formula can be used to give the correct result of $x \approx 2.1234$.

Example 5 Change of Base Formula

AL Write $\dfrac{\log_{10} 20}{\log_{10} 3}$ without any subscripts. $\dfrac{\log 20}{\log 3}$

OL Explain how we can use the change of base formula to use our calculator to find any logarithm of any base. We can turn $\log_b x$ into $\dfrac{\log x}{\log b}$.

BL Explain how you can use the change of base formula to graph $y = \log_5 x$. You can put $\dfrac{\log x}{\log 5}$ into Y1.

Need Another Example?

Express $\log_5 140$ in terms of common logarithms. Then round to the nearest ten-thousandth.

$\log_5 140 = \dfrac{\log_{10} 140}{\log_{10} 5}$; $\log_5 140 \approx 3.0704$

Practice

Formative Assessment Use Exercises 1–15 to assess students' understanding of the concepts in this lesson.

The Practice and Problem Solving exercises assess the content taught in the lesson. The Preparing for Assessment page is meant to be used as preparation for end-of-course assessments.

Teaching the Mathematical Practices

Sense-Making Mathematically proficient students start by explaining to themselves the meaning of a problem and looking for entry points to its solution. They analyze givens, constraints, relationships, and goals, and make conjectures about the form and meaning of the solution and plan a solution pathway rather than simply jumping into a solution attempt.

Extra Practice

See page R6 for extra exercises for students who are approaching level or for on-level students who need additional reinforcement.

Go Online!   eBook

Interactive Student Guide

Use the *Interactive Student Guide* to deepen conceptual understanding.
· Common Logarithms

The Change of Base Formula makes it possible to evaluate a logarithmic expression of any base by translating the expression into one that involves common logarithms.

A.CED.1

Example 5 Change of Base Formula

Express $\log_3 20$ in terms of common logarithms. Then round to the nearest ten-thousandth.

$\log_3 20 = \dfrac{\log_{10} 20}{\log_{10} 3}$  Change of Base Formula

≈ 2.7268 Use a calculator.

▶ **Guided Practice** $\dfrac{\log_{10} 8}{\log_{10} 6} \approx 1.1606$

5. Express $\log_6 8$ in terms of common logarithms. Then round to the nearest ten-thousandth.

> **Go Online!** for a Self-Check Quiz

Check Your Understanding   ○ = Step-by-Step Solutions begin on page R11.

Example 1
A.CED.1
Use a calculator to evaluate each expression to the nearest ten-thousandth.
1. log 5 0.6990   **2.** log 21 1.3222   **3.** log 0.4 −0.3979   **4.** log 0.7 −0.1549

Example 2
6-6
5. SCIENCE The amount of energy E in ergs that an earthquake releases is related to its Richter scale magnitude M by the equation $\log E = 11.8 + 1.5M$. Use the equation to find the amount of energy released by the 1960 Chilean earthquake, which measured 8.5 on the Richter scale. 3.55×10^{24} ergs

Example 3
6-6
Solve each equation. Round to the nearest ten-thousandth.
6. $6^x = 40$ 2.0588   **7.** $2.1^{a+2} = 8.25$ 0.8442   **8.** $7^{x^2} = 20.42$ ±1.2451   **9** $11^{b-3} = 5^b$ 9.1237

Example 4
A.CED.1
Solve each inequality. Round to the nearest ten-thousandth.
10. $5^{4n} > 33$ $\{n \mid n > 0.5431\}$ **11.** $6^{p-1} \le 4^p$ $\{p \mid p \le 4.4190\}$

Example 5
A.CED.1
Express each logarithm in terms of common logarithms. Then approximate its value to the nearest ten-thousandth. 12–15. See margin.
12. $\log_3 7$ **13.** $\log_4 23$ **14.** $\log_9 13$ **15.** $\log_2 5$

Practice and Problem Solving   Extra Practice is on page R6.

Example 1
A.CED.1
Use a calculator to evaluate each expression to the nearest ten-thousandth.
16. log 3 0.4771   **17.** log 11 1.0414   **18.** log 3.2 0.5051
19. log 8.2 0.9138   **20.** log 0.9 −0.0458   **21.** log 0.04 −1.3979

Example 2
A.CED.1
22. **SENSE-MAKING** Loretta had a new muffler installed on her car. The noise level of the engine dropped from 85 decibels to 73 decibels.
a. How many times the minimum intensity of sound detectable by the human ear was the car with the old muffler, if m is defined to be 1? about 316,227,766 times
b. How many times the minimum intensity of sound detectable by the human ear is the car with the new muffler? Find the percent of decrease of the intensity of the sound with the new muffler. about 19,952,623 times; about 93.7%

Differentiated Homework Options

Levels	AL Basic	OL Core	BL Advanced
Exercises	16–38, 72–84	17–39 odd, 40, 41–69 odd, 70–84	39–84
2-Day Option	17–37 odd, 77–84	16–38, 77–84	
	16–38 even, 72–76	39–76	

You can use ALEKS to provide additional remediation support with personalized instruction and practice.

Additional Answers

12. $\dfrac{\log 7}{\log 3} \approx 1.7712$

13. $\dfrac{\log 23}{\log 4} \approx 2.2618$

14. $\dfrac{\log 13}{\log 9} \approx 1.1674$

15. $\dfrac{\log 5}{\log 2} \approx 2.3219$

Example 3
A.CED.1

Solve each equation. Round to the nearest ten-thousandth.

23. $8^x = 40$ 1.7740

24. $5^x = 55$ 2.4899

25. $2.9^{a-4} = 8.1$ 5.9647

26. $9^{b-1} = 7^b$ 8.7429

27. $13^{x^2} = 33.3$ ± 1.1691

28. $15^{x^2} = 110$ ± 1.3175

Example 4
A.CED.1

Solve each inequality. Round to the nearest ten-thousandth.

29. $6^{3n} > 36$ $\{n \mid n > 0.6667\}$

30. $2^{4x} \le 20$ $\{x \mid x \le 1.0805\}$

31. $3^{y-1} \le 4^y$ $\{y \mid y \ge -3.8188\}$

32. $5^{p-2} \ge 2^p$ $\{p \mid p \ge 3.5129\}$

Example 5
A.CED.1

Express each logarithm in terms of common logarithms. Then approximate its value to the nearest ten-thousandth.

33. $\log_7 18$ $\dfrac{\log 18}{\log 7} \approx 1.4854$

34. $\log_5 31$ $\dfrac{\log 31}{\log 5} \approx 2.1337$

35. $\log_2 16$ $\dfrac{\log 16}{\log 2} = 4$

36. $\log_4 9$ $\dfrac{\log 9}{\log 4} \approx 1.5850$

37. $\log_3 11$ $\dfrac{\log 11}{\log 3} \approx 2.1827$

38. $\log_6 33$ $\dfrac{\log 33}{\log 6} \approx 1.9514$

 39 PETS The number n of pet owners in thousands after t years can be modeled by $n = 35[\log_4 (t + 2)]$. Let $t = 0$ represent 2000. Use the Change of Base Formula to answer the following questions.

   a. How many pet owners were there in 2010?   62,737 owners

   b. In what year are there 80,000 pet owners? Does your answer seem reasonable?   See margin.

40. (MP) PRECISION Five years ago the grizzly bear population in a certain national park was 325. Today it is 450. Studies show that the park can support a population of 750.

   a. What is the average annual rate of growth in the population if the grizzly bears reproduce once a year?   0.067 or 6.7%

   b. How many more years will it take to reach the maximum population if the population growth continues at the same average rate? Does your answer seem reasonable?   See margin.

Solve each equation or inequality. Round to the nearest ten-thousandth.

41. $3^x = 40$ 3.3578

42. $5^{3p} = 15$ 0.5609

43. $4^{n+2} = 14.5$ -0.0710

44. $8^{z-4} = 6.3$ 4.8851

45. $7.4^{n-3} = 32.5$ 4.7393

46. $3.1^{y-5} = 9.2$ 6.9615

47. $5^x \ge 42$ $\{x \mid x \ge 2.3223\}$

48. $9^{2a} < 120$ $\{a \mid a < 1.0894\}$

49. $3^{4x} \le 72$ $\{x \mid x \le 0.9732\}$

50. $7^{2n} > 52^{4n+3}$ $\{n \mid n < -0.9950\}$

51. $6^p \le 13^{5-p}$ $\{p \mid p \le 2.9437\}$

52. $2^{y+3} \ge 8^{3y}$ $\{y \mid y \le 0.3750\}$

Express each logarithm in terms of common logarithms. Then approximate its value to the nearest ten-thousandth.

53. $\log_4 12$ $\dfrac{\log 12}{\log 4} \approx 1.7925$

54. $\log_3 21$ $\dfrac{\log 21}{\log 3} \approx 2.7712$

55. $\log_5 (2.7)^2$ $\dfrac{\log 7.29}{\log 5} \approx 1.2343$

56. $\log_7 \sqrt{5}$ $\dfrac{\log \sqrt{5}}{\log 7} \approx 0.4135$

57. MUSIC A musical cent is a unit in a logarithmic scale of relative pitch or intervals. One octave is equal to 1200 cents. The formula $n = 1200\left(\log_2 \dfrac{a}{b}\right)$ can be used to determine the difference in cents between two notes with frequencies a and b. Find the interval in cents when the frequency changes from 443 Hertz (Hz) to 415 Hz.   113.03 cents

GRAPHING CALCULATOR Graph $f(x) = \log x$ and the transformation graph, $g(x)$. Determine the effects on each of the key attributes of the graph of $f(x)$.   See Ch. 6 Answer Appendix.

58. $g(x) = 2f(x)$

59. $g(x) = -4f(x)$

60. $g(x) = f(x) + 3$

61. $g(x) = f(x) - 5$

62. $g(x) = f(x + 2)$

63. $g(x) = f(x - 6)$

Additional Answers

39b. 2022; Sample answer: This seems reasonable as it indicates that the number of pet owners increases at a realistic rate.

40b. 8 yr; Sample answer: The time seems slightly unreasonable as there would be a population growth of 300 bears in 8 years.

Teaching the Mathematical Practices

Critique Arguments Mathematically proficient students are also able to compare the effectiveness of two plausible arguments, distinguish correct logic or reasoning from that which is flawed, and—if there is a flaw in an argument—explain what it is.

Watch Out!

Error Analysis For Exercise 72, point out to students that in the Power Property of Logarithms, $\log_b m^p = p \log_b m$, the p represents the entire exponent of m.

Additional Answers

73.

$\log_{\sqrt{a}} 3 = \log_a x$	Original equation
$\dfrac{\log_a 3}{\log_a \sqrt{a}} = \log_a x$	Change of Base Formula
$\dfrac{\log_a 3}{\frac{1}{2}} = \log_a x$	$\sqrt{a} = a^{\frac{1}{2}}$
$2 \log_a 3 = \log_a x$	Multiply numerator and denominator by 2.
$\log_a 3^2 = \log_a x$	Power Property of Logarithms
$3^2 = x$	Property of Equality for Logarithmic Functions
$9 = x$	Simplify.

75. $\log_3 27 = 3$ and $\log_{27} 3 = \dfrac{1}{3}$; Conjecture:

$\log_a b = \dfrac{1}{\log_b a}$

Proof:

$\log_a b \overset{?}{=} \dfrac{1}{\log_b a}$	Original statement
$\dfrac{\log_b b}{\log_b a} \overset{?}{=} \dfrac{1}{\log_b a}$	Change of Base Formula
$\dfrac{1}{\log_b a} = \dfrac{1}{\log_b a}$	Inverse Property of Exponents and Logarithms

---

Solve each equation. Round to the nearest ten-thousandth.

64. $10^{x^2} = 60$ ± 1.3335

65. $4^{x^2-3} = 16$ $\approx \pm 2.2361$

66. $9^{6y-2} = 3^{3y+1}$ 0.5556

67. $8^{2x-4} = 4^{x+1}$ 3.5

68. $16^x = \sqrt{4^{x+3}}$ 1

69. $2^y = \sqrt{3^{y-1}}$ -3.8188

70. ENVIRONMENTAL SCIENCE An environmental engineer is testing drinking water wells in coastal communities for pollution, specifically unsafe levels of arsenic. The safe standard for arsenic is 0.025 parts per million (ppm). Also, the pH of the arsenic level should be less than 9.5. The formula for hydrogen ion concentration is pH $= -\log H$. (*Hint*: 1 kilogram of water occupies approximately 1 liter. 1 ppm = 1 mg/kg.)

   a. Suppose the hydrogen ion concentration of a well is 1.25×10^{-11}. Should the environmental engineer be worried about too high an arsenic content? yes; 10.9 > 9.5

   b. The environmental engineer finds 1 milligram of arsenic in a 3-liter sample, is the well safe? no

   c. What is the hydrogen ion concentration that meets the troublesome pH level of 9.5? 3.16×10^{-10}

71. MULTIPLE REPRESENTATIONS In this problem, you will solve the exponential equation $4^x = 13$. **a. The solution is between 1.8 and 1.9.**

   a. Tabular Enter the function $y = 4^x$ into a graphing calculator, create a table of values for the function, and scroll through the table to find x when $y = 13$.

   b. Graphical Graph $y = 4^x$ and $y = 13$ on the same screen. Use the **intersect** feature to find the point of intersection. (1.85, 13)

   c. Numerical Solve the equation algebraically. Do all of the methods produce the same result? Explain why or why not. Yes; all methods produce the solution of 1.85. They all should produce the same result because you are starting with the same equation. If they do not, then an error was made. A.CED.1

H.O.T. Problems   Use Higher-Order Thinking Skills

76. Logarithms are exponents. To solve logarithmic equations, write each side of the equation using exponents and solve by using the Inverse Property of Exponents and Logarithms. To solve exponential equations, use the Property of Equality for Logarithmic Functions and the Power Property of Logarithms.

72. CRITIQUE ARGUMENTS Sam and Rosamaria are solving $4^{3p} = 10$. Is either of them correct? Explain your reasoning.

Sam	Rosamaria
$4^{3p} = 10$	$4^{3p} = 10$
$\log 4^{3p} = \log 10$	$\log 4^{3p} = \log 10$
$p \log 4 = \log 10$	$3p \log 4 = \log 10$
$p = \dfrac{\log 10}{\log 4}$	$p = \dfrac{\log 10}{3 \log 4}$

Rosamaria; Sam forgot to bring the 3 down from the exponent when he took the log of each side.

73. CHALLENGE Solve $\log_{\sqrt{a}} 3 = \log_a x$ for x and explain each step. See margin.

74. REASONING Write $\dfrac{\log_5 9}{\log_5 3}$ as a single logarithm. $\dfrac{\log_5 9}{\log_5 3} = \log_3 9$

75. PROOF Find the values of $\log_3 27$ and $\log_{27} 3$. Make and prove a conjecture about the relationship between $\log_a b$ and $\log_b a$. See margin.

76. WRITING IN MATH Explain how exponents and logarithms are related. Include examples like how to solve a logarithmic equation using exponents and how to solve an exponential equation using logarithms.

Standards for Mathematical Practice

Emphasis On	Exercises
1 Make sense of problems and persevere solving them.	1–4, 16–21, 84
4 Model with mathematics.	5–15, 23–83

Preparing for Assessment

77. Let $p^{2x} = q^{16x}$ for all nonzero values of x. What is the value of p in terms of q? 4 A.CED.1 **B**

- ○ A q^4
- ○ B q^8
- ○ C q^{14}
- ○ D q^{32}

78. Let $\log_k m = 3.6$ and $\log_k n = 0.9$. What is the value of $\log_m n$? 4 A.CED.1 **A**

- ○ A 0.25
- ○ B 2.7
- ○ C 3.24
- ○ D 4

79. What is the value of x if $5^{2-x} = 20$? 4 A.CED.1 **B**

- ○ A −1.560
- ○ B 0.139
- ○ C 0.500
- ○ D 0.602

80. What is the solution set of $6^{4-x} > 11$? 4 A.CED.1 **C**

- ○ A $\{x \mid x < -2.6617\}$
- ○ B $\{x \mid x > -2.6617\}$
- ○ C $\{x \mid x < 2.6617\}$
- ○ D $\{x \mid x > 2.6617\}$

81. If $5^{x+2} = 9$, what is the value of x? 4 A.CED.1 **B**

- ○ A −1.26751
- ○ B −0.63479
- ○ C 0.63479
- ○ D 1.26751

82. If $7^{2x+2} = 9^x$, what is the value of x to the nearest thousandth? 4 A.CED.1

$$-2.297$$

83. MULTI-STEP The population of a town $f(t)$ in thousands after t years can be represented by the function $f(t) = 5[\log_6 (t + 5)]$. Let $t = 0$ represent 2010. 4 A.CED.1

a. What was the population of the town in the year 2010?

$$4491$$

b. What was the population of the town in the year 2025?

$$8360$$

c. In what year does the town have 12,000 people? Does this year seem reasonable? Explain. c–e. See Ch. 6 Answer Appendix.

d. What is the domain of the function? Explain your reasoning.

e. What is the range of the function? Explain your reasoning.

f. What was the population of the town in the year 2007?

$$1934$$

84. Solve each equation or inequality. Round to the nearest ten-thousandths. 1 A.CED.1

a. $2^x = 30$ $\quad x = 4.9069$

b. $4^{\frac{x}{2}} > 7$ $\quad x > 2.8074$

c. $\dfrac{5}{12^x} = 100$ $\quad x = -1.2056$

d. $\dfrac{2^x}{3} = \dfrac{5^x}{4}$ $\quad x \geq 0.3140$

Differentiated Instruction OL BL

Extension Remind students that the formula $A = P\left(1 + \dfrac{r}{n}\right)^{nt}$ can be used to find the final amount of an investment using compound interest. Have them use logarithms to find the number of years t that it will take for an investment of $5000 to grow to $8000 at an interest rate of 5% compounded monthly. about 9.4 years

79.

A	Subtracted the log 2 instead of 2
B	CORRECT
C	Solved $2 - x = 4$
D	Subtracted logarithms instead of dividing

80.

| A | Made a sign error when solving |
| B | Made a sign error when solving and did not reverse the inequality |

| C | CORRECT |
| D | Did not reverse the inequality when dividing by −1 |

81.

A	Incorrect calculation with the log
B	CORRECT
C	Made a sign error when solving
D	Made a sign error when solving and incorrect calculation with the log

---

Assess

Yesterday's News Have students write how knowing the properties of logarithms has helped them with solving the exponential equations and inequalities in today's lesson.

Preparing for Assessment

Exercises 77–84 require students to use the skills they will need on standardized assessments. Each exercise is dual-coded with content.

Dual Coding		
Items	Content Standards	MP Mathematical Practices
77	A.CED.1	4
78	A.CED.1	4
79	A.CED.1	4
80	A.CED.1	4
81	A.CED.1	4
82	A.CED.1	4
83	A.CED.1	4
84	A.CED.1	1

Diagnose Student Errors

Survey student responses for each item. Class trends may indicate common errors and misconceptions.

77.

A	Squared the two powers instead of dividing
B	CORRECT
C	Added the two powers instead of dividing
D	Multiplied the powers instead of dividing

78.

A	CORRECT
B	Subtracted the logarithms instead of dividing
C	Multiplied the logarithms instead of dividing
D	Found $\log_n m$

Go Online!

Self-Check Quiz

Students can use *Self-Check Quizzes* to check their understanding of this lesson. You can also give the *Chapter Quiz*, which covers the content in Lessons 6-6 and 6-7.

LESSON 6-8

Natural Logarithms

SUGGESTED PACING (DAYS)

90 min.	0.75
45 min.	1.5

Instruction

Track Your Progress

Objectives

1 Evaluate expressions involving the natural base and natural logarithm.

2 Solve exponential equations and inequalities using natural logarithms.

Mathematical Background

A logarithm with base e is called a natural logarithm, written as $\log_e x$ or $\ln x$. The natural logarithmic function, $y = \ln x$, is the inverse of the natural base exponential function $y = e^x$.

THEN	NOW	NEXT
F.IF.7e: Graph exponential and logarithmic functions, showing intercepts and end behavior, and trigonometric functions, showing period, midline, and amplitude. **A.SSE.2** Use the structure of an expression to identify ways to rewrite it.	**A.CED.1:** Create equations and inequalities in one variable and use them to solve problems.	**F.IF.8b** Use the properties of exponents to interpret expressions for exponential functions. For example, identify percent rate of change in functions such as $y = (1.02)^t$, $y = (0.97)^t$, $y = (1.01)12^t$, $y = (1.2)^t/10$, and classify them as representing exponential growth or decay. **F.LE.4:** For exponential models, express as a logarithm the solution to $ab^{ct} = d$ where a, c, and d are numbers and the base b is 2, 10, or e; evaluate the logarithm using technology.

Go Online! All of these resources and more are available at connectED.mcgraw-hill.com

Use Self-Check Quiz to assess students' understanding of the concepts in this lesson.

Personal Tutors (for every example) let students hear real teachers solve problems. Students can pause and repeat as many times as necessary.

Chapter Project allows students to create and customize a project as a nontraditional method of assessment.

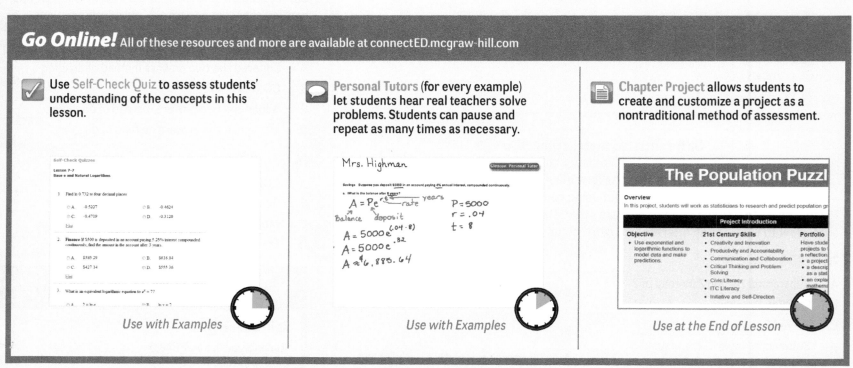

Use with Examples   *Use with Examples*   *Use at the End of Lesson*

OER Using Open Educational Resources

Extension To connect the key concepts that they've learned in this chapter, have students read one or more articles in the series *Logarithms: The Early History of a Familiar Function* on **maa.org.** *Use as an extension*

Differentiate Your Resources

Extra Practice Additional practice or homework; Skills Practice is best for approaching-level students and Practice is best for on-level and beyond-level students

Skills Practice

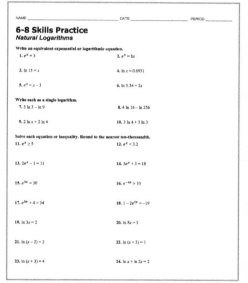

Practice

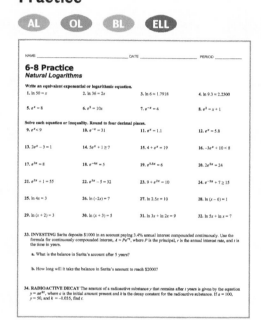

Word Problem Practice

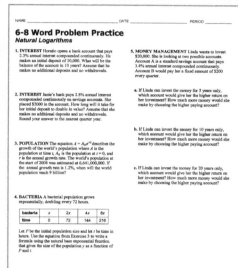

Intervention Reteaching and vocabulary activities that can be used with struggling or absent students and as ELL support

Study Guide and Intervention

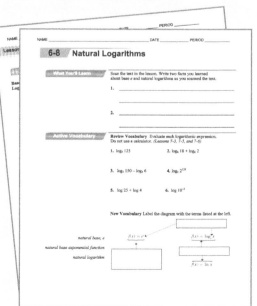

Study Notebook

Extension Activities that can be used to extend lesson concepts

Enrichment

Launch

Have students read the Why? section of the lesson. Ask:

- What other arch-shaped structures have you seen? Sample answers: arches in fast-food advertising, arch bridges

- What other special numbers have you studied? π

Teach

Ask the scaffolded questions for each example to build conceptual understanding for students at all levels.

1 Base *e* and Natural Logarithms

Example 1 Write Equivalent Expressions

AL What is the domain and range of $f(x) = e^x$? domain: $(-\infty, \infty)$ range: $(0, \infty)$

OL Solve $y = e^{\ln 5}$ for y. $y = 5$

BL How can you use your calculator to solve $\ln x = 5$? Convert it to the exponential equation $x = e^5$ and put e^5 into your calculator.

Need Another Example?

Write each exponential equation in logarithmic form.
a. $e^x = 23$ $\ln 23 = x$
b. $e^4 = x$ $\ln x = 4$

Go Online!

Interactive Whiteboard

Use the *eLesson* or *Lesson Presentation* to present this lesson.

Natural Logarithms

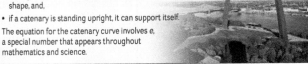

:: Then	:: Now	:: Why?
• You worked with common logarithms.	**1** Evaluate expressions involving the natural base and natural logarithm. **2** Solve exponential equations and inequalities using natural logarithms.	• The St. Louis Gateway Arch in Missouri is in the form of an inverted catenary curve. A catenary curve directs the force of its weight along itself, so that: • if a rope or chain is hanging, it is pulled into that shape, and, • if a catenary is standing upright, it can support itself. The equation for the catenary curve involves *e*, a special number that appears throughout mathematics and science.

New Vocabulary
natural base, *e*
natural base exponential function
natural logarithm

MP Mathematical Practices
4 Model with mathematics.

Content Standards
A.CED.1 Create equations and inequalities in one variable and use them to solve problems.

1 Base *e* and Natural Logarithms Like π and $\sqrt{2}$, the number *e* is an irrational number. The value of *e* is 2.71828… . It is referred to as the **natural base, *e***. An exponential function with base *e* is called a **natural base exponential function**.

🔑 Key Concept Natural Base Functions

The function $f(x) = e^x$ is used to model continuous exponential growth. The function $f(x) = e^{-x}$ is used to model continuous exponential decay.

The inverse of a natural base exponential function is called the **natural logarithm**. This logarithm can be written as $\log_e x$, but is more often abbreviated as $\ln x$.

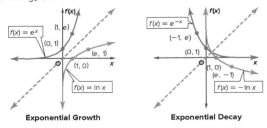

Exponential Growth **Exponential Decay**

You can write an equivalent base *e* exponential equation for a natural logarithmic equation by using the fact that $\ln x = \log_e x$.

$$\ln 4 = x \;\rightarrow\; \log_e 4 = x \;\rightarrow\; e^x = 4$$

A.CED.1

Example 1 Write Equivalent Expressions

Write each exponential equation in logarithmic form.

a. $e^x = 8$
$e^x = 8 \;\rightarrow\; \log_e 8 = x$
$\ln 8 = x$

b. $e^5 = x$
$e^5 = x \;\rightarrow\; \log_e x = 5$
$\ln x = 5$

Guided Practice

1A. $e^x = 9$ $\ln 9 = x$

1B. $e^7 = x$ $\ln x = 7$

MP Mathematical Practices Strategies

Model with mathematics. To help students understand the concepts of natural logarithms, ask the following questions:

- If you put $2000 into an account with an annual interest rate of 3%, how much money would be in the account after 5 years? $2323.69

- If you took this amount and put it into a different account for another 5 years at the same rate, how much would you have in the account? $2699.72

- How much money would be in the account if you had left the money in the original account for the full 10 years? $2699.72

- Mathematically, why are these values the same? Using the formula $A = A_0 e^{rt}$ and the property of the exponentials, we see that $A(10) = A_0 e^{10r} = (A_0 e^{5r})e^{5r}$, which is the formula for 5 years of interest added to an investment which started with value A_0 and had 5 years of interest.

You can also write an equivalent natural logarithm equation for a natural base e exponential equation.

$$e^x = 12 \quad \rightarrow \quad \log_e 12 = x \quad \rightarrow \quad \ln 12 = x$$

A.CED.1

Example 2   Write Equivalent Expressions

Write each logarithmic equation in exponential form.

a. $\ln x \approx 0.7741$

$\ln x \approx 0.7741 \quad \rightarrow \quad \log_e x = 0.7741$

$\qquad\qquad\qquad\qquad x \approx e^{0.7741}$

b. $\ln 10 = x$

$\ln 10 = x \quad \rightarrow \quad \log_e 10 = x$

$\qquad\qquad\qquad\qquad 10 = e^x$

Guided Practice

2A. $\ln x \approx 2.1438$ $x = e^{2.1438}$

2B. $\ln 18 = x$ $18 = e^x$

The properties of logarithms you learned in Lesson 6-6 also apply to the natural logarithms. The logarithmic expressions below can be simplified into a single logarithmic term.

A.CED.1

Example 3   Simplify Expressions with e and the Natural Log

> **Study Tip**
>
> Simplifying When you simplify logarithmic expressions, verify that the logarithm contains no operations and no powers.

Write each expression as a single logarithm.

a. $3 \ln 10 - \ln 8$

$3 \ln 10 - \ln 8 = \ln 10^3 - \ln 8$	Power Property of Logarithms
$\qquad\qquad = \ln \dfrac{10^3}{8}$	Quotient Property of Logarithms
$\qquad\qquad = \ln 125$	Simplify
$\qquad\qquad = \ln 5^3$	$5^3 = 125$
$\qquad\qquad = 3 \ln 5$	Power Property of Logarithms

CHECK Use a calculator to verify the solution.

KEYSTROKES: 3 [LN] 10 [)] [−] [LN] 8 [)] [ENTER] 4.828313737

KEYSTROKES: 3 [LN] 5 [)] [ENTER] 4.828313737 ✓

b. $\ln 40 + 2 \ln \frac{1}{2} + \ln x$

$\ln 40 + 2 \ln \frac{1}{2} + \ln x = \ln 40 + \ln \frac{1}{4} + \ln x$	Power Property of Logarithms
$\qquad\qquad = \ln \left(40 \cdot \frac{1}{4} \cdot x\right)$	Product Property of Logarithms
$\qquad\qquad = \ln 10x$	Simplify

Guided Practice

3A. $6 \ln 8 - 2 \ln 4$ $14 \ln 2$

3B. $2 \ln 5 + 4 \ln 2 + \ln 5y$ $\ln 2000y$

Because the natural base and natural log are inverse functions, they can be used to *undo* or eliminate each other.

$$e^{\ln x} = x \qquad\qquad\qquad \ln e^x = x$$

Example 2 Write Equivalent Expressions

AL What is the domain and range of $f(x) = \ln x$? domain: $(0, \infty)$ range: $(-\infty, \infty)$

OL To what is $\ln e^{5x}$ equal? $5x$

BL Use the change of base formula to convert $\ln 10$ into an expression involving the common log. $\dfrac{1}{\log e}$

Need Another Example?

Write each logarithmic equation in exponential form.
a. $\ln x \approx 1.2528$ $x \approx e^{1.2528}$
b. $\ln 25 = x$ $25 = e^x$

Example 3 Simplify Expressions with e and the Natural Log

AL What rules can we use to write the expression as a single logarithm? Product Property of Logarithms, Quotient Property of Logarithms, Power Property of Logarithms

OL Write out the order of steps you take to write an expression as a single logarithm. Use the Power Property of Logarithms to write the product of a constant with a logarithm as an exponent on the argument of the logarithm. Then from left to right use the Product Property or the Quotient Property to write the logarithms as a single logarithm with multiplication or division.

BL Describe another way you can use your calculator to check your solution. Graph 3 ln 10 − ln 8 and 3 ln 5 and make sure they are the same horizontal line.

Need Another Example?

Write each expression as a single logarithm.
a. $4 \ln 3 + \ln 6$ ln 486
b. $2 \ln 3 + \ln 4 + \ln y$ ln 36y

> **Watch Out!**
>
> Common Misconceptions Stress that e is a constant like π and not a variable like x or y.

2 Equations and Inequalities with *e* and In

Example 4 Solve Base *e* Equations

AL **How can you check the solution?** Substitute −0.3466 in for *x* and see if the left hand side of the equation is close to 3.

OL **Why is In $e^{-2x} = -2x$?** If In $e^{-2x} = -2x$, then In $e = -2x$.

BL **Could you solve this equation using another log? How would you solve it?** You could use any base logarithm. $x = \dfrac{\log_b 2}{-2\log_b e}$

Need Another Example?
Solve $3e^{-2x} + 4 = 10$. Round to the nearest ten-thousandth. $x \approx -0.3466$

Example 5 Solve Natural Log Equations and Inequalities

AL **What is the domain of In $(x - 8)^4$?** $(-\infty, 8) \cup (8, \infty)$

OL **What property allows you to write $e^{\ln (x - 8)^4} < e^4$ in part b?** Property of Inequality for Exponential Functions

BL **Why is $e^{\ln 4x} = 4x$?** If $y = e^{\ln 4x}$, then In $4x = \ln y$. If we convert it to a logarithmic equation, then $y = 4x$ by the Property of Equality for Logarithms.

Need Another Example?
Solve each equation or inequality. Round to the nearest ten-thousandth.
a. $2 \ln 5x = 6$ 4.0171
b. $\ln (3x + 1)^2 > 8$ $\{x \mid x > 17.8661\}$

---

2 **Equations and Inequalities with *e* and In** Equations and inequalities involving base *e* are easier to solve by using natural logarithms rather than by using common logarithms, because In $e = 1$.

A.CED.1

| **Example 4** | Solve Base *e* Equations |

Solve $4e^{-2x} - 5 = 3$. Round to the nearest ten-thousandth.

$$4e^{-2x} - 5 = 3 \qquad \text{Original equation}$$
$$4e^{-2x} = 8 \qquad \text{Add 5 to each side.}$$
$$e^{-2x} = 2 \qquad \text{Divide each side by 4.}$$
$$\ln e^{-2x} = \ln 2 \qquad \text{Property of Equality for Logarithms}$$
$$-2x = \ln 2 \qquad \ln e^x = x$$
$$x = \frac{\ln 2}{-2} \qquad \text{Divide each side by −2.}$$
$$x \approx -0.3466 \qquad \text{Use a calculator.}$$

KEYSTROKES: [LN] 2 [ )] [÷] −2 [ENTER] −.34657359

> **Study Tip**
> **Tools** Most calculators have an e^x and LN key for evaluating natural base and natural log expressions.

> **Guided Practice**
> Solve each equation. Round to the nearest ten-thousandth.
> **4A.** $3e^{4x} - 12 = 15$ 0.5493
> **4B.** $4e^{-x} + 8 = 17$ −0.8109

Just like the natural logarithm can be used to eliminate e^x, the natural base exponential function can eliminate In x.

A.CED.1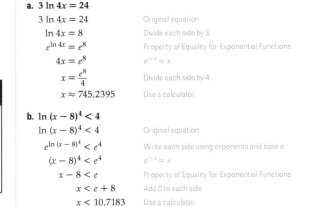

| **Example 5** | Solve Natural Log Equations and Inequalities |

Solve each equation or inequality. Round to the nearest ten-thousandth.

a. $3 \ln 4x = 24$

$$3 \ln 4x = 24 \qquad \text{Original equation}$$
$$\ln 4x = 8 \qquad \text{Divide each side by 3.}$$
$$e^{\ln 4x} = e^8 \qquad \text{Property of Equality for Exponential Functions}$$
$$4x = e^8 \qquad e^{\ln x} = x$$
$$x = \frac{e^8}{4} \qquad \text{Divide each side by 4.}$$
$$x \approx 745.2395 \qquad \text{Use a calculator.}$$

b. $\ln (x - 8)^4 < 4$

$$\ln (x - 8)^4 < 4 \qquad \text{Original equation}$$
$$e^{\ln (x - 8)^4} < e^4 \qquad \text{Write each side using exponents and base } e.$$
$$(x - 8)^4 < e^4 \qquad e^{\ln x} = x$$
$$x - 8 < e \qquad \text{Property of Equality for Exponential Functions}$$
$$x < e + 8 \qquad \text{Add 8 to each side.}$$
$$x < 10.7183 \qquad \text{Use a calculator.}$$

> **Guided Practice**
> Solve each equation or inequality. Round to the nearest ten-thousandth.
> **5A.** $5 \ln 6x = 8$ 0.8255
> **5B.** $\ln (2x - 3)^3 > 6$ $\{x \mid x > 5.1945\}$

Go Online!

Natural logarithms are widely used in problems arising in everyday life, society, and the workplace. Got a question about logarithms? Send a message to your teacher in ConnectED.

Differentiated Instruction **AL**

IF some students mistakenly think that an equation like $4e^{-2x} - 5 = 3$ contains two variables,

THEN point out that the letter *e* represents a constant, just as π does. Both *e* and π are irrational numbers, which cannot be expressed exactly with numerals. To help students avoid this confusion, have them highlight the variables in the equation with a marker.

Interest that is compounded continuously can be found using e or the natural logarithm.

> **Key Concept** Continuously Compounded Interest
>
> The formula for continuously compounded interest can be presented in exponential or logarithmic form, where A is the amount in the account after t years, P is the principal amount invested, and r is the annual interest rate.
>
> **Exponential** $A = Pe^{rt}$ **Logarithmic** $\ln \frac{A}{P} = rt$

A.CED.1

Real-World Example 6 Solve Base e Equations and Inequalities

FINANCIAL LITERACY When Angelina was born, her grandparents deposited $3000 into a college savings account paying 4% interest compounded continuously.

a. Assuming there are no deposits or withdrawals from the account, what will the balance be after 10 years?

$A = Pe^{rt}$ Continuous Compounding Exponential Formula

$= 3000e^{(0.04)(10)}$ $P = 3000, r = 0.04,$ and $t = 10$

$= 3000e^{0.4}$ Simplify.

≈ 4475.47 Use a calculator.

The balance will be $4475.47.

b. How long will it take the balance to reach at least $10,000?

$\ln \frac{A}{P} < rt$ Continuous Compounding Logarithmic Formula

$\ln \frac{10{,}000}{3000} < 0.04t$ $P = 3000, r = 0.04,$ and $A = 10{,}000$

$\ln \frac{10}{3} < 0.04t$ Simplify fraction.

$\frac{\ln \frac{10}{3}}{0.04} < t$ Divide each side by 0.04.

$30.099 < t$ Use a calculator.

It will take about 30 years to reach at least $10,000.

c. If her grandparents want Angelina to have $10,000 after 18 years, how much would they need to invest?

$10{,}000 = Pe^{(0.04)18}$ $A = 10{,}000, r = 0.04,$ and $t = 18$

$\frac{10{,}000}{e^{0.72}} = P$ Divide each side by $e^{0.72}$

$4867.52 \approx P$ Use a calculator.

They need to invest $4867.52.

Study Tip

Rounding In order to avoid any errors due to rounding, do not round until the very end of your calculations.

▶ **Guided Practice**

6. Use the information in Example 6 to answer the following.
   a. If they invested $8000 at 3.75% interest compounded continuously, how much money would be in the account in 30 years? $24,641.73
   b. If they invested $10,000 at 6% interest compounded continuously, then how long would it take the balance to reach at least $30,000? about 19 years
   c. If Angelina's grandparents found an account that paid 5% compounded continuously and wanted her to have $30,000 after 18 years, how much would they need to deposit? $12,197.09

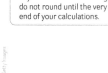

Real-World Link

The average cost of tuition, room, and board at four-year public colleges in Texas is about $16,232 per year.

Source: College for Texans

Example 6 Solve Base e Equations and Inequalities

AL What do the variables A, r, and t represent? A is the amount in the account, r is the annual interest rate of the account, and t is the amount of time the money has been in the account.

OL How old will Angelina be when the amount in the account has doubled? 17

BL Use a table to determine how much Angelina will have after her 18th birthday if her grandparents add an additional $500 in the account each year on her birthday. $19,081.86

Need Another Example?

Savings Suppose you deposit $700 into an account paying 3% annual interest, compounded continuously.
a. What is the balance after 8 years? $889.87
b. How long will it take for the balance in your account to reach at least $1200? about 18 years
c. How much would have to be deposited in order to reach a balance of $1500 after 12 years? $1046.51

Practice

Formative Assessment Use Exercises 1–17 to assess students' understanding of the concepts in this lesson.

The Practice and Problem Solving exercises assess the content taught in the lesson. The Preparing for Assessment page is meant to be used as preparation for end-of-course assessments.

Teaching the Mathematical Practices

Sense-Making Mathematically proficient students start by explaining to themselves the meaning of a problem and looking for entry points to its solution. They analyze givens, constraints, relationships, and goals, and make conjectures about the form and meaning of the solution and plan a solution pathway rather than simply jumping into a solution attempt.

Levels of Complexity Chart

The levels of the exercises progress from 1 to 3, with Level 1 indicating the lowest level of complexity.

Exercises	18–50	51–59, 67–76	60–66
Level 3			●
Level 2		○	
Level 1	●		

Extra Practice

See page R6 for extra exercises for students who are approaching level or for on-level students who need additional reinforcement.

Teaching Tip

Reasonableness Encourage students to evaluate each solution to logarithmic equations and inequalities for reasonableness.

Go Online!   eBook

Interactive Student Guide

Use the *Interactive Student Guide* to deepen conceptual understanding.
· Base *e* and Natural Logarithms

Check Your Understanding   = Step-by-Step Solutions begin on page R11.

 Go Online! for a Self-Check Quiz

Examples 1–2
A.CED.1
Write an equivalent exponential or logarithmic function.

1. $e^x = 30$ $\ln 30 = x$ **2.** $\ln x = 42$ $e^{42} = x$ **3.** $e^3 = x$ $\ln x = 3$ **4.** $\ln 18 = x$ $e^x = 18$

Example 3
A.CED.1
Write each as a single logarithm.

5. $3 \ln 2 + 2 \ln 4$ $7 \ln 2$ **6.** $5 \ln 3 - 2 \ln 9$ $\ln 3$ **7.** $3 \ln 6 + 2 \ln 9$ $\ln 17496$

Example 4
A.CED.1
Solve each equation. Round to the nearest ten-thousandth.

8. $-3e^x + 9 = 4$ 0.5108   **9.** $3e^{-3x} + 4 = 6$ 0.1352   **10.** $2e^{-x} - 3 = 8$ −1.7047

14. $\{x \mid x > 150.4132\}$

13. $\{x \mid -25.0855 < x < 15.0855, x \neq -5\}$

Example 5
A.CED.1
Solve each equation or inequality. Round to the nearest ten-thousandth.

11. $\ln 3x = 8$ 993.6527   **12.** $-4 \ln 2x = -26$ 332.5708   **13.** $\ln (x + 5)^2 < 6$

14. $\ln (x - 2)^3 > 15$ **15.** $e^x > 29$ $\{x \mid x > 3.3673\}$ **16.** $5 + e^{-x} > 14$ $\{x \mid x < -2.1972\}$

Example 6
A.CED.1
17. TECHNOLOGY A virus is spreading through a computer network according to the formula $v(t) = 30e^{0.1t}$, where v is the number of computers infected and t is the time in minutes.

   a. How many computers will be infected after 1.5 hours? about 243,092

   b. How long will it take the virus to infect 10,000 computers? about 58 min

   c. How many computers would need to be infected initially for the virus to spread to 1 million computers in 2 hours? about 7

Practice and Problem Solving   Extra Practice is on page R6.

Examples 1–2
A.CED.1
Write an equivalent exponential or logarithmic function.   **20.** $0.25 = e^x$ **21.** $5.4 = e^x$

18. $e^{-x} = 8$ $\ln 8 = -x$ **19.** $e^{-5x} = 0.1$ $\ln 0.1 = -5x$ **20.** $\ln 0.25 = x$ **21.** $\ln 5.4 = x$

22. $e^{x-3} = 2$ $\ln 2 = x - 3$ **23.** $\ln (x + 4) = 36$ $e^{36} = x + 4$ **24.** $e^{-2} = x^6$ $-2 = 6 \ln x$ **25.** $\ln e^x = 7$ $e^7 = e^x$

Example 3
A.CED.1
Write each as a single logarithm.

26. $\ln 125 - 2 \ln 5$ $\ln 5$ **27.** $3 \ln 10 + 2 \ln 100$ $7 \ln 10$ **28.** $4 \ln \frac{1}{3} - 6 \ln \frac{1}{9}$ $-8 \ln \frac{1}{3}$

29. $7 \ln \frac{1}{2} + 5 \ln 2$ $-2 \ln 2$ **30.** $8 \ln x - 4 \ln 5$ $\ln \frac{x^8}{625}$ **31.** $3 \ln x^2 + 4 \ln 3$ $\ln 81x^6$

Example 4
A.CED.1
Solve each equation. Round to the nearest ten-thousandth.

32. $6e^x - 3 = 35$ 1.8458   **33.** $4e^x + 2 = 180$ 3.7955   **34.** $3e^{2x} - 5 = -4$ −0.5493

35. $-2e^{3x} + 19 = 3$ 0.6931   **36.** $6e^{4x} + 7 = 4$ no solution   **37.** $-4e^{-x} + 9 = 2$ −0.5596

Examples 5–6
A.CED.1
38. SENSE-MAKING Due to depreciation, the value of a car after t years is given by $v(t) = v_0e^{-0.186t}$, where v_0 is the cost of the car new.

   a. What will the value of a car that cost $28,500 new be in 18 months? about $21,561

   b. When will the value of a car be half of its cost new? about 3.73 yr

   c. If a car were worth $30,000 after 1 year, then how much did it cost new? about $36,133

43. $\{x \mid x < -239.8802$ or $x > 239.8802\}$

Solve each inequality. Round to the nearest ten-thousandth.

44. $\{x \mid 6 < x \leq 26.0855\}$

39. $e^x \leq 8.7$ $\{x \mid x \leq 2.1633\}$ **40.** $e^x \geq 42.1$ $\{x \mid x \geq 3.7400\}$ **41.** $\ln (3x + 4)^3 > 10$ $\{x \mid x > 8.0105\}$

42. $4 \ln x^2 < 72$ $\{x \mid -8103.0839 < x < 8103.0839, x \neq 0\}$ **43.** $\ln (8x^4) > 24$ **44.** $-2 [\ln (x - 6)^{-1}] \leq 6$

GRAPHING CALCULATOR Graph $f(x) = \ln x$ and the transformation graph, $g(x)$. Determine the effects on each of the key attributes of the graph of $f(x)$. See Ch. 6 Answer Appendix.

45. $g(x) = 0.5f(x)$ **46.** $g(x) = -0.25f(x)$ **47.** $g(x) = f(x) + 8$

48. $g(x) = f(x) - 9$ **49.** $g(x) = f(x + 5)$ **50.** $g(x) = f(x - 4)$

Differentiated Homework Options

Levels	AL Basic	OL Core	BL Advanced
Exercises	18–44, 62–76	19–59 odd, 60–71	45–76
2-Day Option	19–43 odd, 67–76	18–44, 67–76	
	18–44 even, 62–66	45–66	

 You can use ALEKS to provide additional remediation support with personalized instruction and practice.

B **51 FINANCIAL LITERACY** Use the logarithmic and exponential formulas for continuously compounded interest.

a. If you deposited $800 in an account paying 4.5% interest compounded continuously, how much money would be in the account in 5 years? $1001.86

b. How long would it take you to double your money? about 15.4 yr

c. If you want to double your money in 9 years, what rate would you need? about 7.7%

d. If you want to open an account that pays 4.75% interest compounded continuously and have $10,000 in the account 12 years after your deposit, how much would you need to deposit? about $5655.25   **61c.** $\ln(-x)$ is a reflection in the y-axis. $-\ln x$ is a reflection in the x-axis. See margin for graph.

Write the expression as a sum or difference of logarithms or multiples of logarithms.

52. $\ln 12x^2$ **53.** $\ln \frac{16}{125}$ **54.** $\ln \sqrt[5]{x^3}$ $\frac{3}{5}\ln x$ **55.** $\ln xy^4z^{-3}$
$\ln 12 + 2\ln x$ $4\ln 2 - 3\ln 5$ $\ln x + 4\ln y - 3\ln z$

Use the natural logarithm to solve each equation.

56. $8^x = 24$ 1.5283   **57.** $3^x = 0.4$ −0.8340   **58.** $2^{3x} = 18$ 1.3900   **59.** $5^{2x} = 38$ 1.1301

C **60.** **MODELING** Newton's Law of Cooling, which can be used to determine how fast an object will cool in given surroundings, is represented by $T(t) = T_s + (T_0 - T_s)e^{-kt}$, where T_0 is the initial temperature of the object, T_s is the temperature of the surroundings, t is the time in minutes, and k is a constant value that depends on the type of object.

a. If a cup of coffee with an initial temperature of 180° is placed in a room with a temperature of 70° and the coffee cools to 140° after 10 minutes, find k.   0.045

b. Use this value of k to determine the temperature of the coffee after 20 minutes. about 114.7°

c. When will the temperature of the coffee reach 75°? about 68 min

61. **MULTIPLE REPRESENTATIONS** In this problem, you will use $f(x) = e^x$ and $g(x) = \ln x$.

a. **Graphical** Graph both functions and their axis of symmetry, $y = x$, for $-5 \le x \le 5$. Then graph $a(x) = e^{-x}$ on the same graph. See margin.

b. **Analytical** The graphs of $a(x)$ and $f(x)$ are reflections in which axis? What function would be a reflection of $f(x)$ in the other axis? y-axis; $a(x) = -e^x$

c. **Graphical** Determine the two functions that are reflections of $g(x)$. Graph these new functions.

d. **Verbal** We know that $f(x)$ and $g(x)$ are inverses. Are any of the other functions that we have graphed inverses as well? Explain your reasoning.
Sample answer: no; These functions are reflections over $y = -x$, which indicates that they are not inverses.   A.CED.1

H.O.T. Problems   Use Higher-Order Thinking Skills

62. **CHALLENGE** Solve $4^x - 2^{x+1} = 15$ for x. 2.3219

63. **PROOF** Prove $\ln ab = \ln a + \ln b$ for natural logarithms. See margin.

64. **REASONING** Determine whether $x > \ln x$ is *sometimes*, *always*, or *never* true. Explain your reasoning. Sample answer: Always; the graph of $y = x$ is always greater than the graph of $y = \ln x$ and the graphs never intersect.

65. **OPEN-ENDED** Express the value 3 using e^x and the natural log. Sample answer: $e^{\ln 3}$

66. **WRITING IN MATH** Explain how the natural log can be used to solve a natural base exponential function. See margin.

Standards for Mathematical Practice	
Emphasis On	**Exercises**
1 Make sense of problems and persevere in solving them.	1–4, 8–16, 18–37, 39–44, 52–59, 62, 68, 69, 71, 73–76
2 Reason abstractly and quantitatively.	61, 63–66, 72
4 Model with mathematics.	17, 38, 51, 60, 67, 70
5 Use appropriate tools strategically.	5–7
7 Look for and make use of structure.	45–50

 Teaching the Mathematical Practices

Modeling Mathematically proficient students can apply the mathematics they know to solve problems arising in everyday life, analyze relationships mathematically to draw conclusions, and interpret their mathematical results in the context of a situation.

Assess

Name the Math Have each student write a natural logarithmic equation and a natural logarithmic inequality. Then have students write out all the steps for solving their problems.

Additional Answers

61a.
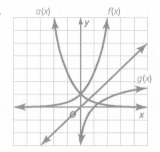

61c.
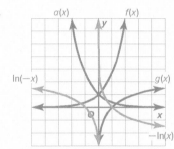

63. Let $p = \ln a$ and $q = \ln b$. That means that $e^p = a$ and $e^q = b$.

$ab = e^p \times e^q$
$ab = e^{p+q}$

$\ln(ab) = (p + q)$
$\ln(ab) = \ln a + \ln b$

66. Sample answer: The natural log and natural base are inverse functions, so taking the natural log of a natural base will *undo* the natural base and make the problem easier to solve.

Go Online!

eSolutions Manual
Create worksheets, answer keys, and solutions handouts for your assignments.

Preparing for Assessment

Exercises 67–76 require students to use the skills they will need on standardized assessments. Each exercise is dual-coded with content standards and mathematical practice standards.

Dual Coding		
Items	Content Standards	Mathematical Practices
67	A.CED.1	2, 4
68	A.SSE.2	1, 8
69	A.CED.1	1
70	A.CED.1	4, 6
71	A.CED.1	1
72	A.SSE.2	2, 5
73	A.CED.1	1, 6
74	A.CED.1	1
75	A.CED.1	1, 2
76	A.CED.1	1

Diagnose Student Errors

Survey student responses for each item. Class trends may indicate common errors and misconceptions.

71.

A	Added 1 to 9 and divided by 5 to both sides and rewrote the equation as $e^{\ln(x)} = e^2$
B	CORRECT
C	Added 1 to 9 and rewrote the equation as $e^{\ln(5x)} > e^{10}$
D	Added 6 to −1 and rewrote the equation as $e^{\ln(5x+5)} > e^{15}$

72.

A	Did not add $\ln e^x + e^x$
B	Misapplied or forgot the Inverse Property of Exponents and Logarithms
C	Believed $e^x \cdot e^x = e^{x^2}$ and forgot the Inverse Property of Exponents and Logarithms
D	CORRECT
E	Believed $e^x \cdot e^x = e^{x^2}$

Go Online!

Self-Check Quiz

Students can use *Self-Check Quizzes* to check their understanding of this lesson.

---

67. MULTI-STEP Population growth can be modeled by exponential growth, using the formula $P(t) = P_0\, e^{rt}$, where P_0 is the initial population size, $P(t)$ is the population at time t (measured in years) after the population size is measured, and r is the rate of growth of the population. 2, 4 A.CED.1

a. The human population growth has an estimated average growth rate of 0.013 (1.3%). If the population in 2016 was measured to be 7.4 billion people, estimate the population we can expect in 2022.

> $P(6) = 8$ billion

b. Estimate the human population of 2000.

> $P(-16) = 6.01$ billion

c. How long ago was the human population half of its value in 2016?

> 53.319 years

d. Estimate the year in which the population was half of the projected population in 2022.

> 1968

68. Simplify the following expression. 1, 8 A.SSE.2

$$\ln 3x^2\, z^4\, \sqrt[3]{y^5}$$

> $\ln(3) + 2\ln(x) + 4\ln(z) + (5/3) * \ln(y)$

69. Solve the inequality: $6\ln(x^3) > 14$. 1 A.CED.1

> $x > e^{\frac{7}{9}}$

70. Which account would have more money in it, given a 5% interest rate with continuously compounded interest: an initial investment of $1000 for 30 years, or an investment of $3000 for 8 years? 4, 6 A.CED.1

The 30-year investment provides a slightly better balance of $4481.69, while the 8-year investment provides $4475.47.

71. What is the approximate value of x in this equation? 1 A.CED.1 **B**

$$\ln(5x-1) + 6 = 15$$

- A 7.39
- B 1620.82
- C 4405.29
- D 653,802.47

72. Let $m \Diamond n = \ln m - \ln n + \ln(mn)$. What is the value of $e^x \Diamond e^x$? 2, 5 A.SSE.2 **D**

- A 0
- B e^{2x}
- C e^{x^2}
- D $2x$
- E x^2

73. What is the value of x, to the nearest thousandth, if $-5^{3x} + 15 = 8$? 1, 6 A.CED.1

> 0.403

74. Which equation is equivalent to $e^{2x} = 14$? 1 A.CED.1 **C**

- A $x = \ln 7$
- B $x = \ln 28$
- C $x = \frac{1}{2}\ln 14$
- D $x = 2\ln 14$

75. Which function generates this table of values? 1, 2 A.CED.1 **B**

x	−2	−1	0	1	2
$f(x)$	1.25	1.5	2	3	5

- A $f(x) = \left(\frac{1}{2}\right)^x + 1$
- B $f(x) = 2^x + 1$
- C $f(x) = 3^x - 1$
- D $f(x) = 3^x$
- E $f(x) = 3^x + 1$

76. Given that $\ln 5 = x$ and $\ln 3 = y$, express in terms of x and y: 1 A.CED.1

- a. $\ln 75$ | $2x + y$ |
- b. $\ln 45$ | $x + 2y$ |
- c. $\ln 15$ | $x + y$ |
- d. $\ln 0.6$ | $y - x$ |
- e. $\ln 0.36$ | $2y - 2x$ |

Differentiated Instruction OL BL

Extension Explain to students that 2! means 2×1, 3! means $3 \times 2 \times 1$, 4! means $4 \times 3 \times 2 \times 1$, and so on. Ask students to use a calculator to find the value of the following series, for as many terms as they want: $1 + \dfrac{1}{1!} + \dfrac{1}{2!} + \dfrac{1}{3!} + \dfrac{1}{4!} + \cdots$.

Discuss with students how the value of this series approaches the value of *e*. Point out that the series expressed to about 20 terms will give the accuracy for *e* that Euler gave. Euler, a Swiss mathematician, studied the number *e* in the 1720s.

74.

A	Divided by 2 before taking the natural logarithm of both sides
B	Isolated x incorrectly
C	CORRECT
D	Multiplied by 2 instead of dividing

75.

A	Forgot the effect of a negative power on the value of the function
B	CORRECT
C	Believed $3^0 = 3$
D	Focused exclusively on $f(1)$
E	Recognized 1 was added to the exponent but chose the wrong base

Solving Logarithmic Equations and Inequalities

Track Your Progress

Objectives

1 Solve logarithmic equations.

2 Solve logarithmic inequalities.

Mathematical Background

When solving logarithmic equations, it is important to remember that a defining characteristic of a logarithmic function is that its domain is the set of all positive numbers. This means that the logarithm of 0 or of a negative number for any base is undefined.

THEN

F.IF.7e Graph exponential and logarithmic functions, showing intercepts and end behavior, and trigonometric functions, showing period, midline, and amplitude.

NOW

A.CED.1 Create equations and inequalities in one variable and use them to solve problems.

NEXT

F.IF.8b Use the properties of exponents to interpret expressions for exponential functions. For example, identify percent rate of change in functions such as $y = (1.02)^t$, $y = (0.97)^t$, $y = (1.01)12^t$, $y = (1.2)^{t}/10$, and classify them as representing exponential growth or decay.

F.LE.4: For exponential models, express as a logarithm the solution to $ab^{ct} = d$ where a, c, and d are numbers and the base b is 2, 10, or e; evaluate the logarithm using technology.

Go Online! All of these resources and more are available at connectED.mcgraw-hill.com

eToolkit contains outstanding tools for enhancing understanding.

Personal Tutors (for every example) let students hear real teachers solve problems. Students can pause and repeat as many times as necessary.

Use **Self-Check Quiz** to assess students' understanding of the concepts in this lesson.

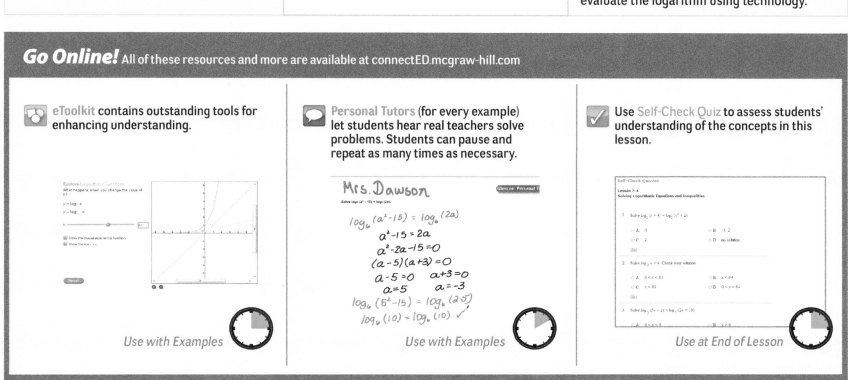

Use with Examples   *Use with Examples*   *Use at End of Lesson*

OER **Using Open Educational Resources**

Tutorials Have students watch the Solving Logarithmic Equations video from the **Kahn Academy** before beginning the lesson or as reinforcement of the key concepts after the lesson is taught. *Use as flipped learning or as an instructional aid*

Differentiate Your Resources

Extra Practice Additional practice or homework; Skills Practice is best for approaching-level students and Practice is best for on-level and beyond-level students

Skills Practice

Practice

Word Problem Practice

Intervention Reteaching and vocabulary activities that can be used with struggling or absent students and as ELL support

Extension Activities that can be used to extend lesson concepts

Study Guide and Intervention

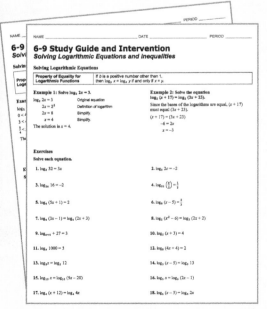

Study Notebook

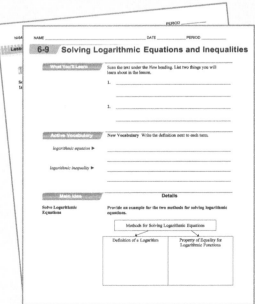

Enrichment

LESSON 9
Solving Logarithmic Equations and Inequalities

::Then
- You evaluated logarithmic expressions.

::Now
 1 Solve logarithmic equations.

 2 Solve logarithmic inequalities.

::Why?
Each year the National Weather Service documents about 1000 tornado touchdowns in the United States. The intensity of a tornado is measured on the Enhanced Fujita scale. Tornadoes are divided into six categories according to their wind speed, path length, path width, and damage caused.

EF-Scale	Wind Speed (mph)	Type of Damage
EF-0	65-85	chimneys, branches
EF-1	86-110	mobile homes overturned
EF-2	111-135	roof torn off
EF-3	136-165	tree uprooted
EF-4	166-200	homes leveled, cars thrown
EF-5	201+	homes thrown

 New Vocabulary
logarithmic equation
logarithmic inequality

Mathematical Practices
4 Model with mathematics.

Content Standards
A.CED.1 Create equations and inequalities in one variable and use them to solve problems.

1 Solve Logarithmic Equations A **logarithmic equation** contains one or more logarithms. You can use the definition of a logarithm to help you solve logarithmic equations.

A.CED.1

Example 1 Solve a Logarithmic Equation

Solve $\log_{36} x = \frac{3}{2}$.

$\log_{36} x = \frac{3}{2}$ Original equation

$x = 36^{\frac{3}{2}}$ Definition of logarithm

$x = (6^2)^{\frac{3}{2}}$ $36 = 6^2$

$x = 6^3$ or 216   Power of a Power

$x^{\frac{3}{2}}$ is $x \cdot x^{\frac{1}{2}}$, so $36^{\frac{3}{2}} = 36 \cdot 6$ or 216. The answer is reasonable.

Guided Practice

Solve each equation.

1A. $\log_9 x = \frac{3}{2}$ 27   **1B.** $\log_{16} x = \frac{5}{2}$ 1024

Key Concept Property of Equality for Logarithmic Functions

Symbols	If b is a positive number other than 1, then $\log_b x = \log_b y$ if and only if $x = y$.
Example	If $\log_5 x = \log_5 8$, then $x = 8$. If $x = 8$, then $\log_5 x = \log_5 8$.

Launch

Have students read the Why? section of the lesson. Ask:
- A tornado with a wind speed of 100 miles per hour is in which category? EF-1
- How many category F-6 tornadoes are known to have occurred? 0

Teach

Ask the scaffolded questions for each example to build conceptual understanding for students at all levels.

1 Solve Logarithmic Equations

Example 1 Solve a Logarithmic Equation

AL What type of equation are you converting the logarithmic equation to? exponential

OL Solve $\log_b x = 1$ for any base b. $x = b$

BL Solve $\log_b b^x = c$ for any values of b and c. $x = c$

Need Another Example?
Solve $\log_8 x = \frac{4}{3}$. 16

 Mathematical Practice Strategies

Reason abstractly and quantitatively. When considering data with very large or very small values, modeling the data using logarithms can be a more useful tool than using polynomials. To help students understand, ask questions like these:

- Which is larger, 100,000,000 or 10,000,000,000? by what factor? 10,000,000,000; 100
- Taking the $\log_{10}$ of both sides, what do we get? How does this relate to the factor? 8, 10; factor is $10^{10-8} = 10^2$
- Consider a math class of 30 students, in a school of 1000 students. What percentage of Earth's population is in this class? What percentage of the population is in the school? Assume that there are 7 billion people total on Earth. ~0.00000000428; ~0.000000142
- Again, take the $\log_{10}$ of both sides, and what do we get? -8.37; -6.85
- Comparing the raw data to the log data, what are the advantages of each? Sample answer: The raw data are more precise, but the log data can be more helpful in determining the relationship between data in terms of factors of their difference instead of the difference in their values.

Go Online!

Interactive Whiteboard

Use the *eLesson* or *Lesson Presentation* to present this lesson.

Example 2 Solve a Logarithmic Equation

AL What must be true about the bases of the logarithmic expressions if you use the Property of Equality for Logarithmic Functions? They must be equal.

OL Why is -1 not a solution to $\log_2 (x^2 - 4) = \log_2 3x$? -1 is not in the domain of either $\log_2 (x^2 - 4)$ or $\log_2 3x$. To be a solution it must be in the domain of both expressions.

BL Write a proof of the statement: "$\log_b x = \log_b y$ implies that $x = y$". If $\log_b x = \log_b y$, then by the definition of logarithms, $b^{\log_b y} = x$. $b^{\log_b y} = y$, so by substitution, $x = y$.

Need Another Example?

Solve $\log_4 x^2 = \log_4 (-6x - 8)$. C
A 4
B 2
C -4 and -2
D no solutions

2 Solve Logarithmic Inequalities

Example 3 Solve a Logarithmic Inequality

AL Write the solution to $\log_3 x > 4$ in interval notation. $(81, +\infty)$

OL Why is the requirement that $x > 0$ in the statement of the Property of Inequality for Logarithmic Functions? Because the domain of $\log_b x$ is $x > 0$

BL Write a Property of Inequality for $0 < b < 1$. If $\log_b x > y$, then $0 < x < b^y$. If $\log_b x < y$, then $x > b^y$.

Need Another Example?

Solve $\log_6 x > 3$. $\{x \mid x > 216\}$

Go Online!

You can use TI-Nspire® technology to solve logarithmic equations. Learn how with the TI-Nspire Worksheet in ConnectED.

Math History Link

Zhang Heng (A.D. 78–139) The earliest known seismograph was invented by Zhang Heng in China in 132 B.C. It was a large brass vessel with a heavy pendulum and several arms that tripped when an earthquake tremor was felt. This helped determine the direction of the quake.

A.CED.1

Example 2 Solve a Logarithmic Equation

Solve $\log_2 (x^2 - 4) = \log_2 3x$.

A -2 B -1 C 2 D 4

Read the Item

You need to find x for the logarithmic equation.

Solve the Item

$\log_2 (x^2 - 4) = \log_2 3x$	Original equation
$x^2 - 4 = 3x$	Property of Equality for Logarithmic Functions
$x^2 - 3x - 4 = 0$	Subtract $3x$ from each side.
$(x - 4)(x + 1) = 0$	Factor.
$x - 4 = 0$ or $x + 1 = 0$	Zero Product Property
$x = 4$ \qquad $x = -1$	Solve each equation.

CHECK Substitute each value into the original equation.

$x = 4$
$\log_2 (4^2 - 4) \stackrel{?}{=} \log_2 3(4)$
$\log_2 12 = \log_2 12$ ✓

$x = -1$
$\log_2 [(-1)^2 - 4] \stackrel{?}{=} \log_2 3(-1)$
$\log_2 (-3) \stackrel{?}{=} \log_2 (-3)$ ✗

The domain of a logarithmic function is $(0, \infty)$, so $\log_2 (-3)$ is undefined and -1 is an extraneous solution. The answer is D.

> **Guided Practice**

2. Solve $\log_3 (x^2 - 15) = \log_3 2x$. C

A -3 B -1 C 5 D 15

---

2 **Solve Logarithmic Inequalities** A **logarithmic inequality** is an inequality that involves logarithms. The following property can be used to solve logarithmic inequalities.

⬦ Key Concept Property of Inequality for Logarithmic Functions

If $b > 1$, $x > 0$, and $\log_b x > y$, then $x > b^y$.

If $b > 1$, $x > 0$, and $\log_b x < y$, then $0 < x < b^y$.

This property also holds true for $\leq$ and $\geq$.

A.CED.1

Example 3 Solve a Logarithmic Inequality

Solve $\log_3 x > 4$.

$\log_3 x > 4$	Original inequality
$x > 3^4$	Property of Inequality for Logarithmic Functions
$x > 81$	Simplify.

> **Guided Practice**

Solve each inequality.

3A. $\log_4 x \geq 3$ $\{x \mid x \geq 64\}$

3B. $\log_2 x < 4$ $\{x \mid 0 < x < 16\}$

Differentiated Instruction **AL** **OL**

IF students need help visualizing the relative locations of the digits in equivalent logarithmic and exponential equations,

THEN have students create colorful posters showing several equivalent exponential and logarithmic equations, such as $2^3 = 8$ and $3 = \log_2 8$. Suggest that students use a different color for each of the digits 2, 3, and 8.

The following property can be used to solve logarithmic inequalities that have logarithms with the same base on each side. Exclude from your solution set values that would result in taking the logarithm of a number less than or equal to zero in the original inequality.

Key Concept Property of Inequality for Logarithmic Functions

Symbols   If $b > 1$, then $\log_b x > \log_b y$ if and only if $x > y$, and $\log_b x < \log_b y$ if and only if $x < y$.

Example   If $\log_6 x > \log_6 35$, then $x > 35$

This property also holds true for ≤ and ≥.

A.CED.1

Example 4   Solve Inequalities with Logarithms on Each Side

Solve $\log_4 (x + 3) > \log_4 (2x + 1)$.

$\log_4 (x + 3) > \log_4 (2x + 1)$ Original inequality

$\qquad x + 3 > 2x + 1$ Property of Inequality for Logarithmic Functions

$\qquad\qquad 2 > x$ Subtract $x + 1$ from each side.

Exclude all values of x for which $x + 3 \le 0$ or $2x + 1 \le 0$. So, $x > -3$, $x > -\frac{1}{2}$, and $x < 2$. The solution set is $\left\{x \mid -\frac{1}{2} < x < 2\right\}$ or $\left(-\frac{1}{2}, 2\right)$.

Guided Practice

4. Solve $\log_5 (2x + 1) \le \log_5 (x + 4)$. Check your solution. $\left\{x \mid -\frac{1}{2} < x \le 3\right\}$

Check Your Understanding   ○ = Step-by-Step Solutions begin on page R11.   **Go Online!** for a Self-Check Quiz

Example 1
A.CED.1
Solve each equation.

1. $\log_8 x = \frac{4}{3}$ 16

2. $\log_{16} x = \frac{3}{4}$ 8

Example 2
A.CED.1
3. **MULTIPLE CHOICE** Solve $\log_5 (x^2 - 10) = \log_5 3x$.   C

   A 10   B 2   C 5   D 2, 5

Example 3
A.CED.1
Solve each inequality.

4. $\log_5 x > 3$ $\{x \mid x > 125\}$

5. $\log_8 x \le -2$ $\left\{x \mid 0 < x \le \frac{1}{64}\right\}$

6. $\log_4 (2x + 5) \le \log_4 (4x - 3)$ $\{x \mid x \ge 4\}$

7. $\log_8 (2x) > \log_8 (6x - 8)$ $\left\{x \mid \frac{4}{3} < x < 2\right\}$

Practice and Problem Solving   Extra Practice is on page R6.

Examples 1–2
A.CED.1
STRUCTURE Solve each equation.

8. $\log_{81} x = \frac{3}{4}$ 27

9. $\log_{25} x = \frac{5}{2}$ 3125

10. $\log_8 \frac{1}{2} = x$ $-\frac{1}{3}$

11. $\log_6 \frac{1}{36} = x$ -2

12. $\log_x 32 = \frac{5}{2}$ 4

13. $\log_x 27 = \frac{3}{2}$ 9

14. $\log_3 (3x + 8) = \log_3 (x^2 + x)$ -2 or 4

15. $\log_{12} (x^2 - 7) = \log_{12} (x + 5)$ 4 or -3

16. $\log_6 (x^2 - 6x) = \log_6 (-8)$ no solution

17. $\log_9 (x^2 - 4x) = \log_9 (3x - 10)$ 5

18. $\log_4 (2x^2 + 1) = \log_4 (10x - 7)$ 1 or 4

19. $\log_7 (x^2 - 4) = \log_7 (-x + 2)$ -3

Differentiated Homework Options			
Levels	**AL** Basic	**OL** Core	**BL** Advanced
Exercises	8–33, 38–48	9–33 odd, 34–48	34–53
2-Day Option	9–33 odd, 45–48	9–33, 45–48	
	8–32 even, 38–44	34–44	

 You can use ALEKS to provide additional remediation support with personalized instruction and practice.

Example 4   Solve Inequalities with Logarithms on Each Side

AL What inequality is $\log_4 (x + 3) > \log_4 (2x + 1)$ reduced to? $x + 3 > 2x + 1$

OL Why must we exclude values of x for which $x + 3 \le 0$ or $2x + 1 \le 0$? Those values are not in the domain of our expressions because the domain of $\log_b x$ is $x > 0$.

BL What property of the graph of $\log_b x$ allows for the Property of Inequality for Logarithmic Equations to be true? The fact that $\log_b x$ is an increasing function when $b > 1$.

Need Another Example?
Solve $\log_7 (2x + 8) > \log_7 (x + 5)$. $\{x \mid x > -3\}$

Practice

Formative Assessment Use Exercises 1–7 to assess students' understanding of the concepts in this lesson.

The Practice and Problem Solving exercises assess the content taught in the lesson. The Preparing for Assessment page is meant to be used as preparation for end-of-course assessments.

Extra Practice

See page R6 for extra exercises for students who are approaching level or for on-level students who need additional reinforcement.

Levels of Complexity Chart			
The levels of the exercises progress from 1 to 3, with Level 1 indicating the lowest level of complexity.			
Exercises	8–33	34–36, 45–53	37–44
Level 3			●
Level 2		●	
Level 1	●		

Go Online!   eBook

Interactive Student Guide   ALGEBRA 2
INTERACTIVE STUDENT GUIDE

Use the *Interactive Student Guide* to deepen conceptual understanding.
· Solving Logarithmic Equations and Inequalities

Teaching the Mathematical Practices

Modeling Mathematically proficient students can apply the mathematics they know to solve problems arising in everyday life, analyze relationships mathematically to draw conclusions, and interpret their mathematical results in the context of a situation.

Follow-Up

Students have explored solving both logarithmic and exponential models.

- **Ask: How are solving logarithmic equations and exponential equations similar and different?** In both cases, you need to simplify the equations so that the variable is isolated on one side of the equations. With logarithmic equations, you will use the definition of logarithm to solve. For exponential equations, you will rewrite the equation so you can utilize the Property of Equality for Exponential Functions to simplify it.

Additional Answer

36b. The graphs are reflections of each other over the x-axis.

36c. 1. The second graph is the same as the first, except it is shifted vertically up 2 units.

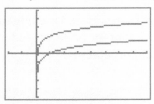

[−2, 8] scl: 1 by [−5, 5] scl: 1

2. The second graph is the same as the first, except it is shifted horizontally to the left 2 units.

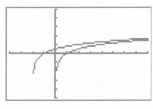

[−4, 8] scl: 1 by [−5, 5] scl: 1

3. Each point on the second graph has a y-coordinate 3 times that of the corresponding point on the first graph.

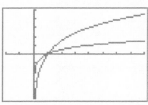

[−2, 8] scl: 1 by [−5, 5] scl: 1

36d. The graphs are reflections of each other over the x-axis. D = {x | x > 0}; R = {all real numbers}

SCIENCE The equation for wind speed w, in miles per hour, near the center of a tornado is $w = 93 \log_{10} d + 65$, where d is the distance in miles that the tornado travels.

20. Write this equation in exponential form. $d = 10^{\frac{w-65}{93}}$

21. In May of 1999, a tornado devastated Oklahoma City with the fastest wind speed ever recorded. If the tornado traveled 525 miles, estimate the wind speed near the center of the tornado. Is your answer reasonable? 318 mph; The wind speed of 318 mph seems high because most tornadoes have wind speeds of about 110 mph.

Solve each inequality.

Examples 3–4
A.CED.1

22. $\log_6 x < -3$ $\left\{x \mid 0 < x < \frac{1}{216}\right\}$

23. $\log_4 x \geq 4$ $\{x \mid x \geq 256\}$

24. $\log_3 x \geq -4$ $\left\{x \mid x \geq \frac{1}{81}\right\}$

(25) $\log_2 x \leq -2$ $\left\{x \mid 0 < x \leq \frac{1}{4}\right\}$

26. $\log_5 x > 2$ $\{x \mid x > 25\}$

27. $\log_7 x < -1$ $\left\{x \mid 0 < x < \frac{1}{7}\right\}$

28. $\log_2 (4x - 6) > \log_2 (2x + 8)$ $\{x \mid x > 7\}$

29. $\log_7 (x + 2) \geq \log_7 (6x - 3)$ $\left\{x \mid \frac{1}{2} < x \leq 1\right\}$

30. $\left\{x \mid \frac{6}{7} < x < 5\right\}$ 30. $\log_3 (7x - 6) < \log_3 (4x + 9)$

31. $\log_5 (12x + 5) \leq \log_5 (8x + 9)$ $\left\{x \mid -\frac{5}{12} < x \leq 1\right\}$

32. $\log_{11} (3x - 24) \geq \log_{11} (-5x - 8)$ $\{x \mid x \geq 2\}$

33. $\log_9 (9x + 4) \leq \log_9 (11x - 12)$ $\{x \mid x \geq 8\}$

34. **MODELING** The magnitude of an earthquake is measured on a logarithmic scale called the Richter scale. The magnitude M is given by $M = \log_{10} x$, where x represents the amplitude of the seismic wave causing ground motion.

34a. 10^3 or 1000 times as great

a. How many times as great is the amplitude caused by an earthquake with a Richter scale rating of 8 as an aftershock with a Richter scale rating of 5?

34b. $10^{0.3}$ or about 2 times as great

b. In 1906, San Francisco was almost completely destroyed by a 7.8 magnitude earthquake. In 1911, an earthquake estimated at magnitude 8.1 occurred along the New Madrid fault in the Mississippi River Valley. How many times greater was the New Madrid earthquake than the San Francisco earthquake?

35. **MUSIC** The first key on a piano keyboard corresponds to a pitch with a frequency of 27.5 cycles per second. With every successive key, going up the black and white keys, the pitch multiplies by a constant. The formula for the frequency of the pitch sounded when the nth note up the keyboard is played is given by $n = 1 + 12 \log_2 \frac{f}{27.5}$.

a. A note has a frequency of 220 cycles per second. How many notes up the piano keyboard is this? 37

b. Another pitch on the keyboard has a frequency of 880 cycles per second. After how many notes up the keyboard will this be found? 61

36a. The shapes of the graphs are the same. The asymptote for each graph is the y-axis and the x-intercept for each graph is 1.

36. **MULTIPLE REPRESENTATIONS** In this problem, you will explore the graphs shown: $y = \log_4 x$ and $y = \log_{\frac{1}{4}} x$.

a. **Analytical** How do the shapes of the graphs compare? How do the asymptotes and the x-intercepts of the graphs compare?

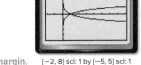

[−2, 8] scl: 1 by [−5, 5] scl: 1

b. **Verbal** Describe the relationship between the graphs. See margin.

c. **Graphical** Use what you know about transformations of graphs to compare and contrast the graph of each function and the graph of $y = \log_4 x$. c, d. See margin.

1. $y = \log_4 x + 2$ 2. $y = \log_4 (x + 2)$ 3. $y = 3 \log_4 x$

d. **Analytical** Describe the relationship between $y = \log_4 x$ and $y = -1(\log_4 x)$. What are a reasonable domain and range for each function?

e. **Analytical** Write an equation for a function for which the graph is the graph of $y = \log_3 x$ translated 4 units left and 1 unit up. $y = \log_3 (x + 4) + 1$

SOUND The relationship between the intensity of sound I and the number of decibels β is $\beta = 10 \log_{10} \left(\frac{I}{10^{-12}} \right)$, where I is the intensity of sound in watts per square meter.

a. Find the number of decibels of a sound with an intensity of 1 watt per square meter. **120**

b. Find the number of decibels of sound with an intensity of 10^{-2} watts per square meter. **100**

c. The intensity of the sound of 1 watt per square meter is 100 times as much as the intensity of 10^{-2} watts per square meter. Why are the decibels of sound not 100 times as great? **See margin.**

40. Sample answer: When $0 < b < 1$, $\log_b x > \log_b y$ if and only if $x < y$. The inequality symbol is switched because a fraction that is less than 1 becomes smaller when it is taken to a greater power.

A.CED.1

H.O.T. Problems Use Higher-Order Thinking Skills

38. Sample answer: Ryan; heather did not need to switch the inequality symbol when raising to a negative power.

38. **CRITIQUE ARGUMENTS** Ryan and Heather are solving $\log_3 x \geq -3$. Is either of them correct? Explain your reasoning.

Ryan	Heather
$\log_3 x \geq -3$	$\log_3 x \geq -3$
$x \geq 3^{-3}$	$x \geq 3^{-3}$
$x \geq \frac{1}{27}$	$0 < x \leq \frac{1}{27}$

39. **CHALLENGE** Find $\log_3 27 + \log_9 27 + \log_{27} 27 + \log_{81} 27 + \log_{243} 27$. $6\frac{17}{20}$

41. The logarithmic function of the form $y = \log_b x$ is the inverse of the exponential function of the form $y = b^x$. The domain of one of the two inverse functions is the range of the other. The range of one of the two inverse functions is the domain of the other.

40. **REASONING** The Property of Inequality for Logarithmic Functions states that when $b > 1$, $\log_b x > \log_b y$ if and only if $x > y$. What is the case for when $0 < b < 1$? Explain your reasoning.

41. **WRITING IN MATH** Explain how the domain and range of logarithmic functions are related to the domain and range of exponential functions.

42. **OPEN-ENDED** Give an example of a logarithmic equation that has no solution. Sample answer: $\log_3 (x + 4) = \log_3 (2x + 12)$

43. **REASONING** Choose the appropriate term. Explain your reasoning. All logarithmic equations are of the form $y = \log_b x$.

a. If the base of a logarithmic equation is greater than 1 and the value of x is between 0 and 1, then the value for y is (_less than_, greater than, equal to) 0.

b. If the base of a logarithmic equation is between 0 and 1 and the value of x is greater than 1, then the value of y is (_less than_, greater than, equal to) 0.

c. There is/are (_no_, one, infinitely many) solution(s) for b in the equation $y = \log_b 0$.

d. There is/are (no, one, _infinitely many_) solution(s) for b in the equation $y = \log_b 1$.

44. **WRITING IN MATH** Explain why any logarithmic function of the form $y = \log_b x$ has an x-intercept of $(1, 0)$ and no y-intercept. See margin.

Standards for Mathematical Practice

Emphasis On	Exercises
1 Make sense of problems and persevere in solving them.	14, 15, 16, 29, 30, 31, 45, 50, 51
2 Reason abstractly and quantitatively.	39, 40, 43, 46–52
3 Construct viable arguments and critique the reasoning of others.	38, 41, 42
4 Model with mathematics.	20, 21, 34, 35, 37, 47
7 Look for and make use of structure.	36, 44

Teaching the Mathematical Practices

Critique Arguments Mathematically proficient students are also able to compare the effectiveness of two plausible arguments, distinguish correct logic or reasoning from that which is flawed, and—if there is a flaw in an argument—explain what it is.

Watch Out!

Error Analysis For Exercise 38, remind students that raising a number to a negative power is not the same as multiplying by −1.

Assess

Name the Math Have students write a step-by-step explanation of the procedure for solving a logarithmic equation such as $\log_8 n = \frac{7}{3}$.

Additional Answers

37c. Sample answer: The power of the logarithm only changes by 2. The power is the answer to the logarithm. That 2 is multiplied by the 10 before the logarithm. So we expect the decibels to change by 20.

44. The y-intercept of the exponential function $y = b^x$ is $(0, 1)$. When the x and y coordinates are switched, the y-intercept is transformed to the x-intercept of $(1, 0)$. There was no x-intercept $(1, 0)$ in the exponential function of the form $y = b^x$. So when the x- and y-coordinates are switched there would be no point on the inverse of $(0, 1)$, and there is no y-intercept.

Go Online!

eSolutions Manual
Create worksheets, answer keys, and solutions handouts for your assignments.

Preparing for Assessment

Exercises 45–53 require students to use the skills they will need on standardized tests. Each exercise is dual-coded with content standards and mathematical practice standards.

Dual Coding		
Items	Content Standards	MP Mathematical Practices
45	A.CED.1	1, 4
46	A.CED.1	2
47	A.CED.1	2, 4
48	A.CED.1	2
49	A.CED.1	2, 4
50	A.CED.1	1, 2
51	A.CED.1	2, 4
52	A.CED.1	1, 2
53	A.CED.1	1, 7

Diagnose Student Errors

Survey student responses for each item. Class trends may indicate common errors and misconceptions.

47.

A	Misunderstood the concept
B	Answered with $I/10^{-12}$ instead of I
C	Did not solve; only chose the proper decibel level
D	CORRECT

49.

A	Solved $x + 8 = 0$
B	CORRECT
C	Solved $x + 8 = 3x$
D	Solved $3x + 4 = 0$

50.

A	Did not check for extraneous solutions
B	Only included the extraneous solution in the set
C	CORRECT
D	Dismissed the valid solution, 8

Go Online!

Self-Check Quiz

Students can use *Self-Check Quizzes* to check their understanding of this lesson. You can also give the *Chapter Quiz*, which covers the content in Lesson 6-9.

45. MULTI-STEP An aftershock is a smaller earthquake that happens in the same area after a large earthquake. A typical aftershock has an amplitude approximately one-thirtieth that of the initial earthquake. MP 1, 4 A.CED.1

The Richter scale reports the magnitude M of an earthquake using the equation $M = \log_{10} x$, where x is the amplitude of the earthquake.

a. Determine the amplitude of an earthquake that measured 7.3 on the Richter scale.

> 19,952,623.15

b. Determine the amplitude of a typical aftershock of an initial earthquake with magnitude 7.3.

> 665,087.4383

c. Determine the magnitude of an aftershock with the amplitude from part b.

> 5.8

46. How many solutions are there to the following problem? MP 2 A.CED.1 B

$$\log_4 (x^2) = \log_4 (x)$$

- A 0
- B 1
- C 2
- D 3

47. If a noise has $ß = 30$ decibels, what is the intensity I of the sound in watts per square meter, where $ß = 10 \log_{10}\left(\frac{I}{10^{-12}}\right)$? MP 2, 4 A.CED.1 D

- A 10^4
- B 10^{-3}
- C 30
- D 10^{-9}

48. What is the value of x? MP 2 A.CED.1 A

$$\log_4 1 = x$$

- A 0
- B 1
- C 4
- D $\frac{1}{4}$

49. What is the value of x in $\log_{19} (x + 8) = \log_{19} (3x + 4)$? MP 2, 4 A.CED.1 B

- A -8
- B 2
- C 4
- D $-\frac{9}{4}$

50. What is the solution set of this equation?

$$\log_3 (x^2 - 16) = \log_3 (6x)$$

MP 1, 2 A.CED.1 C

- A $\{-2, 8\}$
- B $\{-2\}$
- C $\{8\}$
- D ∅

51. The intensity I of normal conversation is about 1×10^{-6} watts per square meter. How many decibels $ß$ is this sound if $ß = 10 \log_{10}\left(\frac{I}{10^{-12}}\right)$? MP 2, 4 A.CED.1 C

- A -6
- B 6
- C 60
- D 600

52. Let $\log_{\frac{1}{64}} m = \frac{3}{2}$. What is the value of m? MP 1, 2 A.CED.1 A

- A $\frac{1}{512}$
- B $\frac{1}{96}$
- C $\frac{3}{128}$
- D $\frac{1}{16}$
- E $\frac{1}{8}$

53. What is the solution set for $\log_7 (2x - 1) < \log_7 (5 - 3x)$? MP 1, 7 A.CED.1 B

- A $\left\{x \mid \frac{1}{2} < x < \frac{4}{5}\right\}$
- B $\left\{x \mid \frac{1}{2} < x < \frac{6}{5}\right\}$
- C $\left\{x \mid \frac{1}{2} < x < \frac{5}{3}\right\}$
- D $\left\{x \mid \frac{4}{5} < x < \frac{5}{3}\right\}$
- E $\left\{x \mid \frac{6}{5} < x < \frac{5}{3}\right\}$

Differentiated Instruction OL BL

Extension Ask students to evaluate $\log_3 9$ and $\log_3 27$. 2, 3 Then ask them to predict the value of $\log_3 (9 \cdot 27)$. After they have made their predictions, ask them to check to see if 3 raised to their predicted values is equal to $9 \cdot 27$, or 243. Have them predict the value of $\log_3 (mn)$. $\log_3 m + \log_3 n$

51.

A	Found the log of 1×10^{-6}
B	Forgot to multiply by the coefficient of 10
C	CORRECT
D	Multiplied by an extra factor of 10

53.

A	Made sign errors when solving $2x - 1 < 5 - 3x$
B	CORRECT
C	Found the interval on which the logarithmic functions would be defined
D	Made sign errors when solving and reversed the inequality
E	Reversed the inequality

EXTEND 6-9

Graphing Technology Lab

Solving Logarithmic Equations and Inequalities

You have solved logarithmic equations algebraically. You can also solve logarithmic equations by graphing or by using a table. The TI-83/84 Plus has $y = \log_{10} x$ as a built-in function. Enter $\boxed{Y=}$ $\boxed{LOG}$ $\boxed{X,T,\theta,n}$ $\boxed{GRAPH}$ to view this graph. To graph logarithmic functions with bases other than 10, you must use the Change of Base Formula, $\log_a n = \dfrac{\log_b n}{\log_b a}$.

Content Standards
A.REI.11 Explain why the x-coordinates of the points where the graphs of the equations $y = f(x)$ and $y = g(x)$ intersect are the solutions of the equation $f(x) = g(x)$; find the solutions approximately, e.g., using technology to graph the functions, make tables of values, or find successive approximations. Include cases where $f(x)$ and/or $g(x)$ are linear, polynomial, rational, absolute value, exponential, and logarithmic functions.

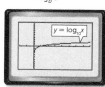

$[-2, 8]$ scl: 1 by $[-10, 10]$ scl: 1

Activity 1

Solve $\log_2 (6x - 8) = \log_3 (20x + 1)$.

Step 1 Graph each side of the equation.

Graph each side of the equation as a separate function. Enter $\log_2 (6x - 8)$ as **Y1** and $\log_3 (20x + 1)$ as **Y2**. Then graph the two equations.

KEYSTROKES: $\boxed{Y=}$ $\boxed{LOG}$ 6 $\boxed{X,T,\theta,n}$ $\boxed{-}$ 8 $\boxed{)}$ $\boxed{\div}$ $\boxed{LOG}$ 2 $\boxed{)}$
$\boxed{ENTER}$ $\boxed{LOG}$ 20 $\boxed{X,T,\theta,n}$ $\boxed{+}$ 1 $\boxed{)}$ $\boxed{\div}$ $\boxed{LOG}$ 3 $\boxed{)}$ $\boxed{GRAPH}$

$[-2, 8]$ scl: 1 by $[-2, 8]$ scl: 1

Step 2 Use the **intersect** feature.

Use the **intersect** feature on the **CALC** menu to approximate the ordered pair of the point at which the curves intersect.

The calculator screen shows that the x-coordinate of the point at which the curves intersect is 4. Therefore, the solution of the equation is 4.

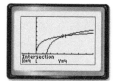

$[-2, 8]$ scl: 1 by $[-2, 8]$ scl: 1

Step 3 Use the **TABLE** feature.

Examine the table to find the x-value for which the y-values for the graphs are equal. At $x = 4$, both functions have a y-value of 4. Thus, the solution of the equation is 4.

You can use a similar procedure to solve logarithmic inequalities using a graphing calculator. (continued on the next page)

Focus

Objective Use a graphing calculator to solve exponential and logarithmic equations and inequalities.

Materials for Each Group

- TI-83/84 Plus or other graphing calculator

Teaching Tip

Point out that the operation **log (** can be used with numbers or expressions. Remind students that it is always a good idea to use closing parentheses at the end of the expression.

Teach ⓔⓛⓛ

Working in Cooperative Groups Put students in groups of two or three, mixing abilities. Then have groups complete Activities 1 and 2 and Exercises 1–4.

Activity 1

- Before discussing the Activity, have students use the Change of Base Formula to express each side of the equation in terms of common logarithms.

- Be sure students understand how to read the table feature to find the x-value for which the y-values are equal.

- Have students substitute the solution into the original equation to verify that it is correct.

Activity 2

- Before discussing Activity 2, have students use the Change of Base Formula to express each side of the inequality in terms of common logarithms.

- Make sure students understand why the inequality needs to be rewritten as a system of inequalities.

- Be sure students understand how to read the table feature to find the x-value for which the y-values are undefined as well as the x-value for which the y-values are equal.

Practice Have students complete Exercises 5–8.

Assess

Formative Assessment
In Exercise 8, check that students record the inequalities in the solution set correctly. In particular, students must include the fact that x must be greater than 0.

From Concrete to Abstract
Have students explain how the solution set for Activity 2 would change if the inequality were $\log_4 (10x + 1) > \log_5 (16 + 6x)$.

Solving Logarithmic Equations and Inequalities *Continued*

Activity 2

Solve $\log_4 (10x + 1) < \log_5 (16 + 6x)$.

Step 1 Enter the inequalities.

Rewrite the problem as a system of inequalities.

The first inequality is $\log_4 (10x + 1) < y$ or $y > \log_4 (10x + 1)$. Because this inequality includes the *greater than* symbol, shade above the curve.

First enter the boundary and then use the arrow and ENTER keys to choose the shade above icon, ▲.

The second inequality is $y < \log_5 (16 + 6x)$. Shade below the curve because this inequality contains *less than*.

KEYSTROKES: [Y=] [◄] [◄] [ENTER] [ENTER] [►] [►] [LOG] 10 [X,T,θ,n] [+] 1 [)] [÷] [LOG] 4 [)] [ENTER] [◄] [◄] [ENTER] [ENTER] [ENTER] [►] [►] [LOG] 16 [+] 6 [X,T,θ,n] [)] [÷] [LOG] 5 [)]

Step 2 Graph the system.

KEYSTROKES: [GRAPH]

The left boundary of the solution set is where the first inequality is undefined. It is undefined for $10x + 1 \le 0$.

$$10x + 1 \le 0$$
$$10x \le -1$$
$$x \le -\frac{1}{10}$$

Use the calculator's **intersect** feature to find the right boundary. You can conclude that the solution set is $\{x \mid -0.1 < x < 1.5\}$.

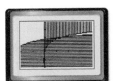

$[-2, 4]$ scl: 1 by $[-2, 4]$ scl: 1

Step 3 Use the **TABLE** feature to check your solution.

Start the table at -0.1 and show x-values in increments of 0.1. Scroll through the table.

KEYSTROKES: [2nd] [TBLSET] -0.1 [ENTER] 0.1 [ENTER] [2nd] [TABLE]

The table confirms the solution of the inequality is $\{x \mid -0.1 < x < 1.5\}$.

Exercises

Solve each equation or inequality. Check your solution.

1. $\log_2 (3x + 2) = \log_3 (12x + 3)$ 0, 2

2. $\log_6 (7x + 1) = \log_4 (4x - 4)$ 5

3. $\log_2 3x = \log_3 (2x + 2)$ about 0.7

4. $\log_{10} (1 - x) = \log_5 (2x + 5)$ about -1.5

5. $\log_4 (9x + 1) > \log_3 (18x - 1)$ $\{x \mid 0.06 < x < 0.17\}$

6. $\log_3 (3x - 5) \ge \log_3 (x + 7)$ $\{x \mid x \ge 6\}$

7. $\log_5 (2x + 1) < \log_4 (3x - 2)$ $\{x \mid x > 2\}$

8. $\log_2 2x \le \log_4 (x + 3)$ $\{x \mid 0 < x \le 1\}$

LESSON 6-10
Using Logarithms to Solve Exponential Problems

SUGGESTED PACING (DAYS)

90 min.	0.5	0.25
45 min.	0.5	0.5
	Instruction	Extend Lab

Track Your Progress

Objectives

1 Use logarithms to solve problems involving exponential growth and decay.

2 Use logarithms to solve problems involving logistic growth.

Mathematical Background

The exponential decay formulas are of the form $y = a(1 - r)^t$, or $y = ae^{-kt}$. The exponential growth formulas are of the form $y = a(1 + r)^t$, or $y = ae^{kt}$. Logarithms can be used to solve problems involving exponential growth and decay.

THEN	NOW	NEXT
A.REI.11 Explain why the x-coordinates of the points where the graphs of the equations $y = f(x)$ and $y = g(x)$ intersect are the solutions of the equation $f(x) = g(x)$; find the solutions approximately, e.g., using technology to graph the functions, make tables of values, or find successive approximations. Include cases where $f(x)$ and/or $g(x)$ are linear, polynomial, rational, absolute value, exponential, and logarithmic functions.*	**F.IF.8.b** Use the properties of exponents to interpret expressions for exponential functions. **F.LE.4** For exponential models, express as a logarithm the solution to $ab^{ct} = d$ where a, c, and d are numbers and the base b is 2, 10 or e; evaluate the logarithm using technology.	**F.BF.1b** Combine standard function types using arithmetic operations.

Go Online! All of these resources and more are available at connectED.mcgraw-hill.com

Graphing Tools are outstanding tools for enhancing understanding.

Personal Tutors (for every example) let students hear real teachers solve problems. Students can pause and repeat as many times as necessary.

Chapter Project allows students to create and customize a project as a nontraditional method of assessment.

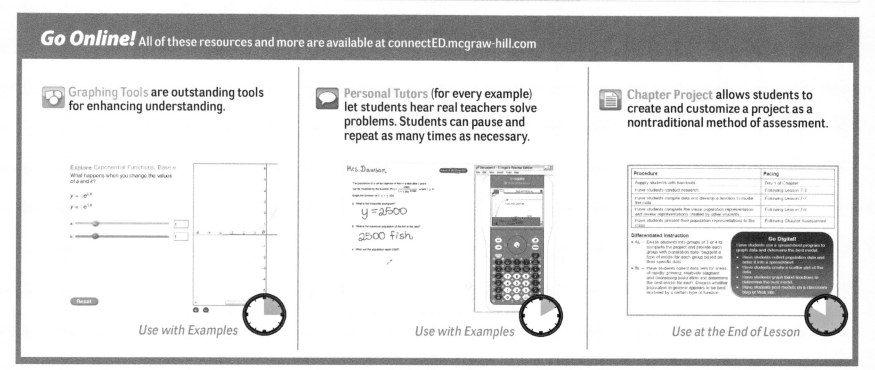

Use with Examples   *Use with Examples*   *Use at the End of Lesson*

OER **Using Open Educational Resources**

Communicating Use **Remind 101** to send text message reminders to parents that Chapter 6 has been completed and there will be a test shortly. This is a free service where parents can opt in to an established group. *Use as reminder*

Go Online!
connectED.mcgraw-hill.com   Worksheets

Differentiate Your Resources

Extra Practice Additional practice or homework; Skills Practice is best for approaching-level students and Practice is best for on-level and beyond-level students

Skills Practice
 AL   **OL**   **ELL**

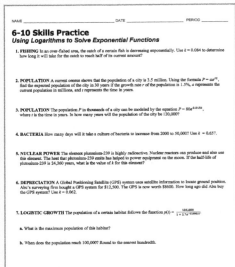

Practice
 AL **OL** **BL** **ELL**

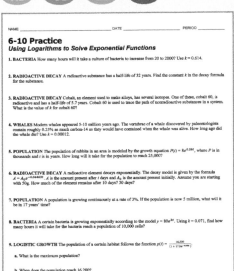

Word Problem Practice
 AL **OL** **BL** **ELL**

Intervention Reteaching and vocabulary activities that can be used with struggling or absent students and as ELL support

Extension Activities that can be used to extend lesson concepts

Study Guide and Intervention
 AL **OL** **ELL**

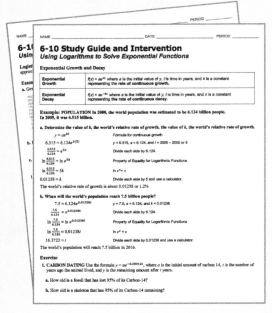

Study Notebook
 AL **OL** **BL** **ELL**

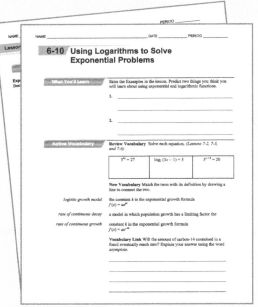

Enrichment
 OL **BL** **ELL**

LESSON 10
Using Logarithms to Solve Exponential Problems

∷Then	∷Now	∷Why?
• You used exponential growth and decay formulas.	**1** Use logarithms to solve problems involving exponential growth and decay. **2** Use logarithms to solve problems involving logistic growth.	• Scientists can use carbon dating to determine the age of an artifact that contained organic material. A living thing exchanges carbon with the environment. When it dies, this carbon exchange stops and the amount of Carbon-14 begins to decrease. This can be modeled using an exponential decay function.

New Vocabulary
rate of continuous growth
rate of continuous decay
logistic growth model

 Mathematical Practices
1 Make sense of problems and persevere in solving them.

Content Standards
F.IF.8.b Use the properties of exponents to interpret expressions for exponential functions.
F.LE.4 For exponential models, express as a logarithm the solution to $ab^{ct} = d$ where a, c, and d are numbers and the base b is 2, 10, or e; evaluate the logarithm using technology.

1 Exponential Growth and Decay
Scientists and researchers frequently use alternate forms of the growth and decay formulas.

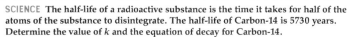

Key Concept Exponential Growth and Decay

Exponential Growth	Exponential Decay
Exponential growth can be modeled by the function $$f(t) = ae^{kt},$$ where a is the initial value, t is time in years, and k is a constant representing the **rate of continuous growth.**	Exponential decay can be modeled by the function $$f(t) = ae^{-kt},$$ where a is the initial value, t is time in years, and k is a constant representing the **rate of continuous decay.**

F.LE.4

Real-World Example 1   Exponential Decay

SCIENCE The half-life of a radioactive substance is the time it takes for half of the atoms of the substance to disintegrate. The half-life of Carbon-14 is 5730 years. Determine the value of k and the equation of decay for Carbon-14.

If a is the initial amount of the substance, then the amount y that remains after 5730 years can be represented by $\frac{1}{2}a$ or $0.5a$.

$y = ae^{-kt}$ Exponential Decay Formula

$0.5a = ae^{-k(5730)}$ $y = 0.5a$ and $t = 5730$

$0.5 = e^{-5730k}$ Divide each side by a

$\ln 0.5 = \ln e^{-5730k}$ Property of Equality for Logarithmic Functions

$\ln 0.5 = -5730k$ $\ln e^x = x$

$\frac{\ln 0.5}{-5730} = k$ Divide each side by -5730

$0.00012 \approx k$ Use a calculator

Thus, the equation for the decay of Carbon-14 is $y = ae^{-0.00012t}$.

> **Guided Practice**   **1.** $k \approx 2.888 \cdot 10^{-5}$
>
> **1.** The half-life of Plutonium-239 is 24,000 years. Determine the value of k.

MP Mathematical Practices Strategies

Make sense of problems and persevere in solving them. Help students maintain oversight of the process of solving exponential growth and decay problems and logistic growth. For example, ask:

• What are the differences between exponential growth and decay? Sample answer: Growth is increasing and decay is decreasing.

• What does k represent in exponential growth and exponential decay? k is the rate of continuous growth (exponential growth); k is the rate of continuous decay (exponential decay).

• What are the differences between the exponential growth model and the logistic growth model? Sample answer: Logistic growth has limiting factors while exponential growth has no limitations.

Launch

Have students read the Why? section of the lesson. Ask:

• In what region is Nicaragua located? Central America

• Why can carbon be used for dating purposes? It decays over time in a measurable pattern.

Teach

Ask the scaffolded questions for each example to build conceptual understanding for students at all levels.

1 Exponential Growth and Decay

Example 1 Exponential Decay

AL What does a represent? The initial amount of the substance.

OL What equation would you use to find the growth constant if a population was growing exponentially and you knew that it took 10 years for the population to double? $2 = e^{10k}$

BL Write an expression to determine k for any substance if you know the half-life.
$$k = \frac{-\ln \frac{1}{2}}{\text{half-life}}$$

Need Another Example?
Geology The half-life of Sodium-22 is 2.6 years. Determine the value of k and the equation of decay for Sodium-22. $k \approx 0.2666$; $y = ae^{-0.2666t}$, where t is given in years

---

Go Online!

Interactive Whiteboard

Use the *eLesson* or *Lesson Presentation* to present this lesson.

Example 2 Carbon Dating

AL What do y, a, and t represent? $a =$ initial amount of Carbon-14, $y =$ amount of carbon remaining, $t =$ years

OL Why is ln used to solve the equation instead of log? Because $\ln e^x = x$, the right side simplifies more.

BL Explain how you could use the common log to solve this problem. Would you get the same answer? Instead of taking ln of both sides you could take log of both sides. Then you would put $\dfrac{\log 0.02}{-0.00012 \log e}$ into your calculator to solve for t. It will give you the same answer.

Need Another Example?

Geology A geologist examining a meteorite estimates that it contains only about 10% as much Sodium-22 as it would have contained when it reached Earth's surface. How long ago did the meteorite reach the surface of Earth?

about 9 years ago

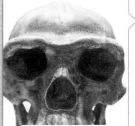

Real-World Link

The oldest modern human fossil, found in Ethiopia, is approximately 160,000 years old.

Source: National Public Radio

Study Tip

Carbon Dating When given a percent or fraction of decay, use an original amount of 1 for a.

Go Online!

Investigate how changing the values in the equation of an exponential function with base e affects the graph by using the eToolkit in ConnectED.

Now that the value of k for Carbon-14 is known, it can be used to date fossils.

F.LE.4

Real-World Example 2   Carbon Dating

SCIENCE A paleontologist examining the bones of a prehistoric animal estimates that they contain 2% as much Carbon-14 as they would have contained when the animal was alive.

a. How long ago did the animal live?

Analyze The formula for the decay of Carbon-14 is $y = ae^{-0.00012t}$. You want to find out how long ago the animal lived.

Formulate Let a be the initial amount of Carbon-14 in the animal's body. The amount y that remains after t years is 2% of a or $0.02a$.

Determine

$y = ae^{-0.00012t}$	Formula for the decay of Carbon-14
$0.02a = ae^{-0.00012t}$	$y = 0.02a$
$0.02 = e^{-0.00012t}$	Divide each side by a.
$\ln 0.02 = \ln e^{-0.00012t}$	Property of Equality for Logarithmic Functions
$\ln 0.02 = -0.00012t$	$\ln e^x = x$
$\dfrac{\ln 0.02}{-0.00012} = t$	Divide each side by -0.00012.
$32,600 \approx t$	Use a calculator.

The animal lived about 32,600 years ago.

Justify Use the formula to find the amount of a sample remaining after 32,600 years. Use an original amount of 1.

$y = ae^{-0.00012t}$	Original equation
$\quad = 1e^{-0.00012(32,600)}$	$a = 1$ and $t = 32,600$
$\quad \approx 0.02$ or 2% ✓	Use a calculator.

Evaluate The half-life of Carbon-14 is 5730 years. 32,600 years is a reasonable answer given the level of Carbon-14 that remains in the bones.

b. If prior research points to the animal being around 20,000 years old, how much Carbon-14 should be in the animal?

$y = ae^{-0.00012t}$	Formula for the decay of Carbon-14
$\quad = 1e^{-0.00012(20,000)}$	$a = 1$ and $t = 20,000$
$\quad = e^{-2.4}$	Simplify.
$\quad = 0.09$ or 9%	Use a calculator.

Guided Practice

2. A specimen that originally contained 42 milligrams of Carbon-14 now contains 8 milligrams. How old is the fossil? about 13,819 yr

The exponential growth equation $y = ae^{kt}$ is identical to the continuously compounded interest formula.

Continuous Compounding	Population Growth
$A = Pe^{rt}$	$y = ae^{kt}$
$P =$ initial amount	$a =$ initial population
$A =$ amount at time t	$y =$ population at time t
$r =$ interest rate	$k =$ rate of continuous growth

Differentiated Instruction   ELL

Logical Learners Have students work in pairs or small groups. Ask them to examine the growth and decay formulas used in Examples 1–3 and to discuss how the equations are related. In particular, ask them to discuss how they can identify which equations are used for exponential decay situations (minus/negative sign) and which are used for exponential growth.

Real-World Example 3   Continuous Exponential Growth

F.LE.4

POPULATION In 2015, the population of the state of Georgia was 10.2 million people. In 2010, it was 9.7 million.

a. Determine the value of k, Georgia's relative rate of growth.

$y = ae^{kt}$ Formula for continuous exponential growth

$10.2 = 9.7e^{k(5)}$ $y = 10.2, a = 9.7,$ and $t = 2015 - 2010$ or 5

$\dfrac{10.2}{9.7} = e^{5k}$ Divide each side by 9.7.

$\ln \dfrac{10.2}{9.7} = \ln e^{5k}$ Property of Equality for Logarithmic Functions

$\ln \dfrac{10.2}{9.7} = 5k$ $\ln e^x = x$

$\dfrac{\ln \frac{10.2}{9.7}}{5} = k$ Divide each side by 5.

$0.01005 = k$ Use a calculator.

Georgia's relative rate of growth is about 0.01005 or about 1%.

b. When will Georgia's population reach 12 million people?

$y = ae^{kt}$ Formula for continuous exponential growth

$12 = 9.7e^{0.01005t}$ $y = 12, a = 9.7,$ and $k = 0.01005$

$1.237 = e^{0.01005t}$ Divide each side by 9.36.

$\ln 1.237 = \ln e^{0.01005t}$ Property of Equality for Logarithmic Functions

$\ln 1.237 = 0.01005t$ $\ln e^x = x$

$\dfrac{\ln 1.237}{0.01005} = t$ Divide each side by 0.01005.

$21.16 \approx t$ Use a calculator.

Georgia's population will reach 12 million people about 21 years after 2010, or in 2031.

c. Ohio's population in 2010 was 11.6 million and can be modeled by $y = 11.6e^{0.004t}$. Determine when Georgia's population will surpass Ohio's.

$9.7e^{0.01005t} > 11.6e^{0.004t}$ Formula for exponential growth

$\ln (9.7e^{0.01005t}) > \ln (11.6e^{0.004t})$ Property of Inequality for Logarithms

$\ln 9.7 + \ln e^{0.01005t} > \ln 11.6 + \ln e^{0.004t}$ Product Property of Logarithms

$\ln 9.7 + 0.01005t > \ln 11.6 + 0.004t$ $\ln e^x = x$

$0.00605t > \ln 11.6 - \ln 9.7$ Subtract $(0.004t + \ln 9.7)$ from each side.

$t > \dfrac{\ln 11.6 - \ln 9.7}{0.00605}$ Divide each side by 0.00605.

$t > 29.57$ Use a calculator.

Georgia's population will surpass Ohio's about 29 years after 2010, or in 2039.

> **Guided Practice**
>
> **3. BIOLOGY** A type of bacteria is growing exponentially according to the model $y = 1000e^{kt}$, where t is the time in minutes.
>
> **A.** If there are 1000 cells initially and 1650 cells after 40 minutes, find the value of k for the bacteria. $k \approx 0.0125$
>
> **B.** Suppose a second type of bacteria is growing exponentially according to the model $y = 50e^{0.0432t}$. Determine how long it will be before the number of cells of this bacteria exceed the number of cells in the other bacteria.   about 97.58 min

Problem-Solving Tip

Use a Formula When dealing with population, it is almost always necessary to use an exponential growth or decay formula.

Example 3   Continuous Exponential Growth

AL **What does t represent?** the number of years since 2010

OL **What is the annual growth rate for the population of Georgia given in the problem?** 0.257%

BL **What must be true about the annual growth rate of a city for its population to surpass the population of another city?** It must be greater.

Need Another Example?

Population In 2007, the population of China was 1.32 billion. In 2000, it was 1.26 billion.

a. Determine the value of k, China's relative rate of growth. 0.0066

b. When will China's population reach 1.5 billion? in 2026

c. India's population in 2007 was 1.13 billion, and can be modeled by $1.13\,e^{0.015t}$. Determine when India's population will surpass China's. (Note: t represents years after 2007.) after 18.6 years, or midway through the year 2025

Differentiated Instruction   OL   BL

Extension Mathematically and scientifically talented students can research the growth rates of different bacteria types. Students can explore how these growth rates are determined, environmental factors that cause them to thrive or inhibit their prosperity, and graph the growth of different types for comparison purposes.

Go Online!

The most up-to-date resources available for your program can be found at connectED.mcgraw-hill.com.

2 Logistic Growth

Example 4 Logistic Growth

AL What values are plugged into the Logistic Growth Function for *a*, *b*, and *c*? $a = 1.66$, $b = 0.048$, $c = 2.0666$.

OL What does the horizontal asymptote represent in this context? It is what the population is limiting to.

BL Why does this function have $y = 2.0666$ as a horizontal asymptote? As $t \to +\infty$, $1.66e^{-0.048t} \to 0$, so the denominator of the expression approaches 1, so the function approaches 2.0666.

Need Another Example?

A city's population in millions is modeled by $f(t) = \dfrac{1.432}{1 + 1.05e^{-0.32t}}$, where *t* is the number of years since 2000.

a. Graph the function.

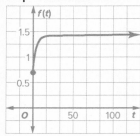

b. What is the horizontal asymptote? $f(t) = 1.432$
c. What will be the maximum population? 1,432,000
d. According to the function, when will the city's population reach 1 million? 2003

e Follow-Up

Students have used exponential and logarithmic function to solve problems.
Ask:
• How can mathematical models help you make good decisions? Sample answer: Mathematical models can be used to compare different options that are available, as well as to predict the impact of an option if it is chosen.

2 Logistic Growth Refer to the equation representing Georgia's population in Example 3. According to the graph at the right, Georgia's population will be about one billion by the year 2130. Does this seem logical?

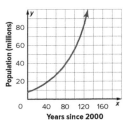

Populations cannot grow infinitely large. There are limitations, such as food supplies, war, living space, diseases, available resources, and so on.

Exponential growth is unrestricted, meaning it will increase without bound. A **logistic growth model**, however, represents growth that has a limiting factor. Logistic models are the most accurate models for representing population growth.

Real-World Link

Phoenix is the fifth largest city in the country and has a population of 1.5 million.

> **Key Concept** Logistic Growth Function
>
> Let *a*, *b*, and *c* be positive constants where $b < 1$. The logistic growth function is represented by $f(t) = \dfrac{c}{1 + ae^{-bt}}$, where *t* represents time.

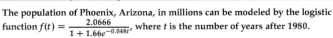

F.LE.4

Real-World Example 4   Logistic Growth

The population of Phoenix, Arizona, in millions can be modeled by the logistic function $f(t) = \dfrac{2.0666}{1 + 1.66e^{-0.048t}}$, where *t* is the number of years after 1980.

a. Graph the function for $0 \le t \le 500$.

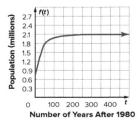

b. What is the horizontal asymptote?
The horizontal asymptote is at $y = 2.0666$.

c. Will the population of Phoenix increase indefinitely? If not, what will be their maximum population?
No. The population will reach a maximum of a little less than 2.0666 million people.

d. According to the function, when will the population of Phoenix reach 1.8 million people?
The graph indicates the population will reach 1.8 million people at $t \approx 50$. Replacing $f(t)$ with 1.8 and solving for *t* in the equation yields $t = 50.35$ years. So, the population of Phoenix will reach 1.8 million people by 2031.

Study Tip

MP Tools To determine where the graph intersects 1.8 on the calculator, graph $y = 1.8$ on the same graph and select *intersection* in the CALC menu.

Guided Practice

4. The population of a certain species of fish in a lake after *t* years can be modeled by the function $P(t) = \dfrac{1880}{1 + 1.42e^{-0.037t}}$, where $t \ge 0$.

   A. Graph the function for $0 \le t \le 500$. See margin.

   B. What is the horizontal asymptote? $y = 1880$

   C. What is the maximum population of the fish in the lake? 1880

   D. When will the population reach 1875? about 170 yr

Additional Answer (Guided Practice)

4A.
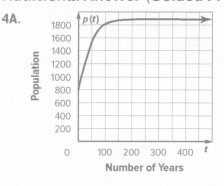

Check Your Understanding

○ = Step-by-Step Solutions begin on page R11.

 Go Online! for a Self-Check Quiz

Examples 1-2
F.LE.4

1. PALEONTOLOGY The half-life of Potassium-40 is about 1.25 billion years.

　a. Determine the value of k and the equation of decay for Potassium-40. 5.545×10^{-10}

　b. A specimen currently contains 36 milligrams of Potassium-40. How long will it take the specimen to decay to only 15 milligrams of Potassium-40? 1,578,843,530 yr

　c. How many milligrams of Potassium-40 will be left after 300 million years? about 30.48 mg

　d. How long will it take Potassium-40 to decay to one eighth of its original amount?
　　3,750,120,003 yr

Example 3
F.LE.4

2. SCIENCE A certain food is dropped on the floor and is growing bacteria exponentially according to the model $y = 2e^{kt}$, where t is the time in seconds. **2b.** about 2.828 cells

　a. If there are 2 cells initially and 8 cells after 20 seconds, find the value of k for the bacteria. $k \approx 0.0693$

　b. The "5-second rule" says that if a person who drops food on the floor eats it within 5 seconds, there will be no harm. How much bacteria is on the food after 5 seconds?

　c. Would you eat food that had been on the floor for 5 seconds? Why or why not? Do you think that the information you obtained in this exercise is reasonable? Explain.

Example 4
F.LE.4

3. ZOOLOGY Suppose the red fox population in a restricted habitat follows the function

　$P(t) = \dfrac{16{,}500}{1 + 18e^{-0.085t}}$, where t represents the time in years.

　a. Graph the function for $0 \le t \le 200$. See margin.

　b. What is the horizontal asymptote? $P(t) = 16{,}500$

　c. What is the maximum population? 16,500

　d. When does the population reach 16,450? about 102 years

2c. Sample answer: Yes; it has not even grown 1 cell in 5 seconds. There are many factors that affect this equation, such as how clean the floor is and what type of food was dropped.

Practice and Problem Solving

Extra Practice is on page R6.

Examples 1-2
F.LE.4

4.  PERSEVERANCE The half-life of Rubidium-87 is about 48.8 billion years.

　a. Determine the value of k and the equation of decay for Rubidium-87. $k \approx 1.42 \times 10^{-11}$

　b. A specimen currently contains 50 milligrams of Rubidium-87. How long will it take the specimen to decay to only 18 milligrams of Rubidium-87? 71,947,270,950 yr

　c. How many milligrams of Rubidium-87 will be left after 800 million years? about 49.4 mg

　d. How long will it take Rubidium-87 to decay to one-sixteenth its original amount? 195.3 billion yr

Example 3
F.LE.4

5 BIOLOGY A certain bacteria is growing exponentially according to the model $y = 80e^{kt}$, where t is the time in minutes.

　a. If there are 80 cells initially and 675 cells after 30 minutes, find the value of k for the bacteria. $k \approx 0.071$

　b. When will the bacteria reach a population of 6000 cells? about 60.8 min

　c. If a second type of bacteria is growing exponentially according to the model $y = 35e^{0.0978t}$, determine how long it will be before the number of cells of this bacteria exceed the number of cells in the other bacteria. about 30.85 min

Example 4
F.LE.4

6. FORESTRY The population of trees in a certain forest follows the function

　$f(t) = \dfrac{18{,}000}{1 + 16e^{-0.084t}}$, where t is the time in years.

　a. Graph the function for $0 \le t \le 100$. See margin on page 450.

　b. When does the population reach 17,500 trees? about 75.33 yr

Differentiated Homework Options

Levels	AL Basic	OL Core	BL Advanced
Exercises	4–6, 14–27	5, 7–27	7–27
2-Day Option	5, 19–27	4–6, 19–27	
	4, 6, 14–18	7–18	

 You can use ALEKS to provide additional remediation support with personalized instruction and practice.

Practice

Formative Assessment Use Exercises 1–3 to assess students' understanding of the concepts in this lesson.

The Practice and Problem Solving exercises assess the content taught in the lesson. The Preparing for Assessment page is meant to be used as preparation for end-of-course assessments.

MP Teaching the Mathematical Practices

Perseverance Mathematically proficient students start by explaining to themselves the meaning of a problem and looking for entry points to its solution. They analyze givens, constraints, relationships, and goals and make conjectures about the form and meaning of the solution and plan a solution pathway rather than simply jumping into a solution attempt.

Levels of Complexity Chart

The levels of the exercises progress from 1 to 3, with Level 1 indicating the lowest level of complexity.

Exercises	4–6	7–11, 19–27	12–18
◆ Level 3			●
◆ Level 2		●	
Level 1	●		

Extra Practice

See page R6 for extra exercises for students who are approaching level or for on-level students who need additional reinforcement.

Additional Answer

3a.

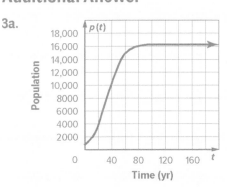

Teaching the Mathematical Practices

Construct Arguments Mathematically proficient students understand and use stated assumptions, definitions, and previously established results in constructing arguments. They make conjectures and build a logical progression of statements to explore the truth of their conjectures. And they are able to analyze situations by breaking them into cases, and can recognize and use counterexamples.

Additional Answers

6a.

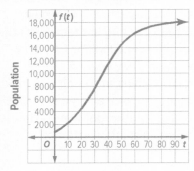

Time (yr)

13a.

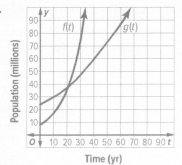

Time (yr)

17. Sample answer: The spread of the flu throughout a small town. The growth of this is limited to the population of the town itself.

18. Sample answer: Exponential functions can be used to model situations that incorporate a percentage of growth or decay for a specific number of times per year. Continuous exponential functions can be used to model situations that incorporate a percentage of growth or decay continuously. Logistic functions can be used to model situations that incorporate a percentage of growth or decay continuously and consist of a limiting factor.

Go Online!

eSolutions Manual

Create worksheets, answer keys, and solutions handouts for your assignments.

---

7 PALEONTOLOGY A paleontologist finds a human bone and determines that the Carbon-14 found in the bone is 85% of that found in living bone tissue. How old is the bone? about 1354 yr old

8. ANTHROPOLOGY An anthropologist has determined that a newly discovered human bone is 8000 years old. How much of the original amount of Carbon-14 is in the bone? about 38%

9. RADIOACTIVE DECAY 100 milligrams of Uranium-238 are stored in a container. If Uranium-238 has a half-life of about 4.47 billion years, after how many years will only 10 milligrams be present? about 14.85 billion yr

10. POPULATION GROWTH The population of the state of Oregon has grown from 3.4 million in 2000 to 3.9 million in 2012.
   a. Write an exponential growth equation of the form $y = ae^{kt}$ for Oregon, where t is the number of years after 2000. $y = 3.4e^{0.011t}$
   b. Use your equation to predict the population of Oregon in 2025. 4.5 million
   c. According to the equation, when will Oregon reach 6 million people? about 2052

11. HALF-LIFE A substance decays 99.9% of its total mass after 200 years. Determine the half-life of the substance. about 20.1 yr

13b.
The graphs intersect at $t = 20.79$. Sample answer: This intersection indicates the point at which both functions determine the same population at the same time.

16.
Sample answer: As $t \to +\infty$, $e^{-t} \to 0$. So, the denominator approaches $1 + 0$ or 1. As the denominator approaches 1, $f(t) \to \frac{c}{1}$ or c. However, because e^{-t} never reaches 0, $f(t)$ can never reach c.

12. LOGISTIC GROWTH The population in millions of the state of Ohio after 1900 can be modeled by $P(t) = \frac{7.85}{1 + 12.19e^{-kt}} + 3.83$, where t is the number of years after 1900 and k is a constant.
   a. If Ohio had a population of 10.7 million in 1970, find the value of k. $k \approx 0.063543$
   b. According to the equation, will the population of Ohio reach 12 million? What is the maximum population? No; about 11.68 million

13. MULTIPLE REPRESENTATIONS In this problem, you will explore population growth. The population growth of a country follows the exponential function $f(t) = 8e^{0.075t}$ or the logistic function $g(t) = \frac{400}{1 + 16e^{-0.025t}}$. The population is measured in millions and t is time in years.
   a. **Graphical** Graph both functions for $0 \le t \le 100$. See margin.
   b. **Analytical** Determine the intersection of the graphs. What is the significance of this intersection?
   c. **Analytical** Which function is a more accurate estimate of the country's population 100 years from now? Explain your reasoning. Sample answer: The logistic function $g(t)$ is a more accurate estimate of the country's population since $f(t)$ will continue to grow exponentially and $g(t)$ considers limitations on population growth such as food supply. F.IF.8.b, F.LE.4

H.O.T. Problems   Use Higher-Order Thinking Skills

14. OPEN-ENDED Give an example of a quantity that grows or decays at a fixed rate. Write a real-world problem involving the rate and solve by using logarithms. Sample answer: money in a bank; See students' work.

15. CHALLENGE Solve $\frac{120,000}{1 + 48e^{-0.015t}} = 24e^{0.055t}$ for t. $t \approx 113.45$

16. CONSTRUCT ARGUMENTS Explain mathematically why $f(t) = \frac{c}{1 + 60e^{-0.5t}}$ approaches, but never reaches the value of c as $t \to +\infty$.

17. OPEN-ENDED Give an example of a quantity that grows logistically and has limitations to growth. Explain why the quantity grows in this manner. See margin.

18. WRITING IN MATH How are exponential, continuous exponential, and logistic functions used to model different real-world situations? See margin.

Standards for Mathematical Practice

Emphasis On	Exercises
1 Make sense of problems and persevere in solving them.	15, 19, 24, 25, 26, 27
2 Reason abstractly and quantitatively.	14, 17, 23
4 Model with mathematics.	1–12, 20–22
6 Attend to precision.	13
7 Look for and make use of structure.	18
8 Look for and express regularity in repeated reasoning.	16

Preparing for Assessment

19. Over the last six years, the population of dolphins in Ocean Bay has increased from 308 to 353. What is the annual relative rate of growth, to the nearest hundredth of a percent? 🔲 1, 4 F.IF.8.b, F.LE.4

> 2.30

20. During its exponential phase, *E. coli* bacteria in a culture increase in number at a rate proportional to the current population. If the growth rate is 2.9% per minute and the current population is 292 million, what will the population be 5.6 minutes from now? 🔲 1, 4 F.IF.8.b, F.LE.4 **B**

- ○ **A** 292 million
- ○ **B** 343.5 million
- ○ **C** 47.4 million
- ○ **D** 33,087 million

21. Solve for $f(3)$ if 🔲 1, 4 F.IF.8.b, F.LE.4 **A**
$$f(t) = \frac{5692}{1 + 9e^{-2t}}$$

- ○ **A** 5,568
- ○ **B** 5,692
- ○ **C** 255,146.3

22. Jeremy has $2500 that he deposits in the bank. The interest rate is 3.2% with continuous compounding. How much money does he have after 10 years? 🔲 1, 4 F.IF.8.b, F.LE.4. **C**

- ○ **A** $2581.29
- ○ **B** $61,331.33
- ○ **C** $3442.82

23. Jada bought a used car for $6000. The value of the car is expected to depreciate at a uniform rate of 30% per year. What will be the approximate value of the car in 3 years? 🔲 1, 4 F.IF.8.b, F.LE.4 **C**

- ○ **A** $600
- ○ **B** $735
- ○ **C** $2060
- ○ **D** $2440

24. Let $p = \log_q r$. What is the value of $\log_q r^2$ in terms of p? 🔲 2 F.IF.8.b, F.LE.4 **B**

- ○ **A** $p + 2$
- ○ **B** $2p$
- ○ **C** p^2
- ○ **D** $\frac{p}{2}$
- ○ **E** $\frac{2}{p}$

25. The formula for decay of Carbon-14 is $y = ae^{-0.00012t}$. After how many years will 3% of the original amount of carbon be left in a sample? 🔲 1, 4 F.IF.8.b, F.LE.4 **C**

- ○ **A** approximately 10,000 years
- ○ **B** approximately 25,000 years
- ○ **C** approximately 30,000 years
- ○ **D** approximately 36,000 years

26. The population of a certain species of squirrel in a forest after t years is modeled by the function $P(t) = \frac{1540}{1 + 1.35e^{-0.07t}}$. What is the maximum population of the squirrels in the forest? 🔲 1, 4 F.IF.8.b, F.LE.4

> 1540

27. MULTI STEP The temperature of an object, T (in degrees Celsius), after t minutes is given by the model $T(t) = 25 + 75e^{kt}$. 🔲 1 F.IF.8.b

a. What is the initial temperature of the object?

> 100 degrees Celsius

b. If the temperature cools to 50 degrees Celsius after 10 minutes, what is the value of k?

> about −0.1099

c. How long will it take for the temperature of the object to reach 30 degrees Celsius? Round your answer to the nearest tenth of a minute.

> 24.6 minutes

Differentiated Instruction 🅞🅛 🅑🅛

Extension *Catenary curves* are curves similar to those formed by a chain hanging between two hooks. Every segment of chain pulls on every other segment of chain, giving the hanging chain its curved shape. The Gateway Arch in St. Louis, Missouri, is an example of an inverted catenary curve.

Ask students to use graphing calculators to graph $y = \dfrac{e^x + e^{-x}}{2}$ to see a model of a catenary curve.

22.

A	Didn't calculate the interest correctly
B	Forgot to divide the interest rate by 100
C	CORRECT

23.

A	Evaluated $6000 - 3(0.3 \cdot 6000)$
B	Used 0.7 instead of 0.3 for k in the uniform decrease formula
C	CORRECT
D	Misused exponential decay function.

24.

A	Misused the Exponent Property of Logarithms
B	CORRECT
C	Substituted p without moving the exponent
D	Misused the Exponent Property of Logarithms
E	Believed this option compensated for squaring r

Assess

Ticket Out the Door Ask students to write questions that can be solved using $y = ae^{kt}$ on one side of an index card. Then have students solve their problems on the reverse sides of the cards.

Preparing for Assessment

Exercises 19–27 require students to use the skills they will need on standardized tests. Each exercise is dual-coded with content standards and mathematical practice standards.

	Dual Coding	
Items	Content Standards	🔲 Mathematical Practices
19	F.IF.8.b, F.LE.4	1, 4
20	F.IF.8.b, F.LE.4	1, 4
21	F.IF.8.b, F.LE.4	1, 4
22	F.IF.8.b, F.LE.4	1, 4
23	F.IF.8.b, F.LE.4	1, 4
24	F.IF.8.b, F.LE.4	2
25	F.IF.8.b, F.LE.4	1, 4
26	F.IF.8.b, F.LE.4	1, 4
27	F.IF.8.b	1

Diagnose Student Errors

Survey student responses for each item. Class trends may indicate common errors and misconceptions.

20.

A	Forgot to use the exponent to find the solution
B	CORRECT
C	Did not use the time when calculating the answer
D	Forgot to put the percent into a decimal

21.

A	CORRECT
B	Did not use the exponent when solving the problem
C	Didn't use the negative sign when calculating the exponential value

Go Online!

Self-Check Quiz

Students can use *Self-Check Quizzes* to check their understanding of this lesson. You can also give the *Chapter Quiz*, which covers the content in Lessons 6-9 and 6-10.

Focus

Objective Use a data collection device to investigate the differences between types of insulated cups and cooling time.

Materials

- TI-83/84 Plus graphing calculator
- data collection device and compatible
- temperature probe
- variety of containers
- very hot water

Teaching Tip

If the device being used is the Calculator-Based Laboratory 2, note the following: The CBL has three keys:

- [TRANSFER] begins transfer of programs of Calculator Software Applications between the CBL and an attached TI graphing calculator.
- [QUICK SET-UP] clears any data stored in CBLs memory.
- [START/STOP] begins sampling for Quick Set-Up. Sampling continues until the number of samples is collected or you press [START/STOP].

Teach ⓔⓛⓛ

Working in Cooperative Groups

Put students in groups of three or four, mixing abilities. Then have groups complete the Activity and Exercises 1–4.

- To set up the calculator and CBL for data collection, select **Collect Data** from the Main Menu. Select **Time Graph** from the **Data Collection** menu. Enter 60 as the time between samples in seconds. Then enter 20 as the number of samples (the CBL will collect data for a total of 1200 seconds). Press ⌈ENTER⌉, then select **Use Time Setup** to continue.

- Discuss with students the **Ymin** and the **Ymax** and the **Yscl** that are appropriate for this lab.

Practice Have groups complete Exercises 5–6.

Cooling

In this lab, you will explore the type of equation that models the change in the temperature of water as it cools under various conditions.

Content Standards
F.BF.1.b Combine standard function types using arithmetic operations.

Set Up the Lab

- Collect a variety of containers, such as a foam cup, a ceramic coffee mug, and an insulated cup.
- Boil water or collect hot water from a tap.
- Choose a container to test and fill with hot water. Place the temperature probe in the cup.
- Connect the temperature probe to your data collection device.

Activity

Step 1 Program the device to collect 20 or more samples in 1 minute intervals.
Step 2 Wait a few seconds for the probe to warm to the temperature of the water.
Step 3 Press the button to begin collecting data.

2. Sample answer: The exponential model fits best. Its correlation coefficient is closest to 1.
3. Sample answer: No; the temperature will approach the temperature of the room but it will not drop below it.

Analyze the Results

1. When the data collection is complete, graph the data in a scatter plot. Use time as the independent variable and temperature as the dependent variable. Write a sentence that describes the points on the graph. Sample answer: It's an exponential pattern.

2. Use the **STAT** menu to find an equation to model the data you collected. Try linear, quadratic, and exponential models. Which model appears to fit the data best? Explain.

3. Would you expect the temperature of the water to drop below the temperature of the room? Explain your reasoning.

4. Use the data collection device to find the temperature of the air in the room. Graph the function $y = t$, where t is the temperature of the room, along with the scatter plot and the model equation. Describe the relationship among the graphs. What is the meaning of the relationship in the context of the experiment?

4. Sample answer: The linear graph of the room temperature is the asymptote for the exponential graph. The temperature of the water will approach the temperature of the room, but it will not go below that temperature.

5. Sample answer: Yes; the thermal container will slow the rate of cooling. This affects the rate of decay in the function.

Make a Conjecture

5. Do you think the results of the experiment would change if you used an insulated container for the water? What part of the function will change, the constant or the rate of decay? Repeat the experiment to verify your conjecture.

6. How might the results of the experiment change if you added ice to the water? What part of the function will change, the constant or the rate of decay? Repeat the experiment to verify your conjecture.

6. Sample answer: The ice will speed the rate of cooling. This affects the rate of decay in the function.

Assess

Formative Assessment

In Exercise 5, check that students have used different containers, but the same initial water temperature. Ask students why it is necessary to begin the second experiment with water at the same temperature as the first experiment.

From Concrete to Abstract

Have students describe the containers that make the best insulators.

CHAPTER 6
Study Guide and Review

 *Go Online!* for Vocabulary Review Games and key vocabulary in 13 languages

Study Guide

Key Concepts

Exponential Functions (Lessons 6-1 and 6-2)

• An exponential function is in the form $y = ab^x$, where $a \neq 0$, $b > 0$ and $b \neq 1$.

• Property of Equality for Exponential Functions: If b is a positive number other than 1, then $b^x = b^y$ if and only if $x = y$.

Geometric Sequences and Series (Lesson 6-3)

• The nth term a_n of a geometric sequence with first term a_1 and common ratio r is given by $a_n = a_1 \cdot r^{n-1}$, where n is any positive integer.

• The sum S_n of the first n terms of a geometric series is given by $S_n = \frac{a_1(1 - r^n)}{1 - r}$ or $S_n = \frac{a_1 - a_1 r^n}{1 - r}$, where $r \neq 1$.

Modeling Data (Lesson 6-5)

• The closer the correlation coefficient r is to 1 or -1, the better the fit.

Logarithms and Logarithmic Functions
(Lessons 6-4, 6-6, 6-7, and 6-9)

• Suppose $b > 0$ and $b \neq 1$. For $x > 0$, there is a number y such that $\log_b x = y$ if and only if $b^y = x$.

• Product Property of Logarithms: $\log_x ab = \log_x a + \log_x b$

• Quotient Property of Logarithms: $\log_x \frac{a}{b} = \log_x a - \log_x b$

• Power Property of Logarithms: $\log_b m^p = p \log_b m$

• The Change of Base Formula: $\log_a n = \frac{\log_b n}{\log_b a}$

Natural Logarithms (Lesson 6-8)

• Because the natural base function and the natural logarithmic function are inverses, these two can be used to "undo" each other.

Using Exponential and Logarithmic Functions (Lesson 6-10)

• Exponential growth can be modeled by the function $f(x) = ae^{kt}$, where k is a constant representing the rate of continuous growth.

• Exponential decay can be modeled by the function $f(x) = ae^{-kt}$, where k is a constant representing the rate of continuous decay.

 **Study Organizer**

Use your Foldable to review the chapter. Working with a partner can be helpful. Ask for clarification of concepts as needed.

Key Vocabulary

asymptote (p. 373)	geometric series (p. 392)
Change of Base Formula (p. 425)	growth factor (p. 375)
	logarithm (p. 397)
common logarithm (p. 423)	logarithmic equation (p. 437)
compound interest (p. 384)	logarithmic function (p. 398)
correlation coefficient (p. 409)	logarithmic inequality (p. 438)
decay factor (p. 376)	logistic growth model (p. 448)
exponential decay (p. 375)	natural base, e (p. 430)
exponential equation (p. 383)	natural base exponential function (p. 430)
exponential function (p. 373)	
exponential growth (p. 373)	natural logarithm (p. 430)
exponential inequality (p. 385)	rate of continuous decay (p. 445)
geometric means (p. 391)	
geometric sequences (p. 390)	rate of continuous growth (p. 445)

Vocabulary Check

Choose a word or term from the list above that best completes each statement or phrase.

1. In $x = b^y$, the variable y is called the _____ of x.   **logarithm**

2. A(n) _____ is an equation in which variables occur as exponents.   **exponential equation**

3. The _____ allows you to write equivalent logarithmic expressions that have different bases.   **change of base formula**

4. The base of the exponential function, $A(t) = a(1 - r)^t$, $1 - r$ is called the _____.   **decay factor**

5. The function $y = \log_b x$, where $b > 0$ and $b \neq 1$, is called a(n) _____.   **logarithmic function**

6. An exponential function with base e is called the _____.   **natural base exponential function**

Concept Check

7. Explain the difference between common logarithms and natural logarithms. **See margin.**

8. Explain how to determine whether an exponential function of the form $f(x) = b^x$ represents growth or decay. **See margin.**

 Study Organizer

A completed Foldable for this chapter should include the Key Concepts related to exponential and logarithmic functions and relations

Key Vocabulary **ELL**

The page reference after each word denotes where that term was first introduced. If students have difficulty answering questions 1–6, remind them that they can use these page references to refresh their memories about the vocabulary terms.

Have students work together to review. Encourage them to summarize their understanding and ask each other questions about exponential and logarithmic functions.

You can use the detailed reports in ALEKS to automatically monitor students' progress and pinpoint remediation needs prior to the chapter test.

Additional Answers

7. Common logarithms have base 10, while natural logarithms have base e.

8. If $b > 1$, then the function represents exponential growth. If $0 < b < 1$, then the function represents exponential decay.

ⓔ Answering the Essential Question

Before answering the Essential Question, have students review their answers to the *Building on the Essential Question* exercises found throughout the chapter.

• How can you make good decisions? (p. 370)

• What factors can affect good decision making? (p. 370)

• How can being financially literate help you to make good decisions? (p. 378)

• How do you make accurate predictions from several different types of data sets? (p. 406)

• What real-world situations or jobs would require the skill to make predictions? (p. 413)

• How are solving logarithmic equations and exponential equations similar and different? (p. 440)

• How can mathematical models help you make good decisions? (p. 448)

Go Online!

Vocabulary Review

Students can use the *Vocabulary Review Games* to check their understanding of the vocabulary terms in this chapter. Students should refer to the *Student-Built Glossary* they have created as they went through the chapter to review important terms. You can also give a *Vocabulary Test* over the content of this chapter.

Lesson-by-Lesson Review

Intervention If the given examples are not sufficient to review the topics covered by the questions, remind students that the lesson references tell them where to review that topic in their textbook.

Two-Day Option Have students complete the Lesson-by-Lesson Review. Then you can use McGraw-Hill eAssessment to customize another review worksheet that practices all the objectives of this chapter or only the objectives on which your students need more help.

Additional Answers

9.

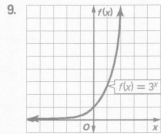

$D = (-\infty, \infty)$ or {all real numbers}
$R = (0, \infty)$ or $\{f(x) \mid f(x) > 0\}$

10.

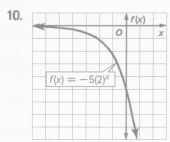

$D = (-\infty, \infty)$ or {all real numbers}
$R = (-\infty, 0)$ or $\{f(x) \mid f(x) < 0\}$

11.

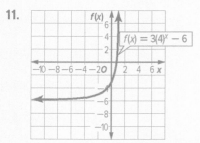

$D = (-\infty, \infty)$ or {all real numbers}
$R = (-6, \infty)$ or $\{f(x) \mid f(x) > -6\}$

12.

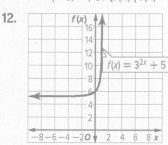

$D = (-\infty, \infty)$ or {all real numbers}
$R = (5, \infty)$ or $\{f(x) \mid f(x) > 5\}$

Lesson-by-Lesson Review

6-1 Graphing Exponential Functions

F.IF.7e, F.BF.3

Graph each function. State the domain and range. **9–14.** See margin.

9. $f(x) = 3^x$
10. $f(x) = -5(2)^x$
11. $f(x) = 3(4)^x - 6$
12. $f(x) = 3^{2x} + 5$
13. $f(x) = 3\left(\frac{1}{4}\right)^{x+3} - 1$
14. $f(x) = \frac{3}{5}\left(\frac{2}{3}\right)^{x-2} + 3$

15. POPULATION A city with a population of 120,000 decreases at a rate of 3% annually.

  a. Write the function that represents this situation. $f(x) = 120,000(0.97)^x$
  b. What will the population be in 10 years? about 88,491

Example 1

Graph $f(x) = -2(3)^x + 1$. State the domain and range.

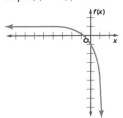

The domain is all real numbers or $D = (-\infty, \infty)$, and the range is all real numbers less than 1 or $R = (-\infty, 1)$.

6-2 Solving Exponential Equations and Inequalities

A.CED.1, F.LE.4

Solve each equation or inequality. **22.** $x > -\frac{2}{3}$

16. $16^x = \frac{1}{64}$ $-\frac{3}{2}$
17. $3^{4x} = 9^{3x+7}$ -7
18. $64^{3n} = 8^{2n-3}$ $-\frac{3}{4}$
19. $8^{3-3y} = 256^{4y}$ $\frac{9}{41}$
20. $5^{1-x} = 125^x$ $\frac{1}{4}$
21. $4^{2x} = \left(\frac{1}{16}\right)^{x+4}$ -2
22. $9^{x-2} > \left(\frac{1}{81}\right)^{x+2}$
23. $27^{3x} \leq 9^{2x-1}$ $x \leq -\frac{2}{5}$
24. $6^x < 36^{x-2}$ $x > 4$
25. $49^{2x-1} \geq 343^{x-1}$ $x \geq -1$

26. BACTERIA A bacteria population started with 5000 bacteria. After 8 hours there were 28,000 in the sample.

  a. Write an exponential function that could be used to model the number of bacteria after x hours if the number of bacteria changes at the same rate. $y = 5000(1.240)^x$
  b. How many bacteria can be expected in the sample after 32 hours? about 4,880,496

Example 2

Solve $4^{3x} = 32^{x-1}$ for x.

$4^{3x} = 32^{x-1}$	Original equation
$(2^2)^{3x} = (2^5)^{x-1}$	Rewrite so each side has the same base
$2^{6x} = 2^{5x-5}$	Power of a Power
$6x = 5x - 5$	Property of Equality for Exponential Functions
$x = -5$	Subtract 5x from each side.

The solution is -5.

13.

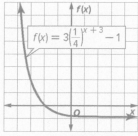

$D = (-\infty, \infty)$ or {all real numbers}
$R = (-1, \infty)$ or $\{f(x) \mid f(x) > -1\}$

14.

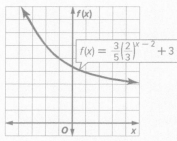

$D = (-\infty, \infty)$ or {all real numbers}
$R = (3, \infty)$ or $\{f(x) \mid f(x) > 3\}$

6-3 Geometric Sequences and Series

A.SSE.4

Find the indicated term for each geometric sequence.

27. $a_1 = 5, r = 2, n = 7$ 320

28. $a_1 = 11, r = 3, n = 3$ 99

29. $a_1 = 128, r = -\frac{1}{2}, n = 5$ 8

30. a_8 for $\frac{1}{8}, \frac{3}{8}, \frac{9}{8} \ldots$ $\frac{2187}{8}$

Find the geometric means in each sequence.

31. $6, \_\_, \_\_, 162$ 18, 54

32. $8, \_\_, \_\_, \_\_, 648$ $\pm 24, 72, \pm 216$

33. $-4, \_\_, \_\_, 108$ 12, -36

34. **SAVINGS** Nolan has a savings account with a current balance of $1500. What would be Nolan's account balance after 4 years if he receives 5% interest annually? $1823.26

Find S_n for each geometric series.

35. $a_1 = 15, r = 2, n = 4$ 225

36. $a_1 = 9, r = 4, n = 6$ 12,285

37. $.5 - 10 + 20 - \cdots$ to 7 terms   215

38. $243 + 81 + 27 + \cdots$ to 5 terms   363

Evaluate the sum of each geometric series.

39. $\sum_{k=1}^{7} 3 \cdot (-2)^{k-1}$ 129

40. $\sum_{k=1}^{8} -1\left(\frac{2}{3}\right)^{k-1}$ $-\frac{6305}{2187}$

Example 3

Find the sixth term of a geometric sequence for which $a_1 = 9$ and $r = 4$.

$a_n = a_1 \cdot r^{n-1}$ Formula for the nth term

$a_6 = 9 \cdot 4^{6-1}$ $n = 6, a_1 = 9, r = 4$

$a_6 = 9216$

The sixth term is 9216.

Example 4

Find two geometric means between 1 and 27.

$a_n = a_1 \cdot r^{n-1}$ Formula for the nth term

$a_4 = 1 \cdot r^{4-1}$ $n = 4$ and $a_1 = 1$

$27 = r^3$ $a_4 = 27$

$3 = r$ Simplify.

The geometric means are 1(3) or 3 and 3(3) or 9.

Example 5

Find the sum of a geometric series for which $a_1 = 3, r = 5$, and $n = 11$.

$S_n = \dfrac{a_1 - a_1 r^n}{1 - r}$ Sum formula

$S_{11} = \dfrac{3 - 3 \cdot 5^{11}}{1 - 5}$ $n = 11, a_1 = 3, r = 5$

$S_{11} = 36,621,093$ Use a calculator.

6-4 Logarithms and Logarithmic Functions

A.SSE.2, F.IF.4, F.IF.7e, F.BF.3

41. Write $\log_2 \frac{1}{16} = -4$ in exponential form. $2^{-4} = \frac{1}{16}$

42. Write $10^2 = 100$ in logarithmic form. $\log_{10} 100 = 2$

Evaluate each expression.

43. $\log_4 256$ 4

44. $\log_2 \frac{1}{8}$ -3

Graph each function.   45, 46. See margin.

45. $f(x) = 2 \log_{10} x + 4$

46. $f(x) = \frac{1}{6} \log_{\frac{1}{3}} (x - 2)$

Example 6

Evaluate $\log_2 64$.

$\log_2 64 = y$ Let the logarithm equal y.

$64 = 2^y$ Definition of logarithm

$2^6 = 2^y$ $64 = 2^6$

$6 = y$ Property of Equality for Exponential Functions

Additional Answers

45.

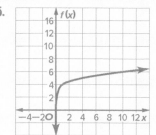

46.

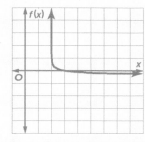

CHAPTER 6
Study Guide and Review *Continued*

6-5 Modeling Data
A.CED.2

Determine the type of function that would best model the data in the table. Then determine the regression equation.

47.

x	2	3	6	10	15
y	6	20	35	50	60

logarithmic; $y = -10.9503 + 26.2680 \ln x$

48.

x	1	2	3	7	8
y	5	20	25	30	50

power; $y = 5.8413x^{1.3185}$

49.

x	1	5	8	10	15
y	1	20	50	200	2000

exponential; $y = 0.8383(1.6997)^x$

Example 7

Determine the type of function that would best model the data in the table. Then determine the regression equation.

x	1	2	3	4	5	6
y	2.4	5	7.8	12.3	20.1	50.8

Use a graphing calculator to determine the correlation coefficient for each regression model.

Linear: $r \approx 0.8682$ Power: $r \approx 0.9550$
Quadratic: $r \approx 0.9438$ Logarithmic: $r \approx 0.7486$
Exponential: $r \approx 0.9918$

The exponential model has the best fit as the value of r is closest to 1. Therefore, the regression equation is $y = 1.4055(1.7653)^x$.

6-6 Properties of Logarithms
A.CED.1

Use $\log_5 16 \approx 1.7227$ and $\log_5 2 \approx 0.4307$ to approximate the value of each expression.

50. $\log_5 8$ 1.2920

51. $\log_5 64$ 2.5841

52. $\log_5 4$ 0.8614

53. $\log_5 \frac{1}{8}$ -1.2921

54. $\log_5 \frac{1}{2}$ -0.4307

Solve each equation. Check your solution.

55. $\log_5 x - \log_5 2 = \log_5 15$ 30

56. $3\log_4 a = \log_4 27$ 3

57. $2\log_3 x + \log_3 3 = \log_3 36$ $2\sqrt{3}$

58. $\log_4 n + \log_4 (n-4) = \log_4 5$ 5

59. $2\log_5 x + 3\log_5 2 = \log_5 10$ $\frac{\sqrt{5}}{2}$

60. $\log_6 8 + \log_6 (n-4) = \log_6 (n+11) - \log_6 2$ 5

61. **SOUND** Use the formula $L = 10 \log_{10} R$, where L is the loudness of a sound and R is the sound's relative intensity, to find out how many times louder 20 people talking would be than one person talking. Suppose one person talks with a loudness of 80 decibels.   1.16 times

---

Example 8

Use $\log_5 16 \approx 1.7227$ and $\log_5 2 \approx 0.4307$ to approximate $\log_5 32$.

$\log_5 32 = \log_5 (16 \cdot 2)$ Replace 32 with 16.

$\quad\quad = \log_5 16 + \log_5 2$ Product Property

$\quad\quad \approx 1.7227 + 0.4307$ Use a calculator.

$\quad\quad \approx 2.1534$

Example 9

Solve $\log_3 3x + \log_3 4 = \log_3 36$.

$\log_3 3x + \log_3 4 = \log_3 36$ Original equation

$\log_3 3x(4) = \log_3 36$ Product Property

$3x(4) = 36$ Definition of logarithm

$12x = 36$ Multiply.

$x = 3$ Divide each side by 12.

Go Online!

Anticipation Guide

Students should complete the *Chapter 6 Anticipation Guide*, and discuss how their responses have changed now that they have completed Chapter 6.

6-7 Common Logarithms

A.CED.1

Solve each equation or inequality. Round to the nearest ten-thousandth.

62. $3^x = 15$ $x \approx 2.4650$ **63.** $6^{x^2} = 28$ $x \approx \pm 1.3637$

64. $8^{m+1} = 30$ **65.** $12^{r-1} = 7r$ $r \approx 4.6102$
      $m \approx 0.6356$

66. $3^{5n} > 24$ **67.** $5^{x+2} \leq 3^x$
    $\{n \mid n > 0.5786\}$ $\{x \mid x \leq -6.3013\}$

68. $2^{3x} < 5^{x-1}$ **69.** $6^{2w-5} \geq 23$
    $\{x \mid x < -3.4243\}$ $\{w \mid w \geq 3.3750\}$

70. SAVINGS You deposited $1000 into an account that pays an annual interest rate r of 5% compounded quarterly. Use $A = P\left(1 + \frac{r}{n}\right)^{nt}$.

   a. How long will it take until you have $1500 in your account?   about 8.2 years

   b. How long it will take for your money to double?   about 13.9 years

Express each logarithm in terms of common logarithms. Then approximate to the nearest ten-thousandth.

71. $\log_8 61$ 1.9769   **72.** $\log_3 42$ 3.4022

73. $\log_5 97$ 2.8424   **74.** $\log_9 150$ 2.2804

75. $\log_7 128$ 2.4935   **76.** $\log_6 295$ 3.1740

Example 10

Solve $5^{3x} > 7^{x+1}$.

$5^{3x} > 7^{x+1}$	Original inequality
$\log 5^{3x} > \log 7^{x+1}$	Property of Inequality
$3x \log 5 > (x+1) \log 7$	Power Property
$3x \log 5 > x \log 7 + \log 7$	Distributive Property
$3x \log 5 - x \log 7 > \log 7$	Subtract $x \log 7$
$x(3 \log 5 - \log 7) > \log 7$	Distributive Property
$x > \dfrac{\log 7}{3 \log 5 - \log 7}$	Divide by $3 \log 5 - \log 7$
$x > 0.6751$	Use a calculator

The solution set is $\{x \mid x > 0.6751\}$.

Example 11

Express $\log_4 15$ in terms of common logarithms. Then round to the nearest ten-thousandth.

$\log_4 15 = \dfrac{\log_{10} 15}{\log_{10} 4}$	Change of Base formula
≈ 1.9534	Use a calculator

6-8 Natural Logarithms

A.CED.1

Solve each equation or inequality. Round to the nearest ten-thousandth.   77–84. See margin.

77. $4e^x - 11 = 17$ **78.** $2e^{-x} + 1 = 15$

79. $\ln 2x = 6$ **80.** $\ln (4x - 1) = 5$

81. $\ln (x + 3)^5 < 5$ **82.** $e^{-x} > 18$

83. $2 + e^x < 9$ **84.** $\ln (x + 1) > 2$

85. SAVINGS If you deposit $2000 in an account paying 6.4% interest compounded continuously, how long will it take for your money to triple? Use $A = Pe^{rt}$.   about 17.2 years

Example 12

Solve $3e^{5x} + 1 = 10$. Round to the nearest ten-thousandth.

$3e^{5x} + 1 = 10$	Original equation
$3e^{5x} = 9$	Subtract 1 from each side
$e^{5x} = 3$	Divide each side by 3
$\ln e^{5x} = \ln 3$	Property of Equality
$5x = \ln 3$	$\ln e^x = x$
$x = \dfrac{\ln 3}{5}$	Divide each side by 5
$x \approx 0.2197$	Use a calculator

Additional Answers

77. 1.9459

78. −1.9459

79. 201.7144

80. 37.3533

81. $\{x \mid -3 < x < -0.2817\}$

82. $\{x \mid x < -2.8904\}$

83. $\{x \mid x < 1.9459\}$

84. $\{x \mid x > 6.3891\}$

Before the Test

Have students complete the Study Notebook Tie it Together activity to review topics and skills presented in the chapter.

Additional Answers (Practice Test)

1.

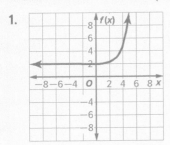

$D = (-\infty, \infty)$ or {all real numbers}
$R = (2, \infty)$ or $\{f(x) \mid f(x) > 2\}$

2.

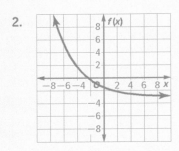

$D = (-\infty, \infty)$ or {all real numbers}
$R = (-3, \infty)$ or $\{f(x) \mid f(x) > -3\}$

6-9 Solving Logarithmic Equations and Inequalities

A.CED.1

Solve each equation or inequality.

86. $\log_4 x = \frac{3}{2}$ 8

87. $\log_2 \frac{1}{64} = x$ −6

88. $\log_4 x < 3$ $\{x \mid 0 < x < 64\}$

89. $\log_5 x < -3$ $\left\{x \mid 0 < x < \frac{1}{125}\right\}$

90. $\log_9 (3x - 1) = \log_9 (4x)$ no solution

91. $\log_2 (x^2 - 18) = \log_2 (-3x)$ −6

92. $\log_3 (3x + 4) \le \log_3 (x - 2)$ no solution

93. EARTHQUAKE The magnitude of an earthquake is measured on a logarithmic scale called the Richter scale. The magnitude M is given by $M = \log_{10} x$, where x represents the amplitude of the seismic wave causing ground motion. How many times as great is the amplitude caused by an earthquake with a Richter scale rating of 10 as an aftershock with a Richter scale rating of 7? 1000

Example 13

Solve $\log_{27} x < \frac{2}{3}$.

$\log_{27} x < \frac{2}{3}$	Original inequality
$x < 27^{\frac{2}{3}}$	Logarithmic to Exponential Inequality
$x < 9$	Simplify.

Example 14

Solve $\log_5 (p^2 - 2) = \log_5 p$.

$\log_5 (p^2 - 2) = \log_5 p$	Original equation
$p^2 - 2 = p$	Property of Equality
$p^2 - p - 2 = 0$	Subtract p from each side.
$(p - 2)(p + 1) = 0$	Factor.
$p - 2 = 0$ or $p + 1 = 0$	Zero Product Property.
$p = 2$ $p = -1$	Solve each equation.

The solution is $p = 2$, since $\log_5 p$ is undefined for $p = -1$.

6-10 Using Logarithms to Solve Exponential Problems

F.IF.8b, F.LE.4

94. CARS Abe bought a used car for $2500. It is expected to depreciate at a rate of 25% per year. What will be the value of the car in 3 years? $1054.69

95. BIOLOGY For a certain strain of bacteria, k is 0.728 when t is measured in days. Using the formula $y = ae^{kt}$, how long will it take 10 bacteria to increase to 675 bacteria? ≈5.8 days

96. POPULATION The population of a city 20 years ago was 24,330. Since then, the population has increased at a steady rate each year. If the population is currently 55,250, find the annual rate of growth for this city. about 4.1%

Example 15

A certain culture of bacteria will grow from 250 to 2000 bacteria in 1.5 hours. Find the constant k for the growth formula. Use $y = ae^{kt}$.

$y = ae^{kt}$	Exponential Growth Formula
$2000 = 250e^{k(1.5)}$	Replace y with 2000, a with 250, and t with 1.5.
$8 = e^{1.5k}$	Divide each side by 250.
$\ln 8 = \ln e^{1.5k}$	Property of Equality
$\ln 8 = 1.5k$	Inverse Property
$\frac{\ln 8}{1.5} = k$	Divide each side by 1.5.
$1.3863 \approx k$	Use a calculator.

Go Online!

ᵉAssessment

Customize and create multiple versions of chapter tests and answer keys that align to your standards. Tests can be delivered on paper or online.

CHAPTER 6
Practice Test

 **Go Online!** for another Chapter Test

Graph each function. State the domain and range.

1. $f(x) = 3^{x-3} + 2$ 1, 2. See margin.

2. $f(x) = 2\left(\frac{3}{4}\right)^{x+1} - 3$

Solve each equation or inequality. Round to the nearest ten-thousandth if necessary.

3. $8^{c+1} = 16^{2c+3}$ $c = -\frac{9}{5}$

4. $9^{x-2} > \left(\frac{1}{27}\right)^x$ $\left\{x \mid x > \frac{4}{5}\right\}$

5. $2^{a+3} = 3^{2a-1}$ $a \approx 2.1130$

6. $\log_2 (x^2 - 7) = \log_2 6x$ $x = 7$

7. $\log_5 x > 2$ $\{x \mid x > 25\}$

8. $\log_3 x + \log_3 (x - 3) = \log_3 4$ $x = 4$

9. $6^{n-1} \le 11^n$ $\{n \mid n \le -2.9560\}$

10. $4e^{2x} - 1 = 5$ $x \approx 0.2027$

11. $\ln (x + 2)^2 > 2$ $\{x \mid x < -4.7183 \text{ or } x > 0.7183, x \ne -2\}$

Use $\log_5 11 \approx 1.4899$ and $\log_5 2 \approx 0.4307$ to approximate the value of each expression.

12. $\log_5 44$ 2.3513

13. $\log_5 \frac{11}{2}$ 1.0592

14. POPULATION The population of a city 10 years ago was 150,000. Since then, the population has increased at a steady rate each year. The population is currently 185,000. a. $y = 185{,}000(1.0212)^x$

  a. Write an exponential function that could be used to model the population after x years if the population changes at the same rate.

  b. What will the population be in 25 years? about 312,566

15. Write $\log_9 27 = \frac{3}{2}$ in exponential form. $9^{\frac{3}{2}} = 27$

16. AGRICULTURE An equation that models the declining number of U.S. farms is $y = 3{,}962{,}520(0.98)^x$, where x is the number of years since 1960 and y is the number of farms. $\{b \mid b < 1\}$

  a. How can you tell that the number is declining?

  b. By what annual rate is the number declining? 2%

  c. Predict when the number of farms will be less than 1 million. in about 2028

17. MULTIPLE CHOICE What is the value of $\log_4 \frac{1}{64}$? A

A -3 C $\frac{1}{3}$

B $-\frac{1}{3}$ D 3

18. MULTIPLE CHOICE What is the next term in the geometric sequence below? C

$$10, \frac{5}{2}, \frac{5}{8}, \frac{5}{32} \cdots$$

A $\frac{5}{8}$ C $\frac{5}{128}$

B $\frac{5}{32}$ D $\frac{5}{256}$

19. Find the three geometric means between 6 and 1536. 24, 96, 384

20. Find the sum of the geometric series for which $a_1 = 15$, $r = \frac{2}{3}$, and $n = 5$. $\frac{1055}{27}$

21. Determine the type of function that would best model the data in the table. Then determine the regression equation.

x	1	5	15	25	60
y	5	50	80	100	120

logarithmic; $y = 4.7710 + 28.4716 \ln x$

22. MULTIPLE CHOICE What is the solution of $\log_4 16 - \log_4 x = \log_4 8$? B

A $\frac{1}{2}$ C 4

B 2 D 8

23. MULTIPLE CHOICE Which function is graphed below? C

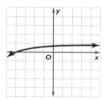

A $y = \log_{10} (x - 5)$ C $y = \log_{10} (x + 5)$

B $y = 5 \log_{10} x$ D $y = -5 \log_{10} x$

$\ln \dfrac{(6^2)(4^3)}{\left(\frac{1}{3}\right)^5}$ or $\ln 559{,}872$

24. Write $2 \ln 6 + 3 \ln 4 - 5 \ln \left(\frac{1}{3}\right)$ as a single logarithm.

Go Online! ✓

Chapter Tests
You can use premade leveled *Chapter Tests* to differentiate assessment for your students. Students can also take self-checking *Chapter Tests* to plan and prepare for chapter assessments.

MC = *multiple-choice questions*
FR = *free-response questions*

Form	Type	Level
1	MC	AL
2A	MC	OL
2B	FR	OL
2C	FR	OL
3	FR	BL
Vocabulary Test		
Extended-Response Test		

RtI Response to Intervention

Use the Intervention Planner to help you determine your Response to Intervention.

Intervention Planner

TIER 1 On Level OL

IF students miss 25% of the exercises or less,

THEN choose a resource:

  SE Lessons 6-1 through 6-10

  Go Online!

  📄 Skills Practice

  📄 Chapter Project

  ✓ Self-Check Quizzes

TIER 2 Strategic Intervention AL
Approaching grade level

IF students miss 50% of the exercises,

THEN ***Go Online!***

  📄 Study Guide and Intervention

  ➕ Extra Examples

  💬 Personal Tutors

  📄 Homework Help

TIER 3 Intensive Intervention
2 or more grades below level

IF students miss 75% of the exercises,

THEN choose a resource:

  Use *Math Triumphs, Alg. 1 MNE*

  Go Online!

  ➕ Extra Examples

  💬 Personal Tutors

  📄 Homework Help

  🔤 Review Vocabulary

Launch

Objective Apply concepts and skills from this chapter in a real-world setting.

Teach

Ask:

- **What is the best way to graph an exponential function by hand?** Sample answer: Make a table of values.

- **In the context of this situation, in the equation $y = ab^x$, what does each variable represent?** Sample answer: a is the starting amount, b is the growth rate, x is time, and y is the population after x amount of years since 1900.

- **How are the data values related?** Sample answer: Each term is 3 times greater than the one before it.

- **In the context of this situation, in the equation $y = ae^{kt}$, what do the variables y, a, k, and t represent?** Sample answer: a is the starting amount, k is the growth rate, t is the time, and y is the population after t years since 1980.

- **What is the first step when performing a regression analysis?** Sample answer: Input the values into a calculator.

- **A perfect fit for the data would have a regression coefficient with what value?** Sample answer: 1

The Performance Task focuses on the following content standards and standards for mathematical practice.

Dual Coding

Parts	Content Standards	Mathematical Practices
A	F.IF.7	1, 4, 6
B	F.BF.1, F.LE.4	1, 4, 6
C	A.SSE.4	1, 4, 6
D	F.BF.1, F.LE.4	1, 4, 6

Go Online!

eBook

Interactive Student Guide

Refer to *Interactive Student Guide* for an additional Performance Task.

ALGEBRA 2
INTERACTIVE STUDENT GUIDE

Preparing for Assessment

Performance Task

Provide a clear solution to each part of the task. Be sure to show all of your work, include all relevant drawings, and justify your answers.

APPLY MATH A United Nations committee commissioned a data analysis company to conduct analyses that will assess the health of various nations' economies, populations, and food supplies.

Part A

The company first does an analysis on the expected food supply for a coastal nation. Due to overfishing and a recent bacterial outbreak, the fish population in this region is expected to decline significantly over the next few years. The company creates the following equation to model the expected decline: $p = 1{,}000{,}000(1 - 0.2)^t$, where p is the fish population and t is the time in years after the study is conducted.

1. Determine the best increments to use on the x- and y-axes to graph the given equation.   See Ch. 6 Answer Appendix.
2. Graph the expected fish population for the next 5 years on a coordinate grid.   See Ch. 6 Answer Appendix.

Part B

The population of a certain country was about 650,000 in 1900. By 2000, the population of that country had increased to about 2.05 million.

3. Write an exponential function of the form $y = ab^x$ that could be used to model the data given above. Let x represent the number of years since 1900. Round the value of b to the nearest ten-thousandth if necessary. $y = 650{,}000(1.1217)^x$
4. Use your equation to estimate the nation's population in 2005. Round your answer to the nearest whole number.   3,666,371

Part C

The company does computer simulations on the infection rate of a certain virus that has an extremely high contagion rate. They run the simulation to see how many people in the populace they can expect to be infected after 1, 2, 3, 4, and 5 days if the virus breaks out. They collect the following data: 20, 60, 180, 540, 1620.

5. Determine the common ratio of the data.   3
6. Write an equation to find the nth term of the sequence. $a_n = 20(3)^{n-1}$
7. Determine how many people can be expected to be infected after 14 days.   31,886,460

Part D

The population of a certain region was about 820,000 in 1980. In 2010, the population had increased to about 1.78 million.

8. Write an exponential function of the form $y = ae^{kt}$ that could be used to model the data given above. Let t represent the number of years since 1980. Round all values to the nearest thousandth if necessary. $y = 820{,}000e^{0.026t}$
9. Using your equation, determine after how many years the population of the region would be 2.5 million. Round your answer to the nearest whole number.   43 years

Levels of Complexity Chart

The levels of the exercises progress from 1 to 3, with Level 1 indicating the lowest level of complexity.

Parts	Level 1	Level 2	Level 3
A		●	
B		●	
C	●		
D			●

Test-Taking Strategy

Example

Read the problem. Identify what you need to know. Then use the information in the problem to solve.

A certain can of soda contains 60 milligrams of caffeine. The caffeine is eliminated from the body at a rate of 15% per hour. What is the half-life of the caffeine? That is, how many hours does it take for half of the caffeine to be eliminated from the body?

A 4 hours   C 4.5 hours

B 4.25 hours   D 4.75 hours

Step 1 How would you normally solve a problem like this?
I would create an exponential decay function and solve it using logarithms.

Step 2 Is a calculator necessary? Would using a calculator be faster? Is there any part that would be quicker to do by hand?
A calculator is necessary to evaluate the log functions.

Step 3 What is the correct answer?
The answer is B.

> **Test-Taking Tip**
> Strategies for Using Technology
> Your calculator can be a useful tool in taking tests. Some problems that you encounter might have steps or computations that require the use of a calculator. A calculator may also help you solve a problem more quickly, especially in cases involving decimals, large numbers, or percents, as in the problem below.

Apply the Strategy

Read the problem. Identify what you need to know. Then use the information in the problem to solve.

Jason recently purchased a new truck for $34,750. The value of the truck decreases by 12% each year. What will the approximate value of the truck be 7 years after Jason purchased it? **D**

A $13,775   C $14,125

B $13,890   D $14,200

Answer the questions below.

a. How would you normally solve a problem like this?   I would create an exponential decay function and solve it using logarithms.

b. Is a calculator necessary? Would using a calculator be faster? Is there any part that would be quicker to do by hand?   A calculator is necessary to evaluate the log functions.

c. What is the correct answer?   **D**

Test-Taking Strategy

Step 1 Read the problem. Determine how you would normally solve this problem.

Step 2 Determine whether a calculator is required. Ask yourself whether a calculator would be faster or if there are parts of the problem that would be quicker to calculate by hand.

Step 3 Solve the problem.

Need Another Example?

A certain city had a population of 4590 in 1800. Since then, the population has grown by approximately 1.05% annually. In what year was the population of this city 37,075? **D**

A   1808
B   1820
C   1939
D   1999

a. How would you normally solve a problem like this? I would create an exponential growth function and solve it using logarithms.

b. Is a calculator necessary? Would using a calculator be faster? Is there any part that would be quicker to do by hand? A calculator is necessary to evaluate the log functions.

c. What is the correct answer? D

Go Online!

The most up-to-date resources available for your program can be found at connectED.mcgraw-hill.com.

Diagnose Student Errors

Survey student responses for each item. Class trends may indicate common errors and misconceptions.

1.	A	Solved $3(x + 2) = 4(x + 3)$
	B	Solved $8(x + 2) = 16(x + 3)$
	C	Solved $16(x + 2) = 8(x + 3)$
	D	CORRECT
3.	A	Misremembered the change of base formula
	B	Misremembered the change of base formula
	C	Misremembered the change of base formula
	D	Used an incorrect change of base formula
	E	CORRECT
5.	A	Misinterpreted the roles of the -4 and the 3 in the equation
	B	Did not recognize the impact of adding in the equation
	C	CORRECT
	D	Confused the domain and range; misinterpreted the roles of the -4 and the 3 in the equation
	E	Confused the domain and range
7.	A	Solved $4x^2 + 25 = 41$
	B	CORRECT
	C	Did not recognize the square of the negative was positive; believed the negative solution must be discarded because the log would not exist
	D	Solved $4x^2 + 25 = 41$ and then dismissed both solutions because they fail when checked
9.	A	Replaced p, q, and 5 with x, y, and 1
	B	Used multiplication instead of addition for the logarithm of a product
	C	Used multiplication instead of addition; did not replace 5 with the logarithm base 5 of 5
	D	CORRECT
	E	Did not replace 5 with the logarithm base 5 of 5
11.	A	Believed $a - bi$ and $-a - bi$ were conjugates
	B	Believed $a - bi$ and $-(a - bi)$ were conjugates
	C	CORRECT
	D	Believed $a - bi$ and $b - ai$ were conjugates
	E	Believed $a - bi$ and $b + ai$ were conjugates

Go Online!

Standardized Test Practice

Students can take self-checking tests in standardized format to plan and prepare for assessments.

CHAPTER 6
Preparing for Assessment
Cumulative Review

Read each question. Then fill in the correct answer on the answer document provided by your teacher or on a sheet of paper.

1. What is the value of x in this equation? D

$$16^{x+2} = 8^{x+3}$$

- ○ A -6
- ○ B -4
- ○ C -1
- ○ D 1

2. Rivka shares a funny meme on a social media site. If six of her friends share the meme, then six of each of their friends share the meme, and the pattern continues, how many times will the picture have been shared, including the time Rivka posted it, after eight rounds of sharing?

> 1,679,616

3. Which expression is equivalent to $\log_7 9$? E

- ○ A $(\log_{10} 9)(\log_{10} 7)$
- ○ B $(\log_{10} 9) + (\log_{10} 7)$
- ○ C $(\log_{10} 7)^{(\log_{10} 9)}$
- ○ D $(\log_{10} 9)^{(\log_{10} 7)}$
- ○ E $\dfrac{\log_{10} 9}{\log_{10} 7}$

4. Which are solutions to the inequality? Select all that apply. A, B, C, D

$$27^{x-1} \le 3^{2x}$$

- ☐ A -2
- ☐ B 0
- ☐ C 1
- ☐ D 3
- ☐ E 9

5. What are the domain and range of the function $y = 2^{x-4} + 3$? C

- ○ A D = {all real numbers}, R = {$y \mid y > -4$}
- ○ B D = {all real numbers}, R = {$y \mid y > 0$}
- ○ C D = {all real numbers}, R = {$y \mid y > 3$}
- ○ D D = {$x \mid x > -4$}, R = {all real numbers}
- ○ E D = {$x \mid x > 3$}, R = {all real numbers}

6.

x	$f(x)$		x	$f(x)$
1	6245		6	6
2	1560		7	1.5
3	390		8	0.25
4	100		9	0.1
5	25		10	0.02

Given the data in the table above, determine whether a linear, quadratic, exponential, or power regression would be most appropriate. Justify your answer.

> Exponential; see margin.

7. What is the solution set of this equation? B

$$\log_5 4x^2 + 2 = \log_5 (41)$$

- ○ A $\{-2, 2\}$
- ○ B $\left\{ -\dfrac{\sqrt{41}}{10}, \dfrac{\sqrt{41}}{10} \right\}$
- ○ C $\left\{ \dfrac{\sqrt{41}}{10} \right\}$
- ○ D ∅

8. Write an equation for the nth term of the geometric sequence below.

$$a_6 = 243; r = -3$$

> $a_n = -1(-3)^{n-1}$

12.	A	Made sign errors when solving and reversed the inequality
	B	Chose the incorrect lower boundary of the solution set
	C	CORRECT
	D	Chose the interval on which the logarithmic functions are defined
	E	Used the wrong relation and chose the incorrect boundary of the solution set
13.	A	Shifted left instead of down
	B	Shifted up instead of down
	C	CORRECT
	D	Graphed $y = \log_4 x - 8$

Additional Answer

6. Sample answer: An exponential regression has regression coefficient of $r = -0.9996$, which is the closest to 1, as opposed to the other options:

Linear: $r = -0.6488$
Quadratic: $r = -0.8756$
Power: $r = -0.9496$

Go Online! for
Standardized
Test Practice

9. Let $\log_5 p = x$ and $\log_5 q = y$. What is the value of $\log_5 \frac{pq}{5}$? **D**

- A xy
- B $xy - 1$
- C $xy - 5$
- D $x + y - 1$
- E $x + y - 5$

10. Leslie invested some money in an account that paid 2% annual interest compounded continuously. After 40 years, her investment was worth $6231.51. To the nearest whole dollar, what was the value of Leslie's original investment?

> 2800

11. Let $6 - 7i$ be a zero of the function $f(x)$. Which value must also be a zero of the function? **C**

- A $-6 - 7i$
- B $-6 + 7i$
- C $6 + 7i$
- D $7 - 6i$
- E $7 + 6i$

12. What is the solution set for the inequality $\log_6 (x - 2) < \log_6 (18 - 3x)$? **C**

- A $\{x \mid -8 < x < 2\}$
- B $\{x \mid -8 < x < 6\}$
- C $\{x \mid 2 < x < 5\}$
- D $\{x \mid 2 < x < 6\}$
- E $\{x \mid 5 < x < 6\}$

13. Which of the following graphs shows the equation $f(x) = \log_8 x - 4$? **C**

- A
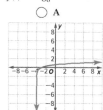

- B
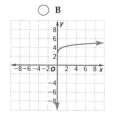

- C
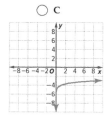

- D
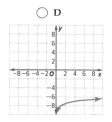

14. Find four geometric means between 0.0625 and 2048.

> 0.5, 4, 32, 256

15. Which equation represents the inverse of the function $y = 0.2^x$? **D**

- A $y = -0.2^x$
- B $y = 2^x$
- C $y = 5^x$
- D $y = \log_{\frac{1}{5}} x$
- E $y = \log_{\frac{1}{2}} x$

Need Extra Help?

If you missed Question...	1	2	3	4	5	6	7	8	9	10	11	12	13	14	15
Go to Lesson...	6-2	6-3	6-7	6-2	6-1	6-5	6-6	6-3	6-8	6-10	3-3	6-9	6-4	6-3	6-4

LS LEARNSMART®

Use LearnSmart as part of your test-preparation plan to measure student topic retention. You can create a student assignment in LearnSmart for additional practice on these topics.

Exercise Question Types	
Question Type	**Exercises**
Multiple-Choice	1, 3, 5, 7, 9, 11–13, 15
Multiple Correct Answers	4
Type Entry: Short Response	2, 6, 8, 10, 14

Formative Assessment

You can use these pages to benchmark student progress.

📄 Standardized Test Practice

Answer Sheet Practice

Have students simulate taking a standardized test by recording their answers on a practice recording sheet.

Homework Option

Get Ready for Chapter 7 Assign students the exercises on p. 466 as homework to assess whether they possess the prerequisite skills needed for the next chapter.

15.	A	Chose an option that used the additive inverse of 0.2
	B	Believed $0.2 = \frac{1}{2}$ and chose an option that used the reciprocal of $\frac{1}{2}$
	C	Chose an option that used the reciprocal of 0.2
	D	CORRECT
	E	Believed $0.2 = \frac{1}{2}$

Go Online!

eAssessment

Customize and create multiple versions of chapter tests and answer keys that align to the standards. Tests can be delivered on paper or online.

Lesson 6-1 (Guided Practice)

1.
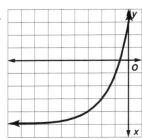

$D = (-\infty, \infty)$ or {all real numbers}; $R = (0, \infty)$ or $\{y \mid y > 0\}$

2A.
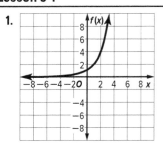

$D = (-\infty, \infty)$, {all real numbers}, or $\{-\infty < x < \infty\}$; $R = (-5, \infty), \{y \mid y > -5\}$, or $\{-5 < x < \infty\}$

2B.
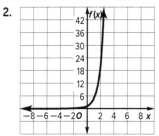

$D = (-\infty, \infty)$, {all real numbers}, or $\{-\infty < x < \infty\}$; $R = (-3, \infty)$ or $\{y \mid y > -3\}$

3.
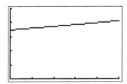

[0, 5] scl: 1 by [0, 300] scl: 60

Lesson 6-1

1.
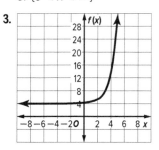

$D = (-\infty, \infty)$, {all real numbers}, or $\{-\infty < x < \infty\}$; $R = (0, \infty), \{f(x) \mid f(x) > 0\}$, or $\{0 < x < \infty\}$

2.
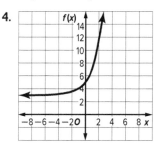

$D = (-\infty, \infty)$, {all real numbers}, or $\{-\infty < x < \infty\}$; $R = (0, \infty), \{f(x) \mid f(x) > 0\}$, or $\{0 < x < \infty\}$

3.
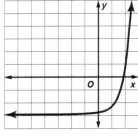

$D = (-\infty, \infty)$, {all real numbers}, or $\{-\infty < x < \infty\}$; $R = (4, \infty), \{f(x) \mid f(x) > 4\}$, or $\{4 < x < \infty\}$

4.

$D = (-\infty, \infty)$, {all real numbers}, or $\{-\infty < x < \infty\}$; $R = (3, \infty), \{f(x) \mid f(x) > 3\}$, or $\{3 < x < \infty\}$

5.
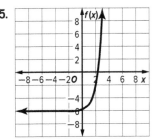

$D = (-\infty, \infty)$, {all real numbers}, or $\{-\infty < x < \infty\}$; $R = (-6, \infty), \{f(x) \mid f(x) > -6\}$, or $\{-6 < x < \infty\}$

6.
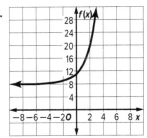

$D = (-\infty, \infty)$, {all real numbers}, or $\{-\infty < x < \infty\}$; $R = (8, \infty), \{f(x) \mid f(x) > 8\}$, or $\{8 < x < \infty\}$

8.
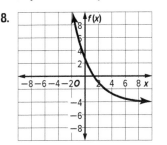

$D = (-\infty, \infty)$, {all real numbers}, or $\{-\infty < x < \infty\}$; $R = (-4, \infty), \{f(x) \mid f(x) > -4\}$, or $\{-4 < x < \infty\}$

9.
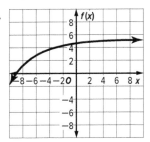

$D = (-\infty, \infty)$, {all real numbers}, or $\{-\infty < x < \infty\}$; $R = (-\infty, 5), \{f(x) \mid f(x) < 5\}$, or $\{-\infty < x < 5\}$

10.
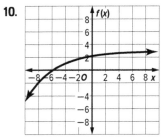

$D = (-\infty, \infty)$, {all real numbers}, or $\{-\infty < x < \infty\}$; $R = (-\infty, 3), \{f(x) \mid f(x) < 3\}$, or $\{-\infty < x < 3\}$

11.
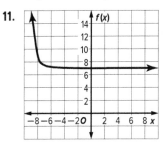

$D = (-\infty, \infty)$, {all real numbers}, or $\{-\infty < x < \infty\}$; $R = (7, \infty), \{f(x) \mid f(x) > 7\}$, or $\{7 < x < \infty\}$

13.
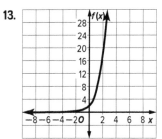

$D = (-\infty, \infty)$, {all real numbers}, or $\{-\infty < x < \infty\}$; $R = (0, \infty), \{f(x) \mid f(x) > 0\}$, or $\{0 < x < \infty\}$

14.
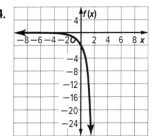

$D = (-\infty, \infty)$, {all real numbers}, or $\{-\infty < x < \infty\}$; $R = (0, \infty), \{f(x) \mid f(x) > 0\}$, or $\{0 < x < \infty\}$

15.
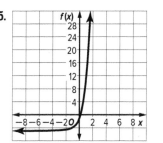

$D = (-\infty, \infty)$, {all real numbers}, or $\{-\infty < x < \infty\}$; $R = (-5, \infty), \{f(x) \mid f(x) > -5\}$, or $\{-5 < x < \infty\}$

16.
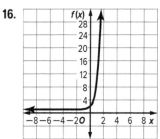

$D = (-\infty, \infty)$, {all real numbers}, or $\{-\infty < x < \infty\}$; $R = (-\infty, 1), \{f(x) \mid f(x) < 1\}$, or $\{-\infty < x < 1\}$

17.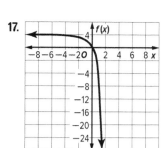

$D = (-\infty, \infty)$, {all real numbers},
or $\{-\infty < x < \infty\}$;
$R = (-\infty, 4)$, $\{f(x) \mid f(x) < 4\}$,
or $\{-\infty < x < 4\}$

18.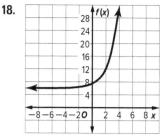

$D = (-\infty, \infty)$, {all real numbers},
or $\{-\infty < x < \infty\}$;
$R = (6, \infty)$, $\{f(x) \mid f(x) > 6\}$,
or $\{6 < x < \infty\}$

19.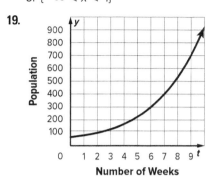

20.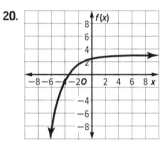

$D = (-\infty, \infty)$ or {all real numbers};
$R = (-\infty, 3)$, $\{f(x) \mid f(x) < 3\}$,
or $\{-\infty < x < 3\}$

21.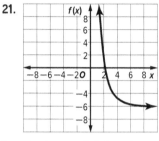

$D = (-\infty, \infty)$, {all real numbers},
or $\{-\infty < x < \infty\}$;
$R = (-6, \infty)$, $\{f(x) \mid f(x) > -6\}$,
or $\{-6 < x < \infty\}$

22.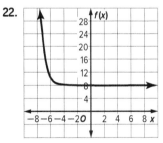

$D = (-\infty, \infty)$ or {all real numbers};
$R = (8, \infty)$, $\{f(x) \mid f(x) > 8\}$,
or $\{8 < x < \infty\}$

23.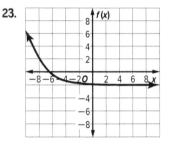

$D = (-\infty, \infty)$, {all real numbers},
or $\{-\infty < x < \infty\}$;
$R = (-2, \infty)$, $\{f(x) \mid f(x) > -2\}$,
or $\{-2 < x < \infty\}$

24.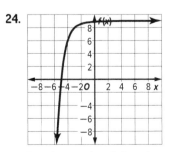

$D = (-\infty, \infty)$ or {all real numbers};
$R = (-\infty, 9)$, $\{f(x) \mid f(x) < 9\}$,
or $\{-\infty < x < 9\}$

25.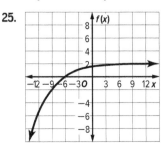

$D = (-\infty, \infty)$, {all real numbers},
or $\{-\infty < x < \infty\}$;
$R = (-\infty, 2)$, $\{f(x) \mid f(x) < 2\}$,
or $\{-\infty < x < 2\}$

33a.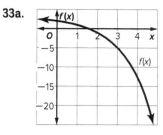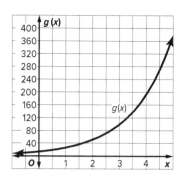

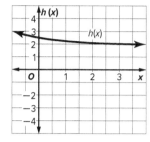

33b. Sample answer: $f(x)$; the graph of $f(x)$ is a reflection along the x-axis and the output values in the table are negative.

33d. Sample answer: $f(x)$ and $g(x)$ are growth and $h(x)$ is decay; The absolute value of the output is increasing for the growth functions and decreasing for the decay function.

Lesson 6-2

30b. Sample answer: Solve $A = 15,000(1 + 0.05)^{18}$ for A. Then multiply by 1.05 three times to determine the tuition cost for all four years. Tuition will be about \$36,099 in 18 years, and \$37,904, \$39,799, and \$41,789 for the next three years, for a total of \$155,591 for four years of college. Solve $155,591 = P(1 + 0.04)^{18}$ for P.

30c. Sample answer: I assumed that there are no additional depostis or withdrawals.

41b.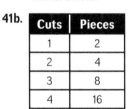

Cuts	Pieces
1	2
2	4
3	8
4	16

42. Sample answer: The more frequently interest is compounded, the higher the account balance becomes.

48.

$27^{2x} \cdot 81^{x+1} = 3^{2x+2} \cdot 9^{4x+1}$	Original equation
$(3^3)^{2x} \cdot (3^4)^{x+1} = 3^{2x+2} \cdot (3^2)^{4x+1}$	$3^2 = 9, 3^3 = 27$, and $3^4 = 81$
$3^{6x} \cdot 3^{4x+4} = 3^{2x+2} \cdot 3^{8x+2}$	Power of a Power
$3^{10x+4} = 3^{10x+4}$	Product of Powers
$10x + 4 = 10x + 4$	Property of Equality for Exponential Functions
$10x = 10x$	Subtract 4 from each side.
$x = x$	Divide each side by 10.

Lesson 6-4

37.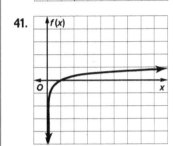

38.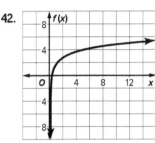

39.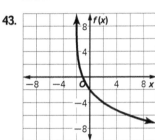

40.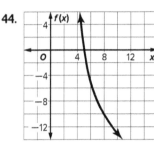

41.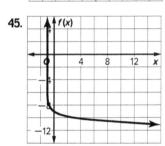

42.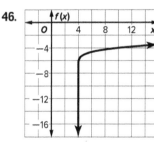

43.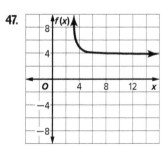

44.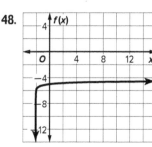

45.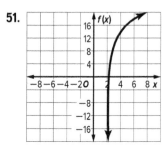

46.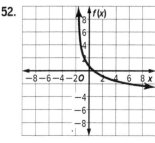

47.

48.

51.

52.

53.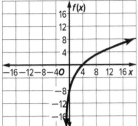

54.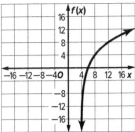

55.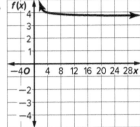

56.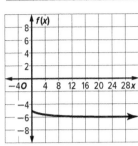

57c.

**Sales versus Money
Spent on Advertising**

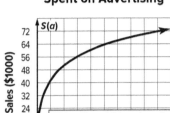

$$S(a) = 10 + 20 \log_4 (a + 1)$$

Advertising ($1000)

60a. The graph of $g(x)$ is a dilation of the graph of $f(x)$, which is stretched.

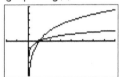

$[-2, 8]$ scl: 1 by $[-10, 10]$ scl: 2

60b. The graph of $g(x)$ is a dilation of the graph of $f(x)$, which is stretched vertically. This graph is also reflected across the x-axis.

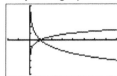

$[-2, 8]$ scl: 1 by $[-10, 10]$ scl: 2

60c. The graph of $g(x)$ is a translation of the graph of $f(x)$, which is translated up 4 units.

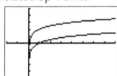

$[-2, 8]$ scl: 1 by $[-10, 10]$ scl: 2

60d. The graph of $g(x)$ is a translation of the graph of $f(x)$, which is translated down 6 units.

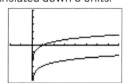

$[-2, 8]$ scl: 1 by $[-10, 10]$ scl: 2

60e. The graph of $g(x)$ is a translation of the graph of $f(x)$, which is translated left 5 units.

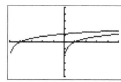

[−5, 5] scl: 1 by [−10, 10] scl: 2

60f. The graph of $g(x)$ is a translation of the graph of $f(x)$, which is translated right 8 units.

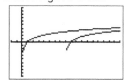

[−2, 18] scl: 1 by [−10, 10] scl: 1

64a–e. Sample answers given.

64a. $\log_2 33{,}554{,}432 = 25$

64b. $\log_4 \frac{1}{64} = -3$

64c. $\log_2 \sqrt{2} = \frac{1}{2}$

64d. $\log_7 1 = 0$

64e. There is no possible solution; this is the empty set.

65. No; Elisa was closer. She should have $-y = 2$ or $y = -2$ instead of $y = 2$. Matthew used the definition of logarithms incorrectly.

66. Sample answer: In $g(x) = a \log_{10}(x - h) + k$, the value of k is a vertical translation and the graph will shift up k units if k is positive and down $|k|$ units if k is negative. The value of h is a horizontal translation and the graph will shift h units to the right if h is positive and $|h|$ units to the left if h is negative. If $a < 0$, the graph will be reflected across the x-axis. If $|a| > 1$, the graph will be expanded vertically and if $0 < |a| < 1$, then the graph will be compressed vertically.

72a.

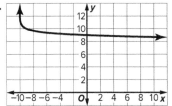

72b. All real numbers greater than or equal to 0

72c. All real numbers greater than or equal to 0

Mid-Chapter Quiz

1.

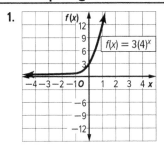

$f(x) = 3(4)^x$

$D = (-\infty, \infty)$ or {all real numbers}; $R = (0, \infty)$ or $\{y \mid y > 0\}$

2.

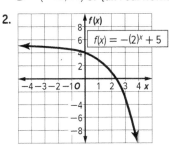

$f(x) = -(2)^x + 5$

$D = (-\infty, \infty)$ or {all real numbers}; $R = (-\infty, 5)$ or $\{f(x) \mid f(x) < 5\}$

3.

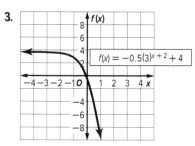

$f(x) = -0.5(3)^{x+2} + 4$

$D = (-\infty, \infty)$ or {all real numbers}; $R = (-\infty, 4)$ or $\{f(x) \mid f(x) < 4\}$

4.

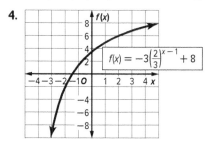

$f(x) = -3\left(\frac{2}{3}\right)^{x-1} + 8$

$D = (-\infty, \infty)$ or {all real numbers}; $R = (-\infty, 8)$ or $\{f(x) \mid f(x) < 8\}$

19.

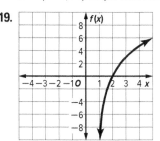

20.

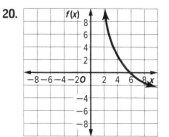

Lesson 6-6

59b. Sample answer: She will earn $1050 × 12 or $12,600 from Social Security. She needs to earn $50,000 − $12,600 or $37,400 from interest. The principal in her CD account will need to be $37,400 ÷ 0.05 or $748,000. Set up the exponential growth equation and solve for t.

$A = P(1 + r)^t$	Exponential Growth
$748{,}000 = 320{,}000(1.15)^t$	$A = 748{,}000, P = 320{,}000$, and $r = 0.15$
$2.3375 = 1.15^t$	Divide each side by 320,000.
$\log 2.3375 = \log 1.15^t$	Property of Equality of Logarithms
$\log 2.3375 = t \log 1.15$	Power Property of Logarithms
$\dfrac{\log 2.3375}{\log 1.15} = t$	Divide each side by log 1.15.
$6.07 \approx t$	Simplify.

59c. Sample answers: The annual increase in her account remains at 15% She needs to pay taxes on Social Security. Each CD will earn 5% interest.

66.

$\log_a \frac{1}{x} = -\log_a x$	Original equation
$\log_a x^{-1} = -\log_a x$	Definition of negative exponents
$\log_a x^{-1} = (-1)\log_a x$	Power Property of Logarithms
$\log_a \frac{1}{x} = -\log_a x$	Simplify.

67.
$$x^{3\log_x 2 - \log_x 5} = x^{\log_x 2^3 - \log_x 5}$$
$$= x^{\log_x 8 - \log_x 5}$$
$$= x^{\log_x \frac{8}{5}}$$
$$= \frac{8}{5}$$

68. Because logarithms are exponents, the properties of logarithms are similar to the properties of exponents. The Product Property states that to multiply two powers that have the same base, add the exponents. Similarly, the logarithm of a product is the sum of the logarithms of its factors. The Quotient Property states that to divide two powers that have the same base, subtract their exponents. Similarly the logarithm of a quotient is the difference of the logarithms of the numerator and the denominator. The Power Property states that to find the power of a power, multiply the exponents. Similarly, the logarithm of a power is the product of the logarithm and the exponent. Answers should include the following.

- Quotient Property: $\log_2\left(\frac{32}{8}\right) = \log_2\left(\frac{2^5}{2^3}\right)$ Replace 32 with 2^5 and 8 with 2^8.

$$= \log_2 2^{(5-3)} \quad \text{Quotient of Powers}$$
$$= 5 - 3 \text{ or } 2 \quad \text{Inverse Property of Exponents and Logarithms}$$

$\log_2 32 - \log_2 8 = \log_2 2^5 - \log_2 2^3$ Replace 32 with 2^5 and 8 with 2^3.

$$= 5 - 3 \text{ or } 2 \quad \text{Inverse Property of Exponents and Logarithms}$$

So, $\log_2\left(\frac{32}{8}\right) = \log_2 32 - \log_2 8$.

Power Property: $\log_3 9^4 = \log_3 (3^2)^4$ Replace 9 with 3^2.

$$= \log_3 3^{(2 \cdot 4)} \quad \text{Power of a Power}$$
$$= 2 \cdot 4 \text{ or } 8 \quad \text{Inverse Property of Exponents and Logarithms}$$

$4 \log_3 9 = (\log_3 9) \cdot 4$ Commutative Property ($\times$)

$$= (\log_3 3^2) \cdot 4 \quad \text{Replace 9 with } 3^2.$$
$$= 2 \cdot 4 \text{ or } 8 \quad \text{Inverse Property of Exponents and Logarithms}$$

So, $\log_3 9^4 = 4 \log_3 9$.

- The Product of Powers Property and Product Property of Logarithms both involve the addition of exponents, because logarithms are exponents.

Lesson 6-7

58. The graph of $g(x)$ is a dilation of the graph of $f(x)$, which is stretched vertically.

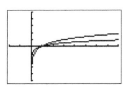

[−2, 8] scl: 1 by [−5, 5] scl: 1

59. The graph of $g(x)$ is a dilation of the graph of $f(x)$, which is stretched vertically. This graph is also reflected across the x-axis.

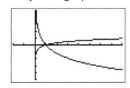

[−2, 8] scl: 1 by [−5, 5] scl: 1

60. The graph of $g(x)$ is a translation of the graph of $f(x)$, which is translated up 3 units.

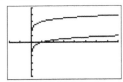

[−2, 8] scl: 1 by [−5, 5] scl: 1

61. The graph of $g(x)$ is a translation of the graph of $f(x)$, which is translated down 5 units.

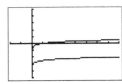

[−2, 8] scl: 1 by [−10, 10] scl: 2

62. The graph of $g(x)$ is a translation of the graph of $f(x)$, which is translated left 2 units.

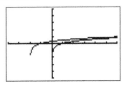

[−4, 6] scl: 1 by [−5, 5] scl: 1

63. The graph of $g(x)$ is a translation of the graph of $f(x)$, which is translated right 6 units.

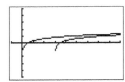

[−2, 18] scl: 2 by [−5, 5] scl: 1

83c. 2079. Yes; Sample answer: The year seems reasonable based on the model of the equation.

83d. The domain is all real numbers greater than −5. These numbers allow for the value inside the parenthesis to be positive.

83e. The range is all real numbers greater than or equal to 0. These numbers allow for a positive or 0 population.

Lesson 6-8

45. The graph of $g(x)$ is a dilation of the graph of $f(x)$, which is compressed vertically by a factor of 0.5.

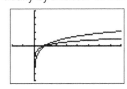

[−2, 8] scl: 1 by [−5, 5] scl: 1

46. The graph of $g(x)$ is a dilation of the graph of $f(x)$, which is compressed vertically by a factor of 0.25. This graph is also reflected across the x-axis.

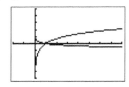

[−2, 8] scl: 1 by [−5, 5] scl: 1

47. The graph of *g(x)* is a translation of the graph of *f(x)*, which is translated up 8 units.

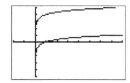

[−2, 8] scl: 1 by [−10, 10] scl: 2

48. The graph of *g(x)* is a translation of the graph of *f(x)*, which is translated down 9 units.

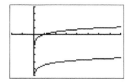

[−2, 8] scl: 1 by [−12, 8] scl: 2

49. The graph of *g(x)* is a translation of the graph of *f(x)*, which is translated left 5 units.

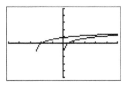

[−10, 10] scl: 2 by [−10, 10] scl: 2

50. The graph of *g(x)* is a translation of the graph of *f(x)*, which is translated right 4 units.

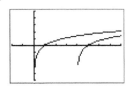

[−2, 8] scl: 1 by [−5, 5] scl: 1

STP

Part A

- The x-axis should be from 0 to 5 by 0.5s. The y-axis should be from 0 to 1,000,000 by 100,000s.

-

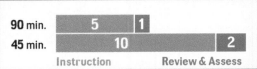
Track Your Progress

This chapter focuses on content from the Arithmetic with Polynomials and Rational Expressions, Creating Equations, and Reasoning with Equations and Inequalities domains.

THEN

A.CED.1 Create equations and inequalities in one variable and use them to solve problems.

F.IF.7e Graph exponential and logarithmic functions, showing intercepts and end behavior, and trigonometric functions, showing period, midline, and amplitude.

F.IF.8b Use the properties of exponents to interpret expressions for exponential functions.

F.LE.4 For exponential models, express as a logarithm the solution to $ab^{ct} = d$ where a, c, and d are numbers and the base b is 2, 10, or e; evaluate the logarithm using technology.

NOW

A.APR.7 Understand that rational expressions form a system analogous to the rational numbers, closed under addition, subtraction, multiplication, and division by a nonzero rational expression; add, subtract, multiply, and divide rational expressions.

A.CED.2 Create equations in two or more variables to represent relationships between quantities; graph equations on coordinate axes with labels and scales.

A.REI.2 Solve simple rational and radical equations in one variable, and give examples showing how extraneous solutions may arise.

NEXT

S.IC.4 Use data from a sample survey to estimate a population mean or proportion; develop a margin of error through the use of simulation models for random sampling.

S.IC.5 Use data from a randomized experiment to compare two treatments; use simulations to decide if differences between parameters are significant.

S.MD.6 Use probabilities to make fair decisions.

S.MD.7 Analyze decisions and strategies using probability concepts.

Standards for Mathematical Practice

All of the Standards for Mathematical Practice will be covered in this chapter. The MP icon notes specific areas of coverage.

Teaching the Mathematical Practices
Help students develop the mathematical practices by asking questions like these.

Questioning Strategies As students approach problems in this chapter, help them develop mathematical practices by asking:

Sense-Making
· What is a rational expression?
· What is the least common multiple of a polynomial?
· What are the properties of reciprocal functions?

Constructing Arguments
· How do you differentiate between a direct variation and joint variation problem?
· How do you differentiate between an inverse variation and combined variation problem?

Reasoning
· How do you simplify complex fractions?
· How do you add and subtract rational expressions?

Modeling
· What does the graph of the transformation of reciprocal functions look like?
· How do you graph rational functions with vertical and horizontal asymptotes?
· How do you graph rational functions with oblique asymptotes and point discontinuity?

Using Tools
· How do you solve rational equations and rational inequalities?

Go Online!

 StudySync:
SMP Modeling Videos

These demonstrate how to apply the Standards for Mathematical Practice to collaborate, discuss, and solve real-world math problems.

Go Online!
connectED.mcgraw-hill.com

 LearnSmart

 The Geometer's Sketchpad

 Vocabulary

 Personal Tutor

 Tools

Calculator Resources

Self-Check Practice

 Animations

Customize Your Chapter

Use the *Plan & Present*, *Assignment Tracker*, and *Assessment* tools in ConnectED to introduce lesson concepts, assign personalized practice, and diagnose areas of student need.

Differentiated Instruction

Throughout the program, look for the icons to find specialized content designed for your students.

- **AL** Approaching Level
- **OL** On Level
- **BL** Beyond Level
- **ELL** English Language Learners

Personalize

Differentiated Resources				
FOR EVERY CHAPTER	**AL**	**OL**	**BL**	**ELL**
✓ Chapter Readiness Quizzes	●	●	◑	●
✓ Chapter Tests	●	●	●	●
✓ Standardized Test Practice	●	●	●	●
Vocabulary Review Games	●	●	●	●
Anticipation Guide (English/Spanish)	●	●	●	●
Student-Built Glossary	●	●	●	●
Chapter Project	●	●	●	●
FOR EVERY LESSON	**AL**	**OL**	**BL**	**ELL**
Personal Tutors (English/Spanish)	●	●	●	●
Graphing Calculator Personal Tutors	●	●	●	●
▷ Step-by-Step Solutions	●	●	◑	●
✓ Self-Check Quizzes	●	●	●	●
5-Minute Check	●	●	●	●
Study Notebook	●	●	●	●
Study Guide and Intervention	●	●	◑	●
Skills Practice	●	◑	◑	●
Practice	◑	●	●	●
Word Problem Practice	●	●	●	◑
Enrichment	◑	●	●	●
✚ Extra Examples	●	◑	◑	◑
Lesson Presentations	●	●	●	●

◑ Aligned to this group   ● Designed for this group

Engage

Featured IWB Resources

Geometer's Sketchpad provides students with a tangible, visual way to learn. *Use with Lessons 7-3 through 7-5.*

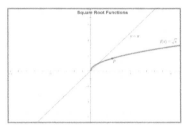

Animations help students make important connections through motion. *Use with Lessons 7-1, 7-4, and 7-6.*

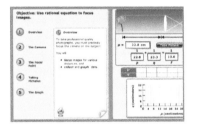

eLessons engage students and help build conceptual understanding of big ideas. *Use with Lessons 7-1 through 7-4.*

Time Management How long will it take to use these resources? Look for the clock in each lesson interleaf.

Introduce the Chapter

Mathematical Background

Rational expressions are ratios of two polynomial expressions. Operations with rational expressions are similar to operations with fractions. The graphs of some rational functions have breaks in continuity and may have vertical and horizontal asymptotes. Rational equations can be solved as polynomial equations once the fractions are eliminated by multiplying by the LCD.

Essential Question

At the end of this chapter, students should be able to answer the Essential Question.

Why are graphs useful? Sample answer: Graphs are useful because they can help you visualize relationships between real-world quantities. They can also be used to estimate function values.

Apply Math to the Real World

TRAVEL In this activity, students will use what they already know about rational functions to explore how these functions can be useful to travelers. Have students complete this activity individually or in small groups. 1

CHAPTER 7
Rational Functions

THEN
You used factoring to solve quadratic equations and you graphed quadratic equations.

NOW
You will:
- Simplify rational expressions.
- Graph rational functions.
- Solve direct, joint, and inverse variation problems.
- Solve rational equations and inequalities.

 WHY

TRAVEL Whatever way you travel, mathematical functions can be used to find distance traveled, time spent traveling, and speed.

Use the Mathematical Practices to complete the activity.

1. **Using Tools** Use the Internet to find the National Park closest to your home. Use a mapping tool to determine the distance to that park. How long will it take to get there? Record the information you learn in a KWL chart.

Distance and Time to National Park		

2. **Applying Math** Set up an equation to find the time it takes to reach the park if your speed is 50 mph.

3. **Discuss** Compare your answer to the one provided by the mapping tool. If there are differences, why do you think that is?

Daniel Grill/Getty Images

ALEKS®

Your Student Success Tool ALEKS is an adaptive, personalized learning environment that identifies precisely what each student knows and is ready to learn—ensuring student success at all levels.

- **Formative Assessment:** Dynamic, detailed reports monitor students' progress toward standards mastery.
- **Automatic Differentiation:** Strengthen prerequisite skills and target individual learning gaps.
- **Personalized Instruction:** Supplement in-class instruction with personalized assessment and learning opportunities.

Go Online!

Chapter Project

Stop and Smell the Roses Students use what they have learned about rational functions to complete a project.

This chapter project addresses financial literacy, as well as several specific skills identified as being essential to student success by the Framework for 21st Century Learning. 1, 3, 4

Go Online to Guide Your Learning

Explore & Explain	Organize

Graphing Tools: Reciprocal Functions

Use the **Graphing Tools: Reciprocal Functions** to explore the graphs of reciprocal functions in Lesson 7-3.

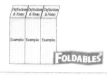

The Geometer's Sketchpad

Use **The Geometer's Sketchpad** to explore graphs of reciprocal and rational functions and to learn more about variation functions in Lesson 7-5.

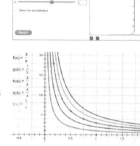

Interactive Student Guide

Before starting the chapter, answer the **Chapter Focus** preview questions. Check your answers as you complete each lesson. At the end of the chapter, try the **Performance Task**.

Foldables

Get organized! Create this **Rational Functions Foldable** before you start the chapter to help you organize your notes about rational functions.

Collaborate

Chapter Project

In the **Stop and Smell the Roses** project, you will use what you have learned about rational functions to complete a project that addresses financial literacy.

Focus

LEARNSMART

Need help studying? Complete the **Polynomial, Rational, and Radical Relationships** and the **Modeling with Functions** domains in LearnSmart to review for the chapter test.

ALEKS

You can use the **Rational Expressions with Functions** topic in ALEKS to explore what you know about relations and functions and what you are ready to learn.*

270

* Ask your teacher if this is part of your program.

Dinah Zike's FOLDABLES

Focus Students write notes about rational functions and relations in this chapter.

Teach Have students make and label their Foldables as illustrated. Students should use the appropriate tab to record their notes and examples for the concepts in each lesson of this chapter.

When to Use It Encourage students to add to their Foldables as they work through the chapter and to use them to review for the chapter test.

Go Online!

Notebooking with Foldables

Save a tree! In this video, you will learn tips for decreasing the amount of paper used when creating notebooks with Foldables. **MP** 5

Get Ready for the Chapter

Response to Intervention

Use the Concept Check results and the Intervention Planner chart to help you determine your Response to Intervention.

Intervention Planner

TIER 1 On Level

IF students miss 25% of the exercises or less,

THEN choose a resource:

Go Online!
- Skills Practice, Chapter 1
- Chapter Project

TIER 2 Approaching Level

IF students miss 50% of the exercises,

THEN choose a resource:

Go Online!
- Study Guide and Intervention, Chapter 1
- Extra Examples
- Personal Tutors
- Homework Help

TIER 3 Intensive Intervention

IF students miss 75% of the exercises,

THEN Use *Math Triumphs, Alg. 2,*

Go Online!
- Extra Examples
- Personal Tutors
- Homework Help
- Review Vocabulary

Additional Answers

1. Sample answer: multiply each side by 8
2. Sample answer: divide each side by 7
3. The GCF of 72 and 77 is 1.
4. the LCD
5. $a = -4$, $b = 16$, $c = -1$
6. Distributive Property
7. Associative Property
8. Line B is the function and line A is its inverse.
9. Write the equation $5(11) = 8p$ and then solve for p.

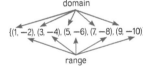 **Go Online!** for Vocabulary Review Games and key vocabulary in 13 languages.

Get Ready for the Chapter

Connecting Concepts	New Vocabulary	

Concept Check

Review the concepts used in this chapter by answering the questions below. 1–9. See margin.

1. What step would you take first to solve $\frac{9}{11} = \frac{7}{8}r$ for r?

2. What step would you take first to solve $\frac{72}{11} = 7r$ for r?

3. How do you know that $\frac{72}{77} = r$ is in simplest form?

4. What do you need to determine in order to simplify the expression $\frac{1}{3} + \frac{3}{4} + \frac{5}{6}$?

5. When applying the Quadratic Formula to $-4x^2 + 16x - 1$, what are the values for a, b, and c?

6. What property would you use to rewrite the expression $\frac{1}{4(4b + 6)}$ without parentheses?

7. Given $0 = 7x^2 + 7x - 9x - 9$, what property can you apply to begin to simplify the equation?

8. In the graph shown, one line is $f(x) = -4x$ and the other is the inverse of that function. Which line is the function and which is its inverse?

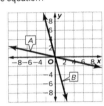

9. How could you use cross products to solve the proportion $\frac{5}{8} = \frac{p}{11}$ for p?

New Vocabulary

English		Español
rational expression	p. 467	expresión racional
complex fraction	p. 470	fracción compleja
reciprocal function	p. 483	función recíproco
hyperbola	p. 483	hipérbola
rational function	p. 491	función racional
vertical asymptote	p. 491	asíntota vertical
horizontal asymptote	p. 491	asíntota horizontal
oblique asymptote	p. 493	asíntota oblicua
point discontinuity	p. 494	discontinuidad evitable
direct variation	p. 500	variación directa
constant of variation	p. 500	constante de variación
joint variation	p. 501	variación conjunta
inverse variation	p. 502	variación inversa
combined variation	p. 503	variación combinada
rational equation	p. 508	ecuación racional
weighted average	p. 510	media ponderada
rational inequality	p. 513	desigualdad racional

Performance Task Preview

You can use the concepts and skills in this chapter to help a car company evaluate the safety and efficiency of its vehicles. Understanding rational functions will help you finish the Performance Task at the end of the chapter.

In this Performance Task you will:
- make sense of problems and persevere in solving them
- model with mathematics
- attend to precision

Review Vocabulary

function *función* a relation in which each element of the domain is paired with exactly one element of the range

domain

$\{(1, -2), (3, -4), (5, -6), (7, -8), (9, -10)\}$

range

least common multiple *mínimo común múltiplo* the least number that is a common multiple of two or more numbers

rational number *número racional* a number expressed in the form $\frac{a}{b}$, where a and b are integers and $b \neq 0$

Key Vocabulary ELL

Introduce the key vocabulary in the chapter using the routine below.

Define An asymptote is a line that a graph approaches.

Example The graph shows asymptotes at $x = -3$ and $f(x) = 2$.

Here, $f(x) = \dfrac{1}{(x + 3) + 2}$. It is less confusing to simply say $y = 2$.

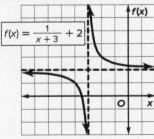

$f(x) = \dfrac{1}{x + 3} + 2$

Ask **What does a vertical asymptote show?** where a function is undefined

LESSON 7-1
Multiplying and Dividing Rational Expressions

SUGGESTED PACING (DAYS)

90 min.	0.75
45 min.	1.5

Instruction

Track Your Progress

Objectives

1 Simplify rational expressions.

2 Simplify complex fractions.

Mathematical Background

The primary skill needed to multiply and divide rational expressions is simplifying. After division is changed to multiplication by the reciprocal of the divisor, and numerators and denominators are multiplied, complete the problem by dividing by the common factors.

THEN

A.APR.1 Understand that polynomials form a system analogous to the integers, namely, they are closed under the operations of addition, subtraction, and multiplication; add, subtract, and multiply polynomials.

NOW

A.APR.7 (+) Understand that rational expressions form a system analogous to the rational numbers, closed under addition, subtraction, multiplication, and division by a nonzero rational expression; add, subtract, multiply, and divide rational expressions.

NEXT

A.CED.2 Create equations in two or more variables to represent relationships between quantities; graph equations on coordinate axes with labels and scales.

F.IF.4 For a function that models a relationship between two quantities, interpret key features of graphs and tables in terms of the quantities, and sketch graphs showing key features given a verbal description of the relationship. Key features include: intercepts; intervals where the function is increasing, decreasing, positive, or negative; relative maximums and minimums; symmetries; end behavior; and periodicity.*

F.BF.3 Identify the effect on the graph of replacing $f(x)$ by $f(x) + k$, $k\,f(x)$, $f(kx)$, and $f(x + k)$ for specific values of k (both positive and negative); find the value of k given the graphs. Experiment with cases and illustrate an explanation of the effects on the graph using technology. Include recognizing even and odd functions from their graphs and algebraic expressions for them.

Go Online! All of these resources and more are available at connectED.mcgraw-hill.com

eLessons utilize the power of your interactive whiteboard in an engaging way. Use **Rational Expressions**, screen 7, to introduce the concepts in this lesson.

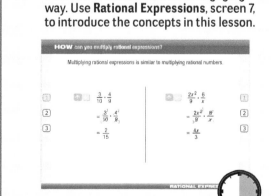

Use at Beginning of Lesson

Animations illustrate key concepts through step-by-step tutorials and videos.

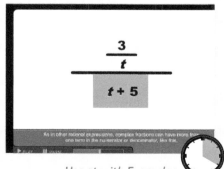

Use at with Examples

Chapter Project allows students to create and customize a project as a nontraditional method of assessment.

Use at the End of Lesson

OER Using Open Educational Resources

Video Conferencing Have absent students view lessons they have missed on **Wetoku**. **Wetoku** offers a way for students to see the teacher's presentation and the class' participation at the same time. It does not require any software installation. All calls are automatically recorded, so it can also be used for struggling students who need the lesson repeated. *Use as remediation*

Differentiate Your Resources

Extra Practice Additional practice or homework; Skills Practice is best for approaching-level students and Practice is best for on-level and beyond-level students

Skills Practice

7-1 Skills Practice
Multiplying and Dividing Rational Expressions

Practice

7-1 Practice
Multiplying and Dividing Rational Expressions

Word Problem Practice

7-1 Word Problem Practice
Multiplying and Dividing Rational Expressions

Intervention Reteaching and vocabulary activities that can be used with struggling or absent students and as ELL support

Study Guide and Intervention

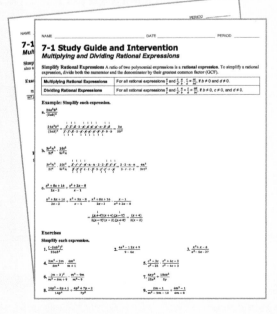

7-1 Study Guide and Intervention
Multiplying and Dividing Rational Expressions

Study Notebook

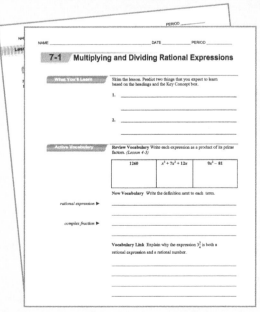

7-1 Multiplying and Dividing Rational Expressions

Extension Activities that can be used to extend lesson concepts

Enrichment

7-1 Enrichment
Dimensional Analysis

LESSON 1
Multiplying and Dividing Rational Expressions

:·Then	**:·Now**	**:·Why?**

- You factored polynomials.

1 Simplify rational expressions.

2 Simplify complex fractions.

If a scuba diver goes to depths greater than 33 feet, the rational function $T(d) = \frac{1700}{d-33}$ gives the maximum time a diver can remain at those depths and still surface at a steady rate with no stops. $T(d)$ represents the dive time in minutes and d represents the depth in feet.

New Vocabulary

rational expression
complex fraction

Mathematical Practices

8 Look for and express regularity in repeated reasoning.

Content Standards
A.APR.7 Understand that rational expressions form a system analogous to the rational numbers, closed under addition, subtraction, multiplication, and division by a nonzero rational expression; add, subtract, multiply, and divide rational expressions.

1 **Simplify Rational Expressions** A ratio of two polynomial expressions such as $\frac{1700}{d-33}$ is called a **rational expression**.

Because variables in algebra often represent real numbers, operations with rational numbers and rational expressions are similar. Just as with reducing fractions, to simplify a rational expression, you divide the numerator and denominator by their greatest common factor (GCF).

$$\frac{8}{12} = \frac{2 \cdot \overset{1}{4}}{3 \cdot \underset{1}{4}} = \frac{2}{3} \qquad \frac{x^2 - 4x + 3}{x^2 - 6x + 5} = \frac{(x-3)(\overset{1}{x-1})}{(x-5)(\underset{1}{x-1})} = \frac{(x-3)}{(x-5)}$$

GCF = 4 | GCF = (x − 1)

A.APR.7

Example 1 Simplify a Rational Expression

a. Simplify $\frac{5x(x^2 + 4x + 3)}{(x-6)(x^2-9)}$.

$$\frac{5x(x^2 + 4x + 3)}{(x-6)(x^2-9)} = \frac{5x(x+3)(x+1)}{(x-6)(x+3)(x-3)}$$ — Factor numerator and denominator.

$$= \frac{5x(x+1)}{(x-6)(x-3)} \cdot \frac{\overset{1}{(x+3)}}{\underset{1}{(x+3)}}$$ — Eliminate common factors.

$$= \frac{5x(x+1)}{(x-6)(x-3)}$$ — Simplify.

b. Under what conditions is this expression undefined?

The original factored denominator is $(x-6)(x+3)(x-3)$. Determine the values that would make the denominator equal to 0. These values are 6, −3, or 3, so the expression is undefined when $x = 6$, −3, or 3.

Guided Practice **1A.** $\frac{4(y+4)}{(y+2)}$; $y \neq 0, -2,$ or 3 **1B.** $\frac{2z(z+4)}{(z-1)}$; $z \neq 1, 2,$ or −5

Simplify each expression. Under what conditions is the expression undefined?

1A. $\frac{4y(y-3)(y+4)}{y(y^2-y-6)}$ | **1B.** $\frac{2z(z+5)(z^2+2z-8)}{(z-1)(z+5)(z-2)}$

Mathematical Practices Strategies

Make sense of problems and persevere in solving them.
Help students apply previous skills and knowledge to the more complicated practice of dividing polynomials. For example, ask:

- State a similarity between dividing polynomials and multiplying fractions. In both cases you cross out common factors.
- How do you keep your work organized? Some students may mention color coding. Others may say that they find it easier to work on unlined paper to reduce the clutter on the page.
- For what values is a quotient of rational expressions undefined? Why is the expression undefined for these values? A quotient of rational expressions is undefined for values of x that make the denominator equal to 0. The quotient is undefined because you cannot divide by zero.

Launch

Have students read the Why? section of the lesson. Ask:

- How can the term *rational expression* help you identify what it means? The word *rational* contains the word *ratio*.
- What ratio is in the function $T(d) = \frac{1700}{d-33}$? $\frac{1700}{d-33}$

Teach

Ask the scaffolded questions for each example to build conceptual understanding for students at all levels.

1 Simplify Rational Expressions

Example 1 Simplify a Rational Expression

AL What does it mean to simplify a rational expression? Cancel any common factors in the numerator and denominator.

OL How do you determine where the expression is undefined? Find the values that make the denominator equal 0.

BL Why can you cancel out common factors and still have an equivalent expression? Because you are really multiplying the rational expression by 1, and anything multiplied by 1 is itself.

Need Another Example?

a. Simplify $\frac{3y(y+7)}{(y+7)(y^2-9)} \cdot \frac{3y}{y^2-9}$

b. Under what conditions is this expression undefined? when $y = -7$, $y = -3$, or $y = 3$

Go Online!

Interactive Whiteboard
Use the *eLesson* or *Lesson Presentation* to present this lesson.

Example 2 Determine Undefined Values

AL How can you verify that a rational expression is undefined at a value? Substitute the value into the denominator and see if the denominator equals 0.

OL Does the numerator matter at all in answering this question? no

BL Write a rational expression that is undefined at $x = 0$, $x = 3$, and $x = -2$. $\dfrac{1}{x(x-3)(x+2)}$

Need Another Example?

For what value(s) is $\dfrac{p^2 + 2p - 3}{p^2 - 2p - 15}$ undefined? B

A 5   C 3, −5

B −3, 5   D 5, 1, −3

Example 3 Simplify Using −1

AL Write the solution without any parentheses.
$\dfrac{-w^2 - wy}{5w + y}$

OL How can you determine when you will be able to simplify a rational expression using −1? When in the numerator and denominator you have one term that is of the form $(a - b)$ and one of the form $(b - a)$.

BL Prove that $(a - b) = -1(b - a)$. By the commutative property of addition, $a - b = -b + a$. We can use the Distributive Property to see that $-b + a = -1(b - a)$. Therefore, $(a - b) = -1(b - a)$.

Need Another Example?

Simplify $\dfrac{a^4 b - 2a^4}{2a^3 - a^3 b}$. $-a$

Example 2   Determine Undefined Values   A.APR.7

For what value(s) is $\dfrac{x^2(x^2 - 5x - 14)}{4x(x^2 + 6x + 8)}$ undefined?

A −2, −4   C 0, −2, −4

B −2, 7   D 0, −2, −4, 7

Read the Item

You want to determine which values of x make the denominator equal to 0.

Study Tip

Eliminating Choices
Sometimes you can save
time by looking at the
possible answers and
eliminating choices.

Solve the Item

With $4x$ in the denominator, x cannot equal 0. So, choices A and B can be eliminated. Next, factor the denominator.

$x^2 + 6x + 8 = (x + 2)(x + 4)$, so the denominator is $4x(x + 2)(x + 4)$.

Because the denominator equals 0 when $x = 0$, −2, and −4, the answer is C.

> **Guided Practice**

2. For what value(s) of x is $\dfrac{x(x^2 + 8x + 12)}{-6(x^2 - 3x - 10)}$ undefined? B

   A 0, 5, −2   B 5, −2   C 0, −2, −6   D 5, −2, −6

Sometimes you can factor out −1 in the numerator or denominator to help simplify a rational expression.

                             A.APR.7

Example 3   Simplify Using −1

Simplify $\dfrac{(4w^2 - 3wy)(w + y)}{(3y - 4w)(5w + y)}$.

$\dfrac{(4w^2 - 3wy)(w + y)}{(3y - 4w)(5w + y)} = \dfrac{w(4w - 3y)(w + y)}{(3y - 4w)(5w + y)}$ Factor.

$= \dfrac{w(-1)(3y \overset{1}{-} 4w)(w + y)}{(3y - 4w)(5w + y)}$ $4w - 3y = -1(3y - 4w)$

$= \dfrac{(-w)(w + y)}{5w + y}$ Simplify.

> **Guided Practice**

Simplify each expression.

3A. $\dfrac{(xz - 4z)}{z^2(4 - x)}$ $-\dfrac{1}{z}$

3B. $\dfrac{ab^2 - 5ab}{(5 + b)(5 - b)}$ $-\dfrac{ab}{(5 + b)}$

The method for multiplying and dividing fractions also works with rational expressions. Remember that to multiply two fractions, you multiply the numerators and multiply the denominators. To divide two fractions, you multiply by the multiplicative inverse, or the reciprocal, of the divisor.

Multiplication

$\dfrac{2}{9} \cdot \dfrac{15}{4} = \dfrac{2 \cdot 3 \cdot 5}{3 \cdot 3 \cdot 2 \cdot 2} = \dfrac{5}{3 \cdot 2} = \dfrac{5}{6}$

Division

$\dfrac{3}{5} \div \dfrac{6}{35} = \dfrac{3}{5} \cdot \dfrac{35}{6} = \dfrac{3 \cdot 5 \cdot 7}{5 \cdot 2 \cdot 3} = \dfrac{7}{2}$

Watch Out!

Common Misconceptions Point out that rational expressions are usually used without specifically excluding those values that make the expression undefined. It is understood that only those values for which the expression has meaning are included.

The following table summarizes the rules for multiplying and dividing rational expressions.

Key Concept

Multiplying Rational Expressions	
Words	To multiply rational expressions, multiply the numerators and multiply the denominators.
Symbols	For all rational expressions $\frac{a}{b}$ and $\frac{c}{d}$ with $b \neq 0$ and $d \neq 0$, $\frac{a}{b} \cdot \frac{c}{d} = \frac{ac}{bd}$.
Dividing Rational Expressions	
Words	To divide rational expressions, multiply by the reciprocal of the divisor.
Symbols	For all rational expressions $\frac{a}{b}$ and $\frac{c}{d}$ with $b \neq 0$, $c \neq 0$, and $d \neq 0$, $\frac{a}{b} \div \frac{c}{d} = \frac{a}{b} \cdot \frac{d}{c} = \frac{ad}{bc}$.

Study Tip

Eliminating Common Factors Be sure to eliminate factors from both the numerator and denominator.

A.APR.7

Example 4   Multiply and Divide Rational Expressions

Simplify each expression.

a. $\dfrac{6c}{5d} \cdot \dfrac{15cd^2}{8a}$

$\dfrac{6c}{5d} \cdot \dfrac{15cd^2}{8a} = \dfrac{2 \cdot 3 \cdot c \cdot 5 \cdot 3 \cdot c \cdot d \cdot d}{5 \cdot d \cdot 2 \cdot 2 \cdot 2 \cdot a}$ Factor.

$= \dfrac{2 \cdot 3 \cdot c \cdot 5 \cdot 3 \cdot c \cdot d \cdot d}{5 \cdot d \cdot 2 \cdot 2 \cdot 2 \cdot a}$ Eliminate common factors.

$= \dfrac{3 \cdot 3 \cdot c \cdot c \cdot d}{2 \cdot 2 \cdot a}$ Simplify.

$= \dfrac{9c^2d}{4a}$ Simplify.

b. $\dfrac{18xy^3}{7a^2b^2} \div \dfrac{12x^2y}{35a^2b}$

$\dfrac{18xy^3}{7a^2b^2} \div \dfrac{12x^2y}{35a^2b} = \dfrac{18xy^3}{7a^2b^2} \cdot \dfrac{35a^2b}{12x^2y}$ Multiply by reciprocal of the divisor.

$= \dfrac{2 \cdot 3 \cdot 3 \cdot x \cdot y \cdot y \cdot y \cdot 5 \cdot 7 \cdot a \cdot a \cdot b}{7 \cdot a \cdot a \cdot b \cdot b \cdot 2 \cdot 2 \cdot 3 \cdot x \cdot x \cdot y}$ Factor.

$= \dfrac{2 \cdot 3 \cdot 3 \cdot x \cdot y \cdot y \cdot y \cdot 5 \cdot 7 \cdot a \cdot a \cdot b}{7 \cdot a \cdot a \cdot b \cdot b \cdot 2 \cdot 2 \cdot 3 \cdot x \cdot x \cdot y}$ Eliminate common factors.

$= \dfrac{3 \cdot 5 \cdot y \cdot y}{2 \cdot b \cdot x}$ Simplify.

$= \dfrac{15y^2}{2bx}$ Simplify.

Guided Practice

4A. $\dfrac{12d^2}{21b} \cdot \dfrac{14b}{8c^2}$ $\dfrac{d^2}{c^2}$

4B. $\dfrac{6y}{15a} \cdot \dfrac{21a}{18y}$ $\dfrac{7}{15}$

4C. $\dfrac{16m}{21a} \div \dfrac{24m}{7a}$ $\dfrac{2}{9}$

4D. $\dfrac{12x^4y^2}{40a^4b^4} \div \dfrac{6x^2y^4}{16a^2x}$ $\dfrac{4x^3}{5a^2b^4y^2}$

Example 4 Multiply and Divide Rational Expressions

AL What is the reciprocal of $\frac{c}{d}$? $\frac{d}{c}$

OL Does it matter if you eliminate common factors first or simplify first? No, but eliminating common factors first makes simplifying easier.

BL Why is dividing the same as multiplying by the reciprocal? Because $\dfrac{1}{\frac{a}{b}} = \dfrac{b}{a}$

Need Another Example?

Simplify each expression.

a. $\dfrac{8x}{21y^3} \cdot \dfrac{7y^2}{16x^3}$ $\dfrac{1}{6x^2y}$

b. $\dfrac{10mk^2}{3c^2d} \div \dfrac{5m^5}{6c^2d^2}$ $\dfrac{4dk^2}{m^4}$

Teaching Tip

Building on Prior Knowledge To help students understand why division is equivalent to multiplying by the reciprocal, discuss simple examples such as this: dividing 18 marbles between two people means that each person gets one half, or 9, of the marbles.

Example 5 Polynomials in the Numerator and Denominator

AL In part **a**, why does the answer have a 1 in the numerator? Because all of the factors in the numerator cancelled with a factor in the denominator, and when you cancel a factor you put a 1 in its place.

OL What would the rational expression simplify to if every factor in the numerator cancelled with a factor in the denominator? 1

BL Give two rational expressions that are multiplicative inverses of each other.

Example: $\dfrac{x^2 + 5}{x^3}$, $\dfrac{x^3}{x^2 + 5}$

Need Another Example?

Simplify each expression.

a. $\dfrac{k-3}{k+1} \cdot \dfrac{1-k^2}{k^2 - 4k + 3}$ -1

b. $\dfrac{2d+6}{d^2 + d - 2} \div \dfrac{d+3}{d^2 + 3d + 2}$ $\dfrac{2(d+1)}{d-1}$

(MP) Teaching the Mathematical Practices

Regularity Mathematically proficient students notice if calculations are repeated, and look both for general methods and for shortcuts. Encourage students to examine all of the factors involved, not just the factors in each expression individually.

Sometimes you must factor the numerator and/or the denominator first before you can simplify a product or a quotient of rational expressions.

A.APR.7

Example 5   Polynomials in the Numerator and Denominator

Simplify each expression.

Study Tip

(MP) **Regularity** When simplifying rational expressions, factors in one polynomial will often reappear in other polynomials. In Example 5a, $x - 8$ appears four times. Use this as a guide when factoring challenging polynomials.

a. $\dfrac{x^2 - 6x - 16}{x^2 - 16x + 64} \cdot \dfrac{x - 8}{x^2 + 5x + 6}$

$\dfrac{x^2 - 6x - 16}{x^2 - 16x + 64} \cdot \dfrac{x - 8}{x^2 + 5x + 6} = \dfrac{(x-8)(x+2)}{(x-8)(x-8)} \cdot \dfrac{x-8}{(x+3)(x+2)}$ Factor.

$= \dfrac{\cancel{(x-8)}\cancel{(x+2)}}{\cancel{(x-8)}(x-8)} \cdot \dfrac{\cancel{x-8}}{(x+3)\cancel{(x+2)}}$ Eliminate common factors.

$= \dfrac{1}{x+3}$ Simplify.

b. $\dfrac{x^2 - 16}{12y + 36} \div \dfrac{x^2 - 12x + 32}{y^2 - 3y - 18}$

$\dfrac{x^2 - 16}{12y + 36} \div \dfrac{x^2 - 12x + 32}{y^2 - 3y - 18} = \dfrac{x^2 - 16}{12y + 36} \cdot \dfrac{y^2 - 3y - 18}{x^2 - 12x + 32}$ Multiply by reciprocal.

$= \dfrac{(x+4)(x-4)}{12(y+3)} \cdot \dfrac{(y-6)(y+3)}{(x-4)(x-8)}$ Factor.

$= \dfrac{(x+4)\cancel{(x-4)}}{12\cancel{(y+3)}} \cdot \dfrac{(y-6)\cancel{(y+3)}}{\cancel{(x-4)}(x-8)}$ Eliminate common factors.

$= \dfrac{(x+4)(y-6)}{12(x-8)}$ Simplify.

▶ **Guided Practice**

5A. $\dfrac{8x^2 - 4x - 40}{x^2 + 2x - 35} \cdot \dfrac{x^2 - 7x + 10}{4x^2 - 16} \cdot \dfrac{2x-5}{x+7}$ **5B.** $\dfrac{x^2 - 9x + 20}{x^2 + 10x + 21} \div \dfrac{x^2 - x - 12}{6x^2 + 60x + 126} \cdot \dfrac{6x - 30}{x+3}$

2 **Simplify Complex Fractions** A **complex fraction** is a rational expression with a numerator and/or denominator that is also a rational expression. The following expressions are complex fractions.

$$\dfrac{\frac{c}{6}}{5d} \qquad \dfrac{\frac{8}{x}}{x-2} \qquad \dfrac{\frac{x-3}{8}}{\frac{x-2}{x+4}} \qquad \dfrac{\frac{4}{a} + 6}{\frac{12}{a} - 3}$$

To simplify a complex fraction, first rewrite it as a division expression.

A.APR.7

Example 6   Simplify Complex Fractions

Simplify each expression.

a. $\dfrac{\frac{a+b}{4}}{\frac{a^2 + b^2}{4}}$

$\dfrac{\frac{a+b}{4}}{\frac{a^2 + b^2}{4}} = \dfrac{a+b}{4} \div \dfrac{a^2 + b^2}{4}$ Express as a division expression.

$= \dfrac{a+b}{4} \cdot \dfrac{4}{a^2 + b^2}$ Multiply by the reciprocal.

$= \dfrac{a+b}{\cancel{4}} \cdot \dfrac{\cancel{4}}{a^2 + b^2}$ or $\dfrac{a+b}{a^2 + b^2}$ Simplify.

Differentiated Instruction **AL** **OL**

IF   students are having difficulty with these problems,

THEN   encourage them to use several steps, writing each one below the previous and keeping each line equivalent to the one above. Caution them to make only one change per step.

Go Online! ▶

Watch and listen to the animation to learn about simplifying complex fractions. Summarize what you hear for a partner, and have them ask you questions to help your understanding. **EL**

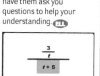

b. $\dfrac{\dfrac{x^2}{x^2-y^2}}{\dfrac{4x}{y-x}}$

$$\dfrac{\dfrac{x^2}{x^2-y^2}}{\dfrac{4x}{y-x}} = \dfrac{x^2}{x^2-y^2} \div \dfrac{4x}{y-x}$$ Express as a division expression

$$= \dfrac{x^2}{x^2-y^2} \cdot \dfrac{y-x}{4x}$$ Multiply by the reciprocal

$$= \dfrac{x \cdot x}{(x+y)(x-y)} \cdot \dfrac{(-1)(x-y)}{4x}$$ Factor

$$= \dfrac{x \cdot x}{(x+y)(x-y)} \cdot \dfrac{(-1)(x-y)}{4x}$$ Eliminate Factors

$$= \dfrac{-x}{4(x+y)}$$ Simplify

> **Guided Practice**

Simplify each expression.

6A. $\dfrac{\dfrac{(x-2)^2}{2(x^2-5x+4)}}{\dfrac{x^2-4}{4x-10}}$ $\dfrac{(2x-5)(x-2)}{(x+2)(x-4)(x-1)}$

6B. $\dfrac{\dfrac{x^2-y^2}{y^2-49}}{\dfrac{y-x}{y+7}}$ $\dfrac{-x-y}{y-7}$

Check Your Understanding   ◯ = Step-by-Step Solutions begin on page R11.

✓ **Go Online!** for a Self-Check Quiz

Example 1
A.APR.7

Simplify each expression. Under what conditions is the expression undefined?

1. $\dfrac{x^2-5x-24}{x^2-64} \cdot \dfrac{x+3}{x+8}$; $x \neq -8$

2. $\dfrac{c+d}{3c^2-3d^2} \cdot \dfrac{1}{3(c-d)}$; $c \neq d$

Example 2
A.APR.7

3. **MULTIPLE CHOICE** Identify all values of x for which $\dfrac{x+7}{x^2-3x-28}$ is undefined. **D**

A $-7, 4$ B $7, 4$ C $4, -7, 7$ D $-4, 7$

Examples 3–6
A.APR.7

Simplify each expression.

4. $\dfrac{y^2+3y-40}{25-y^2} \cdot \dfrac{y+8}{y+5}$

5. $\dfrac{a^2x-b^2x}{by-ay} \cdot \dfrac{-x(a+b)}{y}$

6. $\dfrac{27x}{16y} \cdot \dfrac{8z}{9x} \cdot \dfrac{3z}{2y}$

7. $\dfrac{12y}{13a} \div \dfrac{36x}{26b}$ $\dfrac{2by}{3ax}$

8. $\dfrac{x^2-4x-21}{x^2-6x+8} \cdot \dfrac{x-4}{x^2-2x-35}$ $\dfrac{x+3}{(x-2)(x+5)}$

9. $\dfrac{a^2-b^2}{3a^2-6a+3} \div \dfrac{4a+4b}{a^2-1}$ $\dfrac{(a-b)(a+1)}{12(a-1)}$

10. $\dfrac{\dfrac{a^3b^3}{xy^4}}{\dfrac{a^2b}{x^2y}}$ $\dfrac{ab^2x}{y^3}$

11. $\dfrac{\dfrac{4x}{x+6}}{\dfrac{x^2-3x}{x^2+3x-18}} - 4$

12. **MP** SENSE-MAKING The volume of a shipping container in the shape of a rectangular prism can be represented by the polynomial $6x^3 + 11x^2 + 4x$, where the height is x.

a. Find the length and width of the container if each dimension is a linear polynomial in x. $2x+1, 3x+4$

b. Find the ratio of the three dimensions of the container when $x = 2$. $2:5:10$

c. Will the ratio of the three dimensions be the same for all values of x? no

Volume = $6x^3 + 11x^2 + 4x$

---

2 **Simplify Complex Fractions**

Example 6 Simplify Complex Fractions

AL What does it mean to simplify a complex fraction? Write it as one fraction in lowest terms.

OL What operation is equivalent to the fraction bar? division

BL How do you simplify $\dfrac{\dfrac{a}{b}}{c}$? $\dfrac{a}{bc}$

Need Another Example?

Simplify $\dfrac{\dfrac{x^2}{9x^2-4y^2}}{\dfrac{x^3}{2y-3x}} \cdot \dfrac{-1}{3x^2+2xy}$

Teaching Tip

Reasoning Help students understand why the quotient of $(x-y)$ and $(y-x)$ is -1 by pointing out that these two expressions are opposites (or additive inverses), just like 2 and -2.

Practice

Formative Assessment Use Exercises 1–12 to assess students' understanding of the concepts in this lesson.

The Practice and Problem Solving exercises assess the content taught in the lesson. The Preparing for Assessment page is meant to be used as preparation for end-of-course assessments.

MP Teaching the Mathematical Practices

Sense-Making Mathematically proficient students start by explaining to themselves the meaning of a problem and looking for entry points to its solution. They analyze givens, constraints, relationships, and goals. They check their answers to problems using a different method, and they continually ask themselves, "Does this make sense?"

Go Online!   eBook

Interactive Student Guide

Use the *Interactive Student Guide* to deepen conceptual understanding.
· Multiplying and Dividing Rational Expressions

ALGEBRA 2
INTERACTIVE STUDENT GUIDE

Teaching the Mathematical Practices

Reasoning Mathematically proficient students make sense of quantities and their relationships in problem situations. Quantitative reasoning entails habits of creating a coherent representation of the problem at hand; considering the units involved; attending to the meaning of quantities, not just how to compute them; and knowing and flexibly using different properties of operations and objects.

Extra Practice

See page R7 for extra exercises for students who are approaching level or for on-level students who need additional reinforcement.

Levels of Complexity Chart

The levels of the exercises progress from 1 to 3, with Level 1 indicating the lowest level of complexity.

Exercises	13–39	40–52, 66–70	53–65
Level 3			●
Level 2		○	
Level 1	●		

Practice and Problem Solving — Extra Practice is on page R7.

Example 1
A.APR.7

Simplify each expression. Under what conditions is this expression undefined?

13. $\dfrac{x(x-3)(x+6)}{x^2+x-12} \quad \dfrac{x(x+6)}{x+4}; x \neq -4$

14. $\dfrac{y^2(y^2+3y+2)}{2y(y-4)(y+2)} \quad \dfrac{y(y+1)}{2(y-4)}; y \neq 4$

15. $\dfrac{(x^2-9)(x^2-z^2)}{4(x+z)(x-3)} \quad \dfrac{(x+3)(x-z)}{4}$

16. $\dfrac{(x^2-16x+64)(x+2)}{(x^2-64)(x^2-6x-16)} \quad \dfrac{1}{x+8}; x \neq -8$

17. $\dfrac{x^2(x+2)(x-4)}{6x(x^2+x-20)} \quad \dfrac{x(x+2)}{6(x+5)}; x \neq -5$

18. $\dfrac{3y(y-8)(y^2+2y-24)}{15y^2(y^2-12y+32)} \quad \dfrac{(y+6)}{5y}; y \neq 0$

Example 2
A.APR.7

19. MULTIPLE CHOICE Identify all values of x for which $\dfrac{(x-3)(x+6)}{(x^2-7x+12)(x^2-36)}$ is undefined. **D**

A 3, −6   B 4, 6   C −6, 6   D −6, 3, 4, 6

Example 3
A.APR.7

Simplify each expression.

20. $\dfrac{x^2-5x-14}{28+3x-x^2} \quad \dfrac{x+2}{x+4}$

21. $\dfrac{x^3-9x^2}{x^2-3x-54} \quad \dfrac{x^2}{x+6}$

22. $\dfrac{(x-4)(x^2+2x-48)}{(36-x^2)(x^2+4x-32)} \quad \dfrac{1}{x+6}$

23. $\dfrac{16-c^2}{c^2+c-20} \quad \dfrac{c+4}{c+5}$

24. GEOMETRY The cylinder at the right has a volume of $(x+3)(x^2-3x-18)\pi$ cubic centimeters. Find the height of the cylinder. $x - 6$ cm

2x + 6 cm

Examples 4–6
A.APR.7

Simplify each expression.

25. $\dfrac{3ac^3f^3}{8a^2bcf^4} \cdot \dfrac{12ab^2c}{18ab^3c^2f} \quad \dfrac{c}{4ab^2f^2}$

26. $\dfrac{14xy^2z^3}{21w^4x^2yz} \cdot \dfrac{7wxyz}{12w^2y^3z} \quad \dfrac{7z^2}{18w^5y}$

27. $\dfrac{64a^2b^5}{35b^2c^3f^4} \div \dfrac{12a^4b^3c}{70abcf^2} \quad \dfrac{32b}{3ac^3f^2}$

28. $\dfrac{9x^2yz}{5z^4} \div \dfrac{12x^4y^2}{50xy^4z^2} \quad \dfrac{15y^3}{2xz}$

29. $\dfrac{15a^2b^2}{21ac} \cdot \dfrac{14a^4c^2}{6ab^3} \quad \dfrac{5a^4c}{3b}$

30. $\dfrac{14c^2f^5}{9a^2} \div \dfrac{35cf^4}{18ab^3} \quad \dfrac{4b^3cf}{5a}$

31. $\dfrac{y^2+8y+15}{y-6} \cdot \dfrac{y^2-9y+18}{y^2-9} \quad y+5$

32. $\dfrac{c^2-6c-16}{c^2-d^2} \div \dfrac{c^2-8c}{c+d} \quad \dfrac{c+2}{c(c-d)}$

33. $\dfrac{x^2+9x+20}{8x+16} \cdot \dfrac{4x^2+16x+16}{x^2-25} \quad \dfrac{(x+4)(x+2)}{2(x-5)}$

34. $\dfrac{3a^2+6a+3}{a^2-3a-10} \div \dfrac{12a^2-12}{a^2-4} \quad \dfrac{(a+1)(a-2)}{4(a-5)(a-1)}$

35. $\dfrac{\dfrac{x^2-9}{6x-12}}{\dfrac{x^2+10x+21}{x^2-x-2}} \quad \dfrac{(x-3)(x+1)}{6(x+7)}$

36. $\dfrac{\dfrac{y-x}{z^3}}{\dfrac{x-y}{6z^2}} \quad -\dfrac{6}{z}$

37. $\dfrac{\dfrac{a^2-b^2}{b^3}}{\dfrac{b^2-ab}{a^2}} \quad \dfrac{-a^2(a+b)}{b^4}$

38. $\dfrac{\dfrac{x-y}{a+b}}{\dfrac{x^2-y^2}{b^2-a^2}} \quad \dfrac{b-a}{x+y}$

39. REASONING At the end of her high school soccer career, Ashley had made 33 goals out of 121 attempts.

  a. Write a ratio to represent the ratio of the number of goals made to goals attempted by Ashley at the end of her high school career. $\dfrac{33}{121}$

  b. Suppose Ashley attempted a goals and made m goals during her first year at college. Write a rational expression to represent the ratio of the number of goals made to the number of goals attempted from her first year in college and high school career combined. $\dfrac{33+m}{121+a}$

Differentiated Homework Options

Levels	AL Basic	OL Core	BL Advanced
Exercises	13–39, 59–70	13–39 odd	40–70
2-Day Option	13–39 odd, 66–70	13–39, 66–70	
	14–38 even, 59–65	40–65	

You can use ALEKS to provide additional remediation support with personalized instruction and practice.

Go Online!

eSolutions Manual

Create worksheets, answer keys, and solutions handouts for your assignments.

B **40. GEOMETRY** Parallelogram F has an area of $12x^2 - 10x - 42$ square meters and a height of $4x + 6$ meters. Parallelogram G has an area of $24x^2 + 22x - 10$ square meters and a height of $6x - 2$ meters. Find the area of right triangle H. $\frac{1}{2}(12x^2 - 13x - 35)$ m^2

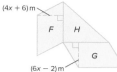

41. POLLUTION The thickness of an oil spill from a ruptured pipe on a rig is modeled by the function $T(x) = \frac{0.4(x^2 - 2x)}{x^3 + x^2 - 6x}$, where T is the thickness of the oil slick in meters and x is the distance from the rupture in meters.

a. Simplify the function. $T(x) = \frac{0.4}{x + 3}$

b. How thick is the slick 100 meters from the rupture? about 3.9 mm thick

Simplify each expression.

42. $\frac{x^2 - 16}{3x^3 + 18x^2 + 24x} \cdot \frac{x^3 - 4x}{2x^2 - 7x - 4} \quad \frac{x - 2}{3(2x + 1)}$

43. $\frac{3x^2 - 17x - 6}{4x^2 - 20x - 24} \div \frac{6x^2 - 7x - 3}{2x^2 - x - 3} \quad \frac{1}{4}$

44. $\frac{9 - x^2}{x^2 - 4x - 21} \cdot \left(\frac{2x^2 + 7x + 3}{2x^2 - 15x + 7}\right)^{-1} \quad \frac{(3 - x)(2x - 1)}{(x + 3)(2x + 1)}$

45. $\left(\frac{2x^2 + 2x - 12}{x^2 + 4x - 5}\right)^{-1} \cdot \frac{2x^3 - 8x}{x^2 - 2x - 35} \quad \frac{x(x + 2)(x - 1)}{(x + 3)(x - 7)}$

46. $\left(\frac{3xy^3z}{2a^2bc^2}\right)^3 \cdot \frac{16a^4b^3c^5}{15x^7yz^3} \quad \frac{18y^8}{5a^2cx^4}$

47. $\frac{15y^3}{4a^2cxz}$

(47) $\frac{20x^2y^6z^{-2}}{3a^3c^2} \cdot \left(\frac{16x^3y^3}{9acz}\right)^{-1}$

48. $\left(\frac{2xy^3}{3abc}\right)^{-2} \div \frac{6a^2b}{x^2y^4} \quad \frac{3bc^2}{8y^2}$

49. $\frac{\dfrac{8x^2 - 10x - 3}{10x^2 + 35x - 20}}{\dfrac{2x^2 + x - 6}{4x^2 + 18x + 8}} \quad \frac{2(4x + 1)(2x + 1)}{5(2x - 1)(x + 2)}$

50. $\frac{\dfrac{2x^2 + 7x - 30}{-6x^2 + 13x + 5}}{\dfrac{4x^2 + 12x - 72}{3x^2 - 11x - 4}} \quad \frac{x - 4}{-4(x - 3)}$

51. $\frac{\dfrac{4x^2 - 1}{3x^3 - 6x^2 - 24x}}{\dfrac{12x^2 + 12x - 9}{-2x^2 + 5x + 12}} \quad \frac{2x + 1}{-9x(x + 2)}$

52. GEOMETRY The area of the base of the rectangular prism at the right is 20 square centimeters.

a. Find the length of $\overline{BC}$ in terms of x. $\frac{20}{x}$

b. If $DC = 3BC$, determine the area of the shaded region in terms of x. $\frac{1200}{x^2}$

c. Determine the volume of the prism in terms of x. $\frac{1200}{x}$

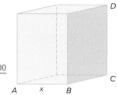

C **Simplify each expression.**

53. $\frac{x^2 + 4x - 32}{2x^2 + 9x - 5} \cdot \frac{3x^2 - 75}{3x^2 - 11x - 4} \div \frac{6x^2 - 18x - 60}{x^3 - 4x} \quad \frac{x(x - 2)(x + 8)}{2(2x - 1)(3x + 1)}$

54. $\frac{8x^2 + 10x - 3}{3x^2 - 12x - 36} \div \frac{2x^2 - 5x - 12}{3x^2 - 17x - 6} \cdot \frac{4x^2 + 3x - 1}{4x^2 - 40x + 24} \quad \frac{(4x - 1)^2(3x + 1)(x + 1)}{12(x + 2)(x - 4)(x^2 - 10x + 6)}$

55. $\frac{4x^2 - 9x - 9}{3x^2 + 6x - 18} \div \frac{-2x^2 + 5x + 3}{x^2 - 4x - 32} \div \frac{8x^2 + 10x + 3}{6x^2 - 6x - 12} \quad \frac{-2(x - 8)(x + 4)(x - 2)(x + 1)}{(2x + 1)^2(x^2 + 2x - 6)}$

56a. $r_s = \frac{d + 1}{t - \frac{1}{6}}$

56b. $r = \dfrac{\dfrac{d + 1}{t - \frac{1}{6}}}{\div \dfrac{d}{t}} = \dfrac{t(d + 1)}{d\left(t - \frac{1}{6}\right)}$

56c. $d = \dfrac{t}{t - \frac{1}{3}}$

56. MULTI-STEP Mike and Scott rode their bikes home from the park. As soon as Mike arrived, he sent Scott a text saying, "Home." Scott responded with, "Wow! I've been home for ten minutes." Shocked, Mike replies, "How did you do that? You live a mile farther away than I do!" Scott retorts, "Apparently, I ride my bike twice as fast as you!"

a. Write a rational expression representing Scott's average speed in miles per hour in terms of Mike's distance d and time t.

b. Find a rational expression describing the ratio of Scott's speed to Mike's speed.

c. Assuming Scott's statement is true, write a rational expression for Mike's distance in terms of his time.

d. If it took Mike 24 minutes to get home, how far did he ride? 6 miles

Exercise Alert

Formula For Exercise 52, students will need to know the formula for the volume of a rectangular prism, $V = Bh$, where B is the area of the base, and h is the height.

Assess

Crystal Ball Ask students to describe how they think their practice with multiplying and simplifying rational expressions will help them when they add or subtract rational expressions.

Additional Answers

58b.

x	0	1	2	3
f(x)	−1	0	1	2
g(x)	−1	0	1	2

x	4	5	6	7
f(x)	ERR	4	5	6
g(x)	3	4	5	6

x	8	9	10
f(x)	7	8	9
g(x)	7	8	9

58d.

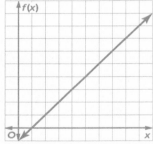

The graphs appear to be the same on the graphing calculator. But $f(x)$ is undefined for $f(4)$, and $g(4) = 3$.

62. $\dfrac{x+1}{\sqrt{x+3}}$ does not belong with the other three. The other three expressions are rational expressions. Since the denominator of $\dfrac{x+1}{\sqrt{x+3}}$ is not a polynomial, $\dfrac{x+1}{\sqrt{x+3}}$ is not a rational expression.

63. Sample answer: Sometimes; with a denominator like $x^2 + 2$, in which the denominator cannot equal 0, the rational expression can be defined for all values of x.

65. Sample answer: When the original expression was simplified, a factor of x was taken out of the denominator. If x were to equal 0, then this expression would be undefined. So, the simplified expression is also undefined for x = 0.

---

57 **TRAINS** Trying to get into a train yard one evening, all of the trains are backed up for 2 miles along a system of tracks. Assume that each car occupies an average of 75 feet of space on a track and that the train yard has 5 tracks.

  a. Write an expression that could be used to determine the number of train cars involved in the backup.   **57a.** $5 \text{ tracks} \cdot \dfrac{2 \text{ miles}}{1 \text{ track}} \cdot \dfrac{5280 \text{ feet}}{1 \text{ mile}} \cdot \dfrac{1 \text{ car}}{75 \text{ feet}}$

  b. How many train cars are involved in the backup?   **704**

  c. Suppose that there are 8 attendants doing safety checks on each car, and it takes each vehicle an average of 45 seconds for each check. Approximately how many hours will it take for all the vehicles in the backup to exit?   **70.4 h**

58. MULTIPLE REPRESENTATIONS In this problem, you will investigate the graph of a rational function. **b, d. See margin.**

  a. Algebraic Simplify $\dfrac{x^2 - 5x + 4}{x - 4}$. $x - 1$ **58c.** $f(4)$ results in an error because the function is undefined at $x = 4$. $g(4) = 3$

  b. Tabular Let $f(x) = \dfrac{x^2 - 5x + 4}{x - 4}$. Use the expression you wrote in part **a** to write the related function $g(x)$. Use a graphing calculator to make a table for both functions for $0 \le x \le 10$.

  c. Analytical What are $f(4)$ and $g(4)$? Explain the significance of these values.

  d. Graphical Graph the functions on the graphing calculator. Use the TRACE function to investigate each graph, using the ▲ and ▼ keys to switch from one graph to the other. Compare and contrast the graphs.

  e. Verbal What conclusions can you draw about the expressions and the functions? The expressions and functions are equivalent except for $x = 4$.

A.APR.7

H.O.T. Problems   Use Higher-Order Thinking Skills

59. Sample answer: The two expressions are equivalent, except that the rational expression is undefined at $x = 3$.

59. MP REASONING Compare and contrast $\dfrac{(x - 6)(x + 2)(x + 3)}{x + 3}$ and $(x - 6)(x + 2)$.

60. MP CRITIQUE ARGUMENTS Troy and Beverly are simplifying $\dfrac{x+y}{x-y} \div \dfrac{4}{y-x}$. Is either of them correct? Explain your reasoning.

Troy

$$\dfrac{x+y}{x-y} \div \dfrac{4}{y-x} = \dfrac{x-y}{x+y} \cdot \dfrac{4}{y-x}$$
$$= \dfrac{-4}{x+y}$$

Beverly

$$\dfrac{x+y}{x-y} \div \dfrac{4}{y-x} = \dfrac{x+y}{y-x} \cdot \dfrac{y-x}{4}$$
$$= -\dfrac{x+y}{4}$$

61. CHALLENGE Find the expression that makes the following statement true. $x^2 + x - 6$

$$\dfrac{x - 6}{x + 3} \cdot \dfrac{?}{x - 6} = x - 2$$

60. Sample answer: Beverly; Troy's mistake was multiplying by the reciprocal of the dividend instead of the divisor.

62. WHICH ONE DOESN'T BELONG? Identify the expression that does not belong with the other three. Explain your reasoning. **See margin.**

| $\dfrac{1}{x - 1}$ | $\dfrac{x^2 + 3x + 2}{x - 5}$ | $\dfrac{x + 1}{\sqrt{x + 3}}$ | $\dfrac{x^2 + 1}{3}$ |

63. MP REASONING Determine whether the following statement is *sometimes*, *always*, or *never* true. Explain your reasoning. **See margin.**

  A rational function that has a variable in the denominator is defined for all real values of x.

64. OPEN-ENDED Write a rational expression that simplifies to $\dfrac{x - 1}{x + 4}$.   Sample answer: $\dfrac{x^2 - 1}{x^2 + 5x + 4}$

65. WRITING IN MATH The rational expression $\dfrac{x^2 + 3x}{4x}$ is simplified to $\dfrac{x + 3}{4}$. Explain why this new expression is not defined for all values of x. **See margin.**

MP **Standards for Mathematical Practice**

Emphasis On	Exercises
1 Make sense of problems and persevere in solving them.	24, 39, 40, 41, 52, 56, 70
2 Reason abstractly and quantitatively.	13–23, 25–38, 42–51, 53–55, 58, 67–69
3 Construct viable arguments and critique the reasoning of others.	59–65
4 Model with mathematics.	57, 66

Preparing for Assessment

66. The surface area of the cylindrical oatmeal container shown is $(54x^2 - 48x + 8)\pi$ square inches. What is the height of the box h in terms of x? ⊕ 4 A.APR.1 **B**

3x − 2

h

○ **A** 6
○ **B** $6x$
○ **C** $6x^2$
○ **D** $12\pi x$

67. For what values of x is the quotient undefined?
⊕ 2 A.APR.1 **E**

$$\frac{2x^2 - 3x - 9}{x(2x^2 - 5x - 3)} \div \frac{4x^2 + 12x + 9}{x^2(4x^2 + 4x + 1)}$$

○ **A** 0
○ **B** $-\dfrac{3}{2}$
○ **C** $0, -\dfrac{1}{2}$
○ **D** $0, -\dfrac{1}{2}, 3$
○ **E** $0, -\dfrac{3}{2}, -\dfrac{1}{2}, 3$

68. Divide $14x^3 - 15x^2 - 98x + 15$ by $x - 5$. Which option correctly shows the factored dividend? ⊕ 2 A.APR.1 **A**

○ **A** $(7x - 1)(2x + 3)(x - 5)$
○ **B** $(7x + 1)(2x - 3)(x - 5)$
○ **C** $(2x - 1)(7x + 3)(x - 5)$
○ **D** $(2x + 1)(7x - 3)(x - 5)$

69. Divide $3x^3 + 37x^2 + 82x - 72$ by $x + 4$. Which option correctly shows the factored quotient? ⊕ 2 A.APR.1 **D**

○ **A** $(3x + 2)(x + 9)$
○ **B** $(3x - 2)(x - 9)$
○ **C** $(3x + 2)(x - 9)$
○ **D** $(3x - 2)(x + 9)$

70. **MULTI-PART** Consider the figure of the tissue box below. The volume of this box is $2x^3 - 3x^2 - 2x$ cubic inches. ⊕ 1, 2, 4 A.APR.1

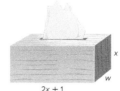

x

w

2x + 1

a. Which rearranged formula for the volume of a rectangular prism would you use to find the width of the box in terms of x? **A**

○ **A** $w = \dfrac{V}{\ell h}$
○ **B** $w = \dfrac{\ell h}{V}$
○ **C** $w = \dfrac{V\ell}{h}$
○ **D** $w = \dfrac{Vh}{\ell}$

b. Complete the work to divide the volume of the box by the length and height to find the width. Which of the expressions below are correct expressions that could be included in this work? Choose all that apply. **A, C**

☐ **A** $\dfrac{2x^3 - 3x^2 - 2x}{(2x + 1)x}$
☐ **B** $\dfrac{x(2x^2 - 3x - 2)}{(2x + 1)}$
☐ **C** $\dfrac{x(2x + 1)(x - 2)}{(2x + 1)x}$
☐ **D** $\dfrac{(2x + 1)(x - 2)}{(2x + 1)x}$
☐ **E** $\dfrac{x(x - 2)}{(2x + 1)}$
☐ **F** $x(x - 2)$

c. What is the correct expression for the width of this tissue box? **A**

○ **A** $x - 2$
○ **B** x
○ **C** $2x + 1$
○ **D** $x + 2$

Preparing for Assessment

Exercises 66–70 require students to use the skills they will need on standardized assessments. The exercises are dual-coded with content standards and mathematical practice standards.

Dual Coding		
Exercises	Content Standards	⊕ Mathematical Practice
66	A.APR.1	4
67	A.APR.1	2
68	A.APR.1	2
69	A.APR.1	2
70	A.APR.1	1, 2, 4

Diagnose Student Errors

Survey student responses for each item. Class trends may indicate common errors and misconceptions.

66.

A	Dropped the variable of the monomial expression
B	CORRECT
C	Factored out the monomial incorrectly
D	Did not factor out 2π

67.

A	Found only one excluded value
B	Found excluded values after division
C	Multiplied instead of divided and found excluded values after multiplication
D	Included a value that makes the numerator 0
E	CORRECT

Differentiated Instruction ⬤ OL ⬤ BL

Extension To prepare students for the next lesson and to build a strong base for future work with rational expressions, give them an expression like $\dfrac{5x^2(x^2 + 3)}{5x(x + 3)}$. Ask them to explain in detail, citing fundamentals from arithmetic, why the fives can be divided out but not the threes. Also explain why the first x^2 and x can be divided to simplify, but not those within parentheses. Students' explanations should mention that common factors of both the numerator and denominator can be divided out, but not terms that are parts of polynomials. Substitution of a number, like 2, for x may help some students reach this realization.

Go Online!

Self-Check Quiz

Students can use *Self-Check Quizzes* to check their understanding of this lesson.

LESSON 7-2
Adding and Subtracting Rational Expressions

SUGGESTED PACING (DAYS)

90 min.	0.75
45 min.	1.5

Instruction

Track Your Progress

Objectives

1 Determine the LCM of polynomials.

2 Add and subtract rational expressions.

Mathematical Background

When rational expressions are given the same denominators in preparation for addition or subtraction, they are equivalent to the original expressions. Multiplying a rational expression by a form of 1 such as $\frac{6x}{6x}$ or $\frac{y-3}{y-3}$ does not change the value of the expression.

THEN	NOW	NEXT

A.APR.1 Understand that polynomials form a system analogous to the integers, namely, they are closed under the operations of addition, subtraction, and multiplication; add, subtract, and multiply polynomials.

A.APR.7 (+) Understand that rational expressions form a system analogous to the rational numbers, closed under addition, subtraction, multiplication, and division by a nonzero rational expression; add, subtract, multiply, and divide rational expressions.

A.CED.2 Create equations in two or more variables to represent relationships between quantities; graph equations on coordinate axes with labels and scales.

F.IF.4 For a function that models a relationship between two quantities, interpret key features of graphs and tables in terms of the quantities, and sketch graphs showing key features given a verbal description of the relationship. Key features include: intercepts; intervals where the function is increasing, decreasing, positive, or negative; relative maximums and minimums; symmetries; end behavior; and periodicity.*

F.BF.3 Identify the effect on the graph of replacing $f(x)$ by $f(x) + k$, $k f(x)$, $f(kx)$, and $f(x + k)$ for specific values of k (both positive and negative); find the value of k given the graphs. Experiment with cases and illustrate an explanation of the effects on the graph using technology. Include recognizing even and odd functions from their graphs and algebraic expressions for them.

Go Online! All of these resources and more are available at connectED.mcgraw-hill.com

eLessons utilize the power of your interactive whiteboard in an engaging way. Use **Rational Expressions**, screen 11, to introduce the concepts in this lesson.

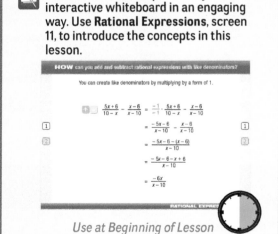

Use at Beginning of Lesson

Personal Tutors (for every example) let students hear real teachers solve problems. Students can pause and repeat as many times as necessary.

Use with Examples

Use **Self-Check Quiz** to assess students' understanding of the concepts in this lesson.

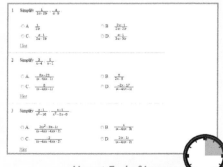

Use at End of Lesson

OER Using Open Educational Resources

Classroom Management Visit **LearnBoost** for software that can help with classroom management. This site includes grade books, lesson plans, and progress reports. *Use as an organizational tool*

Differentiate Your Resources

Extra Practice Additional practice or homework; Skills Practice is best for approaching-level students and Practice is best for on-level and beyond-level students

Skills Practice

NAME _____ DATE _____ PERIOD _____

7-2 Skills Practice
Adding and Subtracting Rational Expressions

Find the LCM of each set of polynomials.

1. $12c, 6c^2d$
2. $18a^2bc^3, 24b^2c^2$
3. $2x - 6, x - 3$
4. $5a, a - 1$
5. $t^2 - 25, t + 5$
6. $x^2 - 3x - 4, x + 1$

Simplify each expression.

7. $\frac{5}{y} + \frac{6}{y}$
8. $\frac{9}{2pr} + \frac{5}{apbr}$
9. $\frac{3x-7}{3} + 4$
10. $\frac{1}{m^2p} - \frac{2}{p}$
11. $\frac{18}{3xy} - \frac{3}{5yz}$
12. $\frac{7}{4gh} + \frac{2}{4h^3}$
13. $\frac{2}{a+2} - \frac{3}{2a}$
14. $\frac{5}{3b+d} - \frac{2}{3bd}$
15. $\frac{3}{w-3} - \frac{2}{w^2-9}$
16. $\frac{3d}{3-a} + \frac{2}{a-3}$
17. $\frac{8}{k-n} - \frac{8}{n-k}$
18. $\frac{2}{x-4} + \frac{2x+4}{x+1}$
19. $\frac{1}{x^2+2x+1} + \frac{4}{x+1}$
20. $\frac{2x+1}{x-5} - \frac{4}{x^2-5x+10}$
21. $\frac{n}{n-3} + \frac{2n+2}{n^2-2n-3}$
22. $\frac{3}{y^2+y-12} - \frac{2}{y^2+4y+8}$

Practice

NAME _____ DATE _____ PERIOD _____

7-2 Practice
Adding and Subtracting Rational Expressions

Find the LCM of each set of polynomials.

1. x^2y, xy^3
2. a^2b^3c, abc^2
3. $x + 1, x + 3$
4. $g - 1, g^2 + 3g - 4$
5. $2r + 2, r^3 + r, r + 1$
6. $3, 4w + 2, 4w^2 - 1$
7. $x^2 + 2x - 8, x + 4$
8. $x^2 - x - 6, x^2 + 6x + 8$
9. $d^2 + 6d + 9, 2(d^2 - 9)$

Simplify each expression.

10. $\frac{5}{x} - \frac{7}{6x}$
11. $\frac{5}{13x^2y} - \frac{1}{3x^2y^2}$
12. $\frac{1}{6c^2d} + \frac{9}{4cd^3}$
13. $\frac{4m}{3mn} + 2$
14. $2x - 5 - \frac{x-8}{x+4}$
15. $\frac{4}{a-3} + \frac{3}{x-3}$
16. $\frac{16}{x^2-16} + \frac{2}{x+4}$
17. $\frac{2-3m}{m-9} + \frac{4m-5}{9-m}$
18. $\frac{y-5}{y^2-3y-10} + \frac{y}{y^2+y-2}$
19. $\frac{6}{2x-11} - \frac{10}{2x^2-4x-12}$
20. $\frac{2p-3}{p^2-3p+6} - \frac{6}{p^3-6}$
21. $\frac{1}{5m} - \frac{3}{m} + \frac{7}{13m}$
22. $\frac{2a}{a-3} + \frac{2a}{a^2+3} - \frac{4}{a^2-5}$
23. $\frac{2x}{a^2-y}$
24. $\frac{r+4}{\frac{r^2}{x} + \frac{4r+2}{y}}$

25. **GEOMETRY** The expressions $\frac{3x}{x}$, $\frac{28}{x+4}$, and $\frac{10}{x-4}$ represent the lengths of the sides of a triangle. Write a simplified expression for the perimeter of the triangle.

26. **KAYAKING** Mai is kayaking on a river that has a current of 2 miles per hour. If r represents her rate in calm water, then $r + 2$ represents her rate with the current, and $r - 2$ represents her rate against the current. Mai kayaks 2 miles downstream and then back to her starting point. Use the formula for time, $t = \frac{d}{r}$, where d is the distance, to write a simplified expression for the total time it takes Mai to complete the trip.

Word Problem Practice

NAME _____ DATE _____ PERIOD _____

7-2 Word Problem Practice
Adding and Subtracting Rational Expressions

1. **SQUARES** Susan's favorite perfect square is s^2 and Travis' is t^2, where s and t are whole numbers. What perfect square is guaranteed to be divisible by both Susan's and Travis' favorite perfect squares regardless of their specific value?

2. **ELECTRIC POTENTIAL** The electrical potential function between two electrons is given by a formula that has the form $\frac{1}{r} + \frac{1}{1-r}$. Simplify this expression.

3. **TRAPEZOIDS** The cross section of a stand consists of two trapezoids stacked one on top of the other.

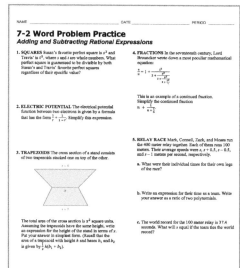

The total area of the cross section is x^2 square units. Assuming the trapezoids have the same height, write an expression for the height of the stand in terms of x. Put your answer in simplest form. (Recall that the area of a trapezoid with height h and bases b_1 and b_2 is given by $\frac{1}{2}h(b_1 + b_2)$.)

4. **FRACTIONS** In the seventeenth century, Lord Brouncker wrote down a most peculiar mathematical equation:

$$\frac{4}{\pi} = 1 + \cfrac{1^2}{2 + \cfrac{3^2}{2 + \cfrac{5^2}{2 + \cdots}}}$$

This is an example of a continued fraction. Simplify the continued fraction $n + \cfrac{1}{n + \frac{1}{n}}$.

5. **RELAY RACE** Mark, Cornell, Zack, and Moses run the 400 meter relay together. Each of them runs 100 meters. Their average speeds were s, $s + 0.5$, $s - 0.5$, and $s - 1$ meters per second, respectively.

a. What were their individual times for their own legs of the race?

b. Write an expression for their time as a team. Write your answer as a ratio of two polynomials.

c. The world record for the 100 meter relay is 37.4 seconds. What will s equal if the team ties the world record?

Intervention Reteaching and vocabulary activities that can be used with struggling or absent students and as ELL support

Study Guide and Intervention

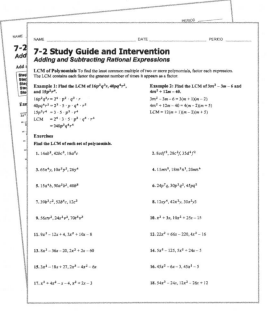

NAME _____ DATE _____ PERIOD _____

7-2 Study Guide and Intervention
Adding and Subtracting Rational Expressions

LCM of Polynomials To find the least common multiple of two or more polynomials, factor each expression. The LCM contains each factor the greatest number of times it appears as a factor.

Example 1: Find the LCM of $16p^2q^3r$, $40pq^4r^2$, and $15p^3r^4$.
$16p^2q^3r = 2^4 \cdot p^2 \cdot q^3 \cdot r$
$40pq^4r^2 = 2^3 \cdot 5 \cdot p \cdot q^4 \cdot r^2$
$15p^3r^4 = 3 \cdot 5 \cdot p^3 \cdot q^4 \cdot r^4$
LCM $= 2^4 \cdot 3 \cdot 5 \cdot p^3 \cdot q^4 \cdot r^4$
$= 240p^3q^4r^4$

Example 2: Find the LCM of $3m^2 - 3m - 6$ and $4m^2 + 12m - 40$.
$3m^2 - 3m - 6 = 3(m + 1)(m - 2)$
$4m^2 + 12m - 40 = 4(m - 2)(m + 5)$
LCM $= 12(m + 1)(m - 2)(m + 5)$

Exercises

Find the LCM of each set of polynomials.

1. $14ab^2, 42bc^3, 18a^2c$
2. $8cd^3, 28c^2f, 35d^4f^2$
3. $65x^3y, 10x^2y^2, 26y^4$
4. $11mn^2, 18m^2n^3, 20mn^4$
5. $15a^4b, 50a^2b^2, 40b^8$
6. $24p^7q, 30p^3q^2, 45pq^3$
7. $39b^2c^2, 52b^4c, 12c^3$
8. $12xy^4, 42x^2y, 30x^4y3$
9. $56stv^2, 24t^2v^2, 70t^2v^3$
10. $x^2 + 3x, 10x^2 + 25x - 15$
11. $9x^2 - 12x + 4, 3x^2 + 10x - 8$
12. $22x^2 + 66x - 220, 4x^2 - 16$
13. $8x^2 - 36x - 20, 2x^2 + 2x - 60$
14. $5x^2 - 125, 5x^2 + 24x - 5$
15. $3x^2 - 18x + 27, 2x^2 - 4x^2 - 6x$
16. $45x^2 - 6x - 3, 45x^2 - 5$
17. $x^2 + 4x^2 - x - 4, x^2 + 2x - 3$
18. $54x^2 - 24x, 12x^2 - 26x + 12$

Study Notebook

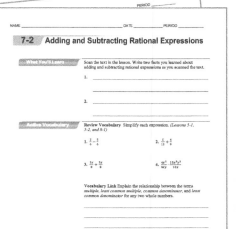

NAME _____ DATE _____ PERIOD _____

7-2   Adding and Subtracting Rational Expressions

What You'll Learn   Scan the text in the lesson. Write two facts you learned about adding and subtracting rational expressions as you scanned the text.

1. _____

2. _____

Active Vocabulary   Review Vocabulary Simplify each expression. (Lessons 5-1, 5-2, and 8-1)

1. $\frac{2}{6} - \frac{5}{6}$
2. $\frac{1}{13} + \frac{4}{5}$
3. $\frac{3x}{6} + \frac{5x}{6}$
4. $\frac{4x^2}{6xy} - \frac{15x^3y^2}{10x}$

Vocabulary Link Explain the relationship between the terms *multiple, least common multiple, common denominator,* and *least common denominator* for any two whole numbers.

_____
_____
_____
_____
_____

Extension Activities that can be used to extend lesson concepts

Enrichment

NAME _____ DATE _____ PERIOD _____

7-2 Enrichment
Zeno's Paradox

The Greek philosopher Zeno of Elea (born sometime between 495 and 480 B.C.) proposed four paradoxes to challenge the notions of space and time. Zeno's first paradox works like this:

Suppose you are on your way to school. Assume you are able to cover half of the remaining distance each minute that you walk. You leave your house at 7:45 A.M. After the first minute, you are half of the way to school. In the next minute you cover half of the remaining distance to school, and at 7:47 A.M. you are three-quarters of the way to school. This pattern continues each minute. At what time will you arrive at school? Before 8:00 A.M.? Before lunch?

Since space is infinitely divisible, we can repeat this pattern forever. Thus, on the way to school you must reach an infinite number of 'midpoints' in a finite time. This is impossible, so you can never reach your goal. In general, according to Zeno anyone who wants to move from one point to another must meet these requirements, and motion is impossible. Therefore, what we perceive as motion is merely an illusion.

Addition of fractions can be defined by $\frac{a}{b} + \frac{c}{d} = \frac{ad + bc}{bd}$, similarly for subtraction.

Assume your house is one mile from school. At 7:46 A.M., you have walked half of a mile, so you have left $1 - \frac{1}{2}$, or $\frac{1}{2}$ a mile. At 7:47 A.M. you only have $1 - \frac{1}{2} - \frac{1}{4}$ of a mile to go.

To determine how far you have walked and how far away from the school you are at 7:48 A.M., add the distances walked each minute, $\frac{1}{2} + \frac{1}{4} + \frac{1}{8} = \frac{7}{8}$ of a mile so far and you still have $1 - \frac{1}{2} - \frac{1}{8}$ of a mile to go.

1. Determine how far you have walked and how far away from the school you are at 7:50 A.M.

2. Suppose instead of covering one-half the distance to school each minute, you cover three-quarters of the distance remaining to school each minute, now will you be able to make it to school on time? Determine how far you still have left to go at 7:47 A.M.

3. Suppose that instead of covering one-half or three-quarters of the distance to school each minute, you cover $\frac{1}{x+n}$ of the distance remaining, where x is a whole number greater than 2. What is your distance from school at 7:47 A.M.?

Launch

Have students read the Why? section of the lesson.
Ask:
- Can $\frac{s_0}{s_0 - v}$ be factored to $s_0\left(\frac{1}{1-v}\right)$? Explain. **No; s_0 is not a common factor.**

- To multiply P_0 times $\frac{s_0}{s_0 - v}$, do you need a common denominator? **no**

Teach

Ask the scaffolded questions for each example to build conceptual understanding for students at all levels.

1 LCM of Polynomials

Example 1 LCM of Monomials and Polynomials

AL What does LCM stand for? *Least Common Multiple*

OL Explain the relationship between LCM and LCD. *The least common denominator (LCD) is the least common multiple (LCM) of the denominators of two or more rational expressions.*

BL What is the smallest an LCM could be? *the largest of the terms you are finding the LCM of*

Need Another Example?

Find the LCM of each set of polynomials.
a. $15a^2bc^3$, $16b^5c^2$, and $20a^3c^6$ $240a^3b^5c^6$
b. $x^3 - x^2 - 2x$ and $x^2 - 4x + 4$ $x(x+1)(x-2)^2$

Go Online!

Interactive Whiteboard

Use the *eLesson* or *Lesson Presentation* to present this lesson.

Adding and Subtracting Rational Expressions

:: Then	:: Now	:: Why?
• You added and subtracted polynomial expressions.	**1** Determine the LCM of polynomials. **2** Add and subtract rational expressions.	• As a fire engine moves toward a person, the pitch of the siren sounds higher to that person than it would if the fire engine were at rest. This is because the sound waves are compressed closer together, referred to as the *Doppler effect*. The Doppler effect can be represented by the rational expression $P_0\left(\frac{s_0}{s_0 - v}\right)$, where P_0 is the actual pitch of the siren, v is the speed of the fire truck, and s_0 is the speed of sound in air.

MP Mathematical Practices
3 Construct viable arguments and critique the reasoning of others.

Content Standards
A.APR.7 Understand that rational expressions form a system analogous to the rational numbers, closed under addition, subtraction, multiplication, and division by a nonzero rational expression; add, subtract, multiply, and divide rational expressions.

1 LCM of Polynomials Just as with rational numbers in fractional form, to add or subtract two rational expressions that have unlike denominators, you must first find the least common denominator (LCD). The LCD is the least common multiple (LCM) of the denominators.

To find the LCM of two or more numbers or polynomials, factor them. The LCM contains each factor the greatest number of times it appears as a factor.

Numbers	Polynomials
$\frac{5}{6} + \frac{4}{9}$	$\frac{3}{x^2 - 3x + 2} + \frac{5}{2x^2 - 2}$
LCM of 6 and 9	**LCM of $x^2 - 3x + 2$ and $2x^2 - 2$**
$6 = 2 \cdot 3$	$x^2 - 3x + 2 = (x-1)(x-2)$
$9 = 3 \cdot 3$	$2x^2 - 2 = 2 \cdot (x-1)(x+1)$
LCM $= 2 \cdot 3 \cdot 3$ or 18	LCM $= 2(x-1)(x-2)(x+1)$

A.APR.7

Example 1 LCM of Monomials and Polynomials

Find the LCM of each set of polynomials.

a. $6xy$, $15x^2$, and $9xy^4$

$6xy = 2 \cdot 3 \cdot x \cdot y$ — Factor the first monomial.
$15x^2 = 3 \cdot 5 \cdot x^2$ — Factor the second monomial.
$9xy^4 = 3 \cdot 3 \cdot x \cdot y^4$ — Factor the third monomial.
LCM $= 2 \cdot 3 \cdot 3 \cdot 5 \cdot x^2 \cdot y^4$ — Use each factor the greatest number of times it appears.
$= 90x^2y^4$ — Then simplify.

b. $y^4 + 8y^3 + 15y^2$ and $y^2 - 3y - 40$

$y^4 + 8y^3 + 15y^2 = y^2(y+5)(y+3)$ — Factor the first polynomial.
$y^2 - 3y - 40 = (y+5)(y-8)$ — Factor the second polynomial.
LCM $= y^2(y+5)(y+3)(y-8)$ — Use each factor the greatest number of times it appears as a factor.

Guided Practice

1A. $12a^2b$, $15abc$, $8b^3c^4$ $120a^2b^3c^4$

1B. $4a^2 - 12a - 16$ and $a^3 - 9a^2 + 20a$ $4a(a-4)(a-5)(a+1)$

MP Mathematical Practices Strategies

Look for and express regularity in repeated reasoning.
Help students remember how to find lowest common multiples, and how to add and subtract fractions by showing some examples with real numbers. Choose one or two examples and work through them. For example, ask:

- What is the lowest common denominator for $\frac{1}{2}$, $\frac{3}{8}$, and $\frac{3}{4}$? **8**

- Rewrite the fractions using the LCD so that they can be more easily added together. $\frac{4}{8}$, $\frac{3}{8}$, and $\frac{6}{8}$

- Add the fractions. $\frac{13}{8}$

- What is the next step? *To simplify it to lowest terms and/or put it into a mixed fraction, or $1\frac{5}{8}$.*

2 **Add and Subtract Rational Expressions** As with fractions, rational expressions must have common denominators in order to be added or subtracted.

Key Concept

Adding Rational Expressions

Words	To add rational expressions, find the least common denominator (LCD). Rewrite each expression with the LCD. Then add.
Symbols	For all $\frac{a}{b}$ and $\frac{c}{d}$, with $b \neq 0$ and $d \neq 0$, $\frac{a}{b} + \frac{c}{d} = \frac{ad}{bd} + \frac{bc}{bd} = \frac{ad + bc}{bd}$.

Subtracting Rational Expressions

Words	To subtract rational expressions, find the least common denominator (LCD). Rewrite each expression with the LCD. Then subtract.
Symbols	For all $\frac{a}{b}$ and $\frac{c}{d}$, with $b \neq 0$ and $d \neq 0$, $\frac{a}{b} - \frac{c}{d} = \frac{ad}{bd} - \frac{bc}{bd} = \frac{ad - bc}{bd}$.

A.APR.7

Example 2 Monomial Denominators

Simplify each expression.

a. $\dfrac{3y}{2x} + \dfrac{2z}{5x}$

$\dfrac{3y}{2x} + \dfrac{2z}{5x} = \dfrac{3y}{2x} \cdot \dfrac{5}{5} + \dfrac{2z}{5x} \cdot \dfrac{2}{2}$

$= \dfrac{15y}{10x} + \dfrac{4z}{10x}$

$= \dfrac{15y + 4z}{10x}$

b. $\dfrac{4}{x^2 - 4x + 3} + \dfrac{x}{x^2 + 5x - 6}$

$\dfrac{4}{x^2 - 4x + 3} + \dfrac{x}{x^2 + 5x - 6}$

$= \dfrac{4}{(x - 3)(x - 1)} + \dfrac{x}{(x - 1)(x + 6)}$

$= \dfrac{4}{(x - 3)(x - 1)} \cdot \dfrac{x + 6}{x + 6} + \dfrac{x}{(x - 1)(x + 6)} \cdot \dfrac{x - 3}{x - 3}$

$= \dfrac{4x + 24}{(x - 3)(x - 1)(x + 6)} + \dfrac{x^2 - 3x}{(x - 3)(x - 1)(x + 6)}$

$= \dfrac{x^2 + x + 24}{(x - 3)(x - 1)(x + 6)}$

Guided Practice

2A. $\dfrac{6}{7y} - \dfrac{2c}{3y}$ $\dfrac{18 - 14c}{21y}$

2B. $\dfrac{a}{a^2 + 7a + 12} - \dfrac{3}{a^2 + 9a + 20}$ $\dfrac{a^2 + 2a - 9}{(a + 4)(a + 3)(a + 5)}$

The LCD is also used to combine rational expressions with polynomial denominators.

A.APR.7

Example 3 Polynomial Denominators

Simplify $\dfrac{5}{6x - 18} - \dfrac{x - 1}{4x^2 - 14x + 6}$.

$\dfrac{5}{6x - 18} - \dfrac{x - 1}{4x^2 - 14x + 6} = \dfrac{5}{6(x - 3)} - \dfrac{x - 1}{2(2x - 1)(x - 3)}$ Factor denominators.

$= \dfrac{5(2x - 1)}{6(x - 3)(2x - 1)} - \dfrac{(x - 1)(3)}{2(2x - 1)(x - 3)(3)}$ Multiply by missing factors.

$= \dfrac{10x - 5 - 3x + 3}{6(x - 3)(2x - 1)}$ Subtract numerators.

$= \dfrac{7x - 2}{6(x - 3)(2x - 1)}$ Simplify.

Study Tip

Precision After you add or subtract rational expressions, it is possible the resulting expression can be further simplified.

Guided Practice

Simplify each expression.

3A. $\dfrac{x - 1}{x^2 - x - 6} - \dfrac{4}{5x + 10}$ $\dfrac{x + 7}{5(x + 2)(x - 3)}$

3B. $\dfrac{x - 8}{4x^2 + 21x + 5} + \dfrac{6}{12x + 3}$ $\dfrac{3x + 2}{(4x + 1)(x + 5)}$

Watch Out!

Preventing Errors Have students discuss the differences between procedures for adding and multiplying fractions. It is important that they see why common denominators are required for addition but not for multiplication.

---

2 Add and Subtract Rational Expressions

Example 2 Monomial Denominators

AL Why can't we combine $12y^3$ and $5x^2z$? They are not like terms.

OL What do you with do with the numerators when you add two rational expressions? We add them and then put the sum over the common denominator.

BL If you are adding $\frac{a}{b}$ and $\frac{c}{d}$, does bd always have to be the denominator? No, the LCM of b and d will be the denominator. The LCM could be less than bd.

Need Another Example?

Simplify $\dfrac{5a^2}{6b} + \dfrac{9}{14a^2b^2}$. $\dfrac{35a^4b + 27}{42a^2b^2}$

Example 3 Polynomial Denominators

AL How do you find the LCD of polynomials? Factor the polynomials and then take each of the factors that show up at least once in the polynomials.

OL For what values is the solution undefined? 3 and $\frac{1}{2}$

BL Explain why after the numerators are subtracted the numerator is $10x - 5 - 3x + 3$ and not $10x - 5 - 3x - 3$. When subtracting we must distribute the negative sign.

Need Another Example?

Simplify $\dfrac{x + 10}{3x - 15} - \dfrac{3x + 15}{6x - 30}$. $-\dfrac{1}{6}$

Example 4 Complex Fractions with Different LCDs

(AL) **What is this example asking you to do?** Simplify the expression to one fraction with no common terms in the numerator and denominator.

(OL) **What is the LCD of $1 + \frac{1}{a}$? why?** a, because the denominator of 1 is 1, so we are finding the LCM of 1 and a which is a.

(BL) **Why would $\dfrac{\frac{x+1}{x}}{\frac{y-x}{x}}$ simplify to $\dfrac{x+1}{y-x}$?** When multiplying by the reciprocal there would be a factor of x in the numerator and denominator that would cancel out.

Need Another Example?

Simplify $\dfrac{\frac{1}{a}+\frac{1}{b}}{\frac{1}{b}-1}$. $\dfrac{a+b}{a(1-b)}$ or $\dfrac{a+b}{a-ab}$

Example 5 Complex Fractions with Same LCDs

(AL) **Do complex fractions have anything to do with complex numbers?** no

(OL) **How do you know when you are finished simplifying a complex fraction?** when you only have one fraction and it is in lowest terms

(BL) **Explain the differences in methods shown in Example 4 and Example 5.** In Example 4 the LCD of the numerator and denominators were found separately and used to simplify. In Example 5 the LCD was found of the numerator and the denominator and then the expression was simplified.

Need Another Example?

Simplify $\dfrac{\frac{2}{x}-1}{\frac{1}{y}-\frac{3}{x}}$. $\dfrac{2y-xy}{x-3y}$

One way to simplify a complex fraction is to simplify the numerator and the denominator separately, and then simplify the resulting expressions.

A.APR.7

Example 4   Complex Fractions with Different LCDs

> **Study Tip**
> Undefined Terms Remember that there are restrictions on variables in the denominator.

Simplify $\dfrac{1+\frac{1}{x}}{1-\frac{x}{y}}$.

$\dfrac{1+\frac{1}{x}}{1-\frac{x}{y}} = \dfrac{\frac{x}{x}+\frac{1}{x}}{\frac{y}{y}-\frac{x}{y}}$ The LCD of the numerator is x.
The LCD of the denominator is y.

$= \dfrac{\frac{x+1}{x}}{\frac{y-x}{y}}$ Simplify the numerator and denominator.

$= \dfrac{x+1}{x} \div \dfrac{y-x}{y}$ Write as a division expression.

$= \dfrac{x+1}{x} \cdot \dfrac{y}{y-x}$ Multiply by the reciprocal of the divisor.

$= \dfrac{xy+y}{xy-x^2}$ Simplify.

▷ **Guided Practice**

Simplify each expression.

4A. $\dfrac{1-\frac{y}{x}}{\frac{1}{y}+\frac{1}{x}}$ $\dfrac{xy-y^2}{x+y}$

4B. $\dfrac{\frac{c}{d}-\frac{d}{c}}{\frac{d}{c}+2}$ $\dfrac{c^2-d^2}{d^2+2cd}$

Another method of simplifying complex fractions is to find the LCD of all of the denominators. Then, the denominators are all eliminated by multiplying by the LCD.

A.APR.7

Example 5   Complex Fractions with Same LCDs

Simplify $\dfrac{1+\frac{1}{x}}{1-\frac{x}{y}}$.

$\dfrac{1+\frac{1}{x}}{1-\frac{x}{y}} = \dfrac{\left(1+\frac{1}{x}\right)}{\left(1-\frac{x}{y}\right)} \cdot \dfrac{xy}{xy}$ The LCD of all of the denominators is xy.
 Multiply by $\frac{xy}{xy}$.

$= \dfrac{xy+y}{xy-x^2}$ Distribute xy.

Notice that the same problem is solved in Examples 4 and 5 using different methods, but both produce the same answer. So, how you solve problems similar to these is left up to your own discretion.

▷ **Guided Practice**

Simplify each expression.

5A. $\dfrac{1+\frac{2}{x}}{\frac{3}{y}-\frac{4}{x}}$ $\dfrac{xy+2y}{3x-4y}$

5B. $\dfrac{\frac{1}{d}-\frac{d}{c}}{\frac{1}{c}+6}$ $\dfrac{c-d^2}{d+6cd}$

5C. $\dfrac{\frac{1}{y}+\frac{1}{x}}{\frac{1}{y}-\frac{1}{x}}$ $\dfrac{x+y}{x-y}$

5D. $\dfrac{\frac{a}{b}+1}{1-\frac{b}{a}}$ $\dfrac{a(a+b)}{b(a-b)}$

Differentiated Instruction (AL) (OL) (ELL)

IF students have difficulty adding and subtracting rational expressions,

THEN have students work with a partner, one in the role of a coach, the other in the role of an athlete. The athlete works a problem, using steps and explaining the thinking while the coach listens and watches for errors, correcting as necessary. Then the partners exchange roles.

Check Your Understanding

 ◯ = Step-by-Step Solutions begin on page R11.

Go Online! for a Self-Check Quiz

Example 1
A.APR.7

Find the LCM of each set of polynomials.

1. $16x$, $8x^2y^3$, $5x^3y$ $80x^3y^3$

2. $7a^2$, $9ab^3$, $21abc^4$ $63a^2b^3c^4$

3. $3y^2 - 9y$, $y^2 - 8y + 15$ $3y(y-3)(y-5)$

4. $x^3 - 6x^2 - 16x$, $x^2 - 4$ $x(x+2)(x-2)(x-8)$

Examples 2–3
A.APR.7

Simplify each expression.

5. $\dfrac{12y}{5x} + \dfrac{5x}{4y^3}$ $\dfrac{48y^4 + 25x^2}{20xy^3}$

6. $\dfrac{5}{6b} + \dfrac{3b^2}{14a}$ $\dfrac{35a + 9b^3}{42ab}$

7. $\dfrac{7b}{12a} - \dfrac{1}{18a}$ $\dfrac{21b - 2}{36a}$

8. $\dfrac{y^2}{8c^2d^2} - \dfrac{3x}{14c^4d}$ $\dfrac{7c^2y^2 - 12dx}{56c^4d^2}$

9. $\dfrac{4x}{x^2 + 9x + 18} + \dfrac{5}{x + 6}$ $\dfrac{9x + 15}{(x+3)(x+6)}$

10. $\dfrac{8}{y - 3} + \dfrac{2y - 5}{y^2 - 12y + 27}$ $\dfrac{10y - 77}{(y-3)(y-9)}$

11. $\dfrac{4}{3x + 6} - \dfrac{x + 1}{x^2 - 4}$ $\dfrac{x - 11}{3(x+2)(x-2)}$

12. $\dfrac{3a + 2}{a^2 - 16} - \dfrac{7}{6a + 24}$ $\dfrac{11a + 40}{6(a+4)(a-4)}$

13. GEOMETRY Find the perimeter of the rectangle.

$\dfrac{14x - 10}{(x+1)(x-2)}$ $\dfrac{3}{x-2}$ $\dfrac{4}{x+1}$

Examples 4–5
A.APR.7

Simplify each expression.

14. $\dfrac{4 + \frac{2}{x}}{3 - \frac{2}{x}}$ $\dfrac{4x + 2}{3x - 2}$

15. $\dfrac{6 + \frac{4}{y}}{2 + \frac{6}{y}}$ $\dfrac{3y + 2}{y + 3}$

16. $\dfrac{\frac{3}{x} + \frac{2}{y}}{1 + \frac{4}{y}}$ $\dfrac{3y + 2x}{xy + 4x}$

17. $\dfrac{\frac{2}{b} + \frac{5}{a}}{\frac{3}{a} - \frac{8}{b}}$ $\dfrac{2a + 5b}{3b - 8a}$

Practice and Problem Solving

Extra Practice is on page R7.

Example 1
A.APR.7

Find the LCM of each set of polynomials.

18. $24cd$, $40a^2c^3d^4$, $15abd^3$ $120a^2bc^3d^4$

19. $4x^2y^3$, $18xy^4$, $10xz^2$ $180x^2y^4z^2$

20. $x^2 - 9x + 20$, $x^2 + x - 30$ $(x-4)(x-5)(x+6)$

21. $6x^2 + 21x - 12$, $4x^2 + 22x + 24$ $6(x+4)(2x-1)(2x+3)$

Examples 2–3 Ⓜ **PERSEVERANCE** Simplify each expression.
A.APR.7

22. $\dfrac{5a}{24cf^4} + \dfrac{a}{36bc^4f^3}$ $\dfrac{15abc^3 + 2af}{72bc^4f^4}$

23. $\dfrac{4b}{15x^3y^2} - \dfrac{3b}{35x^2y^4z}$ $\dfrac{28by^2z - 9bx}{105x^3y^4z}$

24. $\dfrac{5b}{6a} + \dfrac{3b}{10a^2} + \dfrac{2}{ab^2}$ $\dfrac{25ab^3 + 9b^3 + 60a}{30a^2b^2}$

25. $\dfrac{4}{3x} + \dfrac{8}{x^3} + \dfrac{2}{5xy}$ $\dfrac{20x^2y + 120y + 6x^2}{15x^3y}$

26. $\dfrac{8}{3y} + \dfrac{2}{9} - \dfrac{3}{10y^2}$ $\dfrac{240y + 20y^2 - 27}{90y^2}$

27. $\dfrac{1}{16a} + \dfrac{5}{12b} - \dfrac{9}{10b^3}$ $\dfrac{15b^3 + 100ab^2 - 24a^2}{240ab^3}$

28. $\dfrac{8}{x^2 - 6x - 16} + \dfrac{9}{x^2 - 3x - 40}$ $\dfrac{17x + 58}{(x-8)(x+2)(x+5)}$

㉙ $\dfrac{6}{y^2 - 2y - 35} + \dfrac{4}{y^2 + 9y + 20}$ $\dfrac{10y - 4}{(y-7)(y+5)(y+4)}$

30. $\dfrac{12}{3y^2 - 10y - 8} - \dfrac{3}{y^2 - 6y + 8}$ $\dfrac{3y - 30}{(3y+2)(y-4)(y-2)}$

31. $\dfrac{6}{2x^2 + 11x - 6} - \dfrac{8}{x^2 + 3x - 18}$ $\dfrac{-10x - 10}{(2x-1)(x+6)(x-3)}$

32. $\dfrac{2x}{4x^2 + 9x + 2} + \dfrac{3}{2x^2 - 8x - 24}$ $\dfrac{4x^2 - 12x + 3}{2(x-6)(4x+1)(x+2)}$

33. $\dfrac{4x}{3x^2 + 3x - 18} - \dfrac{2x}{2x^2 + 11x + 15}$ $\dfrac{2x^2 + 32x}{3(x-2)(x+3)(2x+5)}$

34. BIOLOGY After a person eats something, the pH or acid level A of his or her mouth can be determined by the formula $A = \dfrac{20.4t}{t^2 + 36} + 6.5$, where t is the number of minutes that have elapsed since the food was eaten.

a. Simplify the equation. $A = \dfrac{6.5t^2 + 20.4t + 234}{t^2 + 36}$

b. What would the acid level be after 30 minutes? ≈ 7.2

Differentiated Homework Options

Levels	ⒶⓁ Basic	ⓄⓁ Core	ⒷⓁ Advanced
Exercises	18–39, 63–71	19–39 odd, 40, 41–51 odd, 53–55, 63, 67–71	40–71
2-Day Option	19–39 odd, 67–71	18–39, 67–71	
	18–38 even, 63–66	40–66	

 You can use ALEKS to provide additional remediation support with personalized instruction and practice.

Practice

Formative Assessment Use Exercises 1–17 to assess students' understanding of the concepts in this lesson.

The Practice and Problem Solving exercises assess the content taught in the lesson. The Preparing for Assessment page is meant to be used as preparation for end-of-course assessments.

 Ⓜ **Teaching the Mathematical Practices**

Perseverance Mathematically proficient students start by explaining to themselves the meaning of a problem and looking for entry points to its solution. They analyze givens, constraints, relationships, and goals and make conjectures about the form and meaning of the solution and plan a solution pathway rather than simply jumping into a solution attempt.

Extra Practice

See page R7 for extra exercises for students who are approaching level or for on-level students who need additional reinforcement.

Levels of Complexity Chart

The levels of the exercises progress from 1 to 3, with Level 1 indicating the lowest level of complexity.

Exercises	18–39	40–62, 67–71	63–66
Ⓒ Level 3			●
Ⓑ Level 2		●	
Level 1	●		

Go Online!   eBook

Interactive Student Guide

Use the *Interactive Student Guide* to deepen conceptual understanding.

· Adding and Subtracting Rational Expressions

ALGEBRA 2
INTERACTIVE STUDENT GUIDE

Exercise Alert

Formulas For Exercise 35, students will need to know the formula for the area of a triangle, $A = \frac{1}{2}bh$.

Teaching the Mathematical Practices

Modeling Mathematically proficient students can apply the mathematics they know to solve problems arising in everyday life, analyze relationships mathematically to draw conclusions, and interpret their mathematical results in the context of a situation.

---

35. GEOMETRY Both triangles in the figure at the right are equilateral. If the area of the smaller triangle is 200 square centimeters and the area of the larger triangle is 300 square centimeters, find the minimum distance from A to B in terms of x and y and simplify. $\dfrac{1000x + 800y}{x(x + 2y)}$

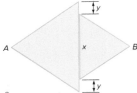

Examples 4–5 Simplify each expression.

A.APR.7

36. $\dfrac{\frac{2}{x-3} + \frac{3x}{x^2-9}}{\frac{3}{x+3} - \frac{4x}{x^2-9}}$ $\dfrac{5x+6}{-x-9}$

37. $\dfrac{\frac{4}{x+5} + \frac{9}{x-6}}{\frac{5}{x-6} - \frac{8}{x+5}}$ $\dfrac{13x+21}{-3x+73}$

38. $\dfrac{\frac{5}{x+6} - \frac{2x}{2x-1}}{\frac{x}{2x-1} + \frac{4}{x+6}}$ $\dfrac{-2x^2-2x-5}{x^2+14x-4}$

39. $\dfrac{\frac{8}{x-9} - \frac{x}{3x+2}}{\frac{3}{3x+2} + \frac{4x}{x-9}}$ $\dfrac{-x^2+33x+16}{12x^2+11x-27}$

40. OIL PRODUCTION Managers of an oil company have estimated that oil will be pumped from a certain well at a rate based on the function $R(x) = \dfrac{20}{x} + \dfrac{200x}{3x^2+20}$, where $R(x)$ is the rate of production in thousands of barrels per year x years after pumping begins.

a. Simplify $R(x)$. $R(x) = \dfrac{260x^2 + 400}{3x^3 + 20x}$

b. At what rate will oil be pumping from the well in 50 years? about 1730 barrels/yr

Find the LCM of each set of polynomials.

41. $12xy^4, 14x^4y^2, 5xyz^3, 15x^5y^3$ $420x^5y^4z^3$

42. $-6abc^2, 18a^2b^2, 15a^4c, 8b^3$ $-360a^4b^3c^2$

43. $x^2 - 3x - 28, 2x^2 + 9x + 4, x^2 - 16$ $(x+4)(x-4)(2x+1)(x-7)$

44. $x^2 - 5x - 24, x^2 - 9, 3x^2 + 8x - 3$ $(x+3)(x-3)(x-8)(3x-1)$

Simplify each expression.

45. $\dfrac{1}{12a} + 6 - \dfrac{3}{5a^2}$ $\dfrac{360a^2 + 5a - 36}{60a^2}$

46. $\dfrac{5}{16y^2} - 4 - \dfrac{8}{3x^2y}$ $\dfrac{15x^2 - 192x^2y^2 - 128y}{48x^2y^2}$

47. $\dfrac{5}{6x^2 + 46x - 16} + \dfrac{2}{6x^2 + 57x + 72}$

48. $\dfrac{1}{8x^2 - 20x - 12} + \dfrac{4}{6x^2 + 27x + 12}$

49. $\dfrac{x^2 + y^2}{x^2 - y^2} + \dfrac{y}{x+y} - \dfrac{x}{x-y}$ 0

50. $\dfrac{x^2 + x}{x^2 - 9x + 8} + \dfrac{4}{x-1} - \dfrac{3}{x-8}$ $\dfrac{x^2 + 2x - 29}{x^2 - 9x + 8}$

47. $\dfrac{42x + 41}{6(3x-1)(x+8)(2x+3)}$

51. $\dfrac{\frac{2}{a-1} + \frac{3}{a-4}}{\frac{6}{a^2-5a+4}}$ $\dfrac{5a-11}{6}$

52. $\dfrac{\frac{1}{x} + \frac{1}{y}}{\left(\frac{1}{x} - \frac{1}{y}\right)(x+y)}$ $\dfrac{1}{y-x}$

48. $\dfrac{19x - 36}{12(2x+1)(x-3)(x+4)}$

53. GEOMETRY The length of one rectangular room is $\dfrac{x^2 - 16}{x - 5}$. The length of a similar rectangular room is expressed as $\dfrac{x+4}{x^2 - 25}$. What is the scale factor of the lengths of the two rooms? Write in simplest form. $(x-4)(x+5)$

54. MODELING Cameron is taking a 20-mile kayaking trip. He travels half the distance at one rate. The rest of the distance he travels 2 miles per hour slower.

a. If x represents the faster pace in miles per hour, write an expression that represents the time spent at that pace. $\dfrac{10}{x}$

b. Write an expression for the amount of time spent at the slower pace. $\dfrac{10}{x-2}$

c. Write an expression for the amount of time Cameron needed to complete the trip. $\dfrac{20(x-1)}{x(x-2)}$

Find the slope of the line that passes through each pair of points.

55. $A\left(\frac{2}{p}, \frac{1}{2}\right)$ and $B\left(\frac{1}{3}, \frac{3}{p}\right)$ $-\dfrac{3}{2}$

56. $C\left(\frac{1}{4}, \frac{4}{q}\right)$ and $D\left(\frac{5}{q}, \frac{1}{5}\right)$ $-\dfrac{4}{5}$

57. $E\left(\frac{7}{w}, \frac{1}{7}\right)$ and $F\left(\frac{1}{7}, \frac{7}{w}\right)$ -1

58. $G\left(\frac{6}{n}, \frac{1}{6}\right)$ and $H\left(\frac{1}{6}, \frac{6}{n}\right)$ -1

Differentiated Instruction BL

Extension *Partial fraction decomposition* is a useful algebraic skill in more advanced mathematics courses, including calculus. Give students a rational expression like $\dfrac{5x + 3}{x(x + 1)}$, and ask what kinds of simpler fractions might be added to yield this expression. If they can identify x and $x + 1$ as denominators for such "partial fractions," ask them to try and find values of A and B that will make the equation $\dfrac{A}{x} + \dfrac{B}{x + 1} = \dfrac{5x + 3}{x(x + 1)}$ a true statement for any value of x. $A = 3$ and $B = 2$

Go Online! e

eSolutions Manual

Create worksheets, answer keys, and solutions handouts for your assignments.

59b. Sample answer: When the object is 70 mm away, y needs to be 0, which is impossible.

59. PHOTOGRAPHY The focal length of a lens establishes the field of view of the camera. The shorter the focal length is, the larger the field of view. For a camera with a fixed focal length of 70 millimeters to focus on an object x millimeters from the lens, the film must be placed a distance y from the lens. This is represented by $\frac{1}{x} + \frac{1}{y} = \frac{1}{70}$.

  a. Express y as a function of x. $y = \frac{70x}{x - 70}$

  b. What happens to the focusing distance when the object is 70 millimeters away?

60. PHARMACOLOGY Two drugs are administered to a patient. The concentrations in the bloodstream of each are given by $f(t) = \frac{2t}{3t^2 + 9t + 6}$ and $g(t) = \frac{3t}{2t^2 + 6t + 4}$ where t is the time, in hours, after the drugs are administered.

  a. Add the two functions together to determine a function for the total concentration of drugs in the patient's bloodstream. $h(t) = \frac{13t}{6t^2 + 18t + 12}$

  b. What is the concentration of drugs after 8 hours? about 0.19

 61. DOPPLER EFFECT Refer to the application at the beginning of the lesson. George is equidistant from two fire engines traveling toward him from opposite directions.

  a. Let x be the speed of the faster fire engine and y be the speed of the slower fire engine. Write and simplify a rational expression representing the difference in pitch between the two sirens according to George. **61a.** $\frac{P_0 s_0 x - P_0 s_0 y}{(s_0 - x)(s_0 - y)}$

  b. If one is traveling at 45 meters per second and the other is traveling at 70 meters per second, what is the difference in their pitches according to George? The speed of sound in air is 332 meters per second, and both engines have a siren with a pitch of 500 Hz. about 55.2 Hz

62. MULTIPLE REPRESENTATIONS Consider the following sets of rational functions. See Ch. 7 Answer Appendix.

$af(x) = \frac{a}{x}$ for $a = \{-2, -1, -0.5, 0.5, 2, 4\}$ $g(bx) = \frac{1}{bx}$ for $b = \{-2, -1, -0.5, 0.5, 2, 4\}$

$h(x - c) = \frac{1}{x - c}$ for $c = \{-4, -2, -0.5, 0.5, 2, 4\}$ $k(x) + d = \frac{1}{x} + d$ for $d = \{-4, -2, -0.5, 0.5, 2, 4\}$

  a. Graphical Graph each set of functions on a graphing calculator.

  b. Verbal Compare and contrast the graphs of the functions in each set.

  c. Tabular Choose two functions from any set. Find the slope between consecutive points on the graphs.

  d. Verbal Describe how the slopes of the two sections of a rational function graph are related.

A.APR.7

H.O.T. Problems   Use Higher-Order Thinking Skills

63. CHALLENGE Simplify $\frac{5x^{-2} - \frac{x + 1}{x}}{\frac{4}{3 - x^{-1}} + 6x^{-1}} \cdot \frac{-3x^3 - 2x^2 + 16x - 5}{4x^3 + 18x^2 - 6x}$

64. CONSTRUCT ARGUMENTS The sum of any two rational numbers is always a rational number. So, the set of rational numbers is said to be closed under addition. Determine whether the set of rational expressions is closed under addition, subtraction, multiplication, and division by a nonzero rational expression. Justify your reasoning. See margin.

65. OPEN-ENDED Write three monomials with an LCM of $180a^4b^6c$. Sample answer: $20a^4b^2c$, $15ab^6$, $9abc$

66. WRITING IN MATH Write a how-to manual for adding rational expressions that have unlike denominators. How does this compare to adding rational numbers? See margin.

Construct Arguments Mathematically proficient students understand and use stated assumptions, definitions, and previously established results in constructing arguments. They make conjectures and build a logical progression of statements to explore the truth of their conjectures. And they are able to analyze situations by breaking them into cases, and can recognize and use counterexamples.

Assess

Yesterday's News Ask students to describe how yesterday's lesson on multiplication and division of rational expressions helped prepare them for today's lesson on addition and subtraction of rational expressions.

Additional Answers

64. Sample answer: The set of rational expressions is closed under all of these operations because the sum, difference, product, and quotient of two rational expressions is a rational expression.

66. Sample answer: First, factor the denominators of all of the expressions. Find the LCD of the denominators. Convert each expression so they all have the LCD. Add or subtract the numerators. Then simplify. It is the same.

Preparing for Assessment

Exercises 67–71 require students to use the skills they will need on standardized assessments. The exercises are dual-coded with content standards and mathematical practice standards.

Dual Coding		
Exercise	Content Objective	Mathematical Practice
67	A.APR.7	4
68	A.APR.7	6
69	A.APR.7	1
70	A.APR.7	1
71	A.APR.7	4

Diagnose Student Errors

Survey student responses for each item. Class trends may indicate common errors and misconceptions.

67.

A	Chose the least speed as the average
B	Chose a speed between the two speeds
C	CORRECT
D	Found the mean of the two speeds

68.

A	Chose the excluded value for x
B	CORRECT
C	Assumed x cannot be negative
D	Assumed x cannot be positive
E	Assumed x can be any real number

69.

A	Added only 2 sides of the triangle
B	CORRECT
C	Added the terms in the numerators without multiplying by $x - 1$
D	Did not distribute the 12
E	Did not add the $15x$ term

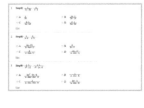

Go Online!

Self-Check Quizzes

Students can use *Self-Check Quizzes* to check their understanding of this lesson. You can also give the *Chapter Quiz*, which covers the content in Lessons 7-1 and 7-2.

Preparing for Assessment

67. Kim drove to work at an average speed of 48 miles per hour. On the ride home, her average speed was 36 miles per hour because of traffic. Which is closest to her average speed for the entire commute? 4 A.APR.7 C

○ A   36 mph
○ B   38 mph
○ C   41 mph
○ D   42 mph

68. What is the domain of the function shown? 6
A.APR.7 B

$$f(x) = 3 + \frac{2}{x^2}$$

○ A $\{x \mid x = 0\}$
○ B $\{x \mid x \neq 0\}$
○ C $\{x \mid x > 0\}$
○ D $\{x \mid x < 0\}$
○ E   all real numbers

69. What is the perimeter of the triangle? 1 A.APR.7 B

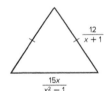

○ A $\frac{27x - 12}{x^2 - 1}$
○ B $\frac{39x - 24}{x^2 - 1}$
○ C $\frac{15x + 24}{x^2 - 1}$
○ D $\frac{27x - 2}{x^2 - 1}$
○ E $\frac{24x - 24}{x^2 - 1}$

70. What is the sum of $\frac{6x - 2}{3x^2 + 2x - 1} + \frac{7}{x + 1} + \frac{2}{x}$? 1
A.APR.7 C

○ A $\frac{7}{x + 1} + \frac{1}{2x}$
○ B $\frac{18x}{x + 1}$
○ C $\frac{11x + 2}{x^2 + x}$
○ D $\frac{11}{2x + 1}$

71. MULTI-STEP Jason hiked up a mountain trail at an average rate of 2 miles per hour. He hiked down the same trail at a rate of 3 miles per hour. 4 A.APR.7

a. Let d represent the distance from the bottom of the trail to the top of the mountain (the end of the trail). Which expression shows the total time Jason spent hiking the trail? C

○ A d
○ B $\frac{d}{2 + 3}$
○ C $\frac{d}{2} + \frac{d}{3}$
○ D $2d + 3d$

b. Let d represent the distance from the bottom of the trail to the top of the mountain (the end of the trail). Which expression shows Jason's average rate for the whole hike? C

○ A dt
○ B $\frac{12d}{2d + 4d}$
○ C $\frac{2d}{\frac{d}{2} + \frac{d}{3}}$
○ D $\frac{5d}{12d}$

c. What was Jason's average rate for the whole hike? B

○ A   1 mph
○ B   2.4 mph
○ C   2.5 mph
○ D   5 mph

LESSON 7-3

Graphing Reciprocal Functions

SUGGESTED PACING (DAYS)

90 min.	0.75
45 min.	1.5

Instruction

Track Your Progress

Objectives

1 Determine properties of reciprocal functions.

2 Graph transformations of reciprocal functions.

Mathematical Background

A reciprocal function has an equation of the form $f(x) = \frac{1}{a(x)}$ where $a(x)$ is a linear function and $a(x) \neq 0$. Reciprocal functions may have breaks in continuity for values that are excluded from the domain, and some may have an asymptote, a line that the graph of the function approaches.

THEN	NOW	NEXT
A.APR.7 (+) Understand that rational expressions form a system analogous to the rational numbers, closed under addition, subtraction, multiplication, and division by a nonzero rational expression; add, subtract, multiply, and divide rational expressions.	**A.CED.2** Create equations in two or more variables to represent relationships between quantities; graph equations on coordinate axes with labels and scales. **F.IF.4** For a function that models a relationship between two quantities, interpret key features of graphs and tables in terms of the quantities, and sketch graphs showing key features given a verbal description of the relationship. Key features include: intercepts; intervals where the function is increasing, decreasing, positive, or negative; relative maxima and minima; symmetries; end behavior; and periodicity.* **F.BF.3** Identify the effect on the graph of replacing $f(x)$ by $f(x) + k$, $k\,f(x)$, $f(kx)$, and $f(x + k)$ for specific values of k (both positive and negative); find the value of k given the graphs. Experiment with cases and illustrate an explanation of the effects on the graph using technology. Include recognizing even and odd functions from their graphs and algebraic expressions for them.	**F.IF.9** Compare properties of two functions each represented in a different way (algebraically, graphically, numerically in tables, or by verbal descriptions). For example, given a graph of one quadratic function and an algebraic expression for another, say which has the larger maximum.

Go Online! All of these resources and more are available at connectED.mcgraw-hill.com

eLessons utilize the power of your interactive whiteboard in an engaging way. Use **Rational Functions**, screen 3, to introduce the concepts in this lesson.

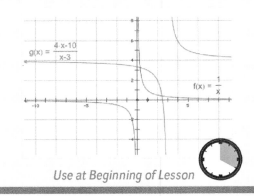

Use at Beginning of Lesson

Use **The Geometer's Sketchpad** to explore the graphs of reciprocal functions.

Use at Beginning of Lesson

Graphing Tools are outstanding tools for enhancing understanding.

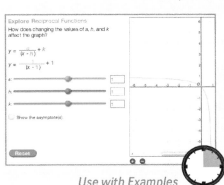

Use with Examples

OER **Using Open Educational Resources**

Publishing Have students work in groups to create a book using **Story Jumper** about reciprocal functions to reinforce their knowledge of the key concepts in this lesson. *Use as homework*

Go Online!
connectED.mcgraw-hill.com
Worksheets

Differentiate Your Resources

Extra Practice Additional practice or homework; Skills Practice is best for approaching-level students and Practice is best for on-level and beyond-level students

Skills Practice
AL OL ELL

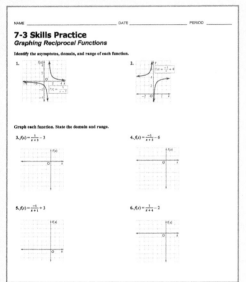

7-3 Skills Practice
Graphing Reciprocal Functions

Identify the asymptotes, domain, and range of each function.

Graph each function. State the domain and range.

Practice
AL OL BL ELL

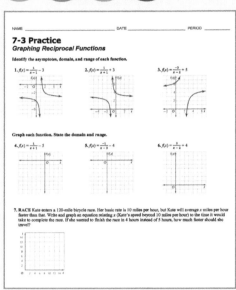

7-3 Practice
Graphing Reciprocal Functions

Word Problem Practice
AL OL BL ELL

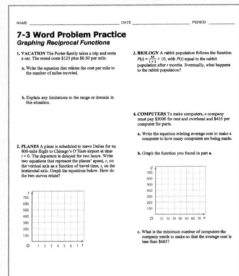

7-3 Word Problem Practice
Graphing Reciprocal Functions

Intervention Reteaching and vocabulary activities that can be used with struggling or absent students and as ELL support

Study Guide and Intervention
AL OL ELL

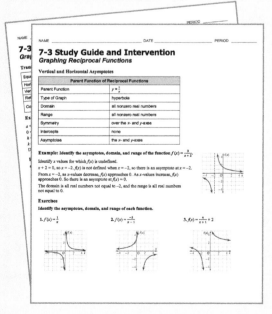

Study Notebook
AL OL BL ELL

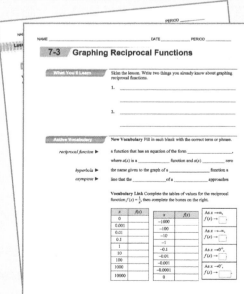

Extension Activities that can be used to extend lesson concepts

Enrichment
OL BL ELL

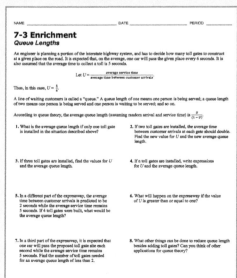

7-3 Enrichment
Queue Lengths

483B | Lesson 7-3 | Graphing Reciprocal Functions

LESSON 3

Graphing Reciprocal Functions

::Then	::Now	::Why?
● You graphed polynomial functions.	**1** Determine properties of reciprocal functions. **2** Graph transformations of reciprocal functions.	● The sophomore class is renting an indoor trampoline park for a class party. The cost of renting the facility is $900, which is shared equally among all students who attend. If c represents the cost to each student and n represents the number of students, then $c = \frac{900}{n}$.

 New Vocabulary
reciprocal function
hyperbola

 Mathematical Practices
2 Reason abstractly and quantitatively.

Content Standards
A.CED.2 Create equations in two or more variables to represent relationships between quantities; graph equations on coordinate axes with labels and scales.
F.IF.4 For a function that models a relationship between two quantities, interpret key features of graphs and tables in terms of the quantities, and sketch graphs showing key features given a verbal description of the relationship.
F.BF.3 Identify the effect on the graph of replacing $f(x)$ by $f(x) + k$, $k f(x)$, $f(kx)$, and $f(x + k)$ for specific values of k (both positive and negative); find the value of k given the graphs. Experiment with cases and illustrate an explanation of the effects on the graph using technology. Include recognizing even and odd functions from their graphs and algebraic expressions for them.

1 **Vertical and Horizontal Asymptotes** The function $c = \frac{5000}{n}$ is a reciprocal function. A **reciprocal function** has an equation of the form $f(x) = \frac{1}{a(x)}$, where $a(x)$ is a linear function and $a(x) \neq 0$.

 Key Concept Parent Function of Reciprocal Functions

Parent function:	$f(x) = \frac{1}{x}$
Type of graph:	**hyperbola**
Domain and range:	all nonzero real numbers, $(-\infty, 0) \cup (0, +\infty)$
Asymptotes:	$x = 0$ and $f(x) = 0$
Intercepts:	none
Not defined:	$x = 0$

The domain of a reciprocal function is limited to values for which the denominator is nonzero.

Functions:	$f(x) = \frac{-3}{x + 2}$	$g(x) = \frac{4}{x - 5}$	$h(x) = \frac{3}{x}$
Not defined at:	$x = -2$	$x = 5$	$x = 0$

F.IF.4

Example 1 Limitations on Domain

Determine the value of x for which $f(x) = \frac{3}{2x + 5}$ is not defined.

Find the value for which the denominator of the expression equals 0.

$\frac{3}{2x + 5} \rightarrow 2x + 5 = 0$

$x = -\frac{5}{2}$ The function is undefined for $x = -\frac{5}{2}$.

▶ **Guided Practice**

Determine the value of x for which each function is not defined.

1A. $f(x) = \frac{2}{x - 1}$ $x = 1$ **1B.** $f(x) = \frac{7}{3x + 2}$ $x = -\frac{2}{3}$

 Mathematical Practices Strategies

Reason abstractly and quantitatively.
Help students understand how to graph reciprocal functions and write their equations. For example, ask:

• A rational expression cannot have a denominator equal to 0. How does this help determine the asymptote of a hyperbola? The asymptote occurs where the undefined value is.

• What is the domain of a hyperbola? The domain cannot contain undefined values but otherwise approaches infinity in both directions.

• What kind of function has a hyperola as its reciprocal? a linear function

• Is it possible for a hyperbola to have more than one undefined value? **Explain.** No; linear functions only have one expression for x (no exponents).

Launch

Have students read the Why? section of the lesson. Ask:

• On what does the cost to each student depend? the number of students participating

• What happens to the value of c as the value of n increases? It decreases.

• The variable n must be greater than or equal to what whole number? 1

• The variable n must be less than or equal to what number? the number of students in the sophomore class

Teach

Ask the scaffolded questions for each example to build conceptual understanding for students at all levels.

1 **Vertical and Horizontal Asymptotes**

Example 1 Limitations on Domain

AL Why is $f(x)$ undefined at $x = -\frac{5}{2}$? $-\frac{5}{2}$ makes the denominator equal to 0.

OL What is the domain of $f(x)$? $\left(-\infty, -\frac{5}{2}\right) \cup \left(-\frac{5}{2}, +\infty\right)$

BL Write a rule about how many places a reciprocal function will be undefined. A reciprocal function will be undefined at one place where the denominator is equal to 0.

Need Another Example?
Determine the values of x for which $f(x) = \frac{2}{3x + 24}$ is not defined. $x = -8$

 Go Online!

Interactive Whiteboard
Use the *eLesson* or *Lesson Presentation* to present this lesson.

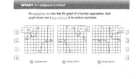

Example 2 Determine Properties of Reciprocal Functions

AL What is the end behavior of $f(x) = \dfrac{2}{x-3}$?
$y \to 0$ as $x \to -\infty$, $y \to 0$ as $x \to \infty$.

OL Why are the equations of vertical asymptotes written in the form $x = a$? Vertical asymptotes are vertical lines and the equations of vertical lines are of the form $x = a$.

BL How many asymptotes will a reciprocal function have? $1 +$ number of unique zeros of the denominator

Need Another Example?

Identify the asymptotes, domain, and range of each function.

a.

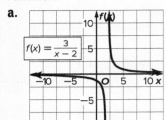

$f(x) = \dfrac{3}{x-2}$

asymptotes: $x = 2$, $f(x) = 0$; $D = \{x \,|\, x \neq 2\}$ or $(-\infty, 2) \cup (2, +\infty)$; $R = \{f(x) \,|\, f(x) \neq 0\}$ or $(-\infty, 0) \cup (0, +\infty)$

b.

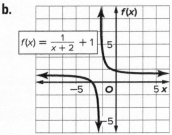

$f(x) = \dfrac{1}{x+2} + 1$

asymptotes: $x = -2$, $f(x) = 1$; $D = \{x \,|\, x \neq -2\}$ or $(-\infty, -2) \cup (-2, +\infty)$; $R = \{f(x) \,|\, f(x) \neq 1\}$ or $(-\infty, 1) \cup (1, +\infty)$

The graphs of reciprocal functions have breaks in continuity for excluded values. They have asymptotes which are lines that the graph of the function approaches.

F.IF.4

Study Tip

Structure Vertical asymptotes show where a function is undefined, while horizontal asymptotes show the end behavior of a graph.

Go Online!

Investigate how changing the values in the equation of a reciprocal function affects the graph by using the eToolkit in ConnectED.

Example 2   Determine Properties of Reciprocal Functions

Identify the asymptotes, domain, and range of each function.

a.

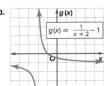

$f(x) = \dfrac{2}{x-3}$

Identify x-values for which $f(x)$ is undefined.
$$x - 3 = 0$$
$$x = 3$$

$f(x)$ is not defined when $x = 3$. So there is an asymptote at $x = 3$.

From $x = 3$, as x-values decrease, $f(x)$-values approach 0, and as x-values increase, $f(x)$-values approach 0. So there is an asymptote at $f(x) = 0$.

The domain is all real numbers not equal to 3 and the range is all real numbers not equal to 0. This is represented as $D = \{x \,|\, x \neq 3\}$ or $(-\infty, 3) \cup (3, +\infty)$ and $R = \{f(x) \,|\, f(x) \neq 0\}$ or $(-\infty, 0) \cup (0, +\infty)$.

b.

$g(x) = \dfrac{1}{x+2} - 1$

Identify x-values for which $g(x)$ is undefined.
$$x + 2 = 0$$
$$x = -2$$

$g(x)$ is not defined when $x = -2$. So there is an asymptote at $x = -2$.

From $x = -2$, as x-values decrease, $g(x)$-values approach -1, and as x-values increase, $g(x)$-values approach -1. So there is an asymptote at $g(x) = -1$.

The domain is all real numbers not equal to -2. The range is all real numbers not equal to -1. This is represented as $D = \{x \,|\, x \neq -2\}$ or $(-\infty, -2) \cup (-2, +\infty)$ and $R = \{g(x) \,|\, g(x) \neq -1\}$ or $(-\infty, -1) \cup (-1, +\infty)$.

▶ **Guided Practice**

$x = 3$, $f(x) = -2$;
$D = \{x \,|\, x \neq 3\}$;
$R = \{f(x) \,|\, f(x) \neq -2\}$

$x = -1$, $g(x) = 5$;
$D = \{x \,|\, x \neq -1\}$;
$R = \{g(x) \,|\, g(x) \neq 5\}$

2A.

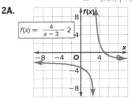

$f(x) = \dfrac{4}{x-3} - 2$

2B.

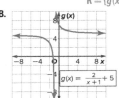

$g(x) = \dfrac{2}{x+1} + 5$

Differentiated Instruction **OL** **BL**

Extension Tell students the following:

- A function is called an **even** function if $f(-x) = f(x)$ for all values of x in its domain. The graph of an even function is symmetric with respect to the y-axis.

- A function is called an **odd** function if $f(-x) = -f(x)$ for all values of x in its domain. The graph of an odd function is symmetric with respect to the origin.

Ask students whether $f(x) = \dfrac{1}{x}$ is an odd or even function. odd

2 Transformations of Reciprocal Functions
The same techniques used to transform the graphs of other functions you have studied can be applied to the graphs of reciprocal functions.

⬦ Key Concept Transformations of Reciprocal Functions

$$f(x) = \frac{a}{b(x-h)} + k$$

h—Horizontal Translation	k—Vertical Translation
h units right if h is positive	k units up if k is positive
|h| units left if h is negative	|k| units down if k is negative
The *vertical* asymptote is at x = h.	The *horizontal* asymptote is at f(x) = k.

a—Orientation and Shape	b—Orientation and Shape
If a < 0, the graph is reflected in the x-axis.	If b < 0, the graph is reflected in the y-axis.
If |a| > 1, the graph is stretched vertically.	If |b| > 1, the graph is compressed horizontally.
If 0 < |a| < 1, the graph is compressed vertically.	If 0 < |b| < 1, the graph is stretched horizontally.

Study Tip

Asymptotes The asymptotes of a reciprocal function move with the graph of the function and intersect at (h, k).

F.BF.3

Example 3 Graph Transformations

Graph each function. State the domain and range.

a. $f(x) = \dfrac{1}{\frac{1}{2}x - 2} + 2$

This represents a transformation of the graph of $f(x) = \frac{1}{x}$.

$b = \frac{1}{2}$: The graph is stretched horizontally.

$h = 4$: The graph is translated 4 units right because $\frac{1}{2}x - 2 = \frac{1}{2}(x - 4)$. There is an asymptote at $x = 4$.

$k = 2$: The graph is translated 2 units up. There is an asymptote at $f(x) = 2$.

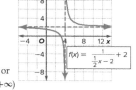

Domain: $\{x \mid x \neq 4\}$ or $(-\infty, 4) \cup (4, +\infty)$

Range: $\{f(x) \mid f(x) \neq 2\}$ or $(-\infty, 2) \cup (2, +\infty)$

b. $f(x) = \dfrac{-3}{x + 1} - 4$

This represents a transformation of the graph of $f(x) = \frac{1}{x}$.

$a = -3$: The graph is stretched vertically and reflected across the x-axis.

$h = -1$: The graph is translated 1 unit left. There is an asymptote at $x = -1$.

$k = -4$: The graph is translated 4 units down. There is an asymptote at $f(x) = -4$.

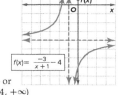

Domain: $\{x \mid x \neq -1\}$ or $(-\infty, -1) \cup (-1, +\infty)$

Range: $\{f(x) \mid f(x) \neq -4\}$ or $(-\infty, -4) \cup (-4, +\infty)$

▶ **Guided Practice 3A, 3B.** See Ch. 7 Answer Appendix.

3A. $f(x) = \dfrac{-2}{x + 4} + 1$ **3B.** $g(x) = \dfrac{1}{-3x + 3} - 2$

Watch Out!

Preventing Errors Suggest that students choose a large unit on their grid paper and estimate point coordinates to the nearest tenth. Point out that they may not be able to see the shape of the graph as a whole unless they use a graphing calculator or computer program.

---

connectED.mcgraw-hill.com 485

Teaching the Mathematical Practices
MP

Structure Mathematically proficient students look closely to discern a pattern or structure. Encourage students to compare the equations of asymptotes to the continuity and behavior of a function.

2 Transformations of Reciprocal Functions

Example 3 Graph Transformations

AL If $f(x) = \dfrac{3}{x - 5} - 7$, what are a, h, and k? a = 3, h = 5, and k = -7

OL Write an expression where $\frac{1}{x}$ has been reflected across the x-axis and shifted up 4 units. $-\dfrac{1}{x} + 4$

BL How does a horizontal transformation change the asymptotes of a function? It does not change the horizontal asymptote. It would move any vertical asymptotes over by that many units.

Need Another Example?
Graph each function. State the domain and range.

a. $f(x) = -\dfrac{1}{x + 1} + 3$

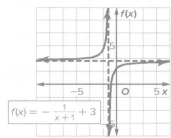

D = $\{x \mid x \neq -1\}$;
R = $\{f(x) \mid f(x) \neq 3\}$

b. $f(x) = \dfrac{-4}{x - 2} - 1$

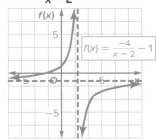

D = $\{x \mid x \neq 2\}$;
R = $\{f(x) \mid f(x) \neq -1\}$

Example 4 Write Equations

AL What do r, t, and d represent? r is the rate (speed) of the airplane, t is the time it takes to complete the flight, and d is the distance of the flight.

OL How long will the trip take if the speed of the airplane is 570 mph? approximately 15.8 hours

BL What inequality would you set up to find what speed the airplane would need to travel at to make the trip in less than 10 hours? $10 < \dfrac{9000}{r}$

Need Another Example?

Commuting A commuter train has a nonstop service from one city to another, a distance of about 25 miles.

a. Write an equation to represent the travel time between these two cities as a function of rail speed. Then graph the equation. $t = \dfrac{25}{r}$

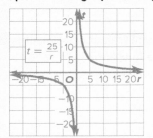

b. Explain any limitations to the range and domain in this situation. The range and domain are limited to all real numbers greater than 0 because negative values do not make sense. There will be further restrictions to the domain because the train has minimum and maximum speeds at which it can travel.

Go Online!

eBook

Interactive Student Guide

ALGEBRA 2
INTERACTIVE STUDENT GUIDE

Use the *Interactive Student Guide* to deepen conceptual understanding.
· Graphing Reciprocal Functions

A.CED.2, F.IF.4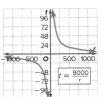

Real-World Example 4 Write Equations

TRAVEL Philip is taking a nonstop flight to Hyderabad, India, for a business trip. A one-way trip is about 9000 miles.

a. Write an equation to represent the travel time to Hyderabad as a function of flight speed. Then graph the equation.

Solve $rt = d$ for t.

$rt = d$ Original formula

$t = \dfrac{d}{r}$ Divide each side by r

$t = \dfrac{9000}{r}$ $d = 9000$

Graph $t = \dfrac{9000}{r}$.

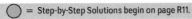

b. Explain any limitations to the range or domain in this situation.

In this situation, the range and domain are limited to all real numbers greater than zero because negative values do not make sense. There will be further restrictions to the domain because the aircraft has minimum and maximum speeds at which it can travel.

▶ **Guided Practice**   4. See Ch. 7 Answer Appendix.

4. HOMECOMING DANCE The junior and senior class officers are sponsoring a homecoming dance. The total cost for the facilities and catering is $45 per person plus a $2500 deposit. Write and graph an equation to represent the average cost per person. Then explain any limitations to the domain and range.

Check Your Understanding   ◯ = Step-by-Step Solutions begin on page R11.   ✓ **Go Online!** for a Self-Check Quiz

Examples 1–2   Identify the asymptotes, domain, and range of each function.
F.IF.4

1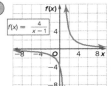
$f(x) = \dfrac{4}{x-1}$
$x = 1$, $f(x) = 0$;
$D = \{x \mid x \neq 1\}$;
$R = \{f(x) \mid f(x) \neq 0\}$

2.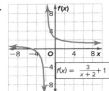
$f(x) = \dfrac{3}{x+2} + 1$
$x = -2$, $f(x) = 1$;
$D = \{x \mid x \neq -2\}$;
$R = \{f(x) \mid f(x) \neq 1\}$

Example 3   Graph each function. State the domain and range.   3–5. See Ch. 7 Answer Appendix.
F.BF.3

3. $f(x) = \dfrac{5}{x}$ **4.** $f(x) = \dfrac{2}{x+3}$ **5.** $f(x) = \dfrac{-1}{x-2} + 4$

Example 4   **6.** **MP** SENSE-MAKING A group of friends plans to get their youth group leader a gift
A.CED.2,   certificate for a day at a spa. The certificate costs $150.
F.IF.4

     a. If c represents the cost for each friend and f represents the number of friends, write an equation to represent the cost to each friend as a function of how many friends give. $c = \dfrac{150}{f}$

     b. Graph the function. See Ch. 7 Answer Appendix.

     c. Explain any limitations to the range or domain in this situation.

     c. In this situation, the range and domain are limited to all real numbers greater than zero because negative values do not make sense.

Differentiated Instruction   **AL** **OL** **ELL**

Visual/Spatial Learners Have students graph one of the functions from the lesson on a large sheet of poster board to clearly show how the graph approaches but never reaches an asymptote. Encourage students to use a variety of colored markers.

Practice and Problem Solving

Extra Practice is on page R7.

Examples 1–2
F.IF.4

Identify the asymptotes, domain, and range of each function.

7.
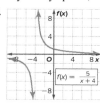
$x = -4, f(x) = 0;$
$D = \{x \mid x \neq -4\};$
$R = \{f(x) \mid f(x) \neq 0\}$
$f(x) = \frac{5}{x+4}$

8.
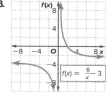
$x = 0, f(x) = -3;$
$D = \{x \mid x \neq 0\};$
$R = \{f(x) \mid f(x) \neq -3\}$
$f(x) = \frac{6}{x} - 3$

9.
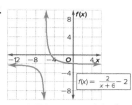
$f(x) = \frac{2}{x+6} - 2$
$x = -6, f(x) = -2; D = \{x \mid x \neq -6\};$
$R = \{f(x) \mid f(x) \neq -2\}$

10.
$f(x) = \frac{-3}{x-1} + 5$
$x = 1, f(x) = 5;$
$D = \{x \mid x \neq 1\};$
$R = \{f(x) \mid f(x) \neq 5\}$

Example 3
F.BF.3

Graph each function. State the domain and range. 11–22. See Ch. 7 Answer Appendix.

11. $f(x) = \frac{3}{x}$

12. $f(x) = \frac{-4}{x+2}$

13. $f(x) = \frac{2}{x-6}$

14. $f(x) = \frac{6}{x} - 5$

15. $f(x) = \frac{2}{x} + 3$

16. $f(x) = \frac{8}{x}$

17. $f(x) = \frac{-2}{x-5}$

18. $f(x) = \frac{3}{x-7} - 8$

19. $f(x) = \frac{9}{x+3} + 6$

20. $f(x) = \frac{8}{-x+3}$

21. $f(x) = \frac{-6}{x+4} - 2$

22. $f(x) = \frac{-5}{\frac{1}{4}x - 2} + 2$

Example 4
A.CED.2,
F.IF.4

23. CYCLING Marina's New Year's resolution is to ride her bike 5000 miles.

a. If m represents the mileage Marina rides each day and d represents the number of days, write an equation to represent the mileage each day as a function of the number of days that she rides. $m = \frac{5000}{d}$

b. Graph the function. See Ch. 7 Answer Appendix.

c. If she rides her bike every day of the year, how many miles should she ride each day to meet her goal? 13.7 mi

24. MODELING Parker has 200 grams of an unknown liquid. Knowing the density will help him discover what type of liquid this is.

a. Density of a liquid is found by dividing the mass by the volume. Write an equation to represent the density of this unknown as a function of volume. $d = \frac{200}{v}$

b. Graph the function. See Ch. 7 Answer Appendix.

c. From the graph, identify the asymptotes, domain, and range of the function.
$v = 0, d = 0; D = \{v \mid v \neq 0\}; R = \{d \mid d \neq 0\}$

Graph each function. State the domain and range. 25–28. See Ch. 7 Answer Appendix.

25. $f(x) = \frac{3}{2x-4}$

26. $f(x) = \frac{5}{3x}$

27. $f(x) = \frac{2}{4x+1}$

28. $f(x) = \frac{1}{2x+3}$

Differentiated Homework Options

Levels	AL Basic	OL Core	BL Advanced
Exercises	7–24, 38–46	7–35 odd, 36–46	25–46
2-Day Option	7–23 odd, 43, 44	7–24, 43, 44	
	8–24 even, 38–42, 45, 46	25–42, 45, 46	

 You can use ALEKS to provide additional remediation support with personalized instruction and practice.

Practice

Formative Assessment Use Exercises 1–6 to assess students' understanding of the concepts in this lesson.

The Practice and Problem Solving exercises assess the content taught in the lesson. The Preparing for Assessment page is meant to be used as preparation for end-of-course assessments.

Extra Practice

See page R7 for extra exercises for students who are approaching level or for on-level students who need additional reinforcement.

Exercise Alert

Grid Paper For Exercises 3–6, 11–22, 25–35, and 41, students will need grid paper.

Levels of Complexity Chart

The levels of the exercises progress from 1 to 3, with Level 1 indicating the lowest level of complexity.

Exercises	7–24	25–35, 43–46	36–42
Level 3			●
Level 2		●	
Level 1	●		

Go Online!

eSolutions Manual
Create worksheets, answer keys, and solutions handouts for your assignments.

MP **Teaching the Mathematical Practices**

Construct Arguments Mathematically proficient students understand and use stated assumptions, definitions, and previously established results in constructing arguments. They make conjectures and build a logical progression of statements to explore the truth of their conjectures. And they are able to analyze situations by breaking them into cases, and can recognize and use counterexamples.

Assess

Crystal Ball Ask students to tell how they think their practice with graphing reciprocal functions will help them when they graph rational functions.

---

 29 BASEBALL The distance from the pitcher's mound to home plate is 60.5 feet.

a. If r represents the speed of the pitch and t represents the time it takes the ball to get to the plate, write an equation to represent the speed as a function of time. $r = \frac{60.5}{t}$

b. Graph the function. See Ch. 7 Answer Appendix.

c. If a two-seam fastball reaches the plate in 0.48 second, what was its speed? about 126 ft/s

30–35. See Ch. 7 Answer Appendix.

Graph each function. State the domain and range, and identify the asymptotes.

30. $f(x) = \frac{-3}{x+7} - 1$ **31.** $f(x) = \frac{-4}{x+2} - 5$ **32.** $f(x) = \frac{6}{x-1} + 2$

33. $f(x) = \frac{2}{x-4} + 3$ **34.** $f(x) = \frac{-7}{x-8} - 9$ **35.** $f(x) = \frac{-6}{x-7} - 8$

36. FINANCIAL LITERACY Lawanda's car went 440 miles on one tank of gas.

a. If g represents the number of miles to the gallon that the car gets and t represents the size of the gas tank, write an equation to represent the miles to the gallon as a function of tank size. $g = \frac{440}{t}$

b. Graph the function. See Ch. 7 Answer Appendix.

c. How many miles does the car get per gallon if it has a 15-gallon tank? $29\frac{1}{3}$ mi/gal

37. MULTIPLE REPRESENTATIONS In this problem you will investigate the similarities and differences between power functions with positive and negative exponents.

a. **TABULAR** Make a table of values for $a(x) = x^2$, $b(x) = x^{-2}$, $c(x) = x^3$, and $d(x) = x^{-3}$.

b. **GRAPHICAL** Graph $a(x)$ and $b(x)$ on the same coordinate plane.

c. **VERBAL** Compare the domain, range, end behavior, and behavior at $x = 0$ for $a(x)$ and $b(x)$.   37. a–f. See Ch. 7 Answer Appendix.

d. **GRAPHICAL** Graph $c(x)$ and $d(x)$ on the same coordinate plane.

e. **VERBAL** Compare the domain, range, end behavior, and behavior at $x = 0$ for $c(x)$ and $d(x)$.

f. **ANALYTICAL** What conclusions can you make about the similarities and differences between power functions with positive and negative exponents?

A.CED.2, F.IF.4, F.BF.3

H.O.T. Problems   Use Higher-Order Thinking Skills

38. OPEN-ENDED Write a reciprocal function for which the graph has a vertical asymptote at $x = -4$ and a horizontal asymptote at $f(x) = 6$.   Sample answer: $f(x) = \frac{1}{x+4} + 6$

39. **MP** REASONING Compare and contrast the graphs of each pair of equations.

a. $y = \frac{1}{x}$ and $y - 7 = \frac{1}{x}$ **b.** $y = \frac{1}{x}$ and $y = 4\left(\frac{1}{x}\right)$ **c.** $y = \frac{1}{x}$ and $y = \frac{1}{x+5}$

d. Without making a table of values, use what you observed in parts **a–c** to sketch a graph of $y - 7 = 4\left(\frac{1}{x+5}\right)$.   **a–d.** See Ch. 7 Answer Appendix.

40. **MP** CONSTRUCT ARGUMENTS Find the function that does not belong. Explain.

$f(x) = \frac{3}{x+1}$	$g(x) = \frac{x+2}{x^2+1}$	$h(x) = \frac{5}{x^2+2x+1}$	$j(x) = \frac{20}{x-7}$

Sample answer: $g(x)$; all other choices have unknowns only in the denominator.

41. CHALLENGE Write two different reciprocal functions with graphs having the same vertical and horizontal asymptotes. Then graph the functions. See Ch. 7 Answer Appendix.

42. WRITING IN MATH Refer to the beginning of the lesson. Explain how rational functions can be used to represent shared costs. Explain why only part of the graph is meaningful in the context of the problem. See Ch. 7 Answer Appendix.

MP **Mathematical Process Standards**

Emphasis On	Exercises
2 Reason abstractly and quantitatively.	7–22, 25–28, 43–46
3 Construct viable arguments and critique the reasoning of others.	37–42
4 Model with mathematics.	23, 24, 29, 36
6 Attend to precision.	30–36

Preparing for Assessment

43. What is the domain of the function? 2 A.CED.2
$$f(x) = \frac{-3}{x-1} - 2 \qquad \text{D}$$

- ○ **A** $D = \{x \mid x \neq 0\}$
- ○ **B** $D = \{x \mid x \neq -1\}$
- ○ **C** $D = \{x \mid x \neq -2\}$
- ○ **D** $D = \{x \mid x \neq 1\}$

44. For which reciprocal function are the x- and y-values for the asymptotes the same? 2 A.CED.2
　　　　　　　　　　　　　　　　　　　　　　D

- ○ **A** $f(x) = \frac{-4}{x-4}$
- ○ **B** $f(x) = \frac{-4}{x-3} - 4$
- ○ **C** $f(x) = \frac{4}{x} + 4$
- ○ **D** $f(x) = \frac{1}{x+4} - 4$
- ○ **E** $f(x) = \frac{1}{x+4} + 4$

45. **MULTI-STEP** Consider the graph of the function $f(x) = \frac{-9}{x+9}$. 2 A.CED.2

a. Which statements about this function are true? Choose all that apply. B, C, F

- ☐ **A** It has an asymptote at $x = 9$.
- ☐ **B** It has an asymptote at $f(x) = 0$.
- ☐ **C** It has an asymptote at $x = -9$.
- ☐ **D** Its transformation from $f(x) = \frac{1}{x}$ includes a translation 9 units down.
- ☐ **E** Its transformation from $f(x) = \frac{1}{x}$ includes a translation 9 units up.
- ☐ **F** Its transformation from $f(x) = \frac{1}{x}$ includes a translation 9 units left.

b. What is the value of the function at $x = -1$? B

- ○ **A** undefined
- ○ **B** negative
- ○ **C** positive
- ○ **D** infinity

46. **MULTI-STEP** Consider the graph of the function $f(x) = \frac{-1}{x-1} + 1$. 2 A.CED.2

a. Which statements about this function are true? Choose all that apply. B, C, E, F

- ☐ **A** It has an asymptote at $x = -1$.
- ☐ **B** It has an asymptote at $x = 1$.
- ☐ **C** It has an asymptote at $f(x) = 1$.
- ☐ **D** Its transformation from $f(x) = \frac{1}{x}$ includes a translation 1 unit down.
- ☐ **E** Its transformation from $f(x) = \frac{1}{x}$ includes a translation 1 unit up.
- ☐ **F** $f(x) = \frac{-8}{x-1} + 1$ has the same asymptotes but is stretched taller.

b. What is the value of the function at $x = -1$? D

- ○ **A** undefined
- ○ **B** infinity
- ○ **C** -1
- ○ **D** 1.5

c. Which is the graph of the function? A

○ **A** 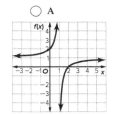　　○ **B**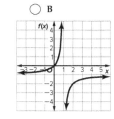

○ **C** 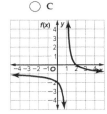　　○ **D**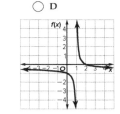

Preparing for Assessment

Exercises 43–46 require students to use the skills they will need on standardized assessments. The exercises are dual-coded with content standards and mathematical practice standards.

Dual Coding		
Exercise	Content Objective	Mathematical Practice
43	A.CED.2	2
44	A.CED.2	2
45	A.CED.2	2
46	A.CED.2	2

Diagnose Student Errors

Survey student responses for each item. Class trends may indicate common errors and misconceptions.

43.

A	Chose domain for function with a denominator of x
B	Chose domain for function with a denominator of $x + 1$
C	Chose the value of k
D	CORRECT

44.

A	Chose function with a horizontal asymptote of 0 and vertical asymptote of 4
B	Chose function with a horizontal asymptote of -4 and vertical asymptote of 3
C	Chose function with a horizontal asymptote of 4 and vertical asymptote of 0
D	CORRECT
E	Chose function with a horizontal asymptote of 4 and vertical asymptote of -4

RtI Response to Intervention

Use the Intervention Planner to help you determine your Response to Intervention.

Intervention Planner

TIER 1 **On Level** OL

IF students miss 25% of the exercises or less,

THEN choose a resource:

SE Lessons 7-1, 7-2, and 7-3

Go Online!

 Skills Practice

Chapter Project

Self-Check Quizzes

TIER 2 **Strategic Intervention** AL
Approaching grade level

IF students miss 50% of the exercises,

THEN *Go Online!*

Study Guide and Intervention

Extra Examples

Personal Tutors

Homework Help

TIER 3 **Intensive Intervention**
2 or more grades below level

IF students miss 75% of the exercises,

THEN choose a resource:

Use *Math Triumphs, Alg. 2*

Go Online!

Extra Examples

Personal Tutors

Homework Help

$^{a}_{bc}$ Review Vocabulary

Go Online!

ᵉAssessment

You can use the premade Mid-Chapter Test to assess students' progress in the first half of the chapter. Customize and create multiple versions of your Mid-Chapter Quiz and answer keys that align to your standards. Tests can be delivered on paper or online.

CHAPTER 7
Mid-Chapter Quiz
Lessons 7-1 through 7-3

Simplify each expression. (Lesson 7-1)

1. $\dfrac{2x^2y^5}{7x^3yz} \cdot \dfrac{14xyz^2}{18x^4y}$ $\dfrac{2y^4z}{9x^4}$

2. $\dfrac{24a^4b^6}{35ab^3} \div \dfrac{12abc}{7a^2c}$ $\dfrac{2a^4b^2}{5}$

3. $\dfrac{3x-3}{x^2+x-2} \cdot \dfrac{4x+8}{6x+18}$ $\dfrac{2}{x+3}$

4. $\dfrac{(m+2)(m+5)}{3}$ $\dfrac{m^2+3m+2}{9} \div \dfrac{m+1}{3m+15}$

5. $\dfrac{\frac{r^2+3r}{r+1}}{\frac{3r}{3r+3}}$ $r+3$

6. $\dfrac{\frac{2y}{y^2-4}}{\frac{3}{y^2-4y+4}}$ $\dfrac{2y(y-2)}{3(y+2)}$

7. **MULTIPLE CHOICE** For all $r \neq \pm 2$, $\dfrac{r^2+6r+8}{r^2-4} =$ ___. (Lesson 7-1) **B**

  A $\dfrac{r-2}{r+4}$ C $\dfrac{r+2}{r-4}$

  B $\dfrac{r+4}{r-2}$ D $\dfrac{r+4}{r+2}$

8. **MULTIPLE CHOICE** Identify all values of x for which $\dfrac{x^2-16}{(x^2-6x-27)(x+1)}$ is undefined. (Lesson 7-1) **C**

  A $-3, -1$ C $-3, -1, 9$

  B $3, 1, -9$ D -1

9. What is the LCM of $x^2 - x$ and $3 - 3x$? (Lesson 7-2)
  $-3x(x-1)$

Simplify each expression. (Lesson 7-2)

10. $\dfrac{2x}{4x^2y} + \dfrac{x}{3xy^3}$ $\dfrac{6xy^2+4x^2}{12x^2y^3}$

11. $\dfrac{3}{4m} + \dfrac{2}{3mn^2} - \dfrac{4}{n}$ $\dfrac{9n^2+8-48mn}{12mn^2}$

12. $\dfrac{6}{r^2-3r-18} - \dfrac{1}{r^2+r-6}$ $\dfrac{5r-6}{(r-6)(r+3)(r-2)}$

13. $\dfrac{3x+6}{x+y} + \dfrac{6}{-x-y}$ $\dfrac{3x}{x+y}$

14. $\dfrac{x-4}{x^2-3x-4} + \dfrac{x+1}{2x-8}$ $\dfrac{x^2+4x-7}{2(x-4)(x+1)}$

15. Determine the perimeter of the rectangle. (Lesson 7-2)

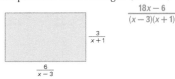

$\dfrac{18x-6}{(x-3)(x+1)}$

16. **TRAVEL** Lucita is going to a beach 100 miles away. She travels half the distance at one rate. The rest of the distance, she travels 15 miles per hour slower. (Lesson 7-2)

  a. If x represents the faster pace in miles per hour, write an expression that represents the time spent at that pace. $\dfrac{50}{x}h$

  b. Write an expression for the amount of time spent at the slower pace. $\dfrac{50}{x-15}h$

  c. Write an expression for the amount of time Lucita needs to complete the trip. $\dfrac{100(x-75)}{x(x-15)}$

Identify the asymptotes, domain, and range of each function. (Lesson 7-3)

17. 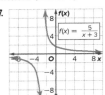 $x=-3; f(x)=0;$
$D=\{x \mid x \neq -3\},$
$R=\{f(x) \mid f(x) \neq 0\}$

18. 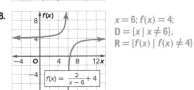 $x=6; f(x)=4;$
$D=\{x \mid x \neq 6\},$
$R=\{f(x) \mid f(x) \neq 4\}$

Graph each reciprocal function. State the domain and range. (Lesson 7-3) **19–24. See Ch. 7 Answer Appendix.**

19. $f(x) = \dfrac{4}{x}$

20. $f(x) = \dfrac{1}{3x}$

21. $f(x) = \dfrac{6}{x-1}$

22. $f(x) = \dfrac{-2}{x} + 4$

23. $f(x) = \dfrac{3}{x+2} - 5$

24. $f(x) = -\dfrac{1}{x-3} + 2$

25. **SANDWICHES** A group makes 45 sandwiches to take on a picnic. The number of sandwiches a person can eat depends on how many people go on the trip. (Lesson 7-3)

  a. Write a function to represent the numbers of sandwiches that can be taken on the picnic.. $f(x) = \dfrac{45}{x}$

  b. Graph the function.   **See Ch. 7 Answer Appendix.**

  c. What mathematical practice did you use to solve this problem?   **See students' work.**

Foldables Study Organizer

Dinah Zike's FOLDABLES®

Before students complete the Mid-Chapter Quiz, encourage them to review the information for Lessons 7-1 through 7-3 in their Foldables. Give students time to ask any questions they have about the rational expressions and reciprocal functions.

ALEKS can be used as a formative assessment tool to target learning gaps for those who are struggling, while providing enhanced learning for those who have mastered the concepts.

Graphing Rational Functions

SUGGESTED PACING (DAYS)

| 90 min. | 0.75 | 0.5 | |
| 45 min. | | 1.5 | 0.5 |

Instruction   Extend Lab

Track Your Progress

Objectives

1 Graph rational functions with vertical and horizontal asymptotes.

2 Graph rational functions with oblique asymptotes and point discontinuity.

Mathematical Background

A rational function has an equation of the form $f(x) = \dfrac{a(x)}{b(x)}$, where $a(x)$ and $b(x)$ are polynomial functions and $b(x) \neq 0$. Some graphs of rational functions have breaks in continuity.

THEN

F.BF.3 Identify the effect on the graph of replacing $f(x)$ by $f(x) + k$, $k\,f(x)$, $f(kx)$, and $f(x + k)$ for specific values of k (both positive and negative); find the value of k given the graphs. Experiment with cases and illustrate an explanation of the effects on the graph using technology. Include recognizing even and odd functions from their graphs and algebraic expressions for them.

NOW

A.CED.2 Create equations in two or more variables to represent relationships between quantities; graph equations on coordinate axes with labels and scales.

F.IF.4 For a function that models a relationship between two quantities, interpret key features of graphs and tables in terms of the quantities, and sketch graphs showing key features given a verbal description of the relationship.

NEXT

A.CED.1 Create equations and inequalities in one variable and use them to solve problems. Include equations arising from linear and quadratic functions, and simple rational and exponential functions.

Go Online! All of these resources and more are available at connectED.mcgraw-hill.com

eLessons utilize the power of your interactive whiteboard in an engaging way. Use **Rational Functions**, screen 12, to introduce the concepts in this lesson.

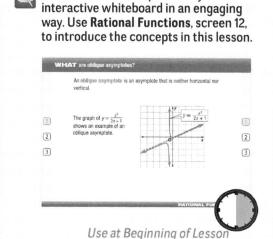

Use at Beginning of Lesson

Use **The Geometer's Sketchpad** to explore graphs of rational functions.

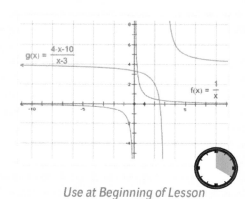

Use at Beginning of Lesson

Animations illustrate key concepts through step-by-step tutorials and videos.

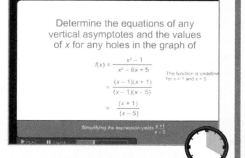

Use at with Examples

OER **Using Open Educational Resources**

Lesson Creation Educators can create their own games, quizzes, activities, and diagrams on **ClassTools.net**. Visit this site to create an activity on graphing rational functions using premade templates. This resource is free, reliable, and will save your work for an entire school year. *Use as a planning resource*

Go Online!

connectED.mcgraw-hill.com   Worksheets

Differentiate Your Resources

Extra Practice Additional practice or homework; Skills Practice is best for approaching-level students and Practice is best for on-level and beyond-level students

Skills Practice

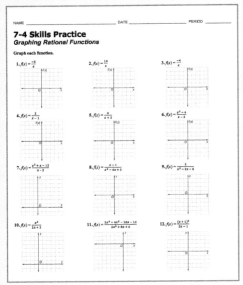

Practice

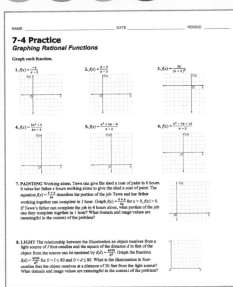

Word Problem Practice

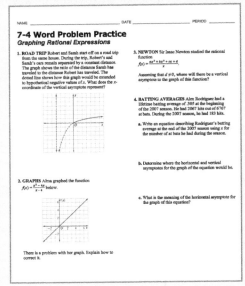

Intervention Reteaching and vocabulary activities that can be used with struggling or absent students and as ELL support

Extension Activities that can be used to extend lesson concepts

Study Guide and Intervention

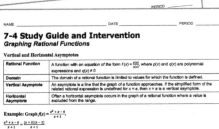

Study Notebook

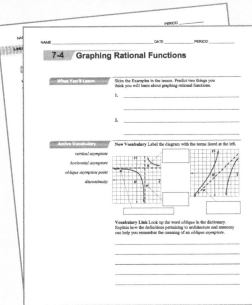

Enrichment

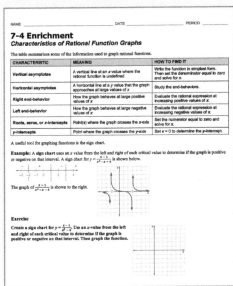

LESSON 4

Graphing Rational Functions

:Then	:Now	:Why?
• You graphed reciprocal functions.	**1** Graph rational functions with vertical and horizontal asymptotes. **2** Graph rational functions with oblique asymptotes and point discontinuity.	• Regina bought a season pass to a water park for $146. She plans on paying for one meal in the park every time she visits. The park claims that meals on average cost $12.49. The rational function $W(m) = \frac{12.49m + 146}{m}$ can be used to determine the average cost $W(m)$ for visiting the park m times.

 New Vocabulary
rational function
vertical asymptote
horizontal asymptote
oblique asymptote
point discontinuity

MP Mathematical Practices
7 Look for and make use of structure.

Content Standards
A.CED.2 Create equations in two or more variables to represent relationships between quantities; graph equations on coordinate axes with labels and scales.
F.IF.4 For a function that models a relationship between two quantities, interpret key features of graphs and tables in terms of the quantities, and sketch graphs given a verbal description of the relationship.

1 **Vertical and Horizontal Asymptotes** A **rational function** has an equation of the form $f(x) = \frac{a(x)}{b(x)}$, where $a(x)$ and $b(x)$ are polynomial functions and $b(x) \neq 0$.

In order to graph a rational function, it is helpful to locate the zeros and asymptotes. A zero of a rational function $f(x) = \frac{a(x)}{b(x)}$ occurs at every value of x for which $a(x) = 0$.

🔑 Key Concept Vertical and Horizontal Asymptotes

Words   If $f(x) = \frac{a(x)}{b(x)}$, $a(x)$ and $b(x)$ are polynomial functions with no common factors other than 1, and $b(x) \neq 0$, then:
- $f(x)$ has a **vertical asymptote** whenever $b(x) = 0$.
- $f(x)$ has at most one **horizontal asymptote.**
  - If the degree of $a(x)$ is greater than the degree of $b(x)$, there is no horizontal asymptote.
  - If the degree of $a(x)$ is less than the degree of $b(x)$, the horizontal asymptote is the line $y = 0$.
  - If the degree of $a(x)$ equals the degree of $b(x)$, the horizontal asymptote is the line $y = \frac{\text{leading coefficient of } a(x)}{\text{leading coefficient of } b(x)}$.

Examples

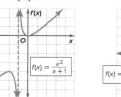

No horizontal asymptote

$f(x) = \frac{x^2}{x+1}$

Vertical asymptote:
$x = -1$

One horizontal asymptote

$f(x) = \frac{3}{x^2 - 1}$

Vertical asymptotes:
$x = -1, x = 1$
Horizontal asymptote:
$f(x) = 0$

$f(x) = \frac{2x + 1}{x - 3}$

Vertical asymptote:
$x = 3$
Horizontal asymptote:
$f(x) = 2$

MP Mathematical Practices Strategies

Look for and make use of structure.

Help students maintain oversight of the process of graphing rational functions while attending to details of asymptotes and point discontinuity. For example, ask:

- Can the vertical asymptote cross the *y*-axis? no
- Can the horizontal asymptote cross the *x*-axis? yes
- How do you find the zeros of the problem? Set the numerator to 0 and solve.
- How do you find the asymptote? Set the denominator to 0 and solve.

Launch

Have students read the Why? section of the lesson. Ask:
- On what does the average cost for visiting the water park depend? the number of visits
- What happens to the value of $W(m)$ as the value of m increases? It decreases and approaches 12.49.
- Can $m = 0$? no
- Can $W(m)$ ever equal 0? Yes, the function has a zero at $m = -\frac{146}{12.49}$, or about -11.69. However, negative values are not realistic to the context of the situation.

Go Online!

Interactive Whiteboard
Use the *eLesson* or *Lesson Presentation* to present this lesson.

Teach

Ask the scaffolded questions for each example to build conceptual understanding for students at all levels.

1 Vertical and Horizontal Asymptotes

Example 1 Graph with No Horizontal Asymptote

AL **Why doesn't this function have a horizontal asymptote?** The degree of the numerator is greater than the degree of the denominator.

OL **How does the domain of this function relate to the vertical asymptotes?** Vertical asymptotes happen at places that are not in the domain.

BL **What are the possibilities for the end behavior of a rational function that doesn't have a horizontal asymptote?** $y \to +\infty$ as $x \to -\infty$, $y \to +\infty$ as $x \to +\infty$; $y \to -\infty$ as $x \to -\infty$, $y \to -\infty$ as $x \to +\infty$; $y \to -\infty$ as $x \to -\infty$, $y \to +\infty$ as $x \to +\infty$; $y \to +\infty$ as $x \to -\infty$, $y \to -\infty$ as $x \to +\infty$

Need Another Example?

Graph $f(x) = \dfrac{x^3}{x+1}$.

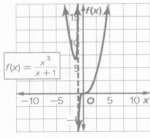

Watch Out!

Zeros vs. Vertical Asymptotes Zeros of rational functions occur at the values that make the numerator equal to zero. Vertical asymptotes occur at the values that make the denominator equal to zero.

Study Tip

Tools The TABLE feature of a graphing calculator can be used to calculate decimal values for x and y.

The asymptotes of a rational function can be used to draw the graph of the function. Additionally, the asymptotes can be used to divide a graph into regions to find ordered pairs on the graph.

F.IF.4

Example 1   Graph with No Horizontal Asymptote

Graph $f(x) = \dfrac{x^3}{x-1}$.

Step 1 Find the zeros.

$x^3 = 0$ Set $a(x) = 0$

$x = 0$ Take the cube root of each side.

There is a zero at $x = 0$.

Step 2 Draw the asymptotes.

Find the vertical asymptote.

$x - 1 = 0$ Set $b(x) = 0$.

$x = 1$ Add 1 to each side.

There is a vertical asymptote at $x = 1$.

The degree of the numerator is greater than the degree of the denominator. So, there is no horizontal asymptote.
$D = \{x \mid x \neq 1\}$ and $R = \{$all real numbers$\}$.

Step 3 Draw the graph.

Use a table to find ordered pairs on the graph. Then connect the points.

x	$f(x)$
−3	6.75
−2	2.67
−1	0.5
0	0
0.5	−0.25
1.5	6.75
2	8
3	13.5

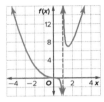

Guided Practice

Graph each function.   1A, 1B. See margin.

1A. $f(x) = \dfrac{x^2 - x - 6}{x + 1}$

1B. $f(x) = \dfrac{(x+1)^3}{(x+2)^2}$

In the real world, sometimes values on the graph of a rational function are not meaningful. In the graph at the right, x-values such as time, distance, and number of people cannot be negative in the context of the problem. So, you do not even need to consider that portion of the graph.

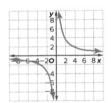

Additional Answers (Guided Practice)

1A.

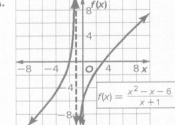

1B.

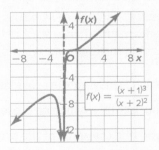

Go Online!

The most up-to-date resources available for your program can be found at connectED.mcgraw-hill.com.

F.IF.4

Real-World Example 2   Use Graphs of Rational Functions

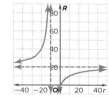

AVERAGE SPEED A boat traveled upstream at r_1 miles per hour. During the return trip to its original starting point, the boat traveled at r_2 miles per hour. The average speed for the entire trip R is given by the formula

$$R = \frac{2r_1 r_2}{r_1 + r_2}.$$

a. Let r_1 be the independent variable, and let R be the dependent variable. Draw the graph if $r_2 = 10$ miles per hour.

The function is $R = \dfrac{2r_1(10)}{r_1 + (10)}$ or $R = \dfrac{20r_1}{r_1 + 10}$.

The vertical asymptote is $r_1 = -10$.

Graph the vertical asymptote and the function.

Notice that the horizontal asymptote is $R = 20$.

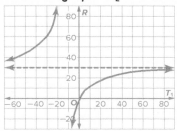

b. What is the R-intercept of the graph?

The R-intercept is 0.

c. What domain and range values are meaningful in the context of the problem?

In the problem context, speeds are nonnegative values. Therefore, only values of r_1 greater than or equal to 0 and values of R between 0 and 20 are meaningful.

Guided Practice

2. SALARIES A company uses the formula $S(x) = \dfrac{45x + 25}{x + 1}$ to determine the salary in thousands of dollars of an employee during his xth year. Graph $S(x)$. What domain and range values are meaningful in the context of the problem? What is the meaning of the horizontal asymptote for the graph? See margin.

2 **Oblique Asymptotes and Point Discontinuity** An **oblique asymptote**, sometimes called a *slant asymptote*, is an asymptote that is neither horizontal nor vertical.

Key Concept Oblique Asymptotes

Words	If $f(x) = \dfrac{a(x)}{b(x)}$, $a(x)$ and $b(x)$ are polynomial functions with no common factors other than 1 and $b(x) \neq 0$, then $f(x)$ has an oblique asymptote if the degree of $a(x)$ minus the degree of $b(x)$ equals 1. The equation of the asymptote is $f(x) = \dfrac{a(x)}{b(x)}$ with no remainder.
Example	$f(x) = \dfrac{x^4 + 3x^3}{x^3 - 1}$ Vertical asymptote: $x = 1$ Oblique asymptote: $f(x) = x + 3$

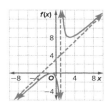

Real-World Career

U.S. Coast Guard Boatswain's Mate

The most versatile member of the U.S. Coast Guard's operational team is the boatswain's mate. BMs are capable of performing almost any task. Training for BMs is accomplished through 12 weeks of intensive training.

Example 2 Use Graphs of Rational Functions

AL What will the average speed of the entire trip be if the boat traveled upstream at 7 mph?
8.23 mph

OL What does the horizontal asymptote represent in this context? the fastest average speed the boat can approach

BL Other than speeds being nonnegative, what other practical limitations would be on the domain of R? the speed of the boat

Need Another Example?

Average Speed Use the situation and formula given in Example 2.

a. Draw the graph if $r_2 = 15$ miles per hour.

b. What is the R-intercept of the graph? The R-intercept is 0.

c. What domain and range values are meaningful in the context of the problem? Values of r_1 greater than or equal to 0 and values of R between 0 and 30 are meaningful.

Additional Answer (Guided Practice)

2. The number of years worked must be greater than or equal to 0, and salary values must be between 25 and 45. The asymptote represents a salary cap of $45,000.

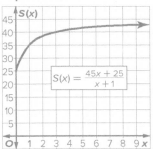

Differentiated Instruction BL

Extension Challenge students to explain the rules for finding horizontal and oblique asymptotes. While the chapter shows students how to find them, the explanation of why the rules work can be left to high ability students. Scaffold the task by having students examine graphs of varying degrees in the numerator and denominator and look for general patterns in the asymptotes

2 Oblique Asymptotes and Point Discontinuity

Example 3 Determine Oblique Asymptotes

AL Why does $f(x)$ have an oblique asymptote? The degree of the numerator is one higher than the degree of the denominator.

OL How could you use synthetic division to find the oblique asymptote? Put $\frac{1}{2}$ in the box and perform synthetic division on $x^2 + 4x + 4$.

BL What is the end behavior of a function with an oblique asymptote with positive slope? Negative slope? Positive slope: $y \rightarrow -\infty$ as $x \rightarrow -\infty$, $y \rightarrow \infty$ as $x \rightarrow \infty$, Negative slope: $y \rightarrow \infty$ as $x \rightarrow -\infty$, $y \rightarrow -\infty$ as $x \rightarrow \infty$.

Need Another Example?

Graph $f(x) = \dfrac{x^2}{x+1}$.

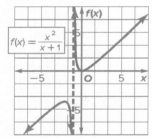

Additional Answers (Guided Practice)

3A.

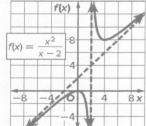

3B.

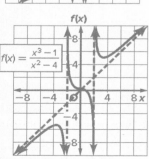

Study Tip

Oblique Asymptotes Oblique asymptotes occur for rational functions that have a numerator polynomial that is one degree higher than the denominator polynomial.

Example 3   Determine Oblique Asymptotes

F.IF.4

Graph $f(x) = \dfrac{x^2 + 4x + 4}{2x - 1}$.

Step 1 Find the zeros.

$$x^2 + 4x + 4 = 0 \qquad \text{Set } a(x) = 0.$$
$$(x + 2)^2 = 0 \qquad \text{Factor.}$$
$$x + 2 = 0 \qquad \text{Take the square root of each side.}$$
$$x = -2 \qquad \text{Subtract 2 from each side.}$$

There is a zero at $x = -2$.

Step 2 Find the asymptotes.

$$2x - 1 = 0 \qquad \text{Set } b(x) = 0.$$
$$2x = 1 \qquad \text{Add 1 to each side.}$$
$$x = \frac{1}{2} \qquad \text{Divide each side by 2.}$$

There is a vertical asymptote at $x = \frac{1}{2}$.

The degree of the numerator is greater than the degree of the denominator, so there is no horizontal asymptote.

The difference between the degree of the numerator and the degree of the denominator is 1, so there is an oblique asymptote.

Divide the numerator by the denominator to determine the equation of the oblique asymptote.

The equation of the asymptote is the quotient excluding any remainder.

$$
\begin{array}{r}
\frac{1}{2}x + \frac{9}{4} \\
2x - 1 \overline{)\, x^2 + 4x + 4} \\
\underline{(-)x^2 - \tfrac{1}{2}x} \\
\frac{9}{2}x + 4 \\
\underline{(-)\tfrac{9}{2}x - \tfrac{9}{4}} \\
\frac{25}{4}
\end{array}
$$

Thus, the oblique asymptote is the line $f(x) = \frac{1}{2}x + \frac{9}{4}$.

Step 3 Draw the asymptotes, and then use a table of values to graph the function.

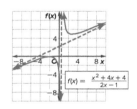

Go Online!

Explore the graphs of rational functions using *The Geometer's Sketchpad®* activity in ConnectED.

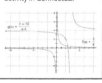

▸ **Guided Practice**

Graph each function.   3A, 3B. See margin.

3A. $f(x) = \dfrac{x^2}{x - 2}$ **3B.** $f(x) = \dfrac{x^3 - 1}{x^2 - 4}$

In some cases, graphs of rational functions may have **point discontinuity**, which looks like a hole in the graph. This is because the function is undefined at that point.

Differentiated Instruction **OL** **ELL**

Verbal/Linguistic Learners Have students write a list of tips to help someone draw the graphs of rational functions.

Key Concept Point Discontinuity

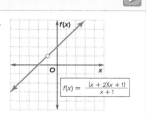

Words If $f(x) = \frac{a(x)}{b(x)}$, $b(x) \neq 0$, and $x - c$ is a factor of both $a(x)$ and $b(x)$, then there is a point discontinuity at $x = c$.

Example $f(x) = \frac{(x+2)(x+1)}{x+1}$
$= x + 2; x \neq -1$

F.IF.4

Watch Out!

Holes Remember that a common factor in the numerator and denominator can signal a point discontinuity.

Example 4   Graph with Point Discontinuity

Graph $f(x) = \frac{x^2 - 16}{x - 4}$.

Notice that $\frac{x^2 - 16}{x - 4} = \frac{(x+4)(x-4)}{x-4}$ or $x + 4$.

Therefore, the graph of $f(x) = \frac{x^2 - 16}{x - 4}$ is the graph of $f(x) = x + 4$ with a point discontinuity at $x = 4$.

$D = \{x \mid x \neq 4\}$ and $R = \{f(x) \mid f(x) \neq 8\}$

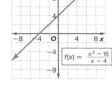

▶ **Guided Practice**   4A, 4B. See margin.

Graph each function.

4A. $f(x) = \frac{x^2 + 4x - 5}{x + 5}$

4B. $f(x) = \frac{x^3 + 2x^2 - 9x - 18}{x^2 - 9}$

Check Your Understanding   ◯ = Step-by-Step Solutions begin on page R11.   ✓ **Go Online!** for a Self-Check Quiz

Example 1
F.IF.4

Graph each function. 1, 2. See Ch. 7 Answer Appendix.

1. $f(x) = \frac{x^4 - 2}{x^2 - 1}$

2. $f(x) = \frac{x^3}{x + 2}$

Example 2
F.IF.4

3. 🔠 REASONING Eduardo is a kicker for his high school football team. So far this season, he has made 7 out of 11 field goals. He would like to improve his field goal percentage. If he can make x consecutive field goals, his field goal percentage can be determined using the function $P(x) = \frac{7 + x}{11 + x}$.

  a. Graph the function.   See Ch. 7 Answer Appendix.

  b. What part of the graph is meaningful in the context of this problem?   the part in the first quadrant

  c. Describe the meaning of the intercept of the vertical axis.   3c. It represents his original field goal percentage of 63.6%.

  d. What is the equation of the horizontal asymptote? Explain its meaning with respect to Eduardo's field goal percentage. $y = 1$; this represents 100% which he cannot achieve because he has already missed 4 field goals.

Examples 3-4
F.IF.4

Graph each function.   4–7. See Ch. 7 Answer Appendix.

4. $f(x) = \frac{6x^2 - 3x + 2}{x}$

⑤ $f(x) = \frac{x^2 + 8x + 20}{x + 2}$

6. $f(x) = \frac{x^2 - 4x - 5}{x + 1}$

7. $f(x) = \frac{x^2 + x - 12}{x + 4}$

Additional Answers (Guided Practice)

4A.

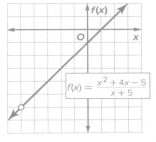

$f(x) = \frac{x^2 + 4x - 5}{x + 5}$

4B.

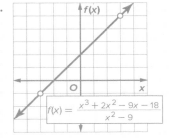

$f(x) = \frac{x^3 + 2x^2 - 9x - 18}{x^2 - 9}$

Example 4 Graph with Point Discontinuity

AL When does a point discontinuity occur? when the numerator and denominator of a rational function have a common factor

OL How can you find the y-coordinate of a point discontinuity? Plug the x-coordinate of the point discontinuity into the factored form of the rational expression.

BL Write a function that has a point discontinuity on the y-axis. Example: $\frac{x}{x(x + 1)}$

Need Another Example?

Graph $f(x) = \frac{x^2 - 4}{x - 2}$.

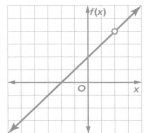

Practice

Formative Assessment Use Exercises 1–7 to assess students' understanding of the concepts in this lesson.

The Practice and Problem Solving exercises assess the content taught in the lesson. The Preparing for Assessment page is meant to be used as preparation for end-of-year assessments.

Exercise Alert

For Exercises 1–26 and 28–42, students will need grid paper.

🅜🅟 Teaching the Mathematical Practices

Reasoning Mathematically proficient students make sense of quantities and their relationships in problem situations. Quantitative reasoning entails habits of creating a coherent representation of the problem at hand; considering the units involved; attending to the meaning of quantities, not just how to compute them; and knowing and flexibly using different properties of operations and objects.

Extra Practice

See page R7 for extra exercises for students who are approaching level or for on-level students who need additional reinforcement.

Additional Answers

8.

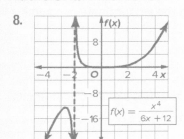

$$f(x) = \frac{x^4}{6x + 12}$$

9.

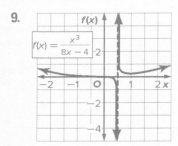

$$f(x) = \frac{x^3}{8x - 4}$$

10.

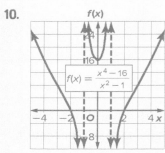

$$f(x) = \frac{x^4 - 16}{x^2 - 1}$$

11.

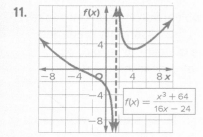

$$f(x) = \frac{x^3 + 64}{16x - 24}$$

Go Online! **eBook**

Interactive Student Guide

Use the *Interactive Student Guide* to deepen conceptual understanding.
· Graphing Rational Functions

ALGEBRA 2
INTERACTIVE STUDENT GUIDE

Practice and Problem Solving Extra Practice is on page R7.

Example 1
F.IF.4

Graph each function. 8–11. See margin.

8. $f(x) = \dfrac{x^4}{6x + 12}$

9. $f(x) = \dfrac{x^3}{8x - 4}$

10. $f(x) = \dfrac{x^4 - 16}{x^2 - 1}$

11. $f(x) = \dfrac{x^3 + 64}{16x - 24}$

12a. $c(t) = \dfrac{9.5t + 75}{t - 15}$
See margin for graph.

Example 2
F.IF.4

12. SCHOOL SPIRIT As president of Student Council, Brandy is getting T-shirts made for a pep rally. Each T-shirt costs $9.50, and there is a set-up fee of $75. The student council plans to sell the shirts, but each of the 15 council members will get one for free.

a. Write a function for the average cost of a T-shirt to be sold. Graph the function.

b. What is the average cost if 200 shirts are ordered? if 500 shirts are ordered? $10.68; $9.95

c. How many T-shirts must be ordered to bring the average cost under $9.75? more than 885

Examples 3–4
F.IF.4

Graph each function. 13–26. See Ch. 7 Answer Appendix.

13. $f(x) = \dfrac{x}{x + 2}$

14. $f(x) = \dfrac{5}{(x - 1)(x + 4)}$

(15) $f(x) = \dfrac{4}{(x - 2)^2}$

16. $f(x) = \dfrac{x - 3}{x + 1}$

17. $f(x) = \dfrac{1}{(x + 4)^2}$

18. $f(x) = \dfrac{2x}{(x + 2)(x - 5)}$

19. $f(x) = \dfrac{(x - 4)^2}{x + 2}$

20. $f(x) = \dfrac{(x + 3)^2}{x - 5}$

21. $f(x) = \dfrac{x^3 + 1}{x^2 - 4}$

22. $f(x) = \dfrac{4x^3}{2x^2 + x - 1}$

23. $f(x) = \dfrac{3x^2 + 8}{2x - 1}$

24. $f(x) = \dfrac{2x^2 + 5}{3x + 4}$

25. $f(x) = \dfrac{x^4 - 2x^2 + 1}{x^3 + 2}$

26. $f(x) = \dfrac{x^4 - x^2 - 12}{x^3 - 6}$

27. PERSEVERANCE The graph of a certain rational function has 2 branches, passes through the point (0, 0), has no horizontal asymptotes, and a vertical asymptote of $x = -1$.

a. Find an equation of the rational function. Sample answer: $f(x) = \dfrac{x^2}{x + 1}$

b. How did you develop a plan for solving this problem? b–c. See Ch. 7 Answer Appendix.

c. What assumptions did you make in your solution process?

Example 4
F.IF.4

Graph each function. 28–35. See Ch. 7 Answer Appendix.

28. $f(x) = \dfrac{x^2 - 2x - 8}{x - 4}$

29. $f(x) = \dfrac{x^2 + 4x - 12}{x - 2}$

30. $f(x) = \dfrac{x^2 - 25}{x + 5}$

31. $f(x) = \dfrac{x^2 - 64}{x - 8}$

32. $f(x) = \dfrac{(x - 4)(x^2 - 4)}{x^2 - 6x + 8}$

33. $f(x) = \dfrac{(x + 5)(x^2 + 2x - 3)}{x^2 + 8x + 15}$

34. $f(x) = \dfrac{3x^4 + 6x^3 + 3x^2}{x^2 + 2x + 1}$

35. $f(x) = \dfrac{2x^4 + 10x^3 + 12x^2}{x^2 + 5x + 6}$

Differentiated Homework Options

Levels	**AL** Basic	**OL** Core	**BL** Advanced
Exercises	8–35, 42–51	9–35 odd, 36–39, 41–51	36–51
2-Day Option	9–35 odd, 47–51	8–35, 47–51	
	8–34 even, 42–46	36–46	

 You can use ALEKS to provide additional remediation support with personalized instruction and practice.

37c. Sample answer: The number of months and the average cost cannot have negative values.

 36. BUSINESS Liam purchased a riding lawn mower for $4500 and mows the lawns of local businesses. Each time he mows a lawn, he incurs a cost of $50 for gas and maintenance.

a. Write and graph the rational function representing his average cost per customer as a function of the number of lawns. **See Ch. 7 Answer Appendix.**

b. What are the asymptotes of the graph? $x = 0$ and $f(x) = 50$

c. Why is the first quadrant in the graph the only relevant quadrant?

d. How many total lawns does Liam need to mow for his average cost per lawn to be less than $80? **150**

36c. Sample answer: The number of lawns and the average cost cannot be negative.

37 FINANCIAL LITERACY Kristina bought a new smartphone with a data plan. The phone cost $150, and her monthly usage charge is $30 plus $10 for the data plan.

a. Write and graph the rational function representing her average monthly cost as a function of the number of months Kristina uses the phone. **See Ch. 7 Answer Appendix.**

b. What are the asymptotes of the graph? $x = 0$ and $f(x) = 40$

c. Why is the first quadrant in the graph the only relevant quadrant?

d. After how many months will the average monthly charge be $45? **30**

38. MP SENSE-MAKING Alana plays softball for Centerville High School. So far this season she has gotten a hit 4 out of 12 times at bat. She is determined to improve her batting average. If she can get x consecutive hits, her batting average can be determined using $B(x) = \frac{4+x}{12+x}$. **c. It represents her original batting average of .333.**

38d. $y = 1$; This represents 100%, which she can never achieve because she has already missed getting a hit 8 times.

a. Graph the function. **See Chapter 7 Answer Appendix.**

b. What part of the graph is meaningful in the context of the problem? **the part in the first quadrant**

c. Describe the meaning of the intercept of the vertical axis.

d. What is the equation of the horizontal asymptote? Explain its meaning with respect to Alana's batting average.

Graph each function. 39–41. See Chapter 7 Answer Appendix.

39. $f(x) = \frac{x+1}{x^2+6x+5}$ **40.** $f(x) = \frac{x^2-10x-24}{x+2}$ **41.** $f(x) = \frac{6x^2+4x+2}{x+2}$

A.CED.2, F.IF.4,

H.O.T. Problems   Use Higher-Order Thinking Skills

42. OPEN-ENDED Sketch the graph of a rational function with a horizontal asymptote $y = 1$ and a vertical asymptote $x = -2$. **See Chapter 7 Answer Appendix.**

43. CHALLENGE Compare and contrast $g(x) = \frac{x^2-1}{x(x^2-2)}$ and $f(x)$ shown at the right. **See margin.**

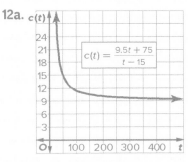

44. MP REASONING What is the difference between the graphs of $f(x) = x - 2$ and $g(x) = \frac{(x+3)(x-2)}{x+3}$? **The graph of $g(x)$ has a hole in it at -3.**

45. PROOF A rational function has an equation of the form $f(x) = \frac{a(x)}{b(x)}$, where $a(x)$ and $b(x)$ are polynomial functions and $b(x) \neq 0$. Show that $f(x) = \frac{x}{a-b} + c$ is a rational function. **45.** $f(x) = \frac{x}{a-b} + \frac{c(a-b)}{(a-b)} = \frac{x+ca-cb}{a-b}$

46. WRITING IN MATH How can factoring be used to determine the vertical asymptotes or point discontinuity of a rational function? **See Ch. 7 Answer Appendix.**

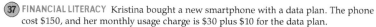
MP Standards for Mathematical Practice

Emphasis On	Exercises
1 Make sense of problems and persevere in solving them.	27
2 Reason abstractly and quantitatively.	1, 2, 4–11, 13–26, 28–35, 38–42
4 Model with mathematics.	12, 36, 37
7 Look for and make use of structure.	3, 43, 46–51
8 Look for and express regularity in repeated reasoning.	44, 45

Levels of Complexity Chart

The levels of the exercises progress from 1 to 3, with Level 1 indicating the lowest level of complexity.

Exercises	8–35	36, 37, 47–51	38–46
▶ Level 3			●
B Level 2		○	
Level 1	●		

e Follow-Up

Students have explored rational expressions and functions.

Ask:

- How are the properties of a rational function reflected in its graph? Sample answer: Vertical asymptotes occur at values that make the denominator 0; horizontal asymptotes occur when the degree of the numerator is less than or equal to the degree of the denominator; oblique asymptotes occur when the degrees of the numerator and denominator differ by 1; holes occur when the numerator and denominator have a common factor that has a zero.

Assess

Name the Math Have students write their own examples of rational functions and graph them, showing any asymptotes or discontinuities.

Additonal Answers

12a.
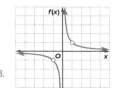

43. Similarities: Both have vertical asymptotes at $x = 0$. Both approach 0 as x approaches $-\infty$ and approach 0 as x approaches ∞. Differences: $f(x)$ has holes at $x = 1$ and $x = -1$, while $g(x)$ has vertical asymptotes at $x = \sqrt{2}$ and $x = -\sqrt{2}$. $f(x)$ has no zeros, but $g(x)$ has zeros at $x = 1$ and $x = -1$.

Preparing for Assessment

Exercises 47–51 require students to use the skills they will need on standardized assessments. The exercises are dual-coded with content standards and mathematical practice standards.

Dual Coding		
Exercises	Content Standards	Mathematical Practices
47	F.BF.4	7
48	F.BF.4, A.CED.2	7
49	F.BF.4	7
50	F.BF.4	7
51	F.BF.4	7

Diagnose Student Errors

Survey student responses for each item. Class trends may indicate common errors and misconceptions.

47.

A	The degree of the numerator is 3 and the degree of the denominator is 1.
B	The degree of the numerator is 5 and the degree of the denominator is 3.
C	The degree of the numerator is 2 and the degree of the denominator is 2.
D	**CORRECT**

48.

A	**CORRECT**
B	Chose a linear function of the line without discontinuity
C	Chose a linear function of the line without discontinuity, but with a negative slope
D	Incorrectly chose the linear denominator

Go Online!

Self-Check Quiz

Students can use *Self-Check Quizzes* to check their understanding of this lesson.

Preparing for Assessment

47. Which function has a graph with an oblique asymptote? 7 F.BF.4 **D**

  ○ A $f(x) = \dfrac{(x-1)^3}{x+4}$

  ○ B $f(x) = \dfrac{(x-1)^3(x-2)^2}{(x-4)^3}$

  ○ C $f(x) = \dfrac{(2x-1)^2}{x^2+4}$

  ○ D $f(x) = \dfrac{x(x+3)^2}{x^2+2}$

48. Which function is graphed below? 7 F.BF.4, A.CED.2 **A**

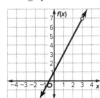

  ○ A $f(x) = \dfrac{2x^2 - 5x - 3}{x - 3}$

  ○ B $f(x) = 2x + 1$

  ○ C $f(x) = -2x + 1$

  ○ D $f(x) = \dfrac{2x^2 - 5x - 3}{x + 3}$

  ○ E $f(x) = \dfrac{2x - 5}{1}$

  ○ F $f(x) = -2x + 5$

49. A group of friends are attending a ball game in another state. Each person spends $25 for a ticket and they share the transportation costs of $150 equally. Which function represents the total cost for each person? 7 F.BF.4 **C**

  ○ A $f(x) = 150x + 25$

  ○ B $f(x) = 25x + 150$

  ○ C $f(x) = \dfrac{150}{x} + 25$

  ○ D $f(x) = \dfrac{150}{x} + \dfrac{25}{x}$

  ○ E $f(x) = \dfrac{150}{x} - 25$

  ○ F $f(x) = \dfrac{25}{x} + 150$

50. Which of the following functions is its own inverse? 7 F.BF.4 **B**

  ○ A $f(x) = 2x^2 - 7x - 15$

  ○ B $f(x) = \dfrac{7}{x}$

  ○ C $f(x) = 7x^2 + 2$

  ○ D $f(x) = \dfrac{x}{7}$

51. Consider the function $f(x) = \dfrac{1}{6x^2 - 12x - 18}$. 7 F.BF.4

  a. Where are the discontinuities? **C**

  ○ A $x = -3, x = -1$

  ○ B $x = -3, x = 1$

  ○ C $x = 3, x = -1$

  ○ D $x = 3, x = 1$

  b. How many horizontal asymptotes does the function have? **A**

  ○ A 1

  ○ B 2

  ○ C 3

  ○ D none

  c. What are the vertical asymptotes? **B**

  ○ A $x = 3, x = 1$

  ○ B $x = 3, x = -1$

  ○ C $x = -3, x = -1$

  ○ D $x = -3, x = 1$

  d. How many x-intercepts are there? **D**

  ○ A 1

  ○ B 2

  ○ C 3

  ○ D none

Differentiated Instruction OL BL

Extension Draw a straight line on the board, but leave a hole in the line for some integral value of *x*. For example, draw the line representing $y = x - 2$, but leave a hole at the point $(1, -1)$. Ask students to write possible rational functions that could be described by the graph. Many answers are possible, as long as $x = 1$ is an excluded value and the simplified form of the function is $f(x) = x - 2$.

Sample answer: $f(x) = \dfrac{(x-1)(x-2)}{x-1}$ or $f(x) = \dfrac{x^2 - 3x + 2}{x - 1}$.

49.

A	Assumed this is a linear function where 150 is the slope and 25 is the y-intercept
B	Assumed this is a linear function where 25 is the slope and 150 is the y-intercept
C	**CORRECT**
D	Wrote a function in which each person also shared the $25 cost
E	Subtracted the amount of the ticket instead of adding

EXTEND 7-4

Graphing Technology Lab
Graphing Rational Functions

A TI-83/84 Plus graphing calculator can be used to explore graphs of rational functions. These graphs have some features that never appear in the graphs of polynomial functions.

Content Standards
A.CED.2 Create equations in two or more variables to represent relationships between quantities; graph equations on coordinate axes with labels and scales.

F.IF.4 For a function that models a relationship between two quantities, interpret key features of graphs and tables in terms of the quantities, and sketch graphs showing key features given a verbal description of the relationship.

Activity 1   Graph with Asymptotes

Work cooperatively. Graph $y = \frac{8x - 5}{2x}$ in the standard viewing window. Find the equations of any asymptotes. State the domain and range of the function.

Step 1 Enter the equation in the **Y=** list, and then graph.

Step 2 Examine the graph.

By looking at the equation, we can determine that if $x = 0$, the function is undefined. The equation of the vertical asymptote is $x = 0$. Notice what happens to the y-values as x grows larger and as x gets smaller. The y-values approach 4. So, the equation for the horizontal asymptote is $y = 4$. The domain is $\{x \mid x \neq 0\}$, and the range is all real numbers.

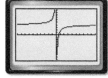

[−10, 10] scl: 1 by [−10, 10] scl: 1

Activity 2   Graph with Point Discontinuity

Work cooperatively. Graph $y = \frac{x^2 - 16}{x + 4}$ in the window [−5, 4.4] by [−10, 2] with scale factors of 1.

Step 1 Because the function is not continuous, put the calculator in dot mode.

Step 2 Examine the graph.

This graph looks like a line. Because the denominator has a factor of $x + 4$ and there is no vertical asymptote at $x = -4$, the graph must have a break in continuity at $x = -4$. Therefore, the function is undefined and has point discontinuity at $x = -4$.

If you **TRACE** along the graph, when you come to $x = -4$, you will see that there is no corresponding y-value.

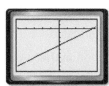

[−5, 4.4] scl: 1 by [−10, 2] scl: 1

Exercises

Work cooperatively. Use a graphing calculator to graph each function. Write the x-coordinates of any points of discontinuity and/or the equations of any asymptotes. State the domain and range.

1. $f(x) = \frac{1}{x}$ $x = 0, y = 0; D = \{x \mid x \neq 0\}, R = \{y \mid y \neq 0\}$

2. $f(x) = \frac{x}{x + 2}$ $x = -2, y = 1; D = \{x \mid x \neq -2\},$ $R = \{y \mid y \neq 1\}$

3. $f(x) = \frac{2}{x - 4}$ $x = 4, y = 0; D = \{x \mid x \neq 4\},$ $R = \{y \mid y \neq 0\}$

4. $f(x) = \frac{2x}{3x - 6}$ $x = 2, y = \frac{2}{3}; D = \{x \mid x \neq 2\},$ $R = \left\{y \mid y \neq \frac{2}{3}\right\}$

5. $f(x) = \frac{4x + 2}{x - 1}$ $x = 1, y = 4; D = \{x \mid x \neq 1\},$ $R = \{y \mid y \neq 4\}$

6. $f(x) = \frac{x^2 - 9}{x + 3}$ point discontinuity at $x = -3$; $D = \{x \mid x \neq -3\}, R = \{$all real numbers$\}$

Launch

Objective Use a graphing calculator to explore the graphs of rational functions.

Materials for Each Group

- TI-83/84 Plus or other graphing calculator

Teaching Tip Students may find it instructive to experiment with the graph style feature found to the left of the equation in the function editor (Y= list). They can begin by using the usual line style, indicated by a backslash icon. The path style line, represented with the icon -0, slows down the graphing so students can follow it more easily.

Teach

Working in Cooperative Groups Have students work in pairs to complete Activities 1 and 2. **ELL**

- Students may need to trace a graph beyond the current window in order to identify asymptotes precisely.

- A calculator graphing in Connected mode may seem to graph a vertical asymptote when it is merely connecting two sequential pixels on a graph. Switching to Dot mode will eliminate this possibility.

Practice Have students complete Exercises 1–6.

Practice

Formative Assessment

Use Exercises 1–6 to assess whether students can use a graphing calculator to explore the graphs of rational functions.

From Concrete to Abstract

Exercise 5 asks students to apply their findings to sketching the graph of a general rational function.

Extending the Concept

In preparation for the study of inversre variation, ask students to graph $y = \frac{1}{x}$, $y = \frac{2}{x}$, $y = \frac{3}{x}$, and $y = \frac{4}{x}$ on the same screen and describe their similarities. The coordinate axes are asymptotes for the graphs of all four functions.

Go Online!

eLesson

You can use the eLesson on Rational Functions to demonstrate graphing rational functions.

Variation Functions

Track Your Progress

Objectives

1 Recognize and solve direct and joint variation problems.

2 Recognize and solve inverse and combined variation problems.

Mathematical Background

The type of variation present can sometimes be identified from a table of values for x and y. If the quotient $\frac{x}{y}$ has a constant value, y varies directly as x. If the product xy has a constant value, y varies inversely as x.

THEN	NOW	NEXT
F.BF.3 Identify the effect on the graph of replacing $f(x)$ by $f(x) + k$, $k\,f(x)$, $f(kx)$, and $f(x + k)$ for specific values of k (both positive and negative); find the value of k given the graphs. Experiment with cases and illustrate an explanation of the effects on the graph using technology. Include recognizing even and odd functions from their graphs and algebraic expressions for them.	**A.CED.2** Create equations in two or more variables to represent relationships between quantities; graph equations on coordinate axes with labels and scales. **F.IF.4** For a function that models a relationship between two quantities, interpret key features of graphs and tables in terms of the quantities, and sketch graphs showing key features given a verbal description of the relationship.	**A.CED.1** Create equations and inequalities in one variable and use them to solve problems. Include equations arising from linear and quadratic functions, and simple rational and exponential functions. **A.REI.2** Solve simple rational and radical equations in one variable, and give examples showing how extraneous solutions may arise.

Go Online! All of these resources and more are available at connectED.mcgraw-hill.com

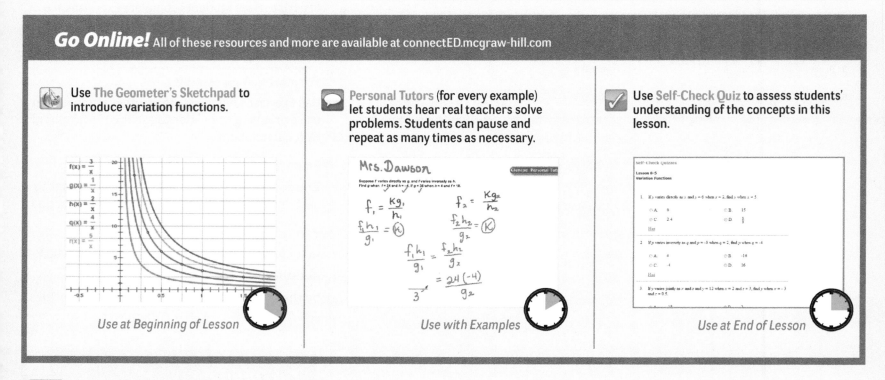

Use The Geometer's Sketchpad to introduce variation functions.

Use at Beginning of Lesson

Personal Tutors (for every example) let students hear real teachers solve problems. Students can pause and repeat as many times as necessary.

Use with Examples

Use Self-Check Quiz to assess students' understanding of the concepts in this lesson.

Use at End of Lesson

OER Using Open Educational Resources

Publishing Have students make an interactive poster or manual explaining variation functions on **glogster**. Glogster allows students to insert text, images, photos, audio, videos, and special effects. Students can trade with each other and critique the manual or poster on understandability and creativity. Ask for volunteers to share with the class. *Use as homework or classwork*

Go Online!
connectED.mcgraw-hill.com
Worksheets

Differentiate Your Resources

Extra Practice Additional practice or homework; Skills Practice is best for approaching-level students and Practice is best for on-level and beyond-level students

Skills Practice

NAME _____ DATE _____ PERIOD _____

7-5 Skills Practice
Variation Functions

State whether each equation represents a *direct, joint, inverse,* or *combined* variation. Then name the constant of variation.

1. $c = 12m$ 2. $p = \frac{4}{q}$ 3. $A = \frac{1}{2}bh$

4. $rw = 15$ 5. $y = 2rgt$ 6. $f = 5280m$

7. $y = 0.2d$ 8. $vz = -25$ 9. $z = 16rh$

10. $R = \frac{6}{w}$ 11. $b = \frac{1}{3}g$ 12. $C = 2\pi r$

13. If y varies directly as x and $y = 35$ when $x = 7$, find y when $x = 11$.

14. If y varies directly as x and $y = 360$ when $x = 180$, find y when $x = 270$.

15. If y varies directly as x and $y = 540$ when $x = 10$, find x when $y = 1080$.

16. If y varies directly as x and $y = 12$ when $x = 72$, find x when $y = 9$.

17. If y varies jointly as x and z and $y = 18$ when $x = 2$ and $z = 3$, find y when x is 5 and z is 6.

18. If y varies jointly as x and z and $y = -16$ when $x = 4$ and $z = 2$, find y when x is -1 and z is 7.

19. If y varies jointly as x and z and $y = 120$ when $x = 4$ and $z = 6$, find y when x is 3 and z is 2.

20. If y varies inversely as x and $y = 2$ when $x = 2$, find y when $x = 1$.

21. If y varies inversely as x and $y = 6$ when $x = 5$, find y when $x = 10$.

22. If y varies inversely as x and $y = 3$ when $x = 14$, find x when $y = 6$.

23. If y varies directly as z and inversely as x and $y = 27$ and $x = -3$ when $x = 2$, find x when $y = 9$ and $z = 5$.

24. If y varies directly as z and inversely as x and $y = -15$ and $x = 5$ when $x = 5$, find y when $z = -36$ and $x = -3$.

Practice

NAME _____ DATE _____ PERIOD _____

7-5 Practice
Variation Functions

State whether each equation represents a *direct, joint, inverse,* or *combined* variation. Then name the constant of variation.

1. $u = 8wz$ 2. $p = 4s$ 3. $L = \frac{5}{k}$ 4. $xy = 4.5$

5. $\frac{c}{d} = x$ 6. $2d = mn$ 7. $\frac{1.23}{g} = h$ 8. $y = \frac{3}{4x}$

9. If y varies directly as x and $y = 8$ when $x = 2$, find x when $y = 6$.

10. If y varies directly as x and $y = -16$ when $x = 6$, find x when $y = -4$.

11. If y varies directly as x and $y = 132$ when $x = 11$, find y when $x = 33$.

12. If y varies directly as x and $y = 7$ when $x = 1.5$, find y when $x = 4$.

13. If y varies jointly as x and z and $y = 24$ when $x = 2$ and $z = 1$, find y when x is 12 and z is 2.

14. If y varies jointly as x and z and $y = 60$ when $x = 3$ and $z = 4$, find y when x is 6 and z is 8.

15. If y varies jointly as x and z and $y = 12$ when $x = -2$ and $z = 3$, find y when x is 4 and z is -1.

16. If y varies inversely as x and $y = 16$ when $x = 4$, find y when $x = 3$.

17. If y varies inversely as x and $y = 3$ when $x = 5$, find x when $y = 2.5$.

18. If y varies directly as z and inversely as x and $y = -18$ and $x = 3$ when $x = 6$, find y when $x = 5$ and $z = -5$.

19. If y varies directly as z and inversely as x and $y = 10$ and $x = 5$ when $x = 12.5$, find x when $y = 37.5$ and $z = 1.2$.

20. **GASES** The volume V of a gas varies inversely as its pressure P. If $V = 80$ cubic centimeters when $P = 2000$ millimeters of mercury, find V when $P = 320$ millimeters of mercury.

21. **SPRINGS** The length S that a spring will stretch varies directly with the weight F that is attached to the spring. If a spring stretches 20 inches with 25 pounds attached, how far will it stretch with 15 pounds attached?

22. **GEOMETRY** The area A of a trapezoid varies jointly as its height and the sum of its bases. If the area is 480 square meters when the height is 20 meters and the bases are 28 meters and 20 meters, what is the area of a trapezoid when its height is 8 meters and its bases are 10 meters and 15 meters?

Word Problem Practice

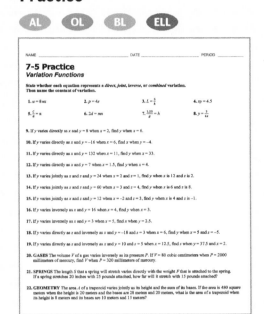

NAME _____ DATE _____ PERIOD _____

7-5 Word Problem Practice
Variation Functions

1. **DIVING** The height that a diver leaps above a diving board varies directly with the amount that the tip of the diving board dips below its normal level. If a diver leaps 44 inches above the diving board when the diving board tip dips 12 inches, how high will the diver leap above the diving board if the tip dips 18 inches?

2. **PARKING LOT DESIGN** As a general rule, the number of parking spaces in a parking lot for a movie theater complex varies directly with the number of theaters in the complex. A typical complex has 30 parking spaces for each theater. A businessman wants to build a new cinema complex on a lot that has enough space for 210 parking spaces. How many theaters should the businessman build in his complex?

3. **RENT** An apartment rents for m dollars per month. If n students share the rent equally, how much would each student have to pay? How does the cost per student vary with the number of students? If 2 students have to pay $700 each, how much money would each student have to pay if there were 5 students sharing the rent?

4. **PAINTING** The cost of painting a wall varies directly with the area of the wall. Write a formula for the cost of painting a rectangular wall with dimensions ℓ by w. With respect to ℓ and w, does the cost vary directly, jointly, or inversely?

5. **HYDROGEN** The cost of a hydrogen storage tank varies directly with the volume of the tank. A laboratory wants to purchase a storage tank shaped like a block with dimensions L by W by H.

   a. Fill in the missing space in the following table from a brochure of various tank sizes.

Hydrogen Tank Dimensions (inches)			Cost
L	**W**	**H**	
36	36	36	
18		24	$150
24	24	24	$800

   b. The hydrogen tank must fit on a shelf that has a fixed height and depth. How does the cost of the hydrogen storage tank vary with the width of tank with fixed depth and height?

   c. How much would a spherical tank of radius 24 inches cost? (Recall that the volume of a sphere is given by $\frac{4}{3}\pi r^3$, where r is the radius.)

Intervention Reteaching and vocabulary activities that can be used with struggling or absent students and as ELL support

Study Guide and Intervention

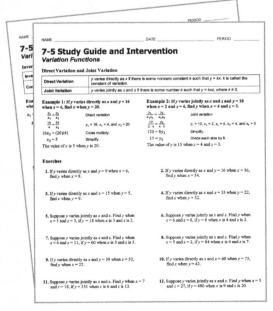

NAME _____ DATE _____ PERIOD _____

7-5 Study Guide and Intervention
Variation Functions

Direct Variation and Joint Variation

Direct Variation	y varies directly as x if there is some nonzero constant k such that $y = kx$. k is called the constant of variation.
Joint Variation	y varies jointly as x and z if there is some number k such that $y = kxz$, where $k \neq 0$.

Example 1: If y varies directly as x and $y = 16$ when $x = 4$, find x when $y = 20$.

$\frac{y_1}{x_1} = \frac{y_2}{x_2}$ Direct variation

$\frac{16}{4} = \frac{20}{x_2}$

$16x_2 = (20)(4)$ Cross multiply.

$x_2 = 5$ Simplify.

The value of x is 5 when $y = 20$.

Example 2: If y varies jointly as x and z and $y = 10$ when $x = 2$ and $z = 4$, find y when $x = 4$ and $z = 3$.

$\frac{y_1}{x_1 z_1} = \frac{y_2}{x_2 z_2}$ Joint variation

$\frac{10}{2 \cdot 4} = \frac{y_2}{4 \cdot 3}$ $y = 10, x_1 = 2, z_1 = 4, x_2 = 4,$ and $z_2 = 3$

$120 = 8y_2$ Simplify.

$15 = y_2$ Divide each side by 8.

The value of y is 15 when $x = 4$ and $z = 3$.

Exercises

1. If y varies directly as x and $y = 9$ when $x = 6$, find y when $x = 8$.

2. If y varies directly as x and $y = 16$ when $x = 36$, find y when $x = 54$.

3. If y varies directly as x and $y = 15$ when $x = 5$, find y when $x = 9$.

4. If y varies directly as x and $y = 33$ when $x = 22$, find x when $y = 32$.

5. Suppose y varies jointly as x and z. Find y when $x = 3$ and $z = 2$, if $y = 18$ when x is 3 and z is 2.

6. Suppose y varies jointly as x and z. Find y when $x = 6$ and $z = 8$, if $y = 6$ when x is 4 and z is 2.

7. Suppose y varies jointly as x and z. Find y when $x = 4$ and $z = 11$, if $y = 60$ when x is 3 and z is 5.

8. Suppose y varies jointly as x and z. Find y when $x = 5$ and $z = 2$, if $y = 84$ when x is 4 and z is 7.

9. If y varies directly as x and $y = 39$ when $x = 52$, find y when $x = 42$.

10. If y varies directly as x and $y = 60$ when $x = 75$, find x when $y = 42$.

11. Suppose y varies jointly as x and z. Find y when $x = 7$ and $z = 18$, if $y = 351$ when x is 6 and z is 13.

12. Suppose y varies jointly as x and z. Find y when $x = 5$ and $z = 27$, if $y = 480$ when x is 9 and z is 20.

Study Notebook

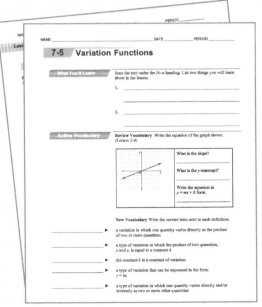

NAME _____ DATE _____ PERIOD _____

7-5   Variation Functions

What You'll Learn   Scan the text under the *Now* heading. List two things you will learn about in the lesson.

1. _____

2. _____

Active Vocabulary   **Review Vocabulary** Write the equation of the graph shown. *(Lesson 2-4)*

What is the slope?

What is the y-intercept?

Write the equation in $y = mx + b$ form.

New Vocabulary Write the correct term next to each definition.

▶ _____ a variation in which one quantity varies directly as the product of two or more quantities

▶ _____ a type of variation in which the product of two quantities, x and y, is equal to a constant k

▶ _____ the constant k in a constant of variation

▶ _____ a type of variation that can be expressed in the form $y = kx$

▶ _____ a type of variation in which one quantity varies directly and/or inversely as two or more other quantities

Extension Activities that can be used to extend lesson concepts

Enrichment

NAME _____ DATE _____ PERIOD _____

7-5 Enrichment
Geosynchronous Satellites

Satellites circling Earth are almost as common as the cell phones that depend on them. A geosynchronous satellite is one that maintains the same position above the Earth at all times. Geosynchronous satellites are used in cell phone communications, transmitting signals from towers on Earth and to each other.

The speed at which they travel is very important. If the speed is too low, the satellite will be forced back down to Earth due to Earth's gravity. However, if it is too fast, it will overcome gravity's force and escape into space, never to return. Newton's second law of motion says that force on an object is equal to mass times acceleration or $F = ma$. It is also well known that the net gravitational force between two objects is inversely proportional to the square of the distance between them. Therefore, there are two variables on which the force depends: speed and height above Earth.

In particular, Newton's second law, $F = ma$, shows that force varies directly with acceleration, where m is the constant taking the place of k.

Exercises

1. Show that the net gravitational force providing a satellite with acceleration is inversely proportional to the square of the distance between them by expressing this variation as an equation.

2. Use your equation from Exercise 1 and equate it with Newton's formula above to determine how the satellite's acceleration varies with its height above Earth.

3. Determine how the speed of a geosynchronous satellite varies with its height above Earth by using the fact that speed is equal to distance divided by time and the path of the satellite is circular.

Launch

Have students read the Why? section of the lesson. Ask:

- If the height of the ramp increases, what happens to the length of the top of the ramp? It increases.

- If h decreases, what happens to ℓ? It decreases.

- Which value remains constant? the ratio $\frac{\ell}{h}$, which equals 1.5

Variation Functions

::Then	::Now	::Why?

Then

 You wrote and graphed linear equations.

Now

1 Recognize and solve direct and joint variation problems.

2 Recognize and solve inverse and combined variation problems.

Why?

While building skateboard ramps, Yu determined that the best ramps were the ones in which the length of the top of the ramp was 1.5 times as long as the height of the ramp.

As shown in the table, the length of the top of the ramp depends on the height of a ramp. The length increases as the height increases, but the ratio remains the same, or is *constant*.

The equation $\frac{\ell}{h} = 1.5$ can be written as $\ell = 1.5h$.

The length *varies directly* with the height of the ramp.

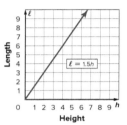

Length (ℓ)	Height (h)	Ratio $\frac{\ell}{h}$
3	2	1.5
6	4	1.5
9	6	1.5
12	8	1.5

 New Vocabulary

direct variation
constant of variation
joint variation
inverse variation
combined variation

MP Mathematical Practices

1 Make sense of problems and persevere in solving them.
2 Reason abstractly and quantitatively.
4 Model with mathematics.

Content Standards
A.CED.2 Create equations in two or more variables to represent relationships between quantities; graph equations on coordinate axes with labels and scales.
F.IF.4 For a function that models a relationship between two quantities, interpret key features of graphs and tables in terms of the quantities, and sketch graphs showing key features given a verbal description of the relationship.

Direct Variation and Joint Variation The relationship given by $\ell = 1.5h$ is an example of direct variation. A **direct variation** can be expressed in the form $y = kx$. In this equation, k is called the **constant of variation**.

Notice that the graph of $\ell = 1.5h$ is a straight line through the origin. A direct variation is a special case of an equation written in slope-intercept form, $y = mx + b$. When $m = k$ and $b = 0$, $y = mx + b$ becomes $y = kx$. So the slope of a direct variation equation is its constant of variation.

To express a direct variation, we say that y varies directly as x. In other words, as x increases, y increases or decreases at a constant rate.

Length / *Height*

$\ell = 1.5h$

Key Concept Direct Variation

Words	y varies directly as x if there is some nonzero constant k such that $y = kx$. k is called the *constant of variation*.
Example	If $y = 3x$ and $x = 7$, then $y = 3(7)$ or 21.

If you know that y varies directly as x and one set of values, you can use a proportion to find the other set of corresponding values.

$$y_1 = kx_1 \qquad \text{and} \qquad y_2 = kx_2$$
$$\frac{y_1}{x_1} = k \qquad\qquad\qquad \frac{y_2}{x_2} = k \qquad \text{Therefore, } \frac{y_1}{x_1} = \frac{y_2}{x_2}.$$

Using the properties of equality, you can find many other proportions that relate these same x- and y-values.

MP Mathematical Practices Strategies

Look for and make sense of problems and persevere in solving them.
Help students maintain oversight of the process of variation of equations. For example, ask:

- **How can a graph illustrate the relationship between two values?** A graph shows how y changes in relation to x. For example, if the graph is an upward line, you know that x increases as y increases.

- **What variable must stay the same to create a direct variation?** the constant of variation, k

- **What is the first step in solving a proportion?** cross-multiplying

---

Example 1 Direct Variation

A.CED.2

If y varies directly as x and $y = 15$ when $x = -5$, find y when $x = 7$.

Use a proportion that relates the values.

$\dfrac{y_1}{x_1} = \dfrac{y_2}{x_2}$ Direct variation

$\dfrac{15}{-5} = \dfrac{y_2}{7}$ $y_1 = 15, x_1 = -5$ and $x_2 = 7$

$15(7) = -5(y_2)$ Cross multiply.

$105 = -5y_2$ Simplify.

$-21 = y_2$ Divide each side by -5.

Guided Practice

1. If r varies directly as t and $r = -20$ when $t = 4$, find r when $t = -6$. **30**

Another type of variation is joint variation. **Joint variation** occurs when one quantity varies directly as the product of two or more other quantities.

> **StudyTip**
> Joint Variation Some mathematicians consider joint variation a special type of combined variation.

Key Concept Joint Variation

Words	y varies jointly as x and z if there is some nonzero constant k such that $y = kxz$.
Example	If $y = 5xz, x = 6$, and $z = -2$, then $y = 5(6)(-2)$ or -60.

If you know that y varies jointly as x and z and one set of values, you can use a proportion to find the other set of corresponding values.

$y_1 = kx_1z_1$ and $y_2 = kx_2z_2$

$\dfrac{y_1}{x_1z_1} = k$ $\dfrac{y_2}{x_2z_2} = k$ Therefore, $\dfrac{y_1}{x_1z_1} = \dfrac{y_2}{x_2z_2}$.

Example 2 Joint Variation

A.CED.2

Suppose y varies jointly as x and z. Find y when $x = 9$ and $z = 2$, if $y = 20$ when $z = 3$ and $x = 5$.

Use a proportion that relates the values.

$\dfrac{y_1}{x_1z_1} = \dfrac{y_2}{x_2z_2}$ Joint variation

$\dfrac{20}{5(3)} = \dfrac{y_2}{9(2)}$ $y_1 = 20, x_1 = 5, z_1 = 3, x_2 = 9$ and $z_2 = 2$

$20(9)(2) = 5(3)(y_2)$ Cross multiply.

$360 = 15y_2$ Simplify.

$24 = y_2$ Divide each side by 15.

Guided Practice

2. Suppose r varies jointly as v and t. Find r when $v = 2$ and $t = 8$, if $r = 70$ when $v = 10$ and $t = 4$. **28**

Teach

Ask the scaffolded questions for each example to build conceptual understanding for students at all levels.

1 Direct Variation and Joint Variation

Example 1 Direct Variation

AL What type of function do you use to model direct variation? linear

OL What does the constant of variation represent of the graph of direct variation? the slope

BL If x varies direction as y, what happens to y if x gets larger? It also gets larger.

Need Another Example?
If y varies directly as x and $y = -15$ when $x = 5$, find y when $x = 3$. -9

Example 2 Joint Variation

AL If y varies jointly with x and z, how many of the variables do we need to solve for k? all 3

OL If x and z both increase, what happens to y? It increases.

BL If y varies jointly with x and z, what must be true if $y = 0$? Either $x = 0$ or $z = 0$.

Need Another Example?
Suppose y varies jointly as x and z. Find y when $x = 10$ and $z = 5$, if $y = 12$ when $z = 8$ and $x = 3$. 25

> **Watch Out!**
> Preventing Errors Discuss with students how to write variation equations that include a constant of variation.

Differentiated Instruction ELL

Beginning Read the lesson opener or an Example aloud one sentence at a time. At the end of each sentence, ask students to say a word or short phrase that describes an important piece of information from the sentence. Model recording the information in preparation for solving the problem. Have students use your model to record information in their notes.

Intermediate Slowly read the lesson opener or an Example aloud. After each sentence or two, pause and ask volunteers to identify an important piece of information. Have students write the important idea in their notes.

Advanced Tell students to listen without taking notes while you read aloud. After you have finished, have students write down what they remember from your reading. Have students work in small groups to compare their notes. Then have each group discuss the problem and its solution.

Advanced High Have students practice active listening as you read aloud by taking notes. Then have students work in pairs to summarize the information and solve the problem. Have pairs share with the class.

2 Inverse Variation and Combined Variation

Example 3 Inverse Variation

AL What type of function do you use to represent inverse variation? reciprocal

OL What does the end behavior of *y* mean in the context of inverse variation? As *x* increases toward infinity or decreases toward negative infinity, *y* gets closer to 0.

BL Why is the condition that $x \neq 0$ and $y \neq 0$ in the statement for inverse variation? $x \neq 0$ is because we cannot divide by 0. The condition $y \neq 0$ is set because there is no value that will satisfy the equation $0 = \frac{k}{x}$.

Need Another Example?
If *r* varies inversely as *t* and $r = -6$ when $t = 2$, find *r* when $t = -7$. $\frac{12}{7}$

Teaching Tips
Alternate Method In Example 3, students may wish to solve the problem by using the equation $a_1 b_1 = a_2 b_2$.

Sense-Making Help students understand the difference between the two types of variation by using the example of speed, distance, and driving time. When driving at a given rate, the distance increases as driving time increases (direct). However, for a given distance, the time needed decreases as speed increases (inverse).

---

2 Inverse Variation and Combined Variation Another type of variation is inverse variation. If two quantities *x* and *y* show **inverse variation**, their product is equal to a constant *k*.

Inverse variation is often described as one quantity increasing while the other quantity is decreasing. For example, speed and time for a fixed distance vary inversely with each other; the faster you go, the less time it takes you to get there.

> **Key Concept** Inverse Variation
>
> | Words | *y* varies inversely as *x* if there is some nonzero constant *k* such that $xy = k$ or $y = \frac{k}{x}$, where $x \neq 0$ and $y \neq 0$. |
> |---|---|
> | Example | If $xy = 2$, and $x = 6$, then $y = \frac{2}{6}$ or $\frac{1}{3}$. |

Study Tip

Reasoning You can identify the type of variation by looking at a table of values for *x* and *y*. If the quotient $\frac{y}{x}$ has a constant value, *y* varies directly as *x*. If the product *xy* has a constant value, *y* varies inversely as *x*.

Suppose *y* varies inversely as *x* such that $xy = 6$ or $y = \frac{6}{x}$. The graph of this equation is shown at the right. Because *k* is a positive value, as the values of *x* increase, the values of *y* decrease.

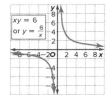

Notice that the graph of an inverse variation is a reciprocal function.

A proportion can be used with inverse variation to solve problems in which some quantities are known. The following proportion is only one of several that can be formed.

$$x_1 y_1 = k \text{ and } x_2 y_2 = k$$

$x_1 y_1 = x_2 y_2$	Substitution Property of Equality
$\dfrac{x_1}{y_2} = \dfrac{x_2}{y_1}$	Divide each side by $y_1 y_2$

A.CED.2

Go Online!

Investigate how variation functions can be used to model astronauts' weights on other planets in a Spreadsheet Activity from the Resources in ConnectED.

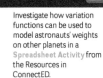

Example 3 Inverse Variation

If *a* varies inversely as *b* and $a = 28$ when $b = -2$, find *a* when $b = -10$.

Use a proportion that relates the values.

$\dfrac{a_1}{b_2} = \dfrac{a_2}{b_1}$	Inverse variation
$\dfrac{28}{-10} = \dfrac{a_2}{-2}$	$a_1 = 28$, $b_1 = -2$, and $b_2 = -10$
$28(-2) = -10(a_2)$	Cross multiply.
$-56 = -10(a_2)$	Simplify.
$5\frac{3}{5} = a_2$	Divide each side by -10.

Guided Practice

3. If *x* varies inversely as *y* and $x = 24$ when $y = 4$, find *x* when $y = 12$. **8**

Inverse variation is often used in real-world situations.

A.CED.2

Real-World Example 4   Write and Solve an Inverse Variation

MUSIC The length of a violin string varies inversely as the frequency of its vibrations. A violin string 10 inches long vibrates at a frequency of 512 cycles per second. Find the frequency of an 8-inch violin string.

Let $v_1 = 10$, $f_1 = 512$, and $v_2 = 8$. Solve for f_2.

$$v_1 f_1 = v_2 f_2 \qquad \text{Original equation}$$
$$10 \cdot 512 = 8 \cdot f_2 \qquad v_1 = 10, f_1 = 512, \text{ and } v_2 = 8$$
$$\frac{5120}{8} = f_2 \qquad \text{Divide each side by 8}$$
$$640 = f_2 \qquad \text{Simplify}$$

The 8-inch violin string vibrates at a frequency of 640 cycles per second.

▶ **Guided Practice**

4. The apparent length of an object is inversely proportional to one's distance from the object. Earth is about 93 million miles from the Sun. Jupiter is about 483.6 million miles from the Sun. Find how many times as large the diameter of the Sun would appear on Earth as on Jupiter. **5.2 times as large**

Another type of variation is combined variation. **Combined variation** occurs when one quantity varies directly and/or inversely as two or more other quantities.

If you know that y varies directly as x, y varies inversely as z, and one set of values, you can use a proportion to find the other set of corresponding values.

$$y_1 = \frac{kx_1}{z_1} \qquad \text{and} \qquad y_2 = \frac{kx_2}{z_2}$$
$$\frac{y_1 z_1}{x_1} = k \qquad\qquad \frac{y_2 z_2}{x_2} = k \qquad \text{Therefore, } \frac{y_1 z_1}{x_1} = \frac{y_2 z_2}{x_2}.$$

A.CED.2

Example 5   Combined Variation

Suppose f varies directly as g, and f varies inversely as h. Find g when $f = 18$ and $h = -3$, if $g = 24$ when $h = 2$ and $f = 6$.

First set up a correct proportion for the information given.

$$f_1 = \frac{kg_1}{h_1} \text{ and } f_2 = \frac{kg_2}{h_2} \qquad \text{g varies directly as f, so g goes in the numerator. h varies inversely as f, so h goes in the denominator.}$$

$$k = \frac{f_1 h_1}{g_1} \text{ and } k = \frac{f_2 h_2}{g_2} \qquad \text{Solve for k.}$$

$$\frac{f_1 h_1}{g_1} = \frac{f_2 h_2}{g_2} \qquad \text{Set the two proportions equal to each other.}$$

$$\frac{6(2)}{24} = \frac{18(-3)}{g_2} \qquad \text{$f_1 = 6$, $g_1 = 24$, $h_1 = 2$, $f_2 = 18$, and $h_2 = -3$}$$

$$24(18)(-3) = 6(2)(g_2) \qquad \text{Cross multiply.}$$

$$-1296 = 12g_2 \qquad \text{Simplify.}$$

$$-108 = g_2 \qquad \text{Divide each side by 12.}$$

When $f = 18$ and $h = -3$, the value of g is -108.

▶ **Guided Practice**

5. Suppose p varies directly as r, and p varies inversely as t. Find t when $r = 10$ and $p = -5$, if $t = 20$ when $p = 4$ and $r = 2$. **−80**

Real-World Link

When you pluck a string, it vibrates back and forth. This causes mechanical energy to travel through the air in waves. The number of times per second these waves hit our ear is called the *frequency*. The more waves per second, the higher the pitch.

Study Tip

Combined Variation
Quantities that vary directly appear in the numerator. Quantities that vary inversely appear in the denominator.

Example 4   Write and Solve an Inverse Variation

AL What do v and f represent? v is the length of the violin string and f is the frequency.

OL What restrictions should be put on v and f? They both must be positive and nonzero.

BL What equation would represent this inverse variation? $v = \frac{5120}{f}$

Need Another Example?

Space Use the information from Guided Practice and the fact that Venus is about 67 million miles away from the Sun. How much larger would the diameter of the Sun appear on Venus than on Earth? about 1.39 times as large

Example 5   Combined Variation

AL How do you solve the correct proportion? cross-multiplication

OL Write a rule about how many of the variables you need to solve for the others. number of variables − 1

BL What equation would represent this situation? $f = \frac{g}{2h}$

Need Another Example?

Suppose f varies directly as g, and f varies inversely as h. Find g when $f = 6$ and $h = -5$, if $g = 18$ when $h = 3$ and $f = 5$. −36

Practice

Formative Assessment Use Exercises 1–6 to assess students' understanding of the concepts in this lesson.

The Practice and Problem Solving exercises assess the content taught in the lesson. The Preparing for Assessment page is meant to be used as preparation for end-of-year assessments.

Teaching the Mathematical Practices

Modeling Mathematically proficient students can apply the mathematics they know to solve problems arising in everyday life, analyze relationships mathematically to draw conclusions, and interpret their mathematical results in the context of a situation.

Extra Practice

See page R7 for extra exercises for students who are approaching level or for on-level students who need additional reinforcement.

Levels of Complexity Chart

The levels of the exercises progress from 1 to 3, with Level 1 indicating the lowest level of complexity.

Exercises	7–24	25–46, 53–57	47–52
Level 3			●
Level 2		●	
Level 1	●		

Check Your Understanding ○ = Step-by-Step Solutions begin on page R11.

Examples 1–3
A.CED.2

1. If y varies directly as x and $y = 12$ when $x = 8$, find y when $x = 14$. 21

2. Suppose y varies jointly as x and z. Find y when $x = 9$ and $z = -3$, if $y = -50$ when z is 5 and x is -10. -27

3. If y varies inversely as x and $y = -18$ when $x = 16$, find x when $y = 9$. -32

Example 4
A.CED.2

4. TRAVEL A map of Texas is scaled so that 2 inches represents 30 miles. How far apart are El Paso and Odessa if they are 20 inches apart on the map? 300 mi

Example 5
A.CED.2

5. Suppose a varies directly as b, and a varies inversely as c. Find b when $a = 8$ and $c = -3$, if $b = 16$ when $c = 2$ and $a = 4$. -48

6. Suppose d varies directly as f, and d varies inversely as g. Find g when $d = 6$ and $f = -7$, if $g = 12$ when $d = 9$ and $f = 3$. -42

Practice and Problem Solving

Extra Practice is on page R7.

Example 1
A.CED.2

If x varies directly as y, find x when $y = 8$.

7. $x = 6$ when $y = 32$ 1.5

8. $x = 11$ when $y = -3$ $-\frac{88}{3}$

9. $x = 14$ when $y = -2$ -56

10. $x = -4$ when $y = 10$ -3.2

11. MOON Astronaut Neil Armstrong, the first man on the Moon, weighed 360 pounds on Earth with all his equipment on, but weighed only 60 pounds on the Moon. Write an equation that relates weight on the Moon m with weight on Earth w. $m = \frac{1}{6}w$

Example 2
A.CED.2

Suppose a varies jointly as b and c. Find a when $b = 4$ and $c = -3$.

12. $a = -96$ when $b = 3$ and $c = -8$ -48

13. $a = -60$ when $b = -5$ and $c = 4$ -36

14. $a = -108$ when $b = 2$ and $c = 9$ 72

15. $a = 24$ when $b = 8$ and $c = 12$ -3

16. **MODELING** According to the A.C. Nielsen Company, the average American watches about 5 hours of television per day.

a. Write an equation to represent the average number of hours spent watching television by m household members during a period of d days. $t = 5md$

b. Assume that members of your household watch the same amount of television each day as the average American. How many hours of television would the members of your household watch in a week? Sample answer for four household members: 140 hours

Example 3
A.CED.2

If f varies inversely as g, find f when $g = -6$.

17. $f = 15$ when $g = 9$ -22.5

18. $f = 4$ when $g = 28$ $-\frac{56}{3}$

19. $f = -12$ when $g = 19$ 38

20. $f = 0.6$ when $g = -21$ 2.1

21. COMMUNITY SERVICE Every year students at West High School collect canned goods for a local food pantry. They plan to distribute flyers to homes in the community asking for donations. Last year, 12 students were able to distribute 1000 flyers in four hours.

a. Write an equation that relates the number of students s to the amount of time t it takes to distribute 1000 flyers. $s = \frac{48}{t}$

b. How long would it take 15 students to hand out the same number of flyers this year? 3.2 hours

---

Differentiated Homework Options

Levels	AL Basic	OL Core	BL Advanced
Exercises	7–24, 48–57	7–43 odd, 47, 49, 53–57	25–57
2-Day Option	7–23 odd, 53–57	7–24, 53–57	
	8–24 even, 48–52	25–52	

You can use ALEKS to provide additional remediation support with personalized instruction and practice.

Example 4
A.CED.2

22. BIRDS When a group of snow geese migrate, the distance that they fly varies directly with the amount of time they are in the air.

a. A group of snow geese migrated 375 miles in 7.5 hours. Write a direct variation equation that represents this situation. $d = 50t$

b. Every year, geese migrate 3000 miles from their winter home in the southwest United States to their summer home in the Canadian Arctic. Estimate the number of hours of flying time that it takes for the geese to migrate. **60 hours**

Example 5
A.CED.2

23. Suppose a varies directly as b, and a varies inversely as c. Find b when $a = 5$ and $c = -4$, if $b = 12$ when $c = 3$ and $a = 8$. **−10**

24. Suppose x varies directly as y, and x varies inversely as z. Find z when $x = 10$ and $y = -7$, if $z = 20$ when $x = 6$ and $y = 14$. **−6**

 Determine whether each relation shows *direct* or *inverse* variation, or *neither*.

25.

x	y
4	12
8	24
16	48
32	96

26.

x	y
8	2
4	4
−2	−8
−8	−2

27.

x	y
2	4
3	9
4	16
5	25

25. direct
26. inverse
27. neither

28. If y varies inversely as x and $y = 6$ when $x = 19$, find y when $x = 2$. **57**

29 If x varies inversely as y and $x = 16$ when $y = 5$, find x when $y = 20$. **4**

30. Suppose a varies directly as b, and a varies inversely as c. Find b when $a = 7$ and $c = -8$, if $b = 15$ when $c = 2$ and $a = 4$. **−105**

31. Suppose x varies directly as y, and x varies inversely as z. Find z when $x = 8$ and $y = -6$, if $z = 26$ when $x = 8$ and $y = 13$. **−12**

State whether each equation represents a *direct*, *joint*, *inverse*, or *combined* variation. Then name the constant of variation.

32. $\frac{x}{y} = 2.75$ **33.** $fg = -2$ **34.** $a = 3bc$ **35.** $10 = \frac{xy^2}{z}$

36. $y = -11x$ **37.** $\frac{n}{p} = 4$ **38.** $9n = pr$ **39.** $-2y = z$

40. $a = 27b$ direct; 27 **41.** $c = \frac{7}{d}$ inverse; 7 **42.** $-10 = gh$ **43.** $m = 20cd$
                                                        inverse; −10   joint; 20

32. direct; 2.75
33. inverse; −2
34. joint; 3
35. combined; 10
36. direct; −11
37. direct; 4
38. combined; 9
39. direct; −2

44. **PRECISION** The volume of a gas v varies inversely as the pressure p and directly as the temperature t.

a. Write an equation to represent the volume of a gas in terms of pressure and temperature. Is your equation a *direct*, *joint*, *inverse*, or *combined* variation? $v = \frac{kt}{p}$; combined

b. A certain gas has a volume of 8 liters, a temperature of 275 Kelvin, and a pressure of 1.25 atmospheres. If the gas is compressed to a volume of 6 liters and is heated to 300 Kelvin, what will the new pressure be? approximately 1.82 atmospheres or $\frac{20}{11}$ atm

c. If the volume stays the same, but the pressure drops by half, then what must have happened to the temperature? It dropped by half.

45. VACATION The time it takes the Levensteins to reach Lake Tahoe varies inversely with their average rate of speed.

a. If they are 800 miles away, write and graph an equation relating their travel time to their average rate of speed. See margin.

b. What minimum average speed will allow them to arrive within 18 hours? $44.\overline{4}$ mph

MP **Teaching the Mathematical Practices**

Precision Mathematically proficient students try to use clear definitions in their reasoning, calculate accurately and efficiently, and make explicit use of definitions.

ℯ Follow-Up

Students have explored variation functions.
Ask:

● How can analyzing a rational function algebraically and graphically help you to see the "whole picture?" Sample answer: An algebraic analysis can help you to determine points of discontinuity that may not be clear or noticeable when viewing the graph of the function. A graphical analysis can help you to see the asymptotes and end behavior of the function.

Additional Answer

45a.
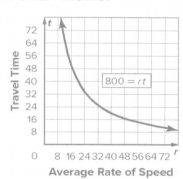

Differentiated Instruction AL OL

IF students have difficulty keeping the formulas for the different variation functions straight,

THEN have students write the formulas using different colors for each variable and another color for k. For example: direct variation: $y = kx$; joint variation: $y = kxz$; and inverse variation: $y = \frac{k}{x}$.

Teaching the Mathematical Practices

Critique Arguements Mathematically proficient students are also able to compare the effectiveness of two plausible arguments, distinguish correct logic or reasoning from that which is flawed, and—if there is a flaw in an argument—explain what it is.

Watch Out!

Preventing Misconceptions Make sure that students understand the essential differences between direct and inverse variation. For example, ask them if the number of candles on a birthday cake varies directly or indirectly with the age of the birthday person. Then ask how the time the candle has burned varies with the length of the candle remaining.

Assess

Ticket Out the Door On a small slip of paper, have each student write an equation for some variation in daily life (time spent studying, hours of sleep, etc.). When leaving the classroom, have each student tell what kind of variation the equation exhibits.

Additional Answer

53d.

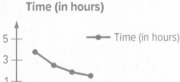

Time (in hours)

---

46. **VIDEO** The maximum number of videos that a smartphone can store depends on the lengths of the videos and the quality of the files. A movie will take up more space on the phone than a television show.

   a. If a certain phone has 64 gigabytes of storage space, write a function that represents the number of files the player can hold as a function of the average size of the files. $f(x) = \frac{64}{x}$

   b. Is your function a *direct, joint, inverse,* or *combined* variation? **inverse**

   c. Suppose the average file size for a movie is 0.75 gigabyte and the average size for a television show is 0.25 gigabyte. Determine how many more files the phone can hold if they are television shows than if they are movies. **171**

▷ 47. **GRAVITY** According to the Law of Universal Gravitation, the attractive force F in newtons between any two bodies in the universe is directly proportional to the product of the masses m_1 and m_2 in kilograms of the two bodies and inversely proportional to the square of the distance d in meters between the bodies. That is, $F = \frac{Gm_1m_2}{d^2}$. G is the universal gravitational constant. Its value is 6.67×10^{-11} Nm^2/kg^2.

   a. The distance between Earth and the Moon is about 3.84×10^8 meters. The mass of the Moon is 7.36×10^{22} kilograms. The mass of Earth is 5.97×10^{24} kilograms. What is the gravitational force that the Moon and Earth exert upon each other? **about 2×10^{20} newtons**

   b. The distance between Earth and the Sun is about 1.5×10^{11} meters. The mass of the Sun is about 1.99×10^{30} kilograms. What is the gravitational force that the Sun and Earth exert upon each other? **about 3.5×10^{22} newtons**

   c. Find the gravitational force exerted on each other by two 1000-kilogram iron balls at a distance of 0.1 meter apart. **6.67×10^{-3} newtons**

A.CED.2, F.IF.4

H.O.T. Problems   Use Higher-Order Thinking Skills

50. Sample answer: Every joint variation is a combined variation because there are two *combined* direct variations. However, a combined variation can have a combination of a direct and an inverse variation, so it cannot be considered as a joint variation.

52. Sample answer: Inverse and some types of combined variation functions cannot have a value of 0 in the domain because division by zero is undefined.

48. **CRITIQUE ARGUMENTS** Jamil and Savannah are setting up a proportion to begin solving the combined variation in which z varies directly as x and z varies inversely as y. Who has set up the correct proportion? Explain your reasoning.

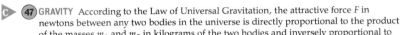

Jamil

$z_1 = \frac{kx_1}{y_1}$ and $z_2 = \frac{kx_2}{y_2}$

$k = \frac{z_1 y_1}{x_1}$ and $k = \frac{z_2 y_2}{x_2}$

$\frac{z_1 y_1}{x_1} = \frac{z_2 y_2}{x_2}$

Savannah

$z_1 = \frac{kx_1}{y_1}$ and $z_2 = \frac{kx_2}{y_2}$

$k = \frac{z_1 x_1}{y_1}$ and $k = \frac{z_2 x_2}{y_2}$

$\frac{z_1 x_1}{y_1} = \frac{z_2 x_2}{y_2}$

Jamil; Savannah multiplied when she should have divided and divided when she should have multiplied.

49. **CHALLENGE** If a varies inversely as b, c varies jointly as b and f, and f varies directly as g, how are a and g related? **a and g are directly related.**

50. **REASONING** Explain why some mathematicians consider every joint variation a combined variation, but not every combined variation a joint variation.

51. **OPEN-ENDED** Describe three real-life quantities that vary jointly with each other. **Sample answer: The force of an object varies jointly as its mass and acceleration.**

52. **WRITING IN MATH** Determine the type(s) of variation(s) for which 0 cannot be one of the values. Explain your reasoning.

Standards for Mathematical Practice

Emphasis On	Exercises
1 Make sense of problems and persevere in solving them.	16
2 Reason abstractly and quantitatively.	1–10, 12–15, 17–20, 23–43, 50, 57
3 Construct viable arguments and critique the reasoning of others.	48
4 Model with mathematics.	11, 21, 22, 44–47, 53–56
7 Look for and make use of structure.	49, 51

Preparing for Assessment

53. MULTI-STEP The time t it takes to paint a house varies inversely as the number of people p painting it. 🔵 4 A.CED.2, F.IF.4

a. Which of the following equations accurately represents the relationship between t and p? **A**

- ○ **A** $t_1 p_1 = t_2 p_2$
- ○ **B** $\dfrac{t_1}{p_1} = \dfrac{t_2}{p_2}$
- ○ **C** $\dfrac{t_1}{t_2} = \dfrac{p_1}{p_2}$
- ○ **D** $t_1 p_2 = p_1 t_2$

b. If 5 people can paint the house in 7.5 hours, how many people are needed to get the house painted in no more than 4.5 hours? Round up to the nearest whole number. **D**

- ○ **A** 3 people
- ○ **B** 5 people
- ○ **C** 7 people
- ○ **D** 9 people

c. Finish the chart to show the change in time t as the number of painters p increases.

Number of Painters (p)	Time to Paint the House (t)
10	3.75
15	2.5
20	1.875
25	1.5

d. Draw a graph to illustrate the time to paint the house in terms of the number of painters.
See margin.

54. Julia earns d total dollars for the number of hours h that she works. What type of variation represents the relationship between d and h? 🔵 4 A.CED.2 **B**

- ○ **A** combined
- ○ **B** direct
- ○ **C** inverse
- ○ **D** joint

55. Which relation shows an inverse variation?
🔵 4 A.CED.2 **D**

- ○ **A**

x	y
3	12
3.5	14
7.5	30
8.5	34

- ○ **B**

x	y
3	7.5
2.5	6.25
9	18
12.2	24.4

- ○ **C**

x	y
2	3
3	2
1	12
4	12

- ○ **D**

x	y
4.5	8
12	3
18	2
6	6

56. The variable x varies directly as y, and x varies inversely as z. What is the value of y when $x = 20$ and $z = -2$, if $y = 30$ when $z = 3$ and $x = 5$? 🔵 4 A.CED.2 **B**

- ○ **A** -180
- ○ **B** -80
- ○ **C** -5
- ○ **D** 5

57. The total number of hours h that a construction job will take to reach completion is related to the average daily number of workers w that are on the job site throughout the whole project. What type of variation represents the relationship between h and w? 🔵 2 A.CED.2 **C**

- ○ **A** combined
- ○ **B** direct
- ○ **C** inverse
- ○ **D** joint

Differentiated Instruction

Extension Write the equation $y = kx^3$ on the board. Ask students to describe the kind of variation modeled by this equation. Have them describe what happens to the value of y when the value of x is doubled, tripled, halved, etc. *For this equation, y varies directly as the cube of x. When x is doubled, y is multiplied by 8. When x is tripled, y is multiplied by 27. When x is halved, y is divided by 8.*

55.

A	Chose direct variation
B	Chose direct variation, but only examined first 2 rows
C	Looked only at the first two rows in the table
D	CORRECT

56.

A	Wrote and solved $\dfrac{3 \times 30}{5} = \dfrac{-2 \times y_2}{20}$
B	CORRECT
C	Wrote and solved $\dfrac{5 \times 30}{3} = \dfrac{20 \times y_2}{-2}$
D	Wrote and solved $\dfrac{5 \times 30}{3} = \dfrac{20 \times y_2}{2}$

Preparing for Assessment

Exercises 53–57 require students to use the skills they will need on assessments. The exercises are dual-coded with content.

Dual Coding		
Exercises	Content Standards	🔵 Mathematical Practices
53	A.CED.2; F.IF.4	4
54	A.CED.2	4
55	A.CED.2	4
56	A.CED.2	4
57	A.CED.2	2

Diagnose Student Errors

Survey student responses for each item. Class trends may indicate common errors and misconceptions.

53a.

A	CORRECT
B	Student misinterpreted the function as a direct variation.
C	Student mixed up variables and compared time to time, rather than comparing time to hours to paint.
D	Student misinterpreted the function as a direct variation.

53b.

A	Student misinterpreted the function as a direct variation.
B	Student possibly misread question or did not know what was being asked.
C	Student misinterpreted the function as a direct variation.
D	CORRECT

54.

A	Chose incorrect variation
B	CORRECT
C	Chose incorrect variation
D	Chose incorrect variation

Go Online!

Quizzes

Students can use *Self-Check Quizzes* to check their understanding of this lesson. You can also give the *Chapter Quiz*, which covers the content in Lessons 7-4 and 7-5.

Track Your Progress

Objectives

1 Solve rational equations.

2 Solve rational inequalities.

Mathematical Background

During the solution process, a rational equation is usually transformed into another type of equation. Solving a rational equation may require solving a related linear, quadratic, or other type of equation.

THEN	NOW	NEXT
A.CED.2 Create equations in two or more variables to represent relationships between quantities; graph equations on coordinate axes with labels and scales.	**A.CED.1** Create equations and inequalities in one variable and use them to solve problems. Include equations arising from linear and quadratic functions, and simple rational and exponential functions. **A.RE1.2** Solve simple rational and radical equations in one variable, and give examples showing how extraneous solutions may arise.	**A.REI.11** Explain why the x-coordinates of the points where the graphs of the equations $y = f(x)$ and $y = g(x)$ intersect are the solutions of the equation $f(x) = g(x)$; find the solutions approximately, e.g., using technology to graph the functions, make tables of values, or find successive approximations. Include cases where $f(x)$ and/or $g(x)$ are linear, polynomial, rational, absolute value, exponential, and logarithmic functions.*

Go Online! All of these resources and more are available at connectED.mcgraw-hill.com

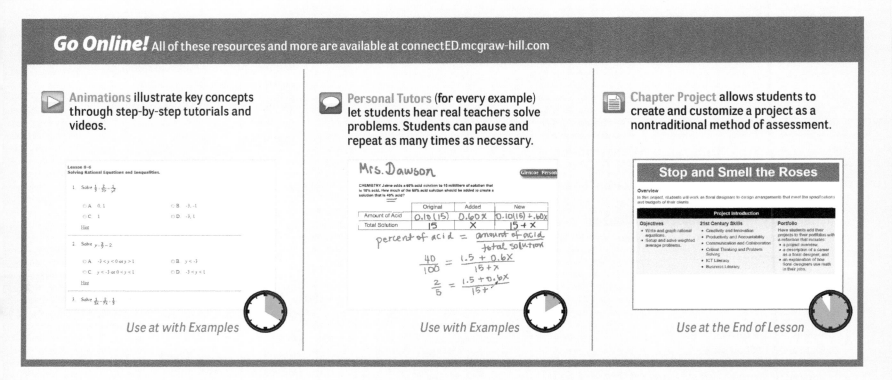

Animations illustrate key concepts through step-by-step tutorials and videos.

Use at with Examples

Personal Tutors (for every example) let students hear real teachers solve problems. Students can pause and repeat as many times as necessary.

Use with Examples

Chapter Project allows students to create and customize a project as a nontraditional method of assessment.

Use at the End of Lesson

OER Using Open Educational Resources

Video Sharing Have students work in groups to create a video lesson on **Knowmia Teach** demonstrating what they have learned about rational functions and equations. Then post students' videos online so students can review them before taking the assessment. *Use as homework*

Differentiate Your Resources

Extra Practice Additional practice or homework; Skills Practice is best for approaching-level students and Practice is best for on-level and beyond-level students

Skills Practice

7-6 Skills Practice
Solving Rational Equations and Inequalities

Practice

7-6 Practice
Solving Rational Equations and Inequalities

Word Problem Practice

7-6 Word Problem Practice
Solving Rational Equations and Inequalities

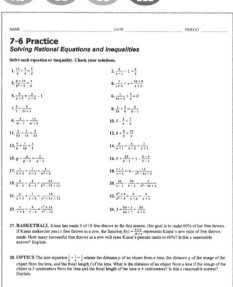

Intervention Reteaching and vocabulary activities that can be used with struggling or absent students and as ELL support

Study Guide and Intervention

7-6 Study Guide and Intervention
Solving Rational Equations and Inequalities

Study Notebook

7-6 Solving Rational Equations and Inequalities

Extension Activities that can be used to extend lesson concepts

Enrichment

7-6 Enrichment
Asymptotes in Three-Dimensions

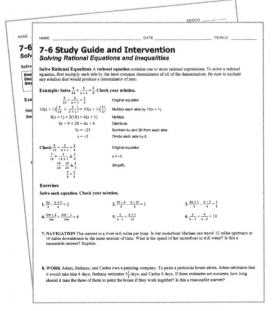

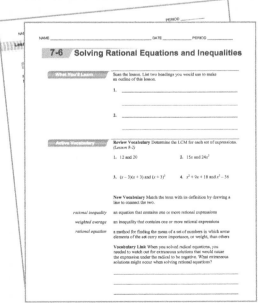

Launch

Have students read the Why? section of the lesson. Ask:

- What is the cost per visit if a member visits 10 times in a given month? $7

- If a member visits the club 10 times in a given month, how much does the monthly membership fee add to the cost of each visit? The fee adds $2 for each visit.

- If a member visits x times in a given month, how much does the monthly membership fee add to the cost of each visit? The fee adds $\frac{20}{x}$ dollars for each visit.

Teach

Ask the scaffolded questions for each example to build conceptual understanding for students at all levels.

1 Solve Rational Equations

Example 1 Solve a Rational Equation

AL What values make the equation undefined? -3

OL Why is the LCD $18(x + 3)$? $18(x + 3)$ is a multiple of $x + 3$, 6, and 18 and is the smallest expression that is a multiple.

BL Explain how you could use a calculator to solve the expression. Let $Y1 = \frac{4}{x + 3} + \frac{5}{6}$ and $Y2 = \frac{23}{18}$ and find the intersection point.

Need Another Example?
Solve $\frac{5}{24} + \frac{2}{3 - x} = \frac{1}{4}$. Check your solution. $x = -45$

Go Online!

Interactive Whiteboard

Use the eLesson or Lesson Presentation to present this lesson.

LESSON 6
Solving Rational Equations and Inequalities

:: Then	:: Now	:: Why?
• You simplified rational expressions.	**1** Solve rational equations. **2** Solve rational inequalities.	• A gaming club charges $20 per month for membership. Members also have to pay $5 each time they visit the club. If a member visits the club x times in one month, then the charge for that month will be $20 + 5x$. The actual cost per visit will be $\frac{20 + 5x}{x}$. To determine how many visits are needed for the cost per visit to be $6, you would need to solve the equation $\frac{20 + 5x}{x} = 6$.

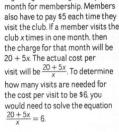

New Vocabulary
rational equation
weighted average
rational inequality

MP Mathematical Practices
6 Attend to precision.

Content Standards
A.CED.1 Create equations and inequalities in one variable and use them to solve problems.
A.REI.2 Solve simple rational and radical equations in one variable, and give examples showing how extraneous solutions may arise.

1 Solve Rational Equations Equations that contain one or more rational expressions are called **rational equations**. These equations are often easier to solve once the fractions are eliminated. You can eliminate the fractions by multiplying each side by the least common denominator (LCD).

A.REI.2

Example 1 Solve a Rational Equation

Solve $\frac{4}{x + 3} + \frac{5}{6} = \frac{23}{18}$. Check your solution.

First estimate. Then, you can later determine if the answer is reasonable. $\frac{23}{18}$ is about $1\frac{1}{3}$, so $\frac{4}{x + 3}$ is about $1\frac{1}{3} - \frac{5}{6}$ or $\frac{1}{2}$. Cross multiply to solve for x. So, x is about 5.

The LCD for the terms is $18(x + 3)$.

$$\frac{4}{x + 3} + \frac{5}{6} = \frac{23}{18} \qquad \text{Original equation}$$

$$18(x + 3)\left(\frac{4}{x + 3}\right) + 18(x + 3)\left(\frac{5}{6}\right) = 18(x + 3)\frac{23}{18} \qquad \text{Multiply by LCD.}$$

$$18(x + 3)\left(\frac{4}{x + 3}\right) + 18(x + 3)\left(\frac{5}{6}\right) = 18(x + 3)\left(\frac{23}{18}\right) \qquad \text{Divide common factors.}$$

$$72 + 15x + 45 = 23x + 69 \qquad \text{Multiply.}$$

$$15x + 117 = 23x + 69 \qquad \text{Simplify.}$$

$$48 = 8x \qquad \text{Subtract 15x and 69.}$$

$$6 = x \qquad \text{Divide.}$$

This answer is close to the estimate, so the answer is reasonable.

Guided Practice
Solve each equation. Check your solution.
1A. $\frac{2}{x + 3} + \frac{3}{2} = \frac{19}{10}$ 2   **1B.** $\frac{7}{12} + \frac{9}{x - 4} = \frac{55}{48}$ 20

MP Mathematical Practices Strategies

Look for and attend to precision.
Help students maintain oversight of the process of solving rational equations and rational inequalities. For example, ask:

- **How do you solve a rational equation or rational inequality?** Find the lowest common denominator and then eliminate the denominator and solve the equation.

- **What is an extraneous solution?** any answer that results in a denominator of 0

- **What is an important step in solving rational equations or rational inequalities that model real world situations?** Sample answer: Check your answers to make sure they make sense in the context of the question.

Multiplying each side of an equation by the LCD of rational expressions can yield results that are not solutions of the original equation. These are extraneous solutions.

A.REI.2

Example 2 Solve a Rational Equation

Solve $\dfrac{2x}{x+5} - \dfrac{x^2-x-10}{x^2+8x+15} = \dfrac{3}{x+3}$. Check your solution.

The LCD for the terms is $(x+3)(x+5)$.

$$\dfrac{2x}{x+5} - \dfrac{x^2-x-10}{x^2+8x+15} = \dfrac{3}{x+3} \qquad \text{Original equation}$$

$$\dfrac{(x+3)(x+5)(2x)}{x+5} - \dfrac{(x+3)(x+5)(x^2-x-10)}{x^2+8x+15} = \dfrac{(x+3)(x+5)3}{x+3} \qquad \text{Multiply by LCD.}$$

$$\dfrac{(x+3)(\overset{1}{\cancel{x+5}})(2x)}{\underset{1}{\cancel{x+5}}} - \dfrac{(\overset{1}{\cancel{x+3}})(\overset{1}{\cancel{x+5}})(x^2-x-10)}{\cancel{x^2+8x+15}} = \dfrac{(x+5)(\overset{1}{\cancel{x+3}})3}{\underset{1}{\cancel{x+3}}} \qquad \text{Divide common factors.}$$

$$(x+3)(2x) - (x^2-x-10) = 3(x+5) \qquad \text{Simplify.}$$

$$2x^2 + 6x - x^2 + x + 10 = 3x + 15 \qquad \text{Distribute.}$$

$$x^2 + 7x + 10 = 3x + 15 \qquad \text{Simplify.}$$

$$x^2 + 4x - 5 = 0 \qquad \text{Subtract } 3x + 15.$$

$$(x+5)(x-1) = 0 \qquad \text{Factor.}$$

$$x + 5 = 0 \qquad \text{or} \qquad x - 1 = 0 \qquad \text{Zero Product Property}$$

$$x = -5 \qquad\qquad x = 1$$

CHECK Try $x = -5$.

$$\dfrac{2x}{x+5} - \dfrac{x^2-x-10}{x^2+8x+15} = \dfrac{3}{x+3}$$

$$\dfrac{2(-5)}{-5+5} - \dfrac{(-5)^2-(-5)-10}{(-5)^2+8(-5)+15} \overset{?}{=} \dfrac{3}{-5+3}$$

$$\dfrac{-10}{0} - \dfrac{25+5-10}{25-40+15} \neq -\dfrac{3}{2} \quad \times$$

Try $x = 1$.

$$\dfrac{2x}{x+5} - \dfrac{x^2-x-10}{x^2+8x+15} = \dfrac{3}{x+3}$$

$$\dfrac{2(1)}{1+5} - \dfrac{1^2-1-10}{1^2+8(1)+15} \overset{?}{=} \dfrac{3}{1+3}$$

$$\dfrac{2}{6} - \dfrac{-10}{24} \overset{?}{=} \dfrac{3}{4}$$

$$\dfrac{8}{24} + \dfrac{10}{24} \overset{?}{=} \dfrac{3}{4}$$

$$\dfrac{3}{4} = \dfrac{3}{4} \quad \checkmark$$

When solving a rational equation, any possible solution that results in a zero in the denominator must be excluded from your list of solutions.

Because $x = -5$ results in a zero in the denominator, it is extraneous. Eliminate -5 from the list of solutions. The solution is 1.

Guided Practice

2A. $\dfrac{5}{y-2} + 2 = \dfrac{17}{6}$ 8

2B. $\dfrac{2}{z+1} - \dfrac{1}{z-1} = \dfrac{-2}{z^2-1}$ no solutions

2C. $\dfrac{7n}{3n+3} - \dfrac{5}{4n-4} = \dfrac{3n}{2n+2}$ $-\dfrac{1}{2}, 3$

2D. $\dfrac{1}{p-2} = \dfrac{2p+1}{p^2+2p-8} + \dfrac{2}{p+4}$ $\dfrac{7}{3}$

Math History Link

Brook Taylor (1685–1731) English mathematician Taylor developed a theorem used in calculus known as Taylor's Theorem that relies on the remainders after computations with rational expressions.

Reading Math Tip ELL

Extraneous Everyday use—something extraneous is not essential, it is unnecessary; Math meaning— extraneous solutions arise in solving an equation, but do not satisfy the original equation.

Example 2 Solve a Rational Equation

AL How do you know from the beginning that 5 cannot be a solution? It makes the equation undefined.

OL Explain how to find the LCD of these expressions. Factor the denominators to be $x + 5$, $(x + 3)(x + 5)$, and $x + 3$. We need both a factor of $x + 3$ and $x + 5$ in our LCD, so the LCD is $(x + 3)(x + 5) = x^2 + 8x + 15$.

BL Explain how finding the points that make an expression undefined can help you easily detect extraneous solutions. If a solution makes the expression undefined it is definitely extraneous, so if you find the points that make the expression undefined if any of these show up as solutions you know they must be extraneous.

Need Another Example?

Solve $\dfrac{p^2-p-5}{p+1} = \dfrac{p^2-7}{p-1} + p$. Check your solution. $p = -3, -2, 2$

Watch Out!

Preventing Misconceptions Remind students that a possible solution must always be checked in the original equation, rather than in any of the steps of the solution.

Go Online!

The most up-to-date resources available for your program can be found at connectED.mcgraw-hill.com.

connectED.mcgraw-hill.com **509**

Example 3 Mixture Problem

AL How does $\frac{60}{100}$ represent 60%? % means "part of 100", so 60% is 60 out of 100 which is represented as $\frac{60}{100}$.

OL What is another method you could use to solve $\frac{60}{100} = \frac{1.8 + .7x}{12 + x}$? cross-multiply

BL Explain how a mixture problem is a weighted average. The percentage of acid in the solution is the weight.

Need Another Example?

Brine Aaron adds an 80% brine (salt and water) solution to 16 ounces of solution that is 10% brine. How much of the solution should be added to create a solution that is 50% brine? $21\frac{1}{3}$ ounces

MP Teaching the Mathematical Practices

Modeling Mathematically proficient students are able to identify important quantities in a practical situation and map their relationships using such tools as diagrams, two-way tables, graphs, flowcharts and formulas. Encourage students to use various representations to better analyze functions.

The **weighted average** is a method for finding the mean of a set of numbers in which some elements of the set carry more importance, or weight, than others. Many real-world problems involving weighted averages can be solved by using rational equations.

A.CED.1, A.REI.2

Real-World Example 3 Mixture Problem

CHEMISTRY How much of a 70% acid solution should Mia add to 12 milliliters of a solution that is 15% acid to create a solution that is 60% acid?

Analyze Mia needs to know how much of a solution needs to be added to an original solution to create a new solution.

Formulate Each solution has a certain percentage that is acid. The percentage of acid in the final solution must equal the amount of acid divided by the total solution. Estimate the amount of 70% solution by seeing that the second solution is about five times as acidic. So, add five times the amount of the first solution, or 60 mL.

	Original	Added	New
Amount of Acid	0.15(12)	0.7(x)	0.15(12) + 0.7x
Total Solution	12	x	12 + x

Determine

$$\frac{\text{percent}}{100} = \frac{\text{amount of acid}}{\text{total solution}}$$ Write a proportion.

$$\frac{60}{100} = \frac{0.15(12) + 0.7x}{12 + x}$$ Substitute.

$$\frac{60}{100} = \frac{1.8 + 0.7x}{12 + x}$$ Simplify numerator.

$$100(12 + x)\frac{60}{100} = 100(12 + x)\frac{1.8 + 0.7x}{12 + x}$$ LCD is 100(12 + x). Multiply by LCD.

$$\underset{1}{100}(12 + x)\underset{1}{\frac{60}{100}} = \underset{1}{100}(12 + \underset{1}{x})\frac{1.8 + 0.7x}{12 + x}$$ Divide common factors.

$$(12 + x)60 = 100(1.8 + 0.7x)$$ Simplify.

$$720 + 60x = 180 + 70x$$ Distribute.

$$540 = 10x$$ Subtract 60x and 180.

$$54 = x$$ Divide by 10.

Justify $\frac{60}{100} = \frac{0.15(12) + 0.7x}{12 + x}$ Original equation

$$\frac{60}{100} \overset{?}{=} \frac{0.15(12) + 0.7(54)}{12 + 54}$$ x = 54

$$\frac{60}{100} \overset{?}{=} 39.6$$ Simplify.

$$0.6 = 0.6 \checkmark$$ Simplify.

Mia needs to add 54 milliliters of the 70% acid solution.

Evaluate The acidity of each solution is the "weight" that must be considered in solving this problem. The answer is close to our estimate and seems reasonable.

▸ **Guided Practice**

3. How much of a 65% fruit juice solution must Justin add to 15 milliliters of a drink that is 10% fruit juice to create a fruit punch that is 35% fruit juice? 12.5 mL

Study Tip

MP Modeling Tables like the one in Example 3 are useful in organizing and solving mixture, work, weighted average, and distance problems.

Go Online!

Many students find mixture, distance, and work problems challenging. Watch the Personal Tutor describe how to solve these problems with a partner. Then try describing how to solve a problem for them. Have them ask questions to help your understanding. **BL**

The formula relating distance, rate, and time, $d = rt$, can be used to solve rational equations. However, it can also be represented by $r = \frac{d}{t}$ and $t = \frac{d}{r}$.

A.CED.1, A.REI.2

Real-World Example 4   Distance Problem

ROWING Sandra's canoeing rate in still water is 6 miles per hour. It takes Sandra 3 hours to travel 10 miles round trip. Assuming a constant rate of speed, determine the rate of the current.

Analyze You know her speed in still water and the time it takes her to travel 5 miles with the current and 5 miles against it. You need to determine the speed of the current.

Formulate The formula that relates distance, rate, and time is $d = rt$, or $t = \frac{d}{r}$. If the current is 6 miles per hour or more, Sandra makes no progress against the current. The current must be greater than 0 miles per hour, so $0 \le r \le 6$.

Time with the Current	Time Against the Current	Total Time
$\frac{5}{6+r}$	$\frac{5}{6-r}$	3 hours

Determine

$$\frac{5}{6+r} + \frac{5}{6-r} = 3 \qquad \text{Write the equation}$$

$$(6+r)(6-r)\frac{5}{6+r} + (6+r)(6-r)\frac{5}{6-r} = (6+r)(6-r)3 \qquad \begin{array}{l}\text{LCD} = (6+r)(6-r)\\ \text{Multiply by LCD.}\end{array}$$

$$(6+r)(6-r)\frac{5}{6+r} + (6+r)(6-r)\frac{5}{6-r} = (6+r)(6-r)3 \qquad \text{Divide common factors.}$$

$$(6-r)5 + (6+r)5 = (36 - r^2)3 \qquad \text{Simplify.}$$

$$30 - 5r + 30 + 5r = 108 - 3r^2 \qquad \text{Distribute.}$$

$$60 = 108 - 3r^2 \qquad \text{Simplify.}$$

$$0 = -3r^2 + 48 \qquad \text{Subtract 10}r$$

$$0 = -3(r+4)(r-4) \qquad \text{Factor.}$$

$$0 = (r+4)(r-4) \qquad \text{Divide each side by } -3$$

$$r = 4 \text{ or } -4 \qquad \text{Zero Product Property}$$

Because speed cannot be negative, the speed of the current is 4 miles per hour.

Justify $\dfrac{5}{6+r} + \dfrac{5}{6-r} = 3 \qquad \text{Original equation}$

$\dfrac{5}{6+4} + \dfrac{5}{6-4} \stackrel{?}{=} 3 \qquad r = 4$

$\dfrac{5}{10} + \dfrac{5}{2} \stackrel{?}{=} 3 \qquad \text{Simplify.}$

$\dfrac{1}{2} + \dfrac{5}{2} = \dfrac{6}{2} \checkmark \qquad \text{Simplify.}$

Evaluate It takes Sandra one half hour to row with the current. It take her two and one half hours to row against the current. The answer is reasonable.

Guided Practice

4. FLYING The speed of the wind is 20 miles per hour. If it takes a plane 7 hours to fly 2368 miles round trip, determine the plane's speed in still air. 339.5 mph

---

Example 4 Distance Problem

AL What do r, d, and t represent in this situation? r is the rate of the current, d is the distance the boat is traveling, and t is time.

OL Explain why the rate against the current is $6 - r$. The current slows Sandra down, so to find the boat's speed against the current we have to take her speed in still water and subtract the current.

BL Write the equation that would find r if Sandra travels 4 miles upstream in the same amount of time she travels 7 miles downstream. $\dfrac{7}{6+r} = \dfrac{4}{6-r}$

Need Another Example?

Swimming Lilia swims for 5 hours in a stream that has a current of 1 mile per hour. She leaves her dock, swims upstream for 2 miles and then swims back to her dock. What is her swimming speed in still water? about 1.5 mi/h

Watch Out!

Avoiding Misconceptions In Example 4, make sure that students understand the difference between the solutions of the quadratic equation (of which there are two) and the solution of the problem (of which there is only one).

Preventing Errors When solving rational equations, students often forget that all terms in the equation must be multiplied by the LCD. Remind them to do so in order to obtain equivalent equations.

Example 5 Work Problems

AL **What does j represent?** the amount of time it would take the junior class working alone

OL **Before solving, what values of j would make sense as an answer?** values that are greater than 0 days

BL **Write a general equation to determine how long it takes two people working together to complete a job. Let x represent the amount of time it takes the first person alone, y represent the time it would take the second person alone, and z the amount of time together.** $\frac{1}{x} + \frac{1}{y} = \frac{1}{z}$

Need Another Example?

Mowing Lawns Wuyi and Uima mow lawns together. Wuyi working alone could complete a particular job in 4.5 hours, and Uima could complete it alone in 3.7 hours. How long does it take to complete the job when they work together? about 2 hours

---

Watch Out!

Preventing Errors Suggest that when a problem involves completing part of a job while working together, students think about what part of the work gets done in one day, or one year, or one unit of the time.

Teaching Tip

Alternative Method The equation in Example 5 could also be solved by first subtracting $\frac{1}{24}$ from each side. Then find the cross products and solve.

Real-world problems that involve work can often be solved using rational equations.

A.CED.1, A.REI.2

Real-World Example 5 Work Problems

COMMUNITY SERVICE Every year, the junior and senior classes at Hillcrest High School build a house for the community. If it takes the senior class 24 days to complete a house and 18 days if they work with the junior class, how long would it take the junior class to complete a house if they worked alone?

Understand We are given how long it takes the senior class working alone and when the classes work together. We need to determine how long it would take the junior class by themselves.

Plan The senior class can complete 1 house in 24 days, so their rate is $\frac{1}{24}$ of a house per day.

The rate for the junior class is $\frac{1}{j}$.

The combined rate for both classes is $\frac{1}{18}$.

Senior Rate	Junior Rate	Combined Rate
$\frac{1}{24}$	$\frac{1}{j}$	$\frac{1}{18}$

Solve

$\frac{1}{24} + \frac{1}{j} = \frac{1}{18}$ Write the equation.

$72j\frac{1}{24} + 72j\frac{1}{j} = 72j\frac{1}{18}$ LCD = 72j
Multiply by LCD.

$\overset{3}{\cancel{72j}}\frac{1}{24} + 72j\frac{1}{\cancel{j}} = \overset{4}{\cancel{72j}}\frac{1}{18}$ Divide common factors.

$3j + 72 = 4j$ Distribute.

$72 = j$ Subtract 3j.

Check Two methods are possible.

Method 1 Substitute values.

$\frac{1}{24} + \frac{1}{j} = \frac{1}{18}$ Original equation

$\frac{1}{24} + \frac{1}{72} \overset{?}{=} \frac{1}{18}$ j = 72

$\frac{3}{72} + \frac{1}{72} \overset{?}{=} \frac{4}{72}$ LCD = 72

$\frac{4}{72} = \frac{4}{72}$ ✓ Simplify.

Method 2 Use a calculator.

It would take the junior class 72 days to complete the house by themselves.

▶ **Guided Practice**

5A. Working together, it took Anthony and Travis 6 hours to mow the lawns of all their clients together last week. The previous week it took Travis 10 hours to do it alone. How long will it take Anthony if he mows by himself this week? 15 hours

5B. Noah and Owen build birdhouses together. If Noah can build a particular house in 6 days and Owen can build the same house in 5 days, how long would it take the two of them if they work together? $2\frac{8}{11}$ days

---

Differentiated Instruction **OL** **BL** **ELL**

Logical Learners Have students think about the difference between "pure" mathematics, such as solving an equation, and "applied" mathematics, such as solving a real-world problem. Ask them to list some ways in which these two are alike and some ways in which they are different.

2 Solve Rational Inequalities To solve **rational inequalities**, which are inequalities that contain one or more rational expressions, follow these steps.

🔑 Key Concept Solving Rational Inequalities

Step 1 State the excluded values. These are the values for which the denominator is 0.

Step 2 Solve the related equation.

Step 3 Use the values determined from the previous steps to divide a number line into intervals.

Step 4 Test a value in each interval to determine which intervals contain values that satisfy the inequality.

A.REI.2

Example 6   Solve a Rational Inequality

Solve $\dfrac{x}{3} - \dfrac{1}{x-2} < \dfrac{x+1}{4}$.

Step 1 The excluded value for this inequality is 2.

Step 2 Solve the related equation.

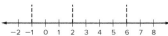

$$\frac{x}{3} - \frac{1}{x-2} = \frac{x+1}{4} \qquad \text{Related equation}$$

$$\overset{4}{12}(x-2)\frac{x}{3} - 12(x-2)\frac{1}{x-2} = \overset{3}{12}(x-2)\frac{x+1}{4} \qquad \text{LCD is 12}(x-2). \text{ Multiply by LCD.}$$

$$4x^2 - 8x - 12 = 3x^2 - 3x - 6 \qquad \text{Distribute.}$$

$$x^2 - 5x - 6 = 0 \qquad \text{Subtract } 3x^2 - 3x - 6.$$

$$(x-6)(x+1) = 0 \qquad \text{Factor.}$$

$$x = 6 \text{ or } -1 \qquad \text{Zero Product Property}$$

Step 3 Draw vertical lines at the excluded value and at the solutions to separate the number line into intervals.

number line marked from -2 to 8 with vertical dashed lines at -1, 2, and 6

> **Study Tip**
> Rational Inequalities It is possible that none or all of the intervals will produce a true statement.

Step 4 Now test a sample value in each interval to determine whether the values in the interval satisfy the inequality.

Test $x = -3$. 　　 Test $x = 0$. 　　 Test $x = 4$. 　　 Test $x = 8$.

$$\frac{-3}{3} - \frac{1}{-3-2} \overset{?}{<} \frac{-3+1}{4} \qquad \frac{0}{3} - \frac{1}{0-2} \overset{?}{<} \frac{0+1}{4} \qquad \frac{4}{3} - \frac{1}{4-2} \overset{?}{<} \frac{4+1}{4} \qquad \frac{8}{3} - \frac{1}{8-2} \overset{?}{<} \frac{8+1}{4}$$

$$-1 + \frac{1}{5} \overset{?}{<} -\frac{2}{4} \qquad 0 + \frac{1}{2} \overset{?}{<} \frac{1}{4} \qquad \frac{4}{3} - \frac{1}{2} \overset{?}{<} \frac{5}{4} \qquad \frac{32}{12} - \frac{2}{12} \overset{?}{<} \frac{27}{12}$$

$$-\frac{4}{5} < -\frac{1}{2} \checkmark \qquad \frac{1}{2} \not< \frac{1}{4} \qquad \frac{5}{6} < \frac{5}{4} \checkmark \qquad \frac{30}{12} \not< \frac{27}{12}$$

The statement is true for $x = -3$ and $x = 4$. Therefore, the solution is $x < -1$ or $2 < x < 6$.

▸ **Guided Practice** Solve each inequality.

6A. $\dfrac{5}{x} + \dfrac{6}{5x} > \dfrac{2}{3}$ 　$0 < x < 9.3$

6B. $\dfrac{4}{3x} + \dfrac{7}{x} < \dfrac{5}{9}$ 　$x < 0$ or $x > 15$

2 Solve Rational Inequalities

Example 6 Solve a Rational Inequality

AL **What values do you place on the number line?** the values that make the expression undefined and the values that solve the related equation.

OL **How do you choose values to test?** values that are in each section on the number line that are easy to plug into the inequality

BL **Explain the process of determining the solution.** After testing values in each section of the number line, the solution is the union of all of the intervals that satisfy the inequality.

Need Another Example?
Solve $\dfrac{1}{3k} + \dfrac{2}{9k} < \dfrac{2}{3}$. $k < 0$ or $k > \dfrac{5}{6}$

> **Watch Out!**
> **Preventing Errors** Suggest that students also verify whether the boundary indicated by the solution of the equation is or is not in the solution set of the inequality.

Practice

Formative Assessment Use Exercises 1–15 to assess students' understanding of the concepts in this lesson.

The Practice and Problem Solving exercises assess the content taught in the lesson. The Preparing for Assessment page is meant to be used as preparation for end-of-year assessments.

Teaching the Mathematical Practices

Structure Mathematically proficient students look closely to discern a pattern or structure. They can see complicated things, such as some algebraic expressions, as single objects or as being composed of several objects.

Extra Practice

See page R7 for extra exercises for students who are approaching level or for on-level students who need additional reinforcement.

Levels of Complexity Chart

The levels of the exercises progress from 1 to 3, with Level 1 indicating the lowest level of complexity.

Exercises	16–30	31–33, 40–45	34–39
▶ Level 3			●
▶ Level 2		●	
Level 1	●		

Exercise Alert

Grid Paper For Exercise 33, students will need grid paper.

Go Online!   eBook

Interactive Student Guide

Use the *Interactive Student Guide* to deepen conceptual understanding.

· Modeling: Rational Functions
· Solving Rational Equations and Inequalities

Check Your Understanding   ○ = Step-by-Step Solutions begin on page R11.

 Go Online! for a Self-Check Quiz

Examples 1–2 **Solve each equation. Check your solution.**
A.REI.2

1. $\frac{4}{7} + \frac{3}{x-3} = \frac{53}{56}$ 11

2. $\frac{7}{3} - \frac{3}{x-5} = \frac{19}{12}$ 9

3. $\frac{10}{2x+1} + \frac{4}{3} = 2$ 7

4. $\frac{11}{4} - \frac{5}{y+3} = \frac{23}{12}$ 3

⑤ $\frac{8}{x-5} - \frac{9}{x-4} = \frac{5}{x^2-9x+20}$ 8

6. $\frac{14}{x+3} + \frac{10}{x-2} = \frac{122}{x^2+x-6}$ 5

7. $\frac{14}{x-8} - \frac{5}{x-6} = \frac{82}{x^2-14x+48}$ 14

8. $\frac{5}{x+2} - \frac{3}{x-2} = \frac{12}{x^2-4}$ 14

Example 3
A.CED.1
A.REI.2

9. **STRUCTURE** Sara has 10 pounds of dried fruit selling for $6.25 per pound. She wants to know how many pounds of mixed nuts selling for $4.50 per pound she needs to make a trail mix selling for $5 per pound.

   a. Let m = the number of pounds of mixed nuts. Complete the following table and estimate an answer.   between 20 and 30 pounds

	Pounds	Price per Pound	Total Price
Dried Fruit	10	$6.25	6.25(10)
Mixed Nuts	m	$4.50	4.5m
Trail Mix	10 + m	$5.00	5(10 + m)

   b. Write a rational equation using the last column of the table. $62.5 + 4.5m = 50 + 5m$

   c. Solve the equation to determine how many pounds of mixed nuts are needed. Is your answer reasonable? Why or why not?   25; yes, the answer falls within the estimate

Example 4
A.CED.1,
A.REI.2

10. **DISTANCE** Alicia's average speed riding her bike is 11.5 miles per hour. She takes a round trip of 40 miles. It takes her 1 hour and 20 minutes with the wind and 2 hours and 30 minutes against the wind.

   a. Write an expression for Alicia's time with the wind. $\frac{20}{11.5+x}$

   b. Write an expression for Alicia's time against the wind. $\frac{20}{11.5-x}$

   c. How long does it take to complete the trip?   3 h and 50 min

   d. Write and solve the rational equation to determine the speed of the wind. $\frac{20}{11.5+x} + \frac{20}{11.5-x} = \frac{23}{6}$; 3.5 mph

Example 5
A.CED.1,
A.REI.2

11. **WORK** Kendal and Chandi wax cars. Kendal can wax a particular car in 60 minutes and Chandi can wax the same car in 80 minutes. They plan on waxing the same car together and want to know how long it will take. How much will Kendal and Chandi complete individually in

   a. 1 minute?   Kendal: $\frac{1}{60}$, Chandi: $\frac{1}{80}$

   b. x minutes?   Kendal: $\frac{x}{60}$, Chandi: $\frac{x}{80}$

   c. Write a rational equation representing Kendal and Chandi working together on the car and estimate an answer. $\frac{x}{60} + \frac{x}{80} = 1$; between 30 and 40 mins

   d. Solve the equation to determine how long it will take them to finish the car. Is your answer reasonable? Why or why not?   about 34.3 min; yes, the answer falls within the estimate

Example 6
A.REI.2

Solve each inequality. Check your solutions.

12. $\frac{3}{5x} + \frac{1}{6x} > \frac{2}{3}$ $0 < x < 1.15$

13. $\frac{1}{4c} + \frac{1}{9c} < \frac{1}{2}$ $c < 0$, or $\frac{13}{18} < c$

14. $\frac{4}{3y} + \frac{2}{5y} < \frac{3}{2}$ $y > \frac{52}{45}$, or $y < 0$

15. $\frac{1}{3b} + \frac{1}{4b} < \frac{1}{5}$ $b < 0$, or $\frac{35}{12} < b$

Differentiated Homework Options

Levels	AL Basic	OL Core	BL Advanced
Exercises	16–30, 36–45	17–29 odd, 31–37 odd, 40–45	31–45
2-Day Option	17–29 odd, 40–45	16–30, 40–45	
	16–30 even, 36–39	31–39	

 You can use ALEKS to provide additional remediation support with personalized instruction and practice.

Practice and Problem Solving

Extra Practice is on page R7.

Examples 1–2
A.REI.2

Solve each equation. Check your solutions.

16. $\dfrac{9}{x-7} - \dfrac{7}{x-6} = \dfrac{13}{x^2-13x+42}$ 9

17. $\dfrac{13}{y+3} - \dfrac{12}{y+4} = \dfrac{18}{y^2+7y+12}$ 2

18. $\dfrac{14}{x-2} - \dfrac{18}{x+1} = \dfrac{22}{x^2-x-2}$ 7

19. $\dfrac{11}{a+2} - \dfrac{10}{a+5} = \dfrac{36}{a^2+7a+10}$ 1

20. $\dfrac{x}{2x-1} + \dfrac{3}{x+4} = \dfrac{21}{2x^2+7x-4}$ −12, 2

21. $\dfrac{2}{y-5} + \dfrac{y-1}{2y+1} = \dfrac{2}{2y^2-9y-5}$ ∅

Examples 3–5
A.CED.1,
A.REI.2

22. CHEMISTRY How many milliliters of a 20% acid solution must be added to 40 milliliters of a 75% acid solution to create a 30% acid solution?   180 mL

23 GROCERIES Ellen bought 3 pounds of bananas for $0.90 per pound. How many pounds of apples costing $1.25 per pound must she purchase so that the total cost for fruit is $1 per pound?   1.2 lb

24. BUILDING Bryan's volunteer group can build a garage in 12 hours. Sequoia's group can build it in 16 hours. How long would it take them if they worked together?   about 6.86 hours

Example 6
A.REI.2

Solve each inequality. Check your solutions.

25. $3 - \dfrac{4}{x} > \dfrac{5}{4x}$ x < 0 or x > 1.75

26. $\dfrac{5}{3a} - \dfrac{3}{4a} > \dfrac{5}{6}$ 0 < a < 1.1

27. $\dfrac{x-2}{x+2} + \dfrac{1}{x-2} > \dfrac{x-4}{x-2}$ x < −2, or 2 < x < 14

28. $\dfrac{3}{4} - \dfrac{1}{x-3} > \dfrac{x}{x+4}$ −4 < x < 3

29. $\dfrac{x}{5} + \dfrac{2}{3} < \dfrac{3}{x-4}$ x < −5 or 4 < x < $\dfrac{17}{3}$

30. $\dfrac{x}{x+2} + \dfrac{1}{x-1} < \dfrac{3}{2}$ x > 2, −2 < x < 1, x < −5

B **31.** AIR TRAVEL It takes a plane 20 hours to fly to its destination against the wind. The return trip takes 16 hours. If the plane's average speed in still air is 500 miles per hour, what is the average speed of the wind during the flight?   55.56 mph

32. FINANCIAL LITERACY Judie wants to invest $10,000 in two different accounts. The risky account could earn 9% interest, while the other account earns 5% interest. She wants to earn $750 interest for the year. Of tables, graphs, or equations, choose the best representation needed and determine how much should be invested in each account. $6250 at 9% and $3750 at 5%

33. MULTIPLE REPRESENTATIONS Consider $\dfrac{2}{x-3} + \dfrac{1}{x} = \dfrac{x-1}{x-3}$.

  a. Algebraic Solve the equation for x. Were any values of x extraneous?   1; yes; 3

  b. Graphical Graph $y_1 = \dfrac{2}{x-3} + \dfrac{1}{x}$ and $y_2 = \dfrac{x-1}{x-3}$ on the same graph for 0 < x < 5.   See margin.

  c. Analytical For what value(s) of x do they intersect? Do they intersect where x is extraneous for the original equation?   1; no

  d. Verbal Use this knowledge to describe how you can use a graph to determine whether an apparent solution of a rational equation is extraneous.   Graph both sides of the equation. Where the graphs intersect, there is a solution. If they do not, then the possible solution is extraneous.

C Solve each equation. Check your solutions.

34. $\dfrac{2}{y+3} - \dfrac{3}{4-y} = \dfrac{2y-2}{y^2-y-12}$ −1

35. $\dfrac{2}{y+2} - \dfrac{y}{2-y} = \dfrac{y^2+4}{y^2-4}$ ∅

A.CED.1, A.REI.2

H.O.T. Problems   Use Higher-Order Thinking Skills

36. OPEN-ENDED Give an example of a rational equation that can be solved by multiplying each side of the equation by $4(x+3)(x-4)$.   See margin.

37. CHALLENGE Solve $\dfrac{1 + \frac{9}{x} + \frac{20}{x^2}}{1 - \frac{25}{x^2}} = \dfrac{x+4}{x-5}$.   all real numbers except 5, −5, 0

38. TOOLS While using the table feature on the graphing calculator to explore $f(x) = \dfrac{1}{x^2-x-6}$, the values −2 and 3 say "**ERROR.**" Explain its meaning.   See margin.

See margin. **39.** WRITING IN MATH Why should you check solutions of rational equations and inequalities?

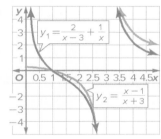
MP Standards for Mathematical Practice

Emphasis On	Exercises
1 Make sense of problems and persevere in solving them.	37, 40–45
2 Reason abstractly and quantitatively.	1–9, 12–21, 25–30, 34–36
4 Model with mathematics.	10, 11, 22–24, 31, 32
5 Use appropriate tools strategically.	38
6 Attend to precision.	33, 40–45
7 Look for and make use of structure.	39

Preparing for Assessment

Exercises 40–45 require students to use the skills they will need on assessments. The exercises are dual-coded with content.

Dual Coding		
Items	Content Standards	Mathematical Practices
40	A.CED.1	1, 6
41	A.CED.1	1, 6
42	A.CED.1	1, 6
43	A.CED.1	1, 6
44	A.CED.1	1, 6
45	A.CED.1	1, 6

Diagnose Student Errors

Survey student responses for each item. Class trends may indicate common errors and misconceptions.

40.

A	Found the difference between the time it takes each person to water the plants
B	CORRECT
C	Divided 800 by 50 instead of 57
D	Found the average of the two times

41.

A	Chose a number from the problem stem
B	Chose the percent of the mixture that is raisins
C	Added the percent of raisins in each mix
D	CORRECT

Preparing for Assessment

40. Jason can water all the plants at the botanical garden in 32 minutes. Celia can water them in 25 minutes. If they work together, about how long will it take for them to water the plants? 🔘 1, 6
A.CED.1   B

- ◯ A   7 min
- ◯ B   14 min
- ◯ C   16 min
- ◯ D   29 min

41. Thirty-two ounces of trail mix containing 25% raisins were mixed with 18 ounces of trail mix containing 15% raisins. About what percent of the new mixture is *not* raisins? 🔘 1, 6 A.CED.1   D

- ◯ A   18%
- ◯ B   21%
- ◯ C   40%
- ◯ D   79%

42. John is preparing to paint his garage. If it would take 3 workers 8 hours to apply 2 coats of paint, how long would it take 4 workers to apply 1 coat of paint? 🔘 1, 6 A.CED.1   A

- ◯ A   3 hours
- ◯ B   4 hours
- ◯ C   6 hours
- ◯ D   7 hours

43. A team of runners is planning to do a 27-mile relay as part of their ongoing training. The first runner will do 8 miles, the second runner will do 9 miles, and the third runner will do 10 miles. If the first runner averages 10 mph, the second runner averages 7 mph, and the third runner averages 9 mph, how long will the relay take in all?
🔘 1, 6 A.CED.1   D

- ◯ A   87.145 minutes
- ◯ B   136.97 minutes
- ◯ C   168.55 minutes
- ◯ D   191.81 minutes

44. MULTI-STEP Manuel's engineering team is surveying a triangular parcel of land. The team measures its perimeter using a measuring wheel. Each of his three team members uses the wheel to measure a side and then passes it to the next member. Each person's average walking speed and the measure of their side is shown in the table.

April	3 mph	1567 ft
Matt	3.5 mph	1125 ft
Henry	3.2 mph	920 ft

Manuel walks at an average speed of 4 mph. How much faster would the measurements have been taken if Manuel had taken them all himself? 🔘 1, 6
A.CED.1   B

- ◯ A   108 seconds faster
- ◯ B   155 seconds faster
- ◯ C   615 seconds faster
- ◯ D   771 seconds faster

45. MULTI-STEP Dara is paddling a kayak on a river. It takes her a total of 6 hours to paddle 4 miles upstream and four miles downstream. When the water is still, Dara can paddle at an average speed of 2 miles per hour. Let r represent the average rate of the current. 🔘 1, 6 A.CED.1

a. Fill in the table below to organize the equation:

	Distance (mi)	Avg Speed (mph)	Time (h)
With Current	4	$2 + r$	$\frac{4}{(2 + r)}$
Against Current	4	$2 - r$	$\frac{4}{2 - r}$

b. Write an equation to represent the average rate of the current. $\frac{4}{(2 + r)} + \frac{4}{2 - r} = 6$

c. What LCD can you use to simplify the equation? $(2 + r)(2 - r)$

d. To the nearest hundredth, what is the average rate of the current?   B

- ◯ A   About 0.76 miles per hour
- ◯ B   About 1.15 miles per hour
- ◯ C   About 1.78 miles per hour
- ◯ D   About 2.21 miles per hour

Differentiated Instruction   OL   BL

Extension The equation $\frac{1}{f} = \frac{1}{d} + \frac{1}{i}$ is sometimes called the "lens equation." It shows how the focal length f of a lens is related to the distances d and i from the lens to an object and to the object's image as seen through the lens, respectively. Ask students to find the focal length of a lens if, for a particular object, d is measured to be 20 cm and i is 12 cm. Ask them to solve the lens equation for f in terms of d and i. 7.5 cm; $f = \frac{di}{d + i}$

EXTEND 7-6

Graphing Technology Lab
Solving Rational Equations and Inequalities

You can use a TI-83/84 Plus graphing calculator to solve rational equations by graphing or by using the table feature. Graph both sides of the equation, and locate the point(s) of intersection.

Mathematical Practices
MP 5 Use appropriate tools strategically.

Content Standards
A.REI.2 Solve simple rational and radical equations in one variable, and give examples showing how extraneous solutions may arise.
A.REI.11 Explain why the x-coordinates of the points where the graphs of the equations $y = f(x)$ and $y = g(x)$ intersect are the solutions of the equation $f(x) = g(x)$; find the solutions approximately, e.g., using technology to graph the functions, make tables of values, or find successive approximations. Include cases where $f(x)$ and/or $g(x)$ are linear, polynomial, rational, absolute value, exponential, and logarithmic functions.

Activity 1   Rational Equation

Work cooperatively. Solve $\dfrac{4}{x+1} = \dfrac{3}{2}$.

Step 1 Graph each side of the equation.

Graph each side of the equation as a separate function. Enter $\dfrac{4}{x+1}$ as **Y1** and $\dfrac{3}{2}$ as **Y2**. Then graph the two equations in the standard viewing window.

KEYSTROKES: [Y=] 4 [÷] [(] [X,T,θ,n] [+] 1 [)] [ENTER] 3 [÷] 2 [ZOOM] 6

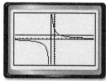

[−10, 10] scl: 1 by [−10, 10] scl: 1

Because the calculator is in connected mode, a vertical line may appear connecting the two branches of the hyperbola. This line is not part of the graph.

Step 3 Use the **TABLE** feature.

Verify the solution using the **TABLE** feature. Set up the table to show x-values in increments of $\frac{1}{3}$.

KEYSTROKES: [2nd] [TBLSET] 0 [ENTER] 1 [÷] 3 [ENTER] [2nd] [TABLE]

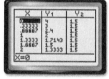

The table displays x-values and corresponding y-values for each graph. At $x = 1\frac{2}{3}$, both functions have a y-value of 1.5. Thus, the solution of the equation is $1\frac{2}{3}$.

Step 2 Use the **intersect** feature.

The **intersect** feature on the **CALC** menu allows you to approximate the ordered pair of the point at which the graphs cross.

KEYSTROKES: [2nd] [CALC] 5

Select one graph and press [ENTER]. Select the other graph, press [ENTER], and press [ENTER] again.

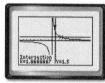

[−10, 10] scl: 1 by [−10, 10] scl: 1

The solution is $1\frac{2}{3}$.

(continued on the next page)

Launch

Objective Use a graphing calculator to solve rational equations by graphing or by using the table feature.

Materials for Each Group

• TI-83/84 Plus or other graphing calculator

Teaching Tip When students enter functions in the function editor, (Y = list), they should use parentheses around any numerator or denominator that is not a single number or variable.

Teach

Working in Cooperative Groups Have students work in pairs so they can help each other correct keystroke errors. Have pairs complete Activities 1 and 2 and Exercises 1 and 6. **ELL**

Activity 1

• When using the intersect feature in Step 2, students should press [ENTER] to select the graph of each function. They will then be prompted for a **Guess?**. They should move the cursor close to an estimated point of intersection before pressing [ENTER] a third time.

• As an alternative to using the table feature in Step 3, students can verify the solution for each function directly on the home screen. Entering **Y1** (5/3) and **Y2** (5/3) will both produce results of 1.5.

Go Online!

Graphing Calculators

Students can use the Graphing Calculator Personal Tutors to review the use of the graphing calculator to represent functions. They can also use the Other Calculator Keystrokes, which cover lab content for students with calculators other than the TI-84 Plus.

Activity 2

- In Activity 2, because the graph of the inequality $y > 9$ begins at the top of the standard window screen, students may want to change the window for y to $[-10, 20]$.

Practice Have students complete Exercises 2–5 and 7–9.

Assess

Formative Assessment

Use Exercise 2 to assess whether students comprehend how to find the intersection of the graphs of the left and right sides of a rational equation.

From Concrete to Abstract

Ask students to solve the equation $\frac{x}{2} = \frac{8}{x}$. They will find that the graphs of $y = \frac{x}{2}$ and $y = \frac{8}{x}$ intersect in two places. It is especially important in a case like this that they use the **intersect** feature correctly. In response to **Guess?**, they must first select one of the points of intersection, then repeat the **intersect** process and select the other point of intersection. The two solutions are $x = 4$ and $x = -4$.

EXTEND 7-6

Graphing Technology Lab

Solving Rational Equations and Inequalities *Continued*

You can use a similar procedure to solve rational inequalities using a graphing calculator.

Activity 2 Rational Inequality

Work cooperatively. Solve $\frac{3}{x} + \frac{7}{x} > 9$.

Step 1 Enter the inequalities.

Rewrite the problem as a system of inequalities.

The first inequality is $\frac{3}{x} + \frac{7}{x} > y$ or $y < \frac{3}{x} + \frac{7}{x}$. Because this inequality includes the *less than* symbol, shade below the curve. First enter the boundary and then use the arrow and ENTER keys to choose the shade below icon, ▖.

The second inequality is $y > 9$. Shade above the curve since this inequality contains *greater than*.

KEYSTROKES: [Y=] [◄] [◄] [ENTER] [ENTER] [ENTER] [▶] [▶] 3 [÷] [X,T,θ,n] [+] 7 [÷] [X,T,θ,n] [ENTER] [◄] [◄] [ENTER] [ENTER] [▶] [▶] 9

Step 2 Graph the system.

KEYSTROKES: [GRAPH]

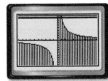

[−10, 10] scl: 1 by [−10, 10] scl: 1

The solution set of the original inequality is the set of x-values of the points in the region where the shadings overlap. Using the calculator's **intersect** feature, you can conclude that the solution set is $\left\{ x \mid 0 < x < 1\frac{1}{9} \right\}$.

Step 3 Use the **TABLE** feature.

Verify using the **TABLE** feature. Set up the table to show x-values in increments of $\frac{1}{9}$.

KEYSTROKES: [2nd] [TBLSET] 0 [ENTER] 1 [÷] 9 [ENTER] [2nd] [TABLE]

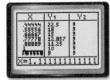

Scroll through the table. Notice that for x-values greater than 0 and less than $1\frac{1}{9}$, $Y_1 > Y_2$. This confirms that the solution of the inequality is $\left\{ x \mid 0 < x < 1\frac{1}{9} \right\}$.

Exercises

Work cooperatively. Solve each equation or inequality.

1. $\frac{1}{x} + \frac{1}{2} = \frac{2}{x}$ 2

2. $\frac{1}{x-4} = \frac{2}{x-2}$ 6

3. $\frac{4}{x} = \frac{6}{x^2}$ 1.5

4. $\frac{1}{1-x} = 1 - \frac{x}{x-1}$ $\{x \mid x \neq 1\}$

5. $\frac{1}{x+4} = \frac{2}{x^2+3x-4} - \frac{1}{1-x}$ ∅

6. $\frac{1}{x} + \frac{1}{2x} > 5$ $\{x \mid 0 < x < 0.3\}$

7. $\frac{1}{x-1} + \frac{2}{x} < 0$ $\left\{ x \mid x < 0 \text{ or } \frac{2}{3} < x < 1 \right\}$

8. $1 + \frac{5}{x-1} \leq 0$ $\{x \mid -4 \leq x < 1\}$

9. $2 + \frac{1}{x-1} \geq 0$ $\{x \mid x \leq 0.5 \text{ or } x > 1\}$

CHAPTER 7
Study Guide and Review

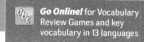 *Go Online!* for Vocabulary Review Games and key vocabulary in 13 languages

Study Guide

Key Concepts

Rational Expressions (Lessons 7-1 and 7-2)

- Multiplying and dividing rational expressions is similar to multiplying and dividing fractions.

- To simplify complex fractions, simplify the numerator and the denominator separately, and then simplify the resulting expression.

Reciprocal and Rational Functions (Lessons 7-3 and 7-4)

- A reciprocal function is of the form $f(x) = \frac{1}{a(x)}$, where $a(x)$ is a linear function and $a(x) \neq 0$.

- A rational function is of the form $\frac{a(x)}{b(x)}$, where $a(x)$ and $b(x)$ are polynomial functions and $b(x) \neq 0$.

Direct, Joint, and Inverse Variation (Lesson 7-5)

- Direct variation: There is a nonzero number k such that $y = kx$.

- Joint variation: There is a nonzero number k such that $y = kxz$.

- Inverse variation: There is a nonzero constant k such that $xy = k$ or $y = \frac{k}{x}$, where $x \neq 0$ and $y \neq 0$.

Rational Equations and Inequalities (Lesson 7-6)

- Eliminate fractions in rational equations by multiplying each side of the equation by the LCD.

- Possible solutions of a rational equation must exclude values that result in zero in the denominator.

 **Study Organizer**

Use your Foldable to review the chapter. Working with a partner can be helpful. Ask for clarification of concepts as needed.

Key Vocabulary

combined variation (p. 503)	point discontinuity (p. 494)
complex fraction (p. 470)	rational equation (p. 508)
constant of variation (p. 500)	rational expression (p. 467)
direct variation (p. 500)	rational function (p. 491)
horizontal asymptote (p. 491)	rational inequality (p. 513)
hyperbola (p. 483)	reciprocal function (p. 483)
inverse variation (p. 502)	vertical asymptote (p. 491)
joint variation (p. 501)	weighted average (p. 510)
oblique asymptote (p. 493)	

Vocabulary Check

Choose a term from the list above that best completes each statement or phrase.

1. A(n) _____ is a rational expression whose numerator and/or denominator contains a rational expression.
   complex fraction

2. A(n) _____ asymptote is a linear asymptote that is neither horizontal nor vertical. **oblique**

3. Equations that contain one or more rational expressions are called _____ **rational equations**

4. The graph of $y = \frac{x}{x+2}$ has a(n) _____ at $x = -2$.
   vertical asymptote

5. _____ occurs when one quantity varies directly as the product of two or more other quantities. **Joint variation**

6. A ratio of two polynomial expressions is called a(n) _____ **rational expression**

7. _____ looks like a hole in a graph because the graph is undefined at that point. **Point discontinuity**

8. _____ occurs when one quantity varies directly **Combined variation** and/or inversely as two or more other quantities.

Concept Check

See margin.

9. Explain how to algebraically determine the horizontal asymptotes of the graph of a rational function $f(x) = \frac{a(x)}{b(x)}$.

10. Explain the difference between direct and inverse variation.
    See margin.

 Answering the Essential Question

Before answering the Essential Question, have students review their answers to the *Building on the Essential Question* exercises found throughout the chapter.

- Why are graphs useful? (p. 464)

- How are the properties of a rational function reflected in its graph? (p. 497)

- How can analyzing a rational function algebraically *and* graphically help you see the "whole picture"? (p. 505)

FOLDABLES **Study Organizer**

A completed Foldable for this chapter should include the Key Concepts related to rational functions and relations.

Key Vocabulary **ELL**

The page reference after each word denotes where that term was first introduced. If students have difficulty answering questions 1–8, remind them that they can use these page references to refresh their memories about the vocabulary terms.

Have students work with a partner to complete the Vocabulary Check. Encourage them to reference and compare their notes from Chapter 7.

You can use the detailed reports in ALEKS to automatically monitor students' progress and pinpoint remediation needs prior to the chapter test.

Additional Answers

9. If the degree of $a(x)$ is greater than the degree of $b(x)$, there is no horizontal asymptote. If the degree of $a(x)$ is less than the degree of $b(x)$, the horizontal asymptote is $y = 0$. If the degree of $a(x)$ is equal to the degree of $b(x)$, the horizontal asymptote is $y = \frac{\text{leading coefficient of } a(x)}{\text{leading coefficient of } b(x)}$.

10. For direct variation, there is a nonzero number k such that $y = kx$. For inverse variation, there is a nonzero number k such that $y = \frac{k}{x}$, where $x \neq 0$ and $y \neq 0$.

Go Online!

Vocabulary Review

Students can use the *Vocabulary Review Games* to check their understanding of the vocabulary terms in this chapter. Students should refer to the *Student-Built Glossary* they have created as they went through the chapter to review important terms. You can also give a *Vocabulary Test* over the content of this chapter.

Lesson-by-Lesson Review

Intervention If the given examples are not sufficient to review the topics covered by the questions, remind students that the lesson references tell them where to review that topic in their textbook.

Two-Day Option Have students complete the Lesson-by-Lesson Review. Then you can use McGraw-Hill eAssessment to customize another review worksheet that practices all the objectives of this chapter or only the objectives on which your students need more help.

Additional Answers

24.

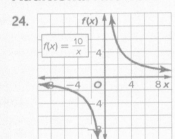

$$f(x) = \frac{10}{x}$$

$D = \{x \mid x \neq 0\}, R = \{f(x) \mid f(x) \neq 0\}$

25.

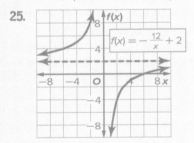

$$f(x) = -\frac{12}{x} + 2$$

$D = \{x \mid x \neq 0\}, R = \{f(x) \mid f(x) \neq 2\}$

26.

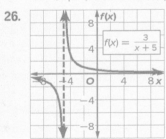

$$f(x) = \frac{3}{x+5}$$

$D = \{x \mid x \neq -5\}, R = \{f(x) \mid f(x) \neq 0\}$

27.

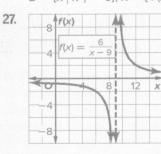

$$f(x) = \frac{6}{x-9}$$

$D = \{x \mid x \neq 9\}, R = \{f(x) \mid f(x) \neq 0\}$

Lesson-by-Lesson Review

7-1 Multiplying and Dividing Rational Expressions
A.APR.7

Simplify each expression.

11. $\dfrac{-16xy}{27z} \cdot \dfrac{15z^3}{8x^2} \cdot \dfrac{10yz^2}{9x}$

12. $\dfrac{x^2 - 2x - 8}{x^2 + x - 12} \cdot \dfrac{x^2 + 2x - 15}{x^2 + 7x + 10} \quad \dfrac{x-4}{x+4}$

13. $\dfrac{x^2 - 1}{x^2 - 4} \cdot \dfrac{x^2 - 5x - 14}{x^2 - 6x - 7} \quad \dfrac{x-1}{x-2}$

14. $\dfrac{x+y}{15x} \div \dfrac{x^2 - y^2}{3x^2} \quad \dfrac{x}{5(x-y)}$

15. $\dfrac{\frac{x^2 + 3x - 18}{x+4}}{\frac{x^2 + 7x + 6}{x+4}} \quad \dfrac{x-3}{x+1}$

16. GEOMETRY A triangle has an area of $3x^2 + 9x - 54$ square centimeters. If the height of the triangle is $x + 6$ centimeters, find the length of the base. $6x - 18$ cm

Example 1

Simplify $\dfrac{4a}{3b} \cdot \dfrac{9b^4}{2a^2}$.

$\dfrac{4a}{3b} \cdot \dfrac{9b^4}{2a^2} = \dfrac{2 \cdot 2 \cdot a \cdot 3 \cdot 3 \cdot b \cdot b \cdot b \cdot b}{3 \cdot b \cdot 2 \cdot a \cdot a}$

$= \dfrac{6b^3}{a}$

Example 2

Simplify $\dfrac{r^2 + 5r}{2r} \div \dfrac{r^2 - 25}{6r - 12}$.

$\dfrac{r^2 + 5r}{2r} \div \dfrac{r^2 - 25}{6r - 12} = \dfrac{r^2 + 5r}{2r} \cdot \dfrac{6r - 12}{r^2 - 25}$

$= \dfrac{r(r+5)}{2r} \cdot \dfrac{6(r-2)}{(r+5)(r-5)}$

$= \dfrac{3(r-2)}{r-5}$

7-2 Adding and Subtracting Rational Expressions
A.APR.7

Simplify each expression.

17. $\dfrac{9}{4ab} + \dfrac{5a}{6b^2} \quad \dfrac{27b + 10a^2}{12ab^2}$

18. $\dfrac{3}{4x - 8} - \dfrac{x-1}{x^2 - 4} \quad \dfrac{-x+10}{4(x-2)(x+2)}$

19. $\dfrac{y}{2x} + \dfrac{4y}{3x^2} - \dfrac{5}{6xy^2} \quad \dfrac{3xy^3 + 8y^3 - 5x}{6x^2y^2}$

20. $\dfrac{2}{x^2 - 3x - 10} - \dfrac{6}{x^2 - 8x + 15} \quad \dfrac{-4x - 18}{(x-5)(x+2)(x-3)}$

21. $\dfrac{3}{3x^2 + 2x - 8} + \dfrac{4x}{2x^2 + 6x + 4} \quad \dfrac{12x^2 - 10x + 6}{2(x+2)(3x-4)(x+1)}$

22. $\dfrac{\frac{3}{2x+3} - \frac{x}{x+1}}{\frac{2x}{x+1} + \frac{5}{2x+3}} \quad \dfrac{-2x^2 + 3}{4x^2 + 11x + 5}$

23. GEOMETRY What is the perimeter of the rectangle?

$$\frac{10x + 20}{(x+6)(x+1)}$$

$\dfrac{1}{x+1}$

$\dfrac{4}{x+6}$

Example 3

Simplify $\dfrac{3a}{a^2 - 4} - \dfrac{2}{a - 2}$.

$\dfrac{3a}{a^2 - 4} - \dfrac{2}{a - 2} = \dfrac{3a}{(a-2)(a+2)} - \dfrac{2}{a-2}$

$= \dfrac{3a}{(a-2)(a+2)} - \dfrac{2(a+2)}{(a-2)(a+2)}$

$= \dfrac{3a - 2(a+2)}{(a-2)(a+2)}$ Subtract numerators.

$= \dfrac{3a - 2a - 4}{(a-2)(a+2)}$ Distributive Property

$= \dfrac{a - 4}{(a-2)(a+2)}$ Simplify.

28.

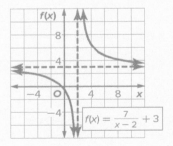

$$f(x) = \frac{7}{x-2} + 3$$

$D = \{x \mid x \neq 2\}, R = \{f(x) \mid f(x) \neq 3\}$

29.

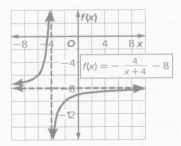

$$f(x) = -\frac{4}{x+4} - 8$$

$D = \{x \mid x \neq -4\}, R = \{f(x) \mid f(x) \neq -8\}$

7-3 Graphing Reciprocal Functions

A.CED.2, F.IF.4, F.BF.3

Graph each function. State the domain and range.

24. $f(x) = \dfrac{10}{x}$

25. $f(x) = -\dfrac{12}{x} + 2$

26. $f(x) = \dfrac{3}{x+5}$

27. $f(x) = \dfrac{6}{x-9}$

28. $f(x) = \dfrac{7}{x-2} + 3$

29. $f(x) = -\dfrac{4}{x+4} - 8$

24–29. See margin.

30. CONSERVATION The student council is planting 28 trees for a service project. The number of trees each person plants depends on the number of student council members.

  a. Write a function to represent this situation.

  b. Graph the function.   See margin. $f(x) = \dfrac{28}{x}$

Example 4

Graph $f(x) = \dfrac{3}{x+2} - 1$. State the domain and range.

$a = 3$:   The graph is stretched vertically.
$h = -2$:   The graph is translated 2 units left. There is an asymptote at $x = -2$.
$k = -1$:   The graph is translated 1 unit down. There is an asymptote at $f(x) = -1$.

Domain: $\{x \mid x \neq -2\}$,
Range: $\{f(x) \mid f(x) \neq -1\}$

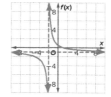

7-4 Graphing Rational Functions

A.CED.2, F.IF.4, F.BF.3

Determine the equations of any vertical asymptotes and the values of x for any holes in the graph of each rational function.

31. $f(x) = \dfrac{3}{x^2 + 4x}$ $x = -4, x = 0$

32. $f(x) = \dfrac{x+2}{x^2 + 6x + 8}$ $x = -4$; hole: $x = -2$

33. $f(x) = \dfrac{x^2 - 9}{x^2 - 5x - 24}$ $x = 8$; hole: $x = -3$

Graph each rational function.   34–37. See margin.

34. $f(x) = \dfrac{x+2}{(x+5)^2}$

35. $f(x) = \dfrac{x}{x+1}$

36. $f(x) = \dfrac{x^2 + 4x + 4}{x+2}$

37. $f(x) = \dfrac{x-1}{x^2 + 5x + 6}$

38. FUNDRAISERS Adelle is selling cookies for a fundraiser. Out of the first 15 houses, she sold cookies to 10 of them. Suppose Adelle goes to x more houses and sells cookies to all of them. The percentage of houses that she sold to out of the total houses can be determined using $P(x) = \dfrac{10 + x}{15 + x}$.

  a. Graph the function.   See margin.

  b. What domain and range values are meaningful in the context of the problem?
     $D = \{x \geq 0\}, R = \{0 \leq P(x) \leq 1.0\}$

Example 5

Determine the equation of any vertical asymptotes and the values of x for any holes in the graph of $f(x) = \dfrac{x^2 - 1}{x^2 + 2x - 3}$.

$$\dfrac{x^2 - 1}{x^2 + 2x - 3} = \dfrac{(x-1)(x+1)}{(x-1)(x+3)}$$

The function is undefined for $x = 1$ and $x = -3$.

Because $\dfrac{(x-1)(x+1)}{(x-1)(x+3)} = \dfrac{x+1}{x+3}$, $x = -3$ is a vertical asymptote, and $x = 1$ represents a hole in the graph.

Example 6

Graph $f(x) = \dfrac{1}{6x(x-1)}$.

The function is undefined for $x = 0$ and $x = 1$. Because $\dfrac{1}{6x(x-1)}$ is in simplest form, $x = 0$ and $x = 1$ are vertical asymptotes. Draw the two asymptotes and sketch the graph.

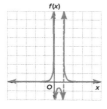

Additional Answers

30b.

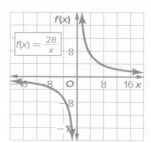

34.

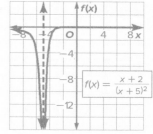

35.

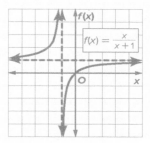

36.

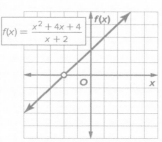

37.

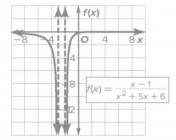

38a.

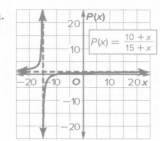

Go Online!

Anticipation Guide

Students should complete the *Chapter 7 Anticipation Guide*, and discuss how their responses have changed now that they have completed Chapter 7.

Before the Test

Have students complete the Study Notebook Tie it Together activity to review topics and skills presented in the chapter.

7-5 Variation Functions

A.CED.2, F.IF.4

39. If a varies directly as b and $b = 18$ when $a = 27$, find a when $b = 10$. $a = 15$

40. If y varies inversely as x and $y = 15$ when $x = 3.5$, find y when $x = -5$. $y = -10.5$

41. If y varies inversely as x and $y = -3$ when $x = 9$, find y when $x = 81$. $y = -\dfrac{1}{3}$

42. If y varies jointly as x and z, and $x = 8$ and $z = 3$ when $y = 72$, find y when $x = -2$ and $z = -5$. $y = 30$

43. If y varies jointly as x and z, and $y = 18$ when $x = 6$ and $z = 15$, find y when $x = 12$ and $z = 4$. $y = \dfrac{48}{5}$

44. **JOBS** Lisa's earnings vary directly with how many hours she babysits. If she earns \$68 for 8 hours of babysitting, find her earnings after 5 hours of babysitting.   \$42.50

Example 7

If y varies inversely as x and $x = 24$ when $y = -8$, find x when $y = 15$.

$$\frac{x_1}{y_2} = \frac{x_2}{y_1} \qquad \text{Inverse variation}$$

$$\frac{24}{15} = \frac{x_2}{-8} \qquad x_1 = 24, \, y_1 = -8, \, y_2 = 15$$

$$24(-8) = 15(x_2) \qquad \text{Cross multiply}$$

$$-192 = 15x_2 \qquad \text{Simplify}$$

$$-12\tfrac{4}{5} = x_2 \qquad \text{Divide each side by 15.}$$

When $y = 15$, the value of x is $-12\tfrac{4}{5}$.

7-6 Solving Rational Equations and Inequalities

A.CED.1, A.REI.2

Solve each equation or inequality. Check your solutions.

45. $\dfrac{1}{3} + \dfrac{4}{x-2} = 6$ $x = \dfrac{46}{17}$

46. $\dfrac{6}{x+5} - \dfrac{3}{x-3} = \dfrac{6}{x^2+2x-15}$ $x = 13$

47. $\dfrac{2}{x^2-9} = \dfrac{3}{x^2-2x-3}$ $x = -7$

48. $\dfrac{4}{2x-3} + \dfrac{x}{x+1} = \dfrac{-8x}{2x^2-x-3}$ $x = -\dfrac{1}{2}, -4$

49. $\dfrac{x}{x+4} - \dfrac{28}{x^2+x-12} = \dfrac{1}{x-3}$ $x = 8$

50. $\dfrac{x}{2} + \dfrac{1}{x-1} < \dfrac{x}{4}$ $x < 1$

51. $\dfrac{1}{2x} - \dfrac{4}{5x} > \dfrac{1}{3}$ $-\dfrac{9}{10} < x < 0$

52. **YARD WORK** Lana can plant a garden in 3 hours. Milo can plant the same garden in 4 hours. How long will it take them if they work together? $1\tfrac{5}{7}$ h

Example 8

Solve $\dfrac{3}{x+2} + \dfrac{1}{x} = 0$.

The LCD is $x(x+2)$.

$$\frac{3}{x+2} + \frac{1}{x} = 0$$

$$x(x+2)\left(\frac{3}{x+2} + \frac{1}{x}\right) = x(x+2)(0)$$

$$x(x+2)\left(\frac{3}{x+2}\right) + x(x+2)\left(\frac{1}{x}\right) = 0$$

$$3(x) + 1(x+2) = 0$$

$$3x + x + 2 = 0$$

$$4x + 2 = 0$$

$$4x = -2$$

$$x = -\frac{1}{2}$$

Go Online!

ᵉAssessment

Customize and create multiple versions of chapter tests and answer keys that align to your standards. Tests can be delivered on paper or online.

CHAPTER 7
Practice Test

 **Go Online!** for another Chapter Test

Simplify each expression. 5–7. See margin.

1. $\dfrac{r^2 + rt}{2r} \div \dfrac{r + t}{16r^2}$ $8r^2$

2. $\dfrac{m^2 - 4}{3m^2} \cdot \dfrac{6m}{2 - m}$ $-\dfrac{2(m+2)}{m}$

3. $\dfrac{m^2 + m - 6}{n^2 - 9} \div \dfrac{m - 2}{n + 3}$ $\dfrac{m+3}{n-3}$

4. $\dfrac{\frac{x^2 + 4x + 3}{x^2 - 2x - 15}}{\frac{x^2 - 1}{x^2 - x - 20}}$ $\dfrac{x+4}{x-1}$

5. $\dfrac{x + 4}{6x + 3} + \dfrac{1}{2x + 1}$

6. $\dfrac{x}{x^2 - 1} - \dfrac{3}{2x + 2}$

7. $\dfrac{1}{y} + \dfrac{2}{7} - \dfrac{3}{2y^2}$

8. $\dfrac{2 + \frac{1}{x}}{5 - \frac{1}{x}}$ $\dfrac{2x+1}{5x-1}$

9. Identify the asymptotes, domain, and range of the function graphed. **See margin.**

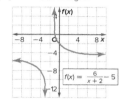

$f(x) = \dfrac{6}{x+2} - 5$

10. **MULTIPLE CHOICE** What is the equation for the vertical asymptote of the rational function $f(x) = \dfrac{x + 1}{x^2 + 3x + 2}$? **A**

   A $x = -2$

   B $x = -1$

   C $x = 1$

   D $x = 2$

Graph each function. 11–16. See Ch. 7 Answer Appendix.

11. $f(x) = -\dfrac{8}{x} - 9$

12. $f(x) = \dfrac{2}{x + 4}$

13. $f(x) = \dfrac{3}{x - 1} + 8$

14. $f(x) = \dfrac{5x}{x + 1}$

15. $f(x) = \dfrac{x}{x - 5}$

16. $f(x) = \dfrac{x^2 + 5x - 6}{x - 1}$

17. Determine the equations of any vertical asymptotes and the values of x for any holes in the graph of the function $f(x) = \dfrac{x + 5}{x^2 - 2x - 35}$. **vertical asymptote: $x = 7$; hole: $x = -5$**

18. Determine the equations of any oblique asymptotes in the graph of the function $f(x) = \dfrac{x^2 + x - 5}{x + 3}$. $f(x) = x - 2$

Solve each equation or inequality.

19. $\dfrac{-1}{x + 4} = 6 - \dfrac{x}{x + 4}$ $x = -5$

20. $\dfrac{1}{3} = \dfrac{5}{m + 3} + \dfrac{8}{21}$ $m = -108$

21. $7 + \dfrac{2}{x} < -\dfrac{5}{x}$ $-1 < x < 0$

22. $r + \dfrac{6}{r} - 5 = 0$ $r = 2, 3$

23. $\dfrac{6}{7} - \dfrac{3m}{2m - 1} = \dfrac{11}{7}$ $m = \dfrac{5}{31}$

24. $\dfrac{r + 2}{3r} = \dfrac{r + 4}{r - 2} - \dfrac{2}{3}$ $r = -\dfrac{1}{4}$

25. If y varies inversely as x and $y = 18$ when $x = -\dfrac{1}{2}$, find x when $y = -10$. $\dfrac{9}{10}$

26. If m varies directly as n and $m = 24$ when $n = -3$, find n when $m = 30$. $\dfrac{15}{4}$

27. Suppose r varies jointly as s and t. If $s = 20$ when $r = 140$ and $t = -5$, find s when $r = 7$ and $t = 2.5$. -2

28. **BICYCLING** When Susan rides her bike, the distance that she travels varies directly with the amount of time she is biking. Suppose she bikes 50 miles in 2.5 hours. At this rate, how many hours would it take her to bike 80 miles? **4 hours**

29. **PAINTING** Peter can paint a house in 10 hours. Melanie can paint the same house in 9 hours. How long would it take if they worked together? **about 4.7 hours**

30. **MULTIPLE CHOICE** How many liters of a 25% acid solution must be added to 30 liters of an 80% acid solution to create a 50% acid solution? **C**

   A 18

   B 30

   C 36

   D 66

31. What is the volume of the rectangular prism?

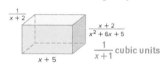
 cubic units

Go Online!

Chapter Tests

You can use premade leveled *Chapter Tests* to differentiate assessment for your students. Students can also take self-checking *Chapter Tests* to plan and prepare for chapter assessments.

MC = multiple-choice questions
FR = free-response questions

Form	Type	Level
1	MC	AL
2A	MC	OL
2B	FR	OL
2C	FR	OL
3	FR	BL
Vocabulary Test		
Extended-Response Test		

RtI Response to Intervention

Use the Intervention Planner to help you determine your Response to Intervention.

Intervention Planner

TIER 1 **On Level** OL

IF students miss 25% of the exercises or less,

THEN choose a resource:

   SE Lessons 7-1 through 7-6

   Go Online!

   📄 Skills Practice

   📄 Chapter Project

   ✓ Self-Check Quizzes

TIER 2 **Strategic Intervention** AL
Approaching grade level

IF students miss 50% of the exercises,

THEN **Go Online!**

   📄 Study Guide and Intervention

   ➕ Extra Examples

   💬 Personal Tutors

   📄 Homework Help

TIER 3 **Intensive Intervention**
2 or more grades below level

IF students miss 75% of the exercises,

THEN choose a resource:

   Use *Math Triumphs, Alg. 2*

   Go Online!

   ➕ Extra Examples

   💬 Personal Tutors

   📄 Homework Help

   🔤 Review Vocabulary

Additional Answers

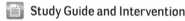

5. $\dfrac{x + 7}{3(2x + 1)}$

6. $\dfrac{-x + 3}{2(x - 1)(x + 1)}$

7. $\dfrac{4y^2 + 14y - 21}{14y^2}$

9. $x = -2$; $f(x) = -5$; $D = \{x \mid x \neq -2\}$, $R = \{f(x) \mid f(x) \neq -5\}$

Launch

Objective Apply concepts and skills from this chapter in a real-world setting.

Teach

Ask:

- For number 1, think about it logically first. If the SUV gets 320 miles out of one gallon, and you knew how many gallons were in the tank, what operation would you perform? Sample answer: I would divide 320 by the number of gallons in tank.

- What is the best way to graph the function in number 1? Sample answer: Make a table of values.

- For Exercise 4, if the company conducted 10 more successful tests, or if $x = 10$, what would be the ratio be? Sample answer: $\frac{15}{18}$

- Which values in the graph in Exercise 4 make sense in this context? Sample answer: Only the positive values make sense.

- For Exercise 8, what are the degrees of the numerator and denominator of the function you created in number 4? Sample answer: Both are 1.

- For Exercises 9–12, which have only two variables? What options does that eliminate? Sample answer: Numbers 9 and 12 have only two variables. It eliminates joint and combined variations.

The Performance Task focuses on the following content standards and standards for mathematical practice.

Dual Coding

Parts	Content Standards	ⓂⓅ Mathematical Practices
A	A.CED.2, F.IF.7	1, 4
B	A.CED.2, F.IF.7	1, 4
C	A.CED.2	1, 6

Go Online!   eBook

Interactive Student Guide

Refer to *Interactive Student Guide* for an additional Performance Task.

ALGEBRA 2
INTERACTIVE STUDENT GUIDE

CHAPTER 7
Preparing for Assessment

Performance Task

Provide a clear solution to each part of the task. Be sure to show all of your work, include all relevant drawings, and justify your answers.

MANUFACTURING A car company evaluates the safety and efficiency of its cars to ensure it meets government standards and consumer expectations.

Part A

EFFICIENCY One of the company's SUVs gets an average of 320 miles out of a single tank of gas.

1. Write a function for y in terms of x, where y represents the number of miles per gallon the SUV gets and x represents the number of gallons of gas in one tank. $y = \frac{320}{x}$

2. **Model** Make a graph of the function you wrote. See margin

3. Determine the number of miles per gallon the SUV gets if the tank holds 20 gallons of gas. 16 mpg

Part B

SAFETY The company is dealing with a defective airbag issue. When conducting crash tests, the rear passenger side airbags deploy every 5 out of 8 times. This is called the crash test success ratio. After the company resolves an issue, the air bags deploy in x consecutive crash tests.

4. **Reasoning** Write a function for the car company's crash test success ratio y after x successful crash tests. $y = \frac{5+x}{8+x}$

5. Make a graph of the function you wrote. See margin

6. **Sense-Making** Explain which parts of the graph are meaningful in this situation.
The first quadrant only; there cannot be a negative number of crash tests or successful deployments.

7. Explain the meaning of the y-intercept in this context. It represents the original crash test success ratio (62.5%).

8. Write the equation of the horizontal asymptote and explain the meaning of the horizontal asymptote in this situation.
$y = 1$; 1 represents a crash test success ratio of 100% which can never be obtained because the airbag will always have failed to deploy at least $8 - 5 = 3$ times.

Part C

FINDINGS The company researches gas mileage for one of their vehicles. Determine whether each finding represents direct, inverse, joint, or combined variation.

9. The amount of gas consumed by a car varies based on the age of the vehicle. As a car ages, it consumes a greater amount of gas. direct

10. The amount of gas consumed is affected by the weight of the load in the car and grade of the road on which the car is driving. As the load weight and the absolute value of the grade increases, so does the amount of gas consumed. joint

11. Average miles per gallon is affected by the ingredients in the fuel used. As ethanol content increases, average miles per gallon decreases. As the purity of the petrol used increases, the average miles per gallon increases. combined

12. For the first 5000 miles, the car's engine experiences a "break-in period" in which the car's engine has not reached maximum efficiency. During this period, the more miles that are driven, the less gas is used per mile. inverse

Levels of Complexity Chart

Parts	Level 1	Level 2	Level 3
A		●	
B			●
C	●		

Additional Answers

2.

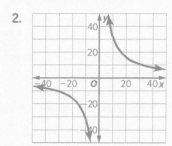

5.

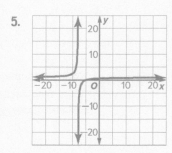

Test-Taking Strategy

Example

Read the problem. Identify what you need to know. Then use the information in the problem to solve.

Solve: $\dfrac{2}{x-3} - \dfrac{4}{x+3} = \dfrac{8}{x^2-9}$.

A -1 C 5

B 1   D 7

Step 1 Are there any answer choices you can eliminate because they are unreasonable?

No, the answer must be a real number, but all of the answer choices are real numbers.

Step 2 Which answer choice should you start with?

Either B or C, because they are in the middle. I'll start with B.

Step 3 Is the answer you chose correct? If not, in which direction do you need to go to find the correct answer?

Choice B was not correct. The result was too small, so I need to go up to find the correct answer. Choice C results in a true statement.

Test-Taking Tip

Strategies for Guessing and Checking

It is very important to pace yourself and keep track of how much time you have when taking a standardized test. If time is running short, or if you are unsure how to solve a problem, the guess-and-check strategy may help you determine the correct answer quickly, especially in a problem where it would be fast and easy to check your answer, such as in the problem below.

Apply the Strategy

Read the problem. Identify what you need to know. Then use the information in the problem to solve.

Solve: $\dfrac{2}{5x} - \dfrac{1}{2x} = -\dfrac{1}{2}$. **B**

A $\dfrac{1}{10}$ C $\dfrac{1}{4}$

B $\dfrac{1}{5}$ D $\dfrac{1}{2}$

Answer the questions below.

a. Are there any answer choices you can eliminate because they are unreasonable? no

b. Which answer choice should you start with? either B or C

c. Is the answer you chose correct? If not, in which direction do you need to go to find the correct answer? B was correct.

Step 1 Read the problem and look over the answer choices. Eliminate any that are unreasonable, such as those that are clearly incorrect, in the improper format, or containing incorrect units.

Step 2 Choose an answer choice and substitute it into the problem. If the answer choices are integers, choose one in the middle. That way, if the answer is incorrect, you know whether to adjust up or down.

Step 3 If the answer choice you have chosen does not satisfy the problem, go on to the next one, until you find the choice that results in a true statement.

Need Another Example?

Suppose y varies jointly as x and z. Find y when $x = 12$ and $z = 7$, if $y = 40$ when $z = 15$ and $x = 28$. A

A 8

B 20

C 80

D 200

a. Are there any answer choices you can eliminate because they are unreasonable? no

b. Which answer choice should you start with? either B or C

c. Is the answer you chose correct? If not, in which direction do you need to go to find the correct answer? B was too large. Choice A is correct.

Go Online!

The most up-to-date resources available for your program can be found at <u>connectED.mcgraw-hill.com</u>.

Diagnose Student Errors

Survey student responses for each item. Class trends may indicate common errors and misconceptions.

1.	A	Chose the height of an infinite number of similar cylinders
	B	Chose the height of 6 similar cylinders
	C	CORRECT
	D	Chose the height of the first 4 cylinders
	E	Chose the height of the first 3 cylinders
2.	A	Considered only the positive square root
	B	CORRECT
	C	Considered only the negative square root
	D	Used the opposite of the denominator in the second fraction
	E	Didn't realize x is not the whole denominator
4.	A	CORRECT
	B	Substituted original slope for y-intercept
	C	Used the positive reciprocal and not the negative reciprocal
	D	Used the opposite slope
	E	Found the equation of a line that is parallel
5.	A	Chose a function whose degree in the denominator is greater than the degree of the numerator
	B	Chose a function whose degree in the denominator is equal to the degree of the numerator
	C	CORRECT
	D	Chose a function whose degree in the denominator is greater than the degree of the numerator
6.	A	Chose the denominator of one of the terms
	B	Used the product of $x + 3$ and $x + 5$
	C	CORRECT
	D	Used the product of $x + 1$ and $x - 5$
7.	A	Did not correctly write or graph the exponential function
	B	Did not correctly write or graph the exponential function
	C	CORRECT
	D	Did not correctly write or graph the exponential function

Go Online!

Standardized Test Practice

Students can take self-checking tests in standardized format to plan and prepare for your state assessments.

Preparing for Assessment
Cumulative Review

Read each question. Then fill in the correct answer on the answer document provided by your teacher or on a sheet of paper.

1. The height of each cylinder is half of the height of the cylinder that it sits upon. The height of the bottom cylinder is 2 inches. How tall is this stack of five cylinders? **C**

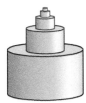

- ○ A 4 in.
- ○ B $3\frac{15}{16}$ in.
- ○ C $3\frac{7}{8}$ in.
- ○ D $3\frac{3}{4}$ in.
- ○ E $3\frac{1}{2}$ in.

2. What is the domain of the function? **B**

$$f(x) = \frac{2x}{x^2 - 1} + \frac{1}{2}$$

- ○ A $D = \{x \mid x \neq 1\}$
- ○ B $D = \{x \mid x \neq -1, 1\}$
- ○ C $D = \{x \mid x \neq -1\}$
- ○ D $D = \{x \mid x \neq -2\}$
- ○ E $D = \{x \mid x \neq 0\}$

3. MULTI-STEP Pilar can build a fence in 12.5 hours. Jackson can build the same fence in 15 hours.

a. If they work together, about how long will it take for them to build the fence? Round to the nearest tenth.

6.8	h

b. What mathematical practice did you use to solve this problem? **See students' work.**

4. Which is an equation of a line perpendicular to the line passing through the points $(-1, 8)$ and $(0, 12)$? **A**

- ○ A $y = -\frac{1}{4}x - 12$
- ○ B $y = -x + 4$
- ○ C $y = \frac{1}{4}x - 48$
- ○ D $y = -4x - 8$
- ○ E $y = 4x - 48$

5. Which function has a graph with no horizontal asymptote? **C**

- ○ A $f(x) = \frac{1}{x}$
- ○ B $f(x) = \frac{x^2 - 1}{x(x + 2)}$
- ○ C $f(x) = \frac{(x + 3)(x - 1)}{2x}$
- ○ D $f(x) = \frac{x^3 - 1}{(x^2 - 1)(x^2 + 1)}$

> **Test-Taking Tip**
> **Question 6** Substitute each of the answer choices for the numerator of the first term. Evaluate the expression on the left to see if it evaluates to $2x + 6$.

6. What expression makes the following a true statement? **C**

$$\frac{?}{x + 1} \cdot \frac{2(x + 1)}{x - 5} = 2x + 6$$

- ○ A $x - 5$
- ○ B $x^2 + 8x + 15$
- ○ C $x^2 - 2x - 15$
- ○ D $x^2 - 4x - 5$

7. Jeremy bought a painting in 2005 for $350. It is estimated that the value of the painting will increase by 12% each year. In what year does the value of the painting exceed $1000? **C**

- ○ A 2011
- ○ B 2013
- ○ C 2015
- ○ D 2017

9.	A	Chose a function with a vertical asymptote at $x = -7$ and a horizontal asymptote at $f(x) = 3$
	B	Chose a function with a vertical asymptote at $x = 3$ and a horizontal asymptote at $f(x) = -7$
	C	Chose a function with a vertical asymptote at $x = -3$ and a horizontal asymptote at $f(x) = 7$
	D	CORRECT
	E	Forgot the sign changes in the equation for the vertical asymptote

11.	A	CORRECT
	B	Chose the wrong direction for vertical translation
	C	Chose the wrong direction for horizontal and vertical transformations
	D	Chose the wrong number of units for horizontal and vertical transformations
	E	Chose the wrong direction and wrong number of units for horizontal and vertical transformations

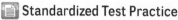

Go Online! for Standardized Test Practice

8. If x varies inversely as y and $x = 15$ when $y = 3$, what is the value of x when $y = 8$?

> 5.625

9. Which function has a graph with a vertical asymptote at $x = 7$ and a horizontal asymptote at $f(x) = -3$? **D**

- ○ **A** $f(x) = \frac{1}{x+7} + 3$
- ○ **B** $f(x) = \frac{1}{x-3} - 7$
- ○ **C** $f(x) = \frac{1}{x+3} + 7$
- ○ **D** $f(x) = \frac{1}{x-7} - 3$
- ○ **E** $f(x) = \frac{1}{x+7} - 3$

10. For what value of x is the function $f(x) = \frac{2x^2 - 7x - 4}{x - 4}$ discontinuous?

> 4

11. The function below is a transformation of the function $f(x) = \frac{1}{x}$. **A**

$$g(x) = \frac{1}{x - 2} + 3$$

Which correctly describes the transformation?

- ○ **A** a horizontal translation 2 units right and a vertical translation 3 units up
- ○ **B** a horizontal translation 2 units right and a vertical translation 3 units down
- ○ **C** a horizontal translation 2 units left and a vertical translation 3 units down
- ○ **D** a horizontal translation 3 units right and a vertical translation 2 units up
- ○ **E** A horizontal translation 3 units left and a vertical translation 2 units down

12. Select all true statements about the function graphed below. **A, D, E**

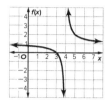

- ☐ **A** The domain of the function is all real numbers except 4.
- ☐ **B** The range of the function is all real numbers.
- ☐ **C** The vertical asymptote is $x = 1$.
- ☐ **D** The horizontal asymptote is $f(x) = 1$.
- ☐ **E** The graph represents a transformation of the function $f(x) = \frac{1}{x}$.

13. Janice spent \$3.85 for 2 pounds of apples and 3 pounds of pears. Lisa paid \$4.94 for 4 pounds of apples and 2 pounds of pears. How much does a pound of apples cost? **D**

- ○ **A** \$0.69
- ○ **B** \$0.77
- ○ **C** \$0.82
- ○ **D** \$0.89
- ○ **E** \$1.58

14. What are the zeros of the function? **D**

$$f(x) = \frac{x^3 + 2x^2 - 35x}{x - 3}$$

- ○ **A** 3
- ○ **B** 0 and 3
- ○ **C** 5 and -7
- ○ **D** 0, 5, and -7
- ○ **E** 3, 5, and -7

Need Extra Help?

If you missed Question...	1	2	3	4	5	6	7	8	9	10	11	12	13	14
Go to Lesson...	6-3	7-2	7-6	1-4	7-4	7-1	6-1	7-5	7-3	7-4	7-4	7-3	1-6	7-4

LS LEARNSMART®

Use LearnSmart as part of your test-preparation plan to measure student topic retention. You can create a student assignment in LearnSmart for additional practice on these topics.

· Rational Expressions
· Solving Rational and Radical Equations
· Graphing Rational, Radical, and Polynomial Functions
· Building Functions

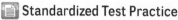

Formative Assessment

You can use these pages to benchmark student progress.

▤ Standardized Test Practice

Test Item Formats

In the Cumulative Review, students will encounter different formats for assessment questions to prepare them for standardized tests.

Exercise Question Types	
Question Type	Exercises
Multiple-Choice	1–2, 4–7, 9, 11, 13–14
Multiple Correct Answers	12
Type Entry: Short Response	3, 8, 10
Type Entry: Extended Response	3

Answer Sheet Practice

Have students simulate taking a standardized test by recording their answers on a practice recording sheet.

Homework Option

Get Ready for Chapter 8 Assign students the exercises on p. 530 as homework to assess whether they possess the prerequisite skills needed for the next chapter.

Go Online!

eAssessment

Customize and create multiple versions of chapter tests and answer keys that align to your standards. Tests can be delivered on paper or online.

62a.

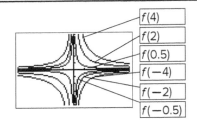

f(4)
f(2)
f(0.5)
f(−4)
f(−2)
f(−0.5)

[−5, 5] scl: 1 by [−5, 5] scl: 1

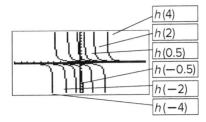

h(4)
h(2)
h(0.5)
h(−0.5)
h(−2)
h(−4)

[−10, 10] scl: 1 by [−10, 10] scl: 1

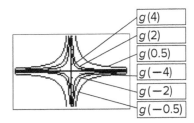

g(4)
g(2)
g(0.5)
g(−4)
g(−2)
g(−0.5)

[−5, 5] scl: 1 by [−5, 5] scl: 1

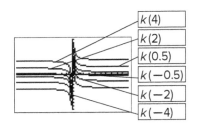

k(4)
k(2)
k(0.5)
k(−0.5)
k(−2)
k(−4)

[−10, 10] scl: 1 by [−10, 10] scl: 1

62b. For $f(x)$ and $g(x)$, when a or b is positive, the curves are in the 1st and 3rd quadrants. When a or b is negative, the curves are in the 2nd and 4th quadrants. For the graph of $k(x)$, as the value of d gets bigger, the graph moves up. For the graph of $h(x)$, as the value of c gets bigger, the graph moves to the right.

62c.

x	−4	−2	1	2	4
$-4f(x) = -4\frac{1}{x}$	1	2	−4	−2	−1
slope		$\frac{1}{2}$	−2	2	$\frac{1}{2}$
x	−4	−2	−1	2	4
$4f(x) = 4\frac{1}{x}$	−1	−2	4	2	1
slope		$-\frac{1}{2}$	6	$-\frac{2}{3}$	$-\frac{1}{2}$

x	−4	−2	1	2	4
$g(-4x) = -4\frac{1}{x}$	1	2	−4	−2	−1
Slope		$\frac{1}{2}$	−2	2	$-\frac{1}{2}$
x	−4	−2	1	2	4
$g(4x) = 4\frac{1}{x}$	−1	−2	4	2	2
slope		$-\frac{1}{2}$	2	−2	$-\frac{1}{2}$

x	−4	−2	1	2	4
$h(x - (-4)) = \frac{1}{x} - (-4)$	3.75	3.5	5	4.5	4.25
slope		−1.25	0.5	−0.5	−0.125
x	−4	−2	1	2	4
$h(x - 4) = \frac{1}{x} - 4$	−4.25	−4.5	−3	−3.5	−3.75
slope		−1.25	0.5	−0.5	−0.125

x	−4	−2	1	2	4
$k(x) + (-4) = \frac{1}{x} + (-4)$	−4.25	−4.5	−3	−3.5	−3.75
slope		−0.125	0.5	−0.5	−0.125
x	−4	−2	1	2	4
$k(x) + 4 = \frac{1}{x} + 4$	3.75	3.5	5	4.5	4.25
slope		−0.125	0.5	−0.5	−0.125

62d. For $f(x)$ and $g(x)$, and $h(x)$ and $k(x)$, the slopes have opposite signs.

Lesson 7-3 (Guided Practice)

3A.

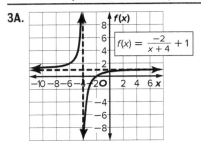

$f(x) = \frac{-2}{x + 4} + 1$

$D = \{x \mid x \neq -4\}; R = \{f(x) \mid f(x) \neq 1\}$

3B.

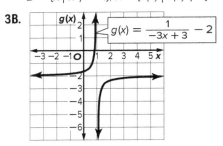

$g(x) = \frac{1}{-3x + 3} - 2$

$D = \{x \mid x \neq 1\}; R = \{g(x) \mid g(x) \neq -2\}$

4. $T = \dfrac{2500 + 45p}{p}$

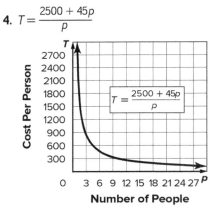

$T = \dfrac{2500 + 45p}{p}$

Cost Per Person (T axis): 300, 600, 900, 1200, 1500, 1800, 2100, 2400, 2700
Number of People (P axis): 0, 3, 6, 9, 12, 15, 18, 21, 24, 27

The domain will never reach zero because someone has to be at the dance. The range will never reach 45 because, even if there are thousands of people there, they will incur some additional cost. Neither value is negative because there cannot be a negative cost or a negative number of people.

Lesson 7-3

3.

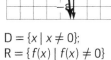

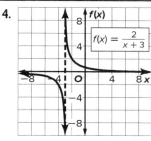

$f(x) = \frac{5}{x}$

$D = \{x \mid x \neq 0\};$
$R = \{f(x) \mid f(x) \neq 0\}$

4.

$f(x) = \frac{2}{x + 3}$

$D = \{x \mid x \neq -3\};$
$R = \{f(x) \mid f(x) \neq 0\}$

5.

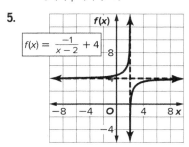

$f(x) = \frac{-1}{x - 2} + 4$

$D = \{x \mid x \neq 2\}; R = \{f(x) \mid f(x) \neq 4\}$

6b.

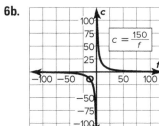

$c = \frac{150}{f}$

11.

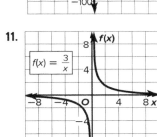

$f(x) = \frac{3}{x}$

$D = \{x \mid x \neq 0\};$
$R = \{f(x) \mid f(x) \neq 0\}$

12.

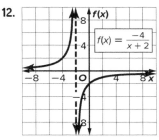

$f(x) = \frac{-4}{x + 2}$

$D = \{x \mid x \neq -2\};$
$R = \{f(x) \mid f(x) \neq 0\}$

13.

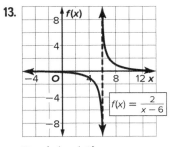

$f(x) = \frac{2}{x - 6}$

$D = \{x \mid x \neq 6\};$
$R = \{f(x) \mid f(x) \neq 0\}$

14.

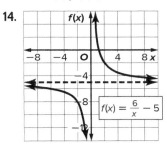

$f(x) = \frac{6}{x} - 5$

$D = \{x \mid x \neq 0\};$
$R = \{f(x) \mid f(x) \neq -5\}$

15.

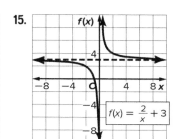

$f(x) = \frac{2}{x} + 3$

$D = \{x \mid x \neq 0\};$
$R = \{f(x) \mid f(x) \neq 3\}$

16.

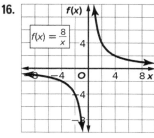

$f(x) = \frac{8}{x}$

$D = \{x \mid x \neq 0\};$
$R = \{f(x) \mid f(x) \neq 0\}$

17.

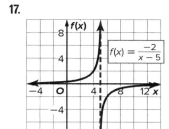

$f(x) = \frac{-2}{x - 5}$

$D = \{x \mid x \neq 5\};$
$R = \{f(x) \mid f(x) \neq 0\}$

18.

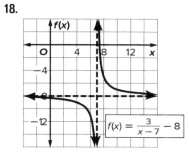

$f(x) = \frac{3}{x - 7} - 8$

$D = \{x \mid x \neq 7\};$
$R = \{f(x) \mid f(x) \neq -8\}$

19.

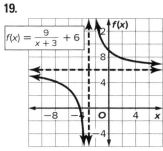

$f(x) = \frac{9}{x + 3} + 6$

$D = \{x \mid x \neq -3\};$
$R = \{f(x) \mid f(x) \neq 6\}$

20.

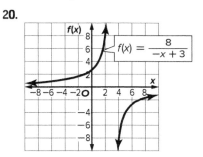

$f(x) = \frac{8}{-x + 3}$

$D = \{x \mid x \neq 3\};$
$R = \{f(x) \mid f(x) \neq 0\}$

21.

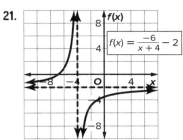

$f(x) = \frac{-6}{x + 4} - 2$

$D = \{x \mid x \neq -4\}; R = \{f(x) \mid f(x) \neq -2\}$

22.

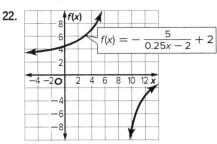

$f(x) = -\frac{5}{0.25x - 2} + 2$

$D = \{x \mid x \neq 8\}; R = \{f(x) \mid f(x) \neq 2\}$

23b.

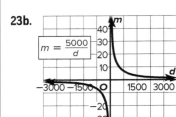

$$m = \frac{5000}{d}$$

24b.

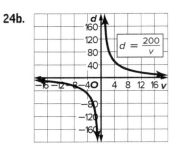

$$d = \frac{200}{v}$$

31.

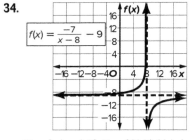

$$f(x) = \frac{-4}{x+2} - 5$$

D = {x | x ≠ −2}; R = {f(x) | f(x) ≠ −5}; x = −2, f(x) = −5

25.

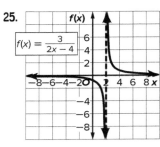

$$f(x) = \frac{3}{2x-4}$$

D = {x | x ≠ 2};
R = {f(x) | f(x) ≠ 0}

26.

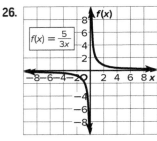

$$f(x) = \frac{5}{3x}$$

D = {x | x ≠ 0};
R = {f(x) | f(x) ≠ 0}

32.

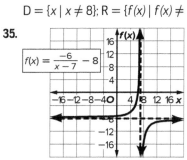

$$f(x) = \frac{6}{x-1} + 2$$

D = {x | x ≠ 1}; R = {f(x) | f(x) ≠ 2}; x = 1, f(x) = 2

27.

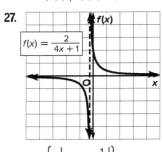

$$f(x) = \frac{2}{4x+1}$$

$$D = \left\{ x \,\middle|\, x \neq -\frac{1}{4} \right\};$$
R = {f(x) | f(x) ≠ 0}

28.

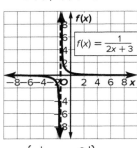

$$f(x) = \frac{1}{2x+3}$$

$$D = \left\{ x \,\middle|\, x \neq -\frac{3}{2} \right\};$$
R = {f(x) | f(x) ≠ 0}

33.

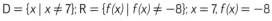

$$f(x) = \frac{2}{x-4} + 3$$

D = {x | x ≠ 4}; R = {f(x) | f(x) ≠ 3}; x = 4, f(x) = 3

29b.

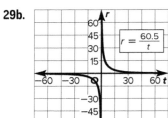

$$r = \frac{60.5}{t}$$

34.

$$f(x) = \frac{-7}{x-8} - 9$$

D = {x | x ≠ 8}; R = {f(x) | f(x) ≠ −9}; x = 8, f(x) = −9

30.

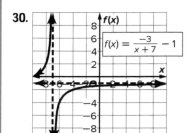

$$f(x) = \frac{-3}{x+7} - 1$$

D = {x | x ≠ −7}; R = {f(x) | f(x) ≠ −1}; x = −7, f(x) = −1

35.

$$f(x) = \frac{-6}{x-7} - 8$$

D = {x | x ≠ 7}; R = {f(x) | f(x) ≠ −8}; x = 7, f(x) = −8

36b.

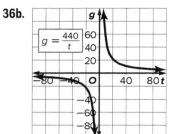

$$g = \frac{440}{t}$$

37a.

x	$a(x) = x^2$	$b(x) = x^{-2}$	$c(x) = x^3$	$d(x) = x^{-3}$
−4	16	$\frac{1}{16}$	−64	$-\frac{1}{64}$
−3	9	$\frac{1}{9}$	−27	$-\frac{1}{27}$
−2	4	$\frac{1}{4}$	−8	$-\frac{1}{8}$
−1	1	1	−1	−1
0	0	undefined	0	undefined
1	1	1	1	1
2	4	$\frac{1}{4}$	8	$\frac{1}{8}$
3	9	$\frac{1}{9}$	27	$\frac{1}{27}$
4	16	$\frac{1}{16}$	64	$\frac{1}{64}$

37b.

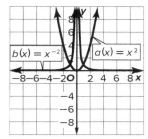

37c. $a(x)$: D = {all real numbers}, R = $\{a(x) \mid a(x) \geq 0\}$; as $x \to -\infty$, $a(x) \to \infty$, as $x \to \infty$, $a(x) \to \infty$; At $x = 0$, $a(x) = 0$, so there is a zero at $x = 0$.

$b(x)$: D = $\{x \mid x \neq 0\}$, R $\{b(x) \mid b(x) > 0\}$; as $x \to -\infty$, $a(x) \to 0$, as $x \to \infty$, $a(x) \to 0$; At $x = 0$, $b(x)$ is undefined, so there is an asymptote at $x = 0$.

37d.

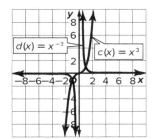

37e. $c(x)$: D = {all real numbers}, R = {all real numbers}; as $x \to -\infty$, $a(x) \to -\infty$, as $x \to \infty$, $a(x) \to \infty$; At $x = 0$, $a(x) = 0$, so there is a zero at $x = 0$.
$d(x)$: D = $\{x \mid x \neq 0\}$, R = $\{b(x) \mid b(x) \neq 0\}$; as $x \to -\infty$, $a(x) \to 0$, as $x \to \infty$, $a(x) \to 0$; At $x = 0$, $b(x)$ is undefined, so there is an asymptote at $x = 0$.

37f. For two power functions $f(x) = ax^n$ and $g(x) = ax^{-n}$, for every x, $f(x)$ and $g(x)$ are reciprocals. The domains are similar except that for $g(x)$, $x \neq 0$. Additionally, wherever $f(x)$ has a zero, $g(x)$ is undefined.

39a. The first graph has a vertical asymptote at $x = 0$ and a horizontal asymptote at $y = 0$. The second graph is translated 7 units up and has a vertical asymptote at $x = 0$ and a horizontal asymptote at $y = 7$.

39b. Both graphs have a vertical asymptote at $x = 0$ and a horizontal asymptote at $y = 0$. The second graph is stretched vertically by a factor of 4.

39c. The first graph has a vertical asymptote at $x = 0$ and a horizontal asymptote at $y = 0$. The second graph is translated 5 units to the left and has a vertical asymptote at $x = -5$ and a horizontal asymptote at $y = 0$.

39d.

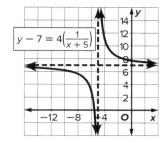

41. Sample answer = $f(x) = \dfrac{2}{x - 3} + 4$ and $g(x) = \dfrac{5}{x - 3} + 4$

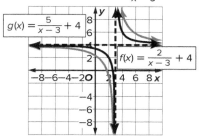

42. A rational function can be used to determine the shared cost of renting the park. The expression that represents the total cost is used as the numerator while the expression representing the total number of students is used for the denominator. Only the first quadrant part of the graph is meaningful in the context of the problem as both the cost and the number of students must be greater than or equal to 0.

Mid-Chapter Quiz

19.

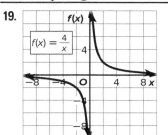

D = $\{x \mid x \neq 0\}$, R = $\{f(x) \mid f(x) \neq 0\}$

20.

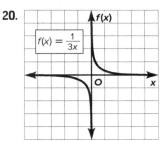

D = $\{x \mid x \neq 0\}$, R = $\{f(x) \mid f(x) \neq 0\}$

21.

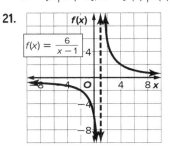

D = $\{x \mid x \neq 1\}$, R = $\{f(x) \mid f(x) \neq 0\}$

22.

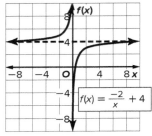

$$f(x) = \frac{-2}{x} + 4$$

$D = \{x \mid x \neq 0\}, R = \{f(x) \mid f(x) \neq 4\}$

23.

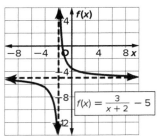

$$f(x) = \frac{3}{x+2} - 5$$

$D = \{x \mid x \neq -2\}, R = \{f(x) \mid f(x) \neq -5\}$

24.

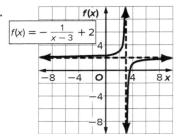

$$f(x) = -\frac{1}{x-3} + 2$$

$D = \{x \mid x \neq 3\}, R = \{f(x) \mid f(x) \neq 2\}$

25b.

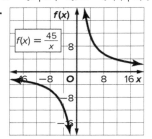

$$f(x) = \frac{45}{x}$$

Lesson 7-4

1.

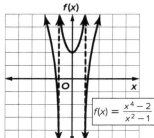

$$f(x) = \frac{x^4 - 2}{x^2 - 1}$$

2.

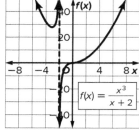

$$f(x) = \frac{x^3}{x+2}$$

3a.

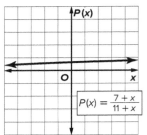

$$P(x) = \frac{7+x}{11+x}$$

4.

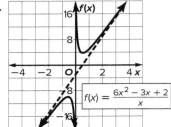

$$f(x) = \frac{6x^2 - 3x + 2}{x}$$

5.

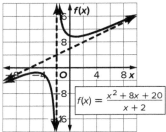

$$f(x) = \frac{x^2 + 8x + 20}{x+2}$$

6.

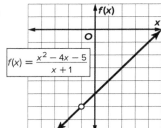

$$f(x) = \frac{x^2 - 4x - 5}{x+1}$$

7.

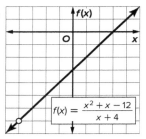

$$f(x) = \frac{x^2 + x - 12}{x+4}$$

13.

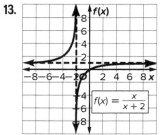

$$f(x) = \frac{x}{x+2}$$

14.

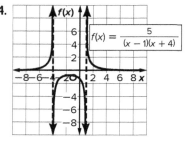

$$f(x) = \frac{5}{(x-1)(x+4)}$$

15.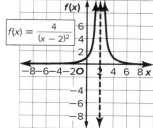

$f(x) = \dfrac{4}{(x-2)^2}$

16.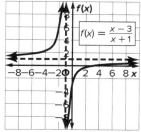

$f(x) = \dfrac{x-3}{x+1}$

17.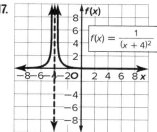

$f(x) = \dfrac{1}{(x+4)^2}$

18.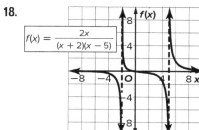

$f(x) = \dfrac{2x}{(x+2)(x-5)}$

19.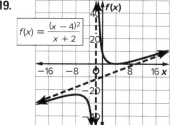

$f(x) = \dfrac{(x-4)^2}{x+2}$

20.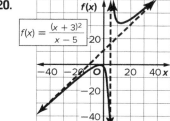

$f(x) = \dfrac{(x+3)^2}{x-5}$

21.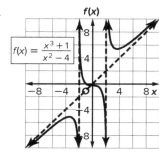

$f(x) = \dfrac{x^3+1}{x^2-4}$

22.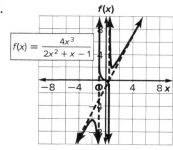

$f(x) = \dfrac{4x^3}{2x^2+x-1}$

23.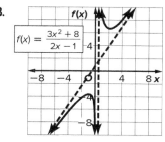

$f(x) = \dfrac{3x^2+8}{2x-1}$

24.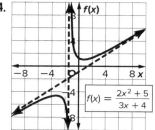

$f(x) = \dfrac{2x^2+5}{3x+4}$

25.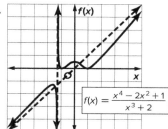

$f(x) = \dfrac{x^4-2x^2+1}{x^3+2}$

26.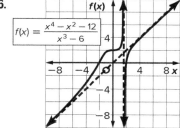

$f(x) = \dfrac{x^4-x^2-12}{x^3-6}$

27b. Sample answer: I took the information I was given and made a visual. Then I set the denominator equal to 0 when x was -1, giving me $x + 1$. Because this is a rational function with no horizontal asymptotes, I made the numerator x^2, meeting all the requirements I was given.

27c. Sample answer: I made the assumption that there were no oblique asymptotes and the graph had no point discontinuity.

28.

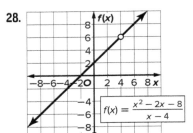

$$f(x) = \frac{x^2 - 2x - 8}{x - 4}$$

29.

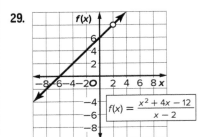

$$f(x) = \frac{x^2 + 4x - 12}{x - 2}$$

30.

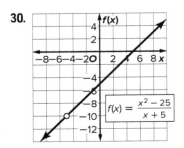

$$f(x) = \frac{x^2 - 25}{x + 5}$$

31.

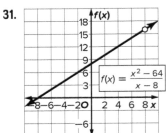

$$f(x) = \frac{x^2 - 64}{x - 8}$$

32.

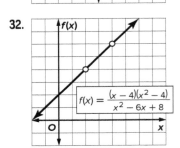

$$f(x) = \frac{(x - 4)(x^2 - 4)}{x^2 - 6x + 8}$$

33.

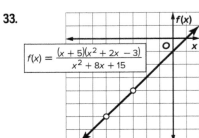

$$f(x) = \frac{(x + 5)(x^2 + 2x - 3)}{x^2 + 8x + 15}$$

34.

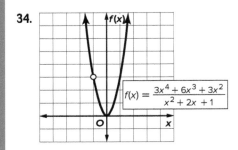

$$f(x) = \frac{3x^4 + 6x^3 + 3x^2}{x^2 + 2x + 1}$$

35.

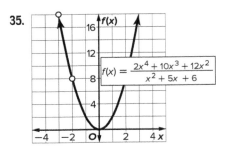

$$f(x) = \frac{2x^4 + 10x^3 + 12x^2}{x^2 + 5x + 6}$$

36a. $f(x) = \frac{4500 + 50x}{x}$

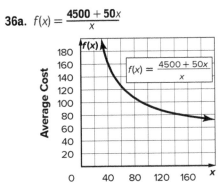

$$f(x) = \frac{4500 + 50x}{x}$$

Average Cost — Number of Lawns

37a. $f(x) = \frac{150 + 40x}{x}$

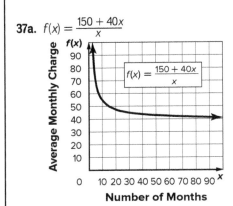

$$f(x) = \frac{150 + 40x}{x}$$

Average Monthly Charge — Number of Months

38a.

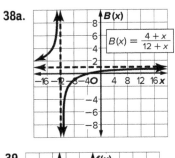

$$B(x) = \frac{4 + x}{12 + x}$$

39.

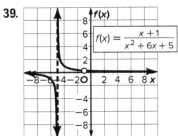

$$f(x) = \frac{x + 1}{x^2 + 6x + 5}$$

40.

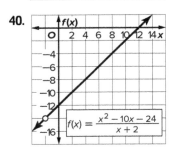

$$f(x) = \frac{x^2 - 10x - 24}{x + 2}$$

41.

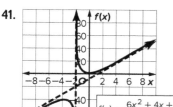
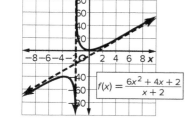

$$f(x) = \frac{6x^2 + 4x + 2}{x + 2}$$

42. Sample graph:

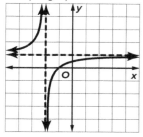

46. Sample answer: By factoring the denominator of a rational function and determining the values that cause each factor to equal zero, you can determine the asymptotes of a rational function. After factoring the numerator and denominator of a rational function, if there is a common factor $x - c$, then there is point discontinuity at $x = c$.

Practice Test

11.

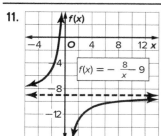

$$f(x) = -\frac{8}{x} - 9$$

12.

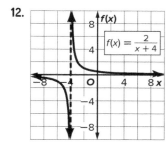

$$f(x) = \frac{2}{x + 4}$$

13.

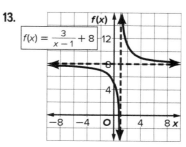

$$f(x) = \frac{3}{x - 1} + 8$$

14.

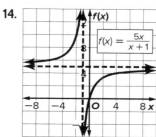

$$f(x) = \frac{5x}{x + 1}$$

15.

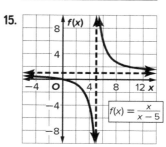

$$f(x) = \frac{x}{x - 5}$$

16.

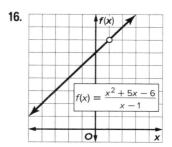

$$f(x) = \frac{x^2 + 5x - 6}{x - 1}$$

SUGGESTED PACING (DAYS)

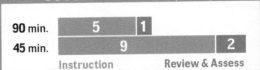

90 min.   5   1

45 min.   9   2

Instruction   Review & Assess

Track Your Progress

This chapter focuses on content from the Making Inferences and Justifying Conclusions and Using Probability to Make Decisions domains.

THEN	NOW	NEXT
A.APR.7 Understand that rational expressions form a system analogous to the rational numbers, closed under addition, subtraction, multiplication, and division by a nonzero rational expression; add, subtract, multiply, and divide rational expressions. **A.CED.2** Create equations in two or more variables to represent relationships between quantities; graph equations on coordinate axes with labels and scales. **A.REI.2** Solve simple rational and radical equations in one variable, and give examples showing how extraneous solutions may arise.	**S.IC.4** Use data from a sample survey to estimate a population mean or proportion; develop a margin of error through the use of simulation models for random sampling. **S.IC.5** Use data from a randomized experiment to compare two treatments; use simulations to decide if differences between parameters are significant. **S.MD.6** Use probabilities to make fair decisions. **S.MD.7** Analyze decisions and strategies using probability concepts.	**F.BF.3** Identify the effect on the graph of replacing $f(x)$ by $f(x) + k$, $k\,f(x)$, $f(kx)$, and $f(x + k)$ for specific values of k (both positive and negative); find the value of k given the graphs. Experiment with cases and illustrate an explanation of the effects on the graph using technology. **F.IF.7e** Graph exponential and logarithmic functions, showing intercepts and end behavior, and trigonometric functions, showing period, midline, and amplitude. **F.TF.1** Understand radian measure of an angle as the length of the arc on the unit circle subtended by the angle.

Standards for Mathematical Practice

All of the Standards for Mathematical Practice will be covered in this chapter. The MP icon notes specific areas of coverage.

 Teaching the Mathematical Practices
Help students develop the mathematical practices by asking questions like these.

Questioning Strategies As students approach problems in this chapter, help them develop mathematical practices by asking:

Sense-Making
· What is the difference between sample surveys, experiments, and observational studies?
· How do the shapes of distributions help you to compare data?

Reasoning
· Can you use data to compare theoretical and experimental probabilities?
· Can you use data from sample surveys to estimate population means or proportions?
· What inferences can you make about population parameters based on random samples of the population?

Modeling
· How can you collect and analyze data by conducting simulations of real-life situations?
· How can you develop margins of error by using simulation models?
· How can the shapes of distributions enable you to select appropriate statistics?

Using Tools
· How do you use the Empirical Rule to analyze normally distributed variables?
· Can you apply the standard normal distribution to z-values?

Go Online!

 StudySync:
SMP Modeling Videos

These demonstrate how to apply the Standards for Mathematical Practice to collaborate, discuss, and solve real-world math problems.

Go Online!

connectED.mcgraw-hill.com

 LearnSmart 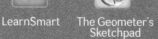 The Geometer's Sketchpad   Vocabulary   Personal Tutor Tools Calculator Resources   Self-Check Practice   Animations

Customize Your Chapter

Use the *Plan & Present*, *Assignment Tracker*, and *Assessment* tools in ConnectED to introduce lesson concepts, assign personalized practice, and diagnose areas of student need.

Differentiated Instruction

Throughout the program, look for the icons to find specialized content designed for your students.

AL	Approaching Level
OL	On Level
BL	Beyond Level
ELL	English Language Learners

Personalize

Differentiated Resources	AL	OL	BL	ELL
FOR EVERY CHAPTER				
✓ Chapter Readiness Quizzes	●	●	◑	●
✓ Chapter Tests	●	●	●	●
✓ Standardized Test Practice	●	●	●	●
Vocabulary Review Games	●	●	◑	●
Anticipation Guide (English/Spanish)	●	●	◑	●
Student-Built Glossary	●	●	◑	●
Chapter Project	◑	●	●	●
FOR EVERY LESSON	AL	OL	BL	ELL
Personal Tutors (English/Spanish)	●	●	◑	●
Graphing Calculator Personal Tutors	●	●	●	●
▷ Step-by-Step Solutions	●	●	◑	●
✓ Self-Check Quizzes	●	●	●	●
5-Minute Check	●	●	●	●
Study Notebook	●	●	●	●
Study Guide and Intervention	●	●		●
Skills Practice	●	◑		●
Practice	◑	●	●	●
Word Problem Practice	●	●	●	◑
Enrichment		●	●	●
✚ Extra Examples	●	◑		◑
Lesson Presentations	●	●	●	●

◑ Aligned to this group   ● Designed for this group

Engage

Featured IWB Resources

 The Geometer's Sketchpad provides students with a tangible, visual way to learn. *Use with Lessons 8-6.*

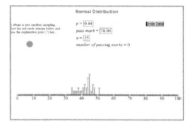

 eLessons engage students and help build conceptual understanding of big ideas. *Use with Lessons 8-1, 8-2, and 8-6.*

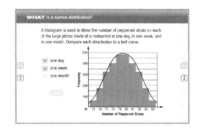

 Animations help students make important connections through motion. *Use with Lessons 8-4 and 8-7.*

 Time Management How long will it take to use these resources? Look for the clock in each lesson interleaf.

Introduce the Chapter

Mathematical Background

A statistic is a measure that describes a characteristic of a sample. The shape of a distribution of data can be symmetric, positively skewed, or negatively skewed. The mean and standard deviation or five-number summary can be used to describe or compare the distribution of sets of data. A probability distribution is a function that maps the sample space to the outcomes in the sample space. The normal distribution is a continuous, symmetric, bell-shaped distribution of a random variable.

Essential Question

At the end of this chapter, students should be able to answer the Essential Question.

How can you effectively evaluate information?
Sample answer: First, determine whether the information source is credible. Then critically analyze the information to determine whether it is useful for the given situation.

How can you use information to make decisions?
Sample answer: You can look for trends, and then make a decision based on what has happened in the past and/or is reflected in the information.

Apply Math to the Real World

EDUCATION In this activity, students will use what they already know about data distribution to consider how studying the distribution of data can aid in assessing how well material was taught or how well an exam assessed the knowledge of students. Have students complete this activity individually or in small groups. 1

Go Online!

Chapter Project

Please Complete This Survey Students use what they have learned about surveys, distributions of data, and normal distributions to complete a project.

This chapter project addresses financial literacy, as well as several specific skills identified as being essential to student success by the Framework for 21st Century Learning. 1, 3, 4

CHAPTER 8
Statistics and Probability

THEN
You calculated weighted averages.

NOW
You will:
- Evaluate surveys, studies, and experiments.
- Create and use graphs of probability distributions.
- Compare sample statistics and population statistics.

WHY

EDUCATION Probability and statistics are used in all facets of education, including determining grades or when teachers weight their grades.

Use the Mathematical Practices to complete the activity.

1. **Using Tools** Use what you already know, your Algebra 1 text, or the Internet to refresh your memory about distributions of data.
2. **Sense Making** This graph represents exam grades. What does this graph tell you? What grade did most students receive?

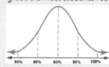

3. **Applying Math** Is the data symmetrically distributed or skewed? Can the mean and standard deviation be used to represent the data set or should the five-number summary be used?
4. **Discuss** What might positively skewed grades say about the exam or the teaching methods used? How can a positively skewed set of data be altered to become symmetrically distributed?

Monkeybusinessimages/Getty Images

⊙ ALEKS®

Your Student Success Tool ALEKS is an adaptive, personalized learning environment that identifies precisely what each student knows and is ready to learn—ensuring student success at all levels.

- **Formative Assessment:** Dynamic, detailed reports monitor students' progress toward standards mastery.
- **Automatic Differentiation:** Strengthen prerequisite skills and target individual learning gaps.
- **Personalized Instruction:** Supplement in-class instruction with personalized assessment and learning opportunities.

Go Online to Guide Your Learning

Explore & Explain	Organize

Coin Toss

Use the **Coin Toss** tool, the **Number Cube**, or the **Spinner** tool to enhance your understanding of statistical experiments and to run simulations in Lesson 8-2.

The Geometer's Sketchpad

Use **The Geometer's Sketchpad** to illustrate the normal distribution of data in Lesson 8-6.

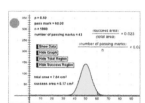

eBook

Interactive Student Guide

Before starting the chapter, answer the **Chapter Focus** preview questions. Check your answers as you complete each lesson. At the end of the chapter, try the **Performance Task**.

Foldables

Get organized! Create this **Statistics and Probability Foldable** before you begin this chapter to help you organize your notes about statistics and probability.

Collaborate

Chapter Project

In the **Please Complete This Survey** project, you will use what you have learned about statistics and probability to complete a project that addresses business literacy.

Focus

 ### LEARNSMART

Need help studying? Complete the **Modeling with Functions** domain in LearnSmart to review for the chapter test.

ALEKS

You can use the **Sequences and Probability** topic in ALEKS to explore what you know about statistics and probability and what you are ready to learn.*

* Ask your teacher if this is part of your program.

Dinah Zike's FOLDABLES

Focus Students write notes about probability and statistics.

Teach Have students make and label their Foldables as illustrated. At the end of each lesson, ask students to write about their experiences with the statistics and probability topics presented in the lessons. Encourage students to explain hat they found interesting or challenging.

What to Use it Encourage students to add to their Foldables as they work through the chapter and to use them to review for the chapter test.

Go Online!

Vocabulary Flashcard Pocketbooks

Help your students take their vocabulary flashcards to the next level! Watch this video to learn how to quickly create a pocketbook that can be used to store vocabulary flashcards. MP 5

Get Ready for the Chapter

RtI Response to Intervention

Use the Concept Check results and the Intervention Planner chart to help you determine your Response to Intervention.

Intervention Planner

TIER 1 On Level **OL**

IF students miss 25% of the exercises or less,

THEN choose a resource:
Go Online!
📄 Chapter Project

TIER 2 Approaching Level **AL**

IF students miss 50% of the exercises,

THEN choose a resource:
Go Online!
📄 Study Guide and Intervention, Chapter 4
➕ Extra Examples
💬 Personal Tutors
📄 Homework Help

TIER 3 Intensive Intervention

IF students miss 75% of the exercises,

THEN Use *Math Triumphs, Alg. 2,*
Go Online!
➕ Extra Examples
💬 Personal Tutors
📄 Homework Help
🔤 Review Vocabulary

Additional Answers

1. the average of the data; add all the values together and divide by the total number of values
2. the value that is halfway in the ordered set
3. the value that occurs most often in the set
4. arrange the data from least to greatest and locate the middle number or average the two middle numbers
5. Multiply the individual probabilities.
6. negatively skewed
7. an outlier in a set of data will not affect the shape of the distribution plot
8. the mean accurately reflects the median of the data

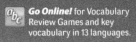
Go Online! for Vocabulary Review Games and key vocabulary in 13 languages.

Get Ready for the Chapter

Connecting Concepts	New Vocabulary		

Concept Check
Review the concepts used in this chapter by answering the questions below. 1–8. See margin.

1. What is the mean of a set of data? How is it determined?
2. What is the median of a set of data?
3. What is the mode of a set of data?
4. How is the median of a set of data determined?
5. If a number cube is rolled and a coin is tossed, how would you determine the probability that the number cube would show a 1 and the coin would land tails up?
6. In the box-and-whisker plot shown, is the data negatively skewed, symmetric, or positively skewed?

50%   50%

7. When plotting data in a distribution chart, how will an outlier affect the shape of the distribution?
8. When a distribution is symmetric, what can be said about the relationship between the mean and the median?

New Vocabulary

English		Español
statistic	p. 531	estadística
parameter	p. 531	parámetro
survey	p. 531	exámenes
experiment	p. 531	experimento
observational study	p. 531	estudio de observación
random sample	p. 531	muestra aleatoria
bias	p. 531	sesgo
experiment	p. 531	experimento
simulation	p. 531	simulación
probability model	p. 538	modelo de probabilidad
theoretical probability	p. 538	probabilidad teórica
experimental probability	p. 538	probabilidad experimental
relative frequency	p. 538	frecuencia relativa
margin of error	p. 546	margen de error muestral
distribution	p. 551	distribución
normal distribution	p. 566	distribución normal
z-value	p. 568	valor de z
standard normal distribution	p. 568	distribución normal estándar

Performance Task Preview
You can use the concepts and skills in this chapter to perform statistical analysis on data for a company that specializes in this service. Understanding statistics and probability will help you finish the Performance Task at the end of the chapter.

In this Performance Task you will:
• make sense of problems and persevere in solving them
• model with mathematics
• attend to precision
• look for and express regularity in repeated reasoning

Review Vocabulary

combination combinación an arrangement or selection of objects in which order is not important

permutation permutación a group of objects or people arranged in a certain order

random arbitrario Unpredictable, or not based on any predetermined characteristics of the population; when a number cube is tossed, a coin is flipped, or a spinner is spun, the outcome is a random event.

Key Vocabulary **ELL**

Introduce the key vocabulary in the chapter using the routine below.

Define Normal distribution is a frequency distribution that often occurs when there is a large number of values in a set of data: about 68% of the values are within one standard deviation of the mean, 95% of the values are within two standard deviations of the mean, and 99% of the values are within three standard deviations.

Example This diagram shows a normal distribution.

Normal Distribution

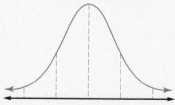

Ask What do you notice about the normal distribution curve? It is shaped like a bell and is symmetrical.

Random Sampling

SUGGESTED PACING (DAYS)

90 min. **0.5**
45 min. **1.0**

Instruction

Track Your Progress

Objectives

1 Distinguish among sample surveys, experiments, and observational studies.

2 Make inferences about population parameters based on random samples of the population.

Mathematical Background

When a census is conducted, data are collected from every member of a population. Therefore, the results are known to be correct. Because a survey investigates only part of a population, the results always involve some uncertainty.

THEN	NOW	NEXT
S.CP.4 Construct and interpret two-way frequency tables of data when two categories are associated with each object being classified. Use the two-way table as a sample space to decide if events are independent and to approximate conditional probabilities. **S.CP.6** Find the conditional probability of *A* given *B* as the fraction of *B*'s outcomes that also belong to *A*, and interpret the answer in terms of the model.	**S.IC.1** Understand statistics as a process for making inferences about population parameters based on a random sample from that population. **S.IC.3** Recognize the purposes of and differences among sample surveys, experiments, and observational studies; explain how randomization relates to each.	**S.IC.2** Decide if a specified model is consistent with results from a given data-generating process, e.g., using simulation. **S.IC.5** Use data from a randomized experiment to compare two treatments; use simulations to decide if differences between parameters are significant.

Go Online! All of these resources and more are available at connectED.mcgraw-hill.com

eLessons utilize the power of your interactive whiteboard in an engaging way. Use **Simulations**, screen 11, to introduce the concepts in this lesson.

Use the **eGlossary** to define observational study, random sample, and other key vocabulary in the lesson.

Chapter Project allows students to create and customize a project as a nontraditional method of assessment.

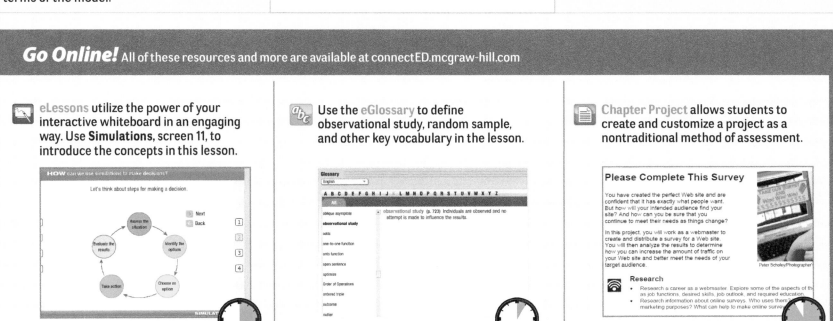

Use at Beginning of Lesson   *Use with Examples*   *Use at End of Lesson*

⊕ Using Open Educational Resources

Lesson Sharing Go to **Do the Math Stats** for ideas about discussion topics and ways for the students to critique statistics in the media. Ask students to bring in articles with statistics to critique in class. Have students work in pairs or small groups to analyze the validity of the article and its parameters. Ask students to turn in written critiques. *Use as group work*

Go Online!
connectED.mcgraw-hill.com

Worksheets

Differentiate Your Resources

Extra Practice Additional practice or homework; Skills Practice is best for approaching-level students and Practice is best for on-level and beyond-level students

Skills Practice

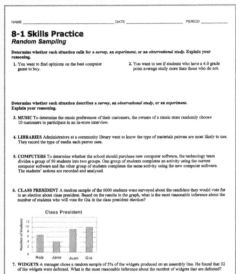

Practice

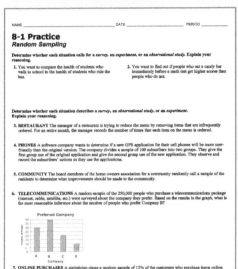

Word Problem Practice

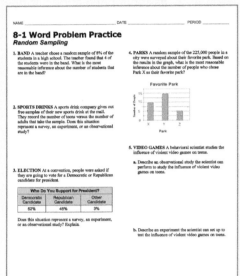

Intervention Reteaching and vocabulary activities that can be used with struggling or absent students and as ELL support

Study Guide and Intervention

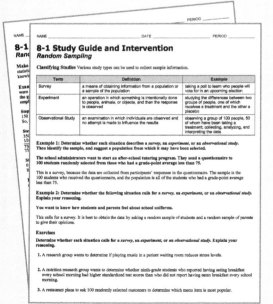

Study Notebook

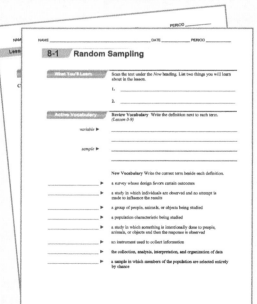

Extension Activities that can be used to extend lesson concepts

Enrichment

LESSON 1
Random Sampling

:: Then	:: Now	:: Why?
• You summarized data using measures of center and measures of variation.	**1** Distinguish among sample surveys, experiments, and observational studies. **2** Make inferences about population parameters based on random samples of the population.	• According to a recent study, 88% of teen cell phone users in the United States send text messages, and one in three teens sends more than 100 texts per day.

New Vocabulary
parameter
statistic
bias
random sample
survey
experiment
observational study

MP Mathematical Practices
1 Make sense of problems and persevere in solving them.
4 Model with mathematics.

Content Standards
S.IC.1 Understand statistics as a process for making inferences about population parameters based on a random sample from that population.
S.IC.3 Recognize the purposes of and differences among sample surveys, experiments, and observational studies; explain how randomization relates to each.

1 Classifying Studies Statistics is the collection, analysis, interpretation, and organization of data. You have used *descriptive statistics* to summarize data using measures like the mean. Due to time and money constraints, it may not be possible to collect data from each member of a population to calculate a population characteristic or **parameter**. In *inferential statistics*, data is collected to make inferences about a population. In these studies, a sample of the population is taken, and a measure called a **statistic** is calculated using the data. The sample statistic, such as the sample mean or sample standard deviation, is then used to make inferences about the population parameter.

The steps in a typical statistical study are shown below.

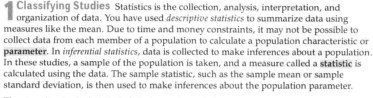

To obtain good information and draw accurate conclusions about a population, it is important to select an *unbiased* sample. A **bias** is an error that results in a misrepresentation of members of a population. A poorly chosen sample can cause biased results. To reduce the possibility of selecting a biased sample, a **random sample** can be taken, in which members of the population are selected entirely by chance.

The following study types can be used to collect sample information.

Key Concept Study Types

Definition	Example
In a **survey**, data are collected from responses given by members of a population regarding their characteristics, behaviors, or opinions.	To determine whether the student body likes the new cafeteria menu, the student council asks a random sample of students for their opinion.
In an **experiment**, the sample is divided into two groups: • an *experimental group* that undergoes a change, and • a *control group* that does not undergo the change. The effect on the experimental group is then compared to the control group.	A restaurant is considering creating meals with chicken instead of beef. They randomly give half of a group of participants meals with chicken and the other half meals with beef. Then they ask how they like the meals.
In an **observational study**, members of a sample are measured or observed without being affected by the study.	Researchers at an electronics company observe a group of teenagers using different laptops and note their reactions.

Launch

Have students read the Why? section of the lesson. Ask:

• **What do you think the purpose of the study was?** Sample answer: to determine the number of teen cell phone users that send texts and the number of texts that these users send per day

• **How do you think the data could have been collected for this study?** Sample answer: Teen cell phone users could have been asked whether they send texts and, if so, how many texts they send per day.

• **According to the study, if 50 teen cell phone users were randomly selected, how many would be texters?** 44

MP Mathematical Practices Strategies

Make sense of problems and persevere in solving them.
Help students understand the meaning of a problem, look for entry points to its solution, and plan a solution pathway. For example, ask:

• **How does the sample differ in an experiment compared to a survey or observational study?** In an experiment, the sample is divided into two groups.

• **Why is it important for a sample to be random?** This ensures that there is no bias and that the sample is representative of the population.

• **How do you find a sample proportion?** Find the ratio of the number of members of the sample with the relevant characteristic to the size of the sample.

• **How do you use a sample proportion to find the corresponding population parameter?** Multiply the sample proportion by the size of the population.

Go Online!

Interactive Whiteboard
Use the *eLesson* or *Lesson Presentation* to present this lesson.

Teach

Ask the scaffolded questions for each example to build conceptual understanding for students at all levels.

1 Classifying Studies

Example 1 Classify Study Types

AL What two groups must you have for an experiment? an experimental group and a control group

OL How could you change part **a** to be a survey? The fan club could be asked which cover they like best.

BL Explain why sometimes an observational study might need to be used instead of an experiment. It might not be practical or moral to conduct the experiment.

Need Another Example?
Determine whether each situation describes a survey, an experiment, or an observational study. Then identify the sample, and suggest a population from which it may have been selected.

a Movies A retro movie theater wants to determine what genre of movies to play during the next year. They plan to poll 50 random area residents and ask them what their favorite movies are. survey; Sample answer: the 50 residents that were polled; population: all potential movie-goers

b Driving A driving school wants to determine the main issue drivers face while taking the driving test. They watch and record 30 random people taking the test. observational study; sample: the people observed while taking the test; population: potential driving school students

Example 2 Choose a Study Type

AL Give two different ways you could conduct a survey. by phone or on paper

OL Why is it important to have a control group in an experiment? You need to compare whether the treatment is having an effect compared to doing nothing.

BL Would a survey always be accurate? No; people might report something inaccurately.

S.IC.3

Example 1 Classify Study Types

Determine whether each situation describes a *survey*, an *experiment*, or an *observational study*. Then identify the sample, and suggest a population from which it may have been selected.

a. MUSIC A band wants to test three designs for an album cover. They randomly select 50 members of their fan club to view the covers while they watch and record their reactions.

This is an observational study, because the band is going to observe the fans without them being affected by the study. The sample is the 50 fans selected, and the population is all potential purchasers of this album.

b. RECYCLING The city council wants to start a recycling program. They send out a questionnaire to 200 random citizens asking what items they would recycle.

This is a survey, because the data are collected from participants' responses in the questionnaire. The sample is the 200 people who received the questionnaire, and the population is all of the citizens of the city.

> **Study Tip**
> **Census** A census is a survey in which each member of a population is questioned. Therefore, when a census is conducted, there is no sample.

> **Guided Practice**

1A. RESEARCH Scientists study the behavior of one group of dogs given a new heartworm treatment and another group of dogs given a false treatment or *placebo*.

1A. experiment; sample: dogs given the heartworm medication; population: all dogs

1B. YEARBOOKS The yearbook committee conducts a study to determine whether students would prefer to have a print yearbook or both print and digital yearbooks.

1B. survey; sample: students participating in the survey; population: the student body

To determine when to use a survey, experiment, or observational study, think about how the data will be obtained and whether or not the participants will be affected by the study.

S.IC.3

Example 2 Choose a Study Type

Determine whether each situation calls for a *survey*, an *experiment*, or an *observational study*. Explain your reasoning.

a. MEDICINE A pharmaceutical company wants to test whether a new medicine is effective.

The treatment will need to be tested on a sample group, which means that the members of the sample will be affected by the study. Therefore, this situation calls for an experiment.

b. ELECTIONS A news organization wants to randomly call citizens to gauge opinions on a presidential election.

This situation calls for a survey because members of the sample population are asked for their opinion.

> **Guided Practice**

2A. RESEARCH A research company wants to study smokers and nonsmokers to determine whether 10 years of smoking affects lung capacity.

2A. Observational study; sample answer: The lung capacities of the participants are observed and compared without them being affected by the study.

2B. PETS A national pet chain wants to know whether customers would pay a small annual fee to participate in a rewards program. They randomly select 200 customers and send them questionnaires.

2B. Survey; sample answer: The data is obtained from opinions given by members of the sample population.

Differentiated Instruction **OL** **ELL**

Verbal/Linguistic Learners Divide students into small groups. Have each group design a survey question and practice asking it in such a way that there is bias built into the tone of voice and facial expression of the questioner. Then have them ask other groups the question and record their answers. As a class, discuss whether the answers corresponded to the bias that the question was designed to elicit.

Need Another Example?
Determine whether each situation calls for a survey, an experiment, or an observational study. Explain your reasoning.

a Video Games A gaming company plans to test whether a new controller is preferable to the old one. A group of teens will be observed while using the controllers to see which one they use the most. observational study; Sample answer: The teens will be observed without being affected by the study.

b Restaurants A restaurant wants to conduct an online study in which they will ask customers whether they were satisfied with their dining experience. survey; Sample answer: The members of the sample will be asked for their opinion.

2 Make Inferences

Make Inferences Once data have been collected using a random sample, you can analyze the data to calculate a sample statistic and make inferences about the population.

S.IC.1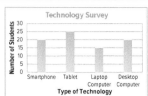

Real-World Example 3   Make an Inference About a Population

TECHNOLOGY A random sample of the 684 students at Sanchez High School were surveyed and asked to name the type of technology they would most like to use for learning a foreign language. Based on the results in the bar graph, what is the most reasonable inference about the number of students at Sanchez High School who would like to use a tablet for learning a foreign language?

Technology Survey

Number of Students (0, 5, 10, 15, 20, 25, 30) vs *Type of Technology* (Smartphone, Tablet, Laptop Computer, Desktop Computer)

Step 1 Determine the number of students in the random sample.

$$20 + 25 + 15 + 20 = 80$$

So, there were 80 students in the random sample.

Step 2 Calculate the sample proportion.

25 students in the random sample chose a tablet as their preferred technology.

$$\frac{25}{80} = 0.3125$$

The sample proportion is 0.3125.

Reading Math 🔵

📐 **Reasoning** To make an inference is to make a logical conclusion from given evidence. When you make an inference, be sure each step of the process can be justified.

Step 3 Use the sample proportion to make an inference about the population.

$$0.3125 \cdot 684 = 213.75$$

It is reasonable to infer that approximately 214 students at the school would most like to use a tablet for learning a foreign language.

▷ **Guided Practice**

3. What is the most reasonable inference about the number of students who would like to use a laptop computer for learning a foreign language? 128

Your inferences might not reflect your population if there is bias in the survey or sample.

S.IC.1

Real-World Example 4   Identify Bias

CLUBS Jamar surveyed a random sample of his school's film club. He found that 30 of the 40 members that he surveyed also play video games. He uses this survey to infer that, at his school of 800 students, approximately 600 students at the school play video games. Identify and explain any bias that might affect the validity of Jamar's inference.

Step 1 Identify the sample and population.

The sample is students in the film club at Jamar's school. The population is all of the students at Jamar's school, because that is the group he makes an inference about.

Step 2 Identify and explain potential bias.

Jamar takes a random sample of his school's film club, but makes an inference about all of the students at the school. Because students with a specific interest in film are in the club, they do not accurately reflect the student population, and thus Jamar's inference might be affected by the bias in his survey.

Need Another Example?

UNIFORMS A university wants to redesign the mascot it uses for each sports team. An administrator surveys a random sample of the football team to see whether the school's athletes want the same mascot with a new design, or a different mascot entirely, and finds that 10 of the 20 respondents want the same mascot. From this, the administrator determines that approximately 1532 of the university's 3064 athletes want the same mascot. Identify and explain any bias that might affect the validity of Jamar's inference. Because the football team is a specific group of athletes, they do not reflect the university's entire athlete population. Thus, the administrator's inference might be affected by the bias in his survey.

2 Make Inferences

Example 3 Make an Inference About a Population

🔴 **AL** How do you find the number of students who were surveyed? Find the height of each bar in the graph and add the heights.

🟢 **OL** What does the ratio $\frac{25}{80}$ represent in this problem? 25 out of the 80 students surveyed would most like to use a tablet for learning a foreign language.

🔵 **BL** What is the sample proportion of students who did not name a tablet as the type of technology they would most like to use for learning a foreign language? $\frac{55}{80}$

Need Another Example?

MUSIC A random sample of the 922 people at a concert were surveyed and asked to name the type of music they listen to most often at home. Based on the results of the bar graph, what is the most reasonable inference about the number of people at the concert who would say that jazz is the type of music they listen to most often at home? 97

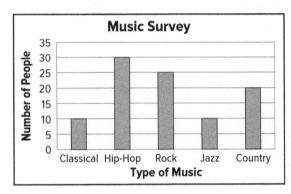

Music Survey

Number of People (0, 5, 10, 15, 20, 25, 30, 35) vs *Type of Music* (Classical, Hip-Hop, Rock, Jazz, Country)

Example 4 Identify Bias

🔴 **AL** Why would bias affect Jamar's inference? Bias means that the population is not accurately reflected by the sample, so inferences on the population will not be correct.

🟢 **OL** Are Jamar's calculations incorrect? No; his calculations are valid, but the calculations do not accurately reflect the intended population because the sample is not representative.

🔵 **BL** How could Jamar fix the bias in his question? He could perform a random sample of the entire student population.

Practice

Formative Assessment Use Exercises 1–6 to assess students' understanding of the concepts in this lesson.

The Practice and Problem Solving exercises assess the content taught in the lesson. The Preparing for Assessment page is meant to be used as preparation for end-of-course assessments.

Teaching the Mathematical Practices

Construct Arguments Mathematically proficient students understand and use stated assumptions, definitions, and previously established results in constructing arguments. They make conjectures and build a logical progression of statements to explore the truth of their conjectures. Remind students that the inferences they make in these exercises should be logically sound and that they should be able to justify each step of their solution process.

Extra Practice

See page R8 for extra exercises for students who are approaching level or for on-level students who need additional reinforcement.

Levels of Complexity Chart

The levels of the exercises progress from 1 to 3, with Level 1 indicating the lowest level of complexity.

Exercises	7–22	23, 24, 31–36	25–30
▶ Level 3			●
▶ Level 2		●	
Level 1	●		

Additional Answer

2. observational study; Sample answer: the participants in the study; population: potential customers

Go Online!   eBook

Interactive Student Guide

Use the *Interactive Student Guide* to deepen conceptual understanding.
· Designing Study

ALGEBRA 2
INTERACTIVE STUDENT GUIDE

▶ **Guided Practice**

4. **CARS** Sumitra wants to estimate how many cars in a mall parking lot have bumper stickers. She drives around the lot and takes random sample of the cars and finds that 30 of the 52 cars she sampled have bumper stickers, and thus determines that approximately 346 of the 600 cars in the whole lot have bumper stickers. Identify and explain any bias that might affect the validity of Sumitra's inference. There is no bias.

Go Online! for a Self-Check Quiz

Check Your Understanding   ○ = Step-by-Step Solutions begin on page R11.

Example 1
S.IC.3

Determine whether each situation describes a *survey*, an *experiment*, or an *observational study*. Then identify the sample, and suggest a population from which it may have been selected.

1. **SCHOOL** A group of high school students is randomly selected and asked to complete the form shown. survey; sample: the students in the study; population: the student body

2. **DESIGN** An advertising company wants to test a new logo design. They randomly select 20 participants and watch them discuss the logo. See margin.

Do you agree with the new lunch rules?
☐ agree
☐ disagree
☐ don't care

Example 2
S.IC.3

CONSTRUCT ARGUMENTS Determine whether each situation calls for a *survey*, an *experiment*, or an *observational study*. Explain your reasoning. 3–4. See Ch. 8 Answer Appendix.

3. **LITERACY** A literacy group wants to determine whether high school students that participated in a recent national reading program had higher standardized test scores than high school students that did not participate in the program.

4. **RETAIL** The research department of a retail company plans to conduct a study to determine whether a dye used on a new T-shirt will begin fading before 50 washes.

Example 3
S.IC.1

5. **EXERCISE** A random sample of the 4240 gym members at Work It! were surveyed about their favorite way to exercise outdoors. Based on the results in the graph, what is the most reasonable inference about the number of gym members whose favorite way to exercise outdoors is cycling? 909

Example 4
S.IC.1

6. **EGGS** An employee at a supermarket wants to see how many eggs in their latest shipment are broken. He checks the top carton in each stack of cartons and sees that 4 of the 72 eggs are broken. From this, he infers that approximately 24 of the 432 total eggs are broken. Identify any bias that might affect the validity of the employee's inference. Because he only checked the top carton in each stack, the eggs he checked do not reflect a random sample of the entire shipment of eggs. Thus, the employee's inference might be affected by bias.

Practice and Problem Solving   Extra Practice is on page R8.

Example 1
S.IC.3

Determine whether each situation describes a *survey*, an *experiment*, or an *observational study*. Then identify the sample, and suggest a population from which it may have been selected. 7–8. See Ch. 8 Answer Appendix.

⑦ **GRADES** A research group randomly selects 80 college students, half of whom took a physics course in high school, and compares their grades in a college physics course.

8. **FOOD** A grocery store conducts an online study in which customers are randomly selected and asked to provide feedback on their shopping experience.

Differentiated Homework Options

Levels	**AL** Basic	**OL** Core	**BL** Advanced
Exercises	7–22, 26–36	7–21 odd, 23, 24, 26–36	23–36
2-Day Option	7–21 odd, 31–36	7–22, 31–36	
	8–22 even, 26–30	23–30	

 You can use **ALEKS** to provide additional remediation support with personalized instruction and practice.

Example 2
S.IC.3

Determine whether each situation calls for a *survey*, an *experiment*, or an *observational study*. Explain your reasoning. 9–12. See margin.

9. FASHION A fashion magazine plans to poll 100 people in the United States to determine whether they would be more likely to buy a subscription if given a free issue.

10. TRAVEL A travel agency randomly calls 250 U.S. citizens and asks them what their favorite vacation destination is.

11. FOOD Chee wants to examine the eating habits of 100 random students at lunch to determine how many students eat in the cafeteria.

12. ENGINEERING An engineer is planning to test 50 metal samples to determine whether a new titanium alloy has a higher strength than a different alloy.

Example 3
S.IC.1

TOOTHPASTE A random sample of a dentist's 847 patients were surveyed about their favorite flavor of toothpaste. Use the results in the graph to determine the most reasonable inference about each of the following.

13. 199   13. the number of patients whose favorite flavor is spearmint

14. 149   14. the number of patients whose favorite flavor is cinnamon

15. 448   15. the number of patients whose favorite flavor is peppermint or spearmint

16. 648   16. the number of patients whose favorite flavor is not spearmint

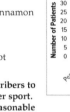

Toothpaste Flavors (bar graph; Number of Patients vs. Flavor: Peppermint, Spearmint, Cinnamon, Raspberry, Other)

WINTER SPORTS A random sample of the 5358 subscribers to a sports blog were asked to name their favorite winter sport. Use the results in the table to determine the most reasonable inference about each of the following.

17. 1674   17. the number of subscribers whose favorite winter sport is ice skating

18. 4395   18. the number of subscribers whose favorite winter sport is not skiing

19. the number of subscribers whose favorite winter sport is hockey or snowboarding   2721

20. the number of subscribers whose favorite winter sport is neither skiing nor hockey   3809

Favorite Winter Sports	
Sport	Frequency
skiing	23
hockey	14
snowboarding	51
ice skating	40

Example 4
S.IC.1

21. CUSTOMERS Gabrielle owns a café. One weekend, she hands out an anonymous survey at random to her customers asking how satisfied they were with their service. Of the 35 respondents, 18 stated they were extremely satisfied. So, she infers that approximately 257 of her average weekend crowd of 500 are extremely satisfied with their experiences. Identify any bias that might affect the validity of the employee's inference. There is no bias.

22. Because Miguel only sampled the modern art section, the paintings do not reflect all of the paintings in the museum. Thus, Miguel's inference might be affected by bias.

22. ART Miguel chose a random sample of 32 paintings in the modern art section and found that 12 were watercolors. Thus, he infers that approximately 171 of the museum's 455 paintings are watercolors. Identify any bias that might affect the validity of the employee's inference.

B 23 REPORTS The graph shown is from a report on the average number of minutes 8- to 18-year-olds in the United States spend on cell phones each day.   a, c–d. See margin.

a. Describe the sample and suggest a population.

b. What type of sample statistic do you think was calculated for this report?   average time

c. Describe the results of the study for each age group.

d. Who do you think would be interested in this type of report? Explain your reasoning.

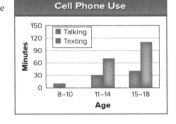

Cell Phone Use (bar graph; Minutes vs. Age: 8–10, 11–14, 15–18; Talking and Texting)

Additional Answers

9. Survey; Sample answer: The data will be obtained from opinions given by members of the sample population.

10. Survey; Sample answer: The data will be obtained from opinions given by members of the sample population.

11. Observational study; Sample answer: The eating habits of the participants will be observed and compared without them being affected by the study.

12. Experiment; Sample answer: Metal samples will need to be tested, which means that the members of the sample will be affected by the study.

23a. Sample answer: the 8- to 18-year-olds surveyed; population: all 8- to 18-year-olds in the United States

23c. Sample answer: The 8- to 10-year-old group talked for about 10 minutes a day and did not text at all. The 11- to 14-year-old group talked for about 30 minutes a day and texted for about 70 minutes a day. The 15- to 18-year-old group talked for about 40 minutes a day and texted for about 110 minutes a day.

23d. Sample answer: A cell phone company might use a report like this to determine which age group to target in their ads.

Go Online!

eSolutions Manual
Create worksheets, answer keys, and solutions handouts for your assignments.

Assess

Ticket Out the Door Have each student explain the main steps for making an inference about a population parameter when data from a random sample is given.

Additional Answers

24a. survey; Sample answer: the 2.4 million people polled, population: all U.S. citizens of voting age in 1936

24b. According to the predicted results, Landon should have won 57% of the popular vote. However, in the actual election, Roosevelt won 60.8% of the popular vote.

24c. Yes; Sample answer: The people sampled could afford magazine subscriptions, automobiles, and telephones, suggesting that they were wealthier than the average American citizen. The sampling method did not represent citizens that could not afford these things, and therefore, was not representative of the entire population.

24. (MP) MODELING In 1936, the *Literary Digest* reported the results of a statistical study used to predict whether Alf Landon or Franklin D. Roosevelt would win the presidential election that year. The sample consisted of 2.4 million Americans, including subscribers to the magazine, registered automobile owners, and telephone users. The results concluded that Landon would win 57% of the popular vote. The actual election results are shown.

ELECTORAL VOTE

POPULAR VOTE
Roosevelt 60.8%
Landon 36.5%

a–c. See margin.

a. Describe the type of study performed, the sample taken, and the population.
b. How do the predicted and actual results compare?
c. Do you think that the survey was biased? Explain your reasoning.

25. MULTIPLE REPRESENTATIONS The results of two experiments concluded that Product A is 70% effective and Product B is 80% effective. **a–d. See Ch. 8 Answer Appendix.**

randInt(0,9)

a. NUMERICAL To simulate the experiment for Product A, use the random number generator on a graphing calculator to generate 30 integers between 0 and 9. Let 0–6 represent an effective outcome and 7–9 represent an ineffective outcome.
b. TABULAR Copy and complete the frequency table shown using the results from part **a.** Then use the data to calculate the probability that Product A was effective. Repeat to find the probability for Product B.

Product A	
Number	Frequency
0–6	
7–9	

c. ANALYTICAL Compare the probabilities that you found in part **b.** Do you think that the difference in the effectiveness of each product is significant enough to justify selecting one product over the other? Explain.
d. LOGICAL Suppose Product B costs twice as much as Product A. Do you think the probability of the product's effectiveness justifies the price difference to a consumer? Explain.

S.IC.1, S.IC.3

H.O.T. Problems Use Higher-Order Thinking Skills

26, 28–30 See Ch. 8 Answer Appendix.

(MP) CONSTRUCT ARGUMENTS Determine whether each statement is *true* or *false*. If false, explain.

26. To save time and money, population parameters are used to estimate sample statistics.

27. Observational studies and experiments can both be used to study cause-and-effect relationships. true

28. OPEN-ENDED Design an observational study. Identify the objective of the study, define the population and sample, collect and organize the data, and calculate a sample statistic.

29 CHALLENGE What factors should be considered when determining whether a given statistical study is reliable?

30. WRITING IN MATH Research each of the following sampling methods. Then describe each method and discuss whether using the method could result in bias.
a. convenience sample   **c.** stratified sample
b. self-selected sample   **d.** systematic sample

(MP) Standards for Mathematical Practice

Emphasis On	Exercises
1 Make sense of problems and persevere in solving them.	25, 30, 32, 34, 35
2 Reason abstractly and quantitatively.	23, 26, 27, 36
4 Model with mathematics.	5, 6, 13–22, 24, 34–36
6 Attend to precision.	28, 29, 31, 33

Go Online!

The most up-to-date resources available for your program can be found at connectED.mcgraw-hill.com.

Preparing for Assessment

31. The principal of a school sends out a questionnaire to a random sample of 60 parents of students at the school and asks them to choose a date for back-to-school night. Which of the following is the best description for this study? ⬤ 6 S.IC.3 **D**

- ◯ **A** census
- ◯ **C** observational study
- ◯ **B** experiment
- ◯ **D** survey

32. Amani takes a random sample of the nuts in his snack mix and finds that 7 of the 11 nuts he chose are cashews. Because the bag has 50 nuts, he infers that approximately 32 of the nuts are cashews. Identify any bias that might affect the validity of the employee's inference. ⬤ 1 S.IC.1 **A**

- ◯ **A** There is no bias because the sample accurately represents the population.
- ◯ **B** There is bias because the sample is not random.
- ◯ **C** There is bias because the sample does not accurately reflect the population, which is all snack mixes.
- ◯ **D** There is bias because he incorrectly calculated the expected number of cashews in his snack mix.

33. Which of these situations call for an observational study? ⬤ 6 S.IC.3 **A, D**

- ☐ **A** A biologist wants to know how long penguins sleep each day.
- ☐ **B** A scientist at a skin-care company wants to know if a new sunscreen is more effective than the current version.
- ☐ **C** The editor of a magazine wants to know if subscribers like the new format of the magazine.
- ☐ **D** A cafeteria employee wants to know if students finish everything on their trays at lunchtime.
- ☐ **E** A veterinarian wants to know if flea powder X works better than flea powder Y.

34. Rich surveyed a random sample of 50 of the 695 employees at his company and found that 32 of them take a bus or train to get to work. Which is the most reasonable inference about the number of employees who take a bus or train to work? ⬤ 1,4 S.IC.1 **C**

- ◯ **A** 222
- ◯ **C** 445
- ◯ **B** 348
- ◯ **D** 1086

35. Morgan surveyed a random sample *s* of registered voters in her town. She asked them for which candidate for mayor they planned to vote. The results are shown in the table. There are *P* registered voters in Morgan's town. ⬤ 1,4 S.IC.1 **D**

Candidate	Frequency
Fernandez	f
Harrison	h
Taylor	t
Wong	w

Which expression can Morgan use to make an inference about the number of registered voters who plan to vote for Fernandez or Harrison?

- ◯ **A** $\frac{f+h}{2} \cdot P$
- ◯ **B** $\frac{f+h}{s}$
- ◯ **C** $\frac{s}{f+h} \cdot P$
- ◯ **D** $\frac{f+h}{s} \cdot P$

36c. The answer to part a would not change since the sample proportion is based only on the survey results; the answer to part b would be doubled since doubling the population doubles the number of subscribers who would choose Spain.

36. MULTI-STEP A random sample of the 952 subscribers to a travel blog were surveyed about the country they would most like to visit. The results are shown in the graph. ⬤ 2,4 S.IC.1

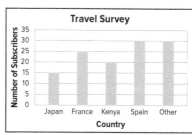

Travel Survey

- **a.** What is the sample proportion of subscribers who did not choose France? $\frac{19}{24}$
- **b.** What is the most reasonable inference about the number of subscribers who would choose Spain as the country they would most like to visit? 238
- **c.** Suppose the number of subscribers to the blog doubles but these survey results remain unchanged. Would your answers to parts **a** and **b** change? If so, explain how. If not, explain why not.

Preparing for Assessment

Exercises 31–36 require students to use the skills they will need on standardized assessments. Each exercise is dual-coded with content standards and mathematical practice standards.

	Dual Coding	
Items	Content Standards	⬤ Mathematical Practices
31	S.IC.3	6
32	S.IC.1	1
33	S.IC.3	6
34	S.IC.1	1, 4
35	S.IC.1	1, 4
36	S.IC.1	2, 4

Diagnose Student Errors

Survey student responses for each item. Class trends may indicate common errors and misconceptions.

31.

A	Did not recognize that not every member of the population is questioned
B	Did not recognize that the sample is not divided into two groups
C	Did not recognize that the members of the sample are not observed
D	CORRECT

33.

A	CORRECT
B	Did not recognize a situation that calls for an experiment
C	Did not recognize a situation that calls for a survey
D	CORRECT
E	Did not recognize a situation that calls for a survey
F	Did not recognize a situation that calls for an experiment

34.

A	Found 32% of 695
B	Found 50% of 695
C	CORRECT
D	Calculated the sample proportion as $\frac{50}{32}$

35.

A	Calculated the average frequency for Fernandez and Harrison instead of the sample proportion
B	Did not multiply the sample proportion by the population
C	Inverted the ratio for the sample proportion
D	CORRECT

Go Online!

Self-Check Quiz

Students can use *Self-Check Quizzes* to check their understanding of this lesson.

Using Statistical Experiments

SUGGESTED PACING (DAYS)

90 min. **0.5**

45 min. **1**

Instruction

Track Your Progress

Objectives

1 Collect and analyze data by conducting simulations of real-life situations.

2 Use data to compare theoretical and experimental probabilities.

Mathematical Background

A *probability model* is a mathematical model used to match a random phenomenon. A *simulation* recreates a situation again and again so that the likelihood of various outcomes can be estimated. Simulations can be constructed with dice, coin tosses, random number tables, and random number generators.

Skills Trace

THEN	NOW	NEXT
S.IC.1 Understand statistics as a process for making inferences about population parameters based on a random sample from that population.	**S.IC.2** Decide if a specified model is consistent with results from a given data-generating process, e.g., using simulation. **S.IC.5** Use data from a randomized experiment to compare two treatments; use simulations to decide if differences between parameters are significant.	**S.IC.4** Use data from a sample survey to estimate a population mean or proportion; develop a margin of error through the use of simulation models for random sampling. **S.IC.6** Evaluate reports based on data.

Go Online! All of these resources and more are available at connectED.mcgraw-hill.com

eLessons utilize the power of your interactive whiteboard in an engaging way. Use Simulations, Screens 2–5, to introduce the concepts in this lesson.

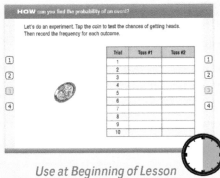

Use at Beginning of Lesson

Personal Tutors (for every example) let students hear real teachers solve problems. Students can pause and repeat as many times as necessary.

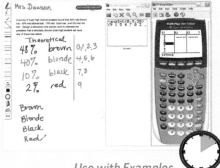

Use with Examples

eToolkit allows students to explore and enhance their understanding of math concepts. Students can use the Spinner, Number Cube, or Coin Toss to simulate a situation.

Use with Examples 1 and 3.

OER **Using Open Educational Resources**

Animations Have students watch the animation on **Math Live** about probability to reinforce their knowledge of simulations. This animation can also be used to introduce students to the concept of simulations. If you are unable to access Math Live, try **e-Learning for Kids** *Use as flipped learning or an instructional aid*

Go Online!
connectED.mcgraw-hill.com

Worksheets

Differentiate Your Resources

Extra Practice Additional practice or homework; Skills Practice is best for approaching-level students and Practice is best for on-level and beyond-level students

Skills Practice

Practice

Word Problem Practice

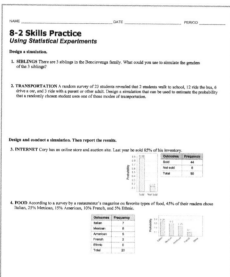

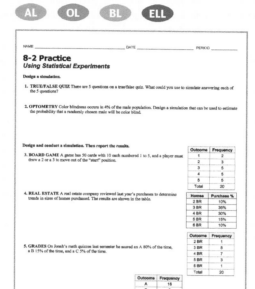

Intervention Reteaching and vocabulary activities that can be used with struggling or absent students and as ELL support

Study Guide and Intervention

Study Notebook

Extension Activities that can be used to extend lesson concepts

Enrichment

Launch

Have students read the Why? section of the lesson. Ask:

- If Alex took 50 kicks, on how many should he score? 33

- How did Alex's actual performance compare with his prediction? Sample answer: He made 70% of his kicks, which agrees with his prediction.

Teach

Ask the scaffolded questions for each example to build conceptual understanding for students at all levels.

1 Simulations

Example 1 Design a Simulation

AL How many possible outcomes could Eloy observe? 3; no defect, weld defect, design defect

OL What tool could we use to simulate these possible outcomes? Sample answer: a spinner

BL Are the three outcomes equally possible? Explain. no; Sample answer: he finds twice as many weld defects as design defects. He finds defects in 3 out of every 20 frames, so 17 out of 20 frames have no defect.

Need Another Example?

Softball Mandy is a pitcher on the softball team. Last season, 70% of her pitches were strikes. Design a simulation to estimate the probability that Mandy's next pitch is a strike. Sample answer: The theoretical probability that her next pitch is a strike is 70%, and the theoretical probability that it is not is 30%. Use a random number generator to generate integers 1 through 10. The integers 1–7 represent a strike, and the integers 8–10 represent not a strike. The simulation will consist of 50 trials.

Go Online!

Interactive Whiteboard

Use the *eLesson* or *Lesson Presentation* to present this lesson.

Using Statistical Experiments

Then	Now	Why?
You calculated simple probability.	**1** Collect and analyze data by conducting simulations of real-life situations. **2** Use data to compare theoretical and experimental probabilities.	Alex has been practicing his penalty kicks. Based on practice, he predicts that he scores on at least 66% of his kicks. To test this, he takes 50 penalty kicks, of which he scores on 35.

New Vocabulary
simulation
probability model
theoretical probability
experimental probability
relative frequency

MP Mathematical Practices
1 Make sense of problems.
2 Reason abstractly and quantitatively.
3 Construct viable arguments.
4 Model with mathematics.

Content Standards
S.IC.2 Decide if a specified model is consistent with results from a given data-generating process, e.g., using simulation.
S.IC.5 Use data from a randomized experiment to compare two treatments; use simulations to decide if differences between parameters are significant.

1 Simulations
An experiment that would be difficult or impractical to perform can be modeled by a **simulation**. In a simulation, a **probability model** to represent the **theoretical probability** is used to recreate a situation in which the **experimental probability** or **relative frequency** of an outcome can be found.

Example Event	Theoretical Probability Ratio of number of favorable outcomes to total number of outcomes	Experimental Probability (Relative Frequency) Ratio of number of outcomes in an experiment to total number of trials
Coin toss lands heads up	$\frac{1}{2}$, or 50%	3 heads land up out of 10 tosses $\frac{3}{10}$, or 30%

S.IC.2

Real-World Example 1   Design a Simulation

QUALITY CONTROL **Eloy** inspects bike frames as they come through the assembly line. From previous observations, he expects to find a weld defect in one out of every 10 frames and a design defect in one out of every 20 frames that he inspects. Design a simulation using Eloy's expectation of defects. Assume that a frame can only have one of the defects.

Step 1 Determine each possible outcome and its theoretical probability. There are three possible outcomes: weld defect, design defect, and no defects. Use Eloy's expectation of defects to calculate the theoretical probability of each outcome.

Possible Outcomes	Theoretical Probability
weld defect	10%
design defect	5%
no defects	85%

Step 2 Describe an appropriate probability model for the situation that accurately represents the theoretical probability of each outcome. We can use the random number generator on a graphing calculator. Assign the integers 0–19 to accurately represent the probability data.

Outcome	Represented by
weld defect	0, 1
design defect	2
no defects	3–19

Step 3 Define what a trial is for the situation, and state the number of trials to be conducted. A trial will represent selecting a frame at random. The simulation can consist of any number of trials. We will use 40.

MP Mathematical Practices Strategies

Use appropriate tools strategically.
Help students understand how to design a simulation, conduct a simulation, and report the findings. For example, ask:

- **How do you design a simulation?** You determine each possible outcome and its theoretical probability. Then you describe an appropriate probability model for the situation and define what a trial is for the simulation. You should also state the number of trials to be conducted.

- **How do you design a simulation that uses random numbers to generate data?** You assign a number to each of the events possible in the trial and then use technology or other means to generate randomly the numbers you need.

- **Why is a bar graph a good way to report the findings after conducting a simulation?** It gives a visual display of the results that is easy to interpret.

▷ **Guided Practice**

1. A survey asked students which method they would choose to travel to school each morning. The results are shown in the circle graph. Design a simulation that can be used to estimate the probability that a randomly chosen student will choose each of the four transportation methods.
See Ch 8. Answer Appendix.

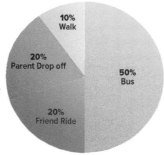

10%
Walk

20%
Parent Drop off

50%
Bus

20%
Friend Ride

Problem-Solving Tip

Use a Simulation
Simulations often provide a safe and efficient problem-solving strategy in situations that otherwise may be costly, dangerous, or impossible to solve using theoretical techniques. Simulations should involve data that are easier to obtain than the actual data you are modeling.

Simulations can also be conducted using number cubes, coin tosses, random number tables, and random number generators, such as those available on graphing calculators.

S.IC.2

Example 2 Design a Simulation by Using Random Numbers

EYE COLOR A survey of East High School students found that 40% had brown eyes, 30% had hazel eyes, 20% had blue eyes, and 10% had green eyes. Design a simulation that can be used to estimate the probability that a randomly chosen student will have one of these eye colors. Assume that the student's eye color will fall into one of these categories.

Step 1 Possible Outcomes | Theoretical Proability
brown eyes → 40%
hazel eyes → 30%
blue eyes → 20%
green eyes → 10%

Step 2 Describe an appropriate probability model for the situation.

Use the random number generator on your calculator. Assign the integers 0–9 to accurately represent the probability data. The actual numbers chosen to represent the outcomes do not matter.

Outcome	Represented by
brown eyes	0, 1, 2, 3
hazel eyes	4, 5, 6
blue eyes	7, 8
green eyes	9

Step 3 A trial will represent selecting a student at random and recording his or her eye color. The simulation will consist of 20 trials.

▷ **Guided Practice**

2. SOCCER Last season, Yao made 18% of his free kicks. Design a simulation using a random number generator that can be used to estimate the probability that he will make his next free kick. **See margin.**

Study Tip

Random Number Generator
To generate a set of random integers on a graphing calculator, press MATH and select randInt(under the PRB menu. Then enter the beginning and ending integer values for your range and the number of integers you want in each trial.

After designing a simulation, you will need to conduct the simulation and report the results. Include both numerical and graphical summaries of the simulation data, as well as an estimate of the probability of the desired outcome.

Example 2 Design a Simulation by Using Random Numbers

AL Could we use different numbers to represent brown eyes? Explain. Yes; Sample answer: As long as you use four numbers to represent brown eyes, it does not matter which ones are used.

OL Do you think experiments become more or less aligned with the theoretical probabilities as you increase the number of experiments? Sample answer: I think they would become more aligned, because with more data you would be able to see the trends more clearly.

BL Do you think random number generators are more likely to be closer to theoretical data than spinners? Explain. Sample answer: No; any well-designed experiment should simulate the theoretical probability.

Need Another Example?

Pizza A survey of Longmeadow High School students found that 30% preferred cheese pizza, 30% preferred pepperoni, 20% preferred peppers and onions, and 20% preferred sausage. Design a simulation that can be used to estimate the probability that a Longmeadow High School student prefers each of these choices. Sample answer: Use a random number generator to generate integers 0 through 9 where 0–2 represents cheese, 3–5 represents pepperoni, 6 and 7 represent peppers and onions, and 8 and 9 represent sausage. A trial will represent selecting a student at random and recording his or her pizza preference. The simulation will consist of 20 trials.

Additional Answer (Guided Practice)

2. Sample answer: The theoretical probability that he will make his next free kick is 18%. I will use a random number generator to generate integers 1 through 50. The integers 1–9 will represent a made free kick and the integers 10–50 will represent a missed free kick. The simulation will consist of 50 trials.

Differentiated Instruction ELL

Beginning To confirm students' understanding of simulations, provide images to highlight each step in the sequence of steps to design a simulation shown in the Concept Summary. Have students copy each term into their notes and write or draw their own image for the step.

Intermediate Use an IWB to project an example and mark text during reading to reinforce steps and concepts. Use different colors to represent the different steps in the sequence.

Advanced Have students write a paragraph summarizing the sequence of steps. They could create a numbered flowchart to help organize their paragraphs.

Advanced High Instruct collaborative groups to create a pamphlet illustrating the sequence of steps to design a simulation. Have students share their pamphlets with the class.

Example 3 Conduct and Evaluate a Simulation

AL Using the theoretical probability percents we determined, how many times would you expect each outcome if we do 40 trials? weld defect: 4 times; design defect: 2 times; no defect: 34 times

OL How close are our results to that of the theoretical probability? Sample answer: they are very close and the design defect is exactly the same.

BL If we did the simulation 80 times, do you think the experimental probability will be closer to or further away from the theoretical probability? Explain. closer to; Sample answer: the more trials we perform in an experiment, the closer the numbers will come to matching the theoretical probability.

Need Another Example?

Softball Conduct the simulation in the Need Another Example? for Example 1. Then report the results.
Sample answer: $P(\text{strike}) = 72\%$, $P(\text{ball}) = 28\%$

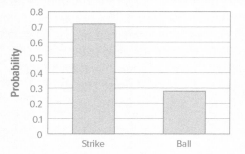

Additional Answer (Guided Practice)

3. Sample answer: $P(\text{Bus}) = 48\%$; $P(\text{Friend}) = 22\%$; $P(\text{Parent Drop Off}) = 18\%$; $P(\text{Walk}) = 12\%$.

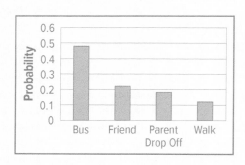

S.IC.2, S.IC.5

Example 3 Conduct and Evaluate a Simulation

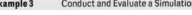

QUALITY CONTROL Refer to the simulation in Example 1. Conduct the simulation and report the results.

Press [MATH] [◄] and select [randInt (]. Then press 0 [,] 19 [,] 40 [)] [ENTER]. Use the left and right arrow buttons to view the results. Make a frequency table and record the results.

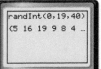
randInt(0,19,40)
{5 16 19 9 8 4 ...

Outcome	Tally	Frequency
weld defect	|||	3
design defect	||	2
no defects	||| ||| ||| ||| ||| ||| |||	35
Total		40

Calculate the experimental probability of finding each type of defect.

Weld defect $= \dfrac{\text{frequency}}{\text{total}} = \dfrac{3}{40}$ or 0.075

Design defect $= \dfrac{\text{frequency}}{\text{total}} = \dfrac{2}{40}$ or 0.05

No defects $= \dfrac{\text{frequency}}{\text{total}} = \dfrac{35}{40}$ or 0.875

The experimental probabilities that a frame will have a weld defect, a design defect, or no defects in this case are 7.5%, 5%, and 87.5%, respectively.

Make a bar graph of these results.

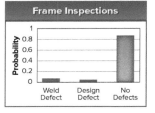

Frame Inspections

Real-World Career

Assemblers Assemblers may work as part of a team, where all members are capable of performing each task. In the automobile manufacturing industry, the median wage for a team assembler in 2013 was $30,554.
Source: Pay Scale

Guided Practice

3. Conduct the simulation in Guided Practice 1. Then report the results. See margin.

Concept Summary Designing, Conducting, and Reporting a Simulation

Designing a Simulation	Conducting a Simulation
Step 1 Determine each possible outcome and its theoretical probability.	**Step 1** Conduct the simulation.
Step 2 Describe an appropriate probability model for the situation.	**Step 2** Report the results using appropriate numerical and graphical summaries, including frequency tables and bar graphs.
Step 3 Define what a trial is for the situation and state the number of trials to be conducted.	**Step 3** Compare the experimental probability results with the theoretical probability.

Differentiated Instruction **AL** **OL**

IF students need more practice with probability simulations,

THEN give each student or pair of students a single number cube. Have them toss the cube 100 times and then create a frequency table to record the number of times each number comes up. Then have them calculate the experimental probability for each number and compare the results to the theoretical probabilities.

Example 4 Conduct and Summarize Data from a Simulation

S.IC.2, S.IC.5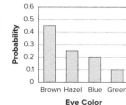

Refer to the simulation in Example 2. **Conduct the simulation and report the results using appropriate numerical and graphical summaries.**

Make a frequency table after using a graphing calculator to conduct the simulation 20 times.

Outcome	Tally	Frequency
brown eyes	ЖΙΙΙΙ	9
hazel eyes	ЖΙ	5
blue eyes	ΙΙΙΙ	4
green eyes	ΙΙ	2

Based on the simulation data, calculate the probability that a randomly selected student will have brown eyes.

$$\frac{\text{number of students with brown eyes}}{\text{total number of students}} = \frac{9}{20} \text{ or } 45\%$$

The experimental probability that a randomly selected student will have brown eyes is 45%. Notice that this is close to the theoretical probability, 40%. So, the experimental probability of not having brown eyes is $1 - 0.45$, or 55%.

Make a bar graph of the results.

Eye Color

▸ **Guided Practice**

4. Use a graphing calculator to conduct the simulation in Example 4 for 30 trials. Report the results using appropriate numerical and graphical summaries. Compare the experimental and theoretical probabilities. **See margin.**

The *Law of Large Numbers* states that as the number of trials of a random process increases, the experimental probability will approach the theoretical probability. The table at the right shows the result of running the simulation again for 40, 80, 150, and 200 trials. Notice the probability that a randomly chosen student having brown eyes is $\frac{81}{200}$, or 40.5%. This experimental probability is very close to the theoretical value of 40%.

Number of Trials	Frequency of Brown Eyes
40	20
80	30
150	58
200	81

Check Your Understanding   ◯ = Step-by-Step Solutions begin on page R11.

✓ **Go Online!** for a Self-Check Quiz

Examples 1, 3
S.IC.2,
S.IC.5

1 GRADES Clara got an A on 80% of her first semester Biology quizzes. Design and conduct a simulation using a geometric model to estimate the probability that she will get an A on a second semester Biology quiz. Report the results using appropriate numerical and graphical summaries. **See Ch. 8 Answer Appendix.**

Age	Frequency
0–7	13
8–12	28
13–17	48
18–23	42
24–31	27
32–45	40
46–64	33
65+	23

Examples 2–3
S.IC.2,
S.IC.5

2. FOOTBALL Rico is a pitcher on the baseball team. Last season, 72% of the pitches he threw were strikes.

a. Design a simulation that can be used to estimate the probability that Rico will throw a strike for his next pitch. **a, b. See Ch. 8 Answer Appendix.**

b. Conduct the simulation, and report the results.

Additional Answer (Guided Practice)

4. Sample answer: $P(\text{brown}) = 43\%$, $P(\text{hazel}) = 27\%$, $P(\text{blue}) = 17\%$, $P(\text{green}) = 13\%$

Outcome	Tally	Frequency
Brown eyes	ЖΙ ЖΙΙΙΙ	13
Hazel eyes	ЖΙΙΙΙ	8
Blue eyes	ЖΙ	5
Green eyes	ΙΙΙΙ	4

The experimental probability that a randomly selected student will have brown eyes is 43%. This is close to the theoretical probability, 40%.

Example 4 Conduct and Summarize Data from a Simulation

AL Why is the simulation data not exactly the same as the theoretical data? Sample answer: A simulation might not necessarily have results that are exactly like the theoretical probability, but if they are well designed, they should be close.

OL If you did a trial with only 10 spins, do you think the simulation results will be closer to the theoretical data? Explain. No; Sample answer: I think a smaller number of spins will result in the data being less aligned with the theoretical data.

BL What do you think would happen to the relationship between the simulation results and the theoretical probabilities as you added more spins? Explain. Sample answer: Because the simulation was designed using the theoretical probabilities, I think adding more spins will make the simulation results closer to the theoretical probabilities.

SOCCER Refer to the simulation in Need Another Example? for Guided Practice 2. Conduct the simulation using 50 trials and report the results, using appropriate numerical and graphical summaries. Compare the experimental and theoretical probabilities. Answers will vary, but should include a frequency table from the simulation and the corresponding bar graph. Sample answer: $P(\text{Made Kick}) = 22\%$, $P(\text{Missed Kick}) = 78\%$. The experimental probability is very close to the theoretical probability of 18%.

Outcome	Tally	Frequency
Made Kick	ЖΙ ЖΙ Ι	11
Missed Kick	ЖΙ ЖΙ ЖΙ ЖΙ ЖΙ ЖΙ ЖΙ ΙΙΙΙ	39

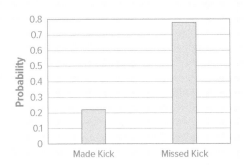

Practice

Formative Assessment Use Exercises 1–4 to assess students' understanding of the concepts in this lesson.

The Practice and Problem Solving exercises assess the content taught in the lesson. The Preparing for Assessment page is meant to be used as preparation for assessment.

Teaching the Mathematical Practices

Modeling Mathematically proficient students can apply the mathematics they know to solve problems arising in everyday life. In Exercises 7–10, encourage students to list the possible outcomes first.

Levels of Complexity Chart

The levels of the exercises progress from 1 to 3, with Level 1 indicating the lowest level of complexity.

Exercises	5–12	13, 14, 21–24	15–20
▷ Level 3			●
▷ Level 2		●	
Level 1	●		

Extra Practice

See page R8 for extra exercises for students who are approaching level or for on-level students who need additional reinforcement.

 Follow-Up

Students have explored simulations.

Ask:

• What should you consider when using the results of a simulation to make a prediction? Sample answers: the design of the simulation, how many trials were used, whether or not the theoretical and experimental probabilities are reasonably close

Additional Answers

11a. Sample answer: The theoretical probability that the player gets a hit is 30%, and the theoretical probability that he does not get a hit is 70%. Use a random number generator to generate integers 1 through 10. The integers 1–3 will represent a hit, and the integers 4–10 will represent the player not getting a hit. The simulation will consist of 50 trials.

---

3. FITNESS The table shows the percent of members participating in four classes offered at a gym. Design and conduct a simulation to estimate the probability that a new gym member will take each class. Report the results using appropriate numerical and graphical summaries. **See Ch. 8 Answer Appendix.**

Class	Sign-Up %
tae kwon do	45%
yoga	30%
swimming	15%
kick-boxing	10%

Examples 3–4
S.IC.2,
S.IC.5
4. Bridget is a member of the bowling club at her school. Last season, she bowled a strike 60% of the time.

   a. Design a simulation to estimate the probability that she will get a strike in the next frame of bowling. **See Ch. 8 Answer Appendix.**

   b. Conduct the simulation and report the results using appropriate numerical and graphical summaries. **See Ch. 8 Answer Appendix.**

   c. Compare the experimental probability to the theoretical probability.
The experimental probability is close to the theoretical probability of 60%.

Practice and Problem Solving   Extra Practice is on page R8.

Examples 1, 3   **Design and conduct a simulation using a probability model. Then report the results.**
S.IC.2,
S.IC.5   **⑤ VIDEO GAMES** Ian works at a video game store. Last year he sold 95% of the new-release video games. **See Ch. 8 Answer Appendix.**

  6. MUSIC Kadisha is listening to music from a playlist set to random. There are 10 songs in the playlist. **See Ch. 8 Answer Appendix.**

Examples 2–4   **MODELING** Design and conduct a simulation using a random number generator. Then
S.IC.2,   **report the results.**
S.IC.5
  7. MOVIES A movie theater reviewed sales from the previous year to determine which genre of movie sold the most tickets. The results are shown at the right. **See Ch. 8 Answer Appendix.**

Genre	Ticket %
drama	40%
mystery	30%
comedy	25%
action	5%

  8. BASEBALL According to a baseball player's on-base percentages, he gets a single 60% of the time, a double 25% of the time, a triple 10% of the time, and a home run 5% of the time. **See Ch. 8 Answer Appendix.**

  9. VACATION According to a survey done by a travel agency, 45% of their clients went on vacation to Europe, 25% went to Asia, 15% went to South America, 10% went to Africa, and 5% went to Australia. **See Ch. 8 Answer Appendix.**

 10. TRANSPORTATION A car dealership's analysis indicated that 35% of the customers purchased a blue car, 30% purchased a red car, 15% purchased a white car, 15% purchased a black car, and 5% purchased any other color. **See Ch. 8 Answer Appendix.**

 11. BATTING AVERAGE In a computer baseball game, a player has a batting average of .300. That is, he gets a hit 300 out of 1000, or 30%, of the times he is at bat.

   a. Design a simulation that can be used to estimate the probability that the player will get a hit at his next at bat. **a–b. See margin.**

   b. Conduct the simulation, and report the results.

 12. JEANS Julie examines the stitching on pairs of jeans that are produced at a manufacturing plant. She expects to find defects in 1 out of every 20 pairs.

   a. Design a simulation that can be used to estimate the probability that the next pair of jeans that Julie examines has a defect. **See margin.**

   b. Conduct the simulation, and report the results. **See Ch. 8 Answer Appendix.**

Differentiated Homework Options

Levels	**AL** Basic	**OL** Core	**BL** Advanced
Exercises	5–12, 16–24	5–11 odd, 13, 14, 17, 19, 21–24	13–24
2-Day Option	5–11 odd, 21–24	5–12, 21–24	
	6–12 even, 16–20	13–20	

You can use ALEKS to provide additional remediation support with personalized instruction and practice.

11b. Sample answer: P(hit) = 28%, P(not a hit) = 72%

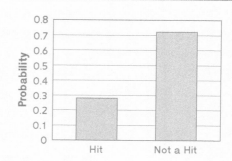

13. FOOD For a promotion, the concession stands at a football stadium are giving away free items. For every tenth customer, a wheel is spun to choose the customer's prize. Each prize is equally likely.

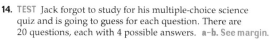

    a. Design a simulation that can be used to estimate the probability that the next spin is one of the five prizes. **a–b. See Chapter 8 Answer Appendix.**

    b. Conduct the simulation, and report the results.

14. TEST Jack forgot to study for his multiple-choice science quiz and is going to guess for each question. There are 20 questions, each with 4 possible answers. **a–b. See margin.**

    a. Design a simulation that can be used to estimate the number of questions that Jack answers correctly.

    b. Conduct the experiment from part **a** five times, and complete the table.

Simulation	Number of Correct Answers
1	
2	
3	
4	
5	

15. MULTIPLE REPRESENTATIONS In this problem, you will investigate expected value. **a–b. See margin. c–f. See Ch. 8 Answer Appendix.**

    a. Concrete Roll two dot cubes 20 times and record the sum of each roll.

    b. Numerical Use the random number generator on a calculator to generate 20 pairs of integers between 1 and 6. Record the sum of each pair.

    c. Tabular Copy and complete the table below using your results from parts **a** and **b**.

15g. Sample answer: They both have the most data points at the middle sums.

Trial	Sum of Roll	Sum of Output from Random Number Generator
1		
2		
...		
20		

    d. Graphical Use a bar graph to graph the number of times each possible sum occurred in the first 5 rolls. Repeat the process for the first 10 rolls and then all 20 outcomes.

    e. Verbal How does the shape of the bar graph change with each additional trial?

    f. Graphical Graph the number of times each possible sum occurred with the random number generator as a bar graph.

    g. Verbal How do the graphs of the cube trial and the random number trial compare?

S.IC.2, S.IC.5

H.O.T. Problems   Use Higher-Order Thinking Skills

16. OPEN-ENDED Describe a situation at your school that could be represented by a simulation. Then design the simulation. **See Ch. 8 Answer Appendix.**

17. JUSTIFY ARGUMENTS An experiment has three equally likely outcomes *A*, *B*, and *C*. Is it possible to use the spinner shown in a simulation to predict the probability of outcome *C*? Explain your reasoning. **See margin.**

120°

18. JUSTIFY ARGUMENTS Can tossing a coin *sometimes*, *always*, or *never* be used to simulate an experiment with two possible outcomes? Explain. **See margin.**

19. WRITING IN MATH What should you consider when using the results of a simulation to make a prediction? **See margin.**

20. WRITING IN MATH How is designing a simulation like solving a word problem?

20. See Ch. 8 Answer Appendix.

19. Sample answer: You should consider the design of the simulation, how many trials were used, and whether the theoretical and experimental probabilities are reasonably close.

Standards for Mathematical Practice

Emphasis On	Exercises
1 Make sense of problems and persevere in solving them.	11, 15
2 Reason abstractly and quantitatively.	19, 20
3 Construct viable arguments and critique the reasoning of others.	4, 8, 13, 17, 18
4 Model with mathematics.	1–3, 5–7, 9, 10, 12, 14, 16

Additional Answers

12a. Sample answer: The theoretical probability that the next pair of jeans has a defect is 5%, and the theoretical probability that the jeans do not have a defect is 95%. Use a random number generator to generate integers 1 through 20. The integer 1 will represent a pair of jeans with a defect, and the integers 2–20 will represent a pair of jeans without a defect. The simulation will consist of 40 trials.

14a. Sample answer: The theoretical probability of Jack answering a question correctly is 25%, and the theoretical probability of Jack answering a question incorrectly is 75%. Use a random number generator to generate the integers 1 through 4. The integer 1 will represent a correct answer, and the integers 2–4 will represent an incorrect answer. The simulation will consist of 20 trials.

14b. Sample answer:

Simulation	Number of Correct Answers
1	4
2	5
3	8
4	5
5	7

15a. Sample answer: 9, 10, 6, 6, 7, 9, 5, 9, 5, 7, 6, 5, 7, 3, 9, 7, 6, 7, 8, 7

15b. Sample answer: 4, 10, 5, 10, 6, 7, 12, 3, 7, 4, 7, 9, 3, 6, 4, 11, 5, 7, 5, 3

17. Yes; Sample answer: If the spinner were going to be divided equally into three outcomes, each sector would measure 120. Since you only want to know the probability of outcome *C*, you can record spins that end in the red area as a success, or the occurrence of outcome *C*, and spins that end in the blue area as a failure, or an outcome of *A* or *B*.

18. Sometimes; Sample answer: Flipping a coin can be used to simulate an experiment with two possible outcomes when both of the outcomes are equally likely. If the probabilities of the occurrence of the two outcomes are different, flipping a coin is not an appropriate simulation.

Go Online!

eSolutions Manual

Create worksheets, answer keys, and solutions handouts for your assignments.

Assess

Ticket Out the Door Have students explain why simulations are sometimes used instead of collecting actual data. Then have them give an example of an instance when a simulation is more practical than collecting actual data.

Preparing for Assessment

Exercises 21–24 require students to use the skills they will need on standardized assessments. Each exercise is dual-coded with content and process standards and mathematical practice standards.

	Dual Coding	
Items	Content Standards	⏱ Mathematical Practices
21	S.IC.2	4
22	S.IC.2, S.IC.5	4
23	S.IC.2, S.IC.5	4
24	S.IC.2, S.IC.5	4

Diagnose Student Errors

Survey student responses for each item. Class trends may indicate common errors and misconceptions.

22c.

A	Used numbers for theoretical probability instead of experimental probability
B	CORRECT
C	Chose percent for not making free throw
D	Used numbers for theoretical probability instead of experimental probability

Additional Answers

21a. Sample answer: Using a spinner with 8 equal regions.

21b. Sample answer: Cat 1 — 0.14, Cat 2 — 0.15, Cat 3 — 0.12, Cat 4 — 0.10, Cat 5 — 0.13, Cat 6 — 0.16, Cat 7 — 0.09, Cat 8 — 0.11

22a. Sample answer: Use a random number generator to generate integers 0 through 100 where

Preparing for Assessment

21. Pilar is playing a board game with eight different categories, each with questions that must be answered correctly to win. ⏱ 4 S.IC.2

  a. Design a simulation that can be used to estimate the probabilities of landing on the eight categories.   a, b. See margin.

  b. Conduct the simulation and report the results using appropriate numerical and graphical summaries.

  c. Use the data to compare the theoretical and experimental probabilities. Which statement is true for the results in part **b**?   A

  - ○ **A** The experimental and theoretical probabilities are close in value.

  - ○ **B** The experimental and theoretical probabilities are not close in value.

  - ○ **C** The experimental and theoretical probabilities are the same in value.

22. MULTI-STEP A basketball player made 62% of her free throws. ⏱ 4 S.IC.2, S.IC.5

  a. Design a simulation using 40 trials to estimate the probability that she will make her next free throw.   See margin.

  b. Which result(s) shown below may be a result of the simulation in part **a**? A, C

  ☐ **A**

Outcome	Tally	Frequency
make free throw	JHT JHT JHT JHT JHT I	26
miss free throw	JHT JHT IIII	14
Total		40

  ☐ **B**

Outcome	Tally	Frequency
1	TNL TNL TNL TNL TNL	25
2	TNL TNL TNL TNL	20
Total		45

  ☐ **C**

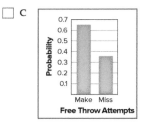

Free Throw Attempts

  c. What is the experimental probability that she will make her next free throw? ⏱ 4 B

  - ○ **A** 70%   ○ **C** 35%

  - ○ **B** 65%   ○ **D** 30%

  d. Use the data to compare the theoretical and experimental probabilities.   See margin.

  e. Explain how to change the simulation so that the experimental probability may be closer to the theoretical probability.   See margin.

23. A national survey showed that 70% of Americans aged 12 and older attended a movie in 2013. Which of the following are the best steps that could be used for a simulation to estimate the probability that the next attendee at a movie theater will be aged 12 or over? Select all that apply. ⏱ 4 S.IC.2, S.IC.5

  ☐ **A** Possible   Moviegoer is 12 or older—70% outcomes: Moviegoer is under 12—30%

  ☐ **B** Possible   Moviegoer is 12 years old—70% outcomes: Moviegoer is under 12—30%

  ☐ **C** Possible   Moviegoer is 12 or older—30% outcomes: Moviegoer is under 12—70%

  ☐ **D** Have theater employees ask attendees their age as they exit the theater. Make a frequency table with categories "12 and over" and "under 12."

  ☐ **E** Send a survey through the mail about the age of movie attendees and ask people to return the survey.

  ☐ **F** Conduct 20 trials.

  ☐ **G** Conduct 200 trials.   A, D, G

24. You roll a dot cube to see how many times you can roll a 1. You generate the following table. ⏱ 4 S.IC.2, S.IC.5

Trials	10	20	50	100
Frequency	3	4	12	18
Experimental Probability	0.3	0.2	0.24	0.18

  a. Describe the simulation that may have generated this table.   a–c. See margin.

  b. As the number of trials increase, what is happening to the experimental probability?

  c. Evaluate the simulation in part **a**. How would you describe the experimental probability of rolling a 1?

Differentiated Instruction BL

Extension Have students work in groups of three or four to design an experiment that has eight possible outcomes. Have them brainstorm ways to simulate the outcome and then select a method and build the simulator. Students should create a frequency table and compare the theoretical and experimental probabilities of the situation. Have students present their results to the class.

0–62 represents making the free throw and 63–100 represents not making the free throw. A trial will represent one free throw attempt.

22d. The experimental probability of 65% is close in value to the theoretical probability of 62%.

22e. Sample answer: Increase the number of trials to a very large number, like 200.

24a. Sample answer: Rolling a die *n* number of times where *n* represents the number of trials.

24b. The experimental probability approaches the theoretical probability.

24c. Sample answer: about 23%

Population Parameters

Track Your Progress

Objectives

1 Use data from sample surveys to estimate population means or proportions.

2 Develop margins of error by using simulation models.

Mathematical Background

Population parameters are used to make inferences based on random samples of a population. Parameters are obtained from collecting and analyzing data from a survey and can be used to estimate important values such as population mean and proportions.

THEN	NOW	NEXT
S.IC.1 Understand statistics as a process for making inferences about population parameters based on a random sample from that population.	**S.IC.4** Use data from a sample survey to estimate a population mean or proportion; develop a margin of error through the use of simulation models for random sampling.	**S.ID.4** Use the mean and standard deviation of a data set to fit it to a normal distribution and to estimate population percentages. Recognize that there are data sets for which such a procedure is not appropriate. Use calculators, spreadsheets, and tables to estimate areas under the normal curve.
S.IC.2 Decide if a specified model is consistent with results from a given data-generating process, e.g., using simulation.		
S.IC.3 Recognize the purposes of and differences among sample surveys, experiments, and observational studies; explain how randomization relates to each.		

Go Online! All of these resources and more are available at connectED.mcgraw-hill.com

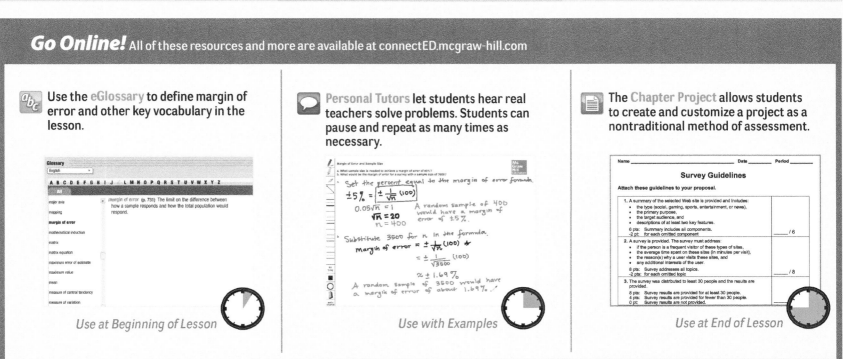

Use the eGlossary to define margin of error and other key vocabulary in the lesson.

Use at Beginning of Lesson

Personal Tutors let students hear real teachers solve problems. Students can pause and repeat as many times as necessary.

Use with Examples

The Chapter Project allows students to create and customize a project as a nontraditional method of assessment.

Use at End of Lesson

OER Using Open Educational Resources

Video Conferencing Have students work in groups to find videos about population mean and estimating the margin of error using **Khan Academy.** *Use as remediation*

Go Online!

connectED.mcgraw-hill.com

Worksheets

Differentiate Your Resources

Extra Practice Additional practice or homework; Skills Practice is best for approaching-level students and Practice is best for on-level and beyond-level students

Skills Practice

Practice

Word Problem Practice

Intervention Reteaching and vocabulary activities that can be used with struggling or absent students and as ELL support

Extension Activities that can be used to extend lesson concepts

Study Guide and Intervention

Study Notebook

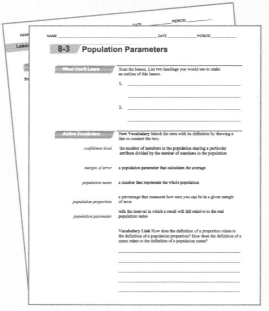

Enrichment

LESSON 3
Population Parameters

:Then	:Now	:Why?
You defined and calculated random samples of a population.	**1** Use data from sample surveys to estimate population means or proportions. **2** Develop margins of error by using simulation models.	A company wants to test out their new product before it launches. To save on cost, they conduct a survey on a sample of people who best represent the point of view of the general population.

New Vocabulary

confidence level
population mean
population parameters
population proportion
margin of error

Mathematical Practices
1 Make sense of problems and persevere in solving them.
4 Model with mathematics.
5 Use appropriate tools strategically.

Content Standards
S.IC.4 Use data from a sample survey to estimate a population mean or proportion; develop a margin of error through the use of simulation models for random sampling.

1 Estimate Population Means or Proportions A population parameter is a number that represents the whole population. These are usually unknown values and must be estimated from given data. The **population mean** is a population parameter that calculates the mean, or average, of the entire population. The population mean can be represented by μ. Another parameter is the **population proportion** p, which is the number of members in the population sharing a particular attribute divided by the number of members in the population.

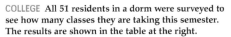
Key Concept Population Mean and Population Proportion

Five students with test scores 100, 100, 91, 84, and 75

Population Mean	**Population Proportion**
average of a sample survey	fraction or percentage of the sample survey that possess a particular trait
Formula	Proportion of test scores higher than 90:
$\mu = \frac{\Sigma x_i}{N}$	$p = \frac{\text{(test scores higher)}}{\text{(number of test)}}$
$= \frac{x_1 + x_2 + x_3 + \dots + x_N}{N}$	$p = \frac{3}{5}$
Σ means "the sum of."	$p = 0.60$
x = all the individual items in the group	
N = the number of items in the group	
Total sum of test scores divided by total number of test scores:	
$\mu = \frac{(100 + 100 + 91 + 84 + 75)}{5}$	
$\mu = \frac{450}{5} = 90$	

S.IC.4

Real-World Example 1 Use Data to Estimate Population Mean

COLLEGE All 51 residents in a dorm were surveyed to see how many classes they are taking this semester. The results are shown in the table at the right.

What is the population mean for the number of classes taken this semester?

Number of Classes	Number of Residents
1	6
2	5
3	26
4	11
5	3

Step 1 Sum the total number of classes taken by the residents. This represents the numerator of the formula Σx_i.

Mathematical Practices Strategies

Reason abstractly and quantitatively.
Help students analyze relationships among data sets in a population survey. For example, ask:

• How do you determine the average response of a population survey? Use the population mean formula.

• What do the population mean and population proportion have in common? They are examples of population parameters which calculate unknown values from a data set.

• Why do you multiply by 100 for the margin of error formula? The margin of error is a percentage, so it should be multiplied by 100 to give a percentage rather than a decimal value.

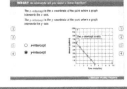

Teaching Tip

Structure Explain that the population mean is calculated in a way similar to the arithmetic mean or average is calculated.

Teaching the Mathematical Practices

Analyze Relationships Mathematically proficient students make sense of population parameters. They recognize that these values are unknown and must be estimated from the data that is given.

Example 2 Use Data to Estimate Population Proportion

AL **How is a proportion calculated?** The total number of members that share a common characteristic is divided by the total number of members.

OL **How can the given data be used to solve part a?** Divide the 90 people who prefer running by the entire population of 250.

BL **If the population proportion is 0.22, what is the number of people that prefer running over hiking?** 55 people

Need Another Example?

Of the 250 guests at a Halloween party, 215 people stated that they thought it was spooky. What is the population proportion of people who were spooked at the Halloween party? The population proportion p is 0.86.

Note that each term x_i represents the number of classes taken by resident i, so there should be 51 values that are summed.

$$\Sigma x_i = (1 \cdot 6) + (2 \cdot 5) + (3 \cdot 26) + (4 \cdot 11) + (5 \cdot 3) = 153$$

Study Tip

Precision Recall that a decimal can be converted to a percentage by multiplying the decimal value by 100.

Step 2 Divide the answer to Step 1 by the number of items in the data set. This is the N part of the population mean formula.

There are 51 residents, so:

$$\frac{153}{51} = 3$$

The population mean is 3 classes per resident.

Guided Practice

What is the population mean for the number of hours worked per day?

Number of Hours	Number of Employees
3	4
4	12
5	21

1. Step 1: $(4 \cdot 3) + (12 \cdot 4)$
   $+ (21 \cdot 5) = 165$

   Step 2: $\frac{165}{37} \approx 4.459459$ The population mean is approximately 4.5 hours.

S.IC.4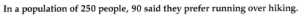

Example 2 Use Data to Estimate Population Proportion

In a population of 250 people, 90 said they prefer running over hiking.

 a. What is the population proportion of people who prefer running over hiking?

 b. What percent of the population prefers running to hiking?

Part a: Divide the number of people who prefer running by the total number of people in the data set. There are 250 people, so:

$$p = \frac{90}{250} = 0.36$$

Part b: A population proportion of 0.36 is equivalent to 36% of the population. Therefore, 36% of the population prefers running to hiking.

Guided Practice

Find the following population proportions.

2A. 105 people prefer running over hiking. 0.42

2B. 55 people prefer hiking over running. 0.22

2C. 140 people prefer hiking over running. 0.56

---

2 **Margin of Error** A survey of a random sample is a valuable tool for generalizing information about a larger population. The accuracy of a sample survey is based on two key factors—the **margin of error** and the confidence level. The margin of error tells you the interval in which the result will fall relative to the real population value. Statisticians have found that for large populations, the margin of error for a random sample of size n can be approximated by the formula, margin of error $= \pm \frac{1}{\sqrt{n}} (100)$.

Study Tip

Sense-Making The margin of error determines how reliable a survey is. The larger the sample size, the smaller the margin of error.

You can use the following formula to calculate the margin of error.

> **⚙ Key Concept** Margin of Error Formula
>
> $$\text{margin of error} = \pm \frac{1}{\sqrt{n}}(100)$$
>
> where n is the sample size

S.IC.4

Example 3   Calculate Margin of Error

A survey is conducted to determine how people will vote for the school presidential candidacy. The result of the 500 students surveyed showed that 62% will vote for Candidate B. Find the margin of error.

$$\frac{1}{\sqrt{n}} \cdot 100 = \frac{1}{\sqrt{500}} \cdot 100 \approx 4.47 \text{ or about } \pm 4.5\%$$

The real proportion is in the range 57.5% to 66.5%.

▸ **Guided Practice**

3. A survey is conducted to determine the most liked color of the rainbow. The results of the 200 people surveyed showed that 44% of people like the color blue the most. Find the margin of error.

S.IC.4

Real-World Example 4   Use Margin of Error to Find Sample Size

RESEARCH You are a member of a research team and are going to run a simulation. The simulation needs to result in an adequate sample size and margin of error for each trial.

a. Find the sample size you should use in your simulation for a margin of error of ±3%.

Substitute ±3% for the margin of error and solve for n in the margin of error formula.

$$\pm 3\% = \pm \frac{1}{\sqrt{n}}(100) \qquad \text{Margin of error formula}$$
$$0.03\sqrt{n} = 1 \qquad \text{Multiply by } \tfrac{\sqrt{n}}{100}$$
$$\sqrt{n} = 33.333 \qquad \text{Divide.}$$
$$n = 1111.11 \qquad \text{Square each side}$$

A sample size of about 1111 would have a margin of error of ±3%.

b. Determine the sample size that has a margin of error of ±10%.

Substitute ±10% for the margin of error and solve for n in the margin of error formula.

$$\pm 10\% = \pm \frac{1}{\sqrt{n}}(100) \qquad \text{Margin of error formula}$$
$$0.1\sqrt{n} = 1 \qquad \text{Multiply by } \tfrac{\sqrt{n}}{100}$$
$$\sqrt{n} = 10 \qquad \text{Divide.}$$
$$n = 100 \qquad \text{Square each side}$$

A sample size of 100 would have a margin of error of ±10%.

▸ **Guided Practice**

4. The finance director decides to conduct a survey with a margin of error of ±4%. Find the appropriate sample size.   625

Sidebar (left margin):

3. $\frac{1}{\sqrt{n}} \times 100$

$= \frac{1}{\sqrt{200}} \times 100 \approx 7.07$ or about ±7.1%.

The real proportion is in the range 36.9% to 51.1%.

Differentiated Homework Options

Levels	**AL** Basic	**OL** Core	**BL** Advanced
Exercises	5–19, 26–39	5–19 odd, 20, 23, 24, 32–39	20–39
2-Day Option	5–19 odd, 32–39	5–19, 32–39	
	6–18 even, 20–23	20–31	

 You can use ALEKS to provide additional remediation support with personalized instruction and practice.

2 Margin of Error

Example 3 Calculate Margin of Error

AL What is n in the margin of error equation? the sample size

OL From where did the range of 57.5% to 66.5% come? added and subtracted the margin of error to the percentage that will vote for Candidate B

BL How would the sample size need to change if the margin of error were to be ±3.2%? It would need to increase.

Need Another Example?

A car company is debating the release of a new feature for drivers. The executive of the company did a poll on 300 drivers and determined that 43% will consider purchasing a new vehicle if that feature came equipped. Find the margin of error.

$$\frac{1}{\sqrt{300}} \times 100 \approx \pm 5.8\%$$

Example 4 Use Margin of Error to Find Sample Size

AL How do you find the sample size when given a margin of error? Sample answer: Plug the value in the formula and solve for n.

OL For a smaller margin of error, should the research team increase or decrease the sample size? increase

BL Will the sample size of 100 be sufficient for the research team? No. The sample size of 100 gives a margin of error of ±10%, which is high.

Need Another Example?

What would be an adequate sample size for a margin of error of ±3.6%? $n = 771.6$; a random sample of about 770 would have a margin of error of ±3.6%.

> **MP Teaching the Mathematical Practices**
>
> **Sense-Making** Explain that when estimating a margin of error, choosing a random sample is important because it is expected that the behavior of the samples will be consistent across different samples.

Practice

Formative Assessment Use Exercises 1–4 to assess students' understanding of the concepts in the lesson.

The Practice and Problem Solving exercises assess the content taught in the lesson.

Extra Practice

See page R8 for extra exercises for students who are approaching level or for on-level students who need additional reinforcement.

Additional Answers

1. $(3 \times 10) + (5 \times 12) + (4 \times 15) + (2 \times 20) +$
    $(2 \times 25) = 240; \frac{240}{16} = 15$

3. Margin of error $= \frac{1}{\sqrt{100}} \times 100 = \pm 10\%$

4. 5% $= \frac{1}{\sqrt{n}} \times 100$
    $0.05\sqrt{n} = 1$
    $\sqrt{n} = 20$
    $n = 400$

20. The sample has a high margin of error of ±7.07%. With a larger sample size, the research center can produce a smaller margin of error which will result in better representation of all teens.

21. Fitness center A: 316.6, Fitness center B: 310.1. The claim is true.

Check Your Understanding   ○ = Step-by-Step Solutions begin on page R11.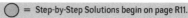

Example 1
S.IC.4
1. **MODELING** Jesse does community service keeping the local park clean. He noticed people who come to the park on a regular basis to play chess for several hours. Jesse surveyed some of the players to find the total number of hours they play each week. Use the table to calculate the population mean. See margin.

Number of Players	Hours per Week
3	10
5	12
4	15
2	20
2	25

Example 2
S.IC.4
2. A survey is conducted to find the most-watched TV network during the times of 5:00 P.M. to 10:00 P.M. Out of a sample of 175 people, 59 viewers are watching the news channel during this time frame. What is the population proportion of people who are watching the news channel? $p = \frac{59}{175} = 0.34$

Example 3
S.IC.4
3. For a sample size of 100 people, what is the margin of error that 53% of consumers will vote "yes" on a new product line from Company A to Z? See margin.

Example 4
S.IC.4
4. A university needs to perform an annual survey and obtain a margin of error of ±5%. What would be the best sample size to reach this goal? See margin.

Practice and Problem Solving   Extra Practice is on page R8.

Example 1
F.BF.3
5. **MANUFACTURING** A clothing manufacturer wants to begin selling their own products. They decide to run a small survey to determine how many items per color they currently manufacture at the warehouse. Find the population mean. $\mu = 52.5$

Color	Number of items
Yellow	13
Red	76
Blue	23
Black	98

Example 2
F.BF.3
For a sample survey of 500 employees, find the following population proportions.

6. 75 prefer the weekend shift $p = 0.15$

7. 270 prefer to stay on site for lunch $p = 0.54$

8. 165 prefer the morning shift $p = 0.33$

9. 300 prefer a flexible schedule $p = 0.60$

Example 3
F.IF.7a,
F.BF.3
Compute the margin of error for each sample size.

10. 100   margin of error $= \pm 10\%$ 11. 300   margin of error $= \pm 5.77\%$

12. 500   margin of error $= \pm 4.47\%$ 13. 755   margin of error $= \pm 3.64\%$

14. 1000   margin of error $= \pm 3.16\%$ 15. 2500   margin of error $= \pm 2\%$

Example 4
S.IC.4
Find the sample size for each margin of error.

16. ±2% $n = 2500$ (17) ±3% $n = 1{,}111$

18. ±6% $n = 278$ 19. ±7% $n = 204$

20. **CONSTRUCT ARGUMENTS** The Pew Research Center recently conducted a survey of a random sample of 200 teens and concluded that 43% of all teens who take their cell phones to school text in class on a daily basis. How accurately did their random sample represent all teens? See margin.

21. **CRITIQUE ARGUMENTS** Fitness center A claims that more of their members burn more calories per day than at the competing fitness center B. A random sample of 10 people from each center results in the following calories burned. See margin.

Levels of Complexity Chart

The levels of the exercises progress from 1 to 3, with Level 1 indicating the lowest level of complexity.

Exercises	5–19	20–23, 32–39	24–31
▶ Level 3			●
▶ Level 2		●	
Level 1	●		

Fitness center A: 391 326 322 297 326 289 293 264 327 331

Fitness center B: 311 304 321 302 307 297 311 313 336 299

Estimate the population mean of calories burned by the members at each fitness center. Does the claim appear to be legitimate?

22. CEREAL A company is conducting a small survey about a new cereal that they want to release next summer. Ten customers produced the following ratings for the cereal after a sample.

Rating	1	2	3	4
Number of Customers	1	3	4	2

a. Estimate the population mean for these ratings.

b. What is the margin of error?

c. What does the margin of error say about the poll?

d. What sample size will produce a margin of 4%?

22. a. $\mu = 2.7$
b. margin of error $= \pm 31.6\%$
c. Sample answer: Due to the large margin of error, the polls cannot be used as the general belief of the entire population.
d. $n = 625$

 23. APPLES A survey of 250 kids was conducted to determine their favorite type of apple between Granny Smith and Golden Delicious apples. It showed that 43% prefer Granny Smith apples. Calculate the margin of error. $\pm 6.3\%$

24. A survey of 1600 customers at a travel agency showed that 66% would prefer to live in a cooler climate. What is the true percent range that those surveyed will choose to live in a cooler climate? With a margin of error of $\pm 2.5\%$, the range is 63.5% to 68.5%.

25. The Department of Education surveyed 20,000 students to determine the graduation rate over a course of five years at a local high school. The results showed that 86% of seniors graduated. Determine the real percentage range of graduated students. With a margin of error of $\pm 0.7\%$, the range is 85.3% to 86.7%.

S.IC.4

H.O.T. Problems   Use Higher-Order Thinking Skills

26. As the sample size increases, the margin of error decreases.

27. Add the sum of all ratings and divide that by 100, the number of candy bars.

26. What happens to the margin of error if the sample size increases?

27. Describe how to calculate the population mean rating for 100 candy bars.

28. The Warriors cross-country team is preparing for a national run. Each team member runs laps around a track and records their time in order to monitor their progress.

28a. $\mu = 170.8$ 28b. $p = \frac{2}{5} = 0.4$

Team Members	Running Time (in secs)
Runner 1	180
Runner 2	172
Runner 3	185
Runner 4	168
Runner 5	149

a. Calculate the mean of the runners' running time.

b. What proportion of the team runs 170 seconds or faster?

29. DOWNTOWN The Mountain Hill High School student council conducted a survey of 300 students and reported that 250 of them hang out in the downtown area on Friday and Saturday nights. Estimate the population proportion of the students who hang out downtown. $p = \frac{250}{300} = 0.83$

30. A survey showed that 250 out of 500 adults prefer coffee over tea. Determine the population proportion of the adults who prefer coffee and estimate the margin of error. See margin.

31. Every attendee at the carnival filled out a short survey before leaving. Out of 1750 people, 75% said they will refer the carnival to a friend next year. What is the population proportion of attendees that will refer the carnival for the following year? $p = 0.75$

Standards for Mathematical Practice

Emphasis On	Exercises
1 Make sense of problems and persevere in solving them.	2, 3, 6–15, 21, 24–26, 29–31, 33–37, 39
4 Model with mathematics.	1, 5, 28, 32
5 Use appropriate tools strategically.	4, 16–20, 22, 23, 27, 38

⊕ Follow-Up

Students have explored the ways to describe a population.

Ask:

● **What strategy could be used to test the accuracy of the size of a random sample?**
Apply the size of the sample n to the margin of error formula.

Assess

Ticket Out the Door Have each student conduct a survey of their choosing, and identify the population mean, proportion, and margin of error.

Additional Answer

30. $p = \frac{250}{500} = 0.5$
Margin of error $= \pm 4.5\%$

Preparing for Assessment

Exercises 32–39 require students to use the skills they will need on assessments. Each exercise is dual-coded with content.

Dual Coding		
Items	Content Standards	🔵 Mathematical Practices
32	S.IC.4	4
33	S.IC.4	1
34	S.IC.4	1
35	S.IC.4	1
36	S.IC.4	1
37	S.IC.4	1
38	S.IC.4	5
39	S.IC.4	1

Diagnose Student Errors

34. Students may make errors by leaving out the square root in the margin of error formula. Remind students of the proper structure of the margin of error formula.

35.

A	CORRECT
B	Population mean is the average.
C	Margin of error is a percentage.
D	All the needed information is provided.

36.

A	Made error when calculating mean
B	CORRECT
C	Made error when calculating mean
D	Made error when calculating mean

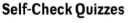
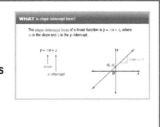
Preparing for Assessment

32. In a fifth grade class, 26 students were asked how many children their parents have. The results are shown in the frequency table. Estimate the mean number of children. 🔵 4 S.IC.4

Children	Frequency
1	5
2	9
3	8
4	4

$$\mu = \frac{(5\times 1)+(9\times 2)+(8\times 3)+(4\times 4)}{26}$$
$$= \frac{63}{26}$$
$$= 2.42$$

33. In Nomax County, 3000 residents are registered to vote. Out of this population of voters, 40% are registered for Party A. What is the population proportion of registered voters who belong to Party A? 🔵 1 S.IC.4 **B**

- ○ **A** $p = 0.3$
- ○ **B** $p = 0.4$
- ○ **C** $p = 0.6$
- ○ **D** $p = 0.12$
- ○ **E** $p = 0.16$
- ○ **F** $p = 0.20$

34. A random sample of middle school students revealed that 15% of children were experiencing reading difficulties. 🔵 1 S.IC.4

a. Find the margin of error if the sample size was 325. margin of error = ±5.5%

b. Find the margin of error if the sample size was 500. margin of error = ±4.5%

35. Every student taking Geology 101 was asked how they felt about the last exam. Twenty percent of the 300 students surveyed found it to be difficult. Which of the following statements below is true about the value 0.20? 🔵 1 S.IC.4 **A**

- ○ **A** It is the population proportion.
- ○ **B** It is the population mean.
- ○ **C** It is the margin of error.
- ○ **D** Is is the confidence interval.
- ○ **E** More information is needed.

36. At a day care, there are 17 children under the age of 4. The ages, in months, are: 19 28 25 24 26 21 20 27 36 35 30 22 23 33 13 34 12. What is the population mean age of children at the day care? 🔵 1 S.IC.4 **B**

- ○ **A** 22
- ○ **B** 25.2
- ○ **C** 26.5
- ○ **D** 30

37. 30 children were given a random amount of marbles. The table represents the number of marbles each child possesses. What information can be gathered from the data? Select all that apply. 🔵 1 S.IC.4 **A, C**

Marbles	Children
2	5
3	19
4	8

- ☐ **A** proportion of children with 3 marbles
- ☐ **B** proportion of children under the age of 10 with marbles
- ☐ **C** mean number of marbles
- ☐ **D** mean number of children without marbles

39c. For better results, Cassy's best option is to survey 473 homeowners with a margin of error of ±4.6%.

38. Explain how to estimate the population mean of the number of times that each player attempts to beat the enemy in Level 5 of a computer game. 🔵 5 S.IC.4
Calculate the total number of times each player played level 5, collectively. Divide that sum by the total number of players.

39. MULTI-STEP Cassy wants to conduct a survey of the amount of water consumed per day for homeowners in her area. Because of the large population, she wants to determine the best sample size with a low margin of error. 🔵 1 S.IC.4

a. What is the margin of error if she surveys 150 homeowners? margin of error = ±8.2%

b. What is the sample size if she wants to accomplish a margin of error of ±4.6%? Round to the nearest whole number. n = 473

c. Use the answers above to determine the better option for Cassy. Explain.

37.

A	CORRECT
B	The table does not include information about the children's ages.
C	CORRECT
D	The table does not include information about children without marbles.

38. Students may incorrectly explain the process of identifying a population mean. Remind students of the proper structure of the margin of error formula.

39. Students may incorrectly apply the margin of error formula. Encourage them to consider the information that is given and identify the unknowns.

Distributions of Data

Track Your Progress

Objectives

1 Use the shapes of distributions to select appropriate statistics.

2 Use the shapes of distributions to compare data.

Mathematical Background

A distribution of data shows the frequency of each possible data value. The shape of a distribution can be determined by looking at its histogram or box-and-whisker plot.

THEN	NOW	NEXT
S.IC.5 Use data from a randomized experiment to compare two treatments; use simulations to decide if differences between parameters are significant.	**S.IC.1** Understand statistics as a process for making inferences about population parameters based on a random sample from that population.	**S.IC.6** Evaluate reports based on data.

Go Online! All of these resources and more are available at connectED.mcgraw-hill.com

Personal Tutors (for every example) let students hear real teachers solve problems. Students can pause and repeat as many times as necessary.

Use with Examples

Use **Self-Check Quiz** to assess students' understanding of the concepts in this lesson.

Use at End of Lesson

Animations illustrate key concepts through step-by-step tutorials and videos.

Use with Examples

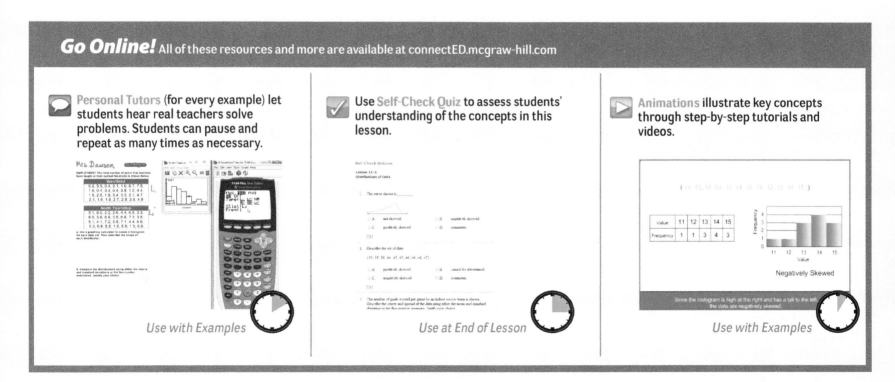

☉ER Using Open Educational Resources

Surveys Have students create a poll on Polldaddy to collect data. Then students can describe the distribution created by that data and use the shape of the distribution to select appropriate statistics. *Use as homework*

Go Online!
connectED.mcgraw-hill.com   Worksheets

Differentiate Your Resources

Extra Practice Additional practice or homework; Skills Practice is best for approaching-level students and Practice is best for on-level and beyond-level students

Skills Practice

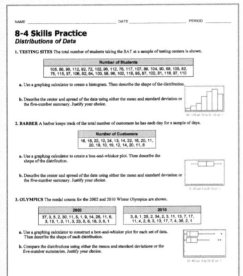

Practice

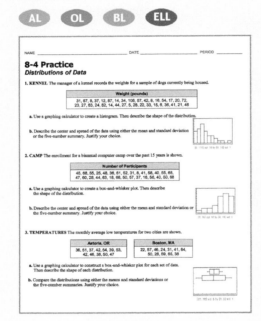

Word Problem Practice

Intervention Reteaching and vocabulary activities that can be used with struggling or absent students and as ELL support

Extension Activities that can be used to extend lesson concepts

Study Guide and Intervention

Study Notebook

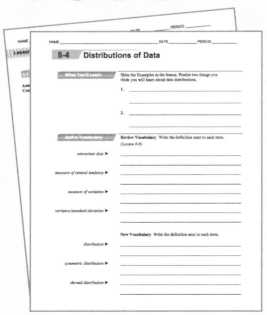

Enrichment

LESSON 4

Distributions of Data

:Then	:Now	:Why?
• You calculated measures of central tendency and variation.	**1** Use the shapes of distributions to select appropriate statistics. **2** Use the shapes of distributions to compare data.	After four games as a reserve player, Craig joined the starting lineup and averaged 18 points per game over the *remaining* games. Craig's scoring average for the *entire* season was less than 18 points per game as a result of the lack of playing time in the first four games.

New Vocabulary

distribution

negatively skewed distribution

symmetric distribution

positively skewed distribution

 Mathematical Practices

1 Make sense of problems and perservere in solving them.

7 Look for and make use of structure.

Content Standards
S.IC.1 Understand statistics as a process for making inferences about population parameters based on a random sample from that population.

1 Analyzing Distributions A **distribution** of data shows the observed or theoretical frequency of each possible data value. In Lesson 0-9, you described distributions of sample data using statistics. You used the mean or median to describe a distribution's center and standard deviation or quartiles to describe its spread. Analyzing the shape of a distribution can help you decide which measure of center or spread best describes a set of data.

The shape of the distribution for a set of data can be seen by drawing a curve over its histogram.

🔑 Key Concept Symmetric and Skewed Distributions

Negatively Skewed Distribution	Symmetric Distribution	Positively Skewed Distribution
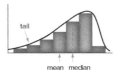		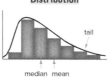
• The mean is less than the median. • The majority of the data are on the right of the mean.	• The mean and median are approximately equal. • The data are evenly distributed on both sides of the mean.	• The mean is greater than the median. • The majority of the data are on the left of the mean.

When a distribution is symmetric, the mean and standard deviation accurately reflect the center and spread of the data. However, when a distribution is skewed, these statistics are not as reliable. Recall that outliers have a strong effect on the mean of a data set, while the median is less affected. Similarly, when a distribution is skewed, the mean lies away from the majority of the data toward the tail. The median is less affected, so it stays near the majority of the data.

When choosing appropriate statistics to represent a set of data, first determine the skewness of the distribution.

• If the distribution is relatively symmetric, the mean and standard deviation can be used.

• If the distribution is skewed or has outliers, use the five-number summary to describe the center and spread of the data.

Launch

Have students read the Why? section of the lesson. Ask:

• **Why would Craig's scoring average for the entire season be less than 18 points per game?** Sample answer: If Craig did not receive much playing time as a reserve player, he probably did not score many points during the first four games. As a result, these lower scores would bring down his average.

• **Is Craig's scoring average for the entire season a good representation of his scoring average? Explain.** Sample answer: No; his scoring average for the entire season is likely lower than it would have been if Craig had been in the starting lineup for the entire season.

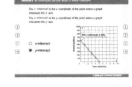

Teach

Ask the scaffolded questions for each example to build conceptual understanding for students at all levels.

1 Analyzing Distributions

Example 1 Describe a Distribution Using a Histogram

AL What values are in the five-number summary? the minimum, Q1, the median, Q3, the maximum

OL By looking at the histogram, what can you determine about the relationship between the median and the mean? The mean is greater than the median.

BL Describe the tails of a distribution that is skewed. If the distribution is negatively skewed there is a longer tail to the left and if the distribution is positively skewed there is a longer tail to the right.

Need Another Example?

Presentations Ms. Shroyer's students each gave a presentation as part of their class project. The length of each presentation is shown in the table below.

Time (minutes)				
17	13	11	17	20
15	23	7	16	10
13	20	12	21	14
17	19	20	18	19

a. Use a graphing calculator to create a histogram. Then describe the shape of the distribution. See bottom margin for graph; negatively skewed

b. Describe the center and spread of the data using either the mean and standard deviation or the five-number summary. Justify your choice. Sample answer: The distribution is skewed, so use the five-number summary. The range is 7 to 23 minutes. The median is 17 minutes, and half of the times are between 13 and 19.5 minutes.

Real-World Link
The first portable computer, the Osborne I, was available for sale in 1981 for $1795. The computer weighed 24 pounds and included a 5-inch display. Laptops can now be purchased for as little as $199 and can weigh as little as 2.3 pounds.
Source: Computer History Museum

1A.

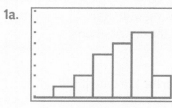

[15, 42] scl: 3 by [0, 8] scl: 1
symmetric

1B. Sample answer: The distribution is symmetric, so use the mean and standard deviation. The mean rainfall was 29 inches with standard deviation of 5.4 inches.

Additional Answer (Additional Example)

1a.

[3, 24] scl: 3 by [0, 8] scl: 1

Real-World Example 1 Describe a Distribution Using a Histogram

LAPTOPS The prices for a random sample of laptops are shown.

Price (dollars)							
723	605	847	410	440	386	572	523
374	915	734	472	420	508	613	659
706	463	470	752	671	618	538	425
811	502	490	552	390	512	389	621

a. Use a graphing calculator to create a histogram. Then describe the shape of the distribution.

First, press STAT ENTER and enter each data value. Then, press 2nd [STAT PLOT] ENTER ENTER and choose ▥. Finally, adjust the window to the dimensions shown.

The majority of the laptops cost between $400 and $700. Some of the laptops are priced significantly higher, forming a tail for the distribution on the right. Therefore, the distribution is positively skewed.

[0, 1000] scl: 100 by [0, 10] scl: 1

b. Describe the center and spread of the data using either the mean and standard deviation or the five-number summary. Justify your choice.

The distribution is skewed, so use the five-number summary to describe the center and spread. Press STAT ▶ ENTER ENTER and scroll down to view the five-number summary.

The prices for this sample range from $374 to $915. The median price is $530.50, and half of the laptops are priced between $451.50 and $665.

1-Var Stats
↑n=32
minX=374
Q1=451.5
Med=530.5
Q3=665
maxX=915

Guided Practice

1. **RAINFALL** The annual rainfall for a region over a 24-year period is shown below.

   A. Use a graphing calculator to create a histogram. Then describe the shape of the distribution.

   B. Describe the center and spread of the data using either the mean and standard deviation or the five-number summary. Justify your choice.

Annual Rainfall (in.)					
27.2	30.2	35.8	26.1	39.3	20.6
28.9	23.0	32.7	26.8	22.7	25.4
29.6	36.8	33.4	28.4	21.9	20.8
24.7	30.6	27.7	31.4	34.9	37.1

A box-and-whisker plot can also be used to identify the shape of a distribution. The position of the line representing the median indicates the center of the data. The "whiskers" show the spread of the data. If one whisker is considerably longer than the other and the median is closer to the shorter whisker, then the distribution is skewed.

Teaching Tip

Skewed Distributions Students may confuse negatively and positively skewed distributions. Remind them that when the tail is on the left of a distribution, the data appear to be going uphill, and since it is harder to go uphill, the distribution is negatively skewed. When the tail is on the right of a distribution, the data appear to be going downhill, and since it is easier to go downhill, the distribution is positively skewed.

Go Online!

You can use the box-and-whisker plot in the eToolkit to create box-and-whisker plots from given data.

Key Concept Box-and-Whisker Plots as Distributions

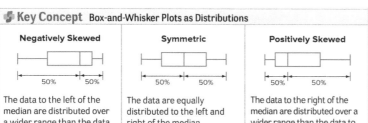

Negatively Skewed	Symmetric	Positively Skewed
50% 50%	50% 50%	50% 50%
The data to the left of the median are distributed over a wider range than the data to the right. The data have a tail to the left.	The data are equally distributed to the left and right of the median.	The data to the right of the median are distributed over a wider range than the data to the left. The data have a tail to the right.

S.IC.1

Example 2   Describe a Distribution Using a Box-and-Whisker Plot

HOMEWORK The students in Mr. Fejis' language arts class found the average number of minutes that they each spent on homework each night.

Minutes per Night					
62	53	46	66	38	45
52	46	73	39	42	56
64	54	48	59	70	60
49	54	48	57	70	33

a. Use a graphing calculator to create a box-and-whisker plot. Then describe the shape of the distribution.

Enter the data as L1. Press [2nd] [STAT PLOT] [ENTER] [ENTER] and choose ⊡⋯. Adjust the window to the dimensions shown.

The lengths of the whiskers are approximately equal, and the median is in the middle of the data. This indicates that the data are equally distributed to the left and right of the median. Thus, the distribution is symmetric.

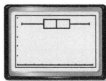

[30, 75] scl: 5 by [0, 5] scl: 1

b. Describe the center and spread of the data using either the mean and standard deviation or the five-number summary. Justify your choice.

The distribution is symmetric, so use the mean and standard deviation to describe the center and spread. The average number of minutes that a student spent on homework each night was 53.5 with standard deviation of about 10.5.

```
1-Var Stats
x̄=53.5
Σx=1284
Σx²=71320
Sx=10.68521937
σx=10.46024219
↓n=24
```

2A.

[500, 750] scl: 25 by [0, 5] scl: 1

negatively skewed

Watch Out!

Standard Deviation Recall from Lesson 0-9 that the formulas for standard deviation for a population σ and for a sample s are slightly different. In Example 2, times for all of the students in Mr. Fejis' class are being analyzed, so use the population standard deviation.

2B. Sample answer: The distribution is skewed, so use the five-number summary. Janet's minutes range from 511 minutes to 695 minutes. The median is 655.5 minutes, and half of the data are between 616.5 and 670.5 minutes.

Guided Practice

2. CELL PHONE Janet reviewed the number of minutes she spent talking to her friends each month for the last two years.

Minutes Used per Month			
582	608	670	620
667	598	671	613
537	511	674	627
638	661	642	641
668	673	680	695
658	653	670	688

A. Use a graphing calculator to create a box-and-whisker plot. Then describe the shape of the distribution.

B. Describe the center and spread of the data using either the mean and standard deviation or the five-number summary. Justify your choice.

Example 2 Describe a Distribution Using a Box-and-Whisker Plot

AL What values are used to plot a box-and-whisker plot? the five-number summary

OL How could the mean compare to the median in this example? They would be close to equal.

BL How do the whiskers on a box-and-whisker plot compare to a histogram? The whiskers will be skewed in the same way as the tails on the histogram.

Need Another Example?

Wages The hourly wages for a random sample of employees of a restaurant are shown in the table.

Wages ($)				
10.50	7.75	8.25	9.50	6.50
7.50	11.25	7.25	6.50	8.25
8.00	7.50	6.75	7.25	9.00
7.50	10.00	7.25	8.00	9.00

a. Use a graphing calculator to create a box-and-whisker plot. Then describe the shape of the distribution.

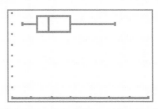

[6, 13] scl: 1 by [0, 8] scl: 1

positively skewed

b. Describe the center and spread of the data using either the mean and standard deviation or the five-number summary. Justify your choice.

Sample answer: The distribution is skewed, so use the five-number summary. The range is $6.50 to $11.25. The median is about $7.88, and half of the data are between $7.25 and $9.00.

Differentiated Instruction AL OL BL ELL

Interpersonal Learners Have students work in pairs to think of examples of data that may have distributions that are symmetric, negatively skewed, or positively skewed.

2 Comparing Distributions

Example 3 Compare Data Using Histograms

AL What types of data sets are easiest to compare? two symmetric data sets

OL If two data sets are both symmetric, how can you determine which set has the larger mean? the histogram that has its highest point more to the right

BL Why would it be important to use the same scales when comparing two data sets? If you use two different scales you could make conclusions about the data that aren't true.

Need Another Example?

Games Tyler and Jordan are working through several brainteasers on the computer. The time in minutes that it took to complete each game is shown.

Tyler (minutes)				
5.1	2.9	3.5	6.0	2.8
3.4	4.1	3.8	4.3	6.1
5.8	5.1	6.0	3.4	4.6
4.4	3.6	4.8	3.1	5.2

Jordan (minutes)				
4.1	5.4	3.8	2.9	6.4
3.1	3.7	5.3	4.5	2.7
3.4	4.3	4.2	5.8	4.7
6.1	5.9	5.4	4.4	3.9

a. Use a graphing calculator to create a histogram for each data set. Then describe the shape of each distribution. See bottom margin for graphs; Tyler, positively skewed; Jordan, symmetric

b. Compare the distributions using either the means and standard deviations or the five-number summaries. Justify your choice. Sample answer: One distribution is symmetric and the other is skewed, so use the five-number summaries. The median for both sets is 4.35 but 50% of Tyler's times occur between 3.45 and 5.15, while 50% of Jordan's times occur between 3.75 and 5.4. The smaller interquartile range for Jordan may suggest that she was slightly more consistent than Tyler.

3A.

3rd Period

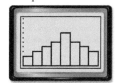

[20, 50] scl: 4 by [0, 10] scl: 1

6th Period

[20, 50] scl: 4 by [0, 10] scl: 1

3rd period, symmetric; 6th period, positively skewed

Study Tip

MP Tools To compare two sets of data, enter one set as L1 and the other as L2. In order to calculate statistics for a set of data in L2, press

STAT ▶ ENTER
2nd [L2] ENTER.

3B. Sample answer: One distribution is symmetric and the other is skewed, so use the five-number summaries. The range for 3rd period is 23, and the range for 6th period is 25. However, the median for 3rd period is 33, and the median for 6th period is 27. The lower quartile for 3rd period is 28. Because this is greater than the median for 6th period, this means that 75% of the speeds for 3rd period are greater than 50% of the speeds for 6th period. Therefore, we can conclude that 3rd period had slightly better typing speeds overall.

---

2 Comparing Distributions To compare two sets of data, first analyze the shape of each distribution. Use the mean and standard deviation to compare two symmetric distributions. Use the five-number summaries to compare two skewed distributions or a symmetric distribution and a skewed distribution.

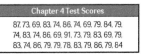
S.IC.1

Example 3	Compare Data Using Histograms

TEST SCORES Test scores from Mrs. Morash's class are shown below.

Chapter 3 Test Scores	Chapter 4 Test Scores
81, 81, 92, 99, 61, 67, 86, 82, 76, 73, 62, 97, 97, 72, 72, 84, 77, 88, 92, 93, 76, 74, 66, 78, 76, 69, 84, 87, 83, 87, 92, 87, 82	87, 73, 69, 83, 74, 86, 74, 69, 79, 84, 79, 74, 83, 74, 86, 69, 91, 73, 79, 83, 69, 79, 83, 74, 86, 79, 79, 78, 83, 79, 86, 79, 84

a. Use a graphing calculator to create a histogram for each data set. Then describe the shape of each distribution.

Chapter 3 Test Scores

[60, 100] scl: 5 by [0, 10] scl: 1

Chapter 4 Test Scores

[60, 100] scl: 5 by [0, 10] scl: 1

Both distributions are symmetric.

b. Compare the distributions using either the means and standard deviations or the five-number summaries. Justify your choice.

The distributions are symmetric, so use the means and standard deviations.

Chapter 3 Test Scores

```
1-Var Stats
x̄=81
Σx=2673
Σx²=219763
Sx=10.07782219
σx=9.923953269
↓n=33
```

Chapter 4 Test Scores

```
1-Var Stats
x̄=79
Σx=2607
Σx²=207085
Sx=5.947688627
σx=5.856878888
↓n=33
```

The Chapter 4 test scores, while lower in average, have a much smaller standard deviation, indicating that the scores are more closely grouped about the mean. Therefore, the mean for the Chapter 4 test scores is a better representation of the data than the mean for the Chapter 3 test scores.

▶ **Guided Practice**

3. TYPING The typing speeds of the students in two classes are shown below.

A. Use a graphing calculator to create a histogram for each data set. Then describe the shape of each distribution.

B. Compare the distributions using either the means and standard deviations or the five-number summaries. Justify your choice.

3rd Period (wpm)	6th Period (wpm)
23, 38, 27, 28, 40, 45, 32, 33, 34, 27, 40, 22, 26, 34, 29, 31, 35, 33, 37, 38, 28, 29, 39, 42	38, 26, 43, 46, 23, 24, 27, 36, 22, 21, 26, 27, 31, 32, 27, 25, 23, 22, 28, 29, 28, 33, 23, 24

Additional Answers (Additional Example)

3a.

Tyler

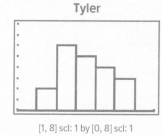

[1, 8] scl: 1 by [0, 8] scl: 1

Jordan

[1, 8] scl: 1 by [0, 8] scl: 1

(MP) Teaching the Mathematical Practices

Tools Mathematically proficient students consider the available tools when solving a mathematical problem. Remind students to select the correct data set when performing calculations.

Box-and-whisker plots can be displayed alongside one another, making them useful for side-by-side comparisons of data.

S.IC.1

Example 4   Compare Data Using Box-and-Whisker Plots

POINTS The points scored per game by a football team for the 2013 and 2014 football seasons are shown.

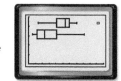

2013							
7	51	24	27	17	35	27	33
28	30	27	21	24	30	14	20

2014							
20	9	3	10	6	14	3	10
3	37	7	21	13	41	20	23

a. Use a graphing calculator to create a box-and-whisker plot for each data set. Then describe the shape of each distribution.

Enter the 2013 scores as **L1**. Graph these data as **Plot1** by pressing [2nd] [STAT PLOT] [ENTER] [ENTER] and choosing ⊡⊶. Enter the 2014 scores as **L2**. Graph these data as **Plot2** by pressing [2nd] [STAT PLOT] [▼] [ENTER] [ENTER] and choosing ⊡⊶. For **Xlist**, enter **L2**. Adjust the window to the dimensions shown.

For the 2013 scores, the left whisker is longer than the right and the median is closer to the right whisker. The distribution is negatively skewed.

[0, 55] scl: 5 by [0, 5] scl: 1

For the 2014 scores, the right whisker is longer than the left and the median is closer to the left whisker. The distribution is positively skewed.

b. Compare the distributions using either the means and standard deviations or the five-number summaries. Justify your choice.

The distributions are skewed, so use the five-number summaries to compare the data.

The lower quartile for the 2013 season and the upper quartile for the 2014 season are both 20.5. This means that 75% of the scores from the 2013 season were greater than 20.5 and 75% of the scores from the 2014 season were less than 20.5.

The minimum of the 2013 season is approximately equal to the lower quartile for the 2014 season. This means that 25% of the scores from the 2014 season are lower than any score achieved in the 2013 season. Therefore, we can conclude that the team scored a significantly higher amount of points during the 2013 season than the 2014 season.

Guided Practice

4. GOLF Robert recorded his golf scores for his sophomore and junior seasons.

A. Use a graphing calculator to create a box-and-whisker plot for each data set. Then describe the shape of each distribution.

B. Compare the distributions using either the means and standard deviations or the five-number summaries. Justify your choice.

Sophomore Season
42, 47, 43, 46, 50, 47, 52, 45, 53, 55, 48, 39, 40, 49, 47, 50

Junior Season
44, 38, 46, 48, 42, 41, 42, 46, 43, 40, 43, 43, 44, 45, 39, 44

4A.

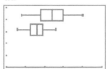

[35, 60] scl: 5 by [0, 5] scl: 1
both symmetric

4B. Sample answer: The distributions are symmetric, so use the means and standard deviations. The mean score for Robert's sophomore season is about 47.1 with standard deviation of about 4.4. The mean score for Robert's junior season is 43 with standard deviation of about 2.6. The lower mean and standard deviation from Robert's junior season indicates that he not only improved, but was also more consistent.

Example 4 Compare Data Using Box-and-Whisker Plots

(AL) Which data set has the larger range? the data set from 2014

(OL) Why is it easier to compare two data sets with box-and-whisker plots? You can plot them together on one graph.

(BL) Will a data set with a larger mean always have a larger maximum value? Use these data sets to support your answer. No, the 2013 data has a larger mean, but the 2014 data set has a larger maximum value.

Need Another Example?

Temperatures The daily high temperatures over a 20-day period for two cities are shown.

Clintonville (°F)				
60	55	68	63	70
59	63	61	54	68
66	57	72	65	67
65	62	58	72	58

Stockton (°F)				
61	63	62	61	64
63	64	60	65	63
63	65	62	60	64
64	63	66	62	66

a. Use a graphing calculator to create a box-and-whisker plot for each data set. Then describe the shape of each distribution.

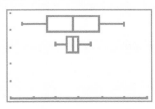

[52, 76] scl: 4 by [0, 5] scl: 1

both symmetric

b. Compare the distributions using either the means and standard deviations or the five-number summaries. Justify your choice. Sample answer: The distributions are symmetric, so use the means and standard deviations. The mean temperature for Clintonville is about 63.15° with standard deviation of about 5.40°. The mean temperature for Stockton is about 63.05° with standard deviation of about 1.76°. The average temperatures for both cities are about the same, but the lower standard deviation for Stockton means that the temperatures there are more consistently near 63° than at Clintonville.

Practice

Formative Assessment Use Exercises 1–4 to assess students' understanding of the concepts in this lesson.

The Practice and Problem Solving exercises assess the content taught in the lesson. The Preparing for Assessment page is meant to be used as preparation for assessment.

Exercise Alert

See page R8 for extra exercises for students who are approaching level or for on-level students who need additional reinforcement.

Levels of Complexity Chart

The levels of the exercises progress from 1 to 3, with Level 1 indicating the lowest level of complexity.

Exercises	5–10	11, 12, 17–22	13–16
Level 3			●
Level 2		●	
Level 1	●		

Check Your Understanding ◯ = Step-by-Step Solutions begin on page R11.

 Go Online! for a Self-Check Quiz

Example 1
S.IC.1

1 **EXERCISE** The amount of time that James ran on a treadmill for the first 24 days of his workout is shown. **a–b.** See margin.

Time (minutes)											
23	10	18	24	13	27	19	7	25	30	15	22
10	28	23	16	29	26	26	22	12	23	16	27

a. Use a graphing calculator to create a histogram. Then describe the shape of the distribution.

b. Describe the center and spread of the data using either the mean and standard deviation or the five-number summary. Justify your choice.

Example 2
S.IC.1

2. **RESTAURANTS** The total number of times that 20 random people either ate at a restaurant or bought fast food in a month are shown. **a–b.** See margin.

Restaurants or Fast Food									
4	7	5	13	3	22	13	6	5	10
7	18	4	16	8	5	15	3	12	6

a. Use a graphing calculator to create a box-and-whisker plot. Then describe the shape of the distribution.

b. Describe the center and spread of the data using either the mean and standard deviation or the five-number summary. Justify your choice.

Example 3
S.IC.1

3. **TOOLS** The total fundraiser sales for the students in two classes are shown. **a–b.** See Ch. 8 Answer Appendix.

Mrs. Johnson's Class (dollars)					
6	14	17	12	38	15
11	12	23	6	14	28
16	13	27	34	25	32
21	24	21	17	16	

Mr. Edmunds' Class (dollars)					
29	38	21	28	24	33
14	19	28	15	30	6
31	23	33	12	38	28
18	34	26	34	24	37

a. Use a graphing calculator to create a histogram for each data set. Then describe the shape of each distribution.

b. Compare the distributions using either the means and standard deviations or the five-number summaries. Justify your choice.

Example 4
S.IC.1

4. **RECYCLING** The weekly totals of recycled paper for the junior and senior classes are shown. **a–b.** See margin.

Junior Class (pounds)					
14	24	8	26	19	38
12	15	12	18	9	24
12	21	9	15	13	28

Senior Class (pounds)					
25	31	35	20	37	27
22	32	24	28	18	32
25	32	22	29	26	35

a. Use a graphing calculator to create a box-and-whisker plot for each data set. Then describe the shape of each distribution.

b. Compare the distributions using either the means and standard deviations or the five-number summaries. Justify your choice.

Differentiated Homework Options

Levels	AL Basic	OL Core	BL Advanced
Exercises	5–10, 13–22	5–9 odd, 11, 12, 13, 17–22	11–22
2-Day Option	5–9 odd, 17–22	5–10, 17–22	
	6–10 even, 13–16	11–16	

 You can use **ALEKS** to provide additional remediation support with personalized instruction and practice.

Go Online!

eBook

Interactive Student Guide

Use the *Interactive Student Guide* to deepen conceptual understanding.
· Distributions of Data

 ALGEBRA 2
INTERACTIVE STUDENT GUIDE

Practice and Problem Solving

Extra Practice is on page R8.

Examples 1–2
S.IC.1

For Exercises 5 and 6, complete each step.

a. Use a graphing calculator to create a histogram and a box-and-whisker plot. Then describe the shape of the distribution.

b. Describe the center and spread of the data using either the mean and standard deviation or the five-number summary. Justify your choice.

5. FANTASY The weekly total points of Kevin's fantasy football team are shown. **a–b. See Ch. 8 Answer Appendix.**

Total Points							
165	140	88	158	101	137	112	127
53	151	120	156	142	179	162	79

6. MOVIES The students in one of Mr. Peterson's classes recorded the number of movies they saw over the past month. **a–b. See Ch. 8 Answer Appendix.**

Movies Seen											
14	11	17	9	6	11	7	8	12	13	10	9
5	11	7	13	9	12	10	9	15	11	13	15

Example 3
S.IC.1

MODELING For Exercises 7 and 8, complete each step.

a. Use a graphing calculator to create a histogram for each data set. Then describe the shape of each distribution.

b. Compare the distributions using either the means and standard deviations or the five-number summaries. Justify your choice. **7–8. See Ch. 8 Answer Appendix.**

7. SAT A group of students took the SAT their sophomore year and again their junior year. Their scores are shown.

Sophomore Year Scores					
1327	1663	1708	1583	1406	1563
1637	1521	1282	1752	1628	1453
1368	1681	1506	1843	1472	1560

Junior Year Scores					
1728	1523	1857	1789	1668	1913
1834	1769	1655	1432	1885	1955
1569	1704	1833	2093	1608	1753

8. INCOME The total incomes for 18 households in two neighboring cities are shown.

Yorkshire (thousands of dollars)					
68	59	61	78	58	66
56	72	86	58	63	53
68	58	74	60	103	64

Applewood (thousands of dollars)					
52	55	60	61	55	65
65	60	45	37	41	71
50	61	65	66	87	55

Example 4
S.IC.1

9 TUITION The annual tuitions for a sample of public and private colleges are shown. Complete each step. **See Ch. 8 Answer Appendix.**

a. Use a graphing calculator to create a box-and-whisker plot for each data set. Then describe the shape of each distribution.

b. Compare the distributions using either the means and standard deviations or the five-number summaries. Justify your choice.

Public Colleges (dollars)					
5760	7304	8230	6248	9064	9794
7155	8736	7344	6640	6960	6869
6283	5978	5760	8480	9211	6207
7630	7328	6664	7462	9152	6558

Private Colleges (dollars)					
26,770	32,665	11,664	10,804	12,297	15,835
14,250	4200	18,000	17,400	20,910	20,670
12,240	6000	16,360	23,600	13120	14,976
6800	30,586	9108	9600	9000	21,450

4a.

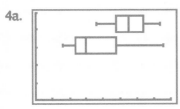

[0, 40] scl: 5 by [0, 5] scl: 1

junior class, positively skewed; senior class, symmetric

4b. Sample answer: One distribution is symmetric and the other is skewed, so use the five-number summaries. The median for the junior class is 15, and the median for the senior class is 27.5. The minimum value for the senior class is 18. This means that all of the weekly totals for the senior class are greater than 50% of the weekly totals for the junior class. Therefore, we can conclude that the senior class's weekly totals were far greater than the junior class's weekly totals.

Additional Answers

1a.

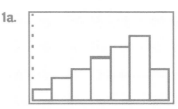

[4, 32] scl: 4 by [0, 8] scl: 1

negatively skewed

1b. Sample answer: The distribution is skewed, so use the five-number summary. The times range from 7 to 30 minutes. The median is 22.5 minutes, and half of the data are between 15.5 and 26 minutes.

2a.

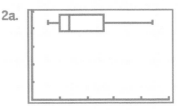

[0, 25] scl: 5 by [0, 5] scl: 1

positively skewed

2b. Sample answer: The distribution is skewed, so use the five-number summary. The data range from 3 to 22 times. The median number is 7 times, and half of the data are between 5 and 13 times.

Assess

Ticket Out the Door Have students generate a set of data, create a histogram using the data, and describe the shape of the distribution. Then have the students describe the center and spread of the data using either the mean and standard deviation or the five-number summary.

Teaching the Mathematical Practices

Construct Arguments Mathematically proficient students understand and use stated assumptions, definitions, and previously established results in constructing arguments. They make conjectures and build a logical progression of statements to explore the truth of their conjectures.

Additional Answers

11a.

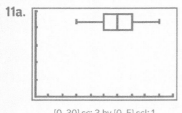

[0, 30] sc: 3 by [0, 5] scl: 1

Sample answer: The distribution is symmetric, so use the mean and standard deviation. The mean of the data is 18 with sample standard deviation of about 5.2 points.

11b.

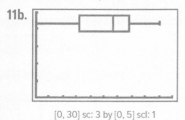

[0, 30] sc: 3 by [0, 5] scl: 1

mean: 14.6; median: 17

11c. Sample answer: Adding the scores from the first four games causes the shape of the distribution to go from being symmetric to being negatively skewed. Therefore, the center and spread should be described using the five-number summary.

13a. Sample answer: mean ≈ 14; median ≈ 10

13b. Sample answer: mean ≈ 20; median ≈ 24

13c. Sample answer: mean ≈ 17; median ≈ 17

---

10. DANCE The total amount of money that a random sample of seniors spent on prom is shown. Complete each step. **See Ch. 8 Answer Appendix.**

  a. Use a graphing calculator to create a box-and-whisker plot for each data set. Then describe the shape of each distribution.

  b. Compare the distributions using either the means and standard deviations or the five-number summaries. Justify your choice.

Boys (dollars)					
253	288	304	283	348	276
322	368	247	404	450	341
291	260	394	302	297	272

Girls (dollars)					
682	533	602	504	635	541
489	703	453	521	472	368
562	426	382	668	352	587

11a–c. See margin.

11. BASKETBALL Refer to the beginning of the lesson. The points that Craig scored in the remaining games are shown.

  a. Use a graphing calculator to create a box-and-whisker plot. Describe the center and spread of the data.

  b. Craig scored 0, 2, 1, and 0 points in the first four games. Use a graphing calculator to create a box-and-whisker plot that includes the new data. Then find the mean and median of the new data set.

  c. What effect does adding the scores from the first four games have on the shape of the distribution and on how you should describe the center and spread? **12a–b. See Ch. 8 Answer Appendix.**

Points Scored			
18	10	18	21
9	25	13	17
17	12	24	19
20	17	27	21

12. SCORES Allison's quiz scores are shown.

  a. Use a graphing calculator to create a box-and-whisker plot. Describe the center and spread.

  b. Allison's teacher allows students to drop their two lowest quiz scores. Use a graphing calculator to create a box-and-whisker plot that reflects this change. Then describe the center and spread of the new data set.

Math Quiz Scores					
83	76	86	82	84	57
86	62	90	96	76	89
76	88	86	86	92	94

S.IC.1

H.O.T. Problems Use Higher-Order Thinking Skills

13. CHALLENGE Approximate the mean and median for each distribution of data. **a–c. See margin.**

a. b. c.

14. CONSTRUCT ARGUMENTS Distributions of data are not always symmetric or skewed. If a distribution has a gap in the middle, like the one shown, two separate clusters of data may result, forming a *bimodal distribution*. How can the center and spread of a bimodal distribution be described? **14. See Ch. 8 Answer Appendix.**

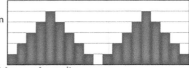

15. OPEN-ENDED Find a real-world data set that appears to represent a symmetric distribution and one that does not. Describe each distribution. Create a visual representation of each set of data. **See Ch. 8 Answer Appendix.**

16. WRITING IN MATH Explain the difference between positively skewed, negatively skewed, and symmetric sets of data, and give an example of each. **See Ch. 8 Answer Appendix.**

Standards for Mathematical Practice

Emphasis On	Exercises
1 Make sense of problems and persevere in solving them.	5, 6, 15, 19, 22
2 Reason abstractly and quantitatively.	13, 16–18, 20, 21
3 Construct viable arguments and critique the reasoning of others.	14
4 Model with mathematics.	7, 8, 10–12

Preparing for Assessment

17. A histogram is used to show the annual snowfall, in inches, for a city over a 20-year period. The center and spread of the data are accurately described as having a mean snowfall of 7.5 inches with a standard deviation of 1.1 inches. Which statement about the data can be assumed from this information? ⓂⓅ 2   S.IC.1   C

- ◯ **A**   The data distribution is flat.
- ◯ **B**   The data distribution is negatively skewed.
- ◯ **C**   The data distribution is symmetric.
- ◯ **D**   The data distribution is skewed to the left.
- ◯ **E**   The mean of the data is much greater than the median.

18. Jane displays attendance data using a box-and-whisker plot. One whisker is much longer than the other. The median is closest to the shorter whisker. Which statement can reasonably be deduced from this information? ⓂⓅ 2   S.IC.1   B

- ◯ **A**   The mean and the median are equal.
- ◯ **B**   The distribution of data is skewed.
- ◯ **C**   The median is less than the mean.
- ◯ **D**   The median is greater than the mean.
- ◯ **E**   The distribution of data is symmetrical.

19. The distributions of exam scores for two high school biology classes are shown in the histograms. Which of the following statements about the histograms is true? ⓂⓅ 1   S.IC.1   E

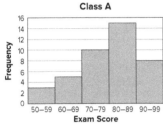

Class A

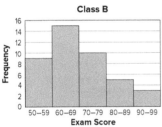

Class B

- ◯ **A**   Most students in class A and in class B scored higher than the mean score.
- ◯ **B**   The data are negatively distributed in class A and symmetrically skewed in class B.
- ◯ **C**   The data are symmetrically distributed in class A and positively skewed in class B.
- ◯ **D**   The mean exam score in class A and in class B is about the same as the corresponding median exam score.
- ◯ **E**   Most students in class A scored higher than the mean score, and most students in class B scored lower than the mean score.

20. A set of data is given as: ⓂⓅ 2   S.IC.1

18, 18, 19, 19, 20, 22, 22, 23, 27, 28, 28, 31, 34, 34, 36.

The box and whisker plot for this data is shown below.

Write down the values of A, B, C, D, and E.

A = [18]   D = [31]

B = [19]   E = [36]

C = [23]

21. MULTI-STEP The Spanish exam results of classes 9A and 9B are given in the box-and-whisker plots below. ⓂⓅ 2   S.IC.1

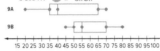

- **a.** Find the maximum and minimum scores in each class. [9A: max 70, min 20   9B: max 80, min 45]
- **b.** About what percent of the students in 9B scored in the same interval as the students in 9A? [75%]
- **c.** About what percent of the students in 9A scored between 65 and 70? [25%]
- **d.** About what percent of the students in 9B scored between 45 and 55? [50%]

22. The distribution of a set of data is given in the box-and-whisker plot below. ⓂⓅ 1   S.IC.1

60   2a   4b   2a + 2b   120

The difference between Q1 and Q3 is 40. Find the median. [80]

Preparing for Assessment

Exercises 17–22 require students to use the skills they will need on standardized assessments. Each exercise is dual-coded with content standards and mathematical practice standards.

	Dual Coding	
Items	**Content Standards**	Ⓜ **Mathematical Practices**
17	S.IC.1	2
18	S.IC.1	2
19	S.IC.1	1
20	S.IC.1	2
21	S.IC.1	2
22	S.IC.1	1

Diagnose Student Errors

Survey student responses for each item. Class trends may indicate common errors and misconceptions.

17.

A	Misinterpreted the center and spread of the data
B	Assumed the mean was less than the median
C	CORRECT
D	Assumed the majority of the data is on the left of the mean
E	Assumed the data were positively skewed

18.

A	Assumed the data was symmetric
B	CORRECT
C	Assumed the data were positively skewed
D	Assumed the data were negatively skewed
E	Assumed the mean and the median were about the same

Go Online!

Self-Check Quiz

Students can use *Self-Check Quizzes* to check their understanding of this lesson. You can also give the *Chapter Quiz*, which covers the content in Lesson 8-4.

RtI Response to Intervention

Use the Intervention Planner to help you determine your Response to Intervention.

Intervention Planner

TIER 1 On Level OL

IF students miss 25% of the exercises or less,

THEN choose a resource:

SE Lessons 8-1, 8-2, 8-3, and 8-4

Go Online!
- 📄 Skills Practice
- 📄 Chapter Project
- ✓ Self-Check Quizzes

TIER 2 Strategic Intervention AL
Approaching grade level

IF students miss 50% of the exercises,

THEN **Go Online!**
- 📄 Study Guide and Intervention
- ➕ Extra Examples
- 💬 Personal Tutors
- 📄 Homework Help

TIER 3 Intensive Intervention
2 or more grades below level

IF students miss 75% of the exercises,

THEN choose a resource:

Use *Math Triumphs, Alg. 2*

Go Online!
- ➕ Extra Examples
- 💬 Personal Tutors
- 📄 Homework Help
- abc Review Vocabulary

Go Online!

eAssessment

You can use the premade Mid-Chapter Test to assess students' progress in the first half of the chapter. Customize and create multiple versions of your Mid-Chapter Quiz and answer keys that align to your standards. Tests can be delivered on paper or online.

Determine whether each situation describes a *survey*, an *experiment*, or an *observational study*. Then identify the sample, and suggest a population from which it may have been selected. (Lesson 8-1)

1–4. See Ch. 8 Answer Appendix.

1. A high school principal wants to test five ideas for a new school mascot. He randomly selects 15 high school students to view pictures of the ideas while he watches and records their reactions.

2. Half of the employees of a grocery store are randomly chosen for an extra hour of lunch break. The managers then compare their attitudes with those of their coworkers.

3. Students want to create a school yearbook. They send out a questionnaire to 100 students asking what they would like to showcase in the yearbook.

4. The producers of a sitcom want to determine if a new character that they are planning to introduce will be well received. They show a clip of the show with the new character to 50 randomly chosen participants and then record the participants' reactions.

5. A random sample of 85 out of 390 students were surveyed and found that 26 of them own a laptop computer. Which is the most reasonable inference about the total number of students who own a laptop computer? (Lesson 8-1)

   | 119 students |

6. Julia chose a random sample of 25% of the green peppers in crate from a farm and found that 4 of them had a crack in them. Which is the most reasonable inference about the total number of peppers with a crack in them in the crate? (Lesson 8-1)
   - ◯ A 4 peppers
   - ◯ B 12 peppers
   - ◯ C 16 peppers   C
   - ◯ D 20 peppers

7. A statistician would like to obtain a margin of error of ±4% in his research study. What is the sample size needed to achieve that margin of error? (Lesson 8-3) **D**
   - ◯ A 25
   - ◯ B 100
   - ◯ C 400
   - ◯ D 625

8. **TRAINING** Aiden and Mark's training times for the 40-meter dash are shown. (Lesson 8-4)
   a–b. See Ch. 8 Answer Appendix.

Aiden's 40-Meter Dash Times (seconds)					
4.84	4.94	4.87	4.78	5.04	4.98
4.83	5.03	4.74	5.15	4.82	4.91
4.62	4.83	4.76	4.93	4.85	4.82
4.76	4.98	4.94	5.05	4.94	5.04
4.86	4.85	4.71	4.66	4.91	4.82

Mark's 40-Meter Dash Times (seconds)					
5.03	4.76	4.69	4.52	4.81	4.78
4.65	4.66	4.83	4.95	4.64	4.76
4.43	4.64	4.50	4.58	4.68	4.65
4.83	4.78	4.71	4.81	4.76	4.84
4.61	4.63	4.33	4.46	4.74	4.63

   a. Use a graphing calculator to create a histogram for each data set. Then describe the shape of each distribution.

   b. Compare the distributions using either the means and standard deviations or the five-number summaries. Justify your choice.

   9a–9d. See margin.

9. **MULTI-STEP** Miguel is taking a 10-question true-false quiz. He is wondering what his chances of passing the test are if he guesses on each question. (Lesson 8-2)

   a. Design a simulation that can be used to estimate the number of correct answers on the quiz.

   b. Conduct the simulation in part **a** five times and report the results using an appropriate summary.

   c. Use the data to compare the theoretical and experimental probabilities. Which statement is true for the results in part **b**? Explain your choice.

      A The experimental and theoretical probabilities are close in value.

      B The experimental and theoretical probabilities are not close in value.

      C The experimental and theoretical probabilities are the same in value.

   d. 🔲 What mathematical practice did you use to solve this problem?

Foldables Study Organizer

Dinah Zike's FOLDABLES

Before students complete the Mid-Chapter Quiz, encourage them to review the information for Lessons 8-1 through 8-4 in their Foldables. Allow students to compare their Foldable with a partner. Encourage them to share what has been helpful to them as they study.

🔲 ALEKS can be used as a formative assessment tool to target learning gaps for those who are struggling, while providing enhanced learning for those who have mastered the concepts.

Additional Answers

9a. Sample answer: Use a six-sided die. Even numbers represent a correct answer, odd numbers incorrect answers. Each simulation will consist of 10 trials. Record the results in a frequency table.

9c. Sample answer: A. The experimental probabilities should be 50%. Each percentage in the simulation was between 40% and 60%, close to 50%

9b.

Simulation	Number of Correct Answers
1	4
2	5
3	5
4	6
5	4

9d. See students' work.

Evaluating Published Data

Track Your Progress

Objectives

1 Evaluate reports based on data.

2 Identify and explain misleading uses of data.

Mathematical Background

Published reports may inadvertently or intentionally present data in a misleading way. Written reports may describe results of studies in terms that do not match the statistical underpinnings of the study. For example, a report may present the result of an observational study as a cause-and-effect relationship. Reports may also present data via inaccurate or misleading data displays. The scales on the axes and other elements of a display can greatly affect the appearance of the data.

THEN	NOW	NEXT
S.IC.4 Use data from a sample survey to estimate a population mean or proportion; develop a margin of error through the use of simulation models for random sampling.	**S.IC.6** Evaluate reports based on data.	**S.ID.4** Use the mean and standard deviation of a data set to fit it to a normal distribution and to estimate population percentages. Recognize that there are data sets for which such a procedure is not appropriate. Use calculators, spreadsheets, and tables to estimate areas under the normal curve.

Go Online! All of these resources and more are available at connectED.mcgraw-hill.com

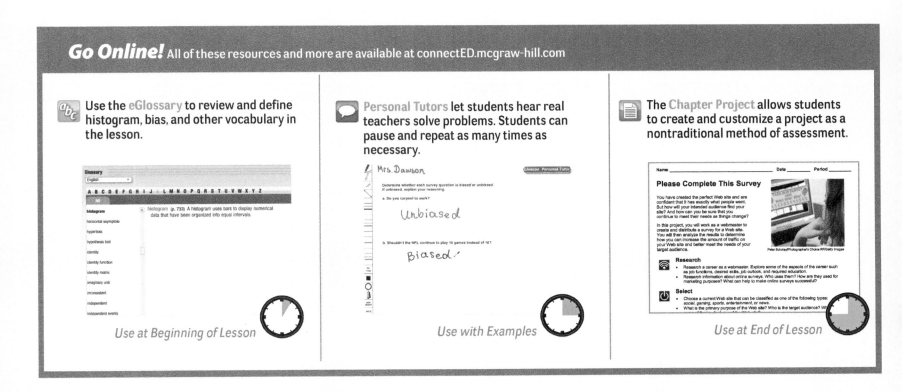

Use the eGlossary to review and define histogram, bias, and other vocabulary in the lesson.

Use at Beginning of Lesson

Personal Tutors let students hear real teachers solve problems. Students can pause and repeat as many times as necessary.

Use with Examples

The Chapter Project allows students to create and customize a project as a nontraditional method of assessment.

Use at End of Lesson

OER Using Open Educational Resources

Publishing Have students work in groups to create a wall on **Padlet** showing examples of misleading data displays. Ask volunteers to share their wall with the class. *Use as homework or classwork*

Go Online!
connectED.mcgraw-hill.com
Worksheets

Differentiate Your Resources

Extra Practice Additional practice or homework; Skills Practice is best for approaching-level students and Practice is best for on-level and beyond-level students

Skills Practice

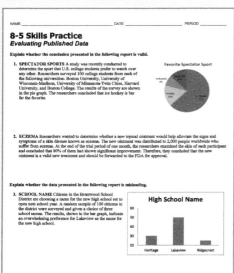

Practice

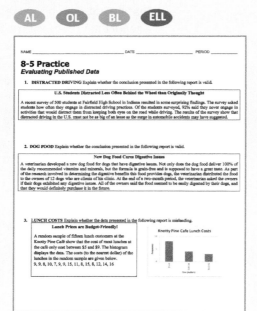

Word Problem Practice

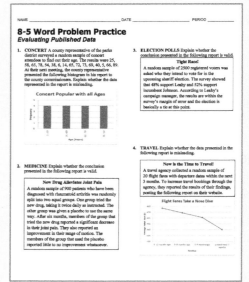

Intervention Reteaching and vocabulary activities that can be used with struggling or absent students and as ELL support

Extension Activities that can be used to extend lesson concepts

Study Guide and Intervention

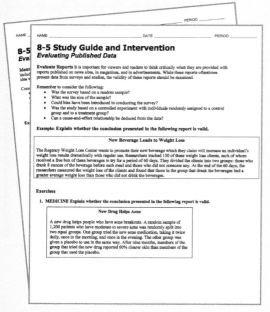

Study Notebook

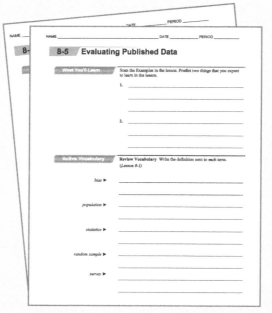

Enrichment

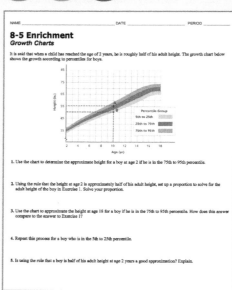

LESSON 5

Evaluating Published Data

Then	Now	Why?
You solved problems about random sampling and statistical experiments.	**1** Evaluate reports based on data. **2** Identify and explain misleading uses of data.	Over the past decade, açai berries have become a popular addition to smoothies and snacks. Some news reports have presented data to promote the health benefits of the berries, while others have pointed out the lack of scientifically controlled studies.

 Mathematical Practices
4 Model with mathematics.
6 Attend to precision.

Content Standards
S.IC.6 Evaluate reports based on data.

1 Evaluate Reports Reports published on news sites, in magazines, and in advertising may present data from surveys or studies. It is important to read the reports critically to determine whether the conclusions presented in the reports are valid. Consider the following.

- Was the survey based on a random sample?
- What was the size of the sample?
- Could bias have been introduced in conducting the survey?
- Was the study based on a controlled experiment with individuals randomly assigned to a control group and to a treatment group?
- Can a cause-and-effect relationship be deduced from the data?

S.IC.6

Real-World Example 1   Evaluate a Report

BLOOD PRESSURE **Explain whether the conclusion presented in the following report is valid.**

Eating Fruit Lowers Blood Pressure

The cafeteria at Midville Hospital offers a bowl of fresh fruit and other items for dessert. Researchers tracked 200 hospital employees who eat lunch at the cafeteria each day. They divided the employees into two groups: those who chose fruit for dessert and those who chose other items for dessert. At the end of six months, the researchers measured the blood pressure of the employees and found that those in the group that chose fruit had a lower average blood pressure than those who chose other desserts.

Was the study based on a controlled experiment with individuals randomly assigned to a control group and to a treatment group? The report is based on the results of an observational study rather than a controlled experiment. This means that it may not be valid to draw a cause-and-effect conclusion from the study.

Can a cause-and-effect relationship be deduced from the data? There may be other reasons why employees in the group that chose fruit had a lower average blood pressure. For example, employees who chose fruit may be more likely to exercise or make other healthy lifestyle choices that affect their blood pressure.

 Mathematical Practices Strategies

Make sense of problems and persevere in solving them.
Help students understand the meaning of a problem, look for entry points to its solution, and analyze givens, constraints, relationships, and goals. For example, ask:

- **What are the characteristics of a random sample?** Each member of the sample has an equal chance of being chosen for the sample.
- **Why is it important to use a random sample when studying a population?** The random sample is representative of the population as a whole. If the random sample is sufficiently large, it is possible to use data from the sample to make predictions about the population.
- **What is the difference between a treatment group and a control group?** The two groups are studied under the same conditions except for one variable. For example, members of a treatment group might receive a medication while members of the control group receive a placebo.
- **What elements of a data display can make it misleading?** Sample answer: The scales on the axes can be chosen to enhance or diminish certain aspects of the data, or the intervals for a histogram can be chosen to group the data in misleading ways.

Launch

Have students read the Why? section of the lesson.
- **What type of data might be presented in a news report to promote the health benefits of the berries?** Sample answer: vitamin content, antioxidant content, data from studies in which participants consumed the berries over a period of time, etc.

- **What is meant by a scientifically controlled study?** This is a study in which participants are randomly assigned to a control group and a treatment group.

Teach

Ask the scaffolded questions for each example to build conceptual understanding for students at all levels.

1 Evaluate Reports

Example 1 Evaluate a Report

AL What is the sample size for this study? 200

OL How do you know that the report is based on an observational study rather than a controlled experiment? The participants were not randomly assigned to a control group and a treatment group.

BL Would it be possible to change the study to make it a controlled experiment? Explain. Sample answer: It would be very difficult to assign participants to a control group and a treatment group so that the only variable that differs between the groups is whether or not the participants eat fruit at lunch.

Need Another Example?
GARDENING Explain whether the conclusion presented in the following report is valid.

Gardening Helps Cholesterol Levels

Researchers tracked 350 randomly-selected adults for one year. The participants were divided into two groups: those who work in a garden at least once a month, and those who do not. At the end of one year, participants were given a blood test and researchers found that the gardeners had healthier cholesterol levels than the nongardeners. Clearly, gardening is good for achieving healthy cholesterol levels.

The report is based on an observational study rather than a controlled experiment. This means that it may not be valid to draw a cause-and-effect conclusion from the study. There may be other reasons why the gardeners had healthier cholesterol levels. For example, they may be more likely to eat fresh fruits and vegetables from the garden.

MP **Teaching the Mathematical Practices**

Tools Mathematically proficient students know that technology can enable them to visualize the results of varying assumptions, explore consequences, and compare predictions with data. Spreadsheets are useful in this lesson because they make it easy to experiment with different displays for the same set of data. For Example 2, you might have students use a spreadsheet to display the data in various ways. Ask students to decide which displays are most misleading and which give the truest picture of the data.

2 Identify Misleading Uses of Data

Example 2 Identify Misleading Uses of Data

AL How many students in the sample chose each mascot? Lions: 12; Pandas: 14; Hawks: 18; Stingrays: 12

OL Based only on the lengths of the bars, what might you conclude about the number of votes for Hawks versus the number of votes for Pandas? The number of votes for Hawks appears to be twice the number of votes for Pandas.

BL Suppose there are 520 students at Redwood High School. What is the best prediction of the total number of students who would choose Hawks as the school's mascot? 167

Need Another Example?

SOCIAL MEDIA Explain whether the data presented in the following report is misleading.

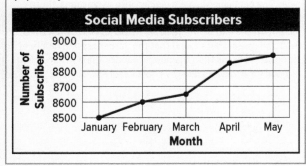

Social Media Site Experiences Fast Growth

The new social media site, FacePal, has grown tremendously in the past few months. As shown in the graph, the number of subscribers has increased rapidly from January to May of this year. The company is excited about its surging popularity.

Social Media Subscribers

The graph is misleading because the y-axis does not begin at 0. This makes it appear as though the number of subscribers has increased greatly over the period shown, but the number of subscribers actually increased by less than 5% from January to May.

---

1. Sample answer: The conclusion may not be valid because the students at Madison High School may not be representative of all students in the United States. Also, students may have hesitated to give a truthful answer to a survey about an unacceptable activity.

Guided Practice

1. TEXTING **Explain whether the conclusion presented in the following report is valid.**

> **U.S. Students Text Less Often in Class Than Originally Thought**
>
> A survey asked 450 students at Madison High School how often they sent one or more text messages while in class. Of the students surveyed, 86% said they never send a text in class. The results of the survey show that texting in class is not as serious a problem as some may have feared.

2 **Identify Misleading Uses of Data** Data from a survey or study can be presented in a variety of ways. In some cases, the data may be presented in a graph so that someone might draw an incorrect conclusion from the data. Consider the following.

- Do the scales on the axes begin at zero? If not, is a break shown on the axes?
- Do the scales on the axes use consistent intervals between tick marks?
- Do the scales on the axes distort differences in the data values?
- Does a histogram group data values appropriately?
- Could the person or organization presenting the data have a reason to create a misleading graph?

S.IC.6

Real-World Example 2   Identify Misleading Uses of Data

MASCOTS **Explain whether the data presented in the following report is misleading.**

> **Hawks: The Overwhelming Choice**
>
> Students at Redwood High School are choosing a new mascot for the school's teams. A random sample of 56 students at the school were surveyed and given a choice of four mascots. The results, shown in the bar graph, indicate an overwhelming preference for Hawks as the new name for the school's teams.

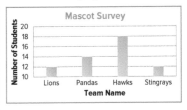

Study Tip
Sense-Making
It can be helpful to compare percentages rather than raw data values. For example, $\frac{18}{56}$ ≈ 32% chose Hawks, but $\frac{14}{56}$ = 25% chose Pandas, which shows that the difference in popularity was not as great as described in the report.

Although Hawks was the most popular team name, the vertical axis of the graph does not begin at zero. This distorts the relative lengths of the bars and makes Hawks appear much more popular than the other choices.

Guided Practice

2. CHARITY **Explain whether the data presented are misleading.**

> **Donations Are Looking Good!**
>
> A random sample of 10 donations received this week at the Gray Foundation show that most donations to the charity are in the $41 to $60 range. The histogram displays the data. The dollar values of the donations in the random sample are given below.
> 42, 21, 35, 41, 43, 20, 43, 25, 41, 42

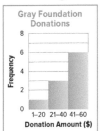

2. Sample answer: Although the histogram is accurate, the intervals on the horizontal axis are misleading because they make it appear that there were many donations of up to $60, while none of the donations were greater than $43.

Differentiated Instruction **OL** **BL**

IF students are successful solving problems about misleading uses of data like the problem in Example 2,

THEN have them work in small groups to find examples of misleading data in online news sources. Ask students to present their findings to the class and have them include suggestions for improving the data displays. You might also ask students to use technology to redraw the data displays in a way that presents the data more clearly and more accurately.

Check Your Understanding

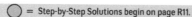 = Step-by-Step Solutions begin on page R11.

Go Online! for a Self-Check Quiz

Example 1
S.IC.6

1. NEWSPAPERS Explain whether the conclusion presented in the report is valid. See margin.

Reading a Newspaper Increases Test Scores

A group of 150 students were divided into two groups: those that read a newspaper at least once a week and those that do not. Test scores in all subject areas were monitored over the course of a school year. At the end of the year, the average test score of students who read a newspaper at least once a week was 6% greater than the average score of students who did not.

Example 2
S.IC.6

Explain whether the data presented in each graph is misleading.   2-3. See margin.

2.

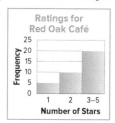

Ratings for Red Oak Café

3.

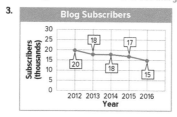

Blog Subscribers

Practice and Problem Solving

Extra Practice is found on page R8.

Example 1
S.IC.6

4. SPORTS Explain whether the conclusion presented in the following report is valid. See margin.

Soccer Is Town's Favorite Sport

A survey of 15 randomly-selected customers at an Elmwood athletic shoe store were asked whether they prefer to watch soccer or another sport. Twenty percent of those surveyed named soccer as their favorite sport to watch. No other sport was named by 20% or more of the participants, so it is safe to conclude that soccer is Elmwood's favorite sport.

5. The conclusion is valid since the study was a controlled experiment with a large sample and individuals randomly assigned to a control group and a treatment group.

5. MEDICINE Explain whether the conclusion presented in the following report is valid.

New Drug Helps Headaches

A random sample of 820 patients who get headaches was randomly split into two equal groups. One group tried the new drug every time they felt a headache starting. The other group was given a placebo to use in the same way. After six months, members of the group that tried the new drug reported 35% fewer headaches than members of the group that used the placebo.

7. Sample answer: The y-axis does not begin at a zero, which makes it appear as if sales revenue is increasing rapidly.

Example 2
S.IC.6

Explain whether the data presented in each graph is misleading.

6. Sample answer: The scale on the y-axis makes it appear as if the losses are generally quite small. A different choice of the scale would give a different impression.

6.

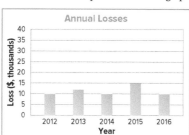

Annual Losses

7.

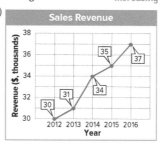

Sales Revenue

Differentiated Homework Options

Levels	AL Basic	OL Core	BL Advanced
Exercises	4–7, 12–21	5, 7, 8–10, 15, 16–21	8–21
2-Day Option	5, 7, 16–21	4–7, 16–21	
	4, 6, 12–15	8–15	

11. The conclusion is valid because the study was a large sample and individuals were randomly selected.

Practice

Formative Assessment Use Exercises 1–3 to assess students' understanding of the concepts in the lesson.

The Practice and Problem Solving exercises assess the content taught in the lesson. The Preparing for Assessment page is meant to be used as preparation for end-of-course assessment.

Additional Answers

1. Sample answer: The conclusion may not be valid because the report is based on an observational study, so a cause-and-effect conclusion cannot be made. There may be other reasons why students who read a newspaper had higher test scores. For example, they may be more likely to read books or may spend more time studying.

2. Sample answer: The intervals are not the same for the horizontal axis.

3. Sample answer: The scale of the y-axis makes it appear as if the number of subscribers was almost constant, although the number of subscribers actually decreased by 25% from 2012 to 2016.

4. The conclusion is not valid since the study was a small sample and individuals were not randomly assigned to a control group or treatment group.

8a. Sample answer:

Frozen Yogurt Sales

8b. Sample answer:

Frozen Yogurt Sales

ALGEBRA 2
INTERACTIVE STUDENT GUIDE

Teaching the Mathematical Practices

Precision Mathematically proficient students calculate carefully and efficiently, and express numerical answers with a degree of precision appropriate for the problem context. Exercise 11 connects the content of this lesson to students' earlier work with margin of error. Remind students to review the formula for margin of error as needed and ask them to express the margin of error in a way that is appropriate for this problem context.

Extra Practice

See page R8 for extra exercises for students who are approaching level or for on-level students who need additional reinforcement.

Assess

Ticket Out the Door Make several copies of misleading data displays. Give one data display to each student. As students leave the room, ask them to explain why the data display might be misleading and have them suggest a way to improve the display so that the data is presented more accurately.

Levels of Complexity Chart

The levels of the exercises progress from 1 to 3, with Level 1 indicating the lowest level of complexity.

Exercises	4–7	8–10, 16–21	11–15
Level 3			●
Level 2		●	
Level 1	●		

Additional Answer

15. Sample answer:

---

9. The conclusion is not valid since 6 out of 5235 people surveyed is approximately 0.1%; the population of the United States is about 320,000,000, and 0.1% of this population is about 320,000. The conclusion that millions of Americans know viral dance moves is not valid.

10a. Sample answer: Because the graphs have different scales on the y-axis, it appears that the Eagles' season attendance is greater than that of the Lions.

12. Sample answer: A random sample could introduce bias if it is not representative of all members of the population. For example, a random sample of high school students in Maine may not be representative of all U.S. high school students.

14. Sample answer: The survey may have missed older voters who do not have a cell phone, and these voters may have been more likely to oppose the new civic center.

8. **SENSE-MAKING** The table shows the number of frozen yogurts sold each month at YoZone. **a, b.** See margin on page 563.

Month	Number Sold
April	1250
May	1360
June	1420
July	1480
August	1390

a. Make a bar graph of the data that YoZone's manager might use to impress the owner of the shop.

b. Make a bar graph of the data that a rival shop owner might use to convince customers that sales are low at YoZone.

9. **TRENDS** Explain whether the conclusion presented in the report is valid.

> **Millions of Americans Know Viral Dance Moves**
>
> A recent study was designed to determine the number of Americans who have learned dance moves from videos that have spread around the internet. Researchers surveyed a random sample of 5235 Americans and found that 6 of them knew dance moves from videos that have gone viral. The researchers concluded that millions of Americans are doing viral dance moves.

10. The graphs show the average attendance at games for two college basketball teams.

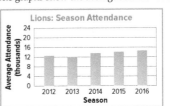
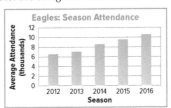

a. Explain why the data presented in the graphs may be misleading.

b. Describe a better way to display the data. Sample answer: Make a double bar graph to display the data.

11. **PRECISION** Explain whether the conclusion presented in the report is valid. See margin on page 563.

> **Mayoral Election Is a Toss-Up**
>
> A random sample of 2050 registered voters were asked who they intend to vote for in the upcoming mayoral election. The survey showed that 53% support Rodriguez and 47% support Baird. According to Baird's campaign manager, the results are within the survey's margin of error and the election is essentially a tie at this point.

S.IC.6

H.O.T. Problems   Use Higher-Order Thinking Skills

12. **WRITING IN MATH** Explain how choosing a random sample for a survey could introduce bias. Provide a specific example.

13. **CRITIQUE ARGUMENTS** Mayumi said that it is never possible to draw a cause-and-effect conclusion from the results of a study. Do you agree or disagree? Explain.

14. **SENSE-MAKING** A news site reported that 56% of registered voters in Ferndale support a proposition to build a new civic center. The survey involved calling a random sample of 1130 registered voters on their cell phones. On election day, the proposition failed to pass. Explain why the conclusion of the report may not have been valid.

15. **CHALLENGE** An employee of a science museum surveyed a random sample of visitors to the museum to find out their age. The results were 22, 30, 11, 17, 12, 50, 69, 24, 8, 16, 27, 51, 26, 60, and 42. Make a histogram of the data that makes it appear as though the museum is equally popular with visitors of all ages. See margin.

13. Disagree; it is possible to draw a cause-and-effect conclusion from a study if the study is a controlled experiment in which individuals are randomly assigned to a control group and to a treatment group.

Standards for Mathematical Practice

Emphasis On	Exercises
1 Make sense of problems and persevere in solving them.	8, 14, 19
3 Construct viable arguments and critique the reasoning of others.	13
4 Model with mathematics.	2, 3, 6–8, 10, 18, 20
6 Attend to precision.	11, 17, 21

Preparing for Assessment

16. The line graph shows the population of Linfield over several years.

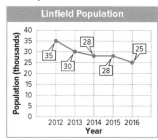

Linfield Population

Which scale on the vertical axis will make it appear as though the population declined more slowly? Assume the graph is drawn at the same overall size. Ⓜ 2 S.IC.6 **A**

- ○ **A** 1 to 60 in increments of 5
- ○ **B** 1 to 20 in increments of 2
- ○ **C** 20 to 40 in increments of 5
- ○ **D** 10 to 40 in increments of 2

17. A group of scientists conducted a study. At the end of the study they were able to make a cause-and-effect conclusion based on the data. Which of the following was most likely *not* a part of their research? ⓂⓅ 6 S.IC.6 **B**

- ○ **A** control group
- ○ **B** observational study
- ○ **C** random sample
- ○ **D** treatment group

18. MULTI-STEP Alana surveyed a random sample of seniors at her high school to determine which city the class should visit for the annual senior trip. The data are shown in the table. ⓂⓅ 4 S.IC.6

City	Number of Votes
Boston	8
New York	12
Philadelphia	10
Washington	16

a. Make a bar graph of the data that might cause someone to think that Washington received many more votes than the other cities.
18a-b. See Ch. 8 Answer Appendix.

b. Make a bar graph of the data that might cause someone to think that the four cities all received about the same number of votes.

19. Consider the report below.

> **Car Washing Improves Performance**
>
> A random sample of 1400 drivers was divided into two groups: those who wash their cars at least once a week and those who do not. After nine months, researchers found that drivers who washed their cars at least once a week had fewer car problems and fewer repairs. So, wash your car to improve its performance.

Which statements about the report are true? ⓂⓅ 1 S.IC.6 **B, E**

- ☐ **A** The conclusion of the report is valid.
- ☐ **B** The report is based on an observational study.
- ☐ **C** The report is based on a controlled experiment.
- ☐ **D** The sample is likely to be biased.
- ☐ **E** The report makes an incorrect cause-and-effect conclusion.
- ☐ **F** The sample is too small to be representative of the population.

20. The table shows the number of tablets sold over four days by a salesperson at a computer store.

Day	Number Sold
Monday	12
Tuesday	15
Wednesday	14
Thursday	12

Which scale on the *y*-axis should the salesperson use to make a bar graph of the data if she wants to make her sales figures appear as impressive as possible? ⓂⓅ 4 S.IC.6 **B**

- ○ **A** 1 to 20 in increments of 5
- ○ **B** 1 to 16 in increments of 2
- ○ **C** 10 to 20 in increments of 2
- ○ **D** 5 to 20 in increments of 5

21. Ming is making a histogram of the data set {2, 2, 3, 4, 5, 6, 7, 8, 8}. Which intervals along the horizontal axis will make it appear that the data are evenly distributed? ⓂⓅ 6 S.IC.6 **D**

- ○ **A** 2–4, 5–7, 8–10
- ○ **B** 1–4, 5–8
- ○ **C** 2–5, 6–9
- ○ **D** 1–3, 4–6, 7–9

20.

A	Chose a reasonable scale, but not the one that gives the appearance of the tallest bars
B	CORRECT
C	Chose a scale that makes the bars appear short
D	Chose a scale that makes the bars appear short

21.

A	Chose intervals that result in three bars of different heights
B	Chose intervals that result in one bar that is taller than the other
C	Chose intervals that result in one bar that is taller than the other
D	CORRECT

Preparing for Assessment

Exercises 16–21 require students to use the skills they will need on assessments. Each exercise is dual-coded with content standards and standards for mathematical practice.

Dual Coding		
Items	Content Standards	Ⓜ Mathematical Practices
16	S.IC.6	2
17	S.IC.6	6
18	S.IC.6	4
19	S.IC.6	1
20	S.IC.6	4
21	S.IC.6	6

Diagnose Student Errors

Survey student responses for each item. Class trends may indicate common errors and misconceptions.

16.

A	CORRECT
B	Chose a scale that does not include the relevant data
C	Chose a scale that makes it appear as though the population declined more rapidly
D	Chose a scale that makes it appear as though the population declined more rapidly

17.

A	Chose an element of a controlled experiment
B	CORRECT
C	Chose an element of a controlled experiment
D	Chose an element of a controlled experiment

19.

A	Did not recognize that there may be other reasons why drivers who washed their cars at least once a week had fewer car problems and fewer repairs
B	CORRECT
C	Did not know the definition of a controlled experiment
D	Did not realize that the sample is a random sample
E	CORRECT
F	Did not know that a random sample of 1400 individuals is generally sufficient

Normal Distributions

Track Your Progress

Objectives

1 Use the Empirical Rule to analyze normally distributed variables.

2 Apply the standard normal distribution and *z*-values.

Mathematical Background

The graphs of all normally distributed variables have essentially the same shape. With appropriate labeling of the mean and the points that are one standard deviation from the mean, the same normal curve can represent any normal distribution.

THEN	NOW	NEXT
S.IC.1 Understand statistics as a process for making inferences about population parameters based on a random sample from that population.	**S.ID.4** Use the mean and standard deviation of a data set to fit it to a normal distribution and to estimate population percentages. Recognize that there are data sets for which such a procedure is not appropriate. Use calculators, spreadsheets, and tables to estimate areas under the normal curve.	**S.MD.6** Use probabilities to make fair decisions.

Go Online! All of these resources and more are available at connectED.mcgraw-hill.com

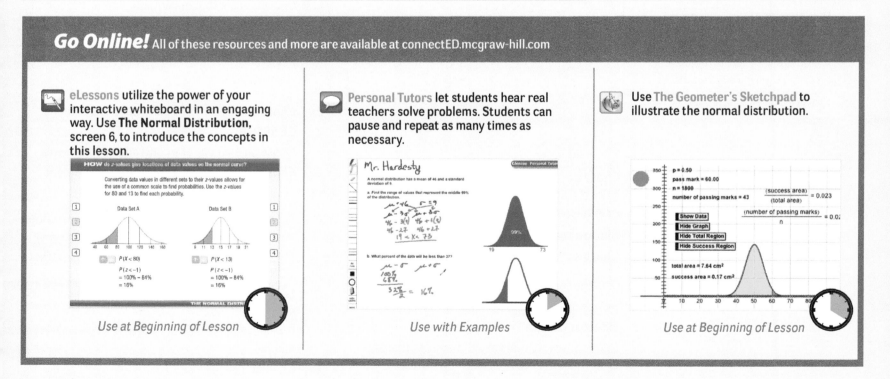

eLessons utilize the power of your interactive whiteboard in an engaging way. Use **The Normal Distribution**, screen 6, to introduce the concepts in this lesson.

Use at Beginning of Lesson

Personal Tutors let students hear real teachers solve problems. Students can pause and repeat as many times as necessary.

Use with Examples

Use **The Geometer's Sketchpad** to illustrate the normal distribution.

Use at Beginning of Lesson

ⓄER Using Open Educational Resources

Games Create a Distributions of Data game on **Super Teacher Tools**. You can create a game in the style of Jeopardy, Who Wants to Be a Millionaire, or a board game. Divide your class into 2–5 teams and use an interactive whiteboard to display the questions. *Use as practice or review*

Go Online!
connectED.mcgraw-hill.com
Worksheets

Differentiate Your Resources

Extra Practice Additional practice or homework; Skills Practice is best for approaching-level students and Practice is best for on-level and beyond-level students

Skills Practice

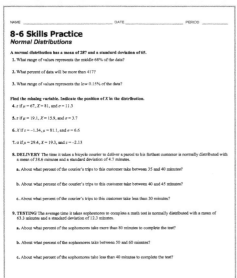

Practice

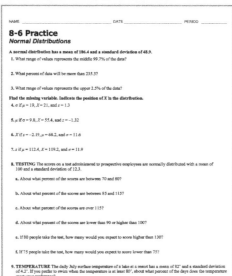

Word Problem Practice

Intervention Reteaching and vocabulary activities that can be used with struggling or absent students and as ELL support

Study Guide and Intervention

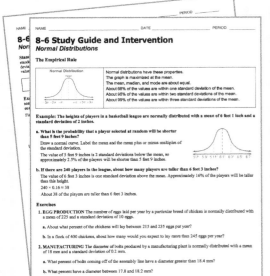

Study Notebook

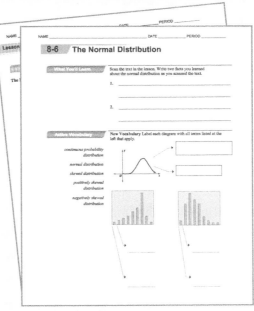

Extension Activities that can be used to extend lesson concepts

Enrichment

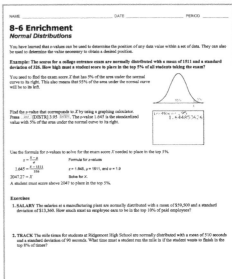

Launch

Have students read the Why? section of the lesson. Ask:

- **What dates are one standard deviation from the mean?** April 11 and May 1

- **What do you think the likelihood is of a Swiss cherry tree flowering before April 1?** about 2.5%

LESSON 6

Normal Distributions

:: Then	:: Now	:: Why?
• You constructed and analyzed discrete probability distributions.	**1** Use the Empirical Rule to analyze normally distributed variables. **2** Apply the standard normal distribution and z-values.	Extensive observations of Swiss cherry trees found that the mean flowering date is April 21 with a standard deviation of about 10 days. Therefore, 95% of the time, a Swiss cherry tree will have a flowering date between April 1 and May 11.

New Vocabulary
normal distribution
Empirical Rule
z-value
standard normal distribution

(MP) Mathematical Practices
6 Attend to precision.
8 Look for and express regularity in repeated reasoning.

Content Standards
S.ID.4 Use the mean and standard deviation of a data set to fit it to a normal distribution and to estimate population percentages. Recognize that there are data sets for which such a procedure is not appropriate. Use calculators, spreadsheets, and tables to estimate areas under the normal curve.

---

1 The Normal Distribution Distributions of mileages of different sample sizes of cars are shown below. As the sample size increases, the distributions become more and more symmetrical and resemble the curve at the right, due to the Law of Large Numbers.

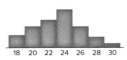
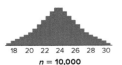
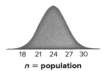

The curve at the right is a **normal distribution**, a continuous, symmetric, bell-shaped distribution of a random variable. It is the most common *continuous probability distribution*. The characteristics of the normal distribution are as follows.

💲 Key Concept The Normal Distribution

- The graph of the curve is continuous, bell-shaped, and symmetric with respect to the mean.
- The mean, median, and mode are equal and located at the center.
- The curve approaches, but never touches, the x-axis.
- The total area under the curve is equal to 1 or 100%.

The area under the normal curve represents the amount of data within a certain interval or the probability that a random data value falls within that interval. The **Empirical Rule** can be used to determine the area under the normal curve at specific intervals.

💲 Key Concept The Empirical Rule

In a normal distribution with mean μ and standard deviation σ,

- approximately 68% of the data fall within 1σ of the mean,
- approximately 95% of the data fall within 2σ of the mean, and
- approximately 99.7% of the data fall within 3σ of the mean.

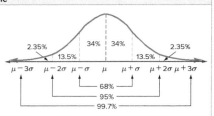

(MP) Mathematical Practices Strategies

Make sense of problems and persevere in solving them. Help students find and interpret probabilities of a normal distribution. For example, ask:

- **How do you find probabilities for a normal distribution?** by finding the areas under the normal distribution curve

- **Why would you determine the z-value of a data value?** to define a data value in terms of the mean and standard deviation of a data set

- **How does the Empirical Rule work?** It gives a set of approximate probabilities for a number of standard deviations from the mean.

Go Online!

Interactive Whiteboard

Use the *eLesson* or *Lesson Presentation* to present this lesson.

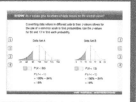

Example 1 Use the Empirical Rule to Analyze Data   S.ID.4

A normal distribution has a mean of 21 and a standard deviation of 4.

a. Find the range of values that represent the middle 68% of the distribution.

The middle 68% of data in a normal distribution is the range from $\mu - \sigma$ to $\mu + \sigma$. Therefore, the range of values in the middle 68% is $17 < X < 25$.

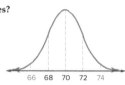

b. What percent of the data will be greater than 29?

29 is 2σ more than μ. 95% of the data fall between $\mu - 2\sigma$ and $\mu + 2\sigma$, so the remaining data values represented by the two tails covers 5% of the distribution. We are only concerned with the upper tail, so 2.5% of the data will be greater than 29.

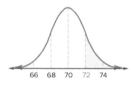

▶ Guided Practice

1. A normal distribution has a mean of 8.2 and a standard deviation of 1.3.

   A. Find the range of values that represent the middle 95% of the distribution. $5.6 < X < 10.8$

   B. What percent of the data will be less than 4.3? 0.15%

Real-World Example 2 Use the Empirical Rule to Analyze a Distribution   S.ID.4

HEIGHTS The heights of 1800 adults are normally distributed with a mean of 70 inches and a standard deviation of 2 inches.

a. About how many adults are between 66 and 74 inches?

66 and 74 are 2σ away from the mean. Therefore, about 95% of the data are between 66 and 74.

Because $1800 \times 95\% = 1710$, we know that about 1710 of the adults are between 66 and 74 inches tall.

b. What is the probability that a random adult is more than 72 inches tall?

From the curve, values greater than 72 are more than 1σ from the mean. 13.5% are between 1σ and 2σ, 2.35% are between 2σ and 3σ, and 0.15% are greater than 3σ.

So, the probability that an adult selected at random has a height greater than 72 inches is $13.5 + 2.35 + 0.15$ or 16%.

▶ Guided Practice

2. NETWORKING SITES The number of friends per member in a sample of 820 members is normally distributed with a mean of 38 and a standard deviation of 12.

   A. About how many members have between 26 and 50 friends? 558

   B. What is the probability that a random member will have more than 14 friends? 97.5%

Differentiated Instruction ELL

Intermediate Before reading, have students take a close look at the visual support. Have them use the graphs in the Key Concepts as they work in pairs to form questions about the normal distribution. After reading, have partners discuss how their ideas changed or stayed the same. Move around the room to monitor progress.

Advanced Have student pairs take turns reading the text to one another. Remind them that they are reading for an audience, and that they need to maintain the audience's interest with their voices. Move around the room, correcting pronunciation as necessary. Ask students what information they learn from the title and how the title makes them approach the text. What do they know about data, and what do they expect to learn? Have them share with the class.

Advanced High Have students write a paragraph explaining and evaluating each photo and graph used in relation to the content of the lesson. How effective was it?

Teach

Ask the scaffolded questions for each example to build conceptual understanding for students at all levels.

1 The Normal Distribution

Example 1 Use the Empirical Rule to Analyze Data

AL What is the median of this distribution? 21

OL What range of values represent the middle 95% of the data? 13 to 29

BL What would the range of values be that represent 100% of the data? $-\infty$ to $+\infty$

Need Another Example?

A normal distribution has a mean of 45.1 and a standard deviation of 9.6.

a. Find the values that represent the middle 99.7% of the distribution. $16.3 < X < 73.9$

b. What percent of the data will be greater than 54.7? 16%

Example 2 Use the Empirical Rule to Analyze a Distribution

AL What is the average height in feet? 5'10"

OL How many adults are taller than 72" tall? 288

BL What would be the cut-off height for the lowest 2.5% of adults? 66 inches

Need Another Example?

Packaging Students counted the number of candies in 100 small packages. They found that the number of candies per package was normally distributed with a mean of 23 candies per package and a standard deviation of 1 piece of candy.

a. About how many packages have between 22 and 24 candies? about 68 packages

b. What is the probability that a package selected at random has more than 25 candies? about 2.5%

2 Standard Normal Distribution

Example 3 Use z-Values to Locate Position

AL Is $\frac{X-\mu}{\sigma} = \frac{\mu-X}{\sigma}$? Why or why not? No, because subtraction is not commutative.

OL What is the z-value of the mean? 0

BL What does it mean about a value if its z-value is negative? The value is below the mean.

Need Another Example?

Find σ if $X = 28.3$, $\mu = 24.6$, and $z = 0.63$. Indicate the position of X in the distribution. 5.87; 0.63 standard deviations greater than the mean

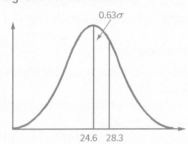

Teaching Tip

Relative Position Like percentiles, z-values can be used to compare the relative positions of two values in two different sets of data.

Additional Answer (Guided Practice)

3. $X \approx 45.0$

Go Online!

Explore the normal distribution using The Geometer's Sketchpad® activity in ConnectED.

2 Standard Normal Distribution The Empirical Rule is only useful for evaluating specific values, such as μ and σ. Once the data set is *standardized*, however, any data value can be evaluated. Data are standardized by converting them to z-values, also known as *z-scores*. The **z-value** represents the number of standard deviations that a given data value is from the mean. Therefore, z-values can be used to determine the position of any data value within a set of data.

🔑 Key Concept Formula for z-Values

The z-value for a data value X in a set of normally distributed data is given by $z = \frac{X-\mu}{\sigma}$, where μ is the mean and σ is the standard deviation.

S.ID.4

Example 3   Use z-Values to Locate Position

Find z if $X = 18$, $\mu = 22$, and $\sigma = 3.1$.
Indicate the position of X in the distribution.

$z = \frac{X-\mu}{\sigma}$ Formula for z values

$= \frac{18-22}{3.1}$ $X = 18, \mu = 22, \sigma = 31$

≈ -1.29 Simplify

The z-value that corresponds to $X = 18$ is approximately -1.29. Therefore, 18 is about 1.29 standard deviations less than the mean of the distribution.

Study Tip
Symmetry The normal distribution is symmetrical, so when you are asked for the middle or outside set of data, the z-values will be opposites.

▶ **Guided Practice**

3. Find X if $\mu = 39$, $\sigma = 8.2$, and $z = 0.73$. Indicate the position of X in the distribution. See margin.

Any combination of mean and standard deviation is possible for a normally distributed set of data. As a result, there are infinitely many normal probability distributions. This makes comparing two individual distributions difficult. Different distributions *can* be compared, however, once they are standardized using z-values. The **standard normal distribution** is a normal distribution with a mean of 0 and a standard deviation of 1.

Study Tip
Standard Normal Distribution The standard normal distribution is the set of all z-values.

🔑 Key Concept Characteristics of the Standard Normal Distribution

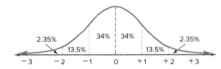

- The total area under the curve is equal to 1 or 100%.
- Almost all of the area is between $z = -3$ and $z = 3$.
- The distribution is symmetric.
- The mean is 0, and the standard deviation is 1.

The standard normal distribution allows us to assign actual areas to the intervals created by z-values. The area under the normal curve corresponds to the proportion of data values in an interval as well as the probability of a random data value falling within the interval. For example, the area between $z = 0$ and $z = 1$ is 0.34. Therefore, the probability of a z-value being in this interval is 34%.

Differentiated Instruction

IF   students need an aid in drawing a normal curve,

THEN   it may be useful to know that the concave side of the curve switches from facing downward to facing upward at points that are one standard deviation from the mean. Drawing a normal curve for each problem can help students with their estimates.

Real-World Example 4   Find Probabilities

S.ID.4

VIDEOS The number of videos uploaded daily to a video sharing site is normally distributed with $\mu = 181{,}099$ videos and $\sigma = 35{,}644$ videos. Find each probability. Then use a graphing calculator to sketch the corresponding area under the curve.

a. $P(180{,}000 < X < 200{,}000)$

The question is asking for the percent of days when between 180,000 and 200,000 videos are uploaded. First, find the corresponding z-values for $X = 180{,}000$ and $X = 200{,}000$.

$$z = \frac{X - \mu}{\sigma} \qquad \text{Formula for } z \text{ values}$$

$$= \frac{180{,}000 - 181{,}099}{35{,}644} \text{ or about } -0.03 \qquad X = 180{,}000, \mu = 181{,}099, \text{ and } \sigma = 35{,}644$$

Use 200,000 to find the other z-value.

$$z = \frac{X - \mu}{\sigma} \qquad \text{Formula for } z \text{ values}$$

$$= \frac{200{,}000 - 181{,}099}{35{,}644} \text{ or about } 0.53 \qquad X = 200{,}000, \mu = 181{,}099, \text{ and } \sigma = 35{,}644$$

The range of z-values that corresponds to $180{,}000 < X < 200{,}000$ is $-0.03 < z < 0.53$. Find the area under the normal curve within this interval.

You can use a graphing calculator to display the area that corresponds to any z-value by selecting [2nd] [DISTR]. Then, under the DRAW menu, select **ShadeNorm(lower z value, upper z value)**. The area between $z = -0.03$ and $z = 0.53$ is about 0.21 as shown in the graph.

[−4, 4] scl: 1 by [0, 0.5] scl: 0.125

Therefore, about 21% of the time, there will be between 180,000 and 200,000 video uploads on a given day.

b. $P(X > 250{,}000)$

$$z = \frac{X - \mu}{\sigma} \qquad \text{Formula for } z \text{ values}$$

$$= \frac{250{,}000 - 181{,}099}{35{,}644} \text{ or about } 1.93 \qquad X = 250{,}000, \mu = 181{,}099, \text{ and } \sigma = 35{,}644$$

Using a graphing calculator, you can find the area between $z = 1.93$ and $z = 4$ to be about 0.027.

Therefore, the probability that more than 250,000 videos will be uploaded is about 2.7%.

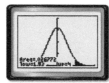

[−4, 4] scl: 1 by [0, 0.5] scl: 0.125

▶ **Guided Practice**

4. **TIRES** The life spans of a certain tread of tire are normally distributed with $\mu = 31{,}066$ miles and $\sigma = 1644$ miles. Find each probability. Then use a graphing calculator to sketch the corresponding area under the curve. **4A–B. See margin.**

A. $P(30{,}000 < X < 32{,}000)$ **B.** $P(X > 35{,}000)$

Another method for calculating the area between two z-values is [2nd] [DISTR] **normalcdf**(*lower z value, upper z value*).

Real-World Link

Video Uploading According to a recent study, 37% of Internet-using students aged 12–17 participate in video chats. 27% record and upload video to the Internet, and 13% stream video to the Internet.

Source: Pew Research Center

Study Tip

Range of z-values The majority of data values are within ±4σ of the mean, so setting the maximum value of z equal to 4 is sufficient in part **b**. Use the window [−4, 4] by [0, 0.5] when using ShadeNorm.

Teaching Tip

Sign Usage In a continuous distribution, there is no difference between $P(X \geq c)$ and $P(X > c)$ because the probability that X is equal to c is zero.

Example 4 Find Probabilities

AL How do you find probabilities of a normal distribution? Find the area under the curve for those values.

OL What is the probability that the number of downloads is not between 180,000 and 200,000? 0.79

BL What value of D would satisfy the equation $P(X < D) = 0.027$? 112,306

Need Another Example?

Health The cholesterol levels for adult males of a specific racial group are normally distributed with a mean of 158.3 and a standard deviation of 6.6. Find each probability. Then use a graphing calculator to sketch the corresponding area under the curve.

a. $P(X > 150)$ 0.90

b. $P(145 < X < 165)$ 0.82

a.

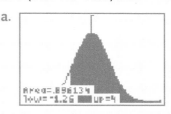

[−4, 4] scl: 1 by [0, 0.5] scl: 0.125

b.

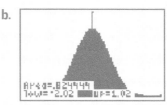

[−4, 4] scl: 1 by [0, 0.5] scl: 0.125

Additional Answers (Guided Practice)

4A. 0.46

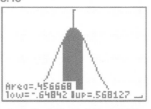

[−4, 4] scl: 1 by [0, 0.5] scl: 0.125

4B. 0.008

[−4, 4] scl: 1 by [0, 0.5] scl: 0.125

Practice

Formative Assessment
Use Exercises 1–8 to assess students' understanding of the concepts in this lesson.

The Practice and Problem Solving exercises assess the content taught in the lesson. The Preparing for Assessment page is meant to be used as preparation for assessment.

(MP) Teaching the Mathematical Practices

Tools When making mathematical models, mathematically proficient students know that technology can enable them to visualize the results of varying assumptions, explore consequences, and compare predictions with data. They are able to use technological tools to explore and deepen their understanding of concepts.

Extra Practice

See page R8 for extra exercises for students who are approaching level or for on-level students who need additional reinforcement.

Levels of Complexity Chart

The levels of the exercises progress from 1 to 3, with Level 1 indicating the lowest level of complexity.

Exercises	9–16	17–20, 27–34	21–26
(C) Level 3			●
(B) Level 2		●	
Level 1	●		

Go Online! for a Self-Check Quiz

Check Your Understanding
○ = Step-by-Step Solutions begin on page R11.

Example 1
S.ID.4

A normal distribution has a mean of 416 and a standard deviation of 55.

1. Find the range of values that represent the middle 99.7% of the distribution. $251 < X < 581$

2. What percent of the data will be less than 361? 16%

Example 2
S.ID.4

3. (MP) **TOOLS** The number of texts sent per day by a sample of 811 teens is normally distributed with a mean of 38 and a standard deviation of 7.
   a. About how many teens sent between 24 and 38 texts? about 386 teens
   b. What is the probability that a teen selected at random sent less than 45 texts? 84%

Example 3
S.ID.4

Find the missing variable. Indicate the position of X in the distribution. 4–7. See Ch. 8 Answer Appendix.

4. z if $\mu = 89$, $X = 81$, and $\sigma = 11.5$
5. z if $\mu = 13.3$, $X = 17.2$, and $\sigma = 1.9$
6. X if $z = -1.38$, $\mu = 68.9$, and $\sigma = 6.6$
7. σ if $\mu = 21.1$, $X = 13.7$, and $z = -2.40$

Example 4
S.ID.4

8. **CONCERTS** The number of concerts attended per year by a sample of 925 teens is normally distributed with a mean of 1.8 and a standard deviation of 0.5. Find each probability. Then use a graphing calculator to sketch the area under each curve.
   a. $P(X < 2)$ 65.5%
   b. $P(1 < X < 3)$ 93.7% a–b. See Ch. 8 Answer Appendix for graphs.

Practice and Problem Solving
Extra Practice is on page R8.

Example 1
S.ID.4

A normal distribution has a mean of 29.3 and a standard deviation of 6.7.

9. Find the range of values that represent the outside 5% of the distribution. $X < 15.9$ or $X > 42.7$

10. What percent of the data will be between 22.6 and 42.7? 81.5%

Example 2
S.ID.4

11. **GYMS** The number of visits to a gym per year by a sample of 522 members is normally distributed with a mean of 88 and a standard deviation of 19.
    a. About how many members went to the gym at least 50 times? about 509 members
    b. What is the probability that a member selected at random went to the gym more than 145 times? 0.15%

Example 3
S.ID.4

Find the missing variable. Indicate the position of X in the distribution. 12–15. See Ch. 8 Answer Appendix.

12. z if $\mu = 3.3$, $X = 3.8$, and $\sigma = 0.2$
13. z if $\mu = 19.9$, $X = 18.7$, and $\sigma = 0.9$
14. μ if $z = -0.92$, $X = 44.2$, and $\sigma = 8.3$
15. X if $\mu = 138.8$, $\sigma = 22.5$, and $z = 1.73$

Example 4
S.ID.4

16. **VENDING** A vending machine dispenses about 8.2 ounces of coffee. The amount varies and is normally distributed with a standard deviation of 0.3 ounce. Find each probability. Then use a graphing calculator to sketch the corresponding area under the curve.
    a. $P(X < 8)$ 25.2%
    b. $P(X > 7.5)$ 99.0% 16a–b. See Ch. 8 Answer Appendix for graphs.

(B) 17 **CAR BATTERIES** The useful life of a certain car battery is normally distributed with a mean of 113,627 miles and a standard deviation of 14,266 miles. The company makes 20,000 batteries a month.
    a. About how many batteries will last between 90,000 and 110,000 miles? about 7000 batteries
    b. About how many batteries will last more than 125,000 miles? about 4200 batteries
    c. What is the probability that if you buy a car battery at random, it will last less than 100,000 miles? about 17.0%

18. **FOOD** The shelf life of a particular snack chip is normally distributed with a mean of 173.3 days and a standard deviation of 23.6 days.
    a. About what percent of the product lasts between 150 and 200 days? 71.0%
    b. About what percent of the product lasts more than 225 days? 1.4%
    c. What range of values represents the outside 5% of the distribution? $X > 220.5$ or $X < 126.1$

Differentiated Homework Options

Levels	(AL) Basic	(OL) Core	(BL) Advanced
Exercises	9–16, 21–34	9–15 odd, 17, 19, 21, 27–34	17–34
2-Day Option	9–15 odd, 27–34	9–16, 27–34	
	10–16 even, 21–26	17–26	

 You can use ALEKS to provide additional remediation support with personalized instruction and practice.

Go Online!
eBook

Interactive Student Guide

Use the *Interactive Student Guide* to deepen conceptual understanding.
· Applying the Normal Distribution
· The Normal Distribution

19. FINANCIAL LITERACY The insurance industry uses various factors including age, type of car driven, and driving record to determine an individual's insurance rate. Suppose insurance rates for a sample population are normally distributed.

19a. between $714 and $944

a. If the mean annual cost per person is $829 and the standard deviation is $115, what is the range of rates you would expect the middle 68% of the population to pay annually?

b. If 900 people were sampled, how many would you expect to pay more than $1000 annually? 62

c. Where on the distribution would you expect a person with several traffic citations to lie? Explain your reasoning. c–d. See Ch. 8 Answer Appendix.

d. How do you think auto insurance companies use each factor to calculate an individual's insurance rate?

20. STANDARDIZED TESTS Nikki took three national standardized tests and scored an 86 on all three. The table shows the mean and standard deviation of each test.

	Math	Science	Social Studies
μ	76	81	72
σ	9.7	6.2	11.6

a. Calculate the z-values that correspond to her score on each test.

b. What is the probability of a student scoring an 86 or *lower* on each test?

c. On which test was Nikki's standardized score the highest? Explain your reasoning.
20a–c. See Ch. 8 Answer Appendix.

H.O.T. Problems   Use Higher-Order Thinking Skills

21. ERROR ANALYSIS A set of normally distributed tree diameters have mean 11.5 centimeters, standard deviation 2.5, and range from 3.6 to 19.8. Monica and Hiroko are to find the range that represents the middle 68% of the data. Is either of them correct? Explain.
Sample answer: Hiroko; Monica's solution would work with a uniform distribution.

Monica	Hiroko
The data span 16.2 cm. 68% of 16.2 is about 11 cm. Center this 11-cm range around the mean of 11.5 cm. This 68% group will range from about 6 cm to about 17 cm.	The middle 68% span from $\mu + \sigma$ to $\mu - \sigma$. So we move 2.5 cm below 11.5 and then 2.5 cm above 11.5. The 68% group will range from 9 cm to 14 cm.

22. CHALLENGE A selection of tablets has an average battery life of 8.2 hours with a standard deviation of 0.7 hour. Eight of the tablets have a battery life greater than 9.3 hours. If the sample is normally distributed, how many tablets are in the selection? about 138

23. REASONING The term *six sigma process* comes from the notion that if one has six standard deviations between the mean of a process and the nearest specification limit, there will be practically no items that fail to meet the specifications. Is this a true assumption? Explain. See margin.

24. REASONING *True* or *false*: According to the Empirical Rule, in a normal distribution, most of the data will fall within one standard deviation of the mean. Explain. See margin.

25. OPEN-ENDED Find a set of real-world data that appears to be normally distributed. Calculate the range of values that represent the middle 68%, the middle 95%, and the middle 99.7% of the distribution. See Ch. 8 Answer Appendix.

26. WRITING IN MATH Describe the relationship between the z-value, the position of an interval of X in the normal distribution, the area under the normal curve, and the probability of the interval occurring. Use an example to explain your reasoning. See Ch. 8 Answer Appendix.

Standards for Mathematical Practice	
Emphasis On	**Exercises**
1 Make sense of problems and persevere in solving them.	9, 10, 12–15, 27, 29, 31
2 Reason abstractly and quantitatively.	22–24
3 Construct viable arguments and critique the reasoning of others.	21
4 Model with mathematics.	11, 16–20
5 Use appropriate tools strategically.	25, 26
6 Attend to precision.	28, 30, 32–34

Assess

Name the Math
Ask students to explain the difference between a normal distribution and the standard normal distribution. **ELL**

Watch Out!
Error Analysis For Exercise 21, remind students that the middle 68% of any normal distribution is within one standard deviation of the mean.

Follow-Up
Students have explored binomial distributions. Ask

- **Can statistics lie?** Sample answer: Statistics can "lie" when they are manipulated and then used to influence the intended audiences' beliefs and behaviors.

Additional Answers

23. Sample answer: True; according to the Empirical Rule, 99% of the data lie within 3 standard deviations of the mean. Therefore, only 1% will fall outside of three sigma. An infinitesimally small amount will fall outside of six-sigma.

24. Sample answer: True; according to the Empirical Rule, 68% of the data lie within 1 standard deviation of the mean.

Go Online!

Self-Check Quiz
Students can use *Self-Check Quizzes* to check their understanding of this lesson. You can also give the *Chapter Quiz*, which covers the content of Lessons 8–5 and 8–6.

Preparing for Assessment

Exercises 27–34 require students to use the skills they will need on standardized assessments. Each exercise is dual-coded with content standards and mathematical practice standards.

Dual Coding		
Items	Content Standards	Mathematical Practices
27	S.ID.4	1
28	S.ID.4	6
29	S.ID.4	1
30	S.ID.4	6
31	S.ID.4	1
32	S.ID.4	6
33	S.ID.4	6
34	S.ID.4	6

Diagnose Student Errors

Survey student responses for each item. Class trends may indicate common errors and misconceptions.

27.

A	Multiplied the number of candles by 13.5%
B	Multiplied the number of candles by 34%
C	CORRECT
D	Computed 2σ away from the mean
E	Computed 3σ away from the mean

28.

A	CORRECT
B	Used 1σ away from the mean
C	Used 3σ away from the mean
D	Used incorrect inequality symbols
E	Used 1σ away from the mean and flipped inequality symbols

29.

A	Used 3σ away from the mean
B	Used 2σ away from the mean
C	Used half of the total trees
D	CORRECT
E	Computed 2σ away from the mean

31.

A	Multiplied the number of workers by one fifth
B	Multiplied the number of workers by one third
C	CORRECT
D	Computed 2σ away from the mean
E	Computed 3σ away from the mean

27. The production costs of 600 candles are normally distributed. The mean cost is $80 and the standard deviation is $12. About how many candles have a production cost between $68 and $92? ⓜ 1 S.ID.4 **C**

- A 81
- B 204
- C 408
- D 570
- E 598

28. A normal distribution has a mean of 18 and a standard deviation of 2.5. What is the range of values that represent the outside 5% of the distribution? ⓜ 6 S.ID.4 **A**

- A $X < 13$ or $X > 23$
- B $X < 15.5$ or $X > 20.5$
- C $X < 10.5$ or $X > 25.5$
- D $X > 13$ or $X < 23$
- E $X > 15.5$ or $X < 20.5$

29. The heights of 1000 trees are normally distributed. The mean height is 30 feet and the standard deviation is 5 feet. How many trees are between 25 and 35 feet? ⓜ 1 S.ID.4 **D**

- A 47
- B 135
- C 500
- D 680
- E 950

30. **MULTI-STEP** The time taken for a student to complete a task is normally distributed with a mean of 20 minutes and a standard deviation of 2.2 minutes. ⓜ 6 S.ID.4

  a. A student is selected at random. Find the probability that the student completes the task in less than 21.8 minutes.

    | 0.7934 |

  b. The probability that a student takes between k and 21.8 minutes is 0.3. Find the value of k.

    | 19.96 |

The hourly wages of 2400 construction workers are normally distributed with a mean of $36 per hour and a standard deviation of $12 per hour. Use this information for Questions 31–32.

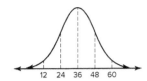

12 24 36 48 60

31. About how many construction workers earn between $24 per hour and $48 per hour? ⓜ 1 S.ID.4 **C**

- A 480
- B 800
- C 1632
- D 2280
- E 2393

32. What is the probability that a construction worker selected at random earns an hourly wage of less than $24 per hour? ⓜ 6 S.ID.4 **D**

- A 0.15%
- B 2.35%
- C 13.5%
- D 16.0%
- E 34.0%

33. A random variable has the normal distribution with mean 82.0 and standard deviation 4.8. Find the probabilities that it will take on a value. ⓜ 6 S.ID.4

  a. less than 89.2 | 0.9332 |

  b. greater than 78.4 | 0.7734 |

  c. between 83.2 and 88.0 | 0.2956 |

  d. between 73.6 and 90.4 | 0.9199 |

34. Find z if: ⓜ 6 S.ID.4

  a. The standard-normal-curve area between 0 and z is 0.4484. | 1.63 |

  b. The standard-normal-curve area to the left of z is 0.9868. | 2.22 |

Differentiated Instruction ⒷⓁ

Extension Explain to students that they can use a graphing calculator to identify the range of z-values that corresponds with any area under the normal curve. For example, to find the range of z-values that corresponds to the middle 50% of data, they need to find the z-values that correspond to the data between 25% and 75% in the distribution. Select ⏚2nd⏚ [DISTR] 3 for the InvNorm(feature. Then enter 0.25 ⏚)⏚. Do the same for 0.75 ⏚)⏚.

32.

A	Assumed the wage fell more than 3σ from the mean
B	Assumed the wage fell between 2σ and 3σ from the mean
C	Assumed the wage fell between 1σ and 2σ from the mean
D	CORRECT
E	Assumed the wage fell 1σ away from the mean

Using Probability to Make Decisions

SUGGESTED PACING (DAYS)

90 min.   **0.5**
45 min.   **1.0**

Instruction

Track Your Progress

Objectives

1 Use probability to make fair decisions.

2 Analyze decisions by using probability concepts.

Mathematical Background

A fair decision is a decision in which all of the possible options have the same probability of being chosen. To determine whether a decision is fair, students can use a variety of probability concepts, including the probability of independent events and conditional probability. Analyzing a decision sometimes depends on a cost-benefit analysis, and this can be accomplished by calculating the average or expected value of a variable, such as the number of prize tickets someone can expect to win with each spin of a wheel.

THEN	NOW	NEXT
S.ID.4 Use the mean and standard deviation of a data set to fit it to a normal distribution and to estimate population percentages. Recognize that there are data sets for which such a procedure is not appropriate. Use calculators, spreadsheets, and tables to estimate areas under the normal curve.	**S.MD.6** Use probabilities to make fair decisions (e.g., drawing by lots, using a random number generator). **S.MD.7** Analyze decisions and strategies using probability concepts (e.g., product testing, medical testing, pulling a hockey goalie at the end of a game).	**F.TF.1** Understand radian measure of an angle as the length of the arc on the unit circle subtended by the angle.

Go Online! All of these resources and more are available at connectED.mcgraw-hill.com

Use the **eGlossary** to review and define conditional probability and other vocabulary in the lesson.

Animations illustrate key concepts through step-by-step tutorials and videos.

The **Chapter Project** allows students to create and customize a project as a nontraditional method of assessment.

Use at Beginning of Lesson   *Use at with Examples*   *Use at the End of Lesson*

OER Using Open Educational Resources

Communicating This lesson builds on several probability concepts. Text the class reminders of definitions or formulas as needed. **Class Parrot** offers an easy, safe, and free way to text messages to an entire class of students. *Use as a reminder tool*

Go Online!
connectED.mcgraw-hill.com Worksheets

Differentiate Your Resources

Extra Practice Additional practice or homework; Skills Practice is best for approaching-level students and Practice is best for on-level and beyond-level students

Skills Practice
 AL OL ELL

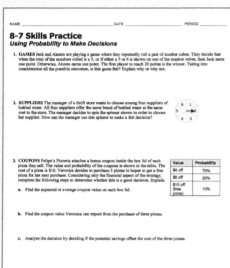

Practice
OL AL BL ELL

Word Problem Practice
AL OL BL ELL

Intervention Reteaching and vocabulary activities that can be used with struggling or absent students and as ELL support

Extension Activities that can be used to extend lesson concepts

Study Guide and Intervention
AL OL ELL

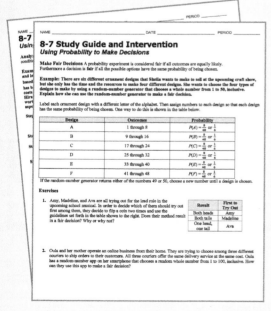

Study Notebook
AL OL BL ELL

Enrichment
OL BL ELL

LESSON 7

Using Probability to Make Decisions

::Then	::Now	::Why?
● You calculated probabilities of dependent and independent events.	1 Use probability to make fair decisions. 2 Analyze decisions by using probability concepts.	● A veterinarian knows that 1% of all dogs are infected with a new virus that has spread around the country. A test for the virus is 98% accurate. Is it a good decision to treat every dog that tests positive for the virus? You can use probability to analyze this situation.

 New Vocabulary
fair

MP **Mathematical Practices**
4 Model with mathematics.
5 Use appropriate tools strategically.

Content Standards
S.MD.6 Use probabilities to make fair decisions (e.g., drawing by lots, using a random number generator).
S.MD.7 Analyze decisions and strategies using probability concepts (e.g., product testing, medical testing, pulling a hockey goalie at the end of a game).

1 Make Fair Decisions A probability experiment is considered fair if all outcomes are equally likely. For example, flipping a coin is a fair experiment. This assumes that the coin is equally likely to land heads up or tails up. Such a coin is called a fair coin. A decision is **fair** if all of the possible options have the same probability of being chosen.

S.MD.6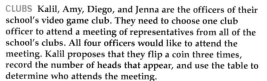

Real-World Example 1   Make a Fair Decision

CLUBS Kalil, Amy, Diego, and Jenna are the officers of their school's video game club. They need to choose one club officer to attend a meeting of representatives from all of the school's clubs. All four officers would like to attend the meeting. Kalil proposes that they flip a coin three times, record the number of heads that appear, and use the table to determine who attends the meeting.

Result	Officer
No heads	Kalil
1 head	Amy
2 heads	Diego
3 heads	Jenna

a. Does Kalil's method result in a fair decision? Why or why not?

Make a table to show all of the possible outcomes. For each outcome, determine which officer would attend the meeting. Then determine the probability that each officer attends the meeting.

Because the four options have different probabilities, this is not a fair decision.

Outcome	Officer	Probability
T T T	Kalil	$P(\text{Kalil}) = \frac{1}{8}$
H T T	Amy	$P(\text{Amy}) = \frac{3}{8}$
T H T		
T T H		
H H T	Diego	$P(\text{Diego}) = \frac{3}{8}$
H T H		
T H H		
H H H	Jenna	$P(\text{Jenna}) = \frac{1}{8}$

b. Is it possible to flip a coin three times to make a fair decision in this situation? Explain.

One possibility is that Kalil attends the meeting if there are no heads, Amy attends if the result is heads-tails-tails, Diego attends if the result is heads-heads-tails, and Jenna attends if the result is three heads. In case of any other result, the coin is flipped three more times until an officer is chosen.

This is fair because the probability of each officer attending the meeting at the end of three coin flips is $\frac{1}{8}$.

MP **Mathematical Practices Strategies**

Model with mathematics. Help students apply the mathematics they know to solve probability problems arising in everyday life, identify important quantities in a practical situation, and interpret their mathematical results in the context of the situation. For example, ask:

● How do you find the sample space for a probability experiment? Make a list of all of the possible outcomes for the experiment.

● What must be true about the probabilities for all the outcomes for any event? The sum of all the probabilities must be equal to 1.

● How can you assure that a real-world situation is solved by using a simulation that makes a fair decision? Set up the simulation so that each outcome has the same probability of occurring.

● How do you find the expected value for the number of tickets that an arcade game pays out each time it is played? The expected value is equal to the sum of the products of each possible number of tickets that can be received times the probability of receiving that number of tickets.

Launch

Have students read the Why? section of the lesson.

● **What do you think it means when a test for a virus is called 98% accurate?** If a dog is infected with the virus, the test returns a positive result 98% of the time. If a dog is not infected with the virus, the test returns a negative result 98% of the time.

● **How can you use conditional probability to restate the question in the Why? section of the lesson?** What is the conditional probability that a dog is infected with the virus given that the dog tests positive for the virus?

Teach

Ask the scaffolded questions for each example to build conceptual understanding for students at all levels.

1 Make Fair Decisions

Example 1 Make a Fair Decision

AL When you flip a coin three times, how many outcomes are in the sample space? 8

OL How do you determine the probability that each officer attends the meeting? Divide the number of outcomes that correspond to the officer attending the meeting by the total number of outcomes in the sample space.

BL Instead of flipping one coin three times, suppose the officers flip three coins at the same time. Does this change the results? Explain. This does not change the results because the sample space is the same and the outcomes assigned to each officer are the same.

(continued on the next page)

Teaching the Mathematical Practices

Tools Mathematically proficient students make sound decisions about tools that might be helpful to solve a problem, recognizing both insight to be gained and the tools' limitations. For Example 2, have students do a simulation using the random-number generator on their calculators. Then ask them whether the results matched what they were expecting and if they noticed any advantages or drawbacks to using a random-number generator in this problem-solving situation.

Need Another Example?

TRAVEL Jamilla is driving to her aunt's house. She is trying to decide whether to take Highway 63, Route 102, or Interstate 20. She decides to flip a nickel and a dime at the same time and use the table to determine which road to take.

Result	Road
2 tails	Highway 63
1 head, 1 tail	Route 102
2 heads	Interstate 20

a. **Does Jamilla's method result in a fair decision? Why or why not?** No; P(Highway 63) = P(Interstate 20) = 0.25, but P(Route 102) = 0.5; because the different options have different probabilities, this is not a fair decision.

b. **Is it possible to flip a nickel and a dime at the same time to make a fair decision in this situation? Explain.** Yes; for example, assign Highway 63 and Interstate 20 to results as shown in the table. Assign Route 102 to the result that the nickel lands heads up and the dime lands tails up. If the result is that the nickel lands tails up and the dime lands heads up, flip the coins again until a route is chosen.

Example 2 Make a Fair Decision

AL Why is $P(A) = \frac{300}{1000}$? There are 300 outcomes assigned to Supplier A, and there are 1000 outcomes in the sample space.

OL Why do the probabilities for Suppliers A, B, and C not add up to 1? The probabilities add up to 1 if you include the possibility of a number from 901 to 1000, which has a probability of $\frac{1}{10}$.

BL What is another way the business owner could have assigned the numbers to the suppliers? Sample answer: Supplier A: 1 through 100; Supplier B: 101 through 200;

1. No; P(Yoshio) = $\frac{5}{9}$ and P(Kaitlyn) = $\frac{4}{9}$; Sample answer: Yoshio goes first if the sum is 2 or 4, Kaitlyn goes first if the sum is 3 or 5; if the sum is 6, spin both spinners again until a winner is chosen.

Study Tip

Tools Most problems that ask how to use a tool to make a fair decision will have many possible correct answers. Check that your answer is correct by making sure each possible option has the same probability.

2. Sample answer: Assign numbers 1–100 to supplier A, 101–200 to supplier B, 201–300 to supplier C, 301–400 to supplier D, 401–500 to supplier E, 501–600 to supplier F, and 601–700 to supplier G. If the random-number generator returns a number from 701 to 1000, choose a new number until a supplier is chosen.

Guided Practice

1. **BOARD GAMES** Yoshio and Kaitlyn are playing a board game that uses two spinners. Each spinner has three equal sections that are numbered 1, 2, and 3. To determine who goes first in the game, they decide to spin both spinners. If the sum of the numbers shown is even, Yoshio goes first. If the sum is odd, Kaitlyn goes first. Does this method result in a fair decision? Explain. If it is not fair, explain how they can use the two spinners to make a fair decision.

S.MD.6

Real-World Example 2 Make a Fair Decision

MANUFACTURING The owner of a candle-making business wants to chose among three suppliers of wax. All three suppliers offer the same wax at the same price. The owner has a random-number app on her phone that chooses a random whole number from 1 to 1000, inclusive. How can she use the app to make a fair decision?

Assign numbers to each supplier so that each supplier has the same probability of being chosen. One way to do this is shown in the table.

If the random-number generator returns a number from 901 to 1000, choose a new number until a supplier is chosen.

Supplier	Outcomes	Probability
A	1 through 300	$P(A) = \frac{300}{1000}$ or $\frac{3}{10}$
B	301 through 600	$P(B) = \frac{300}{1000}$ or $\frac{3}{10}$
C	601 through 900	$P(C) = \frac{300}{1000}$ or $\frac{3}{10}$

Guided Practice

2. The owner of the candle-making business also needs to choose among seven suppliers of wicks. All of these suppliers offer the same wicks at the same price. How can the owner use the random-number app to make a fair decision?

2 Analyze Decisions You can use what you know about the probability of dependent and independent events and conditional probability to analyze decisions and strategies.

S.MD.7

Real-World Example 3 Analyze a Decision

PRIZES Every bottle of Citro orange juice has a cap with a number of points printed inside the cap. The probability of getting a cap with various numbers of points is shown in the table. Caps worth a total of 100 points can be redeemed for a prize valued at $7.95. Dalton decides to pay $54 to buy a case of 24 bottles of the juice. He plans to redeem all of the points he gets for as many prizes as possible. Considering only the financial aspect of the strategy, is this a good decision? Explain.

Points	Probability
10	60%
25	30%
40	9%
75	1%

Step 1 Find the expected or average number of points on each bottle cap.

Multiply each possible number of points by its probability and add.

$10(0.6) + 25(0.3) + 40(0.09) + 75(0.01) = 17.85$

Dalton can expect that each bottle cap will be worth an average of 17.85 points.

Step 2 Find the number of points Dalton can expect from a case of juice.

$24(17.85) = 428.4$

Supplier C: 201 through 300; if a number from 301 to 1000 is chosen, choose a new number until a supplier is chosen.

Need Another Example?

BASKETBALL A coach wants to choose 5 players for the starting lineup of a basketball game. There are 8 players available and the coach has a random number generator that chooses a random whole number from 1 to 100, inclusive. How can the coach use the random number generator to choose the players? Sample answer: assign Player A to the numbers 1 through 10, Player B to 11 through 20, and so on, up to Player H, who is assigned to the numbers 71 through 80. Choose numbers until 5 players are chosen. If the random number generator returns a number from 81 through 100, choose a new number until the starting lineup is chosen.

Step 3 Find the value of the prizes Dalton can expect to win from a case of juice.

Dalton can redeem 428 points for 4 prizes.

The value of the prizes is 4($7.95) = $31.80

Step 4 Analyze the decision.

Dalton paid $54 for the case of juice, which is greater than the value of the prizes, so this was not a good decision.

▸ **Guided Practice**

3. No; the expected number of tickets from each spin is 5.7, so Jaycee can expect to win 28.5 tickets with 5 spins. This results in one prize worth $3.25, which is less than the cost of the 5 spins ($5).

3. CARNIVALS A wheel at a carnival has 20 equal sections. One section is worth 50 prize tickets, 5 sections are each worth 10 tickets, and the remaining sections are each worth 1 ticket. Each spin costs $1, and tickets worth a total of 25 points can be redeemed for a prize valued at $3.25. Jaycee plans to spin the wheel 5 times to win as many prizes as possible. Considering only the financial aspect of the strategy, is this a good decision? Explain.

Real-World Example 4   Analyze a Decision   S.MD.7

VETERINARY MEDICINE A veterinarian knows that 1% of the dogs in the country are infected with a virus. There is a test for the virus that is 98% accurate. This means that it returns a positive result 98% of the time if a dog is infected with the virus and a negative result 98% of the time if a dog is not infected. The vet decides to treat any dog that tests positive for the virus. Do you think this a good decision? Explain.

> **Watch Out!**
>
> **Two-Way Tables** When you make a two-way frequency table, the values in the *Totals* row and the values in the *Totals* column must add up to the value in the cell in the bottom right corner.

Make a two-way frequency table to show the data. You can start by assuming a large population of 100,000 dogs.

Enter the total population of 100,000 dogs in the cell in bottom right corner.

Fill in the *Totals* column using the fact that 1% of the dogs have the virus.

Fill in the interior cells using the fact that the test is 98% accurate.

Add to fill in the *Totals* row.

	Tests Positive	Tests Negative	Totals
Has Virus	980	20	1000
Does Not Have Virus	1980	97,020	99,000
Totals	2960	97,040	100,000

Find the conditional probability that a dog that tests positive for the virus actually has the virus. Use the values in the *Tests Positive* column.

P(has virus | tests positive) = $\frac{980}{2960} \approx 33.1\%$

4. No; the probability that a patty that is identified as not the correct weight actually is not the correct weight is about 33.2%, so it may be wasteful to throw out every patty that is identified as not the correct weight.

The probability that a dog that tests positive for the virus is actually infected with the virus is only about 33%. Therefore, it may not be an efficient use of resources to treat every dog that tests positive for the virus and the vet's decision may not be a good one.

▸ **Guided Practice**

4. PRODUCT TESTING A factory produces and packages hamburger patties. It is known that 0.5% of the patties are not the correct weight. A scale at the factory is 99% accurate in identifying whether or not a patty is the correct weight. A manager at the factory decides to throw out any patty that is identified as not the correct weight. Do you think this is a good decision? Explain.

Differentiated Instruction

IF   students are successful solving conditional probability problems like the one in Example 4,

THEN   have them work in small groups to research Bayes's Theorem. Bayes's Theorem is a formula that allows you to calculate $P(A|B)$ when you know $P(B|A)$ and other probabilities related to a situation. Have students present the solution to Example 4 using Bayes's Theorem and ask them to compare it to the solution method that uses a two-way frequency table.

---

2 Analyze Decisions

Example 3 Analyze a Decision

AL **What do you notice about the probabilities in the table?** The sum of the probabilities is 100%.

OL **Why is it reasonable that the expected or average number of points per cap is 17.85?** Most of the caps are printed with 10 points or 25 points, so it makes sense that the average number of points is somewhere between these values.

BL **What would the value of the prize need to be in order for this to be a good decision?** at least $13.51

Need Another Example?

MINIATURE GOLF The last hole at a miniature golf course gives players a chance to win prize tokens based on where their golf ball lands. Based on observations over the past few years, the probability of winning 5 tokens is 80% and the probability of winning 10 tokens is 20%. Tokens worth a total of 30 points can be redeemed for a prize worth $17.25. Each game of miniature golf costs $4. Rima decides to play 10 games and redeem the tokens for as many prizes as possible. Considering only the financial aspect of the strategy, is this a good decision? Explain. No; the expected number of tokens from each game is 6, so Rima can expect to win 60 tokens with 10 games. This results in 2 prizes worth a total of $34.50, which is less than the cost of the 10 games ($40).

Example 4 Analyze a Decision

AL **How is the value 97,020 in the center of the table calculated?** 0.98(99,000) = 97,020

OL **Would the answer to the problem be different if you started with a population other than 100,000? Why or why not?** No; the numbers in the table would be different, but the relevant probability ratio would be the same.

BL **What is the probability that a dog that tests negative for the virus actually does not have the virus?** ≈99.98%

(continued on the next page)

Need Another Example?

WEATHER It is known that there is a 2% chance of rain on any day in October in Elmwood. The weather forecast for October 15 calls for rain, and the weather forecast is 95% accurate. A principal decides to cancel a school picnic for October 15 based on this forecast. Do you think this is a good decision? Explain. No; the probability that it actually rains given a prediction of rain is about 27.9%, so it may not make sense to cancel the picnic based only on this prediction.

Practice

Formative Assessment Use Exercises 1–4 to assess students' understanding of the concepts in the lesson.

The Practice and Problem Solving exercises assess the content taught in the lesson. The Preparing for Assessment page is meant to be used as preparation for end-of-course assessment.

Teaching Tip

Two-Way Frequency Tables The result of Example 4 may seem surprising to students. A close look at the two-way frequency table can help students understand this counterintuitive result. Point out that the relevant part of the table is the *Tests Positive* column. Restricting the population of dogs to those in this column shows that about one third of the dogs who test positive actually have the virus.

Watch Out!

Preventing Errors When students calculate the expected or average number of points on each bottle cap in Example 3, remind them to pay attention to how they write the probabilities as decimals. For example, some students may write 1% as 0.1 rather than 0.01 and this will cause an error that throws off the results in all remaining parts of the solution.

Go Online! eBook

Interactive Student Guide

Use the *Interactive Student Guide* to deepen conceptual understanding.
· Probability Distributions
· Confidence Intervals and Hypothesis Testing
· Making Decisions
· Analyzing Decisions

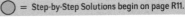 

Go Online! for a Self-Check Quiz

Check Your Understanding   ◯ = Step-by-Step Solutions begin on page R11.

Example 1
S.MD.6

1. SOFTBALL Alexa, Gayle, and Mei are trying to decide which one of them will be captain of their softball team for the coming season. They decide to roll two number cubes and use the sum of the numbers rolled to determine the captain. If the sum is 2 through 5, Alexa is the captain; if the sum is 6 through 8, Gayle is the captain; and if the sum is 9 through 12, Mei is the captain. **1–4. See margin.**

    a. Explain why this method does not result in a fair decision.

    b. Explain how they can use the sum of the numbers rolled to make a fair decision.

Example 2
S.MD.6

2. GARDENING There are eight types of seeds that Derek wants to plant in his garden, but he only has enough space to plant three different types. He wants to choose the three types of seeds to plant by using a random-number generator that chooses a whole number from 1 to 50, inclusive. Explain how he can use the random-number generator to make a fair decision.

Example 3
S.MD.7

3 ARCADE GAMES An arcade game consists of rolling a ball onto a platform with several holes. The player wins tokens based on which hole the ball lands in. The game has been set up to have the probabilities shown in the table. Each roll costs $0.25 and players can redeem 15 tickets for a prize worth $4.95. Latrell decides to play the game 10 times and redeem his tickets for as many prizes as possible. Considering only the financial aspect of the strategy, is this a good decision? Explain.

WIN!!

Tokens	Probability
20	2%
10	10%
5	18%
2	30%
1	40%

Example 4
S.MD.7

4. TRAFFIC ACCIDENTS All of the taxis in a town are yellow, except for 5% that are blue. After a hit-and-run traffic accident, a witness says that the accident was caused by a blue taxi. A detective knows that witnesses in the town are 90% accurate in identifying the color of vehicles involved in accidents. The detective decides to restrict the search to blue taxis. Do you think this is a good decision? Explain.

Practice and Problem Solving   Extra Practice is found on page R8.

Example 1
S.MD.6

Micah and Carrie want to decide who gets the last granola bar in a box. They spin two spinners at the same time and find the sum of the resulting numbers. One spinner has three equal sections labeled 1–3 and the other has four equal sections labeled 1–4. Determine whether each method results in a fair decision. Explain.

5. Micah wins if the sum is odd; Carrie wins if the sum is even.

6. Micah wins if the sum is prime; Carrie wins if the sum is composite.

7. Micah wins if the sum is a multiple of 3; Carrie wins if the sum is not a multiple of 3.

5. Yes; $P(\text{Micah}) = P(\text{Carrie}) = \frac{1}{2}$

6. No; $P(\text{Micah}) = \frac{7}{12}$; $P(\text{Carrie}) = \frac{5}{12}$

7. No; $P(\text{Micah}) = \frac{1}{3}$; $P(\text{Carrie}) = \frac{2}{3}$

Example 2
S.MD.6

8. DRAWING LOTS Drawing lots is a method of using sticks, straws, or other objects to make a fair decision. Sara has 20 straws that are all the same length. She wants to use the straws to decide which two members of the drama club will serve as copresidents. Assuming there are 15 members in the club, describe how Sara can cut some of the straws to make them shorter and then have members choose straws to make a fair decision. **8–9. See margin.**

Example 3
S.MD.7

9. CEREAL Each box of Rice Crunchies cereal is printed so that the inside of the box top shows a number from 1 to 5. The numbers appear with the probabilities shown in the table. Customers can collect the box tops and redeem them for prizes based on the sum of the numbers. A prize worth $18.95 requires a sum of 10 or greater. Ricardo decides to buy 8 boxes of cereal and redeem the box tops for as many prizes as possible. Each box of cereal costs $3.65. Considering only the financial aspect of the strategy, is this a good decision? Explain.

Number	Probability
1	35%
2	29%
3	22%
4	13%
5	1%

Differentiated Homework Options

Levels	**AL** Basic	**OL** Core	**BL** Advanced
Exercises	5–10, 16–26	5–9 odd, 14, 21–26	11–26
2-Day Option	5–9 odd, 21–26	5–10, 21–26	
	6–10 even, 21–26	11–20	

Additional Answers

1a. $P(\text{Alexa}) = \frac{5}{18}$, $P(\text{Gayle}) = \frac{4}{9}$, $P(\text{Mei}) = \frac{5}{18}$; since these probabilities are not equal, the method is not fair.

1b. Sample answer: If the sum is 2 through 5, Alexa is the Captain; if the sum is 6 or 8, Gayle is the captain; if the sum is 9 through 12, Mei is the captain; if the sum is 7, roll again.

2. Sample answer: Assign numbers as follows. Type A: 1–6; Type B: 7–12; Type C: 13–18; Type D: 19–24; Type E: 25–30; Type F: 31–36; Type G: 37–42; Type H: 43–48. Choose numbers until three different seed types are selected, ignoring any result in which 49 or 50 is chosen.

3. Yes; the expected number of tokens from each roll is 3.3, so Latrell can expect to win 33 tickets. These can be redeemed for two prizes worth a total of $9.90, which is more than the cost of the 10 rolls ($2.50).

Example 4
S.MD.7

10. No; the probability that a phone that is identified as defective actually is defective is about 3.1%, so it may be wasteful to discard every phone that is identified as defective by the test.

13. Sample answer: Latoya presents if the number is 5 through 13; David presents if the number is 14 through 22; if the number is 23, choose a new number.

10. QUALITY CONTROL It is known that 0.1% of the smartphones produced at a factory are defective. A quality-control engineer has a quick way of testing the phones that is 97% accurate. The engineer tests some phones and discards any that the test shows to be defective. Do you think this is a good decision? Explain.

TOOLS Latoya and David are biologists. They need to decide which one of them will present their work at a conference. Explain how they can use each of the following tools to make a fair decision.

 11. two number cubes that each have faces numbered 1 through 6

11. Sample answer: Latoya presents if the sum is 6 or less; David presents if the sum is 8 or more; roll again if the sum is 7.

12. a spinner with 5 sections labeled 1–5 and a spinner with 2 sections labeled 1 and 2
Sample answer: Latoya presents if the sum is even; David presents if the sum is odd.

13. a random-number generator that chooses a whole number from 5 to 23, inclusive

14. Visitors to a school fair can spin a wheel to win prize tickets. The sections of the wheel each show a number of tickets according to the probabilities in the table, where the variables represent a probability written as a decimal between 0 and 1.

Tickets	Probability
5	a
10	b
15	c
20	d

a. Mollie spins the wheel x times. Write an expression for the total number of tickets she can expect to win. $x(5a + 10d + 15c + 20d)$

b. Mollie determines that each ticket has a value of t dollars. She decides she will only spin the wheel if the expected dollar value of the tickets after x spins is greater than $12. Write an inequality to represent this decision. $xt(5a + 10d + 15c + 20d) > 12$

(15) SENSE-MAKING A company makes yogurt pops in four flavors: berry, peach, lemon, and cherry. The pops come in boxes of six pops, and, according to the company, the four flavors are produced in equal numbers and are chosen at random for the boxes. **15a–b. See margin.**

a. Brian bought a box of the yogurt pops and was surprised to find that it did not contain any berry pops. He decided that this was very unlikely to happen by chance and wrote an angry email to the company. Do you agree with this decision? Use probability to justify your answer.

b. Suppose Brian had purchased a case of five boxes of the pops and did not get any berry pops. Would you agree with his decision to write the email in this case? Explain.

S.MD.6, S.MD.7

H.O.T. Problems Use Higher-Order Thinking Skills

16–20. See Ch. 8 Answer Appendix.

16. OPEN-ENDED Ellie, Mitchell, and Rosa all want to attend a concert, but they only have two tickets. They have a set of 20 cards that are numbered 1 to 20. Describe one way they can use the cards to make a fair decision about which two of the three friends will get the tickets.

17. CRITIQUE ARGUMENTS A spinner has 7 equal sections that numbered 1 to 7. Dylan said that it is not possible to make a fair decision between two options using this spinner since there is an odd number of outcomes for each spin. Do you agree or disagree? Explain.

18. REASONING A carnival wheel has a 65% chance of landing on "2 tickets" and a 35% chance of landing on "5 tickets." Margo only wants to spin the wheel if the expected value of the prize tickets after one spin is greater than $2. What should be true about the value of each ticket in order for Margo to decide to spin the wheel? Explain.

19. ERROR ANALYSIS Ray said he could use a standard deck of cards to make a fair decision when choosing between three options by assigning one option to the red cards, one option to the clubs, and one option to the spades and choosing a card at random. Explain his error and describe a correct way to use the cards to make a fair decision.

20. WRITING IN MATH Explain how making a fair decision is related to the idea of a fair probability experiment.

Standards for Mathematical Practice

Emphasis On	Exercises
1 Make sense of problems and persevere in solving them.	3, 9, 14, 15, 21, 22, 24
3 Reason abstractly and quantitatively.	14, 17, 18
5 Use appropriate tools strategically.	2, 8, 11–13, 16, 17, 19, 23
7 Look for and make sure of structure.	4, 10, 14

9. No; the expected number on a box top is 2.16, so Ricardo can expect to have a sum of 17.28 with 8 boxes. This results in one prize worth $18.95, which is less than the cost of the 8 boxes ($29.20).

15a. No; the probability that a pop is not berry is 0.75, so the probability that none of the 6 pops is berry is $(0.75)^6 \approx 17.8\%$. This means that approximately

1 out of 5 boxes will not contain any berry pops, so it is not a very unlikely occurrence.

15b. Yes; the probability of no berry pops in a case is $(0.178)^5 \approx 0.02\%$, so this is very unlikely to happen by chance.

Teaching the Mathematical Practices

Structure Mathematically proficient students can see complicated things—such as some algebraic expressions—as single objects or as composed of several objects. Exercise 14 gives students a chance to explore the underlying algebraic structure of a problem. Encourage students to use multiple points of view to describe the expressions they write. For instance, students might see the answer to part **a** as a product of two quantities, but they might also view one of those quantities as a sum.

Extra Practice

See page R8 for extra exercises for students who are approaching level or for on-level students who need additional reinforcement.

Assess

Ticket Out the Door Make several copies of descriptions of different ways to make a decision among two options. (For example: Roll a number cube. Choice A: prime number; Choice B: composite number.) Give one description to each student. As students leave the room, ask them to state whether the method results in a fair decision.

Levels of Complexity Chart

The levels of the exercises progress from 1 to 3, with Level 1 indicating the lowest level of complexity.

Exercises	5–10	11–14, 21–26	15–20
Level 3			●
Level 2		●	
Level 1	●		

Additional Answers

4. No; the probability that a taxi is blue given that a witness said it was blue is only about 32%, so it is probably not a good decision to restrict the search to blue taxis.

8. Sample answer: Cut two of the straws to make them shorter than the others. Arrange those two straws plus 13 full-length straws in a bunch so that it is not clear which straws are the shorter ones. Have all club members choose a straw. The members who choose the two short straws are co-presidents.

Preparing for Assessment

Exercises 21–26 require students to use the skills they will need on assessments. Each exercise is dual-coded with content standards and standards for mathematical practice.

Dual Coding		
Items	Content Standards	(MP) Mathematical Practices
21	S.MD.6	1
22	S.MD.6	1
23	S.MD.6	5
24	S.MD.6	1
25	S.MD.7	4
26	S.MD.7	4

Diagnose Student Errors

Survey student responses for each item. Class trends may indicate common errors and misconceptions.

21.

A	Incorrectly counted the number outcomes assigned to each option
B	CORRECT
C	CORRECT
D	Did not recognize that Option Y should be assigned numbers greater than or equal to 5
E	CORRECT
F	CORRECT

22.

A	Chose the wheel with the greatest probability for some number of tickets
B	Incorrectly calculated the expected number of tickets
C	CORRECT
D	Chose the wheel that offers the greatest possible number of tickets on a spin

23.

A	Did not recognize that there is one more odd number than there are even numbers
B	CORRECT
C	Did not recognize that Tyra is assigned one more number than Connor
D	Did not recognize that there are many more composite numbers than there are prime numbers

Preparing for Assessment

21. Chitra has a spinner with eight equal sections that are numbered 1 through 8. She wants to use the spinner to make a decision between two options, X and Y. Which methods of assigning outcomes to the options result in a fair decision? (MP) 1 S.MD.6 B, C, E, F

- [ ] **A** option X: multiple of 3; option Y: not a multiple of 3
- [ ] **B** option X: prime number; option Y: not a prime number
- [ ] **C** option X: odd number; option Y: even number
- [ ] **D** option X: less than 5; option Y: greater than 5
- [ ] **E** option X: divisible by 2; option Y: not divisible by 2
- [ ] **F** option X: 1, 2, 7, or 8; option Y: any other number

22. A street fair has four different wheels that visitors can spin to win prize tickets. For each wheel, the probability of winning different numbers of tickets is shown in the table. Mario wants to spin the wheel for which he can expect to win the greatest number of tickets on a single spin. Which wheel should he decide to spin? (MP) 1 S.MD.6 C

Wheel A	2 tickets: 80% 4 tickets: 20%
Wheel B	1 ticket: 75% 3 tickets: 15% 5 tickets: 10%
Wheel C	2 tickets: 50% 3 tickets: 35% 4 tickets: 15%
Wheel D	1 ticket: 70% 6 tickets: 30%

- ○ **A** wheel A
- ○ **B** wheel B
- ○ **C** wheel C
- ○ **D** wheel D

23. Tyra and Connor need to decide which one of them will mow the lawn. They decide to use a random-number generator that chooses a whole number between 1 and 99. Which method of assigning numbers results in a fair decision? (MP) 5 S.MD.6 B

- ○ **A** Tyra: odd numbers; Connor: even numbers
- ○ **B** Tyra: less than 50; Connor: greater than 50
- ○ **C** Tyra: 1 through 50; Connor: 51 through 99
- ○ **D** Tyra: prime numbers; Connor: composite numbers

24.

A	Did not recognize that P(Movie) > P(Mall)
B	Did not recognize that P(Movie) < P(Mall)
C	Did not recognize that P(Movie) < P(Mall)
D	CORRECT

25.

A	Incorrectly calculated the cost of the cans and/or the expected value of the prizes
B	Did not understand how to compare costs to evaluate the decision
C	Incorrectly calculated the cost of the cans and/or the expected value of the prizes
D	CORRECT

24. Danielle wants to use the two spinners shown here to decide whether she goes to a movie or goes to the mall. She plans to spin the spinners at the same time and find the sum of the resulting numbers. Which method of assigning outcomes to the options results in a fair decision? (MP) 1 S.MD.6 E

- ○ **A** movie: sum is less than or equal to 7; mall: sum is greater than 7
- ○ **B** movie: sum is even; mall: sum is odd
- ○ **C** movie: sum is odd; mall: sum is even
- ○ **D** movie: sum is 5, 8, or 9; mall: sum is 6 or 7
- ○ **E** movie: sum is less than 7; mall: sum is greater than 7; spin again if the sum equals 7

25. It's a bad decision because the cost of cans is greater than the expected value of the prizes.

25. Every can of Toby's tomato sauce has a number (6, 8, or 10) printed on the inside of the lid. The numbers appear with the following probabilities. 6: 70%, 8: 20%, and 10: 10%. When a customer has lids that total 30 or more, they can be redeemed for a prize worth $5.20. The cans of sauce cost $1.69 each. Miguel decides to buy 10 cans and redeem the lids for as many prizes as possible. Is this a good or bad decision? Explain. (MP) 4 S.MD.7

26c. No; the probability that a cat has the infection, given a positive test result, is only about 14.9%, so it may not be a good use of resources to treat every cat that tests positive.

26. MULTI-STEP There is a 1.5% probability that a cat coming into a shelter will have a particular type of eye infection. A vet has a test that is 92% accurate in detecting the infection. The vet decides to treat every cat at the shelter that tests positive for the infection. (MP) 4 S.MD.7

a. Make a two-way frequency table to represent this situation, assuming a total population of 1,000,000 cats. See margin.

b. Find the probability of an eye infection, given that the test result is positive. ≈ 14.9%

c. Do you think it was a good decision to treat every cat that tests positive? Explain.

Additional Answer

26a.

	Tests Positive	Tests Negative	Totals
Has Infection	13,800	1200	15,000
No Infection	78,800	906,200	985,000
Totals	92,600	907,400	1,000,000

CHAPTER 8
Study Guide and Review

 *Go Online!* for Vocabulary Review Games and key vocabulary in 13 languages

Study Guide

Key Concepts

Random Sampling (Lesson 8-1)
- A survey, an experiment, or an observational study can be used to collect information.
- Samples can be used to make inferences about a population.

Using Statistical Experiments (Lesson 8-2)
- A simulation can be used to predict outcomes of an experiment.
- Data can be used to compare theoretical and experimental probabilities.

Population Parameters (Lesson 8-3)
- The population mean and population proportion can be predicted using samples.
- The margin of error depends on the sample size.

Distributions of Data (Lesson 8-4)
- Use the mean and standard deviation to describe a symmetric distribution.
- Use the five-number summary to describe a skewed distribution.

Evaluating Published Data (Lesson 8-5)
- Analyzing reports is important to discern if the conclusions are valid.

Normal Distributions (Lesson 8-6)
- The graph of a normal distribution is bell-shaped.
- The z-value represents the number of standard deviations that a given data value is from the mean.

Using Probability to Make Decisions (Lesson 8-7)
- Probability experiments can be used to make fair decisions.

FOLDABLES Study Organizer

Use your Foldable to review the chapter. Working with a partner can be helpful. Ask for clarification of concepts as needed.

 Statistics | Probability

Key Vocabulary

bias (p. 531)

confidence level (p. 546)

distribution (p. 551)

Empirical Rule (p. 566)

experiment (p. 531)

experimental probability (p. 538)

fair (p. 573)

margin of error (p. 546)

negatively skewed distribution (p. 551)

normal distribution (p. 566)

observational study (p. 531)

parameter (p. 531)

population mean (p. 545)

population proportion (p. 545)

positively skewed distribution (p. 551)

probability model (p. 538)

random sample (p. 531)

relative frequency (p. 538)

simulation (p. 538)

standard normal distribution (p. 568)

statistic (p. 531)

statistics (p. 531)

survey (p. 531)

symmetric distribution (p. 551)

theoretical probability (p. 538)

z-value (p. 568)

Vocabulary Check

Choose a term from the list above that best completes each statement.

1. A(n) _____ is an error that results in a misrepresentation of members of a population. **bias**

2. In a statistical study, data are collected and used to answer questions about a population characteristic or _____ **parameter**

3. The _____ can be used to determine the area under the normal curve at specific intervals. **Empirical Rule**

4. In a(n) _____ members of a sample are measured or observed without being affected by the study. **observational study**

Concept Check

5. Explain the difference between a negatively skewed and a positively skewed distribution. **5–6. See margin.**

6. Explain how the sample size affects the margin of error.

Answering the Essential Question

Before answering the Essential Question, have students review their answers to the *Building on the Essential Question* exercises found throughout the chapter.

- How can you effectively evaluate information? (p. 528)
- How can you use information to make decisions? (p. 528)
- What should you consider when using the results of a simulation to make a prediction? (p. 542)
- What strategy could be used to test the accuracy of the size of a random sample? (p. 549)
- Can statistics lie? (p. 571)

FOLDABLES Study Organizer

A completed Foldable for this chapter should include the Key Concepts related to statistics and probability.

Allow students to compare their Foldable with a partner. Encourage them to share what has been helpful to them as they study.

Key Vocabulary

The page reference after each word denotes where that term was first introduced. If students have difficulty answering questions 1–6, remind them that they can use these page references to refresh their memories about the vocabulary terms.

Have students work together to review. Encourage them to summarize their understanding and ask each other questions about the content of Chapter 8.

 You can use the detailed reports in ALEKS to automatically monitor students' progress and pinpoint remediation needs prior to the chapter test.

Additional Answers

5. A negatively skewed distribution has most of the data to the right of the mean and a positively skewed distribution has most of the data to the left of the mean.

6. The larger the sample size, the more accurate the statistic is and the smaller the margin of error.

Go Online!

Vocabulary Review

Students can use the *Vocabulary Review Games* to check their understanding of the vocabulary terms in this chapter. Students should refer to the *Student-Built Glossary* they have created as they went through the chapter to review important terms. You can also give a *Vocabulary Test* over the content of this chapter.

Lesson-by-Lesson Review

Intervention If the given examples are not sufficient to review the topics covered by the questions, remind students that the lesson references tell them where to review that topic in their textbook.

Two-Day Option Have students complete the Lesson-by-Lesson Review. Then you can use McGraw-Hill eAssessment to customize another review worksheet that practices all the objectives of this chapter or only the objectives on which your students need more help.

Additional Answers

7. survey; sample: every tenth shopper; population: all potential shoppers

8. observational study; sample: the 25 customers; population: all potential customers

9. survey; sample: every fifth person; population: student body

10a. Sample answer: the theoretical probability that a baseball player gets on base is 30%. Use a spinner with 10 equal sections labeled 1–10. The numbers 1–3 will represent the player getting on base and 4–10 will represent the player not getting on base. The simulation will consist of 40 trials.

10b. Sample answer: P(on base) = 0.35

outcome	Probability
on base	35%
not on base	65%

11a. Sample answer: The theoretical probability that the bus is late is 60%, and the theoretical probability that the bus is not late is 40%. Use a random number generator to generate integers 1 through 5. The integers 1–3 will represent the bus being late, and the integers 4–5 will represent the bus not being late. The simulation will consist of 50 trials.

11b. Sample answer: P(late) = 58%

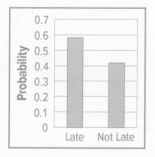

Lesson-by-Lesson Review

8-1 Random Sampling
F.IF.7e, F.BF.3

Determine whether each situation describes a *survey*, an *experiment*, or an *observational study*. Then identify the sample, and suggest a population from which it may have been selected. **7–9. See margin.**

7. **SHOPPING** Every tenth shopper coming out of a store is asked questions about his or her satisfaction with the store.

8. **MILK SHAKE** A fast food restaurant gives 25 of their customers a sample of a new milk shake and employees monitor their reactions as they taste it.

9. **SCHOOL** Every fifth person coming out of a high school is asked what their favorite class is.

Example 1

A dealership wants to test different promotions. They randomly select 100 customers and ask them which promotion they prefer. Does this situation describe a *survey*, an *experiment*, or an *observational study*? Identify the sample, and suggest a population from which it may have been selected.

This is a survey, because the data are collected from participants' responses. The sample is the 100 customers that were selected, and the population is all potential customers.

8-2 Using Statistical Experiments
S.IC.2, S.IC.5

10. **BASEBALL** A baseball player gets on base 30% of the time that he is up to bat. **a–b. See margin.**

   a. Design a simulation that can be used to estimate the probability that the player will get a on base his next bat.

   b. Conduct the simulation and report the results.

11. **SCHOOL BUS** Christy has determined that the school bus is late 60% of the time. **a–b. See margin.**

   a. Design a simulation that can be used to estimate the probability that the school bus is late today.

   b. Conduct the simulation, and report the results.

Example 2

GROUPS Before the random drawing of groups, Dawn has determined that she is 20% likely to get placed in the same group as Sherry. Design a simulation that can be used to estimate the probability of Dawn and Sherry being in the same group.

Step 1 There are two possible outcomes.

Possible Outcomes	Theoretical Probability
grouped together	20%
not grouped together	80%

Step 2 We can use the random number generator on a graphing calculator. Assign the integers 1–5 to accurately represent the probability data.

Outcome	Represented by
grouped together	1
not grouped together	2–5

Step 3 A trial will represent one drawing of groups. The simulation can consist of any number of trials. We will use 20.

8-3 Population Parameters

S.IC.4

In an eleventh grade class, there are 140 total students.

12. In a survey, 35 of the eleventh grade students reported that they share a bedroom with a sibling. What is the population proportion of students who do not share a bedroom? **0.75**

13. In the same survey, the students were asked how many pets they have. The results are shown in the table. What is the population mean number of pets owned by the students?
There is a population mean of 1.5 pets.

Number of Pets	Number of Students
0	26
1	39
2	56
3	17
4	2

14. A sample of 30 of the eleventh grade students is conducted to determine the mean pulse rate of all eleventh graders. What is the margin of error? **about 18%**

Example 3

HOUSING In a new development, 40 new houses are being built. The design plans were analyzed to determine how many bathrooms are in each house. The results are shown in the table. What is the population mean number of bathrooms in the houses?

Number of Bathrooms	Number of Houses
1	3
2	17
3	13
4	6

Step 1 Sum the total number of bathrooms.

$$(1 \times 3) + (2 \times 17) + (3 \times 13) + (4 \times 6) = 100$$

Step 2 Divide the answer in Step 1 by the total number of houses.

$$\frac{100}{40} = 2.5$$

The mean number of bathrooms in the population is 2.5.

Go Online!

Anticipation Guide

Students should complete the *Chapter 8 Anticipation Guide*, and discuss how their responses have changed now that they have completed Chapter 8.

15a.

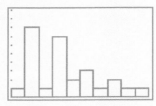

[8, 18] scl: 1 by [0, 10] scl: 1

positively skewed

15b. Sample answer: The distribution is skewed, so use the five-number summary. The values range from 8.7 to 18.1 days. The median is 11.55 days, and half of the data are between 9.55 and 13.3 days.

16a.

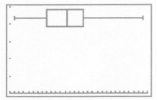

[300, 330] scl: 1 by [0, 5] scl: 1

positively skewed

16b. Sample answer: The distribution is skewed, so use the five-number summary. Kelly's times range from 301 to 329 seconds. The median is 311, and half of the data are between 307 and 316 seconds.

8-4 Distributions of Data

Prep for S.ID4, S.IC.1

15. DOGSLED The Iditarod is a race across Alaska. The table shows the winning times, in days, for recent years.

Iditarod Winning Times

9.1, 9.4, 10.3, 9.3, 9.6, 8.7, 9.5, 9.4, 9.2, 17.3, 15.4, 15.5, 14.2, 12.0, 16.6, 13.5, 13.0, 18.1, 12.4, 11.6, 11.5, 11.3, 11.3, 13.1, 11.2, 11.6, 11.6, 9.7

a–b. See margin.

a. Use a graphing calculator to create a histogram. Then describe the shape of the distribution.

b. Describe the center and spread of the data using either the mean and standard deviation or the five-number summary. Justify your choice.

16. SWIMMING Kelly's practice times in the 400-meter individual medley are shown in the table.

a–b. See margin.

Times in Seconds

301, 311, 320, 308, 312, 307, 303, 305, 309, 308, 304, 302, 311, 313, 313, 316, 314, 306, 329, 326, 319, 310, 306, 309, 320, 318, 315, 318, 314, 309

a. Use a graphing calculator to create a box-and-whisker plot. Then describe the shape of the distribution.

b. Describe the center and spread of the data using either the mean and standard deviation or the five-number summary. Justify your choice.

Example 4

Data collected from a group of sixth graders is shown.

Number of Years Playing an Instrument

2.5, 2.4, 3.1, 2.9, 4.2, 1.3, 2.6, 2.4, 3.3, 1.9, 3.4, 4.8, 2.3, 1.7, 3.2, 2.3, 3.5, 2.2, 3.6, 1.2, 4.4, 2.1, 3.4, 4.5, 1.9, 1.5, 1.4, 0.7, 1.2, 2.5, 1.9, 2.0, 2.4, 2.5, 3.4

a. Use a graphing calculator to create a histogram. Then describe the shape of the distribution.

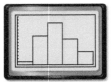

[0, 5] scl: 1 by [0, 15] scl: 1

The distribution is symmetric.

b. Describe the center and spread of the data using either the mean and standard deviation or the five-number summary. Justify your choice.

The distribution is symmetric, so use the mean and standard deviation. The mean number of years is about 2.6 with standard deviation of about 1 year.

8-5 Evaluating Published Data

S.IC.6

17. TEST SCORES Explain whether the conclusion in the following report is valid.

> **Sleeping 8 hours a Night Causes Higher Grades**
>
> A group of students were surveyed. If they slept 8 hours or more per night, they were put into one group, and everyone else in another group. Then, each student's grades were analyzed and it was determined that the group who slept more than 8 hours per night got better grades.

The conclusion may not be valid. The sample may not have been chosen using a simple random sample and may have been too small. This is an observational study, not an experiment. The cause and effect relationship cannot be determined.

Example 5

TEST SCORES Explain whether the conclusion in the following report is valid.

> **We Have the Highest Test Scores**
>
> The first 10 students to arrive at school this morning were surveyed to determine the average test score on the latest history test. The survey showed that the average grade was a 98%.

The conclusion may not be valid. The sample size is too small and the sample was not obtained using a simple random sampling method. Also, students may not be truthful when reporting their grade.

8-6 Normal Distributions

S.ID.4

18. RUNNING TIMES The times in the 40-meter dash for a select group of professional football players are normally distributed with a mean of 4.74 seconds and a standard deviation of 0.13 second.

  a. About what percent of players have times between 4.6 and 4.8 seconds? **53.7%**

  b. About how many of a sample of 800 players will have times below 4.5 seconds? **about 26 players**

19. ATTENDANCE The number of tickets sold at high school basketball games in a particular conference are normally distributed with a mean of 68.7 and a standard deviation of 13.1.

  a. About what percent of the games sell fewer than 75 tickets? **68.5%**

  b. About how many of a sample of 200 games will sell more than 100 tickets? **about 2 games**

20. COMMUTING The number of minutes it takes Phil to commute to work each day are normally distributed with a mean of 18.6 and a standard deviation of 3.5.

  a. About what percent of the time will it take Phil more than 20 minutes to commute to work? **34.5%**

  b. About how many of a sample of 50 days will it take Phil less than 15 minutes to commute to work? **about 8 days**

Example 6

TEST SCORES The midterm test scores for the students in Mrs. Hendrix's classes are normally distributed with a mean of 73.2 and standard deviation of 7.8. About how many test scores are between 70 and 80?

Find the corresponding z-values for $X = 70$ and 80.

$$z = \frac{X - \mu}{\sigma} \qquad\qquad z = \frac{X - \mu}{\sigma}$$
$$= \frac{70 - 73.2}{7.8} \qquad\qquad = \frac{80 - 73.2}{7.8}$$
$$\approx -0.41 \qquad\qquad\qquad \approx 0.87$$

Using a graphing calculator, you can find the area between $-0.41 < z < 0.87$ to be about 0.47.

[−4, 4] scl: 1 by [0, 0.5] scl: 0.125

Additional Answers (Practice Test)

2a.

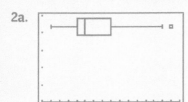

[59, 75] scl: 1 by [0, 5] scl: 1

positively skewed

2b. The distribution is skewed, so use the five-number summary. The values range from 60 to 74. The median is 64, and half of the data are between 63 and 67.

4a. Sample answer: the theoretical probability that an airplane will take off on time is 75%. Use a spinner with 4 equal sections labeled 1–4. The number 1 will represent the plane taking off late and 2–4 will represent the plane taking off on time. The simulation will consist of 20 trials.

4b. Sample answer: $P(\text{late}) = 0.2$

outcome	Probability
on time	80%
late	20%

11. The restaurant can assign each person 12 unique numbers. If the random number generator chooses one of the 40 numbers unassigned, it should be ignored and another number chosen.

Before the Test

Have students complete the Study Notebook Tie it Together activity to review topics and skills presented in the chapter.

8-7 Using Probability to Make Decisions
S.MD.6, S.MD.7

21. SPEECH Of the five members of student council, one must make a speech. To choose who will make the speech, the advisor decided to roll a number cube. He assigns the president numbers 1 and 2 and each of the other positions the other numbers.

  a. Explain why this does not result in a fair decision.

  b. Explain how they can use the results to make a fair decision.

22. MEDICAL TESTS A certain medical test reports errors 0.4% of the time. 8% of people actually have the disease. The treatment for the disease is very risky. The doctor decides to treat everyone who tests positive. Is this a good decision?
No, 4% of people would be treated without needing the treatment.

21a. The President has double the chance of being chosen as everybody else. Everyone should have an equally likely chance of being chosen.

21b. The advisor should assign each member 1 number and if the other number is rolled, ignore it and roll again.

Example 7

MATH COMPETITION A principal must choose one student to represent the school at a mathematics competition. There are 10 eligible students who are all equally qualified. The principal has a random number generator that generates numbers from 1 to 500, inclusive. To determine which student to choose, how can the principal use the random number generator to make a fair decision?

Assign an equal amount of numbers to each of the ten students. 1–50, 51–100, etc. The probability is $\frac{50}{500}$ or $\frac{1}{10}$ for each student.

Example 8

BATTERIES A company determines that 0.8% of their batteries are defective. The machine that tests the batteries is accurate 98% of the time. A quality control specialist decides to destroy all of the batteries that are tested and marked defective. Is this a good decision?

No, the probability that the battery is defective if it tests defective is 28%. Therefore, they would be destroying many batteries that were actually working.

CHAPTER 8
Practice Test

 *Go Online!* for another Chapter Test

1. **BUTTERFLIES** Students in a biology class are learning about the monarch butterfly's life cycle. Each student is given a caterpillar. When a caterpillar turns into chrysalis, it is placed in a glass enclosure with food and a heat lamp and examined.

   a. Determine whether the situation describes a *survey*, an *experiment* or an *observational study*.
   observational study

   b. Identify the sample, and suggest a population from which it was selected. **sample: the monarch butterflies observed in the study; population: all monarch butterflies**

2. **HEIGHTS** The heights of Ms. Joy's dance students are shown.

Height (Inches)				
60	64	62	69	64
63	65	64	66	73
74	63	62	65	64
68	70	66	63	61

   a. Use a graphing calculator to create a box-and-whisker plot. Then describe the shape of the distribution. **See margin.**

   b. Describe the center and spread of the data using either the mean and standard deviation or the five-number summary. Justify your choice.
   See margin.

3. A binomial distribution has a 65% rate of success. There are 15 trials.

   a. What is the probability that there will be exactly 12 successes? **about 11.1%**

   b. What is the probability that there will be at least 10 successes? **about 56.5%**

4. **AIRPLANE** A certain airline's planes take off on time 75% of the time.

   a. Design a simulation that can be used to estimate the probability that the next plane will take off on time. **See margin.**

   b. Conduct the simulation and report the results.
   See margin.

5. **MULTIPLE CHOICE** A sample of 200 nurses is conducted to determine the mean number of hours nurses typically work in one week. Which represents the margin of error? **B**

   ○ A 3.5%   ○ C 10.5%
   ○ B 7%   ○ D 14%

6. A survey was done of every adult in a small town. 8574 people stated that they are satisfied with the condition of the streets. There are 24,503 people in the town. Which can be calculated with this data? **D**

   ○ A sample mean
   ○ B sample proportion
   ○ C population mean
   ○ D population proportion

7. When evaluating a report to determine its validity, which are things to consider? Select all that apply. **A, B, D, E**

   ☐ A Was there bias in survey questions?
   ☐ B Was the sample chosen randomly?
   ☐ C Was the report published?
   ☐ D Was the sample size large enough?
   ☐ E Was it an experiment or an observational study?

8. **WEIGHTS** The weights of 1500 bodybuilders are normally distributed with a mean of 190.6 pounds and a standard deviation of 5.8 pounds.

   a. About how many bodybuilders are between 180 and 190 pounds? **about 638 bodybuilders**

   b. What is the probability that a bodybuilder selected at random has a weight greater than 195 pounds? **22.4%**

   A normal distribution has a mean of 16.4 and a standard deviation of 2.6.

9. Find the range of values that represent the middle 95% of the distribution. **11.2 < X < 21.6**

10. What percent of the data will be less than 19? **84%**

11. **PROMOTION** A restaurant serves lunches to 80 local small businesses employees each month. To say thank you, the restaurant wants to choose one employee to receive a gift card. The restaurant has a random number generator that chooses whole numbers between 1 and 1000, inclusive. To determine which person to choose, how can the restaurant use the random number generator to make a fair decision? **See margin.**

Go Online!

Chapter Tests

You can use premade leveled *Chapter Tests* to differentiate assessment for your students. Students can also take self-checking *Chapter Tests* to plan and prepare for chapter assessments.

MC = multiple-choice questions
FR = free-response questions

Form	Type	Level
1	MC	AL
2A	MC	OL
2B	FR	OL
2C	FR	OL
3	FR	BL
Vocabulary Test		
Extended-Response Test		

RtI Response to Intervention

Use the Intervention Planner to help you determine your Response to Intervention.

Intervention Planner

TIER 1 **On Level** OL

IF students miss 25% of the exercises or less,

THEN choose a resource:

   SE Lessons 8-1, 8-2, 8-3, 8-4, 8-5, and 8-6

   Go Online!
   📄 Skills Practice
   📄 Chapter Project
   ✓ Self-Check Quizzes

TIER 2 **Strategic Intervention** AL
Approaching grade level

IF students miss 50% of the exercises,

THEN *Go Online!*
   📄 Study Guide and Intervention
   ➕ Extra Examples
   💬 Personal Tutors
   📄 Homework Help

TIER 3 **Intensive Intervention**
2 or more grades below level

IF students miss 75% of the exercises,

THEN choose a resource:

   Use *Math Triumphs, Alg. 2*

   Go Online!
   ➕ Extra Examples
   💬 Personal Tutors
   📄 Homework Help
   🔤 Review Vocabulary

Launch

Objective Apply concepts and skills from this chapter in a real-world setting.

Teach

Ask:

- For Part A, what percent of beta testers reported bugs? How can you apply that data to the whole population of consumers? Sample answer: 62%; I can multiply 1,400,000 by 0.62.

- For Part B, how do you calculator the margin of error? Sample answer: Use the formula margin of error $= \pm\frac{1}{\sqrt{n}}(100)$.

- For Part C, what does it mean when a data set is positively skewed? negatively skewed? Sample answer: In data with a positive skew, the majority of the data is to the left of the mean. In data with a negative skew, the majority of the data is to the right of the mean.

- For Part D, in the middle 95%, most of the data will fall within how many standard deviations of the mean? Sample answer: 2 standard deviations

The Performance Task focuses on the following content standards and standards for mathematical practice.

Dual Coding		
Parts	Content Standards	Mathematical Practices
A	S.IC.1, S.IC.3	1, 6, 8
B	S.IC.4	1, 6
C	S.IC.5	1, 4, 8
D	S.ID.4	1, 6

Go Online!   eBook

Interactive Student Guide
Refer to *Interactive Student Guide* for an additional Performance Task.

ALGEBRA 2
INTERACTIVE STUDENT GUIDE

CHAPTER 8

Preparing for Assessment

Performance Task

Provide a clear solution to each part of the task. Be sure to show all of your work, include all relevant drawings, and justify your answers.

TECHNOLOGY A company specializes in statistical analysis. They collect and analyze various types of data.

Part A

A software company wants to get feedback on a new program, so they release a beta version and hire the statistical analysis company to collect data from the beta testers. There are 100 beta testers and of these, 62 report a bug in the software. Based on past sales, the software company expects to sell approximately 1.4 million licenses for the software once they release it.

1. Determine whether the analysis described is a survey, experiment, or observational study. Explain your answer. Survey; the company is "collecting" data and the beta-testers are "reporting" the bugs, which suggests a survey, not an experiment or observational study.

2. **Reason Quantitatively** Determine about how many consumers would be likely to report bugs if the company released the software now. 868,000 people

Part B

The statistical analysis company conducts a survey for a client with 400 respondents.

3. **Use Structure** Determine the margin of error for the survey. Round your answer to the nearest tenth, if necessary. ±5.0%

4. **Use Structure** Determine the number of respondents needed in order to achieve a margin of error of ±2.5%. Round your answer to the nearest whole number, if necessary. 1600 respondents

Part C

A snack company makes two slightly different versions of a product and hires the statistical analysis company to conduct some taste test trials and help them choose which version to sell. The company has two groups of equal numbers of participants, each group getting one version of the snack, and both asked to rate the tastiness of the snack on a scale from 1 to 10. The data is recorded in the two figures shown. Version A: negatively skewed; Version B: positively skewed

5. Determine the skew of each figure.

6. **Make Sense** Determine which version of the snack the company should sell. Justify your answer. The snack company should sell Version A because the data is more densely concentrated above 5, while in Version B, the data is more densely concentrated below 5.

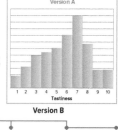
Version A

Tastiness

Version B

Part D

In one data set, the statistical analysis company finds the data has a normal distribution with a mean of 32.5 and a standard deviation of 6.75.

7. **Use Tools** Determine the range of values that represent the middle 95% of the distribution. $19 < X < 46$

8. **Use Tools** Determine the percent of the data that will be greater than 39.25. 15.87%

Levels of Complexity Chart			
Parts	Level 1	Level 2	Level 3
A	●		
B		●	
C		●	●
D		●	

Test-Taking Strategy

Example

Read the problem. Identify what you need to know. Then use the information in the problem to solve.

There are 15 boys and 12 girls in Mrs. Lawrence's homeroom. Suppose a committee is to be made up of 6 randomly selected students. What is the probability that the committee will contain 3 boys and 3 girls? Round your answer to the nearest tenth of a percent.

A 27.2%

B 29.6%

C 31.5%

D 33.8%

> **Test-Taking Tip**
> Strategies for Solving Multi-Step Problems Some problems that you will encounter on standardized tests require you to solve multiple parts in order to come up with the final solution. The question below is such a problem.

Step 1 What are you being asked to solve? What information is given and what do you still need?
The problem asks for the probability of the committee containing 3 boys and 3 girls. I have the number of boys and number of girls. I need to know the probability of 3 boys being chosen and the probability of 3 girls being chosen.

Step 2 What intermediate steps will you take to solve the problem?
I will find the number of possible successes and the number of possible outcomes. Then, I will use them to compute the probability.

Step 3 What is the correct answer?
The answer is D.

Apply the Strategy

Read the problem. Identify what you need to know. Then use the information in the problem to solve.

There are 52 cards in a standard deck. Of these, 4 of the cards are aces. What is the probability of a randomly dealt five-card hand containing exactly one pair of aces? Round your answer to the nearest whole percent.   A

A 4%

B 5%

C 6%

D 7%

Answer the questions below.

a. What are you being asked to solve? What information is given and what do you still need?
a. I need to find the probability of being dealt 2 aces in a hand of 5 cards. I'm given the number of cards in the deck, the number of cards in a hand, and the number of aces. I need the number of possible successes and the number of possible outcomes.

b. What intermediate steps will you take to solve the problem?
I'll calculate the number of possible successes and the number of possible outcomes using combinations.

c. What is the correct answer?   A

Test-Taking Strategy

Step 1 Read the problem. Determine what information is given and what is still needed.

Step 2 List any intermediate steps needed to solve the problem.

Step 3 Solve the problem.

Need Another Example?

A bakery has 8 fudge cakes and 10 yellow cakes, each in a separate, identical box. Anna needs to ice 4 cakes. What is the probability that she will randomly choose 2 fudge cakes and 2 yellow cakes to ice? C

A   about 9.7%

B   about 25.7%

C   about 41.2%

D   about 50%

a. What are you being asked to solve? What information is given and what do you still need? The problem asks for the probability of randomly choosing 2 fudge cakes and 2 yellow cakes. I have the total number of each kind of cake. I need to know the probability of each event separately.

b. What intermediate steps will you take to solve the problem? I'll find the number of possible successes and the number of possible outcomes. Then, I will use them to compute the probability.

c. What is the correct answer? C

Go Online!

The most up-to-date resources available for your program can be found at connectED.mcgraw-hill.com.

Diagnose Student Errors

Survey student responses for each item. Class trends may indicate common errors and misconceptions.

1.	A	Multiplied the numerators by the exponent
	B	Subtracted 4
	C	Divided 4 by -4
	D	Multiplied the exponent by 4 then added $\frac{2}{3}$
	E	CORRECT

2.	A	CORRECT
	B	Added 3 and i as $3i$
	C	Used the quadratic formula incorrectly
	D	Used the quadratic formula incorrectly

3.	A	Chose a solution to first equation only
	B	Chose a solution to second equation only
	C	Chose a solution to first equation only
	D	CORRECT
	E	Chose a solution to third equation only

4.	A	CORRECT
	B	Used mean > median
	C	Used mean < median
	D	Used positively skewed distribution
	E	Used negatively skewed distribution

5.	A	Selected a negatively skewed distribution
	B	Selected a symmetric distribution
	C	Selected a symmetric distribution
	D	CORRECT
	E	Selected a negatively skewed distribution

6.	A	Chose a situation calling for an observational study
	B	CORRECT
	C	CORRECT
	D	Chose a situation calling for a sample survey
	E	Chose a situation calling for a sample survey

7. Student may select the data value found in the middle of the table. Student may compute the range. Student may find the total sum of goals in 7 games.

8. Student may divide 8 games by 2 goals per game. Student may not use correct average computation. Student may incorrectly find the current total sum.

Go Online! ✓

Standardized Test Practice

Students can take self-checking tests in standardized format to plan and prepare for assessments.

CHAPTER 8
Preparing for Assessment
Cumulative Review

Read each question. Then fill in the correct answer on the answer document provided by your teacher or on a sheet of paper.

1. What is the value of the function $f(x) = 4\left(\frac{2}{3}\right)^x$ when $x = -4$? **E**

- ○ A $-\frac{32}{3}$
- ○ B $-\frac{4}{3}$
- ○ C $-\frac{2}{3}$
- ○ D $-16\frac{2}{3}$
- ○ E $20\frac{1}{4}$

> **Test-Taking Tip**
> Question 1 To evaluate an expression that contains a fraction with a negative exponent, take the reciprocal of the fraction and change the sign on the exponent to positive.

2. What is the solution to the equation? **A**

$$x^2 + 6x = -10$$

- ○ A $x = -3 \pm i$
- ○ B $x = \pm 3i$
- ○ C $x = 2$ or 4
- ○ D $x = -2$ or -4
- ○ E $x = -1$ or -3

3. What is the solution of the linear system shown below? **D**

$$x + y - 3z = 6$$
$$2x + y \quad\;\; = 5$$
$$y - 2z = 5$$

- ○ A $\left(\frac{3}{2}, \frac{3}{2}, -1\right)$
- ○ B $(1, 3, 0)$
- ○ C $\left(3, 2, -\frac{1}{3}\right)$
- ○ D $\left(\frac{1}{2}, 4, -\frac{1}{2}\right)$
- ○ E $(0, 3, -1)$

4. Which of the following is a characteristic of a set of data in which the mean and median are approximately equal? **A**

- ○ A The data have a symmetrical distribution.
- ○ B The data have a positively skewed distribution.
- ○ C The data have a negatively skewed distribution.
- ○ D The majority of the data is on the left of the mean.
- ○ E The majority of the data is on the right of the median.

5. Which box-and-whisker plot is the best representation of a positively skewed data distribution? **D**

- ○ A
- ○ B
- ○ C
- ○ D
- ○ E

6. Select all situations that call for an experiment. **B, C**

- ☐ A A teacher wants to examine handwriting skills of her students.
- ☐ B A drug company wants to test whether a new vaccine is effective.
- ☐ C A farmer wants to know which of two fertilizers will make plants grow faster.
- ☐ D A store wants to know whether shoppers would be interested in receiving weekly ads online.
- ☐ E A school district superintendent wants to determine what percentage of residents support the school budget.

9.	A	Chose undefined value for denominator
	B	Chose restriction of undefined value
	C	Chose restricted value
	D	CORRECT
	E	Thought the domain had no restricted value

10.	A	CORRECT
	B	Assumed data could not be described
	C	CORRECT
	D	Assumed that the mean was greater than the median
	E	Assumed data had a mean and median that was about the same

11a. Students may compute the number of students that scored between 80 and 86 points. Students may compute the number of students that scored between 74 and 80 points.

12.	A	Translated the graph down instead of up
	B	Translated the graph horizontally and down instead of up
	C	Translated the graph horizontally instead of vertically
	D	CORRECT
	E	Translated the graph horizontally

Go Online! for Standardized Test Practice

Questions 7–8 refer to the following table, which shows the number of goals scored by the varsity soccer team at Golden Valley High School.

Varsity Soccer Scoring	
Game	Number of Goals
1	5
2	0
3	2
4	1
5	3
6	2
7	0

7. What is the median number of goals scored by the varsity soccer team?

[2] goals

8. The coach wants the soccer team to have an average of 2 goals per game after 8 games played. How many goals does the team need to score in game 8 to achieve this average?

[3] goals

9. What is the domain of the function? **D**

$$f(x) = \frac{3}{x-1} + \frac{1}{2} - \frac{2}{(x-1)^2}$$

- ○ **A** $\{x \mid x = 1\}$
- ○ **B** $\{x \mid x \neq 0\}$
- ○ **C** $\{x \mid x = \pm 1\}$
- ○ **D** $\{x \mid x \neq 1\}$
- ○ **E** all real numbers

10. Select all phrases that describe the graph shown. **A, C**

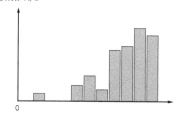

- ☐ **A** not symmetric
- ☐ **B** no correlation
- ☐ **C** negatively skewed
- ☐ **D** positively skewed
- ☐ **E** normally distributed

11. The test scores of 500 students are normally distributed. The mean score is 74 points, and the standard deviation is 6 points.

a. How many more students scored between 74 and 80 points than between 80 and 86 points?

[102] students

b. ⬤ What mathematical practice did you use to solve this problem? **See students' work.**

12. What is the vertex of the function $y = 4 - x^2$? **D**

- ○ **A** $(0, -4)$
- ○ **B** $(-4, -4)$
- ○ **C** $(4, 0)$
- ○ **D** $(0, 4)$
- ○ **E** $(4, 4)$

Need Extra Help?

If you missed Question...	1	2	3	4	5	6	7	8	9	10	11	12
Go to Lesson...	6-2	3-6	1-9	8-4	8-4	8-1	1-1	1-4	7-2	8-4	8-6	3-1

Formative Assessment

You can use these pages to benchmark student progress.

📄 Standardized Test Practice

Test Item Formats

In the Cumulative Review, students will encounter different formats for assessment questions to prepare them for standardized tests.

Exercise Question Types	
Question Type	Exercises
Multiple Choice	1–5, 9, 12
Multiple Correct Answers	6, 10
Type Entry: Short Response	7, 8, 11
Type Entry: Extended Response	11

Answer Sheet Practice

Have students simulate taking a standardized test by recording their answers on a practice recording sheet.

Homework Option

Get Ready for Chapter 9 Assign students the exercises on p. 592 as homework to assess whether they possess the prerequisite skills needed for the next chapter.

LS LEARNSMART®

Use LearnSmart as part of your test-preparation plan to measure student topic retention. You can create a student assignment in LearnSmart for additional practice on these topics.

- · Data Measurements
- · Data Collection Methods
- · Inferences and Conclusions
- · Probability and Decisions

Go Online!

ᵉ Assessment

Customize and create multiple versions of chapter tests and answer keys that align to your standards. Tests can be delivered on paper or online.

Lesson 8-1

3. observation study; sample answer: The scores of the participants are observed and compared without them being affected by the study.

4. experiment; sample answer: A sample of dyed shirts will need to be tested, which means that the members of the sample will be affected by the study.

7. observational study; sample: physics students selected; population: all college students that take a physics course

8. survey; sample: customers that take the online survey; population: all customers

25a. See students' work.

25b. Sample answer for Product A: ≈63.3%

Product A	
Number	**Frequency**
0–6	|||| |||| |||| ||||
7–9	|||| ||||	

Sample answer for Product B: ≈76.7%

Product B	
Number	**Frequency**
0–7	|||| |||| |||| |||| |||
8–9	|||| ||

25c. Sample answer: Yes; the probability that Product B is effective is 13.4% higher than that of Product A.

25d. Sample answer: It depends on what the product is and how it is being used. For example, if the product is a pencil sharpener, then the lower price may be more important than the effectiveness, and therefore may not justify the price difference. However, if the product is a life-saving medicine, the effectiveness may be more important than the price, and therefore may justify the price difference.

26. false; sample answer: A sample statistic is used to estimate a population parameter.

28. Sample answer:

objective: Determine the average amount of time that students spend studying at the library.

population: all students that study at the library

sample: 30 randomly selected students studying at the library during a given week

Study Time (minutes)				
38	16	45	41	63
18	20	17	8	15
41	28	55	19	15
30	11	20	79	24
78	24	26	32	19

mean: ≈26.1 min

29. Sample answer: the sampling method used, the type of sample that was selected, the type of study performed, the survey question(s) that were asked or procedures that were used

30a. Sample answer: In a convenience sample, members are selected based on the convenience of the researcher. One example is handing a survey to shoppers as they walk out of the mall. This method could result in bias if the members of the population who are readily available to be sampled are not representative of the entire population.

30b. Sample answer: In a self-selected sample, members volunteer to be in the sample. This method could result in bias if certain groups of people in the population choose not to volunteer.

30c. Sample answer: In a stratified sample, the population is first divided into similar, nonoverlapping groups, and members are then randomly selected from each group. This method could result in bias if the entire population is not represented when divided into groups or if the members are not randomly selected from each group.

30d. Sample answer: In a systematic sample, a rule is used to select the members. This method could result in bias if the rule does not include everyone in the population.

Lesson 8-2 (Guided Practice)

1. Sample answer: The theoretical probability that a student chooses to ride the bus is 50%, parent drop off 20%, friend ride 20%, and walk 10%. Use a random number generator to generate integers 1–10. The integers 1–5 will represent a bus ride, the integers 6–7 will represent parent drop off, the integers 8–9 will represent a friend ride, and the integer 10 will represent walking. The simulation will consist of 30 trials.

Lesson 8-2

1. Sample answer: Use a spinner that is divided into two sectors, one containing 80% or 288° and the other containing 20% or 72°. Conduct 20 trials and record the results in a frequency table.

Outcome	Frequency
A	17
Below an A	3
Total	20

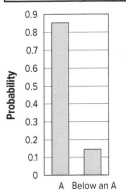

The probability of Clara getting an A on her next quiz is 0.85. The probability of earning any other grade is 1 − 0.85 or 0.15.

2a. Sample answer: P(strike) = 72%, P(ball) = 28%

2b.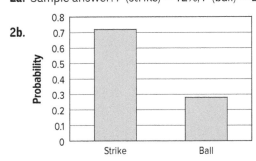

3. Sample answer: Use a random number generator to generate integers 1 through 20 where 1–9 represent tae kwon do, 10–15 represent yoga, 16–18 represent swimming, and 19–20 represent kickboxing. Do 20 trials and record the results in a frequency table.

Outcome	Frequency
tae kwon do	9
yoga	7
swimming	1
kick-boxing	3
Total	20

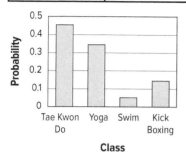

Class

The probability of a customer taking the tae kwon do class is 0.45, yoga is 0.35, swimming is 0.05, and kickboxing is 0.15.

4a. Sample answer: Use a spinner that is divided into two sectors, one containing 60% or 216° and the other containing 40% or 144°. Conduct 20 trials and record the results in a frequency table.

Outcome	Frequency
Strike	13
Not a strike	7
Total	20

4b.

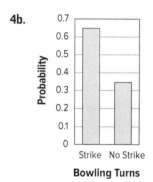

Bowling Turns

The probability of Bridget bowling a strike on her next turn is 0.65. The probability of bowling anything else is 1 − 0.65 or 0.35.

5. Sample answer: Use a spinner that is divided into two sectors, one containing 95% or 342° and the other containing 5% or 18°. Conduct 50 trials and record the results in a frequency table.

Outcome	Frequency
Sale	46
No Sale	4
Total	50

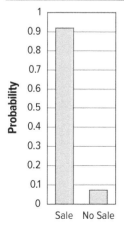

The probability of Ian selling a game is 0.92. The probability of not selling a game is 1 − 0.92 or 0.08.

6. Sample answer: Use a spinner that is divided into 10 equal sectors, each 36°. Conduct 10 trials and record the results in a frequency table.

Outcome	Frequency
Song 1	1
Song 2	0
Song 3	0
Song 4	0
Song 5	1
Song 6	1
Song 7	1
Song 8	2
Song 9	2
Song 10	2
Total	10

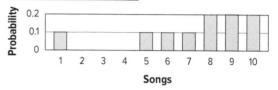

Songs

The probability of hearing song 1 is 0.1, songs 2–4 is 0, songs 5–7 is 0.1, and songs 8–10 is 0.2.

7. Sample answer: Use a random number generator to generate integers 1 through 20, where 1–8 represent drama, 9–14 represent mystery, 15–19 represent comedy, and 20 represents action. Conduct 20 trials and record the results in a frequency table.

Outcome	Frequency
drama	11
mystery	3
comedy	6
action	0
Total	20

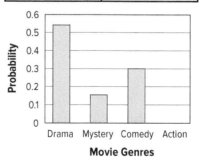

Movie Genres

The probability of a customer choosing a drama is 0.55, choosing a mystery is 0.15, choosing a comedy is 0.3, and choosing an action film is 0.

8. Sample answer: Use a random number generator to generate integers 1 through 20, where 1–12 represent a single, 13–17 represent a double, 18–19 represent a triple, and 20 represents a home run. Conduct 20 trials and record the results in a frequency table.

Outcome	Frequency
single	13
double	4
triple	2
home run	1
Total	20

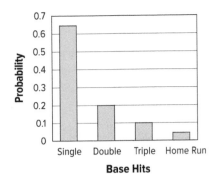

The probability of the baseball player hitting a single is 0.65, a double is 0.2, a triple is 0.1, and a home run is 0.05.

9. Sample answer: Use a random number generator to generate integers 1 through 20 where 1–9 represent Europe, 10–14 represent Asia, 15–17 represent South America, 18–19 represent Africa, and 20 represents Australia. Conduct 20 trials and record the results in a frequency table.

Outcome	Frequency
Europe	7
Asia	6
South America	5
Africa	2
Australia	0
Total	20

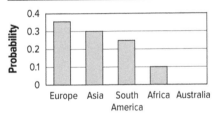

Travel Destinations

The probability of a customer traveling to Europe is 0.35, to Asia is 0.3, to South America is 0.25, to Africa is 0.1, and to Australia is 0.

10. Sample answer: Use a random number generator to generate integers 1 through 20, where 1–7 represent blue, 8–13 represent red, 14–16 represent white, 17–19 represent black, and 20 represents all other colors. Conduct 50 trials and record the results in a frequency table.

Outcome	Frequency
blue	17
red	14
black	7
white	10
other	2
Total	50

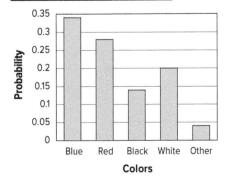

Colors

The probability of a customer buying a blue car is 0.34, buying a red car is 0.28, buying a black car is 0.14, buying a white car is 0.2, and any other color is 0.04.

12b. Sample answer: P(no defect) = 95%, P(defect) = 5%

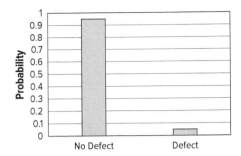

13a. Sample answer: The theoretical probability that a spin results in each prize is 20%. Use a random number generator to generate integers 1 through 5. The integer 1 will represent a hot pretzel, the integer 2 will represent a burger, the integer 3 will represent a large drink, the integer 4 will represent nachos, and the integer 5 will represent a small popcorn. The simulation will consist of 50 trials.

13b. Sample answer: P(hot pretzel) = 20%, P(burger) = 20%, P(large drink) = 28%, P(nachos) = 14%, P(small popcorn) = 18%

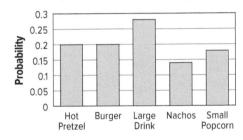

15c. Sample answer:

Trial	Sum of Roll	Sum of Output from Random Number Generator
1	9	4
2	10	10
3	6	5
4	6	10
5	7	6
6	9	7
7	5	12
8	9	3
9	5	7
10	7	4
11	6	7
12	5	9
13	7	3
14	3	6
15	9	4
16	7	11
17	6	5
18	7	7
19	8	5
20	7	3

15d. Sample answer:

Dice — 5 Rolls

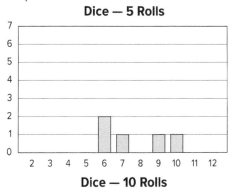

Dice — 10 Rolls

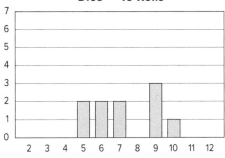

Dice — 20 Rolls

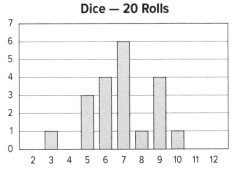

15e. Sample answer: The bar graph has more data points at the middle sums as more trials are added.

15f. Sample answer:

Random Number Generator

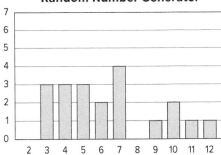

16. Sample answer: A simulation can be designed to estimate the probability that a basketball player will make her next free throw if she made 70% of her free throws the previous season. The theoretical probability that she makes her next free throw is 70%, and the theoretical probability that she misses is 30%. Use a random number generator to generate integers 1 through 10. The integers 1–7 will represent a made free throw, and the integers 8–10 will represent a miss. The simulation will consist of 30 trials.

20. Sample answer: Designing a simulation is like solving a word problem because each entails a series of steps that should lead to a desired outcome--one gives a summary report of data collected, and the other gives a solution of an equation or problem. When you solve a word problem, you start by reading the problem carefully and identifying your variables. Likewise, when designing a simulation, you determine what problem the simulation will help solve, and you start by determining each possible outcome and its theoretical probability. For a word problem, you next describe your plan for solving the problem. Likewise, for the simulation you describe an appropriate model for the situation that accurately represents the theoretical probability of each outcome. For a word problem, you write an equation from your plan and solve it. For a simulation, you describe a trial for the situation, then conduct the trials and summarize the data in a report.

Lesson 8-4

3a. **Mrs. Johnson's Class**

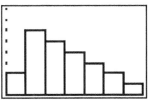

[5, 40] scl: 5 by [0, 8] scl: 1

Mr. Edmunds' Class

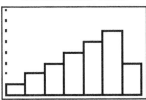

[5, 40] scl: 5 by [0, 8] scl: 1

Mrs. Johnson's class, positively skewed; Mr. Edmunds' class, negatively skewed

3b. Sample answer: The distributions are skewed, so use the five-number summaries. The range for both classes is the same. However, the median for Mrs. Johnson's class is 17 and the median for Mr. Edmunds' class is 28. The lower quartile for Mr. Edmunds' class is 20. Because this is greater than the median for Mrs. Johnson's class, this means that 75% of the data from Mr. Edmunds' class is greater than 50% of the data from Mrs. Johnson's class. Therefore, we can conclude that the students in Mr. Edmunds' class had slightly higher sales overall than the students in Mrs. Johnson's class.

5a.

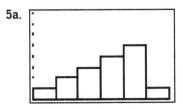

[50, 200] scl: 25 by [0, 8] scl: 1

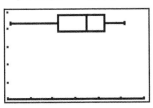

[50, 200] scl: 25 by [0, 5] scl: 1

negatively skewed

5b. Sample answer: The distribution is skewed, so use the five-number summary. The points range from 53 to 179. The median is 138.5 points, and half of the data are between 106.5 and 157 points.

6a.

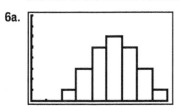

[0, 18] scl: 2 by [0, 8] scl: 1

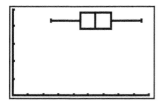

[0, 18] scl: 2 by [0, 5] scl: 1

symmetric

6b. Sample answer: The distribution is symmetric, so use the mean and standard deviation. The mean number of movies watched was about 10.7 with standard deviation of about 3 movies.

7a.

Sophomore Year

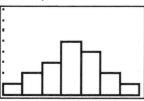

[1200, 1900] scl: 100 by [0, 8] scl: 1

Junior Year

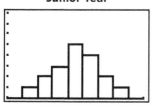

[1300, 2200] scl: 100 by [0, 8] scl: 1

both symmetric

7b. Sample answer: The distributions are symmetric, so use the means and standard deviations. The mean score for sophomore year is about 1552.9 with standard deviation of about 147.2. The mean score for junior year is about 1753.8 with standard deviation of about 159.1. We can conclude that the scores and the variation of the scores from the mean both increased from sophomore year to junior year.

8a.

Yorkshire

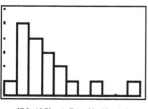

[50, 105] scl: 5 by [0, 6] scl: 1

Applewood

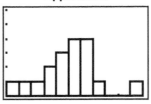

[35, 90] scl: 5 by [0, 6] scl: 1

Yorkshire, positively skewed; Applewood, negatively skewed

8b. Sample answer: The distributions are skewed, so use the five-number summaries. The median for Yorkshire is 63.5, and the median for Applewood is 60. The lower quartile for Applewood is 52, while the minimum for Yorkshire is 53. This means that 25% of the incomes for Applewood are lower than any of the incomes for Yorkshire. Also, the upper 25% of incomes for Yorkshire is between 72 and 103, while the upper 25% of incomes for Applewood is between 65 and 87. We can conclude that the incomes for the households in Yorkshire are greater than the incomes for the households in Applewood.

9a. symmetric

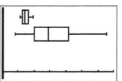

[0, 35,000] scl: 5000 by
[−10, 100] scl: 1

9b. Sample answer: The distributions appear to be symmetric, so use the means and standard deviations. The mean for the public colleges is $7367.38 with standard deviation of about $1160.13. The mean for private colleges is about $15,762.71 with standard deviation of about $7337.14. We can conclude that not only is the average cost of private schools far greater than the average cost of public schools, but the variation of the costs from the mean is also much greater.

10a.

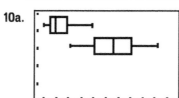

[225, 775] scl: 50 by [0, 5] scl: 1

boys, positively skewed; girls, symmetric

10b. Sample answer: One distribution is skewed and the other is symmetric, so use the five-number summaries. The maximum value for the boys is 450. The lower quartile for the girls is 453. This means that 75% of the data for the girls is higher than any data for the boys. We can conclude that 75% of the girls spent more money on prom than any of the boys.

12a.

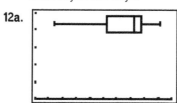

[50, 100] scl: 5 by [0, 5] scl: 1

Sample answer: The distribution is negatively skewed, so use the five-number summary. The scores range from 57 to 96. The median is 86, and half of the data are between 76 and 89.

12b.

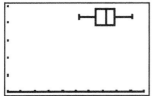

[50, 100] scl: 5 by [0, 5] scl: 1

Sample answer: The distribution is symmetric, so use the mean and standard deviation. The mean is about 85.6 with standard deviation of about 5.9.

14. Sample answer: Because the distribution has two clusters, an overall summary of center and spread would give an inaccurate depiction of the data. Instead, summarize the center and spread of each cluster individually using its mean and standard deviation.

15. Sample answer: The heights of the players on the Pittsburgh Steelers roster appear to represent a normal distribution.

Heights of the Players on the 2009 Pittsburgh Steelers Roster (inches)							
75	74	71	70	74	75	77	72
71	72	70	70	75	78	71	75
77	71	69	70	77	75	74	73
77	71	73	76	76	74	72	75
75	70	70	74	73	76	79	73
71	69	70	77	77	80	75	77
67	74	69	76	77	76		

The mean of the data is about 73.61 in. or 6 ft 1.61 in. The standard deviation is about 2.97 in.

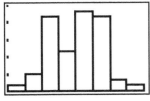

[66, 82] scl: 2 by [0, 15] scl: 3

The birth months of the players do not display central tendency.

Birth Months of the Players on the 2009 Pittsburgh Steelers Roster							
1	12	10	3	11	1	10	5
4	8	9	11	1	1	11	5
8	6	11	4	3	4	8	5
3	7	2	1	11	4	3	2
1	1	6	1	6	8	11	9
3	3	1	6	9	1	9	9
6	5	10	11	11	12		

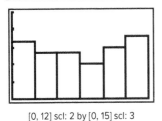

[0, 12] scl: 2 by [0, 15] scl: 3

16. Sample answer: The distribution for a data set is positively skewed when the majority of the data is on the left of the mean and a tail appears to the right of the mean. An example is when the data set includes the height of everyone in an elementary school, most of the data will be on the left side (the students), while a comparatively small amount will be on the right (the teachers and staff). The distribution for a data set is negatively skewed when the majority of the data is on the right of the mean and a tail appears to the left of the mean. An example is when the batting averages of a baseball lineup are listed, most of the data will be at a certain level while the pitcher will typically be much lower. The distribution for a data set is symmetric when the data are evenly distributed on both sides of the mean. An example is when test scores are calculated for an entire state, most of the students will place in the middle, while some will place above or below.

Mid-Chapter Quiz

1. observational study; sample: the 15 high school students selected; population: the student body

2. experiment; sample: the employees that were given an extra hour lunch break; population: all grocery store employees

3. survey; sample: the 100 students who received the questionnaire; population: school body

4. observational study; sample: the 50 participants; population: all potential viewers

8a.

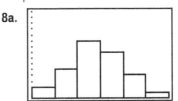

[4.6, 5.2] scl: 0.1 by [0, 15] scl: 1

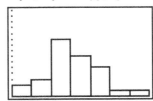

[4.4, 5.1] scl: 0.1 by [0, 15] scl: 1

both symmetric

8b. Sample answer: The distributions are symmetric, so use the means and standard deviations. The mean time for Aiden is about 4.88 with standard deviation of about 0.12. The mean time for Mark is about 4.69 with standard deviation of about 0.15. We can conclude that Aiden's times were higher than Mark's and the variation of Mark's times from the mean is greater than Aiden's.

Lesson 8-5

18a. Sample answer:

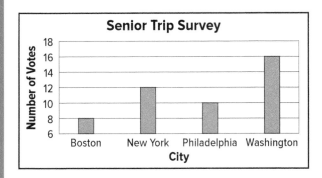

18b. Sample answer:

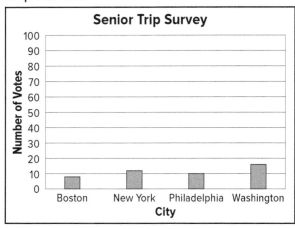

Lesson 8-6

4. −0.70; 0.70 standard deviations less than the mean

5. 2.05; 2.05 standard deviations greater than the mean

6. 59.8; 1.38 standard deviations less than the mean

7. 3.08; 2.40 standard deviations less than the mean

8a.

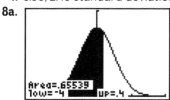

[−4, 4] scl: 1 by [0, .5] scl: 0.125

8b.

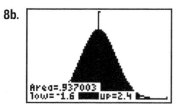

[−4, 4] scl: 1 by [0, .5] scl: 0.125

12. 2.5; 2.5 standard deviations greater than the mean

13. −1.33; 1.33 standard deviations less than the mean

14. 51.8; 0.92 standard deviations less than the mean

15. 177.7; 1.73 standard deviations greater than the mean

16a.

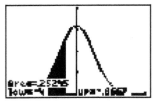

[−4, 4] scl: 1 by [0, .5] scl: 0.125

16b.

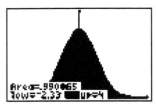

[−4, 4] scl: 1 by [0, .5] scl: 0.125

19c. Sample answer: I would expect people with several traffic citations to lie to the far right of the distribution where insurance costs are highest, because I think insurance companies would charge them more.

19d. Sample answer: As the probability of an accident occurring increases, the more an auto insurance company is going to charge. I think auto insurance companies would charge younger people more than older people because they have not been driving as long. I think they would charge more for expensive cars and sports cars and less for cars that have good safety ratings. I think they would charge a person less if they have a good driving record and more if they have had tickets and accidents.

20a. Math: 1.03; Science: 0.81; Social Studies: 1.21

20b. Math: 84.8%; Science: 79.1%; Social Studies: 88.7%

20c. Social Studies; Sample answer: When the distributions are standardized, Nikki's relative scores on each test are 84.8%, 79.1%, and 88.7%. These standardized scores are Nikki's scores in relation to the population of scores for each individual test. Therefore, her 86% on the Social Studies test was better than 88.7% of the other test takers' scores on that test. This is the highest percentage of the three tests, so Nikki performed the best in Social Studies.

25. Sample answer: The scores per team in each game of the first round of the 2010 NBA playoffs. The mean is 96.56 and the standard deviation is 11.06. The middle 68% of the distribution is $85.50 < X < 107.62$. The middle 95% is $74.44 < X < 118.68$. The middle 99.7% is $63.38 < X < 129.74$.

26. Sample answer: The z-value represents the position of a value X in a normal distribution. A z-value of 1.43 means that the corresponding data value X is 1.43 standard deviations to the right of the mean in the distribution. An interval of all of the values greater than X in the distribution will be represented by the area under the curve from $z = 1.43$ to $z = 4$. This area is equivalent to the probability of the interval occurring (a random data value falling within the interval).

Lesson 8-7

16. Sample answer: Assign Ellie to numbers 1 to 6, Mitchell to numbers 7 to 12, and Rosa to numbers 13 to 18. Shuffle the cards and choose one at random. If the number is 19 or 20, choose again. After the first person has been chosen, continue to choose cards using the same method until a second person has been chosen.

17. Disagree; Sample answer: Assign one option to the outcomes 1–3, assign the other option to the outcomes 4–6, and spin again if the outcome is 7.

18. The value of each ticket should be at least $0.66. The expected number of tickets after each spin is $2(0.65) + 5(0.35) = 3.05$ and $3.05(\$0.66) \approx \$2.01$.

19. $P(\text{red}) = 0.5$, but $P(\text{club}) = P(\text{spade}) = 0.25$, so the probabilities of the options are not all equal. Instead, he could assign the first option to the hearts. If the chosen card is a diamond, he should choose a new card until one of the options is chosen.

20. Sample answer: In a fair probability experiment, each outcome has an equal probability. When making a fair decision, each option should be assigned to one or more outcomes so that the probability of each option has an equal probability.

Notes

Track Your Progress

This chapter focuses on content from the Building Functions, Interpreting Functions, and Trigonometric Functions domains.

THEN	NOW	NEXT
S.IC.4 Use data from a sample survey to estimate a population mean or proportion; develop a margin of error through the use of simulation models for random sampling. **S.IC.5** Use data from a randomized experiment to compare two treatments; use simulations to decide if differences between parameters are significant. **S.MD.6** Use probabilities to make fair decisions. **S.MD.7** Analyze decisions and strategies using probability concepts.	**F.BF.3** Identify the effect on the graph of replacing $f(x)$ by $f(x) + k$, $kf(x)$, $f(kx)$, and $f(x + k)$ for specific values of k (both positive and negative); find the value of k given the graphs. Experiment with cases and illustrate an explanation of the effects on the graph using technology. **F.IF.7e** Graph exponential and logarithmic functions, showing intercepts and end behavior, and trigonometric functions, showing period, midline, and amplitude. **F.TF.1** Understand radian measure of an angle as the length of the arc on the unit circle subtended by the angle.	**F.TF.8** Prove the Pythagorean identity $sin^2(\theta) + cos^2(\theta) = 1$ and use it to calculate trigonometric ratios.

Standards for Mathematical Practice

All of the Standards for Mathematical Practice will be covered in this chapter. The MP icon notes specific areas of coverage.

 Teaching the Mathematical Practices
Help students develop the mathematical practices by asking questions like these.

Questioning Strategies As students approach problems in this chapter, help them develop mathematical practices by asking:

Construct Arguments
· How do you find values of trigonometric functions for acute angles or general angles?
· How do you find values of trigonometric functions by using reference angles?
· How do you find values of trigonometric functions based on the unit circle?

Modeling
· Can you draw and find angles in standard position?
· How do you graph trigonometric functions such as the sine, cosine, and tangent functions?
· How do you graph horizontal translations and vertical translations of trigonometric graphs and find phase shifts?

Using Tools
· Can you use the properties of periodic functions to evaluate trigonometric functions?
· Can you use trigonometric functions to find side lengths and angle measures of right triangles?

Precision
· How do you convert between degree measures and radian measures?

Go Online!

StudySync:
SMP Modeling Videos

These demonstrate how to apply the Standards for Mathematical Practice to collaborate, discuss, and solve real-world math problems.

LearnSmart   The Geometer's Sketchpad   Vocabulary   Personal Tutor Tools   Calculator Resources Self-Check Practice   Animations

Customize Your Chapter

Use the *Plan & Present*, *Assignment Tracker*, and *Assessment* tools in ConnectED to introduce lesson concepts, assign personalized practice, and diagnose areas of student need.

Differentiated Instruction

Throughout the program, look for the icons to find specialized content designed for your students.

AL	Approaching Level
OL	On Level
BL	Beyond Level
ELL	English Language Learners

Personalize

Differentiated Resources

FOR EVERY CHAPTER	AL	OL	BL	ELL
✓ Chapter Readiness Quizzes	●	●	◐	●
✓ Chapter Tests	●	●	●	●
✓ Standardized Test Practice	●	●	●	●
Vocabulary Review Games	●	●	◐	●
Anticipation Guide (English/Spanish)	●	●	◐	●
Student-Built Glossary	●	●	◐	●
Chapter Project	◐	●	●	●

FOR EVERY LESSON	AL	OL	BL	ELL
Personal Tutors (English/Spanish)	●	●	◐	●
Graphing Calculator Personal Tutors	●	●	●	●
▷ Step-by-Step Solutions	●	●	◐	●
✓ Self-Check Quizzes	●	●	●	●
5-Minute Check	●	●	●	●
Study Notebook	●	●	●	●
Study Guide and Intervention	●	●		●
Skills Practice	●	◐		●
Practice	◐	●	●	●
Word Problem Practice	◐	●	●	◐
Enrichment		●	●	●
✚ Extra Examples	●	◐		◐
Lesson Presentations	●	●	●	●

◐ Aligned to this group   ● Designed for this group

Engage

Featured IWB Resources

 Geometer's Sketchpad provides students with a tangible, visual way to learn. *Use with Lessons 9-1 and 9-5.*

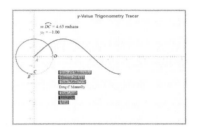

 Animations help students make important connections through motion. *Use with Lesson 9-3, 9-5, and 9-6.*

 eLessons engage students and help build conceptual understanding of big ideas. *Use with Lessons 9-1, 9-2, and 9-5.*

 Time Management How long will it take to use these resources? Look for the clock in each lesson interleaf.

Introduce the Chapter

Mathematical Background

A trigonometric ratio compares the side lengths of a right triangle. Angles can be rotations about the origin and be measured in degrees or radians. When a unit circle is used to generalize the sine and cosine functions, they are known as circular functions.

ℯ Essential Question

At the end of this chapter, students should be able to answer the Essential Question.

What types of real-world problems can be modeled and solved using trigonometry? Sample answer: periodic functions and physical situations involving triangles such as construction and surveying problems

Apply Math to the Real World

WATER SPORTS In this activity, students will use what they already know about right triangles to explore how their basic knowledge of trigonometry can be useful in determining distance traveled in a kayak. Have students complete this activity individually or in small groups.

Go Online! ✓

Chapter Project

To Bee Or Not To Bee Students use what they have learned about trigonometry to complete a project. This chapter project addresses environmental awareness, as well as several specific skills identified as being essential to student success by the Framework for 21st Century Learning.
MP 1, 3, 4

Trigonometric Functions

THEN
You have graphed and analyzed functions.

NOW
You will:
- Find values of trigonometric functions.
- Solve problems by using right triangle trigonometry.
- Graph trigonometric functions.

MP WHY

WATER SPORTS Knowing trigonometric functions has practical applications in water sports. You can use right triangle trigonometry to find the distance a kayak has traveled when paddling against the current.

Use the Mathematical Practices to complete the activity.

1. **Sense-Making** You want to reach a point 100 meters up river, but know the current is too strong to paddle there directly. You can paddle straight out for 50 meters. How would you determine how far you have to paddle diagonally upriver to reach the desired point?

2. **Model with Math** Use the Triangle Special Segments tool to model the problem.

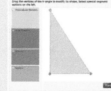

3. **Apply Math** Determine both the distance traveled diagonally upriver as well as the total distance traveled in the kayak.

Michael DeLeong/Blend Images

⊙ ALEKS®

Your Student Success Tool ALEKS is an adaptive, personalized learning environment that identifies precisely what each student knows and is ready to learn—ensuring student success at all levels.

- **Formative Assessment:** Dynamic, detailed reports monitor students' progress toward standards mastery.
- **Automatic Differentiation:** Strengthen prerequisite skills and target individual learning gaps.
- **Personalized Instruction:** Supplement in-class instruction with personalized assessment and learning opportunities.

Go Online to Guide Your Learning

Explore & Explain		Organize

Graphing Tools

Use the **Explore: Trigonometric Functions** tool to enhance your understanding of graphing trigonometric functions in Lesson 9-5.

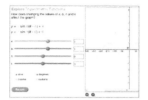

The Geometer's Sketchpad

The Geometer's Sketchpad can be used throughout this chapter to illustrate trigonometric ratios in right angles, to graph and analyze trigonometric functions, and to explore angles, angle measures, transformations of trigonometric graphs, and periodic functions.

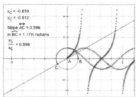

eBook

Interactive Student Guide

Before starting the chapter, answer the **Chapter Focus** preview questions. Check your answers as you complete each lesson. At the end of the chapter, try the **Performance Task**.

Foldables

Get organized! Before you begin this chapter, create this **Trigonometric Functions Foldable** to help you organize your notes about trigonometric functions.

Collaborate

Chapter Project

In the **To Bee or Not to Bee** project, you will use what you have learned about trigonometry to complete a project that addresses environmental awareness.

Focus

LEARNSMART

Need help studying? Complete the **Trigonometric Functions** domain in LearnSmart to review for the chapter test.

ALEKS

You can use the **Trigonometry** topic in ALEKS to explore what you know about statistics and probability and what you are ready to learn.*

*Ask your teacher if this is part of your program.

Dinah Zike's FOLDABLES

Focus Students write notes as they explore trigonometric functions in the lessons of this chapter.

Teach Have students make and label their Foldables as illustrated. Have students use the appropriate tab as they cover each lesson in this chapter. Encourage students to apply what they have learned by writing their own examples as well.

When to Use It Encourage students to add to their Foldables as they work through the chapter and to use them to review for the chapter test.

Go Online!

Extending Vocabulary

Looking for more interesting assessments? Learn strategies for teaching and assessing vocabulary with pocketbooks and notebook Foldables. **MP** 5

Get Ready for the Chapter

RtI Response to Intervention

Use the Concept Check results and the Intervention Planner chart to help you determine your Response to Intervention.

Intervention Planner

TIER 1 On Level OL

IF students miss 25% of the exercises or less,

THEN choose a resource:

Math Triumphs, Alg. 2

Go Online!

 Chapter Project

TIER 2 Approaching Level AL

IF students miss 50% of the exercises,

THEN ***Go Online!***

 Extra Examples

Personal Tutors

Homework Help

TIER 3 Intensive Intervention

IF students miss 75% of the exercises,

THEN choose a resource:

Use *Math Triumphs, Alg. 2*

Go Online!

Extra Examples

Personal Tutors

Homework Help

Review Vocabulary

Additional Answers

1. $a^2 + b^2 = c^2$
2. right triangle
3. yes; they both measure 45°
4. $m^2 + m^2 = 18^2$
5. $x = 12$
6. $y = 12\sqrt{3}$
7. each side will be $\frac{x\sqrt{2}}{2}$
8. the longer side will be $x\sqrt{3}$; the hypotenuse will be $2x$

Get Ready for the Chapter

Connecting Concepts	New Vocabulary		

Concept Check

Review the concepts used in this chapter by answering the questions below. 1–8 See margin.

1. How is the Pythagorean Theorem stated mathematically?
2. What type of triangle does the Pythagorean Theorem apply to?
3. In the triangle shown, can you determine the measures of the two angles that are not the right angle? If so, what are they?
4. If you applied the Pythagorean Theorem to the triangle shown, what would the equation be?
5. In the triangle shown, what is the value of x?
6. In the triangle shown, what is the value of y?
7. Given a 45°–45°–90° special right triangle with a hypotenuse of x, what the values of the other two sides be?
8. Given a 30°–60°–90° special right triangle with the shortest side measuring x, what will the values of the longer side and of the hypotenuse be?

Performance Task Preview

You can use the concepts and skills in the chapter to perform calculations for a safety inspector. Understanding trigonometry and other principles of mathematics will help you finish the Performance Task at the end of the chapter.

In this Performance Task you will:

- make sense of problems and persevere in solving them
- reason abstractly and quantitatively
- construct viable arguments and critique the reasoning of others
- model with mathematics
- attend to precision

New Vocabulary		
English		**Español**
trigonometry	p. 594	trigonometría
trigonometric ratio	p. 594	rázon trigonometric
sine	p. 594	seno
cosine	p. 594	coseno
tangent	p. 594	tangente
cosecant	p. 594	cosecante
secant	p. 594	secante
cotangent	p. 594	cotangente
reciprocal functions	p. 595	funciones recíprocas
angle of elevation	p. 598	ángulo de depresión
angle of depression	p. 598	ángulo de elevación
standard position	p. 604	posición estándar
initial side	p. 604	lado inicial de un ángulo
terminal side	p. 604	lado terminal
coterminal angles	p. 605	ángulos coterminales
radian	p. 606	radián
quadrantal angle	p. 613	ángulo de cuadrante
reference angle	p. 613	ángulo de referencia
unit circle	p. 620	círculo unitario
circular function	p. 620	funciones circulares
periodic function	p. 621	función periódica
cycle	p. 621	ciclo
period	p. 621	período
amplitude	p. 627	amplitud
frequency	p. 628	frecuencia
phase shift	p. 635	desplazamiento de fase
vertical shift	p. 636	cambio vertical
midline	p. 636	linea media

Review Vocabulary

function función a relation in which each element of the domain is paired with exactly one element in the range

inverse function función inversa two functions *f* and *g* are inverse functions if and only if both of their compositions are the identity function

Key Vocabulary ELL

Introduce the key vocabulary in the chapter using the routine below.

Define One radian is the measure of an angle θ in standard position whose rays intercept an arc of length 1 unit on the unit circle.

Example

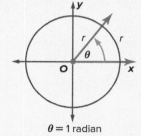

$\theta = 1$ radian

Ask: What does a radian measure? an angle

Trigonometric Functions in Right Triangles

Track Your Progress

Objectives

1 Find values of trigonometric functions.

2 Use trigonometric functions to find side lengths and angle measures of right triangles.

Mathematical Background

Students may be familiar with finding values of trigonometric functions of any angle. However, at this point, the functions have been defined in terms of ratios of the sides of a right triangle. This means that the definitions are only valid for acute angles. The more general case will be considered later when reference angles are introduced.

THEN	NOW	NEXT
G.SRT.4 Prove theorems about triangles. Theorems include: a line parallel to one side of a triangle divides the other two proportionally, and conversely; the Pythagorean Theorem proved using triangle similarity.	**G.SRT.8** Use trigonometric ratios and the Pythagorean Theorem to solve right triangles in applied problems.	**F.TF.1** Understand radian measure of an angle as the length of the arc on the unit circle subtended by the angle.

Go Online! All of these resources and more are available at connectED.mcgraw-hill.com

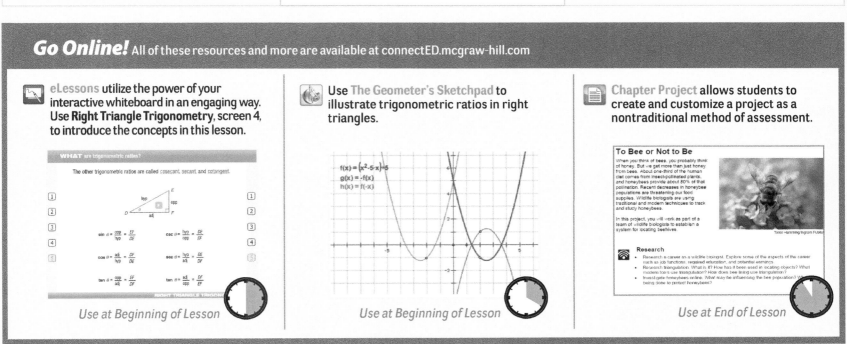

eLessons utilize the power of your interactive whiteboard in an engaging way. Use **Right Triangle Trigonometry**, screen 4, to introduce the concepts in this lesson.

Use at Beginning of Lesson

Use **The Geometer's Sketchpad** to illustrate trigonometric ratios in right triangles.

Use at Beginning of Lesson

Chapter Project allows students to create and customize a project as a nontraditional method of assessment.

Use at End of Lesson

OER Using Open Educational Resources

Mnemonic Devices Have students work in pairs or threes to write a trigonometric ratios song, or a poem, using **Sound Cloud** to record their work. Sound Cloud allows you to create, record, and share audio for free. Have students volunteer to perform, or play the audio, of their mnemonic device in front of the class.
Use as homework

Differentiate Your Resources

Extra Practice Additional practice or homework; Skills Practice is best for approaching-level students and Practice is best for on-level and beyond-level students

Skills Practice

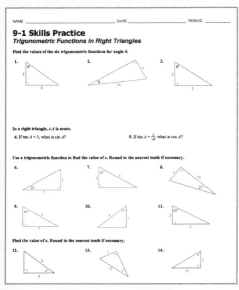

Practice

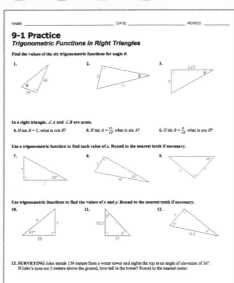

Word Problem Practice

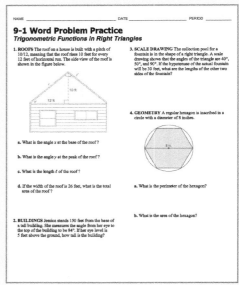

Intervention Reteaching and vocabulary activities that can be used with struggling or absent students and as ELL support

Extension Activities that can be used to extend lesson concepts

Study Guide and Intervention

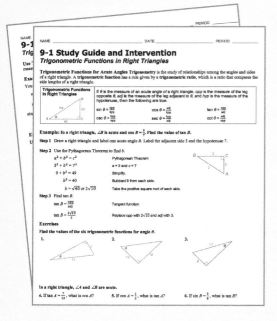

Study Notebook

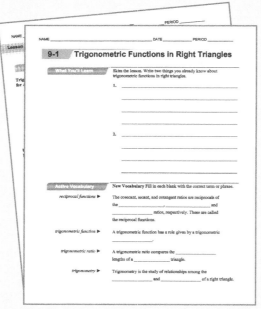

Enrichment

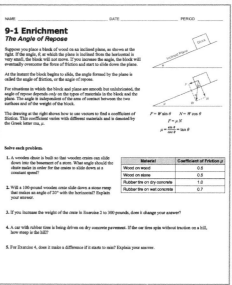

EXPLORE 9-1

Spreadsheet Lab

Investigating Special Right Triangles

You can use a spreadsheet to investigate side measures of special right triangles.

Mathematical Practices
8 Look for and express regularity in repeated reasoning.

Activity 45°-45°-90° Triangle

The legs of a 45°-45°-90° triangle, *a* and *b*, are equal in measure. What patterns do you observe in the ratios of the side measures of these triangles?

Step 1 Enter the indicated formulas in the spreadsheet. The formula uses the Pythagorean Theorem in the form $c = \sqrt{a^2 + b^2}$.

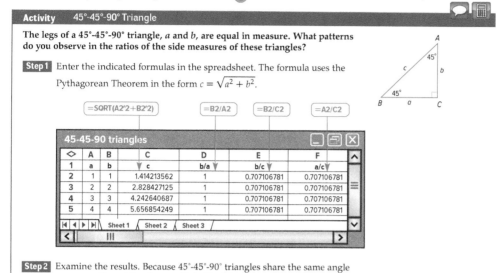

45-45-90 triangles

	A	B	C	D	E	F
	a	b	c	b/a	b/c	a/c
1			=SQRT(A2^2+B2^2)	=B2/A2	=B2/C2	=A2/C2
2	1	1	1.414213562	1	0.707106781	0.707106781
3	2	2	2.828427125	1	0.707106781	0.707106781
4	3	3	4.242640687	1	0.707106781	0.707106781
5	4	4	5.656854249	1	0.707106781	0.707106781

Sheet 1 / Sheet 2 / Sheet 3

Step 2 Examine the results. Because 45°-45°-90° triangles share the same angle measures, these triangles are all similar. The ratios of the sides of these triangles are all the same. The ratios of side *b* to side *a* are 1. The ratios of side *b* to side *c* and of side *a* to side *c* are approximately 0.71.

Model and Analyze

Work cooperatively. Use the spreadsheet below for 30°-60°-90° triangles.

30-60-90 triangles

	A	B	C	D	E	F
	a	b	c	b/a	b/c	a/c
1						
2	1		2			
3	2		4			
4	3		6			
5	4		8			

Sheet 1 / Sheet 2 / Sheet 3

3. All of the ratios of side *b* to side *a* are approximately 1.73. All of the ratios of side *b* to side *c* are approximately 0.87. All of the ratios of side *a* to side *c* are 0.5.

1. Copy and complete the spreadsheet above. **See margin.**

2. Describe the relationship among the 30°-60°-90° triangles with the dimensions given. **The triangles are all similar.**

3. What patterns do you observe in the ratios of the side measures of these triangles?

Additional Answer

1.

	A	B	C	D	E	F
	a	b	c	b/a	b/c	a/c
1	1	1.732050808	2	1.732050808	0.866025404	0.5
2	2	3.464101615	4	1.732050808	0.866025404	0.5
3	3	5.196152423	6	1.732050808	0.866025404	0.5
4	4	6.92820323	8	1.732050808	0.866025404	0.5
5	5	8.660254038	10	1.732050808	0.866025404	0.5
6	6	10.39230485	12	1.732050808	0.866025404	0.5

Launch

Objective Use a spreadsheet to investigate ratios of side lengths in special right triangles.

Materials for Each Student

• spreadsheet program and computer

Teaching Tip

Have students practice their skills in using a spreadsheet by entering the data in the example. Have students format the cells before they start to enter the data by using the Format Cells command. Format columns C, E, and F for numbers with 9 decimal places. Format columns A, B, and D for integers.

Teach

Working in Cooperative Groups Have students
work with partners, mixing abilities so that a student with more knowledge of spreadsheets is paired with one who has less experience. **ELL**

Practice Have students complete the Activity and Exercises 1–3.

Assess

Formative Assessment

Use Exercise 1 to assess whether students comprehend how to enter data into a spreadsheet.

From Concrete to Abstract

Ask students to describe the rule, pattern, or formula for various cells of the spreadsheet.

Extending the Concept

Ask:

• Ask students what formula could have been used in column B for the 45°-45°-90° triangle instead of entering the data? in column C for the 30°-60°-90° triangle? $b = a$; $c = 2a$

• Ask students to compare the 30°-60°-90° triangles. What is the same? What is different? The angle measures are all the same, but the side measures are different. The triangles are similar but not congruent

Launch

Have students read the Why? section of the lesson.
Ask:

- What kind of angle is formed by the tow rope and the horizontal? acute

- Which side of the triangle is opposite the right angle? Which side is opposite the *x*° angle? the hypotenuse *ℓ*; altitude

- If length *ℓ* is constant and *x* increases, how does the altitude change? It increases.

Teach

Ask the scaffolded questions for each example to build conceptual understanding for students at all levels.

1 Trigonometric Functions For Acute Angles

Example 1 Evaluate Trigonometric Functions

AL Which side of a right triangle is always the longest? the hypotenuse

OL How can you verify that this is a right triangle? Verify that the lengths of the sides of the triangle satisfy the Pythagorean Theorem.

BL In a right triangle with acute angle *θ*, explain why sin *θ* is always less than 1. The sine ratio is opposite/hypotenuse. Because the length of the hypotenuse is always the greatest, the sine ratio will always be less than 1.

(continued on the next page)

Go Online!

Interactive Whiteboard

Use the *eLesson* or *Lesson Presentation* to present this lesson.

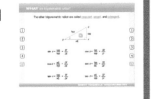

Trigonometric Functions in Right Triangles

··Then	··Now	··Why?
● You used the Pythagorean Theorem to find side lengths of right triangles.	**1** Find values of trigonometric functions for acute angles. **2** Use trigonometric functions to find side lengths and angle measures of right triangles.	● The altitude of a person parasailing depends on the length of the tow rope *ℓ* and the angle the rope makes with the horizontal *x*°. If you know these two values, you can use a ratio to find the altitude of the person parasailing.

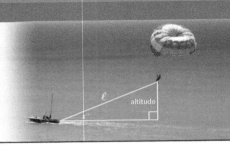

New Vocabulary
trigonometry
trigonometric ratio
trigonometric function
sine
cosine
tangent
cosecant
secant
cotangent
reciprocal functions
inverse sine
inverse cosine
inverse tangent
angle of elevation
angle of depression

MP Mathematical Practices
6 Attend to precision.

1 Trigonometric Functions for Acute Angles Trigonometry is the study of relationships among the angles and sides of a right triangle. A **trigonometric ratio** compares the side lengths of a right triangle. A **trigonometric function** has a rule given by a trigonometric ratio.

The Greek letter *theta θ* is often used to represent the measure of an acute angle in a right triangle. The *hypotenuse*, the *leg opposite θ*, and the *leg adjacent to θ* are used to define the six trigonometric functions.

Key Concept Trigonometric Functions in Right Triangles

Words	If *θ* is the measure of an acute angle of a right triangle, then the following trigonometric functions involving the opposite side *opp*, the adjacent side *adj*, and the hypotenuse *hyp* are true.
Symbols	$\sin (\text{sine}) \; \theta = \dfrac{opp}{hyp}$ $\quad$ $\csc (\text{cosecant}) \; \theta = \dfrac{hyp}{opp}$ $\cos (\text{cosine}) \; \theta = \dfrac{adj}{hyp}$ $\quad$ $\sec (\text{secant}) \; \theta = \dfrac{hyp}{adj}$ $\tan (\text{tangent}) \; \theta = \dfrac{opp}{adj}$ $\quad$ $\cot (\text{cotangent}) \; \theta = \dfrac{adj}{opp}$
Examples	$\sin \theta = \dfrac{4}{5}$ $\quad$ $\cos \theta = \dfrac{3}{5}$ $\quad$ $\tan \theta = \dfrac{4}{3}$ $\csc \theta = \dfrac{5}{4}$ $\quad$ $\sec \theta = \dfrac{5}{3}$ $\quad$ $\cot \theta = \dfrac{3}{4}$

Example 1 Evaluate Trigonometric Functions

Find the values of the six trigonometric functions for angle *θ*.

leg opposite *θ*: *BC* = 8 $\quad$ leg adjacent *θ*: *AC* = 15 $\quad$ hypotenuse: *AB* = 17

$\sin \theta = \dfrac{opp}{hyp} = \dfrac{8}{17}$ $\qquad$ $\cos \theta = \dfrac{adj}{hyp} = \dfrac{15}{17}$ $\qquad$ $\tan \theta = \dfrac{opp}{adj} = \dfrac{8}{15}$

$\csc \theta = \dfrac{hyp}{opp} = \dfrac{17}{8}$ $\qquad$ $\sec \theta = \dfrac{hyp}{adj} = \dfrac{17}{15}$ $\qquad$ $\cot \theta = \dfrac{adj}{opp} = \dfrac{15}{8}$

Guided Practice

1. Find the values of the six trigonometric functions for angle *B*. See margin.

MP Mathematical Practices Strategies

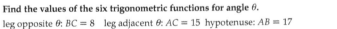

Attend to precision. Help students identify the parts of a right triangle used in the ratios for trigonometric functions. For example, ask:

- **How do you tell which side is the hypotenuse?** It is the longest side and is opposite the right angle.

- **Which is the side opposite *θ*?** It is the side of the triangle that is not one side of the angle.

- **Which is the side adjacent *θ*?** It is the side of the triangle that is not the hypotenuse, but is one side of the angle.

- **What information are you given?** Depending on the diagram, students should identify whether the hypotenuse, adjacent leg, or opposite leg are given.

- **How do you determine which trigonometric function to apply?** Compare the known information to the ratios for each trigonometric function and choose the function that matches the known information.

Notice that the cosecant, secant, and cotangent ratios are reciprocals of the sine, cosine, and tangent ratios, respectively. These are called the **reciprocal functions**.

$$\csc \theta = \frac{1}{\sin \theta} \qquad \sec \theta = \frac{1}{\cos \theta} \qquad \cot \theta = \frac{1}{\tan \theta}$$

The domain of any trigonometric function is the set of all acute angles θ of a right triangle. So, trigonometric functions depend only on the measures of the acute angles, not on the side lengths of a right triangle.

Example 2   Find Trigonometric Ratios

If $\sin B = \frac{5}{8}$, find the exact values of the five remaining trigonometric functions for B.

Step 1 Draw a right triangle and label one acute angle B. Label the opposite side 5 and the hypotenuse 8.

Step 2 Use the Pythagorean Theorem to find a.

$a^2 + b^2 = c^2$	Pythagorean Theorem
$a^2 + 5^2 = 8^2$	$b = 5$ and $c = 8$
$a^2 + 25 = 64$	Simplify.
$a^2 = 39$	Subtract 25 from each side.
$a = \pm\sqrt{39}$	Take the square root of each side.
$a = \sqrt{39}$	Length cannot be negative.

Step 3 Find the other values.

Because $\sin B = \frac{5}{8}$, $\csc B = \frac{\text{hyp}}{\text{opp}}$ or $\frac{8}{5}$.

$\cos B = \frac{\text{adj}}{\text{hyp}} = \frac{\sqrt{39}}{8}$ $\qquad\qquad$ $\sec B = \frac{\text{hyp}}{\text{adj}} = \frac{8}{\sqrt{39}}$ or $\frac{8\sqrt{39}}{39}$

$\tan B = \frac{\text{opp}}{\text{adj}} = \frac{5}{\sqrt{39}}$ or $\frac{5\sqrt{39}}{39}$ $\qquad$ $\cot B = \frac{\text{adj}}{\text{opp}} = \frac{\sqrt{39}}{5}$

> **Guided Practice** **2.** $\sin B = \frac{3\sqrt{58}}{58}$, $\sec B = \frac{\sqrt{58}}{7}$, $\cot B = \frac{7}{3}$, $\csc B = \frac{\sqrt{58}}{3}$, $\cos B = \frac{7\sqrt{58}}{58}$

2. If $\tan B = \frac{3}{7}$, find exact values of the remaining trigonometric fuctions for B.

Angles that measure 30°, 45°, and 60° occur frequently in trigonometry.

Key Concept   Trigonometric Values for Special Angles

30°-60°-90°

$\sin 30° = \frac{1}{2}$ $\qquad$ $\cos 30° = \frac{\sqrt{3}}{2}$ $\qquad$ $\tan 30° = \frac{\sqrt{3}}{3}$

$\sin 60° = \frac{\sqrt{3}}{2}$ $\qquad$ $\cos 60° = \frac{1}{2}$ $\qquad$ $\tan 60° = \sqrt{3}$

45°-45°-90°

$\sin 45° = \frac{\sqrt{2}}{2}$ $\qquad$ $\cos 45° = \frac{\sqrt{2}}{2}$ $\qquad$ $\tan 45° = 1$

Need Another Example?

Find the values of the six trigonometric functions for angle G.

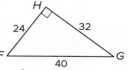

$\sin G = \frac{3}{5}$, $\cos G = \frac{4}{5}$,

$\tan G = \frac{3}{4}$, $\cot G = \frac{4}{3}$,

$\sec G = \frac{5}{4}$, $\csc G = \frac{5}{3}$

Example 2 Find Trigonometric Ratios

AL Why can we use the Pythagorean Theorem to find a? This is a right triangle.

OL Why can't we find sin C? C is a right angle.

BL How do sin A and cos A relate to sin B and cos B? $\sin A = \cos B$ and $\cos A = \sin B$

Need Another Example?

If $\tan A = \frac{5}{3}$, find the exact values of the five remaining trigonometric functions for A.

$\cos A = \frac{3\sqrt{34}}{34}$, $\csc A = \frac{\sqrt{34}}{5}$,

$\cot A = \frac{3}{5}$, $\sec A = \frac{\sqrt{34}}{3}$,

$\sin A = \frac{5\sqrt{34}}{34}$

Teaching Tip

Reasoning The relative lengths of the sides of the two special triangles in the Key Concept box on this page are critical. Have students learn how to recreate these triangles themselves from memory. From the triangles, they can generate the values shown in the Key Concept box.

Additional Answer (Guided Practice)

1. $\sin B = \frac{15}{17}$, $\cos B = \frac{8}{17}$, $\tan B = \frac{15}{8}$, $\csc B = \frac{17}{15}$, $\sec B = \frac{17}{8}$, $\cot B = \frac{8}{15}$

---

2 Use Trigonometric Functions

Example 3 Find a Missing Side Length

AL What is the measure of the unlabeled angle?
60°

OL What is the length of the third side of the triangle? 4

BL What information would change in the problem if we would use tan 30° to solve the problem? the length of the opposite side

Need Another Example?

Use a trigonometric function to find the value of x. Round to the nearest tenth if necessary. 10.4

Example 4 Find a Missing Side Length

AL Does *d* represent the height of the building?
No, *d* is 6 feet less than the height of the building.

OL What does the hypotenuse represent in this situation? the distance between Joel's eyes and the top of the building

BL What would happen to the angle from his eye to the top of the building as Joel walked further from the building? It would decrease.

Need Another Example?

Buildings In Example 4, what is the distance from the top of the building to Joel's eye? about 827 ft

e Essential Question

How are trigonometric functions used to find unknown values?
Trigonometric functions represent ratios of sides of a right triangle. When one of those sides is unknown, or the angle of the acute angle is unknown, a trigonometric ratio is used to form an equation with that unknown, which then can be solved.

---

2 Use Trigonometric Functions You can use trigonometric functions to find missing side lengths and missing angle measures of right triangles.

Study Tip
Choose a Function If the length of the hypotenuse is unknown, then either the sine or cosine function must be used to find the missing measure.

Example 3 Find a Missing Side Length

Use a trigonometric function to find the value of x. Round to the nearest tenth if necessary.

The length of the hypotenuse is 8. The missing measure is for the side adjacent to the 30° angle. Use the cosine function.

$\cos \theta = \dfrac{adj}{hyp}$ Cosine function

$\cos 30° = \dfrac{x}{8}$ Replace θ with 30°, adj with x, and hyp with 8.

$\dfrac{\sqrt{3}}{2} = \dfrac{x}{8}$ $\cos 30° = \dfrac{\sqrt{3}}{2}$

$\dfrac{8\sqrt{3}}{2} = x$ Multiply each side by 8.

$6.9 \approx x$ Use a calculator.

Guided Practice

Use a trigonometric function to find the value of x. Round to the nearest tenth if necessary.

3A.    3B.

You can use a calculator to find the missing side lengths of triangles.

Real-World Link
Inclinometers measure the angle of Earth's magnetic field as well as the pitch and roll of vehicles, sailboats, and airplanes. They are also used for monitoring volcanoes and well drilling.
Source: Science Magazine

Real-World Example 4 Find a Missing Side Length

BUILDINGS To calculate the height of a building, Joel walked 200 feet from the base of the building and used an inclinometer to measure the angle from his eye to the top of the building. If his eye level is at 6 feet, how tall is the building?

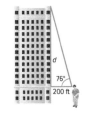

The measured angle is 76°. The side adjacent to the angle is 200 feet. The missing measure is the side opposite the angle. Use the tangent function to find d.

$\tan \theta = \dfrac{opp}{adj}$ Tangent function

$\tan 76° = \dfrac{d}{200}$ Replace θ with 76°, opp with d, and adj with 200.

$200 \tan 76° = d$ Multiply each side by 200.

$802 \approx d$ Use a calculator to simplify: 200 TAN 76 ENTER

Because the inclinometer was 6 feet above the ground, the height of the building is approximately 808 feet.

Guided Practice

4. Use a trigonometric function to find the value of x. Round to the nearest tenth if necessary. 27.0

When solving equations like $3x = -27$, you use the inverse of multiplication to find x. You also can find angle measures by using the inverse of sine, cosine, or tangent.

Key Concept Inverse Trigonometric Ratios

Words	If $\angle A$ is an acute angle and the sine of A is x, then the **inverse sine** of x is the measure of $\angle A$.
Symbols	If $\sin A = x$, then $\sin^{-1} x = m\angle A$.
Example	$\sin A = \frac{1}{2} \rightarrow \sin^{-1}\frac{1}{2} = m\angle A \rightarrow m\angle A = 30°$
Words	If $\angle A$ is an acute angle and the cosine of A is x, then the **inverse cosine** of x is the measure of $\angle A$.
Symbols	If $\cos A = x$, then $\cos^{-1} x = m\angle A$.
Example	$\cos A = \frac{\sqrt{2}}{2} \rightarrow \cos^{-1}\frac{\sqrt{2}}{2} = m\angle A \rightarrow m\angle A = 45°$
Words	If $\angle A$ is an acute angle and the tangent of A is x, then the **inverse tangent** of x is the measure of $\angle A$.
Symbols	If $\tan A = x$, then $\tan^{-1} x = m\angle A$.
Example	$\tan A = \sqrt{3} \rightarrow \tan^{-1}\sqrt{3} = m\angle A \rightarrow m\angle A = 60°$

If you know the sine, cosine, or tangent of an acute angle, you can use a calculator to find the measure of the angle, which is the inverse of the trigonometric ratio.

Example 5 Find a Missing Angle Measure

Find the measure of each angle. Round to the nearest tenth if necessary.

a. $\angle N$

You know the measure of the side opposite $\angle N$ and the measure of the hypotenuse. Use the sine function.

$$\sin N = \frac{6}{10} \qquad \sin\theta = \frac{opp}{hyp}$$

$$\sin^{-1}\frac{6}{10} = m\angle N \qquad \text{Inverse sine}$$

$$36.9° \approx m\angle N \qquad \text{Use a calculator}$$

b. $\angle B$

Use the cosine function.

$$\cos B = \frac{8}{16} \qquad \cos\theta = \frac{adj}{hyp}$$

$$\cos^{-1}\frac{8}{16} = m\angle B \qquad \text{Inverse cosine}$$

$$60° = m\angle B \qquad \text{Use a calculator}$$

Guided Practice Find x. Round to the nearest tenth if necessary.

5A.
28.1

5B.
56.3

Example 5 Find a Missing Angle Measure

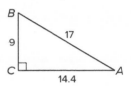

AL What is the measure of angle *L*? 53.1°

OL How could you find side *L* in part **a**? the Pythagorean Theorem

BL Why would you only ever need to use this method once to find all of the angles of a right triangle? You know the right angle and once you have two angles you can use the fact that the three angles of a triangle add up to 180°.

Need Another Example?

Find the measure of each angle. Round to the nearest tenth if necessary.

a. $\angle A$ 32°
b. $\angle B$ 58°

Differentiated Instruction AL OL

Visual/Spatial Learners Have students use a stack of books and a notebook to model a ramp and investigate how steep the ramp needs to be for a toy car to roll down it without being pushed. Have them report their results in terms of the trigonometric functions of a right triangle.

Example 6 Use Angles of Elevation and Depression

AL In part **b**, what is the exact value of sin 60°? $\frac{\sqrt{3}}{2}$

OL Do your answers make sense in these situations? Why? Yes; because we are finding the hypotenuse, the length we find should be longer than the other lengths in the problem.

BL In part **b**, what would tan 60° represent? the ratio of the vertical distance to the horizontal distance

Need Another Example?

a. **Golf** From a camera in a blimp, the apparent distance between the golfer and the hole in Example 6a is the horizontal distance. Find the horizontal distance. 169.4 yd

b. **Roller Coaster** From a blimp, the apparent distance traveled by the roller coaster in Example **6b** is the horizontal distance from the top of the hill to the bottom. Find the horizontal distance. 112.6 ft

Additional Answers

1. $\sin \theta = \frac{4}{5}$; $\cos \theta = \frac{3}{5}$; $\tan \theta = \frac{4}{3}$;
   $\csc \theta = \frac{5}{4}$; $\sec \theta = \frac{5}{3}$; $\cot \theta = \frac{3}{4}$

2. $\sin \theta = \frac{\sqrt{7}}{4}$; $\cos \theta = \frac{3}{4}$;
   $\tan \theta = \frac{\sqrt{7}}{3}$; $\csc \theta = \frac{4\sqrt{7}}{7}$;
   $\sec \theta = \frac{4}{3}$; $\cot \theta = \frac{3\sqrt{7}}{7}$

3. $\sin A = \frac{\sqrt{33}}{7}$, $\tan A = \frac{\sqrt{33}}{4}$,
   $\csc A = \frac{7\sqrt{33}}{33}$, $\sec A = \frac{7}{4}$,
   $\cot A = \frac{4\sqrt{33}}{33}$

4. $\sin A = \frac{20}{29}$, $\cos A = \frac{21}{29}$,
   $\csc A = \frac{29}{20}$, $\sec A = \frac{29}{21}$,
   $\cot A = \frac{21}{20}$

Study Tip
Angles of Elevation and Depression The angle of elevation and the angle of depression are congruent since they are alternate interior angles of parallel lines.

In the figure at the right, the angle formed by the line of sight from the swimmer and a line parallel to the horizon is called the **angle of elevation**. The angle formed by the line of sight from the lifeguard and a line parallel to the horizon is called the **angle of depression**.

Angle of depression
Angle of elevation

Real-World Example 6   Use Angles of Elevation and Depression

a. GOLF A golfer is standing at the tee, looking up to the green on a hill. If the tee is 36 feet lower than the green and the angle of elevation from the tee to the hole is 12°, find the distance from the tee to the hole.

12° 36 ft

Write an equation using a trigonometric function that involves the ratio of the vertical rise (side opposite the 12° angle) and the distance from the tee to the hole (hypotenuse).

$\sin 12° = \frac{36}{x}$ $\sin \theta = \frac{opp}{hyp}$

$x \sin 12° = 36$ Multiply each side by x.

$x = \frac{36}{\sin 12°}$ Divide each side by sin 12°.

$x \approx 173.2$ Use a calculator.

So, the distance from the tee to the hole is about 173.2 feet.

b. ROLLER COASTER The hill of the roller coaster has an *angle of descent*, or an angle of depression, of 60°. Its vertical drop is 195 feet. Estimate the length of the hill.

60°
x
195 ft

Write an equation using a trigonometric function that involves the ratio of the vertical drop (side opposite the 60° angle) and the length of the hill (hypotenuse).

$\sin 60° = \frac{195}{x}$ $\sin \theta = \frac{opp}{hyp}$

$x \sin 60° = 195$ Multiply each side by x.

$x = \frac{195}{\sin 60°}$ Divide each side by sin 60°.

$x \approx 225.2$ Use a calculator.

So, the length of the hill is about 225.2 feet.

Real-World Link
The steepest roller coasters in the world have angles of descent that are close to 90°.

Source: Ultimate Roller Coaster

> **Guided Practice**

6A. MOVING A ramp for unloading a moving truck has an angle of elevation of 32°. If the top of the ramp is 4 feet above the ground, estimate the length of the ramp. about 7.5 ft

MOVE

6B. LADDERS A 14-foot long ladder is placed against a house at an angle of elevation of 72°. How high above the ground is the top of the ladder? about 13.3 ft

Check Your Understanding

 = Step-by-Step Solutions begin on page R11.

Go Online! for a
Self-Check Quiz

Example 1 Find the values of the six trigonometric functions for angle θ. **1-2. See margin.**

1.

2.

Example 2 In a right triangle, $\angle A$ is acute. Find the values of the five remaining trigonometric funtions. **3-4. See margin.**

3. $\cos A = \frac{4}{7}$

4. $\tan A = \frac{20}{21}$

Examples 3-4 Use a trigonometric function to find the value of x. Round to the nearest tenth.

5 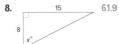 25.4

6. 7.7

7.  8.3

Example 5 Find the value of x. Round to the nearest tenth.

8. 61.9

9. 25.4

10.  68.0

Example 6

11. 🅜🅟 SENSE-MAKING Christian found two trees directly across from each other in a canyon. When he moved 100 feet from the tree on his side (parallel to the edge of the canyon), the angle formed by the tree on his side and the tree on the other side was 70°. Find the distance across the canyon. **about 274.7 ft**

12. LADDERS The recommended angle of elevation for a ladder used in fire fighting is 75°. At what height on a building does a 21-foot ladder reach if the recommended angle of elevation is used? Round to the nearest tenth. **20.3 ft**

Practice and Problem Solving

Extra Practice is on page R9.

Example 1 Find the values of the six trigonometric functions for angle θ. **13-16. See margin.**

13.

14.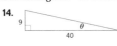

15.

16.

Example 2 In a right triangle, $\angle A$ and $\angle B$ are acute. Find the values of the five remaining trigonometric funtions. **17-20. See margin.**

17. $\tan A = \frac{8}{15}$

18. $\cos A = \frac{3}{10}$

19. $\tan B = 3$

20. $\sin B = \frac{4}{9}$

Differentiated Homework Options

Levels	🔵 AL Basic	🔵 OL Core	🔵 BL Advanced
Exercises	13–36, 58–61	13–45 odd, 58–61	37–61
2-Day Option	13–35 odd, 58–61	13–36, 58–61	
	14–36 even, 54–57	37–57	

 You can use ALEKS to provide additional remediation support with personalized instruction and practice.

Practice

Formative Assessment Use Exercises 1–12 to assess students' understanding of the concepts in this lesson.

The Practice and Problem Solving exercises assess the content taught in the lesson. The Preparing for Assessment page is meant to be used as preparation for end-of-course assessments.

Extra Practice

See page R9 for extra exercises for students who are approaching level or for on-level students who need additional reinforcement.

Additional Answers

13. $\sin \theta = \frac{12}{13}$; $\cos \theta = \frac{5}{13}$; $\tan \theta = \frac{12}{5}$; $\csc \theta = \frac{13}{12}$; $\sec \theta = \frac{13}{5}$; $\cot \theta = \frac{5}{12}$

14. $\sin \theta = \frac{9}{41}$; $\cos \theta = \frac{40}{41}$; $\tan \theta = \frac{9}{40}$; $\csc \theta = \frac{41}{9}$; $\sec \theta = \frac{41}{40}$; $\cot \theta = \frac{40}{9}$

15. $\sin \theta = \frac{\sqrt{51}}{10}$; $\cos \theta = \frac{7}{10}$; $\tan \theta = \frac{\sqrt{51}}{7}$; $\csc \theta = \frac{10\sqrt{51}}{51}$; $\sec \theta = \frac{10}{7}$; $\cot \theta = \frac{7\sqrt{51}}{51}$

16. $\sin \theta = \frac{2\sqrt{13}}{13}$; $\cos \theta = \frac{3\sqrt{13}}{13}$; $\tan \theta = \frac{2}{3}$; $\csc \theta = \frac{\sqrt{13}}{2}$; $\sec \theta = \frac{\sqrt{13}}{3}$; $\cot \theta = \frac{3}{2}$

17. $\sin A = \frac{8}{17}$, $\cos A = \frac{15}{17}$, $\csc A = \frac{17}{8}$, $\sec A = \frac{17}{15}$, $\cot A = \frac{15}{8}$

18. $\sin A = \frac{\sqrt{91}}{10}$, $\tan A = \frac{\sqrt{91}}{3}$, $\csc A = \frac{10\sqrt{91}}{91}$, $\sec A = \frac{10}{3}$, $\cot A = \frac{3\sqrt{91}}{91}$

19. $\sin B = \frac{3\sqrt{10}}{10}$, $\cos B = \frac{\sqrt{10}}{10}$, $\csc B = \frac{\sqrt{10}}{3}$, $\sec B = \sqrt{10}$, $\cot B = \frac{1}{3}$

20. $\cos B = \frac{\sqrt{65}}{9}$, $\tan B = \frac{4\sqrt{65}}{65}$, $\csc B = \frac{9}{4}$, $\sec B = \frac{9\sqrt{65}}{65}$, $\cot B = \frac{\sqrt{65}}{4}$

Teaching Tip

Calculators On a scientific calculator (in contrast to a graphing calculator), the sequence for finding the sine, cosine, or tangent of an angle may be to enter the angle measure, such as 20, first, and then press the SIN , COS , or TAN key

Levels of Complexity Chart

The levels of the exercises progress from 1 to 3, with Level 1 indicating the lowest level of complexity.

Exercises	13–36	37–46, 58–61	47–57
C Level 3			●
B Level 2		●	
Level 1	●		

Additional Answers

46a. Sample answer:

x ft
57°
370 ft

55. True; $\sin \theta = \dfrac{\text{opp}}{\text{hyp}}$ and the values of the opposite side and the hypotenuse of an acute triangle are positive, so the value of the sine function is positive.

56. The value of sin A equals sin C, so $\dfrac{\text{side opp } A}{\text{hyp}} = \dfrac{\text{side opp } C}{\text{hyp}}$. Becaus the hypotenuse is the same, the length of the side opposite angle A must equal the length of the side opposite angle C. Because the two sides have the same measure, the triangle is isosceles.

57. Sample answer: The slope describes the ratio of the vertical rise to the horizontal run of the roof. The vertical rise is opposite the angle that the roof makes with the horizontal. The horizontal run is the adjacent side. So, the tangent of the angle of elevation equals the ratio of the rise to the run, or the slope of the roof; $\theta = 33.7°$.

Go Online!

eSolutions Manual

Create worksheets, answer keys, and solutions handouts for your assignments.

Examples 3–4 Use a trigonometric function to find each value of *x*. Round to the nearest tenth.

21. 9, 12.7, 45°, x

22. 3.6, 64°, 4, x

(23) 10.4, x, 30°, 18

24. 32.9, 22, 48°, x

25. 8.7, 32°, x, 14

26. 70°, 15, x, 5.1

27. PARASAILING Refer to the beginning of the lesson and the figure at the right. Find *a*, the altitude of a person parasailing, if the tow rope is 250 feet long and the angle formed is 32°. Round to the nearest tenth. **132.5 ft**

a, 250 ft, 32°

28. **MP** MODELING Devon wants to build a rope bridge between his treehouse and Cheng's treehouse. Suppose Devon's treehouse is directly behind Cheng's treehouse. At a distance of 20 meters to the left of Devon's treehouse, an angle of 52° is measured between the two treehouses. Find the length of the rope. **25.6 m**

Example 5 Find the value of *x*. Round to the nearest tenth.

29. x°, 30, 10, 5

30. 67.2, 19, 8, x°

31. 36.9, 9, x°, 12

32. 55.2, 4, x°, 7

33. 32.5, x°, 32, 27

34. 23.6, x°, 25, 10

Example 6 **35.** SQUIRRELS Adult flying squirrels can make glides of up to 160 feet. If a flying squirrel glides a horizontal distance of 160 feet and the angle of descent is 9°, find its change in height. **25.3 ft**

36. HANG GLIDING A hang glider climbs at a 20° angle of elevation. Find the change in altitude of the hang glider when it has flown a horizontal distance of 60 feet. **21.8 ft**

B Use trigonometric functions to find the values of *x* and *y*. Round to the nearest tenth.

37. 30.2, x, 46.5°, y

38. y, 50°, x, 71.8

39.  x°, y°, 81, $26\frac{3}{4}$

37. $x = 21.9, y = 20.8$
38. $x = 93.7, y = 60.2$
39. $x = 19.3, y = 70.7$

Solve each equation.

40. $\cos A = \dfrac{3}{19}$ 80.9

41. $\sin N = \dfrac{9}{11}$ 54.9

42. $\tan X = 15$ 86.2

43. $\sin T = 0.35$ 20.5

44. $\tan G = 0.125$ 7.1

45. $\cos Z = 0.98$ 11.5

MP **Teaching the Mathematical Practices**

Modeling Mathematically proficient students can apply the mathematics they know to solve problems arising in everyday life, analyze relationships mathematically to draw conclusions, and interpret their mathematical results in the context of a situation.

46. MONUMENTS The San Jacinto Monument casts a shadow 370 feet long. The angle of elevation from the end of the shadow to the top of the monument is 57°.

   a. Draw and label a right triangle to represent this situation. See margin.
   b. Write a trigonometric function that can be used to find the height of the monument. $\tan 57° = \frac{x}{370}$
   c. Find the value of the function to determine the height of the monument to the nearest tenth. 569.8 ft

 47 NESTS Tabitha's eyes are 5 feet above the ground as she looks up to a bird's nest in a tree. If the angle of elevation is 74.5° and she is standing 12 feet from the tree's base, what is the height of the bird's nest? Round to the nearest tenth of a foot. 48.3 ft

48. RAMPS Two bicycle ramps each cover a horizontal distance of 8 feet. One ramp has a 20° angle of elevation, and the other ramp has a 35° angle of elevation, as shown at the right.

   a. How much taller is the second ramp than the first? Round to the nearest tenth. 2.7 ft
   b. How much longer is the second ramp than the first? Round to the nearest tenth. 1.3 ft

49. FALCONS A falcon at a height of 200 feet sees two mice A and B, as shown in the diagram.

   a. What is the approximate distance z between the falcon and mouse B? about 647.2 ft
   b. How far apart are the two mice? about 239.4 ft

In $\triangle ABC$, $\angle C$ is a right angle. Use the given measurements to find the missing side lengths and missing angle measures of $\triangle ABC$. Round to the nearest tenth if necessary. **50.** $m\angle B = 54°$, $b = 16.5$, $c = 20.4$ **51.** $m\angle A = 59°$, $a = 31.6$, $c = 36.9$

50. $m\angle A = 36°$, $a = 12$

51. $m\angle B = 31°$, $b = 19$

52. $a = 8$, $c = 17$
   $m\angle A = 28.1°$, $m\angle B = 61.9°$, $b = 15$

53. $\tan A = \frac{4}{5}$, $a = 6$
   $m\angle A = 38.7°$, $m\angle B = 51.3°$, $b = 7.5$, $c = 9.6$

H.O.T. Problems   Use Higher-Order Thinking Skills

54. CHALLENGE A line segment has endpoints $A(2, 0)$ and $B(6, 5)$, as shown in the figure at the right. What is the measure of the acute angle θ formed by the line segment and the x-axis? Explain how you found the measure.

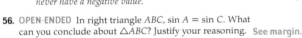

55. CONSTRUCT ARGUMENTS Determine whether the following statement is *true* or *false*. Explain your reasoning. See margin.
   For any acute angle, the sine function will never have a negative value.

56. OPEN-ENDED In right triangle ABC, $\sin A = \sin C$. What can you conclude about $\triangle ABC$? Justify your reasoning. See margin.

57. WRITING IN MATH A roof has a slope of $\frac{2}{3}$. Describe the connection between the slope and the angle of elevation θ that the roof makes with the horizontal. Then use an inverse trigonometric function to find θ. See margin.

54. About 51.3°; if a right triangle is drawn with $\overline{AB}$ as the hypotenuse, then the side opposite angle θ is 5 and the side adjacent angle θ is 4. $\tan A = \frac{5}{4}$, so $A \approx 51.3°$.

Assess

Ticket Out the Door Have students write the minimum information one must have about a right triangle in order to solve it for a missing side or angle value. two of the following: measure of an acute angle, length of a leg, length of the hypotenuse

---

Differentiated Instruction OL BL

Extension Have students compute the value of $\sin^2 x + \cos^2 x$ for some values of x. Have them make a conjecture about the value of the expression and use the definitions of the trigonometric functions to prove their conjecture. The value is always 1.

$$\sin^2 x + \cos^2 x \stackrel{?}{=} 1$$
$$\left(\frac{\text{opp}}{\text{hyp}}\right)^2 + \left(\frac{\text{adj}}{\text{hyp}}\right)^2 \stackrel{?}{=} 1$$
$$\text{opp}^2 + \text{adj}^2 = \text{hyp}^2, \text{ which is true by the Pythagorean Theorem}$$

Preparing for Assessment

Exercises 58–61 require students to use the skills they will need on standardized assessments. Each exercise is dual-coded with content standards and mathematical practice standards.

Dual Coding		
Exercise	Content Objective	MP Mathematical Practices
58	G.SRT.8	6
59	G.SRT.8	6
60	G.SRT.8	6
61	G.SRT.8	6

Diagnose Student Errors

Survey student responses for each item. Class trends may indicate common errors and misconceptions.

58.

A	CORRECT
B	Used adjacent side instead of opposite side
C	Used hypotenuse as adjacent side
D	Misused cosine function
E	Found cosecant of angle

59.

A	Found $-\cos\theta$
B	Found $\sin\theta$
C	CORRECT
D	Found $\sec\theta$
E	Found $\cot\theta$

60.

A	Found the value of $\sin\theta$
B	CORRECT
C	Found the value of $\cos\theta$
D	Found the value of $\cot\theta$
E	Found the value of $\csc\theta$

Go Online!

Self-Check Quiz

Students can use *Self-Check Quizzes* to check their understanding of this lesson.

58. The angle of elevation of a ramp is 17°. The length of the ramp is 5 meters. Which of the following expressions gives the height of the top of the ramp from the ground? MP 6 G.SRT.8 **A**

- A $5\sin 17°$
- B $\dfrac{5}{\sec 17°}$
- C $5\tan 17°$
- D $\dfrac{5}{\cos 17°}$
- E $\dfrac{5}{\sin 17°}$

59. In the first quadrant of a coordinate plane, the angle θ formed by a line segment and the x-axis is described by $\cos\theta = \frac{12}{13}$. Which of the following is the best approximation of $\tan\theta$? MP 6 G.SRT.8 **C**

- A -0.9231
- B 0.3846
- C 0.4167
- D 1.0833
- E 2.4000

60. The measure of an acute angle of a right triangle is given by θ. If $\sin\theta = \frac{8}{17}$, which of the following is the best approximation of $\tan\theta$? MP 6 G.SRT.8 **B**

- A 0.4705
- B 0.5333
- C 0.8824
- D 2.125
- E 1.875

61. **MULTI-STEP** A designer is laying out a section for a mosaic in a garden wall. The measurements of the section are shown on the figure. The width of the base of the triangle is 15 feet and the angle at the top of the section is 32°. Use the image below. MP 6 G.SRT.8

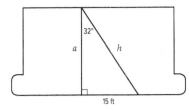

a. Which expression can be used to find the hypotenuse, h? **D**
- A $h = \dfrac{\sin 32°}{15}$
- B $15h = \sin 32°$
- C $h = 15\sin 32°$
- D $h = \dfrac{15}{\sin 32°}$

b. Which expression can be used to find the triangle's height, a? **D**
- A $a = \dfrac{\tan 32°}{15}$
- B $15a = \tan 32°$
- C $a = 15\tan 32°$
- D $a = \dfrac{15}{\tan 32°}$

c. What is the length of the hypotenuse of the triangle? **D**
- A 18.0
- B 20.8
- C 24.0
- D 28.3
- E 49.2

d. What is the height of the triangle? **C**
- A 18.0
- B 20.8
- C 24.0
- D 28.3
- E 49.2

e. Use the Pythagorean Theorem to show that your answers are correct. $28.3^2 \approx 15^2 + 24^2$

MP Standards for Mathematical Practice	
Emphasis On	Exercises
1 Make sense of problems and persevere in solving them.	12, 27, 28, 35, 36, 46–49
3 Construct viable arguments and critique the reasoning of others.	11, 54–57
6 Attend to precision.	1–10, 13–20, 21–26, 29–34, 37–45, 50–53, 58–61

EXTEND 9-1

Geometry Lab
Regular Polygons

You can use central angles of circles to investigate characteristics of regular polygons inscribed in a circle. Recall that a regular polygon is inscribed in a circle if each of its vertices lies on the circle.

Mathematical Practices
MP **5** Use appropriate tools strategically.

Activity Collect the Data

Step 1 Use a compass to draw a circle with a radius of one inch.

Step 2 Inscribe an equilateral triangle inside the circle. To do this, use a protractor to measure three angles of 120° at the center of the circle, because $\frac{360°}{3} = 120°$. Then connect the points where the sides of the angles intersect the circle using a straightedge.

Step 3 The **apothem** of a regular polygon is a segment that is drawn from the center of the polygon perpendicular to a side of the polygon. Use the cosine of angle θ to find the length of an apothem, labeled a in the diagram.

Model and Analyze

Work cooperatively.

1. Make a table like the one shown below and record the length of the apothem of the equilateral triangle. Inscribe each regular polygon named in the table in a circle with radius one inch. Complete the table.

Number of Sides, n	θ	a	Number of Sides, n	θ	a
3	60	0.50	7	25.7	0.90
4	45	0.71	8	22.5	0.92
5	36	0.81	9	20	0.94
6	30	0.87	10	18	0.95

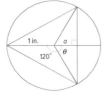

2. What do you notice about the measure of θ as the number of sides of the inscribed polygon increases? The measure of θ decreases.

3. What do you notice about the value of a? The length of the apothem increases as the number of sides increases, approaching 1.

4. **MAKE A CONJECTURE** Suppose you inscribe a 30-sided regular polygon inside a circle. Find the measure of angle θ. 6°

5. Write a formula that gives the measure of angle θ for a polygon with n sides. $\theta = 360 \div 2n$ or $\theta = 180 \div n$

6. Write a formula that gives the length of the apothem of a regular polygon inscribed in a circle with radius one inch. $a = \cos \theta$

7. How would the formula you wrote in Exercise 6 change if the radius was not one inch? See margin.

Ed Imaging

Additional Answer

7. To find the length of the apothem, you need to write the equation $\cos \theta = \dfrac{a}{\text{length of radius}}$. If the radius is 1, then $\cos \theta = a$. If the radius is not 1, then $a = \text{length of radius} \cdot \cos \theta$

Launch

Objective Investigate measures in regular polygons using trigonometry.

Materials for Each Student

- compass
- straightedge
- protractor

Teaching Tip

Have students draw a diameter for each circle. This will make it easier to use the protractor to draw the angles in each circle.

Teach

Working in Cooperative Groups Have students work with partners, mixing abilities so that a student with more experience using drawing tools is paired with one who has less experience. Have students complete the Activity.

Practice Have students complete Exercises 1–7.

Assess

Formative Assessment

Use Exercise 1 to assess whether students understand how to inscribe regular polygons in a circle.

From Concrete to Abstract

Exercises 5–7 require students to determine formulas that express relationships between the number of sides of a regular polygon and the length of the apothem. Students should use the results of Exercise 1 to determine the formulas.

Extending the Concept

Ask:

- What happens to the area of the inscribed regular polygons as n increases? The area increases.

- What value does that area approach? the area of the circle

Angles and Angle Measure

Track Your Progress

Objectives

1 Draw and find angles in standard position.

2 Convert between degree measures and radian measures.

Mathematical Background

An angle on a coordinate plane is in standard position if one ray of the angle (initial side) is placed on the positive *x*-axis and the other ray (terminal side) rotates about the origin. The terminal side rotates counterclockwise to show an angle with a positive measure and rotates clockwise to show an angle with a negative measure.

THEN	NOW	NEXT
G.SRT.8 Use trigonometric ratios and the Pythagorean Theorem to solve right triangles in applied problems.	**F.TF.1** Understand radian measure of an angle as the length of the arc on the unit circle subtended by the angle.	**F.TF.2** Explain how the unit circle in the coordinate plane enables the extension of trigonometric functions to all real numbers, interpreted as radian measures of angles traversed counterclockwise around the unit circle.

Go Online! All of these resources and more are available at connectED.mcgraw-hill.com

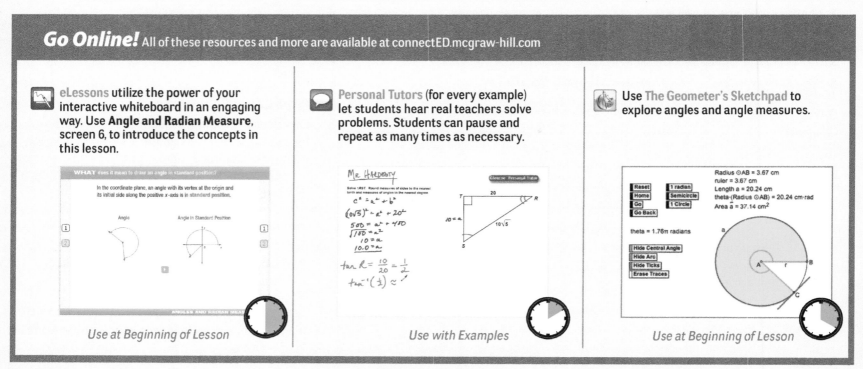

eLessons utilize the power of your interactive whiteboard in an engaging way. Use **Angle and Radian Measure**, screen 6, to introduce the concepts in this lesson.

Use at Beginning of Lesson

Personal Tutors (for every example) let students hear real teachers solve problems. Students can pause and repeat as many times as necessary.

Use with Examples

Use **The Geometer's Sketchpad** to explore angles and angle measures.

Use at Beginning of Lesson

⊙ER Using Open Educational Resources

Tutorials Students can read about degree and radian measure and explore step-by-step solutions of examples on **mathwarehouse.com**. They can also practice converting between degree and radian measure.

Use as flipped learning or as an instructional aid

Differentiate Your Resources

Extra Practice Additional practice or homework; Skills Practice is best for approaching-level students and Practice is best for on-level and beyond-level students

Skills Practice

AL OL ELL

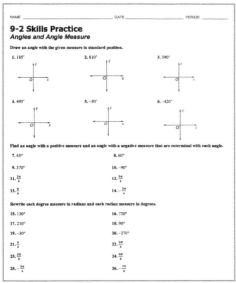

Practice

AL OL BL ELL

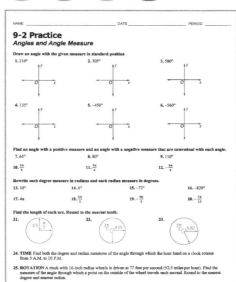

Word Problem Practice

AL OL BL ELL

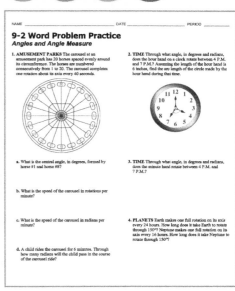

Intervention Reteaching and vocabulary activities that can be used with struggling or absent students and as ELL support

Extension Activities that can be used to extend lesson concepts

Study Guide and Intervention

AL OL ELL

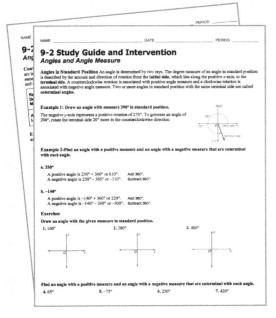

Study Notebook

AL OL BL ELL

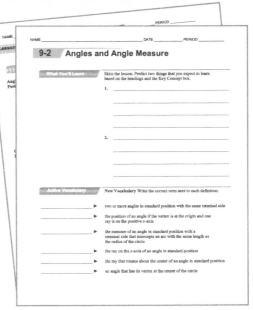

Enrichment

OL BL ELL

Launch

Have students read the Why? section of the lesson. Ask:

- In what direction does the shadow point in the morning? west

- At what time do you get the shortest shadow? noon

- Explain why the shadow moves 15° every hour. The shadow "returns" to any particular position every 24 hours, and the entire circle, divided by 24, is 360° ÷ 24, or 15°.

Teach

Ask the scaffolded questions for each example to build conceptual understanding for students at all levels.

1 Angles in Standard Position

Example 1 Draw an Angle in Standard Position

AL In what quadrant is the terminal side of −40°? Quadrant 4

OL Where would the terminal side of a 270° angle lie? on the negative y-axis

BL What positive angle would have the same terminal side as −40°? 320°

(continued on the next page)

:: Then	:: Now	:: Why?
• You used angles with degree measures.	**1** Draw and find angles in standard position. **2** Convert between degree measures and radian measures.	• A sundial is an instrument that indicates the time of day by the shadow that it casts on a surface marked to show hours or fractions of hours. The shadow moves around the dial 15° every hour.

 New Vocabulary
standard position
initial side
terminal side
coterminal angles
radian
central angle
arc length

 Mathematical Practices
2 Reason abstractly and quantitatively.

Content Standards
F.TF.1 Understand radian measure of an angle as the length of the arc on the unit circle subtended by the angle.

1 **Angles in Standard Position** An angle on the coordinate plane is in **standard position** if the vertex is at the origin and one ray is on the positive x-axis.

- The ray on the x-axis is called the **initial side** of the angle.

- The ray that rotates about the center is called the **terminal side**.

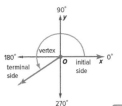

Key Concept Angle Measures

If the measure of an angle is positive, the terminal side is rotated counterclockwise.

If the measure of an angle is negative, the terminal side is rotated clockwise.

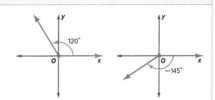

F.TF.1

Example 1 Draw an Angle in Standard Position

Draw an angle with the given measure in standard position.

a. 215°   215° = 180° + 35°

Draw the terminal side of the angle 35° counterclockwise past the negative x-axis.

b. −40°

The angle is negative. Draw the terminal side of the angle 40° clockwise from the positive x-axis.

Guided Practice 1A, 1B. See Ch. 9 Answer Appendix.

1A. 80°   **1B.** −105°

MP Mathematical Practices Strategies

Attend to precision. Help students make sense of the important definitions in this lesson so that they can remember and use them correctly. For example, ask:

- Draw the diagram you see on the first page of this lesson. Where do you always find the initial side of an angle in standard position? along the positive x-axis

- Where do you find the terminal side of the angle? It is the ray on the other side of the angle.

- What is between the initial side and the terminal side of the angle? the vertex of the angle

- Which is the positive direction for an angle in standard position? counterclockwise

- Which is the negative direction for an angle in standard position? clockwise

The terminal side of an angle can make more than one complete rotation. For example, a complete rotation of 360° plus a rotation of 120° forms an angle that measures 360° + 120° or 480°.

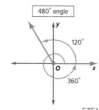

F.TF.1

Real-World Example 2   Draw an Angle in Standard Position

WAKEBOARDING *Wakeboarding* is a combination of surfing, skateboarding, snowboarding, and water skiing. One maneuver involves a 540-degree rotation in the air. Draw an angle in standard position that measures 540°.

540° = 360° + 180°

Draw the terminal side of the angle 180° past the positive *x*-axis.

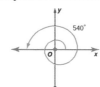

Guided Practice

2. Draw an angle in standard position that measures 600°. See margin.

Two or more angles in standard position with the same terminal side are called **coterminal angles**. For example, angles that measure 60°, 420°, and −300° are coterminal, as shown in the figure at the right.

An angle that is coterminal with another angle can be found by adding or subtracting a multiple of 360°.

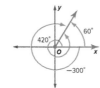

• 60° + 360° = 420°

• 60° − 360° = −300°

F.TF.1

Real-World Link

Wakeboarding is one of the fastest-growing water sports in the United States. Participation has increased more than 100% in recent years.

Source: King of Wake

Reading Math

Angle of Rotation
In trigonometry, an angle is sometimes referred to as an *angle of rotation*.

Example 3   Find Coterminal Angles

Find an angle with a positive measure and an angle with a negative measure that are coterminal with each angle.

a. **130°**

positive angle: 130° + 360° = 490°   Add 360°
negative angle: 130° − 360° = −230°   Subtract 360°

b. **−200°**

positive angle: −200° + 360° = 160°   Add 360°
negative angle: −200° − 360° = −560°   Subtract 360°

Guided Practice

3A. 15° 375°, −345°   **3B.** −45° 315°, −405°

Teaching Tip

Drawing Angles Discuss with students the importance of marking their drawings of angles with directional arrows, and labeling the angle measures.

Additional Answer (Guided Practice)

2.

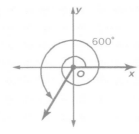

Need Another Example?

Draw an angle with the given measure in standard position.

a. **210°**

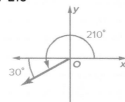

b. **−45°**

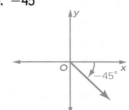

Example 2 Draw an Angle in Standard Position

AL What angle would be two complete rotations? 720°

OL How would we represent this rotation as a negative angle? −540°

BL Describe the position this maneuver ends in compared to its starting position. 180°

Need Another Example?

Diving In a springboard diving competition, a diver made a 900-degree rotation before slicing into the water. Draw an angle in standard position that measures 900°.

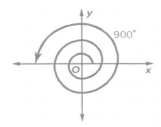

Example 3 Find Coterminal Angles

AL Why do we add or subtract 360° to find coterminal angles? 360° is one full revolution.

OL Give another positive angle that is coterminal to 130°. 850°

BL Write an expression for all angles that are coterminal to 130°. 130° + 360°k where k is any integer

Need Another Example?
Find an angle with a positive measure and an angle with a negative measure that are coterminal with each angle.
a. 210° Sample answers: 570°, −150°
b. −120° Sample answers: 240°, −480°

2 Convert Between Degrees and Radians

Example 4 Convert Between Degrees and Radians

AL How many radians is 2 full revolutions? 4π

OL How many degrees is one radian? 57.3°

BL Give a positive angle in radians that is coterminal to $\frac{5\pi}{2}$. 90°

Need Another Example?
Rewrite the degree measure in radians and the radian measure in degrees.
a. 30° $\frac{\pi}{6}$
b. $-\frac{5\pi}{3}$ −300°

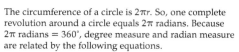

2 Convert Between Degrees and Radians Angles can also be measured in units that are based on arc length. One **radian** is the measure of an angle θ in standard position with a terminal side that intercepts an arc with the same length as the radius of the circle.

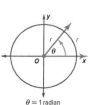

θ = 1 radian

The circumference of a circle is 2πr. So, one complete revolution around a circle equals 2π radians. Because 2π radians = 360°, degree measure and radian measure are related by the following equations.

$$2\pi \text{ radians} = 360° \qquad \pi \text{ radians} = 180°$$

Key Concept Convert Between Degrees and Radians

Degrees to Radians	Radians to Degrees
To convert from degrees to radians, multiply the number of degrees by $\frac{\pi \text{ radians}}{180°}$.	To convert from radians to degrees, multiply the number of radians by $\frac{180°}{\pi \text{ radians}}$.

F.TF.1

Example 4   Convert Between Degrees and Radians

Rewrite the degree measure in radians and the radian measure in degrees.

a. −30°

$$-30° = -30° \cdot \frac{\pi \text{ radians}}{180°}$$

$$= \frac{-30\pi}{180} \text{ or } -\frac{\pi}{6} \text{ radians}$$

b. $\frac{5\pi}{2}$

$$\frac{5\pi}{2} = \frac{5\pi}{2} \text{ radians} \cdot \frac{180°}{\pi \text{ radians}}$$

$$= \frac{900°}{2} \text{ or } 450°$$

Guided Practice

4A. 120° $\frac{2\pi}{3}$

4B. $-\frac{3\pi}{8}$ −67.5°

Concept Summary Degrees and Radians

The diagram shows equivalent degree and radian measures for special angles.

You may find it helpful to memorize the following equivalent degree and radian measures. The other special angles are multiples of these angles.

30° = $\frac{\pi}{6}$ 45° = $\frac{\pi}{4}$

60° = $\frac{\pi}{3}$ 90° = $\frac{\pi}{2}$

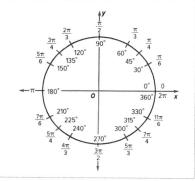

Differentiated Instruction **AL**

IF	students need further practice with coterminal angles,
THEN	have them work with a partner so that one person models an angle with two pencils or yardsticks. The other partner then names a positive and negative angle, less than or more than a full circle, that are each coterminal with the modeled angle.

A **central angle** of a circle is an angle with a vertex at the center of the circle. If you know the measure of a central angle and the radius of the circle, you can find the length of the arc that is intercepted by the angle.

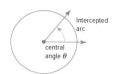

Key Concept Arc Length

Words	For a circle with radius r and central angle θ (in radians), the **arc length** s equals the product of r and θ.	Model

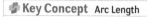

Symbols $s = r\theta$

F.TF.1

Real-World Example 5   Find Arc Length

TRUCKS Monster truck tires have a radius of 33 inches. How far does a monster truck travel in feet after just three fourths of a tire rotation?

Step 1 Find the central angle in radians.

$\theta = \dfrac{3}{4} \cdot 2\pi$ or $\dfrac{3\pi}{2}$ The angle is $\frac{3}{4}$ of a complete rotation.

Step 2 Use the radius and central angle to find the arc length.

$s = r\theta$ Write the formula for arc length.

$= 33 \cdot \dfrac{3\pi}{2}$ Replace r with 33 and θ with $\frac{3\pi}{2}$.

≈ 155.5 in.   Use a calculator to simplify.

≈ 13.0 ft   Divide by 12 to convert to feet.

So, the truck travels about 13 feet after three fourths of a tire rotation.

Watch Out!

Arc Length Remember to write the angle measure in radians, not degrees, when finding arc length. Also, recall that the number of radians in a complete rotation is 2π.

Guided Practice

5. A circle has a diameter of 9 centimeters. Find the arc length if the central angle is 60°. Round to the nearest tenth. **4.7 cm**

Check Your Understanding   ◯ = Step-by-Step Solutions begin on page R11.

Go Online! for a Self-Check Quiz

Examples 1–2   Draw an angle with the given measure in standard position. **1–3. See margin.**
F.TF.1
1. 140°   **2.** −60°   **3.** 390°

Example 3   Find an angle with a positive measure and an angle with a negative measure that are
F.TF.1   coterminal with each angle. **4. Sample answer: 385°, −335°**

4. 25°   **5** 175°   **6.** −100°

5. Sample answer: 535°, −185°
6. Sample answer: 260°, −460°

Example 4   Rewrite each degree measure in radians and each radian measure in degrees.
F.TF.1
7. $\dfrac{\pi}{4}$ **45°**   **8.** 225° $\dfrac{5\pi}{4}$ **9.** −40° $-\dfrac{2\pi}{9}$

Example 5   **10.** **MP** REASONING A tennis player's swing moves along the path of an arc. If the radius
F.TF.1   of the arc's circle is 4 feet and the angle of rotation is 100°, what is the length of the
arc? Round to the nearest tenth. **7.0 ft**

Differentiated Instruction **AL** **OL** **BL** **ELL**

Interpersonal Learners Have students work in small groups to decide on an informal explanation of what a radian is and what coterminal angles are. Then have a reporter from each group share that explanation with the whole class.

Example 5 Find Arc Length

AL What do r and s represent in this situation?
r is the radius of the tire, s is the arc length or the length the tire will travel

OL How far will the truck travel with one full revolution of the tire? approximately 17.3 feet

BL Explain what happens to the length the truck travels if the radius of the tire is larger. It will go farther.

Need Another Example?

Trucks The steering wheel on a monster truck has a radius of 11 inches. How far does a point on the steering wheel travel if the wheel makes four fifths of a rotation? about 55.3 in.

Additional Answers

1.
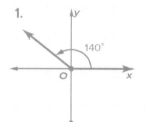

2.
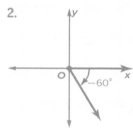

3.
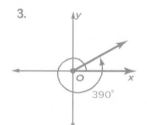

11.
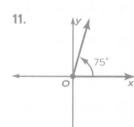

12.
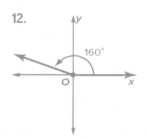

13.
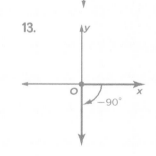

MP **Teaching the Mathematical Practices**

REASONING Mathematically proficient students make sense of quantities and their relationships in problem situations. Quantitative reasoning entails habits of creating a coherent representation of the problem at hand; considering the units involved; attending to the meaning of quantities, not just how to compute them; and knowing and flexibly using different properties of operations and objects.

Practice

Formative Assessment Use Exercises 1–10 to assess students' understanding of the concepts in this lesson.

The Practice and Problem Solving exercises assess the content taught in the lesson. The Preparing for Assessment page is meant to be used as preparation for end-of-course assessments.

Extra Practice

See page R9 for extra exercises for students who are approaching level or for on-level students who need additional reinforcement.

Additional Answers

14.

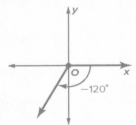

15.

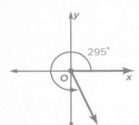

16.

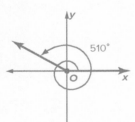

17.

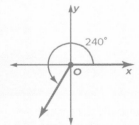

18.

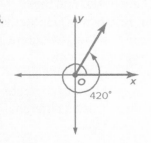

Practice and Problem Solving   Extra Practice is on page R9.

Examples 1–2   Draw an angle with the given measure in standard position.   11–13. See margin on page 607.
F.TF.1

11. 75°   **12.** 160°   **13.** −90°

14. −120°   **15.** 295°   **16.** 510° 14–16. See margin.

17. GYMNASTICS A gymnast on the uneven bars swings to make a 240° angle of rotation.

18. FOOD The lid on a jar of pasta sauce is turned 420° before it comes off. 17, 18. See margin.

Example 3   Find an angle with a positive measure and an angle with a negative measure that are
F.TF.1   coterminal with each angle. 19–24. Sample answers given.

19. 50° 410°, −310°   **20.** 95° 455°, −265°   **21.** 205° 565°, −155°

22. 350° 710°, −10°   **23.** −80° 280°, −440°   **24.** −195° 165°, −555°

Example 4   Rewrite each degree measure in radians and each radian measure in degrees.
F.TF.1

(25) 330° $\frac{11\pi}{6}$ **26.** $\frac{5\pi}{6}$ 150°   **27.** $-\frac{\pi}{3}$ −60°

28. −50° $-\frac{5\pi}{18}$ **29.** 190° $\frac{19\pi}{18}$ **30.** $-\frac{7\pi}{3}$ −420°

Example 5   **31.** SKATEBOARDING The skateboard ramp at the
F.TF.1   right is called a *quarter pipe*. The curved surface
              is determined by the radius of a circle. Find the
              length of the curved part of the ramp. about 12.6 ft

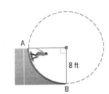

32. RIVERBOATS The paddlewheel of a riverboat has
a diameter of 24 feet. Find the arc length of the
circle made when the paddlewheel rotates 300°. about 62.8 ft

Find the length of each arc. Round to the nearest tenth.

33.   6.7 cm   **34.**   94.2 m

35. CLOCKS How long does it take for the minute hand on a clock to pass through
2.5π radians? 1 h 15 min

36. PERSEVERANCE Refer to the beginning of the lesson. A shadow moves around a
sundial 15° every hour.

a. After how many hours is the angle of rotation of the shadow $\frac{8\pi}{5}$ radians? 19.2 h

b. What is the angle of rotation in radians after 5 hours? $\frac{5\pi}{12}$

c. A sundial has a radius of 8 inches. What is the arc formed by a shadow after
14 hours? Round to the nearest tenth. 29.3 in.

Find an angle with a positive measure and an angle with a negative measure that are
coterminal with each angle. 37–40. Sample answers given.

37. 620° 260°, −100°   **38.** −400° 320°, −40°   **39.** $-\frac{3\pi}{4}$ $\frac{5\pi}{4}$, $\frac{11\pi}{4}$ **40.** $\frac{19\pi}{6}$ $\frac{7\pi}{6}$, $-\frac{5\pi}{6}$

Differentiated Homework Options

Levels	AL Basic	OL Core	BL Advanced
Exercises	11–34, 48–59	11–35 odd, 36, 37–47 odd, 54–59	35–59
2-Day Option	11–33 odd, 54–59	11–34, 54–59	
	12–34 even, 48–53	35–53	

You can use ALEKS to provide additional remediation support with personalized instruction and practice.

41 SWINGS A swing has a 165° angle of rotation.

   a. Draw the angle in standard position. **See margin.**

   b. Write the angle measure in radians. $\frac{11\pi}{12}$

   c. If the chains of the swing are $6\frac{1}{2}$ feet long, what is the length of the arc that the swing makes? Round to the nearest tenth. **18.7 ft**

   d. Describe how the arc length would change if the chain lengths were doubled.

41d. The arc length would double. Because $s = r\theta$, if r is doubled and θ remains unchanged, then the value of s is also doubled.

42b.
$\tan \angle BEA = -\frac{3}{2}$,
$\tan \angle DEC = \frac{4}{3}$

42d. Sample answer: In the coordinate plane, the tangent of the angle in standard position equals the slope of the terminal side of the angle.

42. MULTIPLE REPRESENTATIONS Consider $A(-4, 0)$, $B(-4, 6)$, $C(6, 0)$, and $D(6, 8)$.

   a. Geometric Draw $\triangle EAB$ and $\triangle ECD$ with E at the origin. **See margin.**

   b. Algebraic Find the values of the tangent of $\angle BEA$ and the tangent of $\angle DEC$.

   c. Algebraic Find the slope of $\overline{BE}$ and $\overline{ED}$. slope of $\overline{BE} = -\frac{3}{2}$, slope of $\overline{ED} = \frac{4}{3}$

   d. Verbal What conclusions can you make about the relationship between slope and tangent?

Rewrite each degree measure in radians and each radian measure in degrees.

43. $\frac{21\pi}{8}$ 472.5°   **44.** 124° $\frac{31\pi}{45}$ **45.** $-200°$ $-\frac{10\pi}{9}$ **46.** 5 $\frac{900}{\pi} \approx 286.5°$

47. CAROUSELS At its peak, the carousel of the Sky Screamer at Six Flags Over Texas makes 0.5 revolution per minute. The circle formed by riders sitting in the outside has a radius of 49 feet. The circle formed by riders sitting in the inside has a radius of 47 feet.

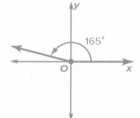

   a. Find the angle θ, in degrees, through which the carousel rotates in one second. **3°**

   b. In one second, what is the difference in arc lengths between the riders sitting in the outside row and the riders sitting in the inside row? **0.1 ft**

F.TF.1

H.O.T. Problems   **Use H**igher-**O**rder **T**hinking Skills

48. ERROR ANALYSIS Tarshia and Alan are writing an expression for the measure of an angle coterminal with the angle shown at the right. Is either of them correct? Explain your reasoning.

Tarshia	Alan
The measure of a coterminal angle is $(x - 360)°$.	The measure of a coterminal angle is $(360 - x)°$.

48. Tarshia; a coterminal can be found by adding a multiple of 360° or by subtracting a multiple of 360°. Alan incorrectly subtracted the original angle measure from 360°.

49. CHALLENGE A line makes an angle of $\frac{\pi}{2}$ radians with the positive x-axis at the point $(2, 0)$. Find an equation for this line. $x = 2$

50. REASONING Express $\frac{1}{8}$ of a revolution in degrees and in radians. Explain your reasoning. $45°$, $\frac{\pi}{4}$; $\frac{1}{8}$ of 360° is 45° and $\frac{1}{8}$ of 2π is $\frac{\pi}{4}$.

51. OPEN-ENDED Draw and label an acute angle in standard position. Find two angles, one positive and one negative, that are coterminal with the angle. **See margin.**

52. REASONING Justify the formula for the length of an arc. **See Ch. 9 Answer Appendix.**

53. WRITING IN MATH Use a circle with radius r to describe what one degree and one radian represent. Then explain how to convert between the measures. **See Ch. 9 Answer Appendix.**

Standards for Mathematical Practice

Emphasis On	Exercises
1 Make sense of problems and persevere in solving them.	10, 17, 18, 31, 32, 35, 41, 42, 59
2 Reason abstractly and quantitatively.	1–9, 11–16, 19–30, 37–40, 43–46, 54, 55, 57, 58
3 Construct viable arguments and critique the reasoning of others.	48–53
4 Model with mathematics.	36, 47
6 Attend to precision.	33, 34, 56

Watch Out!

Error Analysis For Exercise 48, if students think that Alan is correct, point out that if the angle shown is measured clockwise, then the angle measure x is a negative quantity and $(360 - x)°$ will be greater, not less, than 360°.

Levels of Complexity Chart

The levels of the exercises progress from 1 to 3, with Level 1 indicating the lowest level of complexity.

Exercises	11–34	35–47, 54–59	48–53
Level 3			●
Level 2		●	
Level 1	●		

Additional Answers

41a.

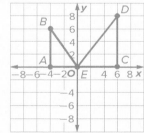

42a.

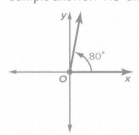

51. Sample answer: 440° and −280°

Go Online!

eSolutions Manual

Create worksheets, answer keys, and solutions handouts for your assignments.

Assess

Yesterday's News Ask students to describe how yesterday's topic of angles in right triangles helped them understand and illustrate today's topic of angles in standard position.

Preparing for Assessment

Exercises 54–59 require students to use the skills they will need on standardized assessments. Each exercise is dual-coded with content standards and mathematical practice standards.

Dual Coding		
Exercise	Content Objective	**MP** Mathematical Practices
54	F.TF.1	2
55	F.TF.1	2
56	F.TF.1	6
57	F.TF.1	2
58	F.TF.1	2
59	F.TF.1	1

Diagnose Student Errors

Survey student responses for each item. Class trends may indicate common errors and misconceptions.

54.

A	Corresponded $\frac{\pi}{2}$ to x-axis
B	Related −90° to the x-axis
C	Divided degrees by 90° to rewrite 120° in radians
D	Added 180° instead of 360° to find coterminal angle
E	CORRECT

55.

A	Rewrote radian measure by multiplying by 360°.
B	Divided 6 by 5 and interpreted quotient as 120°.
C	CORRECT
D	Added 90° to the conversion
E	Misunderstood definition of terminal side

Go Online!

Self-Check Quiz

Students can use *Self-Check Quizzes* to check their understanding of this lesson.

54. Given two angle measures in standard position, which of the following pairs of angles are coterminal? **MP** 2 F.TF.1 **E**

- A   −360° and $\frac{\pi}{2}$
- B   −90° and 4π
- C   120° and $\frac{4\pi}{3}$
- D   180° and 4π
- E   270° and $\frac{7\pi}{2}$

55. Where on a coordinate plane does the terminal side of the angle $\frac{6\pi}{5}$ appear? **MP** 2 F.TF.1 **C**

- A   Quadrant I
- B   Quadrant II
- C   Quadrant III
- D   Quadrant IV
- E   lies on y-axis

56. The shortest distance along circle O from point R to point Q is 800 yards. The radius of circle O is 1200 yards. What is the approximate measure of $\angle ROQ$? **MP** 6 F.TF.1 **C**

- A   less than 20°
- B   between 20° and 30°
- C   between 30° and 40°
- D   between 40° and 50°
- E   greater than 50°

57. MULTI-STEP

a. Which of the following angle measures is coterminal with the angle $\frac{5\pi}{12}$? **MP** 2 F.TF.1 **D**

- A   25°
- B   180°
- C   150°
- D   435°
- E   510°

b. Some students are creating a school tile mosaic to cover an industrial pipe spool's circumference. What would the arc length be for a section of tiles that will cover $\frac{5\pi}{12}$ radians of the spool if the diameter of the spool is 8 feet? **A**

- A   5.2 ft
- B   10.4 ft
- C   3.3 ft
- D   6.7 ft

c. Two other groups of students will split the remaining part of the spool circumference evenly. Calculate the degree measure to which each group is entitled.

$$\frac{24\pi}{12} - \frac{5\pi}{12} = \frac{19\pi}{12} \div 2 = \frac{19\pi}{24} \times \frac{180}{\pi} = 142.5°$$

d. Calculate how many feet of the spool that will be.

$$\frac{142.5}{360} \times 2 \times 3.14 \times 4 = 9.9 \text{ ft}$$

58. a. What is the measure of an angle in degrees if the angle measures 2 radians? **MP** 2 F.TF.1

$$\boxed{\frac{360}{\pi} \text{ degrees}}$$

b. What is the measure of an angle in radians if the angle measures π degrees?

$$\boxed{\frac{\pi}{180} \text{ degrees}}$$

59. Aisha runs 120 yards around a circular racetrack. The radius of the racetrack is 40 yards. **MP** 1 F.TF.1

a. Find the central angle of Aisha's rotation, as she runs the 120 yards. Express your answer in radians.

$$\boxed{3 \text{ radians}}$$

b. Find the central angle of Aisha's rotation, as she runs the 120 yards. Express your answer in degrees.

$$\boxed{\frac{540}{\pi} \text{ degrees}}$$

Differentiated Instruction **OL** **BL**

Extension There are infinitely many angles coterminal with a given angle. Have students write an expression that gives the angle measure of all angles coterminal with an angle of 50°. $50° + k \cdot 360°$, where k is any interger

56.

A	Multiplied arc length by 90° instead of 180°
B	Estimated π
C	CORRECT
D	Used 3 to estimate π and did not round down
E	Swapped arc length and radius measures in formula

EXTEND 9-2

Geometry Lab
Areas of Parallelograms

The area of any triangle can be found using the sine ratios in the triangle. A similar process can be used to find the area of a parallelogram.

Activity

Work cooperatively. Find the area of parallelogram *ABCD*.

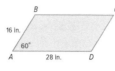

Step 1 Draw diagonal $\overline{BD}$.

$\overline{BD}$ divides the parallelogram into two congruent triangles, $\triangle ABD$ and $\triangle CDB$.

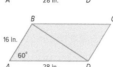

Step 2 Find the area of $\triangle ABD$.

Use the sine ratio to determine the height h from *B* to $\overline{AD}$.

$\sin \theta = \dfrac{\text{opp}}{\text{hyp}}$	Definition of sine
$\sin \theta = \dfrac{h}{AB}$	$h = $ opp, $AB = $ hyp
$AB \sin \theta = h$	Solve for h.

So, $h = AB \sin \theta$.

Area $= \frac{1}{2}bh$	Area of a triangle
$= \frac{1}{2}(AD)(AB) \sin A$	$b = AD, h = AB \sin A$
$= \frac{1}{2}(28)(16) \sin 60°$	$AD = 28, AB = 16, A = 60°$
$= 224\left[\dfrac{\sqrt{3}}{2}\right]$	Multiply and evaluate sin 60°.
$= 112\sqrt{3}$	Simplify.

Step 3 Find the area of $\square ABCD$.

The area of $\square ABCD$ is equal to the sum of the areas of $\triangle ABD$ and $\triangle CDB$. Because $\triangle ABD \cong \triangle CDB$, the areas of $\triangle ABD$ and $\triangle CDB$ are equal. So, the area of $\square ABCD$ equals twice the area of $\triangle ABD$.

$2 \cdot 112\sqrt{3} = 224\sqrt{3}$ or about 387.98 square inches.

Exercises

Work cooperatively. For each of the following:

a. Find the area of each parallelogram.

b. Find the area of each parallelogram when the included angle is half the given measure.

c. Find the area of each parallelogram when the included angle is twice the given measure.

1a. 106.07 m² **1b.** 57.40 m² **1c.** 150 m² **2a.** 22.5 in² **2b.** 11.65 in²
2c. 38.97 in² **3a.** 19,318.52 ft² **3b.** 12,175.23 ft² **3c.** 10,000 ft²

1.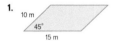

2.

3.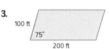

Launch

Objective Use the sine ratio to find the area of a parallelogram.

Teaching Tip
If students use calculators to approximate areas, remind them to make sure that their calculators are in degree mode.

Teach

Working in Cooperative Groups Have students work in pairs, mixing abilities, to complete the Activity.

Have each student in the pair draw a different diagonal of the parallelogram. Then have them compare their area calculations to check their work. **ELL**

Ask:
● What part of the parallelogram is represented by *AB* sin *A*? $\sin A = \dfrac{h}{AB}$, so *AB* sin *A* is the height *h* of the parallelogram.

Practice Have students complete Exercises 1–3.

Assess

Formative Assessment
Use Exercise 2 to assess whether students can calculate the area of a parallelogram given two consecutive sides and the measure of the included angle.

From Concrete to Abstract
Ask students to develop a formula for the area of a parallelogram if the measures of two consecutive sides are *a* and *b*, and one of the angles of the parallelogram is θ. $A = ab \sin \theta$

Trigonometric Functions of General Angles

SUGGESTED PACING (DAYS)

90 min.	0.5	0.5
45 min.	1.0	0.5
	Instruction	Extend Lab

Track Your Progress

Objectives

1 Find values of trigonometric functions for general angles.

2 Find values of trigonometric functions by using reference angles.

Mathematical Background

When a point $P(x, y)$ on the terminal side of the angle θ is known, the value of the six trigonometric functions can be found. Draw a segment from the point perpendicular to the x-axis, forming a right triangle with a leg measuring x units, a leg measuring y units, and a hypotenuse measuring r units, where $r = \sqrt{x^2 + y^2}$. Thus, $\sin \theta = \frac{y}{r}$ and $\cos \theta = \frac{x}{r}$.

THEN	NOW	NEXT
F.TF.1 Understand radian measure of an angle as the length of the arc on the unit circle subtended by the angle.	**F.TF.2** Explain how the unit circle in the coordinate plane enables the extension of trigonometric functions to all real numbers, interpreted as radian measures of angles traversed counterclockwise around the unit circle.	**A.CED.2** Create equations in two or more variables to represent relationships between quantities; graph equations on coordinate axes with labels and scales.

F.IF.4 For a function that models a relationship between two quantities, interpret key features of graphs and tables in terms of the quantities, and sketch graphs showing key features given a verbal description of the relationship. Key features include: intercepts; intervals where the function is increasing, decreasing, positive, or negative; relative maximums and minimums; symmetries; end behavior; and periodicity.

F.IF.7e Graph exponential and logarithmic functions, showing intercepts and end behavior, and trigonometric functions, showing period, midline, and amplitude.

F.TF.5 Choose trigonometric functions to model periodic phenomena with specified amplitude, frequency, and midline.

Go Online! All of these resources and more are available at connectED.mcgraw-hill.com

Animations illustrate key concepts through step-by-step tutorials and videos.

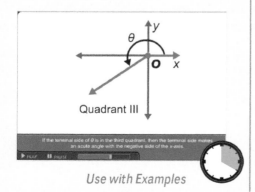

Use with Examples

Personal Tutors (for every example) let students hear real teachers solve problems. Students can pause and repeat as many times as necessary.

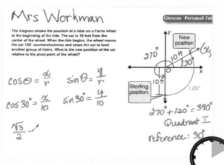

Use with Examples

Use **The Geometer's Sketchpad** to explore trigonometric functions and general angle measures.

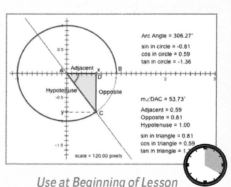

Use at Beginning of Lesson

 Using Open Educational Resources

Publishing Have students work in groups to create a wall on **Padlet** with everything they know about trigonometric functions. Ask for volunteers to share their wall with the class, and have students continue to add to the wall as they progress through the chapter. *Use as homework or classwork*

Go Online!
connectED.mcgraw-hill.com

 Worksheets

Differentiate Your Resources

Extra Practice Additional practice or homework; Skills Practice is best for approaching-level students and Practice is best for on-level and beyond-level students

Skills Practice

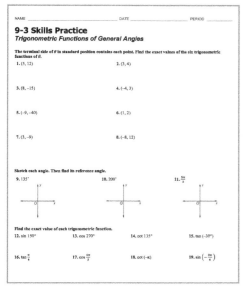

Practice

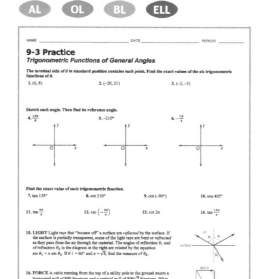

Word Problem Practice

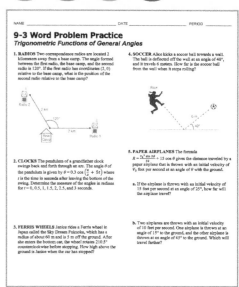

Intervention Reteaching and vocabulary activities that can be used with struggling or absent students and as ELL support

Extension Activities that can be used to extend lesson concepts

Study Guide and Intervention

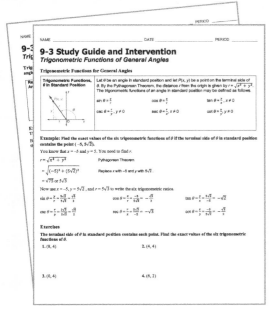

Study Notebook

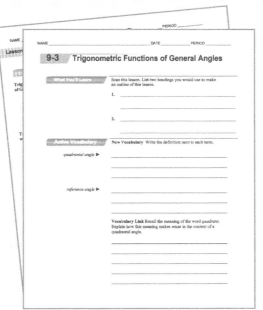

Enrichment

Launch

Have students read the Why? section of the lesson and tell them that the positions of two of the arms are 220° clockwise and 50° counterclockwise from the standard position.

Ask:

- Which quadrant is indicated by "20° clockwise"?
  Quadrant IV

- Which is indicated by "200° counterclockwise"?
  Quadrant III

- How can you describe the position "20° clockwise" in terms of a counterclockwise rotation?
  340° counterclockwise

Teach

Ask the scaffolded questions for each example to build conceptual understanding for students at all levels.

1 Trigonometric Functions for General Angles

Example 1 Evaluate Trigonometric Functions Given a Point

AL What part of the right triangle does r represent? the hypotenuse

OL How is a right triangle formed? A perpendicular line is taken from the point to the x-axis.

BL Explain why sin θ will be negative in the third quadrant. In the third quadrant the y value will be negative and r is always positive, so $\frac{y}{r}$ will be negative.

(continued on the next page)

Go Online!

Interactive Whiteboard

Use the *eLesson* or *Lesson Presentation* to present this lesson.

LESSON 3
Trigonometric Functions of General Angles

Then	Now	Why?
You found values of trigonometric functions for acute angles.	**1** Find values of trigonometric functions for general angles. **2** Find values of trigonometric functions by using reference angles.	In the ride at the right, the cars rotate back and forth about a central point. The positions of the arms supporting the cars can be described using trigonometric angles in standard position, with the central point of the ride at the origin of a coordinate plane.

 New Vocabulary
quadrantal angle
reference angle

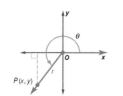 **Mathematical Practices**
6 Attend to precision.

Content Standards
F.TF.1 Understand radian measure of an angle as the length of the arc on the unit circle subtended by the angle.
F.TF.2 Explain how the unit circle in the coordinate plane enables the extension of trigonometric functions to all real numbers, interpreted as radian measures of angles traversed counterclockwise around the unit circle.

1 Trigonometric Functions for General Angles You can find values of trigonometric functions for angles greater than 90° or less than 0°.

> **Key Concept** Trigonometric Functions of General Angles
>
> Let θ be an angle in standard position and let $P(x, y)$ be a point on its terminal side. Using the Pythagorean Theorem, $r = \sqrt{x^2 + y^2}$. The six trigonometric functions of θ are defined below.
>
> $\sin \theta = \dfrac{y}{r}$ $\qquad$ $\cos \theta = \dfrac{x}{r}$ $\qquad$ $\tan \theta = \dfrac{y}{x}, x \neq 0$
>
> $\csc \theta = \dfrac{r}{y}, y \neq 0$ $\qquad$ $\sec \theta = \dfrac{r}{x}, x \neq 0$ $\qquad$ $\cot \theta = \dfrac{x}{y}, y \neq 0$
>
>

Example 1 Evaluate Trigonometric Functions Given a Point

The terminal side of θ in standard position contains the point at $(-3, -4)$. Find the exact values of the six trigonometric functions of θ.

Step 1 Draw the angle, and find the value of r.

$$r = \sqrt{x^2 + y^2}$$
$$= \sqrt{(-3)^2 + (-4)^2}$$
$$= \sqrt{25} \text{ or } 5$$

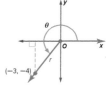

Step 2 Use $x = -3$, $y = -4$, and $r = 5$ to write the six trigonometric ratios.

$\sin \theta = \dfrac{y}{r} = \dfrac{-4}{5}$ or $-\dfrac{4}{5}$ $\qquad$ $\cos \theta = \dfrac{x}{r} = \dfrac{-3}{5}$ or $-\dfrac{3}{5}$ $\qquad$ $\tan \theta = \dfrac{y}{x} = \dfrac{-4}{-3}$ or $\dfrac{4}{3}$

$\csc \theta = \dfrac{r}{y} = \dfrac{5}{-4}$ or $-\dfrac{5}{4}$ $\qquad$ $\sec \theta = \dfrac{r}{x} = \dfrac{5}{-3}$ or $-\dfrac{5}{3}$ $\qquad$ $\cot \theta = \dfrac{x}{y} = \dfrac{-3}{-4}$ or $\dfrac{3}{4}$

▶ **Guided Practice**

1. The terminal side of θ in standard position contains the point at $(-6, 2)$. Find the exact values of the six trigonometric functions of θ. See margin.

MP Mathematical Practices Strategies

Look for and express regularity in repeated reasoning. Help students correlate trigonometric functions in circles with trigonometric functions that they have previously seen in triangles by writing and drawing the two side by side. Doing this freestyle on a blank piece of paper might be best, but you can also give the following guidance.

- What are the base and height of the triangle called in the Pythagorean Theorem for triangles? a and b

- What are the base and height of the triangle called in the Pythagorean Theorem for the radius of circles? x and y

- What is the hypotenuse called in the Pythagorean Theorem for triangles? c

- What is the hypotenuse called in the Pythagorean Theorem for the radius of a circle? r

- Write the Pythagorean Theorem for triangles. $a^2 + b^2 = c^2$

- Write the Pythagorean Theorem for the radii of circles. $x^2 + y^2 = r^2$

If the terminal side of angle θ in standard position lies on the x- or y-axis, the angle is called a **quadrantal angle**.

Key Concept Quadrantal Angles

$\theta = 0°$
or 0 radians

$\theta = 90°$
or $\frac{\pi}{2}$ radians

$\theta = 180°$
or π radians

$\theta = 270°$
or $\frac{3\pi}{2}$ radians

Example 2 Quadrantal Angles

The terminal side of θ in standard position contains the point at $(0, 6)$. Find the values of the six trigonometric functions of θ.

The point at $(0, 6)$ lies on the positive y-axis, so the quadrantal angle θ is $90°$. Use $x = 0$, $y = 6$, and $r = 6$ to write the trigonometric functions.

$\sin \theta = \frac{y}{r} = \frac{6}{6}$ or 1

$\cos \theta = \frac{x}{r} = \frac{0}{6}$ or 0

$\tan \theta = \frac{y}{x} = \frac{6}{0}$ undefined

$\csc \theta = \frac{r}{y} = \frac{6}{6}$ or 1

$\sec \theta = \frac{r}{x} = \frac{6}{0}$ undefined

$\cot \theta = \frac{x}{y} = \frac{0}{6}$ or 0

Guided Practice

2. The terminal side of θ in standard position contains the point at $(-2, 0)$. Find the values of the six trigonometric functions of θ.

2 **Trigonometric Functions with Reference Angles** If θ is a nonquadrantal angle in standard position, its **reference angle** θ' is the acute angle formed by the terminal side of θ and the x-axis. The rules for finding the measures of reference angles for $0° < \theta < 360°$ or $0 < \theta < 2\pi$ are shown below.

Key Concept Reference Angles

Quadrant I	Quadrant II	Quadrant III	Quadrant IV
$\theta' = \theta$	$\theta' = 180° - \theta$ $\theta' = \pi - \theta$	$\theta' = \theta - 180°$ $\theta' = \theta - \pi$	$\theta' = 360° - \theta$ $\theta' = 2\pi - \theta$

Additional Answer (Guided Practice)

1. $\sin \theta = \frac{\sqrt{10}}{10}$, $\cos \theta = -\frac{3\sqrt{10}}{10}$, $\tan\theta = -\frac{1}{3}$,

   $\csc \theta = \sqrt{10}$, $\sec \theta = -\frac{\sqrt{10}}{3}$, $\cot \theta = -3$

2. $\sin \theta = 0$ $\cos \theta = -1$, $\tan \theta = 0$, $\csc \theta = $ undefined, $\sec \theta = -1$, $\cot \theta = $ undefined

---

Need Another Example?
The terminal side of θ in standard position contains the point $(8, -15)$. Find the exact values of the six trigonometric functions of θ.

$\sin \theta = -\frac{15}{17}$; $\cos \theta = \frac{8}{17}$; $\tan \theta = -\frac{15}{8}$; $\sec \theta = \frac{17}{8}$;

$\csc \theta = -\frac{17}{15}$; $\cot \theta = -\frac{8}{15}$

Watch Out!

Unlocking Misconceptions In Example 1, watch for students who think that r must be a negative number any time either x or y has a negative value.

Example 2 Quadrantal Angles

AL List four negative quadrantal angles. $-90°$, $-180°$, $-270°$, $-360°$

OL List the two quadrantal angles where tangent will be undefined. $90°$, $270°$

BL Give the general form of all points that would give θ where $\sin \theta = 0$. $(a, 0)$ where a is any real number

Need Another Example?
The terminal side of θ in standard position contains the point at $(0, -2)$. Find the values of the six trigonometric functions of θ. $\sin \theta = -1$; $\cos \theta = 0$; $\tan \theta$ is undefined; $\csc \theta = -1$; $\sec \theta$ is undefined; $\cot \theta = 0$

Watch Out!

Preventing Errors When finding trigonometric values for an angle, students may find it helpful to sketch an angle in standard position and drop a perpendicular to the x-axis to form a right triangle. Then they can see that, as the angle increases from $90°$ to $180°$, the y value approaches 0, and the values of the other two sides approach each other.

2 Trigonometric Functions with Reference Angles

Example 3 Find Reference Angles

AL What type of angle is a reference angle? acute

OL What is the reference angle for a positive acute angle? The angle is its own reference angle.

BL Explain why the reference angle for an angle less than one full turn in the fourth quadrant is $360° - \theta$. The reference angle is the angle formed by the ray and the positive x-axis. One full revolution is $360°$ and the reference angle is the amount left in one full revolution, so the reference angle is $360° - \theta$.

Need Another Example?

Sketch each angle. Then find its reference angle.
a. 330° 30°

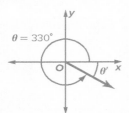

b. $-\dfrac{5\pi}{6}$ $\dfrac{\pi}{6}$

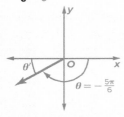

Additional Answers (Guided Practice)

3A. 70°

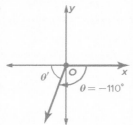

3B. $\dfrac{\pi}{3}$

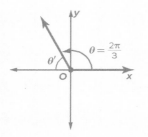

If the measure of θ is greater than 360° or less than 0°, then use a coterminal angle with a positive measure between 0° and 360° to find the reference angle.

Study Tip
Modeling You can refer to the diagram in the Lesson 9-2 Concept Summary to help you sketch angles.

Example 3 Find Reference Angles

Sketch each angle. Then find its reference angle.

a. 210°

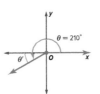

The terminal side of 210° lies in Quadrant III.
$\theta' = \theta - 180°$
$= 210° - 180°$ or 30°

b. $-\dfrac{5\pi}{4}$

coterminal angle: $-\dfrac{5\pi}{4} + 2\pi = \dfrac{3\pi}{4}$

The terminal side of $\dfrac{3\pi}{4}$ lies in Quadrant III.
$\theta' = \pi - \theta$
$= \pi - \dfrac{3\pi}{4}$ or $\dfrac{\pi}{4}$

▶ **Guided Practice** 3A, 3B. See margin.

3A. −110°

3B. $\dfrac{2\pi}{3}$

You can use reference angles to evaluate trigonometric functions for any angle θ. The sign of a function is determined by the quadrant in which the terminal side of θ lies. Use these steps to evaluate a trigonometric function for any angle θ.

✦ Key Concept Evaluate Trigonometric Functions

Step 1 Find the measure of the reference angle θ'.

Step 2 Evaluate the trigonometric function for θ'.

Step 3 Determine the sign of the trigonometric function value. Use the quadrant in which the terminal side of θ lies.

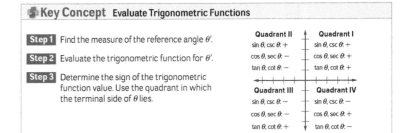

You can use the trigonometric values of angles measuring 30°, 45°, and 60° that you learned in Lesson 9-1.

Trigonometric Values for Special Angles					
Sine	Cosine	Tangent	Cosecant	Secant	Cotangent
$\sin 30° = \dfrac{1}{2}$	$\cos 30° = \dfrac{\sqrt{3}}{2}$	$\tan 30° = \dfrac{\sqrt{3}}{3}$	$\csc 30° = 2$	$\sec 30° = \dfrac{2\sqrt{3}}{3}$	$\cot 30° = \sqrt{3}$
$\sin 45° = \dfrac{\sqrt{2}}{2}$	$\cos 45° = \dfrac{\sqrt{2}}{2}$	$\tan 45° = 1$	$\csc 45° = \sqrt{2}$	$\sec 45° = \sqrt{2}$	$\cot 45° = 1$
$\sin 60° = \dfrac{\sqrt{3}}{2}$	$\cos 60° = \dfrac{1}{2}$	$\tan 60° = \sqrt{3}$	$\csc 60° = \dfrac{2\sqrt{3}}{3}$	$\sec 60° = 2$	$\cot 60° = \dfrac{\sqrt{3}}{3}$

Differentiated Instruction **AL** **OL**

Auditory/Musical Learners Have students work in small groups to create a jingle, song, rap, or short poem to help them remember the trigonometric values for special angles.

Example 4 Use a Reference Angle to Find a Trigonometric Value

Find the exact value of each trigonometric function.

a. cos 240°

The terminal side of 240° lies in Quadrant III.

$\theta' = \theta - 180°$ Find the measure of the reference angle.
$= 240° - 180°$ or $60°$ $\theta = 240°$

$\cos 240° = -\cos 60°$ or $-\dfrac{1}{2}$ The cosine function is
negative in Quadrant III.

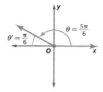

b. $\csc \dfrac{5\pi}{6}$

The terminal side of $\dfrac{5\pi}{6}$ lies in Quadrant II.

$\theta' = \pi - \theta$ Find the measure of the reference angle.
$= \pi - \dfrac{5\pi}{6}$ or $\dfrac{\pi}{6}$ $\theta = \dfrac{5\pi}{6}$

$\csc \dfrac{5\pi}{6} = \csc \dfrac{\pi}{6}$ The cosecant function is
positive in Quadrant II.

$= \csc 30°$ $\dfrac{\pi}{6}$ radians $= 30°$

$= 2$ $\csc 30° = \dfrac{1}{\sin 30}$

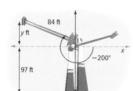

Guided Practice

4A. cos 135° $-\dfrac{\sqrt{2}}{2}$ **4B.** $\tan \dfrac{5\pi}{6}$ $-\dfrac{\sqrt{3}}{3}$

Real-World Example 5   Use Trigonometric Functions

RIDES The swing arms of the ride at the right are 84 feet long and the height of the axis from which the arms swing is 97 feet. What is the total height of the ride at the peak of the arc?

coterminal angle: $-200° + 360° = 160°$

reference angle: $180° - 160° = 20°$

$\sin \theta = \dfrac{y}{r}$ Sine function

$\sin 20° = \dfrac{y}{84}$ $\theta = 20°$ and $r = 84$

$84 \sin 20° = y$ Multiply each side by 84.

$28.7 \approx y$ Use a calculator to solve for y.

Because y is approximately 28.7 feet, the total height of the ride at its peak is $28.7 + 97$ or about 125.7 feet.

Guided Practice   about 106.6 ft

5. RIDES A similar ride that is smaller has swing arms that are 72 feet long. The height of the axis from which the arms swing is 88 feet, and the angle of rotation from the standard position is $-195°$. What is the total height of the ride at the peak of the arc?

Real-World Link

On a swing ride, riders experience weightlessness just like the drop side of a roller coaster. The ride lasts one minute and reaches speeds of 60 miles per hour in both directions.

Source: Cedar Point

Levels of Complexity Chart

The levels of the exercises progress from 1 to 3, with Level 1 indicating the lowest level of complexity.

Exercises	12–32	33–36, 52–56	37–51
Level 3			●
Level 2		●	
Level 1	●		

4. 60°

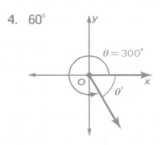

Example 4 Use a Reference Angle to Find a Trigonometric Value

AL In which quadrants is cos θ positive? the first and the fourth

OL Explain why all trigonometric values are positive in the first quadrant. *x*, *y*, and *r* are all positive in the first quadrant.

BL Explain why a trigonomic function and its reciprocal function always have the same sign. If you take the reciprocal of a number you don't change the sign, so these functions will always have the same sign.

Need Another Example?

Find the exact value of each trigonometric function.

a. sin 135° $\dfrac{\sqrt{2}}{2}$

b. $\cot \dfrac{7\pi}{3}$ $\dfrac{\sqrt{3}}{3}$

Example 5 Use Trigonometric Functions

AL What part of the triangle does the swing arm of the ride represent? the hypotenuse

OL Why is the reference angle 20°? The arms swings 20° past the horizontal axis.

BL What would the reference angle of the arm swing be if the ride designers wanted the ride to reach a height of 130 feet? 23.1°

Need Another Example?

Rides Suppose a ride similar to the one in Example 5 has a swing arm 89 feet long. The height of the axis is 99 feet, and the angles are the same. What is the total height of the new ride at the peak of the arc? *y* = 89 sin 20° = 30.4 ft; 30.4 + 99 = 129.4 ft

Additional Answers

1. $\sin \theta = \dfrac{2\sqrt{5}}{5}$, $\cos \theta = \dfrac{\sqrt{5}}{5}$, $\tan \theta = 2$, $\csc \theta = \dfrac{\sqrt{5}}{2}$, $\sec \theta = \sqrt{5}$, $\cot \theta = \dfrac{1}{2}$

2. $\sin \theta = -\dfrac{15}{17}$, $\cos \theta = -\dfrac{8}{17}$, $\tan \theta = \dfrac{15}{8}$, $\csc \theta = -\dfrac{17}{15}$, $\sec \theta = -\dfrac{17}{8}$, $\cot \theta = \dfrac{8}{15}$

3. $\sin \theta = -1$, $\cos \theta = 0$, $\tan \theta =$ undefined, $\csc \theta = -1$, $\sec \theta =$ undefined, $\cot \theta = 0$

Practice

Formative Assessment Use Exercises 1–11 to assess students' understanding of the concepts in this lesson.

The Practice and Problem Solving exercises assess the content taught in the lesson. The Preparing for Assessment page is meant to be used as preparation for end-of-course assessments.

Extra Practice

See page R9 for extra exercises for students who are approaching level or for on-level students who need additional reinforcement.

Additional Answers

5. 65°

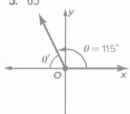

6. $\frac{\pi}{4}$

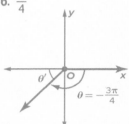

12. $\sin \theta = \frac{12}{13}, \cos \theta = \frac{5}{13}, \tan \theta = \frac{12}{5}, \csc \theta = \frac{13}{12},$
$\sec \theta = \frac{13}{5}, \cot \theta = \frac{5}{12}$

13. $\sin \theta = \frac{4}{5}, \cos \theta = -\frac{3}{5}, \tan \theta = -\frac{4}{3}, \csc \theta = \frac{5}{4},$
$\sec \theta = -\frac{5}{3}, \cot \theta = -\frac{3}{4},$

14. $\sin \theta = 0, \cos \theta = 1, \tan \theta = 0, \csc \theta = $ undefined,
$\sec \theta = 1, \cot \theta = $ undefined

15. $\sin \theta = -1, \cos \theta = 0, \tan \theta = $ undefined,
$\csc \theta = -1, \sec \theta = $ undefined, $\cot \theta = 0$

16. $\sin \theta = -\frac{\sqrt{5}}{5}, \cos \theta = \frac{2\sqrt{5}}{5}, \tan \theta = -\frac{1}{2},$
$\csc \theta = -\sqrt{5}, \sec \theta = \frac{\sqrt{5}}{2}, \cot \theta = -2$

17. $\sin \theta = -\frac{\sqrt{10}}{10}, \cos \theta = -\frac{3\sqrt{10}}{10}, \tan \theta = \frac{1}{3},$
$\csc \theta = -\sqrt{10}, \sec \theta = -\frac{\sqrt{10}}{3}, \cot \theta = 3$

Go Online!

eBook

Interactive Student Guide

Use the *Interactive Student Guide* to deepen conceptual understanding.
· Angles and Angle Measure

Check Your Understanding   = Step-by-Step Solutions begin on page R11.

Go Online! for a Self-Check Quiz

Examples 1–2   The terminal side of θ in standard position contains each point. Find the exact values of the six trigonometric functions of θ. 1–3. See margin on page 615.

1. $(1, 2)$ **2.** $(-8, -15)$ **3.** $(0, -4)$

Example 3   Sketch each angle. Then find its reference angle. 5-6. See margin.

4. 300°
See margin on page 615.   **5.** 115°   **6.** $-\frac{3\pi}{4}$

Example 4   Find the exact value of each trigonometric function.

7. $\sin \frac{3\pi}{4}$ $\frac{\sqrt{2}}{2}$ **8.** $\tan \frac{5\pi}{3}$ $-\sqrt{3}$ **9.** $\sec 120°$ -2 **10.** $\sin 300°$ $-\frac{\sqrt{3}}{2}$

Example 5   **11. ENTERTAINMENT** Alejandra opens her portable DVD player so that it forms a 125° angle. The screen is $5\frac{1}{2}$ inches long.

a. Redraw the diagram so that the angle is in standard position on the coordinate plane. See Ch. 9 Answer Appendix.

b. Find the reference angle. Then write a trigonometric function that can be used to find the distance to the wall d that she can place the DVD player. 55°; $\cos 55° = \frac{d}{5\frac{1}{2}}$

c. Use the function to find the distance. Round to the nearest tenth. 3.2 in.

Practice and Problem Solving

Extra Practice is on page R9.

Examples 1–2   The terminal side of θ in standard position contains each point. Find the exact values of the six trigonometric functions of θ. 12–17. See margin.

12. $(5, 12)$ **13** $(-6, 8)$ **14.** $(3, 0)$

15. $(0, -7)$ **16.** $(4, -2)$ **17.** $(-9, -3)$

Example 3   Sketch each angle. Then find its reference angle. 18–23. See Ch. 9 Answer Appendix.

18. 195°   **19.** 285°   **20.** −250°

21. $\frac{7\pi}{4}$ **22.** $-\frac{\pi}{4}$ **23.** 400°

Example 4   Find the exact value of each trigonometric function.

24. $\sin 210°$ $-\frac{1}{2}$ **25.** $\tan 315°$ -1 **26.** $\cos 150°$ $-\frac{\sqrt{3}}{2}$ **27.** $\csc 225°$ $-\sqrt{2}$

28. $\sin \frac{4\pi}{3}$ $-\frac{\sqrt{3}}{2}$ **29.** $\cos \frac{5\pi}{3}$ $\frac{1}{2}$ **30.** $\cot \frac{5\pi}{4}$ 1 **31.** $\sec \frac{11\pi}{6}$ $\frac{2\sqrt{3}}{3}$

Example 5   **32.** **REASONING** A soccer player x feet from the goalie kicks the ball toward the goal, as shown in the figure. The goalie jumps up and catches the ball 7 feet in the air.

a. Find the reference angle. Then write a trigonometric function that can be used to find how far from the goalie the soccer player was when he kicked the ball. 26°; $\tan 26° = \frac{7}{x}$

b. About how far away from the goalie was the soccer player? about 14.4 ft

Differentiated Homework Options

Levels	AL Basic	OL Core	BL Advanced
Exercises	12–32, 47–56	13–31 odd, 33–36, 37–45 odd, 52–56	33–56
2-Day Option	13–31 odd, 52–56	12–32, 52–56	
	12–32 even, 47–51	33–51	

 You can use ALEKS to provide additional remediation support with personalized instruction and practice.

33 SPRINKLER A sprinkler rotating back and forth shoots water out a distance of 10 feet. From the horizontal position, it rotates 145° before reversing its direction. At a 145° angle, about how far to the left of the sprinkler does the water reach? **about 8.2 ft**

34. MULTI-STEP Seven hikers are stranded at the edge of a cliff. They have determined that the most direct way out is to create a rope long enough to reach their companion in the trees below. The companion estimates the horizontal distance from the cliff to the trees to be about 90 feet and the angle from the safest high point in the tree-line up to the cliff to be between 30 and 40 degrees.

   a. What is the minimum amount of rope needed to reach the tree line? **about 104 ft**

   b. What other factors do you think can cause them to need more rope? Use these factors to determine a new minimum amount of rope needed. Explain your reasoning.

   c. What assumptions did you make? **b-c. See margin.**

35. PHYSICS A rock is shot off the edge of a ravine with a slingshot at an angle of 65° and with an initial velocity of 6 meters per second. The equation that represents the horizontal distance of the rock x is $x = v_0 (\cos \theta)t$, where v_0 is the initial velocity, θ is the angle at which it is shot, and t is the time in seconds. About how far does the rock travel after 4 seconds? **about 10.1 m**

36. FERRIS WHEELS Geri is in line to ride a Ferris wheel that is 212 feet tall and has a radius of about 98 feet. After a person gets on the bottom car, the Ferris wheel rotates 202.5° counterclockwise before stopping. How high above the ground is this car when it has stopped? **about 204.5 ft**

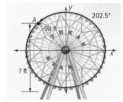

 Suppose θ is an angle in standard position whose terminal side is in the given quadrant. For each function, find the exact values of the remaining five trigonometric functions of θ. **37–40. See margin.**

37. $\sin \theta = \frac{4}{5}$, Quadrant II   38. $\tan \theta = -\frac{2}{3}$, Quadrant IV

39. $\cos \theta = -\frac{8}{17}$, Quadrant III   40. $\cot \theta = -\frac{12}{5}$, Quadrant IV

Find the exact value of each trigonometric function.

41. $\cot 270°$ 0   42. $\csc 180°$ undefined   43. $\sin 570°$ $-\frac{1}{2}$

44. $\tan \left(-\frac{7\pi}{6}\right)$ $-\frac{\sqrt{3}}{3}$ 45. $\cos \left(-\frac{11\pi}{6}\right)$ $\frac{\sqrt{3}}{2}$ 46. $\cot \frac{9\pi}{4}$ 1

H.O.T. Problems   Use Higher-Order Thinking Skills

47. CHALLENGE For an angle θ in standard position, $\sin \theta = \frac{\sqrt{2}}{2}$ and $\tan \theta = -1$. Can the value of θ be 225°? Justify your reasoning.

48. ![MP] CONSTRUCT ARGUMENTS Determine whether $3 \sin 60° = \sin 180°$ is *true* or *false*. Explain your reasoning.

48. False; $3 \sin 60° = 3 \cdot \frac{\sqrt{3}}{2}$ or $\frac{3\sqrt{3}}{2}$ and $\sin 180° = 0.0$

49. ![MP] REASONING Use the sine and cosine functions to explain why $\cot 180°$ is undefined. **See margin.**

50. OPEN-ENDED Identify a negative angle θ for which $\sin \theta > 0$ and $\cos \theta < 0$.

50. Sample answer: $\theta = -200°$

51. WRITING IN MATH Describe the steps for evaluating a trigonometric function for an angle θ that is greater than 90°. Include a description of a reference angle. **See margin.**

Differentiated Instruction OL BL

Extension For an angle with a measure between 0° and 90°, the angle is its own reference angle. Have students write expressions for the measure of the reference angle of θ,

if $90° < \theta < 180°$. $180° - \theta$

if $180° < \theta < 270°$. $\theta - 180°$

if $270° < \theta < 360°$. $360° - \theta$

Additional Answers

34b. Sample answer: I added 15 feet to the rope because the hikers will need to tie the rope around something solid like a tree that likely isn't on the edge of the cliff. I added another 10 feet for the rope to be tied around a tree by the companion. I also added another foot in case some smaller pieces of rope need to be tied together in order to create the final piece of rope. Lastly, I added another 15 feet in case the angle estimate was off and just as a precaution. Therefore, the minimum amount of rope is about 145 feet.

34c. Sample answer: I assumed they could find something within 15 feet of the edge of the cliff to which they could anchor the rope. I assumed that the estimate of the angle of elevation could be inaccurate. I assumed the estimate of the horizontal distance was accurate.

37. $\cos \theta = -\frac{3}{5}$, $\tan \theta = -\frac{4}{3}$, $\csc \theta = \frac{5}{4}$, $\sec \theta = -\frac{5}{3}$, $\cot \theta = -\frac{3}{4}$

38. $\sin \theta = -\frac{2\sqrt{13}}{13}$, $\cos \theta = \frac{3\sqrt{13}}{13}$, $\csc \theta = -\frac{\sqrt{13}}{2}$, $\sec \theta = \frac{\sqrt{13}}{3}$, $\cot \theta = -\frac{3}{2}$

39. $\sin \theta = -\frac{15}{17}$, $\tan \theta = \frac{15}{8}$, $\csc \theta = -\frac{17}{15}$, $\sec \theta = -\frac{17}{8}$, $\cot \theta = \frac{8}{15}$

40. $\sin \theta = -\frac{5}{13}$, $\cos \theta = \frac{12}{13}$, $\csc \theta = -\frac{13}{5}$, $\sec \theta = \frac{13}{12}$, $\tan \theta = -\frac{5}{12}$

47. No; for $\sin \theta = \frac{\sqrt{2}}{2}$ and $\tan \theta = -1$, the reference angle is 45°. However, for $\sin \theta$ to be positive and $\tan \theta$ to be negative, the reference angle must be in the second quadrant. So, the value of θ must be 135° or an angle coterminal with 135°.

49. Sample answer: We know that $\cot \theta = \frac{x}{y}$, $\sin \theta = \frac{y}{r}$, and $\cos \theta = \frac{x}{r}$. Since $\sin 180° = 0$, it must be true that $y = 0$. Thus $\cot 180° = \frac{x}{0}$, which is undefined.

51. Sample answer: First, sketch the angle and determine in which quadrant it is located. Then use the appropriate rule for finding its reference angle θ'. A reference angle is the acute angle formed by the terminal side of θ and the x-axis. Next, find the value of the trigonometric function for θ'. Finally, use the quadrant location to determine the sign of the trigonometric function value of θ.

Assess

Crystal Ball Have students look ahead to Lesson 9-4. Have them write how they think what they learned today will connect with the theme in Lesson 9-4.

Preparing for Assessment

Dual Coding Chart		
Exercise	Content Objective	(MP) Mathematical Practices
52	F.TF.2	2
53	F.TF.2	6
54	F.TF.2	2
55	F.TF.2	2
56	F.TF.2	2

Diagnose Student Errors

Survey student responses for each item. Class trends may indicate common errors and misconceptions.

52.

A	Did not recognize that an angle of $\frac{3\pi}{4}$ terminates in QII and that the sine is positive
B	CORRECT
C	Did not recognize that an angle of $\frac{9\pi}{4}$ terminates in Quadrant I and that the sine is positive
D	Did not recognize that an angle of $\frac{3\pi}{4}$ terminates in Quadrant II and that the sine is positive
E	Did not recognize that one angle terminates in Quadrant I and that the cosine is positive

53.

A	Found cot θ
B	Found $-\tan \theta \cdot \cot \theta$
C	Found tan θ
D	Found $-\tan \theta$
E	CORRECT

54.

A	CORRECT
B	Selected an angle with a negative tangent value
C	Selected an angle with a negative tangent value
D	Selected an angle with a negative sine value
E	Selected an angle with both negative tangent and sine values

Go Online! ✓

Self-Check Quiz

Students can use *Self-Check Quizzes* to check their understanding of this lesson. You can also give the *Chapter Quiz* which covers the content in Lessons 9-1, 9-2, and 9-3.

Preparing for Assessment

52. Which pair of expressions is equal to $\frac{-\sqrt{2}}{2}$? (MP) 2   F.TF.2 B

- ○ A $\sin \frac{3\pi}{4}$ and $\cos \frac{3\pi}{4}$
- ○ B $\sin \frac{5\pi}{4}$ and $\sin \frac{7\pi}{4}$
- ○ C $\cos \frac{3\pi}{4}$ and $\sin \frac{9\pi}{4}$
- ○ D $\cos \frac{5\pi}{4}$ and $\sin \frac{3\pi}{4}$
- ○ E $\cos \frac{\pi}{4}$ and $\cos \frac{3\pi}{4}$

53. The terminal side of θ in standard position contains the point at $(-2, 1)$. What is the exact measure of $(\tan \theta \cdot \cot \theta)$? (MP) 6   F.TF.2 E

- ○ A -2
- ○ B -1
- ○ C $-\frac{1}{2}$
- ○ D $\frac{1}{2}$
- ○ E 1

54. Which angle has a tangent and a sine that are both positive? (MP) 2   F.TF.2 A

- ○ A $25°$
- ○ B $110°$
- ○ C $150°$
- ○ D $225°$
- ○ E $330°$

55. MULTI-STEP

a. When is a trigonometric function undefined? Choose all that apply. (MP) 2   F.TF.2 A, D, E, G

- ☐ A   when it does not exist
- ☐ B   when it is equal to zero
- ☐ C   when it is equal to one
- ☐ D   when it is a cosecant function for an angle with a terminal arm along the x-axis
- ☐ E   when it is a secant function for an angle with a terminal arm along the y-axis
- ☐ F   when it is divided by one
- ☐ G   when it is divided by zero

b. Look at the diagram. The terminal side of the angle in standard position contains the point $(0, -4)$.

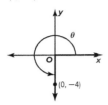

Which of the following trigonometric functions is undefined? E

- ○ A $\sin \theta$
- ○ B $\cos \theta$
- ○ C $\csc \theta$
- ○ D $\cot \theta$
- ○ E $\sec \theta$

c. Which angles are coterminal with the angle shown? Choose all that apply. C, G

- ☐ A $90°$
- ☐ B $730°$
- ☐ C $270°$
- ☐ D $-330°$
- ☐ E $-270°$
- ☐ F $-540°$
- ☐ G $-90°$

56. The terminal side of an angle in standard position contains the point $(-8, 0)$. Which functions are undefined? Choose all that apply. (MP) 2   F.TF.2 C, D

- ☐ A $\sin \theta$
- ☐ B $\cos \theta$
- ☐ C $\csc \theta$
- ☐ D $\cot \theta$
- ☐ E $\sec \theta$
- ☐ F $\tan \theta$

(MP) Standards for Mathematical Practice

Emphasis On	Exercises
2 Reason abstractly and quantitatively.	37–40
3 Construct viable arguments and critique the reasoning of others.	47–51
4 Model with mathematics.	11, 32–36
6 Attend to precision.	1–10, 12–31, 41–46

CHAPTER 9
Mid-Chapter Quiz
Lessons 9-1 through 9-3

Solve $\triangle XYZ$ by using the given measurements. Round measures of sides to the nearest tenth and measures of angles to the nearest degree. (Lesson 9-1)

$X = 25°$,
$y = 34.3$,
$z = 37.9$

$Y = 65°$,
$y = 17.2$,
$z = 18.9$

1. $Y = 65°$, $x = 16$ **2.** $X = 25°$, $x = 8$

3. Find the values of the six trigonometric functions for angle θ. (Lesson 9-1) **See Ch. 9 Answer Appendix.**

4. Draw an angle measuring $-80°$ in standard position. (Lesson 9-2) **See Ch. 9 Answer Appendix.**

Rewrite each degree measure in radians and each radian measure in degrees. (Lesson 9-2)

5. $215°$ $\dfrac{43\pi}{36}$ **6.** $-350°$ $\dfrac{35\pi}{18}$

7. $\dfrac{8\pi}{5}$ $288°$ **8.** $\dfrac{9\pi}{2}$ $810°$

9. MULTIPLE CHOICE What is the length of the arc below rounded to the nearest tenth? (Lesson 9-2) **C**

○ **A** 4.2 cm
○ **B** 17.1 cm
○ **C** 53.9 cm
○ **D** 2638.9 cm

8π⁄7
15 cm

Find the exact value of each trigonometric function. (Lesson 9-3)

10. $\tan \pi$ 0 **11.** $\cos \dfrac{3\pi}{4}$ $-\dfrac{\sqrt{2}}{2}$

The terminal side of θ in standard position contains each point. Find the exact values of the six trigonometric functions of θ. (Lesson 9-3) **12, 13. See Ch. 9 Answer Appendix.**

12. $(0, -5)$ **13.** $(6, 8)$

14. MULTIPLE CHOICE Suppose θ is an angle in standard position with $\cos \theta > 0$. In which quadrant(s) does the terminal side of θ lie? (Lesson 9-3) **D**

○ **A** I
○ **B** II
○ **C** III
○ **D** I and IV

15. MULTI-PART The angle of elevation of a ramp is 11°. The length of the ramp is 14 meters.

a. Which of the following expressions gives the height of the top of the ramp from the ground? **A**

○ **A** 14 sin 11°
○ **B** 14 cos 11°
○ **C** 14 sec 11°
○ **D** 14 csc 11°

b. Which of the following expressions gives the distance on the ground that the ramp covers? **B**

○ **A** 14 sin 11°
○ **B** 14 cos 11°
○ **C** 14 sec 11°
○ **D** 14 csc 11°

16. If both angles are in standard position, which of the following pairs are coterminal? Choose all that apply. **B, D, F**

☐ **A** $180°$ and $\dfrac{\pi}{2}$
☐ **B** $150°$ and $\dfrac{5\pi}{6}$
☐ **C** $210°$ and $\dfrac{3\pi}{2}$
☐ **D** $240°$ and $\dfrac{4\pi}{3}$
☐ **E** $60°$ and $\dfrac{2\pi}{3}$
☐ **F** $315°$ and $\dfrac{-\pi}{4}$

17. Which of the following angles has a cosine value of $\dfrac{-\sqrt{2}}{2}$? **B**

○ **A** 45°   ○ **C** 180°
○ **B** 135°   ○ **D** 315°

RtI **Response to Intervention**

Use the Intervention Planner to help you determine your Response to Intervention.

Intervention Planner

TIER 1 **On Level** OL

IF students miss 25% of the exercises or less,

THEN choose a resource:

SE Lessons 9-1, 9-2, and 9-3

Go Online!
📄 Skills Practice
📄 Chapter Project
✓ Self-Check Quizzes

TIER 2 **Strategic Intervention** AL
Approaching grade level

IF students miss 50% of the exercises,

THEN **Go Online!**
📄 Study Guide and Intervention
➕ Extra Examples
💬 Personal Tutors
📄 Homework Help

TIER 3 **Intensive Intervention**
2 or more grades below level

IF students miss 75% of the exercises,

THEN choose a resource:

Use *Math Triumphs, Alg. 2*

Go Online!
➕ Extra Examples
💬 Personal Tutors
📄 Homework Help
ᵃᵇᵨ Review Vocabulary

Foldables Study Organizer

Dinah Zike's FOLDABLES®

Before students complete the Mid-Chapter Quiz, encourage them to review the information for Lessons 9-1 through 9-3 in their Foldables. Ask students to share the items they have added to their Foldables that have been helpful as they study Chapter 9.

 ALEKS can be used as a formative assessment tool to target learning gaps for those who are struggling, while providing enhanced learning for those who have mastered the concepts.

Go Online!

ᵉAssessment

You can use the premade Mid-Chapter Test to assess students' progress in the first half of the chapter. Customize and create multiple versions of your Mid-Chapter Quiz and answer keys that align to the your standard. Tests can be delivered on paper or online.

LESSON 9-4
Circular and Periodic Functions

SUGGESTED PACING (DAYS)

90 min.	0.75
45 min.	1.0

Instruction

Track Your Progress

Objectives

1 Find values of trigonometric functions based on the unit circle.

2 Use the properties of periodic functions to evaluate trigonometric functions.

Mathematical Background

A unit circle—a circle centered at the origin with a radius of 1 unit—can be used to generalize the sine and cosine functions. If the terminal side of an angle θ in standard position intersects the unit circle at a point P with coordinates (x, y), then $\cos \theta = x$ and $\sin \theta = y$.

THEN	NOW	NEXT

F.TF.1 Understand radian measure of an angle as the length of the arc on the unit circle subtended by the angle.

F.TF.2 Explain how the unit circle in the coordinate plane enables the extension of trigonometric functions to all real numbers, interpreted as radian measures of angles traversed counterclockwise around the unit circle.

A.CED.2 Create equations in two or more variables to represent relationships between quantities; graph equations on coordinate axes with labels and scales.

F.IF.4 For a function that models a relationship between two quantities, interpret key features of graphs and tables in terms of the quantities, and sketch graphs showing key features given a verbal description of the relationship. Key features include: intercepts; intervals where the function is increasing, decreasing, positive, or negative; relative maxima and minima; symmetries; end behavior; and periodicity.*

F.IF.7e Graph exponential and logarithmic functions, showing intercepts and end behavior, and trigonometric functions, showing period, midline, and amplitude.

F.TF.5 Choose trigonometric functions to model periodic phenomena with specified amplitude, frequency, and midline.*

Go Online! All of these resources and more are available at connectED.mcgraw-hill.com

Use **The Geometer's Sketchpad** to explore periodic functions.

Personal Tutors (for every example) let students hear real teachers solve problems. Students can pause and repeat as many times as necessary.

Use **Self-Check Quiz** to assess students' understanding of the concepts in this lesson.

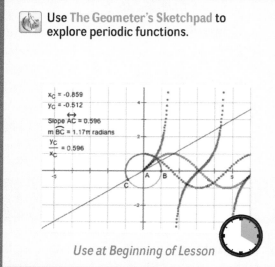

Use at Beginning of Lesson

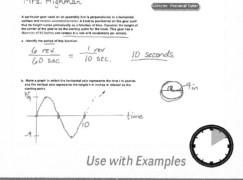

Use with Examples

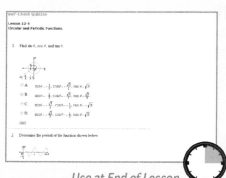

Use at End of Lesson

OER ## Using Open Educational Resources

Communicating Texting is the way most of your students communicate. **Class Parrot** offers an easy, safe, free way to text message entire classes of students.. *Use as a reminder tool*

Go Online!
connectED.mcgraw-hill.com

Worksheets

Differentiate Your Resources

Extra Practice Additional practice or homework; Skills Practice is best for approaching-level students and Practice is best for on-level and beyond-level students

Skills Practice

NAME _____ DATE _____ PERIOD _____

9-4 Skill Practice
Circular and Periodic Functions

The terminal side of angle θ in standard position intersects the unit circle at each point P. Find cos θ and sin θ.

1. $P\left(\frac{3}{5}, \frac{4}{5}\right)$ 2. $P\left(\frac{5}{13}, -\frac{12}{13}\right)$ 3. $P\left(-\frac{9}{41}, -\frac{40}{41}\right)$

4. $P(0, 1)$ 5. $P(-1, 0)$ 6. $P\left(\frac{1}{2}, -\frac{\sqrt{3}}{2}\right)$

Find the exact value of each function.

7. cos 45° 8. sin 210° 9. sin 330°

10. cos 330° 11. cos (−60°) 12. sin (−390°)

13. sin 5π 14. cos 3π 15. sin $\frac{5\pi}{2}$

16. sin $\frac{7\pi}{3}$ 17. cos $\left(-\frac{7\pi}{3}\right)$ 18. cos $\left(-\frac{5\pi}{2}\right)$

Determine the period of each function.

19.

20.

21.

Practice

NAME _____ DATE _____ PERIOD _____

9-4 Practice
Circular and Periodic Functions

The terminal side of angle θ in standard position intersects the unit circle at each point P. Find cos θ and sin θ.

1. $P\left(-\frac{1}{2}, \frac{\sqrt{3}}{2}\right)$ 2. $P\left(\frac{20}{29}, -\frac{21}{29}\right)$ 3. $P(0.8, 0.6)$

4. $P(0, -1)$ 5. $P\left(-\frac{\sqrt{2}}{2}, -\frac{\sqrt{2}}{2}\right)$ 6. $P\left(\frac{\sqrt{3}}{2}, -\frac{1}{2}\right)$

Determine the period of each function.

7.

8.

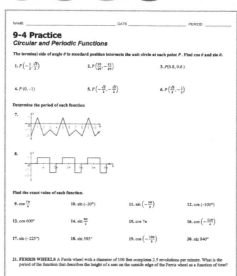

Find the exact value of each function.

9. cos $\frac{7\pi}{4}$ 10. sin (−30°) 11. sin $\left(-\frac{2\pi}{3}\right)$ 12. cos (−330°)

13. cos 600° 14. sin $\frac{8\pi}{3}$ 15. cos 7π 16. cos $\left(-\frac{11\pi}{4}\right)$

17. sin (−225°) 18. sin 585° 19. cos $\left(-\frac{10\pi}{3}\right)$ 20. sin 840°

21. **FERRIS WHEELS** A Ferris wheel with a diameter of 100 feet completes 2.5 revolutions per minute. What is the period of the function that describes the height of a seat on the outside edge of the Ferris wheel as a function of time?

Word Problem Practice

NAME _____ DATE _____ PERIOD _____

9-4 Word Problem Practice
Circular and Periodic Functions

1. **TIRES** A point on the edge of a car tire is marked with paint. As the car moves slowly, the marked point on the tire varies in distance from the surface of the road. The height in inches of the point is given by the expression h = −8 cos t + 8, where t is the time in seconds.

a. What is the maximum height above ground that the point on the tire reaches?

b. What is the minimum height above ground that the point on the tire reaches?

c. How many rotations does the tire make per second?

d. How far does the marked point travel in 30 seconds? How far does the marked point travel in one hour?

2. **GEOMETRY** The temperature T in degrees Fahrenheit of a city t months into the year is approximated by the formula $T = 42 + 30 \sin\left(\frac{\pi}{6}t\right)$.

a. What is the highest monthly temperature for the city?

b. In what month does the highest temperature occur?

c. What is the lowest monthly temperature for the city?

d. In what month does the lowest temperature occur?

3. **THE MOON** The Moon's period of revolution is the number of days it takes for the Moon to revolve around Earth. The period can be determined by graphing the percentage of sunlight reflected by the Moon each day, as seen by an observer on Earth. Use the graph to determine the Moon's period of revolution.

Moon's Orbit

Intervention Reteaching and vocabulary activities that can be used with struggling or absent students and as ELL support

Study Guide and Intervention

NAME _____ DATE _____ PERIOD _____

9-4 Study Guide and Intervention
Circular and Periodic Functions

Circular Functions

Definition of Sine and Cosine	If the terminal side of an angle θ in standard position intersects the unit circle at P(x, y), then cos θ = x and sin θ = y. Therefore, the coordinates of P can be written as P(cos θ, sin θ).

Example: The terminal side of angle θ in standard position intersects the unit circle at $\left(-\frac{5}{6}, \frac{\sqrt{11}}{6}\right)$. Find cos θ and sin θ.

$P\left(-\frac{5}{6}, \frac{\sqrt{11}}{6}\right) = P(\cos\theta, \sin\theta)$, so $\cos\theta = -\frac{5}{6}$ and $\sin\theta = \frac{\sqrt{11}}{6}$.

Exercises

The terminal side of angle θ in standard position intersects the unit circle at each point P. Find cos θ and sin θ.

1. $P\left(-\frac{\sqrt{3}}{2}, \frac{1}{2}\right)$ 2. $P(0, -1)$

3. $P\left(-\frac{5}{13}, \frac{12}{13}\right)$ 4. $P\left(-\frac{4}{5}, -\frac{3}{5}\right)$

5. $P\left(\frac{1}{4}, -\frac{\sqrt{15}}{4}\right)$ 6. $P\left(\frac{\sqrt{3}}{4}, \frac{7}{4}\right)$

7. P is on the terminal side of θ = 45°. 8. P is on the terminal side of θ = 120°.

9. P is on the terminal side of θ = 240°. 10. P is on the terminal side of θ = 330°.

Study Notebook

NAME _____ DATE _____ PERIOD _____

9-4 **Circular and Periodic Functions**

What You'll Learn Skim the Examples in this lesson. Predict two things you think you will learn about circular functions.

1. _____

2. _____

Active Vocabulary New Vocabulary Match the term with its definition by drawing a line to connect the two.

circular function — one complete pattern of a periodic function

cycle — a function that has y-values that repeat at regular intervals

period — a function that is defined using the unit circle

periodic function — a circle with a radius of one unit centered at the origin on the coordinate plane

unit circle — the horizontal length of one cycle of a periodic function

Vocabulary Link Describe how the revolutions of the pedals on a bicycle can be used as a model of a periodic function.

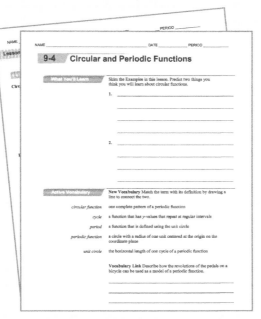

Extension Activities that can be used to extend lesson concepts

Enrichment

NAME _____ DATE _____ PERIOD _____

9-4 Enrichment
Polar Coordinates

Consider an angle in standard position with its vertex at a point O called the *pole*. Its initial side is on a coordinated axis called the *polar axis*. A point P on the terminal side of the angle is named by the *polar coordinates* (r, θ) where r is the directed distance of the point from O and θ is the measure of the angle. Graphs in this system may be drawn on polar coordinate paper such as the kind shown at the right. The polar coordinates of a point are not unique. For example, (3, 30°) names point P as well as (3, 390°). Another name for P is (−3, 210°). Can you see why? The coordinates of the pole are (0, θ) where θ may be any angle.

Example: Draw the graph of the function r = cos θ. Make a table of convenient values for θ and r. Then plot the points.

θ	0°	30°	60°	90°	120°	150°	180°
r	1	$\frac{\sqrt{3}}{2}$	$\frac{1}{2}$	0	$-\frac{1}{2}$	$-\frac{\sqrt{3}}{2}$	−1

Since the period of the cosine function is 180°, values of r for θ > 180° are repeated.

Graph each function by making a table of values and plotting the values on polar coordinate paper.

1. r = 4 2. r = 3 sin θ

3. r = 3 cos 2θ 4. r = 2(1 + cos θ)

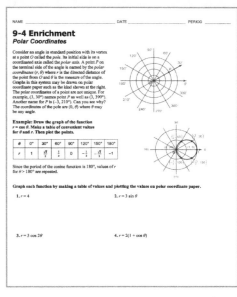

Launch

Have students read the Why? section of the lesson. Ask:

- Use the maximum and minimum heights of the pedal to find the diameter of the circle.
  diameter = maximum − minimum = 18 − 4 = 14 in.

- Use the height of the circle's center to find the circle's diameter. The radius is 18 − 11 = 7 in., or 11 − 4 = 7 in., so the diameter is 2 • 7 = 14 in.

- What is the starting position of the pedal? Explain. The starting position would correspond with $t = 0$ seconds; at that time the pedal is at the top of the diagram.

Teach

Ask the scaffolded questions for each example to build conceptual understanding for students at all levels.

1 Circular Functions

Example 1 Find Sine and Cosine Given a Point on the Unit Circle

AL What angle does the point (0, 1) represent? 90°

OL What is tan θ? $\sqrt{3}$

BL Explain why this definition matches the definition in 9-3. Everything is the same, but here the r value is 1 since we are using the unit circle.

Need Another Example?
The terminal side of angle θ in standard position intersects the unit circle at $P\left(\frac{\sqrt{7}}{4}, \frac{3}{4}\right)$. Find cos θ and sin θ.
$\sin \theta = \frac{3}{4}$, $\cos \theta = \frac{\sqrt{7}}{4}$

Go Online!

Interactive Whiteboard

Use the *eLesson* or *Lesson Presentation* to present this lesson.

LESSON 4

Circular and Periodic Functions

:Then	:Now	:Why?
You evaluated trigonometric functions using reference angles.	**1** Find values of trigonometric functions based on the unit circle.	The pedals on a bicycle rotate as the bike is being ridden. The height of a pedal is a function of time, as shown in the figure at the right.
	2 Use the properties of periodic functions to evaluate trigonometric functions.	Notice that the pedal makes one complete rotation every two seconds.

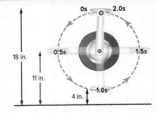

New Vocabulary
unit circle
circular function
periodic function
cycle
period

MP Mathematical Practices
3 Construct viable arguments and critique the reasoning of others.
7 Look for and make use of structure.

Content Standards
F.TF.1 Understand radian measure of an angle as the length of the arc on the unit circle subtended by the angle.
F.TF.2 Explain how the unit circle in the coordinate plane enables the extension of trigonometric functions to all real numbers, interpreted as radian measures of angles traversed counterclockwise around the unit circle.

1 **Circular Functions** A **unit circle** is a circle with a radius of 1 unit centered at the origin on the coordinate plane. You can use a point P on the unit circle to generalize sine and cosine functions.

$$\sin \theta = \frac{y}{r} = \frac{y}{1} \text{ or } y \qquad \cos \theta = \frac{x}{r} = \frac{x}{1} \text{ or } x$$

So, the values of sin θ and cos θ are the y-coordinate and x-coordinate, respectively, of the point where the terminal side of θ intersects the unit circle.

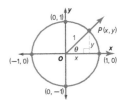

Key Concept Functions on a Unit Circle

Words	If the terminal side of an angle θ in standard position intersects the unit circle at $P(x, y)$, then cos $\theta = x$ and sin $\theta = y$.	Model
Symbols	$P(x, y) = P(\cos \theta, \sin \theta)$	
Example	If $\theta = 120°$, $P(x, y) = P(\cos 120°, \sin 120°)$.	

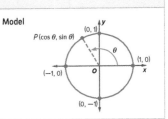

Both cos $\theta = x$ and sin $\theta = y$ are functions of θ. Because they are defined using a unit circle, they are called **circular functions**.

F.TF.1

Example 1   Find Sine and Cosine Given a Point on the Unit Circle

The terminal side of angle θ in standard position intersects the unit circle at $P\left(\frac{1}{2}, \frac{\sqrt{3}}{2}\right)$. Find cos θ and sin θ.

$P\left(\frac{1}{2}, \frac{\sqrt{3}}{2}\right) = P(\cos \theta, \sin \theta)$

$\cos \theta = \frac{1}{2} \qquad \sin \theta = \frac{\sqrt{3}}{2}$

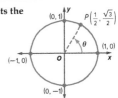

▷ **Guided Practice**

1. The terminal side of angle θ in standard position intersects the unit circle at $P\left(\frac{3}{5}, -\frac{4}{5}\right)$. Find cos θ and sin θ. $\cos \theta = \frac{3}{5}$, $\sin \theta = -\frac{4}{5}$

MP Mathematical Practices Strategies

Use appropriate tools strategically.
Help students become comfortable with circle trigonometry by writing, understanding, and repeatedly using the unit circle.

- Have students copy the unit circle from this lesson. As they work through, ask them to make note of any patterns they see. Ask the following guiding questions:

  - What pattern do you notice in the sine values for 30, 150, 210, and 330 degrees? They're all $\frac{1}{2}$ or $-\frac{1}{2}$.

  - What pattern do you notice in the sine and cosine values for 45, 135, 225, and 315 degrees? They're all $\frac{\sqrt{2}}{2}$ or $-\frac{\sqrt{2}}{2}$.

  - In what quadrant are all values positive? the first quadrant

  - In what quadrant are all values negative? the third quadrant

2 **Periodic Functions** A **periodic function** has y-values that repeat at regular intervals. One complete pattern is a **cycle**, and the horizontal length of one cycle is a **period**.

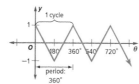

θ	y
0°	1
180°	−1
360°	1
540°	−1
720°	1

The cycle repeats every 360°.

F.TF.1 💬

Example 2   Identify the Period

Determine the period of the function.

The pattern repeats at π, 2π, and so on. So, the period is π.

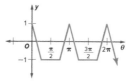

Guided Practice

2. Graph a function with a period of 4. See margin.

The rotations of wheels, pedals, carousels, and objects in space are all periodic.

F.TF.1 💬 👥

Real-World Example 3   Use Trigonometric Functions

CYCLING Refer to the beginning of the lesson. The height of a bicycle pedal varies periodically as a function of time, as shown in the figure.

a. Make a table showing the height of a bicycle pedal at 0, 0.5, 1.0, 1.5, 2.0, 2.5, and 3.0 seconds.

At 0 seconds, the pedal is 18 inches high. At 0.5 second, the pedal is 11 inches high. At 1.0 second, the pedal is 4 inches high, and so on.

Time (s)	Height (in.)
0	18
0.5	11
1.0	4
1.5	11
2.0	18
2.5	11
3.0	4

b. Identify the period of the function.

The period is the time it takes to complete one rotation. So, the period is 2 seconds.

c. Graph the function. Let the horizontal axis represent the time t and the vertical axis represent the height h in inches that the pedal is from the ground.

The maximum height of the pedal is 18 inches, and the minimum height is 4 inches. Because the period of the function is 2 seconds, the pattern of the graph repeats in intervals of 2 seconds.

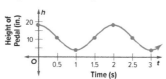

Guided Practice

3. CYCLING Another cyclist pedals the same bike at a rate of 1 revolution per second.

   A. Make a table showing the height of a bicycle pedal at times 0, 0.5, 1.0, 1.5, 2.0, 2.5, and 3.0 seconds.

   B. Identify the period and graph the function. 1; See margin for graph.

3A.

Time (s)	Height of Pedal (in.)
0	18
0.5	4
1.0	18
1.5	4
2.0	18
2.5	4
3.0	18

Additional Answers (Guided Practice)

2.

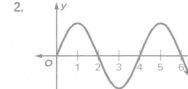

3B.

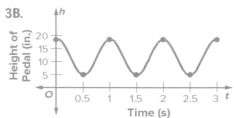

2 Periodic Functions

Example 2 Identify the Period

AL What is the length of two cycles? 2π

OL What is the y-value at $\frac{5\pi}{2}$? -1

BL For what values of this function will the y-values be 1? πk where k is any integer.

Need Another Example?

Determine the period of the function. $\frac{\pi}{3}$

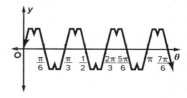

Example 3 Use Trigonometric Functions

AL What will the height be at 4 seconds? 18

OL What is the range of this function? [4, 18]

BL How can you use the range of one period to find the range of the function? The range of one period is the range of the function.

Need Another Example?

Cycling Refer to the beginning of the lesson. The height of a bicycle pedal varies periodically as a function of the time, as shown in the figure.

a. Make a table showing the height of a bicycle pedal at 3.0, 3.5, 4.0, 4.5, and 5.0 seconds.

Time (s)	Height (in.)
3.0	4
3.5	11
4.0	18
4.5	11
5.0	4

b. Identify the period of the function. 2 seconds

c. Graph the function over the interval [3, 5]. Let the horizontal axis represent the time t and the vertical axis represent the height h in inches that the pedal is from the ground.

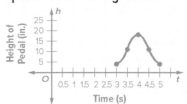

Example 4 Evaluate Trigonometric Expressions

AL What type of angles are 120° and 480°?
coterminal

OL In part **b**, why do you use $\frac{8\pi}{2}$ to find the exact value of $\frac{11\pi}{2}$? $\frac{8\pi}{2} = 2\pi$ which is one period

BL For what values of θ will sin $\theta = 1$? $90° + 360°k$, where k is any integer

Need Another Example?

Find the exact value of each function.

a. cos 690° $\frac{\sqrt{3}}{2}$

b. $\sin\left(-\frac{3\pi}{4}\right)$ $-\frac{\sqrt{2}}{2}$

The exact values of cos θ and sin θ for special angles are shown on the unit circle at the right. The cosine values are the x-coordinates of the points on the unit circle, and the sine values are the y-coordinates.

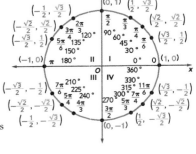

You can use this information to graph the sine and cosine functions. Let the horizontal axis represent the values of θ and the vertical axis represent the values of sin θ or cos θ.

The cycles of the sine and cosine functions repeat every 360°. So, they are periodic functions. The period of each function is 360° or 2π.

Consider the points on the unit circle for $\theta = 45°$, $\theta = 150°$, and $\theta = 270°$.

$(\cos 45°, \sin 45°) = \left(\frac{\sqrt{2}}{2}, \frac{\sqrt{2}}{2}\right)$

$(\cos 150°, \sin 150°) = \left(-\frac{\sqrt{3}}{2}, \frac{1}{2}\right)$

$(\cos 270°, \sin 270°) = (0, -1)$

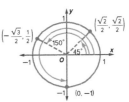

These points can also be shown on the graphs of the sine and cosine functions.

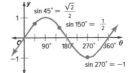 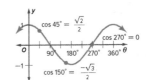

Because the period of the sine and cosine functions is 360°, the values repeat every 360°. So, sin $(x + 360°) = \sin x$, and cos $(x + 360°) = \cos x$.

F.TF.1

Example 4 Evaluate Trigonometric Expressions

Find the exact value of each expression.

a. cos 480°

$\cos 480° = \cos (120° + 360°)$

$= \cos 120°$

$= -\frac{1}{2}$

b. $\sin\frac{11\pi}{4}$

$\sin\frac{11\pi}{4} = \sin\left(\frac{3\pi}{4} + \frac{8\pi}{4}\right)$

$= \sin\frac{3\pi}{4}$

$= \frac{\sqrt{2}}{2}$

▸ **Guided Practice**

4A. $\cos\left(-\frac{3\pi}{4}\right)$ $-\frac{\sqrt{2}}{2}$

4B. sin 420° $\frac{\sqrt{3}}{2}$

Differentiated Instruction **AL** **OL**

Naturalist Learners Have students research various kinds of circular calendars, such as those used by the Maya, to predict the weather and determine the best time for planting crops.

Check Your Understanding

 = Step-by-Step Solutions begin on page R11.

 Go Online! for a Self-Check Quiz

Example 1
F.TF.1

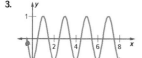 **STRUCTURE** The terminal side of angle θ in standard position intersects the unit circle at each point P. Find $\cos \theta$ and $\sin \theta$.

1. $P\left(\frac{15}{17}, \frac{8}{17}\right)$ $\cos \theta = \frac{15}{17}, \sin \theta = \frac{8}{17}$

2. $P\left(-\frac{\sqrt{2}}{2}, \frac{\sqrt{2}}{2}\right)$ $\cos \theta = -\frac{\sqrt{2}}{2}, \sin \theta = \frac{\sqrt{2}}{2}$

Example 2
F.TF.1

Determine the period of each function.

3. 2

4. 4π

Example 3
F.TF.1

5. SWINGS The height of a swing varies periodically as the function of time. The swing goes forward and reaches its high point of 6 feet. It then goes backward and reaches 6 feet again. Its lowest point is 2 feet. The time it takes to swing from its high point to its low point is 1 second.

a. How long does it take for the swing to go forward and back one time? **4 seconds**

b. Graph the height of the swing h as a function of time t. **See margin.**

Example 4
F.TF.1

Find the exact value of each expression.

6. $\sin \frac{13\pi}{6}$ $\frac{1}{2}$

7. $\sin(-60°)$ $-\frac{\sqrt{3}}{2}$

8. $\cos 540°$ -1

Practice and Problem Solving

Extra Practice is on page R9.

Example 1
F.TF.1

The terminal side of angle θ in standard position intersects the unit circle at each point P. Find $\cos \theta$ and $\sin \theta$.

9. $P\left(\frac{6}{10}, -\frac{8}{10}\right)$ $\cos \theta = \frac{3}{5}, \sin \theta = -\frac{4}{5}$

10. $P\left(-\frac{10}{26}, -\frac{24}{26}\right)$ $\cos \theta = -\frac{5}{13}, \sin \theta = -\frac{12}{13}$

11 $P\left(\frac{\sqrt{3}}{2}, \frac{1}{2}\right)$ $\cos \theta = \frac{\sqrt{3}}{2}, \sin \theta = \frac{1}{2}$

12. $P\left(\frac{\sqrt{6}}{5}, \frac{\sqrt{19}}{5}\right)$ $\cos \theta = \frac{\sqrt{6}}{5}, \sin \theta = \frac{\sqrt{19}}{5}$

Example 2
F.TF.1

Determine the period of each function.

13.  3

14. 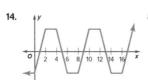 8

15. 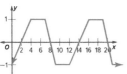 12

16.  6

Differentiated Homework Options

Levels	**AL** Basic	**OL** Core	**BL** Advanced
Exercises	9–25, 37–46	9–25 odd, 26–30 even, 31–35 odd, 37–46	26–46
2-Day Option	9–25 odd, 42–46	9–25, 42–46	
	10–24 even, 37–41	26–41	

 You can use ALEKS to provide additional remediation support with personalized instruction and practice.

Practice

Formative Assessment Use Exercises 1–8 to assess students' understanding of the concepts in this lesson.

The Practice and Problem Solving exercises assess the content taught in the lesson. The Preparing for Assessment page is meant to be used as preparation for end-of-course assessments.

 **Teaching the Mathematical Practices**

Structure Mathematically proficient students look closely to discern a pattern or structure.

Extra Practice

See page R9 for extra exercises for students who are approaching level or for on-level students who need additional reinforcement.

Additional Answer

5b. Sample answer:

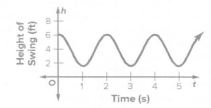

Teaching the Mathematical Practices

Sense-Making Mathematically proficient students start by explaining to themselves the meaning of a problem and looking for entry points to its solution. They analyze givens, constraints, relationships, and goals. They check their answers to problems using a different method, and they continually ask themselves, "Does this make sense?"

Additional Answers

19a.

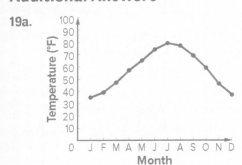

26b. Sample answer:

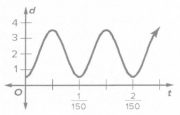

27b. Sample answer:

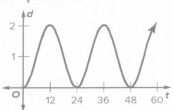

28. Sample answer:

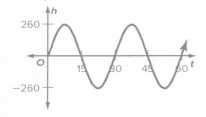

Go Online!

eSolutions Manual

Create worksheets, answer keys, and solutions handouts for your assignments.

Determine the period of each function.

17. 180°

18. 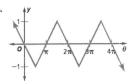 2π

Example 3
F.TF.1

19. WEATHER In a city, the average high temperature for each month is shown in the table.

a. Sketch a graph of the function representing this situation. **See margin.**

b. Describe the period of the function. **12 mo or 1 yr**

Average High Temperatures			
Month	Temperature (°F)	Month	Temperature (°F)
Jan	35	July	79
Feb.	39	Aug.	77
Mar.	47	Sept.	69
Apr.	57	Oct.	59
May	65	Nov.	46
Jun.	74	Dec.	37

Source: The Weather Channel

Example 4
F.TF.1

Find the exact value of each expression.

20. $\sin \frac{7\pi}{3}$ $\frac{\sqrt{3}}{2}$

21. $\cos(-60°)$ $\frac{1}{2}$

22. $\cos 450°$ 0

23. $\sin \frac{11\pi}{4}$ $\frac{\sqrt{2}}{2}$

24. $\sin(-45°)$ $-\frac{\sqrt{2}}{2}$

25. $\cos 570°$ $-\frac{\sqrt{3}}{2}$

26. SENSE-MAKING In the engine at the right, the distance d from the piston to the center of the circle, called the *crankshaft*, is a function of the speed of the piston rod. Point R on the piston rod rotates 150 times per second.

a. Identify the period of the function as a fraction of a second. $\frac{1}{150}$

b. The shortest distance d is 0.5 inch, and the longest distance is 3.5 inches. Sketch a graph of the function. Let the horizontal axis represent the time t. Let the vertical axis represent the distance d. **See margin.**

27 TORNADOES A tornado siren makes 2.5 rotations per minute and the beam of sound has a radius of 1 mile. Ms. Miller's house is 1 mile from the siren. The distance of the sound beam from her house varies periodically as a function of time.

a. Identify the period of the function in seconds. **24 seconds**

b. Sketch a graph of the function. Let the horizontal axis represent the time t from 0 seconds to 60 seconds. Let the vertical axis represent the distance d the sound beam is from Ms. Miller's house at time t. **See margin.**

28. FERRIS WHEEL A Ferris wheel in China has a diameter of approximately 520 feet. The height of a compartment h is a function of time t. It takes about 30 seconds to make one complete revolution. Let the height at the center of the wheel represent the height at time 0. Sketch a graph of the function. **See margin.**

Differentiated Instruction ELL

Intermediate Instruct a small group of students to write a paragraph describing what is happening in the figure illustrating the unit circle. Their paragraphs should describe each part of the diagram in their own words. Ask for volunteers to read their paragraphs. Have students ask for clarification as needed.

30b.

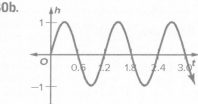

40. Sample answer:

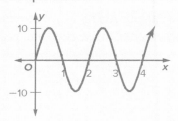

period: 2

29. MULTIPLE REPRESENTATIONS The terminal side of an angle in standard position intersects the unit circle at P, as shown in the figure.

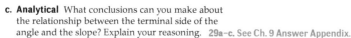

a. Geometric Copy the figure. Draw lines representing 30°, 60°, 150°, 210°, and 315°.

b. Tabular Use a table of values to show the slope of each line to the nearest tenth.

c. Analytical What conclusions can you make about the relationship between the terminal side of the angle and the slope? Explain your reasoning. **29a–c. See Ch. 9 Answer Appendix.**

30. POGO STICK A person is jumping up and down on a pogo stick at a constant rate. The difference between his highest and lowest points is 2 feet. He jumps 50 times per minute.

a. Describe the independent variable and dependent variable of the periodic function that represents this situation. Then state the period of the function in seconds.

b. Sketch a graph of the jumper's change in height in relation to his starting point. Assume that his starting point is halfway between his highest and lowest points. Let the horizontal axis represent the time t in seconds. Let the vertical axis represent the height h. **See margin.**

30a. Sample answer: independent variable: time t in seconds; dependent variable: height h in feet; period: 1.2 seconds

 Find the exact value of each expression.

31. $\cos 45° - \cos 30°$ $\dfrac{\sqrt{2}-\sqrt{3}}{2}$

32. $6(\sin 30°)(\sin 60°)$ $\dfrac{3\sqrt{3}}{2}$

33. $2\sin\dfrac{4\pi}{3} - 3\cos\dfrac{11\pi}{6}$ $-\dfrac{5\sqrt{3}}{2}$

34. $\cos\left(-\dfrac{2\pi}{3}\right) + \dfrac{1}{3}\sin 3\pi$ $-\dfrac{1}{2}$

35. $(\sin 45°)^2 + (\cos 45°)^2$ 1

36. $\dfrac{(\cos 30°)(\cos 150°)}{\sin 315°}$ $\dfrac{3\sqrt{2}}{4}$

37. Benita; Francis incorrectly wrote $\cos\dfrac{-\pi}{3} = -\cos\dfrac{\pi}{3}$.

F.TF.1, F.TF.2

H.O.T. Problems   Use Higher-Order Thinking Skills

37. CRITIQUE ARGUMENTS Francis and Benita are finding the exact value of $\cos\dfrac{-\pi}{3}$. Is either of them correct? Explain your reasoning.

41. The period of a periodic function is the horizontal distance of the part of the graph that is nonrepeating. Each nonrepeating part of the graph is one cycle.

39. Sometimes; the period of a sine curve could be $\dfrac{\pi}{2}$, which is not a multiple of π.

Francis	Benita
$\cos\dfrac{-\pi}{3} = -\cos\dfrac{\pi}{3}$	$\cos\dfrac{-\pi}{3} = \cos\left(-\dfrac{\pi}{3} + 2\pi\right)$
$= -0.5$	$= \cos\dfrac{5\pi}{3}$
	$= 0.5$

38. CHALLENGE A ray has its endpoint at the origin of the coordinate plane, and point $P\left(\dfrac{1}{2}, -\dfrac{\sqrt{3}}{2}\right)$ lies on the ray. Find the angle θ formed by the positive x-axis and the ray.   **−60°**

39. REASONING Is the period of a sine curve *sometimes*, *always*, or *never* a multiple of π? Justify your reasoning.

40. OPEN-ENDED Draw the graph of a periodic function that has a maximum value of 10 and a minimum value of −10. Describe the period of the function.   **See margin.**

41. WRITING IN MATH Explain how to determine the period of a periodic function from its graph. Include a description of a cycle.

MP Teaching the Mathematical Practices

Critique Arguments Mathematically proficient students are also able to compare the effectiveness of two plausible arguments, distinguish correct logic or reasoning from that which is flawed, and—if there is a flaw in an argument—explain what it is.

Assess

Name the Math Ask students to describe how they can find points on the unit circle in each of the quadrants and on each of the axes.

Differentiated Instruction OL BL

Extension Show students the graphs of $y = \cos x$, $y = \cos 2x$, and $y = \cos 3x$. Ask them to state the period of each function. 360°, 180°, 120° Ask them to determine the period of the function $y = \cos kx$. $\dfrac{360°}{k}$

Go Online!

The most up-to-date resources available for your program can be found at connectED.mcgraw-hill.com.

Preparing for Assessment

Dual Coding		
Exercise	Content Objective	Mathematical Practices
42	F.TF.2	6
43	F.TF.1	6
44	F.TF.2	2, 6
45	F.TF.2	2
46	F.TF.1	1

Diagnose Student Errors

Survey student responses for each item. Class trends may indicate common errors and misconceptions.

42.

A	Found the product of $2 \cos \theta \sin \theta$
B	Computed $2 \sin \theta$
C	Forgot to multiply by 2
D	Computed $2 \cos \theta$
E	CORRECT

43.

A	Substituted y-values from unit circle
B	Thought $\cos \frac{4\pi}{3} = -1$
C	CORRECT
D	Used wrong sign on first function
E	Used wrong sign on second function

44.

A	Chose the point of the first peak
B	Selected the amplitude
C	CORRECT
D	Selected an x-axis intersection point
E	Chose the range of the domain

Preparing for Assessment

42. The terminal side of angle θ in standard position intersects the unit circle at $P\left(\frac{\sqrt{3}}{2}, \frac{1}{2}\right)$, as shown in the figure. 6 F.TF.2

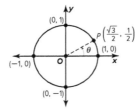

What is two times the sum of $\cos \theta$ and $\sin \theta$? **E**

- A $\frac{\sqrt{3}}{2}$
- B   1
- C $\frac{\sqrt{3}+1}{2}$
- D $\sqrt{3}$
- E $\sqrt{3}+1$

43. What is the exact value of $2 \cos \frac{4\pi}{3} - 4 \cos \frac{\pi}{3}$? 6 F.TF.1 **C**

- A $-3\sqrt{3}$
- B -4
- C -3
- D -1
- E   1

44. What is the period of the function shown in the graph? 2, 6 F.TF.2 **C**

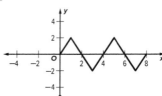

- A   1
- B   2
- C   4
- D   6
- E   8

---

45. MULTI-STEP Consider $\cos \frac{31\pi}{6}$.

a. Which of the following values are equal to $\cos \frac{31\pi}{6}$? Choose all that apply. 2 F.TF.2 **B, E, F**

- ☐ A $\cos\left(\frac{\pi}{6} + \frac{24\pi}{6}\right)$
- ☐ B $\cos\left(\frac{7\pi}{6} + \frac{24\pi}{6}\right)$
- ☐ C $\cos\left(\frac{3\pi}{6} + 5\pi\right)$
- ☐ D $\cos \frac{9\pi}{6}$
- ☐ E $\cos \frac{7\pi}{6}$
- ☐ F $\cos\left(\frac{7\pi}{6} + 4\pi\right)$

b. What is the exact value of $\cos \frac{31\pi}{6}$? **A**

- A $-\frac{\sqrt{3}}{2}$
- B $-\frac{1}{2}$
- C $\frac{-\sqrt{3}+1}{2}$
- D $\frac{1}{2}$
- E $\frac{2\sqrt{3}}{2}$

c. Which facts did you need to use to solve this trigonometry problem? Choose all that apply. **B, D**

- ☐ A   The sine function repeats every 2π radians.
- ☐ B   The cosine function repeats every 2π radians.
- ☐ C   The sine function values are shifted 0.5π to the right of the cosine function values.
- ☐ D $\cos \theta = \cos(\theta + 2\pi)$
- ☐ E $\sin \theta = \sin(\theta + 2\pi)$
- ☐ F   The standard amplitude of the sine function is 1.
- ☐ G   The standard amplitude of the cosine function is 1.

46. The period of a function $f(x)$ is 3 and the period of a function $g(x)$ is 4. What is the fundamental period of the function $h(x) = f(x) + g(x)$? 1 F.TF.1

12

ⓂⓅ Standards for Mathematical Practice

Emphasis On	Exercises
1 Make sense of problems and persevere in solving them.	5, 19, 26, 27, 28, 30
2 Reason abstractly and quantitatively.	1–4, 29
3 Construct viable arguments and critique the reasoning of others.	37–41
5 Use appropriate tools strategically.	31–36
6 Attend to precision.	9–12, 20–25
8 Look for and express regularity in repeated reasoning.	13–18

LESSON 9-5

Graphing Trigonometric Functions

SUGGESTED PACING (DAYS)

	Instruction	Extend Lab
90 min.	0.75	0.25
45 min.	1.0	0.5

Track Your Progress

Objectives

1 Describe and graph the sine, cosine, and tangent functions.

2 Describe and graph other trigonometric functions.

Mathematical Background

The graphs of the sine and cosine functions have amplitude. The graphs of the other trigonometric functions do not have amplitude because there is no maximum or minimum value. The period is the distance along the horizontal axis required for the graph to complete one cycle. The period is easy to determine from the graph.

THEN

F.TF.1 Understand radian measure of an angle as the length of the arc on the unit circle subtended by the angle.

F.TF.2 Explain how the unit circle in the coordinate plane enables the extension of trigonometric functions to all real numbers, interpreted as radian measures of angles traversed counterclockwise around the unit circle.

NOW

A.CED.2 Create equations in two or more variables to represent relationships between quantities; graph equations on coordinate axes with labels and scales.

F.IF.7e Graph exponential and logarithmic functions, showing intercepts and end behavior, and trigonometric functions, showing period, midline, and amplitude.

NEXT

F.IF.7e Graph exponential and logarithmic functions, showing intercepts and end behavior, and trigonometric functions, showing period, midline, and amplitude.

F.TF.5 Choose trigonometric functions to model periodic phenomena with specified amplitude, frequency, and midline.*

Go Online! All of these resources and more are available at connectED.mcgraw-hill.com

eLessons utilize the power of your interactive whiteboard in an engaging way. Use **Trigonometric Functions**, screen 15, to introduce the concepts in this lesson.

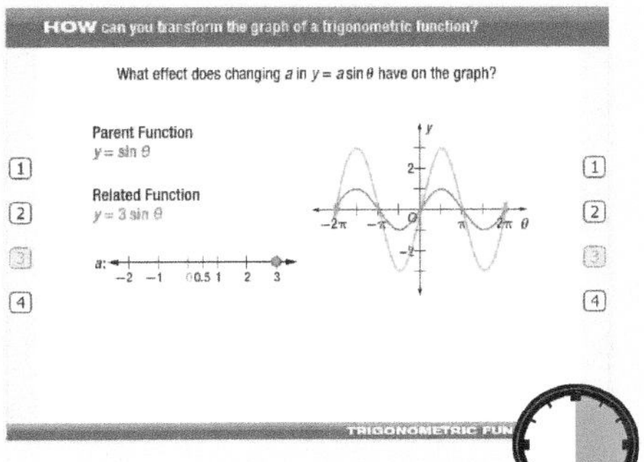

Use at Beginning of Lesson

Use The Geometer's Sketchpad to graph and analyze trigonometric functions.

Use at Beginning of Lesson

Graphing Tools are outstanding tools for enhancing understanding.

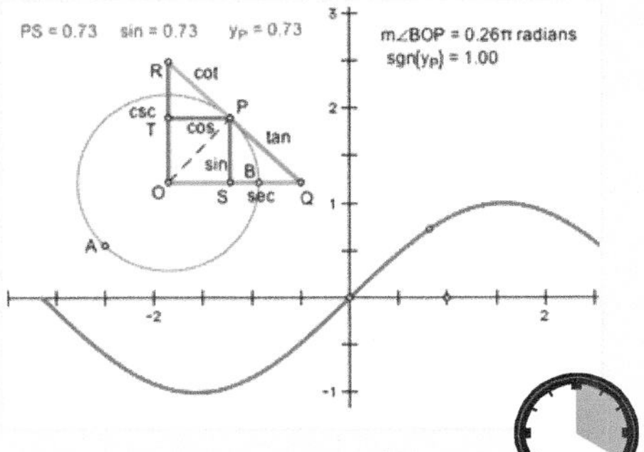

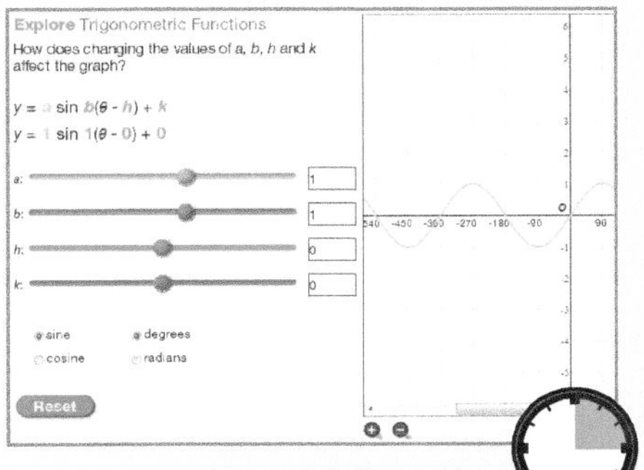

Use with Examples

Using Open Educational Resources

Blogs Have students work in four groups and use **QuadBlogging** to blog about graphing trigonometric functions. Have groups take turns being the main contributor to the blog while the other groups post comments. *Use as homework or review before the assessment*

Go Online!

connectED.mcgraw-hill.com

 Worksheets

Differentiate Your Resources

Extra Practice Additional practice or homework; Skills Practice is best for approaching-level students and Practice is best for on-level and beyond-level students

Skills Practice

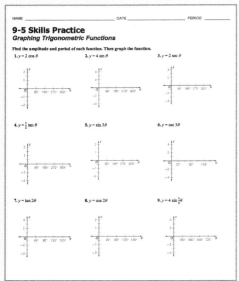

Practice

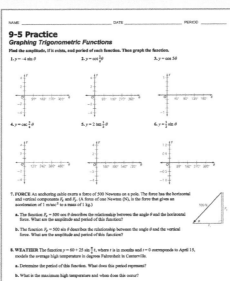

Word Problem Practice

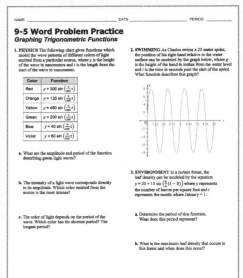

Intervention Reteaching and vocabulary activities that can be used with struggling or absent students and as ELL support

Study Guide and Intervention

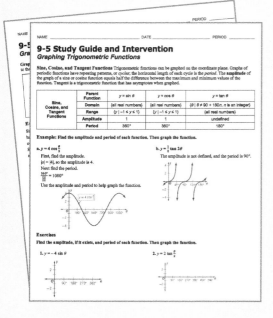

Study Notebook

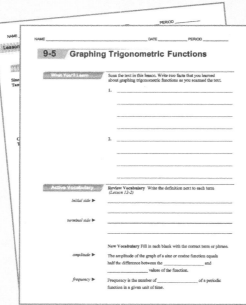

Extension Activities that can be used to extend lesson concepts

Enrichment

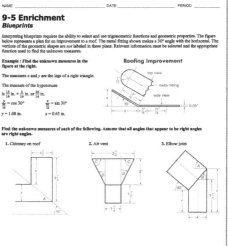

LESSON 5

Graphing Trigonometric Functions

::Then	::Now	::Why?
• You examined periodic functions.	**1** Describe and graph the sine, cosine, and tangent functions. **2** Describe and graph other trigonometric functions.	• Visible light waves have different wavelengths or periods. Red has the longest wavelength and violet has the shortest wavelength.

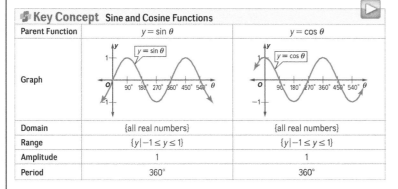

New Vocabulary
amplitude
frequency

MP Mathematical Practices
1 Make sense of problems and persevere in solving them.

Content Standards
A.CED.2 Create equations in two or more variables to represent relationships between quantities; graph equations on coordinate axes with labels and scales.

F.IF.4 For a function that models a relationship between two quantities, interpret key features of graphs and tables in terms of the quantities, and sketch graphs showing key features given a verbal description of the relationship.

F.IF.7.e Graph exponential and logarithmic functions, showing intercepts and end behavior; and trigonometric functions, showing period, midline, and amplitude.

F.TF.5 Choose trigonometric functions to model periodic phenomena with specified amplitude, frequency, and midline.

1 Sine, Cosine, and Tangent Functions Trigonometric functions can also be graphed on the coordinate plane. Recall that graphs of periodic functions have repeating patterns, or *cycles*. The horizontal length of each cycle is the *period*. The **amplitude** of the graph of a sine or cosine function equals half the difference between the maximum and minimum values of the function.

Key Concept Sine and Cosine Functions

Parent Function	$y = \sin \theta$	$y = \cos \theta$
Graph	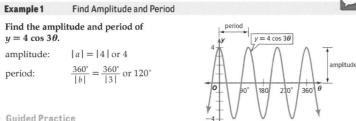	
Domain	{all real numbers}	{all real numbers}
Range	$\{y \mid -1 \le y \le 1\}$	$\{y \mid -1 \le y \le 1\}$
Amplitude	1	1
Period	360°	360°

As with other functions, trigonometric functions can be transformed. For the graphs of $y = a \sin b\theta$ and $y = a \cos b\theta$, the amplitude $= |a|$ and the period $= \dfrac{360°}{|b|}$.

F.IF.4, F.TF.5

Example 1 Find Amplitude and Period

Find the amplitude and period of $y = 4 \cos 3\theta$.

amplitude: $|a| = |4|$ or 4

period: $\dfrac{360°}{|b|} = \dfrac{360°}{|3|}$ or 120°

Guided Practice

Find the amplitude and period of each function.

1A. $y = \cos \frac{1}{2}\theta$ amplitude: 1; period: 720° **1B.** $y = 3 \sin 5\theta$ amplitude: 3; period: 72°

MP Mathematical Practices Strategies

Look for and express regularity in repeated reasoning.

• Use a large paper circle or draw one on the board. Write the degrees on the circle at 90, 180, 270, and 360 degrees. Use a pen to trace around and have students calculate the sine values at each stop. You might stop every 45 or 60 degrees.

measure	sine value
0	0
30	0.5
45	0.707
60	0.866
90	1
120	0.866
180	0
210	−0.500

• As the students call out the sine value at each stop, have them write the values in the data table, and graph each sine value point on a graph on their own note paper. The graph should work out exactly like the one in the Key Concept box.

• Have the students work cooperatively now, to do the same process in groups and generate a cosine graph. One person should trace the circle, and the group should call out the cosine values and watch as each is placed on the graph.

Launch

Have students read the Why? section of the lesson. Ask:

• **How are wavelengths measured in the diagram?** In the diagram, wavelengths are measured as the distance between two sequential high points.

• **What is another way to measure a wavelength?** The wavelengths could be measured as the distance between any two corresponding points of sequential periods.

• **In the diagram, how many violet wavelengths are equivalent to one red wavelength?** between 3 and 4

Teach

Ask the scaffolded questions for each example to build conceptual understanding for students at all levels.

1 Sine, Cosine, and Tangent Functions

Example 1 Find Amplitude and Period

AL What is the period of $\sin \theta$ and $\cos \theta$ in radians? 2π

OL What is the range of this function? $[-4, 4]$

BL What is the range of $y = a \cos \theta$? $[-a, a]$

Need Another Example?
Find the amplitude and period of $y = \sin \frac{1}{3}\theta$.
amplitude: 1; period: 1080°

Mathematical Practices Strategies *Cont.*

• **How are the sine and cosine graphs similar and different?** they have the same period but the values are shifted by $\frac{\pi}{2}$. They have the same amplitude and the same shape. Some students might say the graph is flipped upside down, which is not incorrect, but explain that it is useful to realize that they differ by $\frac{\pi}{2}$ by demonstrating the values as you trace around the circle.

Go Online!

Interactive Whiteboard
Use the *eLesson* or *Lesson Presentation* to present this lesson.

Example 2 Graph Sine and Cosine Functions

AL What is a and b for $y = 3 \cos 2\theta$? $a = 3$ and $b = 2$

OL What are the x-intercepts of $y = a \sin \theta$? $(\pi k, 0)$, where k is any integer.

BL What happens to the period as b gets larger? It gets shorter.

Need Another Example?

Graph each function.

a. $y = \sin 3\theta$

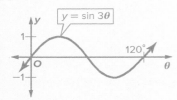

b. $y = \frac{1}{2} \cos \theta$

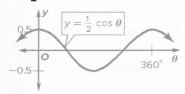

Additional Answers (Guided Practice)

2A.

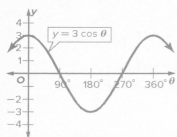

2B.

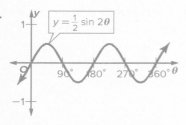

Use the graphs of the parent functions to graph $y = a \sin b\theta$ and $y = a \cos b\theta$. Then use the amplitude and period to draw the appropriate sine and cosine curves. You can also use θ-intercepts to help you graph the functions.

The θ-intercepts of $y = a \sin b\theta$ and $y = a \cos b\theta$ in one cycle are as follows.

$y = a \sin b\theta$	$y = a \cos b\theta$
$(0, 0), \left(\frac{1}{2} \cdot \frac{360°}{b}, 0\right) \left(\frac{360°}{b}, 0\right)$	$\left(\frac{1}{4} \cdot \frac{360°}{b}, 0\right), \left(\frac{3}{4} \cdot \frac{360°}{b}, 0\right)$

F.IF.4, F.TF.5

Example 2   Graph Sine and Cosine Functions

Graph each function.

a. $y = 2 \sin \theta$

Find the amplitude, the period, and the x-intercepts: $a = 2$ and $b = 1$.

amplitude: $|a| = |2|$ or 2   →   The graph is stretched vertically so that the maximum value is 2 and the minimum value is -2.

period: $\frac{360°}{|b|} = \frac{360°}{|1|}$ or $360°$ →   One cycle has a length of $360°$.

x-intercepts: $(0, 0)$

$\left(\frac{1}{2} \cdot \frac{360°}{b}, 0\right) = (180°, 0)$

$\left(\frac{360°}{b}, 0\right) = (360°, 0)$

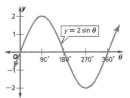

b. $y = \cos 4\theta$

amplitude: $|a| = |1|$ or 1

period: $\frac{360°}{|b|} = \frac{360°}{|4|}$ or $90°$

x-intercepts: $\left(\frac{1}{4} \cdot \frac{360°}{b}, 0\right) = (22.5°, 0)$

$\left(\frac{3}{4} \cdot \frac{360°}{b}, 0\right) = (67.5°, 0)$

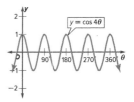

> **Guided Practice** 2A, 2B. See margin.

2A. $y = 3 \cos \theta$ **2B.** $y = \frac{1}{2} \sin 2\theta$

Trigonometric functions are useful for modeling real-world periodic motion such as electromagnetic waves or sound waves. Often these waves are described using *frequency*. **Frequency** is the number of cycles in a given unit of time.

The frequency of the graph of a function is the reciprocal of the period of the function. So, if the period of a function is $\frac{1}{100}$ second, then the frequency is 100 cycles per second.

Study Tip

Periods In $y = a \sin b\theta$ and $y = a \cos b\theta$, b represents the number of cycles in 360°. In Example 1, the 3 in $y = 4 \cos 3\theta$ indicates that there are three cycles in 360°. So, there is one cycle in 120°.

Study Tip

Amplitude The graphs of $y = a \sin b\theta$ and $y = a \cos b\theta$ with amplitude of $|a|$ have maxima at $y = a$ and minima at $y = -a$.

Go Online!

Investigate how changing the values in the equation of a trigonometric function affects the graph by using the eToolkit in ConnectED.

Differentiated Instruction **AL** **OL**

Visual/Spatial Learners Have groups of students make posters showing sketches of the graphs of the six trigonometric functions. Encourage students to color-code the key features of all the graphs, such as period, amplitude, asymptotes, and so on.

Real-World Link

Elephants are able to hear sound coming from up to 5 miles away. Humans can hear sounds with frequencies between 20 Hz and 20,000 Hz.

Source: School for Champions

Study Tip

MP **Reasoning** Note that the amplitude affects the graph along the vertical axis, and the period affects it along the horizontal axis.

Real-World Example 3   Model Periodic Situations F.TF.5

SOUND Sound that has a frequency below the human range is known as *infrasound*. Elephants can hear sounds in the infrasound range, with frequencies as low as 5 hertz (Hz), or 5 cycles per second.

a. Find the period of the function that models the sound waves.

There are 5 cycles per second, and the period is the time it takes for one cycle. So, the period is $\frac{1}{5}$ or 0.2 second.

b. Let the amplitude equal 1 unit. Write a sine equation to represent the sound wave y as a function of time t. Then graph the equation.

period $= \frac{2\pi}{\|b\|}$	Write the relationship between the period and b.
$0.2 = \frac{2\pi}{\|b\|}$	Substitution
$0.2\|b\| = 2\pi$	Multiply each side by $\|b\|$.
$b = 10\pi$	Multiply each side by 5: b is positive.
$y = a \sin b\theta$	Write the general equation for the sine function.
$y = 1 \sin 10\pi t$	$a = 1, b = 10\pi,$ and $\theta = t$
$y = \sin 10\pi t$	Simplify.

$y = \sin 10\pi t$

▸ **Guided Practice**

3. SOUND Humans can hear sounds with frequencies as low as 20 hertz.

   A. Find the period of the function. $\frac{1}{20}$ or 0.05 second

   B. Let the amplitude equal 1 unit. Write a cosine equation to model the sound waves. Then graph the equation. $y = \cos 40\pi t$; See margin for graph.

The tangent function is one of the trigonometric functions whose graphs have asymptotes.

⬥ Key Concept Tangent Functions

Parent Function	$y = \tan \theta$	Graph
Domain	$\{\theta \| \theta \neq 90 + 180n,$ n is an integer$\}$	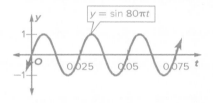
Range	{all real numbers}	
Amplitude	no amplitude	
Period	180°	
θ intercepts in one cycle	$180n$	

For the graph of $y = a \tan b\theta$, the period is $\frac{180°}{\|b\|}$, there is no amplitude, and the asymptotes are odd multiples of $\frac{180°}{2\|b\|}$.

Example 3 Model Periodic Situations

AL What is the domain of this function? all real numbers

OL What would this function be in degrees? $y = \sin (1800t)$

BL Why can you use sine to model this function? Sound is a wave and sine is a wave function.

Need Another Example?

Sound Humans can hear sounds with a frequency of 40 Hz.

a. Find the period of the function that models the sound waves. $\frac{1}{40}$ or 0.025 second

b. Let the amplitude equal 1 unit. Write a sine equation to represent the sound wave y as a function of time t. Then graph the equation. $y = \sin 80\pi t$

Teaching Tip

Graphing You might wish to have students draw their own graphs of the sine, cosine, and tangent functions. Their graphs can be compared to those shown on pages 628 and 630. This immediate feedback can be beneficial in helping students acquire the skills for graphing trigonometric functions, which are more complicated than most of the graphing students have done to this point.

Additional Answer (Guided Practice)

3B.

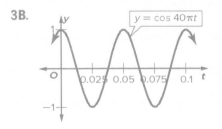

Example 4 Graph Tangent Functions

AL What is a in $y = \tan 2\theta$? 1

OL How many asymptotes will a tangent function have? infinitely many

BL Explain why tangent does not have an amplitude.
because the range of tangent is from $(-\infty, \infty)$

Need Another Example?
Find the period of $y = \tan \frac{1}{2}\theta$. Then graph the function. period: 360°

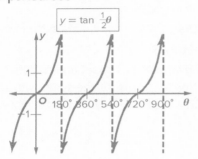

2 Graphs of Other Trigonometric Functions

Example 5 Graph Other Trigonometric Functions

AL List the functions that will have the same period. Sine, cosine, cosecant, and secant all have the same period and tangent and cotangent have the same period.

OL What is $\cos \theta$ everytime that $2 \sec \theta$ is undefined? 0

BL List the asymptotes for $y = A \csc \theta$, $x = 180°k$ where k is any integer.

Need Another Example?
Find the period of $y = 3 \csc \theta$. Then graph the function. period: 360°

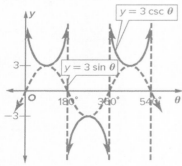

F.IF.4, F.TF.5

Example 4 Graph Tangent Functions

Find the period of $y = \tan 2\theta$. Then graph the function.

period: $\frac{180°}{|b|} = \frac{180°}{|2|}$ or 90°

asymptotes: $\frac{180°}{2|b|} = \frac{180°}{2|2|}$ or 45°

Sketch asymptotes at $-1 \cdot 45°$ or $-45°$,
$1 \cdot 45°$ or 45°, $3 \cdot 45°$ or 135°, and so on.

Use $y = \tan \theta$, but draw one cycle every 90°.

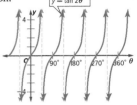

▶ **Guided Practice**

4. Find the period of $y = \frac{1}{2}\tan \theta$. Then graph the function. See margin.

2 Graphs of Other Trigonometric Functions The graphs of the cosecant, secant, and cotangent functions are related to the graphs of the sine, cosine, and tangent functions.

Key Concept Cosecant, Secant, and Cotangent Functions

Parent Function	$y = \csc \theta$	$y = \sec \theta$	$y = \cot \theta$
Graph	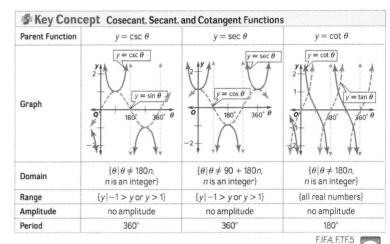		
Domain	$\{\theta \mid \theta \neq 180n,$ n is an integer$\}$	$\{\theta \mid \theta \neq 90 + 180n,$ n is an integer$\}$	$\{\theta \mid \theta \neq 180n,$ n is an integer$\}$
Range	$\{y \mid -1 > y$ or $y > 1\}$	$\{y \mid -1 > y$ or $y > 1\}$	$\{$all real numbers$\}$
Amplitude	no amplitude	no amplitude	no amplitude
Period	360°	360°	180°

F.IF.4, F.TF.5

Example 5 Graph Other Trigonometric Functions

Find the period of $y = 2 \sec \theta$. Then graph the function.

Because $2 \sec \theta$ is a reciprocal of $2 \cos \theta$, the graphs have the same period, 360°. The vertical asymptotes occur at the points where $2 \cos \theta = 0$. So, the asymptotes are at $\theta = 90°$ and $\theta = 270°$.

Sketch $y = 2 \cos \theta$ and use it to graph $y = 2 \sec \theta$.

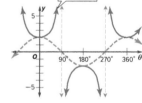

▶ **Guided Practice**

5. Find the period of $y = \csc 2\theta$. Then graph the function. See margin.

Additional Answers (Guided Practice)

4. period: 180°

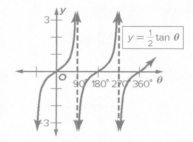

5. period: 180°

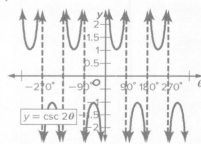

Check Your Understanding = Step-by-Step Solutions begin on page R11.

 **Go Online!** for a Self-Check Quiz

Examples 1–2
F.IF.4, F.TF.5

Find the amplitude and period of each function. Then graph the function.   1–4. See margin.

1. $y = 4 \sin \theta$

2. $y = \sin 3\theta$

3. $y = \cos 2\theta$

4. $y = \frac{1}{2} \cos 3\theta$

Example 3
F.TF.5

5. SPIDERS When an insect gets caught in a spider web, the web vibrates with a frequency of 14 hertz.

   a. Find the period of the function. $\frac{1}{14}$ or about 0.07 second

   b. Let the amplitude equal 1 unit. Write a sine equation to represent the vibration of the web y as a function of time t. Then graph the equation.   See margin.

Examples 4–5
F.IF.4, F.TF.5

Find the period of each function. Then graph the function.   6–8. See Ch. 9 Answer Appendix.

6. $y = 3 \tan \theta$

7. $y = 2 \csc \theta$

8. $y = \cot 2\theta$

Practice and Problem Solving   Extra Practice is on page R9.

Examples 1–2
F.IF.4, F.TF.5

Find the amplitude and period of each function. Then graph the function.

9. $y = 2 \cos \theta$

10. $y = 3 \sin \theta$

11. $y = \sin 2\theta$

12. $y = \cos 3\theta$

13. $y = \cos \frac{1}{2}\theta$

14. $y = \sin 4\theta$

15. $y = \frac{3}{4} \cos \theta$

16. $y = \frac{3}{2} \sin \theta$

17 $y = \frac{1}{2} \sin 2\theta$

18. $y = 4 \cos 2\theta$

19. $y = 3 \cos 2\theta$

20. $y = 5 \sin \frac{2}{3}\theta$ 9–20. See Ch. 9 Answer Appendix.

Example 3
F.TF.5

21. REASONING A boat on Lake Klondike bobs up and down with the waves. The difference between the lowest and highest points of the boat is 8 inches. The boat is at *equilibrium* when it is halfway between the lowest and highest points. Each cycle of the periodic motion lasts 3 seconds.

   a. Write an equation for the motion of the boat. Let h represent the height in inches and let t represent the time in seconds. Assume that the boat is at equilibrium at $t = 0$ seconds. $h = 4 \sin \frac{2}{3}\pi t$

   b. Draw a graph showing the height of the boat as a function of time.   See Ch. 9 Answer Appendix.

22. ELECTRICITY The voltage supplied by an electrical outlet is a periodic function that *oscillates*, or goes up and down, between −165 volts and 165 volts with a frequency of 50 cycles per second.

   a. Write an equation for the voltage V as a function of time t. Assume that at $t = 0$ seconds, the current is 165 volts. $V = 165 \cos 100\pi t$

   b. Graph the function.   See Ch. 9 Answer Appendix.

Examples 4–5
F.TF.5

Find the period of each function. Then graph the function.   23–28. See Ch. 9 Answer Appendix.

23. $y = \tan \frac{1}{2}\theta$

24. $y = 3 \sec \theta$

25. $y = 2 \cot \theta$

26. $y = \csc \frac{1}{2}\theta$

27. $y = 2 \tan \theta$

28. $y = \sec \frac{1}{3}\theta$

Differentiated Homework Options

Levels	AL Basic	OL Core	BL Advanced
Exercises	9–28, 41–48	9–27 odd, 29–31, 33–39 odd, 41–48	29–48
2-Day Option	9–27 odd, 45–48	9–28, 45–48	
	10–28 even, 41–44	29–44	

 You can use ALEKS to provide additional remediation support with personalized instruction and practice.

Practice

Formative Assessment Use Exercises 1–8 to assess students' understanding of the concepts in this lesson.

The Practice and Problem Solving exercises assess the content taught in the lesson. The Preparing for Assessment page is meant to be used as preparation for end-of-course assessments.

Extra Practice

See page R9 for extra exercises for students who are approaching level or for on-level students who need additional reinforcement.

Additional Answers

1. amplitude: 4; period: 360°

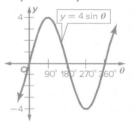

2. amplitude: 1; period: 120°

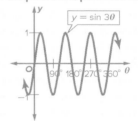

3. amplitude: 1; period: 180°

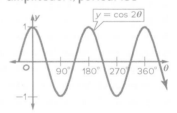

4. amplitude: $\frac{1}{2}$; period: 120°

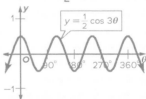

5b. $y = \sin 28\pi t$

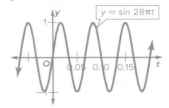

Levels of Complexity Chart

The levels of the exercises progress from 1 to 3, with Level 1 indicating the lowest level of complexity.

Exercises	9–28	29–30, 45–48	31–44
Level 3			●
Level 2		●	
Level 1	●		

Additional Answers

29b.

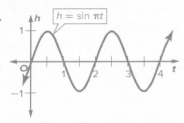

43. Sample answer: $y = 3 \sin 2\theta$

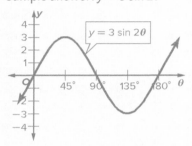

44. Sample answer: Determine the amplitude and period of the function; find and graph any *x*-intercepts, extrema, and asymptotes; use the parent function to sketch the graph.

Assess

Ticket Out the Door Have students write a function of the form $y = a \sin b\theta$ or $y = a \cos b\theta$. Then have them state the amplitude and period of the graph.

B 29 EARTHQUAKES A seismic station detects an earthquake wave that has a frequency of 0.5 hertz and an amplitude of 1 meter. **b. See margin.**

a. Write an equation involving sine to represent the height of the wave *h* as a function of time *t*. Assume that the equilibrium point of the wave, $h = 0$, is halfway between the lowest and highest points. $h = \sin \pi t$

b. Graph the function. Then determine the height of the wave after 20.5 seconds.

41. The domain of $y = a \cos \theta$ is the set of all real numbers. The domain of $y = a \sec \theta$ is the set of all real numbers except the values for which $\cos \theta = 0$. The range of $y = a \cos \theta$ is $-a \le y \le a$. The range of $y = a \sec \theta$ is $y \le -a$ and $y \ge a$.

30. **MP PERSEVERANCE** An object is attached to a spring as shown at the right. It oscillates according to the equation $y = 20 \cos \pi t$, where *y* is the distance in centimeters from its equilibrium position at time *t*.

a. Describe the motion of the object by finding the following: the amplitude in centimeters, the frequency in vibrations per second, and the period in seconds.

b. Find the distance of the object from its equilibrium to its position at $t = \frac{1}{4}$ second.

c. The equation $v = (-20 \text{ cm})(\pi \text{ rad/s}) \cdot \sin (\pi \text{ rad/s} \cdot t)$ represents the velocity *v* of the object at time *t*. Find the velocity at $t = \frac{1}{4}$ second. **about −44.4 cm/s**

30a. amplitude: 20 cm; frequency: 0.5 vibrations per second; period: 2 seconds

about 14.1 cm

31b. The amplitude remains the same. The period decreases because it is the reciprocal of the frequency.

31. PIANOS A piano string vibrates at a frequency of 130 hertz.

C

31a. $y = \cos 260\pi t$; See Ch. 9 Answer Appendix for graph.

a. Write and graph an equation using cosine to model the vibration of the string *y* as a function of time *t*. Let the amplitude equal 1 unit.

b. Suppose the frequency of the vibration doubles. Do the amplitude and period increase, decrease, or remain the same? Explain.

Find the amplitude, if it exists, and period of each function. Then graph the function.

32. $y = 3 \sin \frac{2}{3}\theta$
33. $y = \frac{1}{2} \cos \frac{3}{4}\theta$
34. $y = 2 \tan \frac{1}{2}\theta$
35. $y = 2 \sec \frac{4}{5}\theta$
36. $y = 5 \csc 3\theta$
37. $y = 2 \cot 6\theta$

32–37. See Ch. 9 Answer Appendix.

Identify the period of the graph and write an equation for each function.

42. The graph of $y = \frac{1}{2} \sin \theta$ has an amplitude of $\frac{1}{2}$ and a period of 360°. The graph of $y = \sin \frac{1}{2}\theta$ has an amplitude of 1 and a period of 720°.

38.

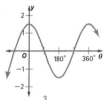

360°; $y = \frac{3}{2} \cos \theta$

39.

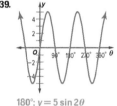

180°; $y = 5 \sin 2\theta$

40.

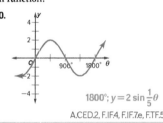

1800°; $y = 2 \sin \frac{1}{5}\theta$

A.CED.2, F.IF.4, F.IF.7e, F.TF.5

H.O.T. Problems Use Higher-Order Thinking Skills

41. **MP CHALLENGE** Describe the domain and range of $y = a \cos \theta$ and $y = a \sec \theta$, where *a* is any positive real number.

42. **MP REASONING** Compare and contrast the graphs of $y = \frac{1}{2} \sin \theta$ and $y = \sin \frac{1}{2}\theta$.

43. **OPEN-ENDED** Write a trigonometric function that has an amplitude of 3 and a period of 180°. Then graph the function. **See margin.**

44. **Q WRITING IN MATH** How can you use the characteristics of a trigonometric function to sketch its graph? **See margin.**

Differentiated Instruction **OL BL**

Extension Ask students to graph $y = \sin (\theta - \pi)$, $y = \sin \left(\theta - \frac{\pi}{2}\right)$, and $y = \sin \left(\theta - \frac{\pi}{4}\right)$. Ask them to describe the effect of the term being subtracted from θ. The value being subtracted from θ determines how far to the right the graph of $y = \sin \theta$ is translated.

Preparing for Assessment

45. Which of the following equations is best represented by the graph shown below? **MP** 2,7 F.TF.5 **D**

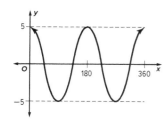

- ○ **A** $y = 2 \cos 5\theta$
- ○ **B** $y = -5 \cos \theta$
- ○ **C** $y = -\cos 2\theta$
- ○ **D** $y = 5 \cos 2\theta$
- ○ **E** $y = 10 \cos \frac{1}{2}\theta$

46. For the graph of the trigonometric function $y = 4 \tan 4\theta$, which of the following is (are) true? **MP** 2,7 F.TF.5 **B**

   I. The amplitude is 4.

   II. The period is 45°.

   III. There are no asymptotes.

- ○ **A** I only
- ○ **B** II only
- ○ **C** I and II only
- ○ **D** I, II, and III
- ○ **E** II and III only

47. What is the amplititude and period of $y = 6 \cos 2\theta$? **MP** 2,7 F.TF.5 **C**

- ○ **A** amplitude 6; period 720°
- ○ **B** amplitude 6; period 360°
- ○ **C** amplitude 6; period 180°

48. **MULTI-STEP** Use this graph of a trigonometric function. **MP** 1,2,8 F.TF.5

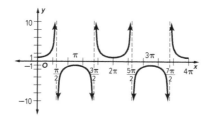

a. At which values of x does the function have a vertical asymptote? Choose all that apply. **B, D, F**

- ☐ **A** π
- ☐ **B** $\frac{\pi}{2}$
- ☐ **C** 2π
- ☐ **D** $\frac{3\pi}{2}$
- ☐ **E** 3π
- ☐ **F** $\frac{5\pi}{2}$

b. What is the period of this function? **B**

- ○ **A** $\frac{5\pi}{2}$
- ○ **B** 2π
- ○ **C** $\frac{\pi}{2}$
- ○ **D** π

c. Which function does this graph represent? **E**

- ○ **A** $y = \sin \theta$
- ○ **B** $y = \cos \theta$
- ○ **C** $y = \cot \theta$
- ○ **D** $y = \tan \theta$
- ○ **E** $y = \sec \theta$

d. What is the reciprocal function of this graph? **B**

- ○ **A** $y = \sin \theta$
- ○ **B** $y = \cos \theta$
- ○ **C** $y = \cot \theta$
- ○ **D** $y = \tan \theta$
- ○ **E** $y = \sec \theta$

Preparing for Assessment

Dual Coding Chart		
Exercise	Content Objective	**MP** Mathematical Practices
45, 46, 47	F.TF.5	2, 7
48	F.TF.5	1, 2, 8

Diagnose Student Errors

Survey student responses for each item. Class trends may indicate common errors and misconceptions.

45.

A	Confused period and amplitude
B	Misinterpreted the amplitude and period
C	Misinterpreted the amplitude
D	CORRECT
E	Doubled the amplitude and halved the period

46.

A	Did not understand that the tangent function does not have amplitude
B	CORRECT
C	Did not understand that the tangent function does not have amplitude
D	Did not understand that tangent functions have asymptotes but not amplitude
E	Did not understand that tangent functions have asymptotes

MP Standards for Mathematical Practice	
Emphasis On	**Exercises**
1 Make sense of problems and persevere in solving them.	5, 21, 22, 29–31
2 Reason abstractly and quantitatively.	1–4, 6–20, 23–28, 32–40
3 Construct viable arguments and critique the reasoning of others.	41–44

Go Online!

Self-Check Quiz

Students can use *Self-Check Quizzes* to check their understanding of this lesson. You can also give the *Chapter Quiz*, which covers the content in Lessons 9-4 and 9-5.

Launch

Objective Use a graphing calculator to explore transformations of the graphs of trigonometric functions.

Materials for Each Student

- TI-83/84 Plus or other graphing calculator

Teaching Tip

To set the calculator for degrees, press [MODE] and move the cursor to highlight **DEGREE** and press [ENTER]. Also, be sure to have students clear the Y= lists before beginning Exercise 1.

Teach

Working in Cooperative Groups Have students work in pairs, mixing abilities, to complete the Activities and Exercises 1 and 2. **ELL**

Ask:

- **How do parentheses change the meanings of the functions in Activities 1 and 2?** In Activity 1, a number is added to the value of sin θ, while in Activity 2, a number is added to θ before finding the sine.

- **How do the parentheses change the graphs in Activities 1 and 2?** The graphs in Activity 1 "move" up or down, while the graphs in Activity 2 "move" right or left.

Practice Have students complete Exercises 3–8.

Assess

Formative Assessment

Use Exercises 1 and 2 to assess whether students can describe the effect of adding a constant to a trigonometric function or to the value of θ in a trigonometric function.

Go Online!

Graphic Calculators

Students can use the *Graphing Calculator Personal Tutors* to review the use of the graphing calculator to solve systems of equations. They can also use the *Other Calculator Keystrokes*, which cover lab content for students with calculators other than the TI–84 Plus.

EXTEND 9-5

Graphing Technology Lab
Trigonometric Graphs

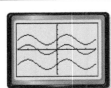

You can use a TI-83/84 Plus graphing calculator to explore transformations of the graphs of trigonometric functions.

Mathematical Practices
MP **5** Use appropriate tools strategically.

Content Standards
A.CED.2 Create equations in two or more variables to represent relationships between quantities; graph equations on coordinate axes with labels and scales.
F.IF.4 For a function that models a relationship between two quantities, interpret key features of graphs and tables in terms of the quantities, and sketch graphs showing key features given a verbal description of the relationship.
F.BF.3 Identify the effect on the graph of replacing f(x) by f(x) + k, k f(x), f(kx), and f(x + k) for specific values of k (both positive and negative); find the value of k given the graphs. Experiment with cases and illustrate an explanation of the effects on the graph using technology.

Activity 1 k in $y = \sin \theta + k$

Work cooperatively. Graph $y = \sin \theta$, $y = \sin \theta + 2$, and $y = \sin \theta - 3$ on the same coordinate plane. Describe any similarities and differences among the graphs.

Set the viewing window to match the window shown at the right. Let Y1 = sin θ, Y2 = sin θ + 2, and Y3 = sin θ − 3.

KEYSTROKES: [Y=] [SIN] [X,T,θ,n] [)] [ENTER]
[SIN] [X,T,θ,n] [)] [+] 2 [ENTER]
[SIN] [X,T,θ,n] [)] [−] 3 [GRAPH]

[−360, 360] scl: 90 by [−5, 5] scl: 1

The graphs have the same shape, but different vertical positions.

Activity 2 h in $y = \sin(\theta - h)$

Work cooperatively. Graph $y = \sin \theta$, $y = \sin(\theta + 45°)$, and $y = \sin(\theta - 90°)$ on the same coordinate plane. Describe any similarities and differences among the graphs.

Let Y1 = sin θ, Y2 = sin (θ + 45), and Y3 = sin (θ − 90).
Be sure to clear the entries from Activity 1.

KEYSTROKES: [Y=] [SIN] [X,T,θ,n] [)] [ENTER]
[SIN] [X,T,θ,n] [+] 45 [)] [ENTER]
[SIN] [X,T,θ,n] [−] 90 [)] [GRAPH]

[−360, 360] scl: 90 by [−5, 5] scl: 1

The graphs have the same shape, but different horizontal positions.

Model and Analyze

Work cooperatively. Repeat the activities for the cosine and tangent functions.

1. What are the domain and range of the functions in Activities 1 and 2? See Ch. 9 Answer Appendix.
2. What is the effect of adding a constant to a trigonometric function?

   2. Sample answer: Adding a constant to a trigonometric function translates the graph of the function vertically.
3. What is the effect of adding a constant to θ in a trigonometric function?

   Sample answer: Adding a constant to θ translates the graph of the function horizontally.

Repeat the activities for each of the following. Describe the relationship between each pair of graphs.

4. $y = \sin \theta + 4$
   $y = \sin(2\theta) + 4$

5. $y = \cos\left(\frac{1}{2}\theta\right)$ 4–7. See Ch. 9 Answer Appendix.
   $y = \cos\frac{1}{2}(\theta + 45°)$

6. $y = 2 \sin \theta$
   $y = 2 \sin \theta - 1$

7. $y = \cos \theta - 3$
   $y = \cos(\theta - 90°) - 3$

8. Write a general equation for the sine, cosine, and tangent functions after changes in amplitude a, period b, horizontal position h, and vertical position k. $y = a \sin(b\theta - h) + k$

From Concrete to Abstract

Ask students to describe how the value of n affects the graphs of $y = \sin \theta + n$ and $y = \sin(\theta + n)$ if n is a positive rational number. The graph of $y = \sin \theta + n$ is n units above $y = \sin \theta$; the graph of $y = \sin(\theta + n)$ is n units left of $y = \sin \theta$.

Track Your Progress

Objectives

1 Graph horizontal translations of trigonometric graphs and find phase shifts.

2 Graph vertical translations of trigonometric graphs.

Mathematical Background

The graphs of the functions $y = a \sin b(\theta - h) + k$, $y = a \cos b(\theta - h) + k$, and $y = a \tan b(\theta - h) + k$ are affected by changing the values of a, b, h, and k. A horizontal translation, or phase shift, is affected by h, and a vertical shift of the horizontal midline is affected by k. The amplitude is determined by the value of $|a|$ and the period is determined by $|b|$.

THEN	NOW	NEXT
A.CED.2 Create equations in two or more variables to represent relationships between quantities; graph equations on coordinate axes with labels and scales.	**F.IF.7e** Graph exponential and logarithmic functions, showing intercepts and end behavior, and trigonometric functions, showing period, midline, and amplitude.	**F.TF.8** Prove the Pythagorean identity $\sin^2(\theta) + \cos^2(\theta) = 1$ and use it to find $\sin(\theta)$, $\cos(\theta)$, and $\tan(\theta)$ given one of $\sin(\theta)$, $\cos(\theta)$, or $\tan(\theta)$ and the quadrant of the angle.
F.IF.7e Graph exponential and logarithmic functions, showing intercepts and end behavior; and trigonometric functions, showing period, midline, and amplitude.	**F.TF.5** Choose trigonometric functions to model periodic phenomena with specified amplitude, frequency, and midline.*	

Go Online! All of these resources and more are available at connectED.mcgraw-hill.com

eLessons utilize the power of your interactive whiteboard in an engaging way. Use **Trigonometric Functions**, screen 16, to introduce the concepts in this lesson.

Animations illustrate key concepts through step-by-step tutorials and videos.

Use **The Geometer's Sketchpad** to explore the transformations of trigonometric graphs.

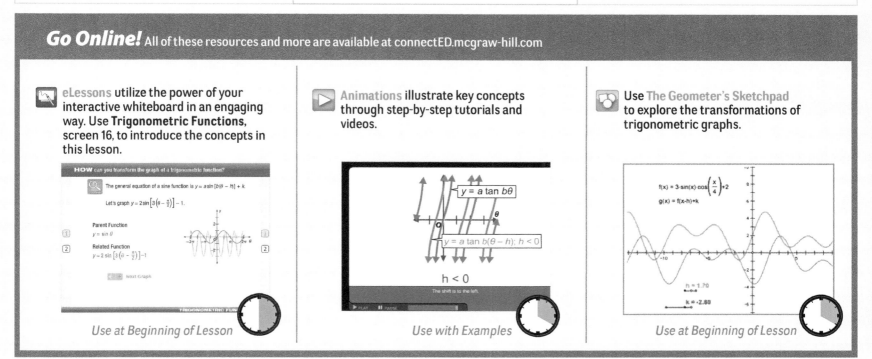

Use at Beginning of Lesson   *Use with Examples*   *Use at Beginning of Lesson*

⊙ER Using Open Educational Resources

Exploration Have students use the Graphical Functions Explorer tool on **mathopenref.com** to explore transformations of trigonometric functions before beginning the lesson. Students can enter up to three equations at once and then manipulate the variables using sliders to discover the effect of changing the equation on the graph. *Use as homework*

Go Online!
connectED.mcgraw-hill.com

Worksheets

Differentiate Your Resources

Extra Practice Additional practice or homework; Skills Practice is best for approaching-level students and Practice is best for on-level and beyond-level students

Skills Practice

NAME _____ DATE _____ PERIOD _____

9-6 Skills Practice
Translations of Trigonometric Graphs

State the amplitude, period, and phase shift for each function. Then graph the function.

1. $y = \sin(\theta + 90°)$
2. $y = \cos(\theta - 45°)$
3. $y = \tan\left(\theta - \frac{\pi}{4}\right)$

State the amplitude, period, vertical shift, and equation of the midline for each function. Then graph the function.

4. $y = \csc\theta - 2$
5. $y = \cos\theta + 1$
6. $y = \sec\theta + 3$

State the amplitude, period, phase shift, and vertical shift of each function. Then graph the function.

7. $y = 2\cos[3(\theta + 45°)] + 2$
8. $y = 3\sin[2(\theta - 90°)] + 2$
9. $y = 4\cot\left[\frac{1}{2}\left(\theta + \frac{\pi}{4}\right)\right] - 2$

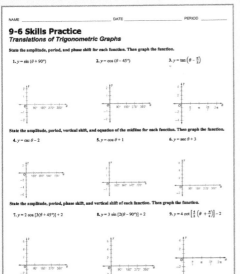

Practice

NAME _____ DATE _____ PERIOD _____

9-6 Practice
Translations of Trigonometric Graphs

State the amplitude, period, phase shift, and vertical shift for each function. Then graph the function.

1. $y = \frac{1}{2}\cos\left(\theta - \frac{\pi}{2}\right)$
2. $y = 2\cos(\theta + 30°) + 3$
3. $y = 3\csc(2\theta + 60°) - 2.5$

4. $y = -3 + 2\sin 2\left(\theta + \frac{\pi}{2}\right)$
5. $y = 3\cos 2(\theta + 45°) + 1$
6. $y = -1 + 4\tan(\theta + \pi)$

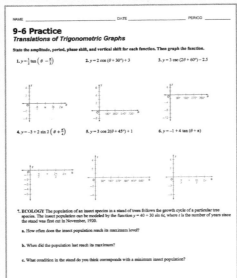

7. **ECOLOGY** The population of an insect species in a stand of trees follows the growth cycle of a particular tree species. The insect population can be modeled by the function $y = 40 + 30\sin 6x$, where x is the number of years since the stand was first cut in November, 1920.

a. How often does the insect population reach its maximum level?

b. When did the population last reach its maximum?

c. What condition in the stand do you think corresponds with a minimum insect population?

Word Problem Practice

NAME _____ DATE _____ PERIOD _____

9-6 Word Problem Practice
Translations of Trigonometric Graphs

1. **CLOCKS** A town hall has a tower with a clock on its face. The center of the clock is 40 feet above street level. The minute hand of the clock has a length of four feet.

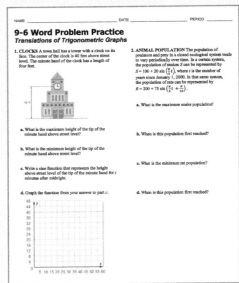

a. What is the maximum height of the tip of the minute hand above street level?

b. What is the minimum height of the tip of the minute hand above street level?

c. Write a sine function that represents the height above street level of the tip of the minute hand for t minutes after midnight.

d. Graph the function from your answer to part c.

2. **ANIMAL POPULATION** The population of predators and prey in a closed ecological system tends to vary periodically over time. In a certain system, the population of snakes S can be represented by $S = 100 + 20\sin\left(\frac{\pi}{6}t\right)$, where t is the number of years since January 1, 2000. In that same system, the population of rats can be represented by $R = 200 + 75\sin\left(\frac{\pi}{6}t + \frac{\pi}{16}\right)$.

a. What is the maximum snake population?

b. When is this population first reached?

c. What is the minimum rat population?

d. When is this population first reached?

Intervention Reteaching and vocabulary activities that can be used with struggling or absent students and as ELL support

Study Guide and Intervention

NAME _____ DATE _____ PERIOD _____

9-6 Study Guide and Intervention
Translations of Trigonometric Graphs

Horizontal Translations When a constant is subtracted from the angle measure in a trigonometric function, a phase shift of the graph results.

| Phase Shift | The phase shift of the graphs of the functions $y = a\sin b(\theta - h)$, $y = a\cos b(\theta - h)$, and $y = a\tan b(\theta - h)$ is h, where $b > 0$. If $h > 0$, the shift is h units to the right. If $h < 0$, the shift is $|h|$ units to the left. |

Example: Find State the amplitude, period, and phase shift for $y = \frac{1}{2}\cos 3\left(\theta - \frac{\pi}{3}\right)$. Then graph the function.

Amplitude: $|a| = \left|\frac{1}{2}\right|$ or $\frac{1}{2}$

Period: $\frac{2\pi}{|b|} = \frac{2\pi}{|3|}$ or $\frac{2\pi}{3}$

Phase Shift: $h = \frac{\pi}{3}$

The phase shift is to the right since $\frac{\pi}{3} > 0$.

Exercises

State the amplitude, period, and phase shift for each function. Then graph the function.

1. $y = 2\sin(\theta + 60°)$
2. $y = \tan\left(\theta - \frac{\pi}{2}\right)$
3. $y = 3\cos(\theta - 45°)$
4. $y = \frac{1}{2}\sin 3\left(\theta - \frac{\pi}{2}\right)$

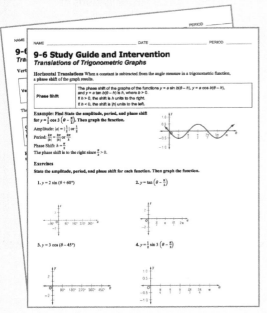

Study Notebook

NAME _____ DATE _____ PERIOD _____

9-6 Translations of Trigonometric Graphs

What You'll Learn Scan the text under the *Now* heading. List two things you will learn about in the lesson.

1. _____

2. _____

Active Vocabulary **New Vocabulary** Label the diagrams with the correct terms.

phase shift

vertical shift

midline

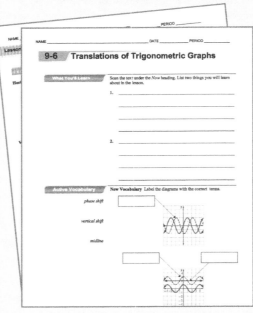

Extension Activities that can be used to extend lesson concepts

Enrichment

NAME _____ DATE _____ PERIOD _____

9-6 Enrichment
Simple Harmonic Motion

Suppose a small object is attached to the end of a spring and then released. The object oscillates up and down in a periodic fashion. The motion of the object is known as simple harmonic motion.

The motion of the object is described by the expression $y = A\sin\sqrt{\frac{k}{m}}t$, where A is how far down the object is pulled to stretch the spring, k is the spring constant, m is the mass of the object in grams, and t is the time in seconds.

1. A 10-gram object is attached to a spring that has a spring constant of 7, and is released after being pulled down 5 inches. Graph the motion of the spring during the first 10 seconds after the object is dropped.

2. How many seconds does it take the object to complete one full oscillation?

3. How does the graph in Exercise 1 change as k increases? As k decreases?

4. How does the graph in Exercise 1 change as m increases? As m decreases?

5. Suppose that while timing the motion of the spring in Exercise 1, the timer was started 1 second early. How would this affect the graph?

LESSON 6

Translations of Trigonometric Graphs

::Then	::Now	::Why?
• You translated exponential functions.	**1** Graph horizontal translations of trigonometric graphs and find phase shifts. **2** Graph vertical translations of trigonometric graphs.	• The graphs at the right represent the waves in a bay during high and low tides. Notice that the shape of the waves does not change.

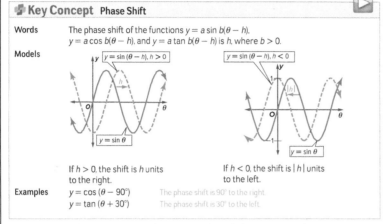

New Vocabulary
phase shift
vertical shift
midline

(MP) Mathematical Practices
4 Model with mathematics.

Content Standards
F.IF.7.e Graph exponential and logarithmic functions, showing intercepts and end behavior, and trigonometric functions, showing period, midline, and amplitude.
F.BF.3 Identify the effect on the graph of replacing $f(x)$ by $f(x) + k$, $k f(x)$, $f(kx)$, and $f(x + k)$ for specific values of k (both positive and negative); find the value of k given the graphs. Experiment with cases and illustrate an explanation of the effects on the graph using technology.
F.TF.5 Choose trigonometric functions to model periodic phenomena with specified amplitude, frequency, and midline.

1 Horizontal Translations Recall that a *translation* occurs when a figure is moved from one location to another on the coordinate plane without changing its orientation. A horizontal translation of a periodic function is called a **phase shift**.

Key Concept   Phase Shift

Words   The phase shift of the functions $y = a \sin b(\theta - h)$, $y = a \cos b(\theta - h)$, and $y = a \tan b(\theta - h)$ is h, where $b > 0$.

Models

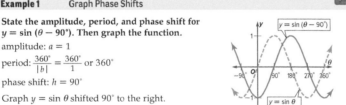

If $h > 0$, the shift is h units to the right.

If $h < 0$, the shift is $|h|$ units to the left.

Examples $y = \cos(\theta - 90°)$ The phase shift is 90° to the right.
$y = \tan(\theta + 30°)$ The phase shift is 30° to the left.

The secant, cosecant, and cotangent can be graphed using the same rules.

F.TF.5

Example 1   Graph Phase Shifts

State the amplitude, period, and phase shift for $y = \sin(\theta - 90°)$. Then graph the function.

amplitude: $a = 1$

period: $\dfrac{360°}{|b|} = \dfrac{360°}{1}$ or 360°

phase shift: $h = 90°$

Graph $y = \sin \theta$ shifted 90° to the right.

> **Guided Practice**

1. State the amplitude, period, and phase shift for $y = 2 \cos(\theta + 45°)$. Then graph the function.   2; 360°; $h = -45°$; See Ch. 9 Answer Appendix for graph.

(MP) Mathematical Practices Strategies

Reason abstractly and quantitatively.
Help students summarize and synthesize the different information about graphs in this lesson by drawing a summary chart or diagram. For example, ask:

• Do you think you will be able to remember and compare the lesson's graphs better by drawing them in a concept diagram, or a chart? Check students' answers.

• Use two colors. Draw a sine graph and its image that results from a phase shift. Write the equation that causes the shift. What are some interesting features of the graph? similar to Example 1

• Use two colors. Draw a sine graph and its image that results from a vertical shift. Write the equation that causes the shift. What are some interesting features of the graph? similar to Example 2

• Use two colors. Draw a sine graph and its image that results from an amplitude shift. Write the equation that causes the shift. What are some interesting features of the graph? similar to Example 3

Launch

Have students read the Why? section of the lesson. Ask:

• **Compare the periods and amplitudes of the waves for high and low tides.** The two waves have the same period and amplitude.

• **What is represented by the horizontal axis?** The horizontal axis represents time measured in seconds.

• **What is represented by the vertical axis?** The vertical axis shows the height above the bottom of the bay, measured in inches.

• **According to the diagram, is the greatest level at low tide greater than the lowest level at high tide?** yes

Teach

Ask the scaffolded questions for each example to build conceptual understanding for students at all levels.

1 Horizontal Translations

Example 1 Graph Phase Shifts

AL What type of shift is a phase shift? horizontal

OL Does a phase shift affect the length of the period? no

BL Explain how $\cos \theta$ is a phase shift of $\sin \theta$. $\cos \theta = \sin(\theta + 90°)$

(Continued on the next page.)

Go Online!

Interactive Whiteboard
Use the *eLesson* or *Lesson Presentation* to present this lesson.

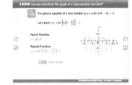

Need Another Example?

State the amplitude, period, and phase shift for $y = 2 \sin(\theta + 20°)$. Then graph the function.
amplitude: 2; period: 360°; phase shift: 20° left

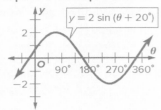

2 Vertical Translations

Example 2 Graph Vertical Translations

AL What is a, b, and k for $y = 2 \sin(3\theta) - 5$?
$a = 2$, $b = 3$, and $k = -5$

OL Does a vertical shift affect the period? no

BL What is the range of $y = a \sin b\theta + k$?
$[-a + k, a + k]$

Need Another Example?

State the amplitude, period, vertical shift, and equation of the midline for $y = \frac{1}{2} \cos \theta + 3$. Then graph the function. amplitude: $\frac{1}{2}$; period: 360° or 2π; vertical shift: 3 units up; midline: $y = 3$

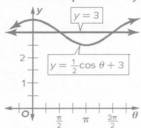

Additional Answer (Guided Practice)

2. no amplitude; 180°; $k = 3$; $y = 3$

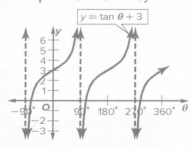

---

2 Vertical Translations Recall that the graph of $y = x^2 + 5$ is the graph of the parent function $y = x^2$ shifted up 5 units. Similarly, graphs of trigonometric functions can be translated vertically through a **vertical shift**.

Key Concept Vertical Shift

Words The vertical shift of the functions $y = a \sin b\theta + k$, $y = a \cos b\theta + k$ and $y = a \tan b\theta + k$ is k.

Models

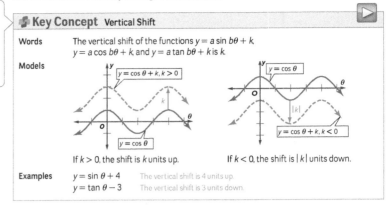

If $k > 0$, the shift is k units up.   If $k < 0$, the shift is $|k|$ units down.

Examples $y = \sin \theta + 4$ The vertical shift is 4 units up.
$y = \tan \theta - 3$ The vertical shift is 3 units down.

The secant, cosecant, and cotangent can be graphed using the same rules.

When a trigonometric function is shifted vertically k units, the line $y = k$ is the new horizontal axis about which the graph oscillates. This line is called the **midline**, and it can be used to help draw vertical translations.

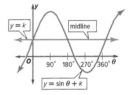

F.BF.3, F.TF.5

Example 2 Graph Vertical Translations

State the amplitude, period, vertical shift, and equation of the midline for $y = \frac{1}{2} \cos \theta - 2$. Then graph the function.

amplitude: $|a| = \frac{1}{2}$

period: $\frac{2\pi}{|b|} = \frac{2\pi}{|1|}$ or 2π

vertical shift: $k = -2$

midline: $y = -2$

To graph $y = \frac{1}{2} \cos \theta - 2$, first draw the midline. Then use it to graph $y = \frac{1}{2} \cos \theta$ shifted 2 units down.

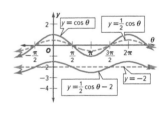

Guided Practice

2. State the amplitude, period, vertical shift, and equation of the midline for $y = \tan \theta + 3$. Then graph the function. **See margin.**

Differentiated Instruction **AL**

IF students struggle with translations of trigonometric graphs,

THEN make coordinate axes with masking tape on the classroom floor. Give students at least 15 feet of rope and have them stand along the x-axis, positioning the rope to model the graph of $y = \sin x$. As you call out equations of functions with graphs being horizontal phase shifts of the graph of $y = \sin x$, students can step left or right to model the translated graph. Similarly, call out functions with graphs being vertical shifts of the graph $y = \sin x$.

You can use the following steps to graph trigonometric functions involving phase shifts and vertical shifts.

Key Concept Graph Trigonometric Functions

$$\underset{\uparrow}{\overset{\text{amplitude}}{\downarrow}} \quad \underset{\uparrow}{\overset{\text{period}}{\downarrow}}$$
$$y = a \sin b(\theta - h) + k$$
$$\underset{\text{phase shift}}{\uparrow} \qquad \underset{\text{vertical shift}}{\uparrow}$$

Step 1 Determine the vertical shift, and graph the midline.

Step 2 Determine the amplitude, if it exists. Use dashed lines to indicate the maximum and minimum values of the function.

Step 3 Determine the period of the function, and graph the appropriate function.

Step 4 Determine the phase shift, and translate the graph accordingly.

F.BF.3, F.TF.5

Example 3 Graph Transformations

State the amplitude, period, phase shift, and vertical shift for
$y = 3 \sin \frac{2}{3}(\theta - \pi) + 4$. Then graph the function.

amplitude: $|a| = 3$

period: $\frac{2\pi}{|b|} = \frac{2\pi}{\left|\frac{2}{3}\right|}$ or 3π The period indicates that the graph will be stretched.

phase shift: $h = \pi$ The graph will shift π to the right.

vertical shift: $k = 4$ The graph will shift 4 units up.

midline: $y = 4$ The graph will oscillate around the line $y = 4$

Step 1 Graph the midline.

Step 2 Because the amplitude is 3, draw dashed lines 3 units above and 3 units below the midline.

Step 3 Graph $y = 3 \sin \frac{2}{3}\theta + 4$ using the midline as a reference.

Step 4 Shift the graph π units to the right.

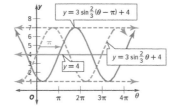

CHECK You can check the accuracy of your transformation by evaluating the function for various values of θ and confirming their location on the graph.

▶ **Guided Practice**

3. State the amplitude, period, phase shift, and vertical shift for $y = 2 \cos \frac{1}{2}\left(\theta + \frac{\pi}{2}\right) - 2$. Then graph the function. See margin.

Go Online! 💬

Follow along with your graphing calculator as you watch a Personal Tutor graph a translated trigonometric function.

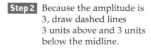
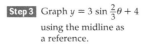
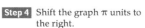

Study Tip

Verifying a Graph After drawing the graph of a trigonometric function, select values of θ and evaluate them in the equation to verify your graph.

Example 3 Graph Transformations

🔵**AL** What are a, b, and k for $y = -7 \sin(4\theta) + 15$? $a = -7$, $b = 4$, and $k = 15$

🔵**OL** What is the domain of any function of the form $y = a \sin b(\theta - h) + k$? all real numbers

🔵**BL** For what values of θ will $y = 3 \csc \frac{2}{3}(\theta - \pi) + 4$ be defined? all real numbers except πn where n is an integer

Need Another Example?

State the amplitude, period, phase shift, and vertical shift for $y = 3 \sin\left[2\left(\theta - \frac{\pi}{2}\right)\right] + 4$. Then graph the function. amplitude: 3; period: π; phase shift: $\frac{\pi}{2}$ to the right; vertical shift: 4 units up

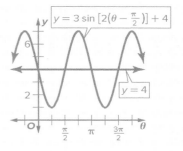

Teaching Tip

Reading Math Make sure students are clear about the difference between the terms *vertical* and *horizontal*. Students can use the words *vertigo* (a sense of dizziness some people experience when looking down from a great height) and *horizon* to help link vertical and horizontal to their relative direction.

Additional Answer (Guided Practice)

3. 2; 4π; $h = -\frac{\pi}{2}$; $k = -2$

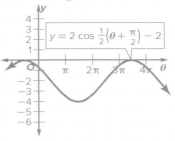

Example 4 Represent Periodic Functions

AL What part of the problem statement helps us determine the period? "6 waves per minute"

OL What is the range of this function? [5, 13]

BL Give the cosine function that would represent the same function. $4 \cos\left(\frac{\pi}{5}t - \frac{\pi}{2}\right) + 9$

Need Another Example?

Wave Pool In Example 4, suppose the height of water oscillates between a maximum of 10 feet and a minimum of 6 feet, and the wave generator pumps 3 waves per minute. Write a sine function that represents the height of the water at time t seconds. Then graph the function. $h = 2 \sin \frac{\pi}{10} t + 8$

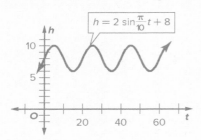

Additional Answer (Guided Practice)

4. $h = 4 \cos \frac{\pi}{6} t + 10$

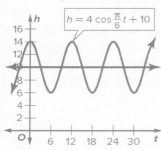

The sine wave occurs often in physics, signal processing, music, electrical engineering, and many other fields.

F.TF.5

Real-World Link
Waves at the Surf and Swim in Garland, Texas reach an apex of 3 feet.

Source: Surf and Swim

Real-World Example 4   Represent Periodic Functions

WAVE POOL The height of water in a wave pool oscillates between a maximum of 13 feet and a minimum of 5 feet. The wave generator pumps 6 waves per minute. Write a sine function that represents the height of the water at time t seconds. Then graph the function.

Step 1 Write the equation for the midline, and determine the vertical shift.

$y = \frac{13 + 5}{2}$ or 9   The midline lies halfway between the maximum and minimum values.

Because the midline is $y = 9$, the vertical shift is $k = 9$.

Step 2 Find the amplitude.

$|a| = |13 - 9|$ or 4   Find the difference between the midline value and the maximum value.

So, $a = 4$.

Step 3 Find the period.

Because there are 6 waves per minute, there is 1 wave every 10 seconds. So, the period is 10 seconds.

$10 = \frac{2\pi}{|b|}$ Period $= \frac{2\pi}{|b|}$

$|b| = \frac{2\pi}{10}$ Solve for $|b|$.

$b = \pm\frac{\pi}{5}$ Simplify.

Watch Out

Parent Functions Often the graph of a trigonometric function can be represented by more than one equation. For example, the graphs of $y = \cos\theta$ and $y = \sin(\theta + 90°)$ are the same.

Step 4 Write an equation for the function.

$h = a \sin b(t - h) + k$ Write the equation for sine relating height h and time t.

$= 4 \sin \frac{\pi}{5}(t - 0) + 9$ Substitution: $a = 4$, $b = \frac{\pi}{5}$, $h = 0$, $k = 9$.

$= 4 \sin \frac{\pi}{5}t + 9$ Simplify.

Then graph the function.

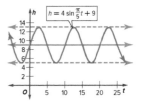

Guided Practice

4. **WAVE POOL** The height of water in a wave pool oscillates between a maximum of 14 feet and a minimum of 6 feet. The wave generator pumps 5 waves per minute. Write a cosine function that represents the height of water at time t seconds. Then graph the function. See margin.

Additional Answers

1. 1; 360°; $h = 180°$

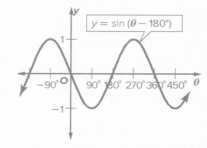

2. no amplitude; 180°; $h = \frac{\pi}{4}$

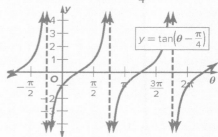

Check Your Understanding

 = Step-by-Step Solutions begin on page R11.

 *Go Online!* for a Self-Check Quiz

Example 1
F.BF.3,
F.TF.5

State the amplitude, period, and phase shift for each function. Then graph the function. **1–4. See margin.**

1. $y = \sin(\theta - 180°)$

2. $y = \tan\left(\theta - \frac{\pi}{4}\right)$

3. $y = \sin\left(\theta - \frac{\pi}{2}\right)$

4. $y = \frac{1}{2}\cos(\theta + 90°)$

Example 2
F.BF.3,
F.TF.5

State the amplitude, period, vertical shift, and equation of the midline for each function. Then graph the function. **5–8. See margin.**

5. $y = \cos\theta + 4$

6. $y = \sin\theta - 2$

7. $y = \frac{1}{2}\tan\theta + 1$

8. $y = \sec\theta - 5$

Example 3
F.BF.3,
F.TF.5

MP REGULARITY State the amplitude, period, phase shift, and vertical shift for each function. Then graph the function. **9–12. See Ch. 9 Answer Appendix.**

9. $y = 2\sin(\theta + 45°) + 1$

10. $y = \cos 3(\theta - \pi) - 4$

11. $y = \frac{1}{4}\tan 2(\theta + 30°) + 3$

12. $y = 4\sin\frac{1}{2}\left(\theta - \frac{\pi}{2}\right) + 5$

Example 4
F.TF.5

13. EXERCISE While doing some moderate physical activity, a person's blood pressure oscillates between a maximum of 130 and a minimum of 90. The person's heart rate is 90 beats per minute. Write a sine function that represents the person's blood pressure P at time t seconds. Then graph the function. **See Ch. 9 Answer Appendix.**

Practice and Problem Solving

Extra Practice is on page R9.

Example 1
F.BF.3,
F.TF.5

State the amplitude, period, and phase shift for each function. Then graph the function.

14. $y = \cos(\theta + 180°)$

15. $y = \tan(\theta - 90°)$

16. $y = \sin(\theta + \pi)$

17. $y = 2\sin\left(\theta + \frac{\pi}{2}\right)$

18. $y = \tan\frac{1}{2}(\theta + 30°)$

19. $y = 3\cos\left(\theta - \frac{\pi}{3}\right)$

14–19. See Ch. 9 Answer Appendix.

Example 2
F.BF.3,
F.TF.5

State the amplitude, period, vertical shift, and equation of the midline for each function. Then graph the function. **20–25. See Ch. 9 Answer Appendix.**

20. $y = \cos\theta + 3$

21. $y = \tan\theta - 1$

22. $y = \tan\theta + \frac{1}{2}$

23 $y = 2\cos\theta - 5$

24. $y = 2\sin\theta - 4$

25. $y = \frac{1}{3}\sin\theta + 7$

Example 3
F.BF.3,
F.TF.5

State the amplitude, period, phase shift, and vertical shift for each function. Then graph the function. **26–33. See Ch. 9 Answer Appendix.**

26. $y = 4\sin(\theta - 60°) - 1$

27. $y = \cos\frac{1}{2}(\theta - 90°) + 2$

28. $y = \tan(\theta + 30°) - 2$

29. $y = 2\tan 2\left(\theta + \frac{\pi}{4}\right) - 5$

30. $y = \frac{1}{2}\sin\left(\theta - \frac{\pi}{2}\right) + 4$

31. $y = \cos 3(\theta - 45°) + \frac{1}{2}$

32. $y = 3 + 5\sin 2(\theta - \pi)$

33. $y = -2 + 3\sin\frac{1}{3}\left(\theta - \frac{\pi}{2}\right)$

Example 4
F.TF.5

34. TIDES The height of the water in a harbor rose to a maximum height of 15 feet at 6:00 P.M. and then dropped to a minimum level of 3 feet by 3:00 A.M. The water level can be modeled by the sine function. Write an equation that represents the height h of the water t hours after noon on the first day. $h = 9 + 6\sin\left[\frac{\pi}{9}(t - 1.5)\right]$

7. no amplitude; 180°; $k = 1$; $y = 1$

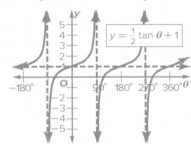

8. $k = -5$; $y = -5$; no amplitude; 360°

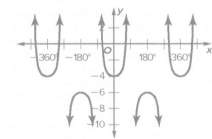

Practice

Formative Assessment Use Exercises 1–13 to assess students' understanding of the concepts in this lesson.

The Practice and Problem Solving exercises assess the content taught in the lesson. The Preparing for Assessment page is meant to be used as preparation for end-of-course assessments.

Extra Practice

See page R9 for extra exercises for students who are approaching level or for on-level students who need additional reinforcement.

Additional Answers

3. $1; 2\pi; h = \frac{\pi}{2}$

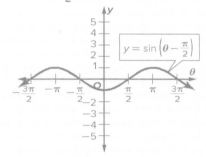

4. $\frac{1}{2}; 360°; h = -90°$

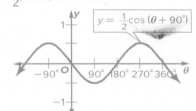

5. $1; 360°; k = 4; y = 4$

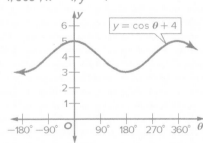

6. $1; 360°; k = -2; y = -2$

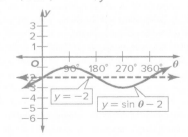

Additional Answers

35.

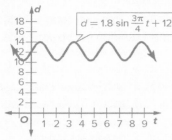

$d = 1.8 \sin \frac{3\pi}{4}t + 12$

min: 10.2 ft; max: 13.8 ft

42c.

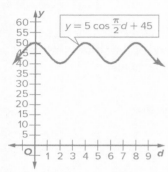

$y = 5 \cos \frac{\pi}{2}d + 45$

60. Sometimes; if the function is shifted vertically, then you also need to know the value of the midline. The maximum value is the value of the midline plus the amplitude. The minimum value is the midline value minus the amplitude.

63. Sample answer: $y = 2 \sin \theta - 3$

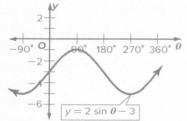

$y = 2 \sin \theta - 3$

B **35. LAKES** A buoy marking the swimming area in a lake oscillates each time a speed boat goes by. Its distance d in feet from the bottom of the lake is given by $d = 1.8 \sin \frac{3\pi}{4}t + 12$, where t is the time in seconds. Graph the function. Describe the minimum and maximum distances of the buoy from the bottom of the lake when a boat passes by. **See margin.**

36. FERRIS WHEEL Suppose a Ferris wheel has a diameter of approximately 100 feet and makes one revolution every 24 seconds. Write an equation for the height of a car h as a function of time t in seconds, assuming at $t = 0$, the car is at its lowest point, 5 feet above ground. $h = -50 \cos \frac{\pi}{12}t + 55$

Write an equation for each translation.

37. $y = \sin x$, 4 units to the right and 3 units up $y = \sin(x - 4) + 3$

38. $y = \cos x$, 5 units to the left and 2 units down $y = \cos(x + 5) - 2$

39. $y = \tan x$, π units to the right and 2.5 units up $y = \tan(x - \pi) + 2.5$

40a. At 1.25 seconds, the height of the rope is 68 inches; at 2.75 seconds, the height of the rope is 2 inches.

40. JUMP ROPE The graph at the right approximates the height of a jump rope h in inches as a function of time t in seconds. A maximum point on the graph is (1.25, 68), and a minimum point is (2.75, 2).

  a. Describe what the maximum and minimum points mean in the context of the situation.

  b. What is the equation for the midline, the amplitude, and the period of the function? $y = 35; 33, 1$

  c. Write an equation for the function. $h = 33 \sin 2\pi t + 35$

41. MULTI-STEP You are designing a carousel with 32 horses. You want only 4 horses to be at their maximum point at any particular time. You want the carousel to do a complete revolution every 20 seconds and the range of motion of each horse to be 5 feet. *See Ch. 9 Answer Appendix.*

  a. Use trigonometry to model the movement of the carousel horses.

  b. Describe each part of your trigonometric function(s).

  c. What assumptions did you make in designing your carousel?

  d. Suppose you also want each horse to be at a different height at any particular point for each rotation so parents will be able to take pictures of their children at different heights without having to move. How do you accomplish this?

42. REASONING During one month, the outside temperature fluctuates between 40°F and 50°F. A cosine curve approximates the change in temperature, with a high of 50°F being reached every four days.

  a. Describe the amplitude, period, and midline of the function that approximates the temperature y on day d. $5; 4; y = 45$

  b. Write a cosine function to estimate the temperature y on day d. $y = 5 \cos \frac{\pi}{2}d + 45$

  c. Sketch a graph of the function. **See margin.**

  d. Estimate the temperature on the 7th day of the month. *about 45°F*

Find a coordinate that represents a maximum for each graph.

43. $y = -2 \cos\left(x - \frac{\pi}{2}\right)$ $\left(\frac{3\pi}{2}, 2\right)$

44. $y = 4 \sin\left(x + \frac{\pi}{3}\right)$ $\left(\frac{\pi}{6}, 4\right)$

45. $y = 3 \tan\left(x + \frac{\pi}{2}\right) + 2$ *no maximum values*

46. $y = -3 \sin\left(x - \frac{\pi}{4}\right) - 4$ $\left(\frac{7\pi}{4}, -1\right)$

Differentiated Homework Options

Levels	AL Basic	OL Core	BL Advanced
Exercises	14–34, 60–70	15–33 odd, 35-37, 39–42, 43–59 odd, 60–70	35–70
2-Day Option	15–33 odd, 65–70	14–34, 65–70	
	14–34 even, 60–64	35–64	

You can use **ALEKS** to provide additional remediation support with personalized instruction and practice.

Go Online!

eBook

Interactive Student Guide

ALGEBRA 2
INTERACTIVE STUDENT GUIDE

Use the *Interactive Student Guide* to deepen conceptual understanding.
• Graphing Trigonometric Functions

Compare each pair of graphs.

47. $y = -\cos 3\theta$ and $y = \sin 3(\theta - 90°)$ The graphs are reflections of each other in the *x*-axis.

48. $y = 2 + 0.5 \tan \theta$ and $y = 2 + 0.5 \tan (\theta + \pi)$ The graphs are identical.

49. $y = 2 \sin \left(\theta - \frac{\pi}{6}\right)$ and $y = -2 \sin \left(\theta + \frac{5\pi}{6}\right)$ The graphs are identical.

 Identify the period of each function. Then write an equation for the graph using the given trigonometric function. **50.** 360°; Sample answer: $y = \sin \theta - 5$

51. 360°;
Sample answer:
$y = 2 \cos (\theta + 90°)$
52. 360°;
Sample answer:
$y = 4 \cos \theta + 1$
53. 180°;
Sample answer:
$y = \sin 2(\theta - 45°) + 3$

50. sine

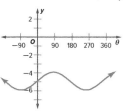

51. cosine

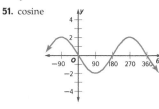

61. The graph of $y = 3 \sin 2\theta + 1$ has an amplitude of 3 rather than an amplitude of 1. It is shifted up 1 unit from the parent graph and is compressed so that it has a period of 180°.

52. cosine

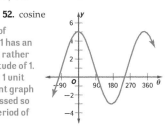

(53) sine

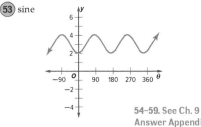

54–59. See Ch. 9 Answer Appendix.

State the period, phase shift, and vertical shift. Then graph the function.

54. $y = \csc (\theta + \pi)$

55. $y = \cot \theta + 6$

56. $y = \cot \left(\theta - \frac{\pi}{6}\right) - 2$

57. $y = \frac{1}{2} \csc 3(\theta - 45°) + 1$

58. $y = 2 \sec \frac{1}{2}(\theta - 90°)$

59. $y = 4 \sec 2\left(\theta + \frac{\pi}{2}\right) - 3$

62. Sample answer: a phase shift 90° left, $y = \sin (\theta + 90°)$; a phase shift 270° right, $y = \sin (\theta - 270°)$

H.O.T. Problems   Use Higher-Order Thinking Skills

60. **CONSTRUCT ARGUMENTS** If you are given the amplitude and period of a cosine function, is it *sometimes*, *always*, or *never* possible to find the maximum and minimum values of the function? Explain your reasoning.   See margin.

61. REASONING Describe how the graph of $y = 3 \sin 2\theta + 1$ is different from $y = \sin \theta$.

64. Sample answer: Infinitely many; any change in amplitude will create a different graph that has the same θ-intercepts.

62. WRITING IN MATH Describe two different phase shifts that will translate the sine curve onto the cosine curve shown at the right. Then write an equation for the new sine curve using each phase shift.

63. OPEN-ENDED Write a periodic function that has an amplitude of 2 and midline at $y = -3$. Then graph the function.   See margin.

64. REASONING How many different sine graphs pass through the origin and all points of the form $(n\pi, 0)$? Explain.

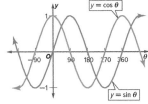

Differentiated Instruction **OL** **BL**

Extension Have students choose a vertical shift, amplitude, period, and phase shift and write a function with these characteristics.

e Follow-Up

Students have explored transformations of trigonometric functions

Ask:

• How are transformations of trigonometric functions similar to transformations of other functions you have studied? Sample answer: Adding or subtracting to any function translates the graph; multiplying a function dilates the graph; multiplying a function by a negative number reflects the graph.

Assess

Name the Math Have students write a function for which the graph involves both a horizontal and a vertical translation of a trigonometric function. Then have them describe the translations.

Preparing for Assessment

Dual Coding

Exercise	Content Objective	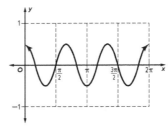 Mathematical Practices
65	F.BF.3	2
66	F.BF.3	2
67	F.BF.3	2
68	F.BF.3	2
69	F.BF.3	2
70	F.BF.3	1

Diagnose Student Errors

Survey student responses for each item. Class trends may indicate common errors and misconceptions.

65.

A	CORRECT
B	Did not consider the amplitude change
C	Misinterpreted the amplitude change
D	Used a vertical stretch instead of a shrink
E	Did not consider the period change and used a vertical stretch instead of a compression

66.

A	Confused amplitude and period
B	Translated in the wrong direction and confused period with translation
C	Confused amplitude with translation; multiplied the period by 2 instead of dividing
D	CORRECT
E	Used a reflection and confused amplitude and period

67.

A	Confused period with amplitude
B	Divided the amplitude by 2
C	CORRECT
D	Included a vertical shift/did not change period
E	Multiplied the period by 2

Go Online!

Self-Check Quiz

Students can use *Self-Check Quizzes* to check their understanding of this lesson.

65. The graph of $y = \cos \theta$ was transformed to produce the graph shown below. ⓂP 2 F.BF.3

What equation best represents the graph? **A**

- ○ **A** $y = \frac{1}{2} \cos 3x$
- ○ **B** $y = \cos 3x$
- ○ **C** $y = -\cos 3x$
- ○ **D** $y = 2 \cos 3x$
- ○ **E** $y = 2 \cos x$

66. Which of the following explains how the graph of $y = \sin \theta$ can be used to produce the graph of $y = \sin 2(\theta - \pi)$? ⓂP 2 F.BF.3 **D**

- ○ **A** Shift right π units and double amplitude.
- ○ **B** Translate left π units and shift up 2 units.
- ○ **C** Move up 2 units and double period.
- ○ **D** Shift right π units and divide period by 2.
- ○ **E** Reflect across x-axis and double amplitude.

67. The period of the graph of $y = \cos \theta$ is reduced by 50%, and the amplitude is doubled. Which function represents this change? ⓂP 2 F.BF.3 **C**

- ○ **A** $y = 0.5 \cos \frac{\theta}{2}$
- ○ **B** $y = 0.5 \cos 2\theta$
- ○ **C** $y = 2 \cos 2\theta$
- ○ **D** $y = 2 \cos \theta + 0.5$
- ○ **E** $y = 2 \cos \frac{\theta}{2}$

68 MULTI-STEP The graph of $y = \cos \theta$ is to be transformed to produce the graph of $y = 4 \cos \theta + 2$.

a. Which of the following transformation(s) of $y = \cos \theta$ need to occur to produce this new function? ⓂP 2 F.BF.3 **D**

   I. The amplitude will increase to 4.

   II. The graph will be stretched horizontally.

   III. The graph will shift 2 units up.

- ○ **A** I only
- ○ **B** II only
- ○ **C** I and II only
- ○ **D** I and III only
- ○ **E** II and III only

b. Which of the transformations would change the period?

   | II |

c. Give an example of a function which uses all three transformations on the sine function.

   Sample answer: $y = 4 \sin(2x) + 2$

69. Look at the list of functions. Write the letters of all the functions that match each description. ⓂP 2 F.BF.3

- **A** $y = 0.5 \cos(0.5\theta) + 2$
- **B** $y = 2 \cos\theta - 2$
- **C** $y = 0.5 \cos(0.5\theta + 0.5\pi) + 2$
- **D** $y = 2 \cos 2\theta - 2$
- **E** $y = 0.5 \cos(0.5\theta - \pi)$

a. amplitude doubled | B, D |

b. a period of 2π | B |

c. shifted π to the left | C |

d. shifted two units up | A, C |

e. a midline on the x-axis | E |

70. The graph of $y = \sin x$ is transformed so that its amplitude becomes 6, its period 3π and its maximum value 4. Find an expression for this new function. ⓂP 1 F.BF.3

Sample answer: $y = 6 \sin\left(\frac{2}{3}x\right) - 2$

ⓂP Standards for Mathematical Practice

Emphasis On	Exercises
1 Make sense of problems and persevere in solving them.	13, 40–42
2 Reason abstractly and quantitatively.	1–12, 14–33, 65–69
3 Construct viable arguments and critique the reasoning of others.	47–49, 60–64
4 Model with mathematics.	34–36
7 Look for and make use of structure.	37–39, 43–46, 50–59

CHAPTER 9
Study Guide and Review

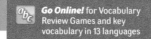 **Go Online!** for Vocabulary Review Games and key vocabulary in 13 languages

Study Guide

Key Concepts

Right Triangle Trigonometry (Lesson 9-1)

- $\sin\theta = \dfrac{opp}{hyp}$, $\cos\theta = \dfrac{adj}{hyp}$, $\tan\theta = \dfrac{opp}{adj}$,

  $\csc\theta = \dfrac{hyp}{opp}$, $\sec\theta = \dfrac{hyp}{adj}$, $\cot\theta = \dfrac{adj}{opp}$

Angle Measures and Trigonometric Functions of General Angles (Lessons 9-2 and 9-3)

- The measure of an angle is determined by the amount of rotation from the initial side to the terminal side.
- You can find the exact values of the six trigonometric functions of θ, given the coordinates of a point $P(x, y)$ on the terminal side of the angle.

Circular and Periodic Functions (Lesson 9-4)

- If the terminal side of an angle θ in standard position intersects the unit circle at $P(x, y)$, then $\cos\theta = x$ and $\sin\theta = y$.

Graphing Trigonometric Functions (Lesson 9-5)

- For trigonometric functions of the form $y = a\sin b\theta$ and $y = a\cos b\theta$, the amplitude is $|a|$, and the period is $\dfrac{360°}{|b|}$ or $\dfrac{2\pi}{b}$.
- The period of $y = a\tan b\theta$ is $\dfrac{180°}{|b|}$ or $\dfrac{\pi}{|b|}$.

Translations of Trigonometric Graphs (Lesson 9-6)

- The same techniques used to transform the graphs of other functions can be applied to the graphs of trigonometric functions
- A horizontal translation of a trigonometric function is called a phase shift.
- The midline can be used to help draw vertical translations of trigonometric functions.

 **Study Organizer**

Use your Foldable to review the chapter. Working with a partner can be helpful. Ask for clarification of concepts as needed.

Key Vocabulary

amplitude (p. 627)	phase shift (p. 635)
angle of depression (p. 598)	quadrantal angle (p. 613)
angle of elevation (p. 598)	radian (p. 606)
central angle (p. 607)	reference angle (p. 613)
circular function (p. 620)	secant (p. 594)
cosecant (p. 594)	sine (p. 594)
cosine (p. 594)	standard position (p. 604)
cotangent (p. 594)	tangent (p. 594)
coterminal angles (p. 605)	terminal side (p. 604)
cycle (p. 621)	trigonometric function (p. 594)
frequency (p. 628)	trigonometric ratio (p. 594)
initial side (p. 604)	trigonometry (p. 794)
midline (p. 636)	unit circle (p. 620)
period (p. 621)	vertical shift (p. 636)
periodic function (p. 621)	

Vocabulary Check

State whether each sentence is *true* or *false*. If *false*, replace the underlined term to make a true sentence.

1. An angle on the coordinate plane is in <u>standard position</u> if the vertex is at the origin and one ray is on the positive *x*-axis. **true**

2. <u>Coterminal angles</u> are angles in standard position that have the same terminal side. **true**

3. A horizontal translation of a periodic function is called a <u>phase shift</u>. **true**

4. The <u>cycle</u> of the graph of a sine or cosine function equals half the difference between the maximum and minimum values of the function. **false; amplitude**

Concept Check

5. When given one trigonometric ratio, explain how to find the other five trigonometric ratios. **See margin.**

6. Explain how points on the unit circle are related to the values of sine and cosine. **See margin.**

Answering the Essential Question

Before answering the Essential Question, have students review their answers to the *Building on the Essential Question* exercises found throughout the chapter.

- What types of real-world problems can be modeled and solved using trigonometry? (p. 590)

- How is trigonometry used to find unknown values? (p. 596)

- How are transformations of trigonometric functions similar to transformations of other functions you have studied? (p. 641)

 Study Organizer

A completed Foldable for this chapter should include the Key Concepts related to trigonometric functions.

Key Vocabulary ELL

The page reference after each word denotes where that term was first introduced. If students have difficulty answering questions 1–4, remind them that they can use these page references to refresh their memories about the vocabulary terms.

Have students work together to determine whether each sentence in the Vocabulary Check is true or false. Have students take turns saying each sentence aloud while the other student listens carefully.

You can use the detailed reports in ALEKS to automatically monitor students' progress and pinpoint remediation needs prior to the chapter test.

Additional Answers

5. Use the given trigonometric ratio to label two sides of a right triangle, and then use the Pythagorean Theorem to find the third side. Apply definitions to determine the other ratios.

6. If the terminal side of an angle θ in standard position intersects the unit circle at $P(x, y)$, then $\cos\theta = x$ and $\sin\theta = y$.

Go Online!

Vocabulary Review

Students can use the *Vocabulary Review Games* to check their understanding of the vocabulary terms in this chapter. Students should refer to the *Student-Built Glossary* they have created as they went through the chapter to review important terms. You can also give a *Vocabulary Test* over the content of this chapter.

Lesson-by-Lesson Review

Intervention If the given examples are not sufficient to review the topics covered by the questions, remind students that the lesson references tell them where to review that topic in their textbook.

Two-Day Option Have students complete the Lesson-by-Lesson Review. Then you can use McGraw-Hill eAssessment to customize another review worksheet that practices all the objectives of this chapter or only the objectives on which your students need more help.

Additional Answers

26. $\sin \theta = \frac{3}{5}, \cos \theta = -\frac{4}{5},$
$\tan \theta = -\frac{3}{4}, \csc \theta = \frac{5}{3},$
$\sec \theta = -\frac{5}{4}, \cot \theta = -\frac{4}{3},$

27. $\sin \theta = \frac{12}{13}, \cos \theta = \frac{5}{13},$
$\tan \theta = \frac{12}{5}, \csc \theta = \frac{13}{12},$
$\sec \theta = \frac{13}{5}, \cot \theta = \frac{5}{12},$

28. $\sin \theta = -\frac{3}{5}, \cos \theta = \frac{4}{5},$
$\tan \theta = -\frac{3}{4}, \csc \theta = -\frac{5}{3},$
$\sec \theta = \frac{5}{4}, \cot \theta = -\frac{4}{3},$

36. amplitude: 4, period: 180°

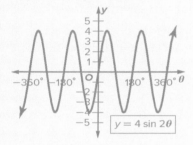
$y = 4 \sin 2\theta$

37. amplitude: 1, period: 720°

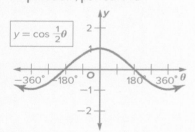
$y = \cos \frac{1}{2}\theta$

Go Online!

Anticipation Guide

Students should complete the *Chapter 9 Anticipation Guide*, and discuss how their responses have changed now that they have completed Chapter 9.

Lesson-by-Lesson Review

9-1 Trigonometric Functions in Right Triangles

Solve △ABC by using the given measurements. Round measures of sides to the nearest tenth and measures of angles to the nearest degree.

7. $c = 12, b = 5$

8. $a = 10, B = 55°$

9. $B = 75°, b = 15$ $A = 15°; a = 4.0; c = 15.5$

10. $B = 45°, c = 16$ $a = 11.3; b = 11.3; A = 45°$

11. $A = 35°, c = 22$ $B = 55°; a = 12.6; b = 18.0$

12. $\sin A = \frac{2}{3}, a = 6$ $A = 42°; B = 48°; b = 6.7; c = 9.0$

13. **TRUCK** The back of a moving truck is 3 feet off the ground. What length does a ramp off the back of the truck need to be in order for the angle of elevation of the ramp to be 20°? **about 8.8 ft**

Example 1

Solve △ABC by using the given measurements. Round measures of sides to the nearest tenth and measures of angles to the nearest degree.

Find b. $a^2 + b^2 = c^2$
$9^2 + b^2 = 16^2$
$b = \sqrt{16^2 - 9^2}$
$b \approx 13.2$

Find A. $\sin A = \frac{9}{16}$

Use a calculator.
To the nearest degree, A = 34°.

Find B. $34° + B \approx 90°$
$B \approx 56°$

Therefore, $b \approx 13.2$, $A \approx 34°$, and $B \approx 56°$.

7. $a = 10.9; A = 65°; B = 25°$
8. $A = 35°; c = 17.4; b = 14.3$

9-2 Angles and Angle Measure   F.TF.1

Rewrite each degree measure in radians and each radian measure in degrees.

14. $215°$ $\frac{43\pi}{36}$

15. $\frac{5\pi}{2}$ $450°$

16. -3π $-540°$

17. $-315°$ $-\frac{7\pi}{4}$

Find one angle with positive measure and one angle with negative measure coterminal with each angle.

18. $265°$ $625°, -95°$

19. $-65°$ $295°, -425°$

20. $\frac{7\pi}{2}$ $\frac{11\pi}{2}, -\frac{\pi}{2}$

21. **BICYCLE** A bicycle tire makes 8 revolutions in one second. The tire has a radius of 15 inches. Find the angle θ in radians through which the tire rotates in one second. $\frac{16}{\pi}$

Example 2

Rewrite 160° in radians.

$160° = 160°\left(\frac{\pi \text{ radians}}{180°}\right)$
$= \frac{160\pi}{180}$ radians or $\frac{8\pi}{9}$

Example 3

Find one angle with positive measure and one angle with negative measure coterminal with 150°.

positive angle:
$150° + 360° = 510°$ Add 360°.

negative angle:
$150° - 360° = -210°$ Subtract 360°.

38. amplitude: not defined, period: 360°

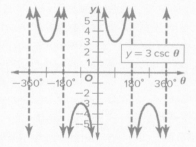
$y = 3 \csc \theta$

39. amplitude: not defined, period: 360°

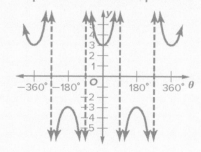

40. amplitude: not defined, period: 90°

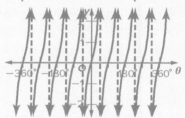

41. amplitude: not defined, period: 720°

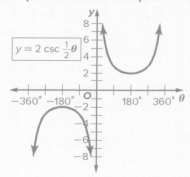
$y = 2 \csc \frac{1}{2}\theta$

Additional Answers

43. vertical shift: up 1, amplitude: 3, period: 180°, phase shift: 90° right

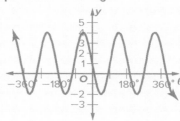

44. vertical shift: down 3, amplitude: undefined, period: 90°, phase shift: 30° right

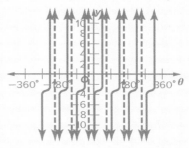

45. vertical shift: up 2, amplitude: not defined, period: $\frac{2\pi}{3}$, phase shift: $\frac{\pi}{2}$ right

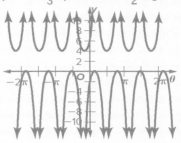

46. vertical shift: down 1, amplitude: $\frac{1}{2}$, period: 4π, phase shift: $\frac{\pi}{4}$ left

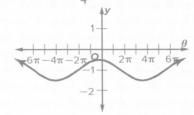

9-3 Trigonometric Functions of General Angles

Find the exact value of each trigonometric function.

22. $\cos 135°$ $-\frac{\sqrt{2}}{2}$ **23.** $\tan 150°$ $-\frac{\sqrt{3}}{3}$

24. $\sin 2\pi$ 0 **25.** $\cos \frac{3\pi}{2}$ 0

The terminal side of θ in standard position contains each point. Find the exact values of the six trigonometric functions of θ.   **26–28. See margin.**

26. $P(-4, 3)$

27. $P(5, 12)$

28. $P(16, -12)$

29. BALL A ball is thrown off the edge of a building at an angle of 70° and with an initial velocity of 5 meters per second. The equation that represents the horizontal distance of the ball x is $x = v_0 (\cos \theta)t$, where v_0 is the initial velocity, θ is the angle at which it is thrown, and t is the time in seconds. About how far will the ball travel in 10 seconds?   **about 17.1 meters**

Example 4

Find the exact value of sin 120°.

Because the terminal side of 120° lies in Quadrant II, the reference angle θ' is 180° − 120° or 60°. The sine function is positive in Quadrant II, so $\sin 120° = \sin 60°$ or $\frac{\sqrt{3}}{2}$.

Example 5

The terminal side of θ in standard position contains the point (6, 5). Find the exact values of the six trigonometric functions of θ.

$\sin \theta = \frac{y}{r}$ or $\frac{5\sqrt{61}}{61}$ $\csc \theta = \frac{r}{y}$ or $\frac{\sqrt{61}}{5}$

$\cos \theta = \frac{x}{r}$ or $\frac{6\sqrt{61}}{61}$ $\sec \theta = \frac{r}{x}$ or $\frac{\sqrt{61}}{6}$

$\tan \theta = \frac{y}{x}$ or $\frac{5}{6}$ $\cot \theta = \frac{x}{y}$ or $\frac{6}{5}$

9-4 Circular and Periodic Functions

             F.TF.1, F.TF.2

Find the exact value of each function.

30. $\cos(-210°)$ $-\frac{\sqrt{3}}{2}$ **31.** $(\cos 45°)(\cos 210°)$ $-\frac{\sqrt{6}}{4}$

32. $\sin\left(-\frac{7\pi}{4}\right)$ $\frac{\sqrt{2}}{2}$ **33.** $\left(\cos \frac{\pi}{2}\right)\left(\sin \frac{\pi}{2}\right)$ 0

34. Determine the period of the function.

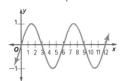

             6

35. A wheel with a diameter of 18 inches completes 4 revolutions in 1 minute. What is the period of the function that describes the height of one spot on the outside edge of the wheel as a function of time?   **15 seconds**

Example 6

Find the exact value of sin 510°.

$\sin 510° = \sin(360° + 150°)$

$= \sin 150°$

$= \frac{1}{2}$

Example 7

Determine the period of the function below.

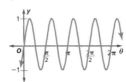

The pattern repeats itself at $\frac{\pi}{2}$, π, and so on. So, the period is $\frac{\pi}{2}$.

---

47. vertical shift: up 2, amplitude: $\frac{1}{3}$, period: 1080°, phase shift: 90° right

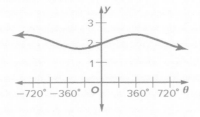

Before the Test

Have students complete the Study Notebook Tie it Together activity to review topics and skills presented in the chapter.

Additional Answers (Practice Test)

16.

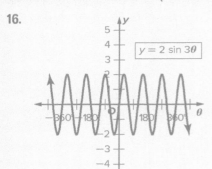

17.

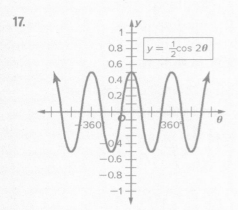

22.

9-5 Graphing Trigonometric Functions

A.CED2, F.IF4, F.IF.7e, F.TF.5

Find the amplitude, if it exists, and period of each function. Then graph the function. **36–41. See margin on page 644.**

36. $y = 4 \sin 2\theta$

37. $y = \cos \frac{1}{2}\theta$

38. $y = 3 \csc \theta$

39. $y = 3 \sec \theta$

40. $y = \tan 2\theta$

41. $y = 2 \csc \frac{1}{2}\theta$

42. When Lauren jumps on a trampoline it vibrates with a frequency of 10 hertz. Let the amplitude equal 5 feet. Write a sine equation to represent the vibration of the trampoline y as a function of time t. $y = 5 \sin 20\pi t$

Example 8

Find the amplitude and period of $y = 2 \cos 4\theta$. Then graph the function.

amplitude: $|a| = |2|$ or 2. The graph is stretched vertically so that the maximum value is 2 and the minimum value is -2.

period:
$\frac{360°}{|b|} = \frac{360°}{|4|}$ or 90°

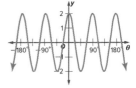

9-6 Translations of Trigonometric Graphs

F.IF.7e, F.BF.3, F.TF.5

State the vertical shift, amplitude, period, and phase shift of each function. Then graph the function. **43–47. See margin on page 645.**

43. $y = 3 \sin [2(\theta - 90°)] + 1$

44. $y = \frac{1}{2} \tan [2(\theta - 30°)] - 3$

45. $y = 2 \sec \left[3\left(\theta - \frac{\pi}{2}\right)\right] + 2$

46. $y = \frac{1}{2} \cos \left[\frac{1}{4}\left(\theta + \frac{\pi}{4}\right)\right] - 1$

47. $y = \frac{1}{3} \sin \left[\frac{1}{3}(\theta - 90°)\right] + 2$

48. The graph below approximates the height y of a rope that two people are twirling as a function of time t in seconds. Write an equation for the function.

$y = 4 \sin 360t + 4$

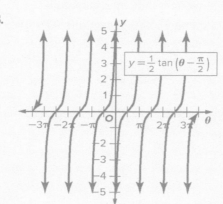

Example 9

State the vertical shift, amplitude, period, and phase shift of $y = 2 \sin \left[3\left(\theta + \frac{\pi}{2}\right)\right] + 4$. Then graph the function.

Identify the values of k, a, b, and h.

$k = 4$, so the vertical shift is 4.

$a = 2$, so the amplitude is 2.

$b = 3$, so the period is $\frac{2\pi}{|3|}$ or $\frac{2\pi}{3}$.

$h = -\frac{\pi}{2}$, so the phase shift is $\frac{\pi}{2}$ to the left.

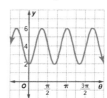

23.

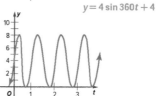

24.

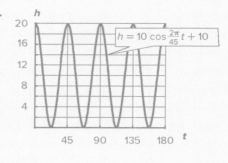

CHAPTER 9
Practice Test

 Go Online! for another Chapter Test

Solve $\triangle ABC$ by using the given measurements. Round measures of sides to the nearest tenth and measures of angles to the nearest degree.

1. $A = 36°$, $c = 9$ $B = 54°$, $a = 5.3$, $b = 7.3$
2. $a = 12$, $A = 58°$ $B = 32°$, $c = 14.2$, $b = 7.5$
3. $B = 85°$, $b = 8$ $A = 5°$, $c = 8.0$, $a = 0.7$
4. $a = 9$, $c = 12$ $b = 7.9$, $B = 41°$, $A = 49°$

Rewrite each degree measure in radians and each radian measure in degrees.

5. $325°$ $\frac{65\pi}{36}$
6. $-175°$ $-\frac{35\pi}{36}$
7. $\frac{9\pi}{4}$ $405°$
8. $-\frac{5\pi}{6}$ $-150°$

Find the exact value of each function. Write angle measures in degrees.

9. $\cos(-90°)$ 0
10. $\sin 585°$ $-\frac{\sqrt{2}}{2}$
11. $\cot \frac{4\pi}{3}$ $\frac{\sqrt{3}}{3}$
12. $\sec\left(-\frac{9\pi}{4}\right)$ $\sqrt{2}$

13. The terminal side of angle θ in standard position intersects the unit circle at point $P\left(\frac{1}{2}, \frac{\sqrt{3}}{2}\right)$. Find $\cos \theta$ and $\sin \theta$. $\cos \theta = \frac{1}{2}$, $\sin \theta = \frac{\sqrt{3}}{2}$

14. **MULTIPLE CHOICE** What angle has a tangent and sine that are both negative? **D**
   A $65°$
   B $120°$
   C $265°$
   D $310°$

15. NAVIGATION Airplanes and ships measure distance in nautical miles. The formula 1 nautical mile = $6077 - 31 \cos 2\theta$ feet, where θ is the latitude in degrees, can be used to find the approximate length of a nautical mile at a certain latitude. Find the length of a nautical mile when the latitude is $120°$. 6092.5 ft

Find the amplitude and period of each function. Then graph the function. 16–17. See margin for graphs.

16. $y = 2 \sin 3\theta$ $2, 120°$
17. $y = \frac{1}{2} \cos 2\theta$ $\frac{1}{2}, 180°$

18. **MULTIPLE CHOICE** What is the period of the function $y = 3 \cot \theta$? **B**
   A $120°$
   B $180°$
   C $360°$
   D $1080°$

Write an equation for each translation.

19. $y = \sin x$, 6 units to the left and 4 units down $y = \sin(x + 6) - 4$
20. $y = \tan x$, $\frac{\pi}{2}$ units to the right and 1 unit up

21. Write an equation for the graph using the given trigonometric function. 20. $y = \tan\left(x - \frac{\pi}{2}\right) + 1$

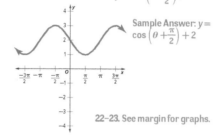

Sample Answer: $y = \cos\left(\theta + \frac{\pi}{2}\right) + 2$

22–23. See margin for graphs.

State the amplitude, period, and phase shift for each function. Then graph the function.

22. $y = \cos(\theta + 180)$ $1, 360°, -180°$
23. $y = \frac{1}{2} \tan\left(\theta - \frac{\pi}{2}\right)$ does not exist, $\pi, \frac{\pi}{2}$

24. WHEELS A water wheel has a diameter of 20 feet. It makes one complete revolution in 45 seconds. Let the height at the top of the wheel represent the height at time 0. Write an equation for the height of point h in the diagram below as a function of time t. Then graph the function. $h = 10 \cos \frac{2\pi}{45} t + 10$; See margin for graph.

RtI Response to Intervention

Use the Intervention Planner to help you determine your Response to Intervention.

Intervention Planner

TIER 1 On Level OL

IF students miss 25% of the exercises or less,

THEN choose a resource:
- SE Lessons 9-1 through 9-6

Go Online!
- Skills Practice
- Chapter Project
- ✓ Self-Check Quizzes

TIER 2 Strategic Intervention AL
Approaching grade level

IF students miss 50% of the exercises,

THEN **Go Online!**
- Study Guide and Intervention
- + Extra Examples
- Personal Tutors
- Homework Help

TIER 3 Intensive Intervention
2 or more grades below level

IF students miss 75% of the exercises,

THEN choose a resource:
Use *Math Triumphs, Alg. 2*

Go Online!
- + Extra Examples
- Personal Tutors
- Homework Help
- Review Vocabulary

Go Online! ✓

Chapter Tests

You can use premade leveled *Chapter Tests* to differentiate assessment for your students. Students can also take self-checking *Chapter Tests* to plan and prepare for chapter assessments.

MC = multiple-choice questions
FR = free-response questions

Form	Type	Level
1	MC	AL
2A	MC	OL
2B	FR	OL
2C	FR	OL
3	FR	BL
Vocabulary Test		
Extended-Response Test		

Launch

Objective Apply concepts and skills from this chapter in a real-world setting.

Teach

Ask:

- How can you organize the information presented in the problem? Sample answer: Draw a quick sketch.

- Which trigonometric function could you use to find the height of the telephone pole in Part A? Sample answer: tangent

- Where is the initial side of an angle in standard position? along the positive *x*-axis

- What is a coterminal angle? an angle that shares the same initial and terminal side with another angle

- What is the period of a function? the time it takes to complete one full rotation

The Performance Task focuses on the following content standards and standards for mathematical practice.

Dual Coding

Parts	Content Standards	^{MP} Mathematical Practices
A	F.TF.8	4, 6
B	F.TF.1	1, 6
C	F.TF.2	1, 4, 6
D	F.TF.4	2, 3

Levels of Complexity Chart

Parts	Level 1	Level 2	Level 3
A		●	
B	●		
C			●
D			●

Go Online!

eBook

Interactive student guide
Refer to *Interactive Student Guide* for an additional Performance Task.

ALGEBRA 2
INTERACTIVE STUDENT GUIDE

Performance Task

Provide a clear solution to each part of the task. Be sure to show all of your work, include all relevant drawings, and justify your answers.

SAFETY INSPECTOR An inspector is often hired to take detailed measurements for government safety commissions, architects, and construction companies. They frequently rely on trigonometry to make their calculations.

Part A
A telephone company needs to perform routine maintenance on some telephone poles. An inspector is taking measurements so they can order materials. The telephone pole is perpendicular to the ground, and a wire runs from the top of the pole diagonally to the ground. The angle formed by the wire and the pole is 20°. The distance from the base of the pole to the point where the wire meets the ground is 12 feet.

1. Find the length of the wire. Round your answer to the nearest tenth, if necessary. 35.1 ft

2. Find the height of the pole. Round your answer to the nearest tenth, if necessary. 33.0 ft

Part B
A safety inspector is evaluating plans for a spiral staircase to make sure it complies with local safety regulations. The inspector needs to draw several angles and make some calculations for the report.

3. Draw a 280° angle in standard position. 3–4. See Ch. 9 Answer Appendix for graphs.

4. Draw a 630° angle in standard position.

5. Find a positive and negative coterminal angle measure for a 110° angle. 470° and −250°

6. Convert 105° to radians. $\frac{7\pi}{12}$

7. Convert $\frac{7\pi}{4}$ to degrees. 315°

Part C
An inspector is reviewing plans for a circular building that will have a domed roof. There are eight evenly-spaced support beams for the dome that form a design similar to the spokes of a wheel. The inspector makes a sketch of the beams where the center is at (0, 0) and four support beams have endpoints at (−36, 0), (36, 0), (0, −36), and (0, 36).

8. Use the unit circle to find the exact locations of the end points of the other four support beams. $(18\sqrt{2}, 18\sqrt{2})$, $(-18\sqrt{2}, 18\sqrt{2})$, $(-18\sqrt{2}, -18\sqrt{2})$, and $(18\sqrt{2}, -18\sqrt{2})$

Part D
Regulations require that the tornado alarm on a large university's campus be audible at every point on the campus within 8 seconds of the alarm being activated. The alarm makes 3 rotations per minute and has a period of 20 seconds. The length of the radius of the sound beam is exactly as long as the farthest place on campus from the alarm. The distance of the sound beam from the edge of campus varies periodically as a function of time.

9. Explain whether the alarm is in compliance with safety regulations. No; the sound will reach its farthest point at exactly halfway through its period, which is at 10 seconds.

Test-Taking Strategy

Example

Read the problem. Identify what you need to know. Then use the information in the problem to solve.

When Molly stands at a distance of 18 feet from the base of a tree, she forms an angle of 57° with the top of the tree. What is the height of the tree to the nearest tenth?

A 27.7 ft   C 29.2 ft

B 28.5 ft   D 30.1 ft

Step 1 Does the problem involve any concepts that may require a calculator?
Yes, it involves a trigonometric function.

Step 2 What steps can you do by hand? How will you use a calculator?
Make a sketch by hand and then set up a trigonometric ratio. Use the calculator to evaluate the trigonometric function.

Step 3 What is the correct answer?
Choice A is correct.

> **Test-Taking Tip**
> Using a Scientific Calculator
> Problems involving *n*th roots, logarithms, and exponential or trigonometric functions may require a scientific calculator for one or more steps.

Apply the Strategy

Read the problem. Identify what you need to know. Then use the information in the problem to solve.

What is the angle of the bike ramp below? D

A   26.3°

B   28.5°

C   30.4°

D   33.6°

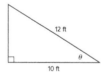

Answer the questions below.

a.   Does the problem involve any concepts that may require a calculator? Yes, it involves a trigonometric function.

b.   What steps can you do by hand? How will you use a calculator? Write the trigonometric equation by hand, and then use the calculator to evaluate it.

c.   What is the correct answer? D

Step 1 Make sure you're familiar with how a scientific calculator works and the types of questions for which one will be necessary.

Step 2 Make a plan. Decide what steps you can do by hand and what calculations require a calculator.

Step 3 If time permits, check your answer.

Need Another Example?

An airplane takes off and climbs at a constant rate. After traveling 800 yards horizontally, the plane has climbed 285 yards vertically. What is the plane's angle of elevation during the takeoff and initial climb? C

A   15.6°

B   19.6°

C   18.4°

D   22.3°

Go Online!

The most up-to-date resources available for your program can be found at connectED.mcgraw-hill.com.

Diagnose Student Errors

Survey student responses for each item. Class trends may indicate common errors and misconceptions.

1. A	CORRECT
B	Used cosecant ratio instead of cosine ratio
C	Used hypotenuse instead of adjacent side
D	Used tangent ratio to write the sine ratio
E	Used cotangent ratio to write cosine ratio

2. Student may use a ratio other than secant. Student may not realize secant is negative in Quadrant III.

3. A	Assumed the radicand of the hypotenuse measure for a 45-45-90 triangle was the measure of one leg
B	Assumed $2x^2 = (2x)^2$
C	Found the length of one leg
D	CORRECT
E	Found the square of one leg

4. A	Used 180° instead of 360° and used radius instead of diameter
B	Used diameter instead of radius
C	CORRECT
D	Found the circumference
E	Used the area formula

5. A	Assumed $\frac{2\pi}{3}$ has the least angle measure
B	CORRECT
C	Assumed π corresponds to 360°
D	Thought that $\frac{3\pi}{4} < \frac{2\pi}{3}$
E	Chose answer where choices are ordered from greatest to least

6. A	Thought given point was y-intercept
B	CORRECT
C	Made an algebra error
D	Made a sign error
E	Used x-coordinate of given point as y-intercept

CHAPTER 9
Preparing for Assessment
Cumulative Review

Read each question. Then fill in the correct answer on the answer document provided by your teacher or on a sheet of paper.

1. A line segment extends from the origin of the coordinate plane $O(0, 0)$ to $B(30, 40)$, as shown in the figure below. Which equation represents the measure of the acute angle θ formed by $\overline{OB}$ and the x-axis? **A**

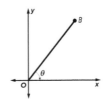

- ○ **A** $\sin \theta = \frac{4}{5}$
- ○ **B** $\cos \theta = \frac{5}{4}$
- ○ **C** $\tan \theta = \frac{4}{5}$
- ○ **D** $\sin \theta = \frac{4}{3}$
- ○ **E** $\cos \theta = \frac{3}{4}$

2. Determine the exact value of csc 210°.

−2

3. The two acute angles of a right triangle have the same measure. The hypotenuse measures 32 inches. What is the sum of the measures of the two legs of the triangle? **D**

- ○ **A** $\sqrt{2}$ in.
- ○ **B** 32 in.
- ○ **C** $16\sqrt{2}$ in.
- ○ **D** $32\sqrt{2}$ in.
- ○ **E** 2048 in.

4. The radius of a children's bicycle tire is 12 inches. What is the arc length of the circle made when the tire rotates 270°? **C**

- ○ **A** 4.5π in.
- ○ **B** 9π in.
- ○ **C** 18π in.
- ○ **D** 24π in.
- ○ **E** 108π in.

Test-Taking Tip

Question 3 Pay attention to the measures given in the question. If diameter is used in a formula, you need to multiply the radius by 2.

5. Which of the following angle measures are ordered from least to greatest? **B**

- ○ **A** $\frac{3\pi}{4}$, 180°, $\frac{2\pi}{3}$, 225°
- ○ **B** $\frac{2\pi}{3}$, $\frac{3\pi}{4}$, 180°, 225°
- ○ **C** 180°, 225°, $\frac{2\pi}{3}$, $\frac{3\pi}{4}$
- ○ **D** $\frac{3\pi}{4}$, $\frac{2\pi}{3}$, 180°, 225°
- ○ **E** 225°, 180°, $\frac{3\pi}{4}$, $\frac{2\pi}{3}$

6. A line with slope of $\frac{3}{4}$ passes through $(-3, -4)$. Which is the equation of the line? **B**

- ○ **A** $y = \frac{3}{4}x - 4$
- ○ **B** $-3x + 4y = -7$
- ○ **C** $y = \frac{3}{4}x + 2$
- ○ **D** $3x - 4y = -7$
- ○ **E** $y = \frac{3}{4}x - 3$

7. Find the exact value of sin 75°.

$\frac{\sqrt{2}+\sqrt{6}}{4}$

8. A	Ignored the coefficient of b in the original binomial
B	Forgot to square the coefficient of b in the original binomial
C	Forgot to square and cube the coefficients of b in the original binomial
D	CORRECT
E	Switched the order

9. A	CORRECT
B	Chose solution to first equation only
C	Chose coefficients in second equation
D	Chose constants from equations
E	Chose solution to second equation only

10. A	CORRECT
B	Selected function that has an amplitude of 2 and a period of 2π
C	Selected function that has an amplitude of 2 and a period of 2π
D	Selected function that has an amplitude of 4 and a period of 2π
E	Selected function that has an amplitude of 1 and a period of π

11. A	Selected a value that result in a horizontal shrink
B	Selected values that result in a horizontal shrink
C	Selected only negative values for W
D	CORRECT
E	Determined that all values for W will result in a horizontal stretch

 **Go Online!** for Standardized Test Practice

8. What are the missing coefficients, from left to right, in this binomial expansion of $(a + 2b)^5$? **D**

$$a^5 + 10a^4b + \underline{\quad} a^3b^2 + \underline{\quad} a^2b^3 + 80ab^4 + 32b^5$$

- ○ **A** 10 and 10
- ○ **B** 10 and 20
- ○ **C** 20 and 10
- ○ **D** 40 and 80
- ○ **E** 80 and 40

9. What is the solution to the system of equations below? **A**

$$x - y = 5$$
$$3x + 5y = 7$$

- ○ **A** $(4, -1)$
- ○ **D** $(5, 7)$
- ○ **B** $(5, 0)$
- ○ **E** $(-1, 2)$
- ○ **C** $(3, 5)$

10. For which of the trigonometric functions is the period and amplitude the same as $y = 2\pi \sin x$? **A**

- ○ **A** $y = -2\pi \cos x$
- ○ **D** $y = 4 \sin x$
- ○ **B** $y = -2 \sin 2\pi x$
- ○ **E** $y = \cos 2x$
- ○ **C** $y = -2 \cos x$

11. For what value(s) of W will the function $y = R \cos (W \cdot x)$ result in a horizontal stretch of $y = \cos x$? **D**

- ○ **A** $W = \pi$
- ○ **D** $0 < W < 1$
- ○ **B** $W > 1$
- ○ **E** $W = $ any real number
- ○ **C** $W < 0$

12. Select all true characteristics of the trigonometric function graphed below. **A, B, C**

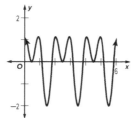

- ☐ **A** The domain of the function is all real numbers.
- ☐ **B** The range of the function is all real numbers between -2 and 1.
- ☐ **C** The minimum value of the function is -2.
- ☐ **D** The function cycles 6 times.
- ☐ **E** The period of the function is 6.

13. The exponential function $f(x) = a(2)^x$ passes through the point $(-1, 6)$. What is the value of a?

12

14. A certain medication has an effective biological half-life of 2 hours. This means that after every 2 hours, the amount of medication remaining in the body is reduced by half.

a. What fraction of an initial dose of the medication has been eliminated after 8 hours? Express your answer as a decimal.

0.9375

b. 🔘 What mathematical practice did you use to solve this problem? See students' work.

Need Extra Help?

If you missed Question...	1	2	3	4	5	6	7	8	9	10	11	12	13	14
Go to Lesson...	9-1	9-3	9-1	9-2	9-2	1-4	10-3	4-2	1-6	9-5	9-4	9-6	6-1	6-3

 LEARNSMART®

Use LearnSmart as part of your test-preparation plan to measure student topic retention. You can create a student assignment in LearnSmart for additional practice on these topics.

- Solve Quadratic Equations in One Variable
- Solve a System of Equations Consisting of a Linear and a Quadratic Equation in Two Variables
- Use Properties of Rational and Irrational Numbers
- Create and interpret Quadratic Models
- Build New Functions From Existing Functions
- Construct and Compare Linear, Quadratic, and Exponential Models and Solve Problems

---

Formative Assessment

You can use these pages to benchmark student progress.

📄 Standardized Test Practice

Test Item Formats

In the Cumulative Review, students will encounter different formats for assessment questions to prepare them for standardized tests.

Question Type	Exercises
Multiple-Choice	1, 3–6, 8–11
Multiple Correct Answers	12
Type Entry: Short Response	2, 7, 13
Type Entry: Extended Response	13

Answer Sheet Practice

Have students simulate taking a standardized test by recording their answers on a practice recording sheet.

Homework Option

Get Ready for Chapter 10 Assign students the exercises on p. 654 as homework to assess whether they possess the prerequisite skills needed for the next chapter.

12.	A	CORRECT
	B	CORRECT
	C	CORRECT
	D	Counted the intersection points where $y = 1$
	E	Selected the last number labeled on the x-axis

13. Student may think $2^{-1} = 2$ or $2^{-1} = 1$. Student may confuse exponentiation with multiplication. Student may think that a negative exponent gives a negative answer.

14a. Student may find the fraction remaining after 8 hours. Student may find the fraction eliminated after 4 hours or 6 hours.

Go Online!

 **eAssessment**

Customize and create multiple versions of chapter tests and answer keys that align to your standards. Tests can be delivered on paper or online.

Lesson 9-2 (Guided Practice)

1A.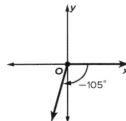

1B.

Lesson 9-2

52. Use a proportion.

$$\frac{\text{measure of the central angle}}{\text{measure of an entire circle}} = \frac{\text{the length of the arc}}{\text{the circumference}}$$

$\frac{\theta}{2\pi} = \frac{s}{2\pi r}$ Substitute.

$2\pi r\theta = 2\pi s$ Find the cross products.

$r\theta = s$ Divide each side by 2π.

53. One degree represents an angle measure that equals $\frac{1}{360}$ rotation around a circle. One radian represents the measure of an angle in standard position that intercepts an arc of length r. To change from degrees to radians, multiply the number of degrees by $\frac{\pi \text{ radians}}{180°}$. To change from radians to degrees, multiply the number of radians by $\frac{180°}{\pi \text{ radians}}$.

Lesson 9-3

11a.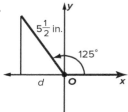

18. 15°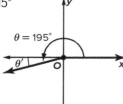

19. 75°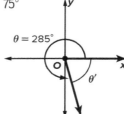

20. 70°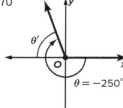

21. $\frac{\pi}{4}$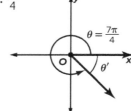

22. $\frac{\pi}{4}$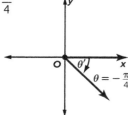

23. 40°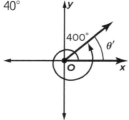

Mid-Chapter Quiz

Additional Answers

3. $\sin \theta = \frac{\sqrt{7}}{4}$, $\cos \theta = \frac{3}{4}$, $\tan \theta = \frac{\sqrt{7}}{3}$, $\csc \theta = \frac{4\sqrt{7}}{7}$, $\sec \theta = \frac{4}{3}$, $\cot \theta = \frac{3\sqrt{7}}{7}$

4.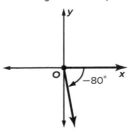

12. $\sin \theta = -1$, $\cos \theta = 0$, $\tan \theta =$ undefined, $\csc \theta = -1$, $\sec \theta =$ undefined, $\cot \theta = 0$

13. $\sin \theta = \frac{4}{5}$, $\cos \theta = \frac{3}{5}$, $\tan \theta = \frac{4}{3}$, $\csc \theta = \frac{5}{4}$, $\sec \theta = \frac{5}{3}$, $\cot \theta = \frac{3}{4}$

Lesson 9-4

29a.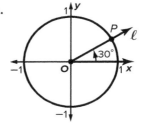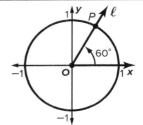

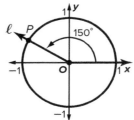 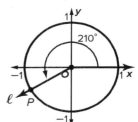

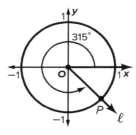

29b.

Angle	Slope
30	0.6
60	1.7
120	−1.7
150	−0.6
210	0.6
315	−1

29c. Sample answer: The slope corresponds to the tangent of the angle.

For $\theta = 120°$, the x-coordinate of P is $-\frac{1}{2}$ and the y-coordinate is $= \frac{\sqrt{3}}{2}$;

slope $= \dfrac{\text{change in } y}{\text{change in } x}$. Because change in $x = -\frac{1}{2}$ and change in $y = \frac{\sqrt{3}}{2}$,

slope $= \dfrac{\sqrt{3}}{2} \div \left(-\dfrac{1}{2}\right) = -\sqrt{3}$ or about -1.7.

Lesson 9-5

6. period: 180°

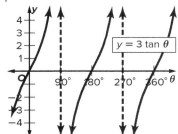

7. period: 360°

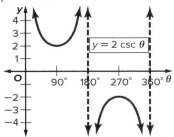

8. period: 90°

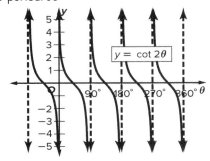

9. amplitude: 2; period: 360°

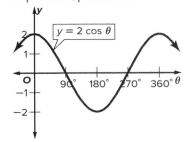

10. amplitude: 3; period: 360°

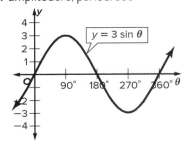

11. amplitude: 1; period: 180°

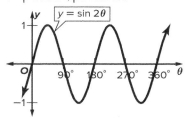

12. amplitude: 1; period: 120°

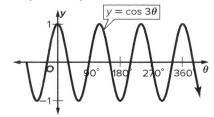

13. amplitude: 1; period: 720°

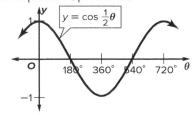

14. amplitude: 1; period: 90°

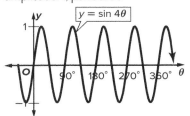

15. amplitude: $\frac{3}{4}$; period: 360°

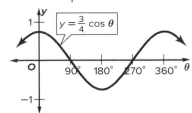

16. amplitude: $\frac{3}{2}$; period: 360°

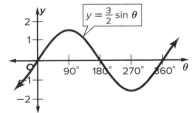

17. amplitude: $\frac{1}{2}$; period: 180°

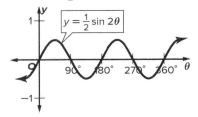

18. amplitude: 4; period: 180°

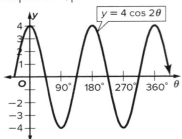

$y = 4 \cos 2\theta$

19. amplitude: 3; period: 180°

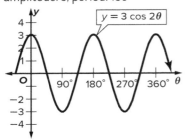

$y = 3 \cos 2\theta$

20. amplitude: 5; period: 540°

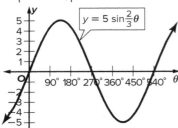

$y = 5 \sin \frac{2}{3}\theta$

21b.

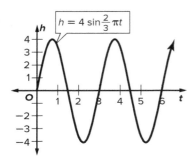

$h = 4 \sin \frac{2}{3}\pi t$

22b.

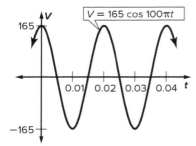

$V = 165 \cos 100\pi t$

23. period: 360°

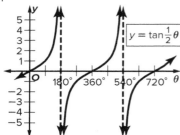

$y = \tan \frac{1}{2}\theta$

24. period: 360°

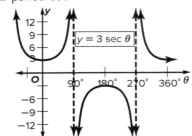

$y = 3 \sec \theta$

25. period: 180°

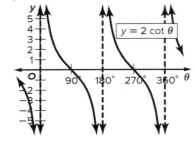

$y = 2 \cot \theta$

26. period: 720°

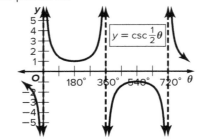

$y = \csc \frac{1}{2}\theta$

27. period: 180°

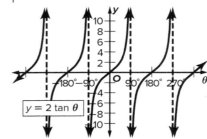

$y = 2 \tan \theta$

28. period: 1080°

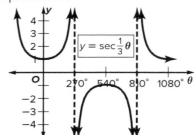

$y = \sec \frac{1}{3}\theta$

31a.

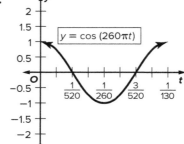

$y = \cos (260\pi t)$

32. amplitude: 3; period: 540°

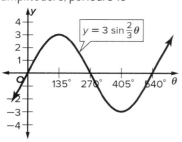
$y = 3 \sin \frac{2}{3}\theta$

33. amplitude: $\frac{1}{2}$; period: 480°

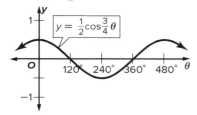
$y = \frac{1}{2}\cos\frac{3}{4}\theta$

34. amplitude: does not exist; period: 360°

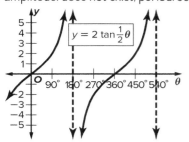
$y = 2 \tan \frac{1}{2}\theta$

35. amplitude: does not exist; period: 450°

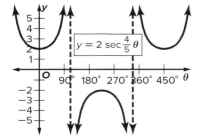
$y = 2 \sec \frac{4}{5}\theta$

36. amplitude: does not exist; period: 120°

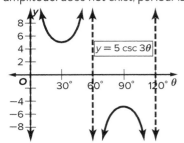
$y = 5 \csc 3\theta$

37. amplitude: does not exist; period: 30°

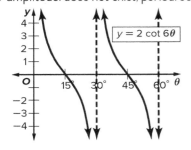
$y = 2 \cot 6\theta$

Extend 9-5

1. **Y1**: D = {all real numbers}, R = {$y \mid -1 \le y \le 1$}, **Y2**: D = {all real numbers}, R = {$y \mid 1 \le y \le 3$}, **Y3**: D = {all real numbers}, R = {$y \mid -4 \le y \le 2$}; **Y1** − **Y3**: D = {all real numbers}, R = {$y \mid -1 \le y \le 1$}

4. Sample answer: The graph of $y = \sin(2\theta) + 4$ goes through two cycles in 360°, and the graph of $y = \sin\theta + 4$ goes through one cycle in 360°.

5. Sample answer: The graph of $y = \cos\frac{1}{2}(\theta + 45°)$ has been translated 45° left from the graph of $y = \cos\left(\frac{1}{2}\theta\right)$.

6. Sample answer: The graph of $y = 2\sin\theta - 1$ has been translated down 1 unit from the graph of $y = 2\sin\theta$.

7. Sample answer: The graph of $y = \cos(\theta - 90°) - 3$ has been translated 90° right from the graph of $y = \cos\theta - 3$.

Lesson 9-6 (Guided Practice)

1.

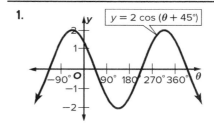
$y = 2\cos(\theta + 45°)$

Lesson 9-6

9. 2; 360°; $h = -45°$; $k = 1$

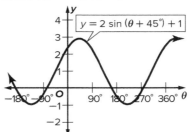
$y = 2\sin(\theta + 45°) + 1$

10. 1; $\frac{2\pi}{3}$; $h = \pi$; $k = -4$

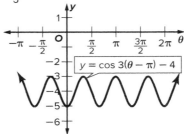
$y = \cos 3(\theta - \pi) - 4$

11. no amplitude; 90°; $h = -30°$; $k = 3$

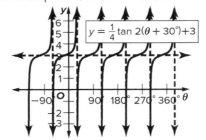
$y = \frac{1}{4}\tan 2(\theta + 30°) + 3$

12. $4; 4\pi; h = \frac{\pi}{2}; k = 5$

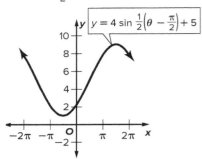

13. $P = 20 \sin 3\pi t + 110$

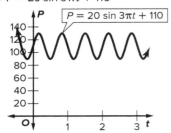

14. $1; 360°; h = -180°$

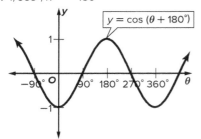

15. no amplitude; 180°; $h = 90°$

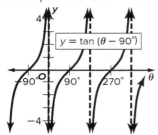

16. $1; 2\pi; h = -\pi$

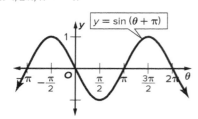

17. $2; 2\pi; h = -\frac{\pi}{2}$

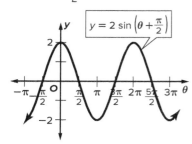

18. no amplitude; 360°; $h = -30°$

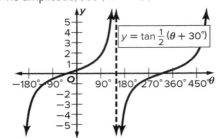

19. $3; 2\pi; h = \frac{\pi}{3}$

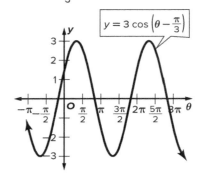

20. $1; 360°; k = 3; y = 3$

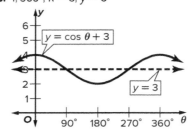

21. no amplitude; 180°; $k = -1; y = -1$

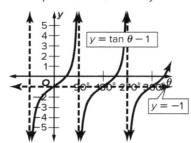

22. no amplitude; 180°; $k = \frac{1}{2}; y = \frac{1}{2}$

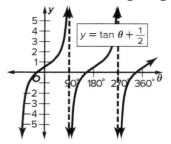

23. $2; 360°; k = -5; y = -5$

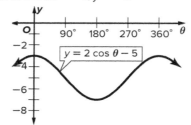

24. 2; $360°$; $k = -4$; $y = -4$

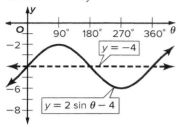

$y = 2 \sin \theta - 4$

25. $\frac{1}{3}$; $360°$; $k = 7$; $y = 7$

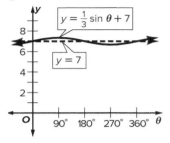

$y = \frac{1}{3} \sin \theta + 7$

$y = 7$

26. 4; $360°$; $h = 60°$; $k = -1$

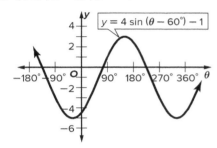

$y = 4 \sin (\theta - 60°) - 1$

27. 1; $720°$; $h = 90°$; $k = 2$

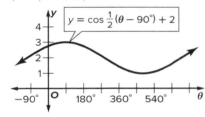

$y = \cos \frac{1}{2}(\theta - 90°) + 2$

28. no amplitude; $180°$; $h = -30°$; $k = -2$

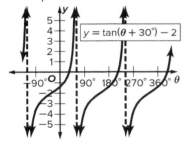

$y = \tan(\theta + 30°) - 2$

29. no amplitude; $\frac{\pi}{2}$; $h = -\frac{\pi}{4}$; $k = -5$

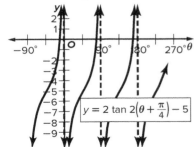

$y = 2 \tan 2\left(\theta + \frac{\pi}{4}\right) - 5$

30. $\frac{1}{2}$; 2π; $h = \frac{\pi}{2}$; $k = 4$

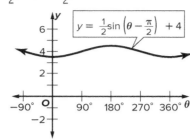

$y = \frac{1}{2}\sin\left(\theta - \frac{\pi}{2}\right) + 4$

31. 1; $120°$; $h = 45°$; $k = \frac{1}{2}$

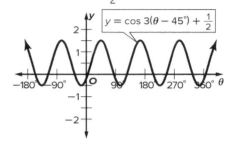

$y = \cos 3(\theta - 45°) + \frac{1}{2}$

32. 5; π; $h = \pi$; $k = 3$

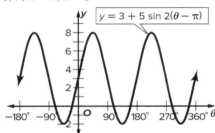

$y = 3 + 5 \sin 2(\theta - \pi)$

33. 3; 6π; $h = \frac{\pi}{2}$; $k = -2$

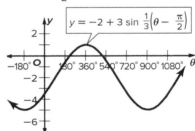

$y = -2 + 3 \sin \frac{1}{3}\left(\theta - \frac{\pi}{2}\right)$

41a. Sample answer: There are 8 groups of 4 horses, with each horse in a different location from other members of its group. Each group is represented by $f(t) = 2.5 \sin\left(\frac{\pi}{10}t - \frac{g\pi}{4}\right) + 2.5$, where g represents a specific group, and $1 \leq g \leq 8$, $g \in N$.

41b. Sample answer: The value 2.5 represents the magnitude, where the bottom of the sine curve, at $f(t) = -2.5$, is when the horse is at its minimum height and the top of the curve, at $f(t) = 2.5$, is when the horse is at its maximum height. The 2.5 vertical shift was added at the end to set the minimum height at 0. There are 8 different maxima, so increments of $\frac{\pi}{4}$ to represent each group. Thus, each group is shifted from the others. The value g represents each of the 8 groups. The carousel does a complete revolution every 20 seconds, so the period is 20 and $b = \frac{2\pi}{20}$ or $\frac{\pi}{10}$.

41c. Sample answer: I assumed that the horses all went up and down at the same speed. I also assumed that there were 8 groups of 4 horses and each group was spaced out vertically by the same increment. This way the entire carousel can be modeled after one trigonometric function. I also assumed that the period of the horses coincided with the period of the carousel.

41d. Sample answer: In order to accomplish this, set the period of the horses so that it is not a multiple of the period of the carousel. Keep the function representing the horses the same, and set the period of the carousel to 23 seconds.

54. $360°$; $h = -\pi$; no vertical shift

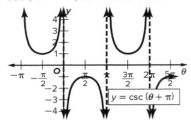

$y = \csc(\theta + \pi)$

55. $180°$; no phase shift; $k = 6$

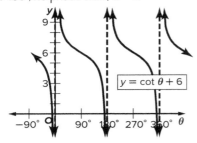

$y = \cot\theta + 6$

56. π; $h = \dfrac{\pi}{6}$; $k = -2$

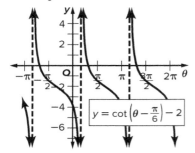

$y = \cot\left(\theta - \dfrac{\pi}{6}\right) - 2$

57. $120°$; $h = 45°$; $k = 1$

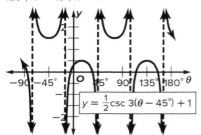

$y = \dfrac{1}{2}\csc 3(\theta - 45°) + 1$

58. $720°$; $h = 90°$; no vertical shift

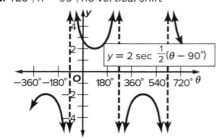

$y = 2\sec\dfrac{1}{2}(\theta - 90°)$

59. π; $h = -\dfrac{\pi}{2}$; $k = -3$

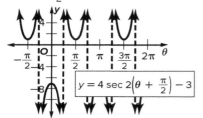

$y = 4\sec 2\left(\theta + \dfrac{\pi}{2}\right) - 3$

Performance Task

3.

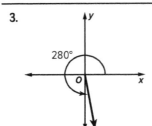

280°

4.

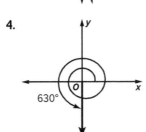

630°

Chapter 9 Answer Appendix

Trigonometric Identities and Equations

Track Your Progress

This chapter focuses on content from the Trigonometric Functions domain.

THEN	NOW	NEXT
F.BF.3 Identify the effect on the graph of replacing $f(x)$ by $f(x) + k$, $k\,f(x)$, $f(kx)$, and $f(x + k)$ for specific values of k (both positive and negative); find the value of k given the graphs. Experiment with cases and illustrate an explanation of the effects on the graph using technology.	**F.TF.8** Prove the Pythagorean identity $sin^2(\theta) + cos^2(\theta) = 1$ and use it to calculate trigonometric ratios.	**A.CED.2** Create equations in two or more variables to represent relationships between quantities; graph equations on coordinate axes with labels and scales.
F.IF.7e Graph exponential and logarithmic functions, showing intercepts and end behavior, and trigonometric functions, showing period, midline, and amplitude.		**F.TF.2** Explain how the unit circle in the coordinate plane enables the extension of trigonometric functions to all real numbers, interpreted as radian measures of angles traversed counterclockwise around the unit circle.
F.TF.1 Understand radian measure of an angle as the length of the arc on the unit circle subtended by the angle.		

Standards for Mathematical Practice

All of the Standards for Mathematical Practice will be covered in this chapter. The MP icon notes specific areas of coverage.

 Teaching the Mathematical Practices
Help students develop the mathematical practices by asking questions like these.

Questioning Strategies As students approach problems in this chapter, help them develop mathematical practices by asking:

Sense-Making
· What is a trigonometric equation and how do you solve one?
· What is an extraneous solution from a trigonometric equation?

Reasoning
· Is it possible to find values of sine and cosine by using sum and difference identities?
· Can you find the values of sine and cosine by using double-angle identities?
· Can you find the values of sine and cosine by using half-angle identities?

Modeling
· Can you verify trigonometric identities by transforming one side of an equation into the form of the other side?
· Can you verify trigonometric identities by transforming each side of an equation into the same form?

Using Tools
· How can you use trigonometric identities to find trigonometric values or to simplify expressions?
· How do you verify trigonometric identities by using sum and difference identities?

Go Online!

 StudySync:
SMP Modeling Videos

These demonstrate how to apply the Standards for Mathematical Practice to collaborate, discuss, and solve real-world math problems.

 LearnSmart

 The Geometer's Sketchpad

 Vocabulary

 Personal Tutor

 Tools

 Calculator Resources

 Self-Check Practice

 Animations

Customize Your Chapter

Use the *Plan & Present*, *Assignment Tracker*, and *Assessment* tools in ConnectED to introduce lesson concepts, assign personalized practice, and diagnose areas of student need.

Differentiated Instruction

Throughout the program, look for the icons to find specialized content designed for your students.

- **AL** Approaching Level
- **OL** On Level
- **BL** Beyond Level
- **ELL** English Language Learners

Personalize

Differentiated Resources				
FOR EVERY CHAPTER	**AL**	**OL**	**BL**	**ELL**
✓ Chapter Readiness Quizzes	●	●	◐	●
✓ Chapter Tests	●	●	●	●
✓ Standardized Test Practice	●	●	●	●
𝑎𝑏𝑐 Vocabulary Review Games	●	●	◐	●
🗎 Anticipation Guide (English/Spanish)	●	●	◐	●
🗎 Student-Built Glossary	●	●	◐	●
🗎 Chapter Project	◐	●	●	●
FOR EVERY LESSON	**AL**	**OL**	**BL**	**ELL**
💬 Personal Tutors (English/Spanish)	●	●	◐	●
💬 Graphing Calculator Personal Tutors	●	●	●	●
▷ Step-by-Step Solutions	●	●	◐	●
✓ Self-Check Quizzes	●	●	●	●
🗎 5-Minute Check	●	●	●	●
🗎 Study Notebook	●	●	●	●
🗎 Study Guide and Intervention	●	●		●
🗎 Skills Practice	●	◐		●
🗎 Practice	◐	●	●	●
🗎 Word Problem Practice	●	●	●	◐
🗎 Enrichment		●	●	●
➕ Extra Examples	●	◐		◐
🖼 Lesson Presentations	●	●	●	●

◐ Aligned to this group ● Designed for this group

Engage

Featured IWB Resources

Geometer's Sketchpad provides students with a tangible, visual way to learn. *Use with Lesson 10-1.*

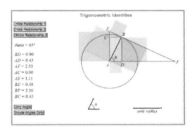

Personal Tutors let you hear a real teacher discuss each step to solving a problem. *Use with Lessons 10-1 to 10-5.*

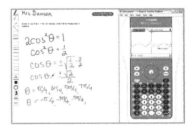

Graphing Calculator Personal Tutors let you hear a real teacher discuss each step to solving a problem. *Use with Explore 10-5 and Lesson 10-5.*

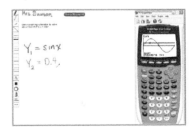

Time Management How long will it take to use these resources? Look for the clock in each lesson interleaf.

Introduce the Chapter

Mathematical Background

A trigonometric identity is an equation involving trigonometric functions that is true for all values of the variable for which every expression in the equation is defined. To verify an identity, transform either one side or both sides of the equation by replacing expressions with equivalent expressions until the two sides are identical. The sum and difference identities and the double- and half-angle identities can be used to verify other trigonometric identities. Identities are often used to solve trigonometric equations.

Ⓔ Essential Question

At the end of this chapter, students should be able to answer the Essential Question.

How can representing the same mathematical concept in different ways be helpful? Sample answer: Depending on the situation, it might be more helpful to use a visual representation such as a graph or diagram. In other situations, it might be more helpful to use a numerical or algebraic representation such as a table of values or equation.

Apply Math to the Real World

ELECTRONICS In this activity, students will use what they already know about trigonometric functions to model electronics. Have students complete this activity individually or in small groups. 1

Go Online! ✓

Chapter Project

Spin the Wheel Students use what they have learned about trigonometric identities to complete a project.

This chapter project addresses business literacy, as well as several specific skills identified as being essential to student success by the Framework for 21st Century Learning. Ⓜ 1, 3, 4

CHAPTER 10

Trigonometric Identities and Equations

THEN
You graphed trigonometric functions and determined the period, amplitude, phase shifts, and vertical shifts.

NOW
You will:
- Use and verify trigonometric identities.
- Use the sum and difference of angles identities.
- Use the double- and half-angle identities.
- Solve trigonometric equations.

Ⓜ WHY

ELECTRONICS Radios, smartphones, and wireless Internet all use waves that can be modeled by trigonometric functions.

Use the Mathematical Practices to complete the activity.

1. Use Tools Use the Internet to learn about the ways in which waves are used for Wi-Fi signals, cell phones, or other wireless devices.

2. Model with Math Use the Pen tool on the Coordinate Grid mat to model the wave you researched. Use the Text tool to label the amplitude, frequency, and so on.

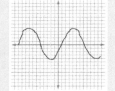

3. Discuss How does the cyclical or periodic behavior of the wave enable it to transmit information to and from our wireless devices?

Ⓢ ALEKS®

Your Student Success Tool ALEKS is an adaptive, personalized learning environment that identifies precisely what each student knows and is ready to learn—ensuring student success at all levels.

- **Formative Assessment:** Dynamic, detailed reports monitor students' progress toward standards mastery.
- **Automatic Differentiation:** Strengthen prerequisite skills and target individual learning gaps.
- **Personalized Instruction:** Supplement in-class instruction with personalized assessment and learning opportunities.

Go Online to Guide Your Learning

Explore & Explain

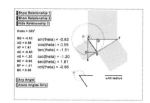 **The Geometer's Sketchpad**

The Geometer's Sketchpad can be used to enhance your understanding of trigonometric identities in Lesson 10-1 and throughout this chapter.

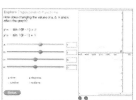 **Graphing Tools**

Use the **Explore: Trigonometric Functions** tool to enhance your understanding of graphing trigonometric functions.

eBook

Interactive Student Guide

Before starting the chapter, answer the **Chapter Focus** preview questions. Check your answers as you complete each lesson. At the end of the chapter, try the **Performance Task**.

Organize

Foldables

Get organized! Before you begin this chapter, create this **Trigonometric Identities and Equations Foldable** to help you organize your notes about trigonometric identities and equations.

Collaborate

Chapter Project

In the **Spin the Wheel** project, you will use what you have learned about trigonometric identities to complete a project that addresses business literacy.

Focus

 LEARNSMART

Need help studying? Complete the **Trigonometric Functions** domain in LearnSmart to review for the chapter test.

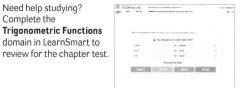

ALEKS

You can use the **Trigonometry** topic in ALEKS to explore what you know about trigonometry and what you are ready to learn.*

* Ask your teacher if this is part of your program.

Dinah Zike's **FOLDABLES**

Focus Students write notes as they explore trigonometric identities and equations in the lessons of this chapter.

Teach Have students make and label their Foldables as illustrated. Have students use the appropriate tab as they cover each lesson in this chapter. Encourage students to apply what they have learned about trigonometric identities by writing their own examples as well.

When to Use It Encourage students to add to their Foldables as they work through the chapter and to use them to review for the chapter test.

Go Online!

Using VKVs to Support Vocabulary Learning

Practice makes perfect! Find out how to use VKVs to incorporate daily vocabulary practice. **MP** 1, 5

Get Ready for the Chapter

RtI Response to Intervention

Use the Concept Check results and the Intervention Planner chart to help you determine your Response to Intervention.

Intervention Planner

TIER 1 On Level OL

IF students miss 25% of the exercises or less,

THEN choose a resource:

Go Online!
📄 Skills Practice, Chapters 3, 4, and 9
📄 Chapter Project

TIER 2 Approaching Level AL

IF students miss 50% of the exercises,

THEN Go Online!
📄 Study Guide and Intervention, Chapters 3, 4, and 9
➕ Extra Examples
💬 Personal Tutors
📄 Homework Help

TIER 3 Intensive Intervention

IF students miss 75% of the exercises,

THEN choose a resource:

use *Math Triumphs, Alg. 2*

Go Online!
➕ Extra Examples
💬 Personal Tutors
📄 Homework Help
🔤 Review Vocabulary

Additional Answers

1. factorable; $-4a(4a - 1)$

2. no; $x(x + 6)(x - 4)$

3. $x^2 + x - 42 = 0$

4. -7 and 6; since a dimension cannot be negative the solution is $x = 6$.

5. $180° - 135°$

6. $\dfrac{\sqrt{2}}{2}$

Get Ready for the Chapter

Go Online! for Vocabulary Review Games and key vocabulary in 13 languages.

Connecting Concepts	New Vocabulary

Concept Check

Review the concepts used in this chapter by answering the questions below. 1–6. See margin.

1. Is the polynomial $-16a^2 + 4a$ factorable or is it prime? If it is factorable, write the expression in its completely factored form.

2. Is the polynomial $x(x^2 + 2x - 24)$ completely factored? If not, write its completely factored form.

3. You are designing a flower bed that has an area of 42 square feet, with the dimensions shown. Write the equation, in polynomial form, to determine the value of x, and hence the length of each side.

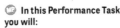

x ft

$x + 1$ ft

4. Solve the polynomial equation in question 3. How do you know which value of x is a reasonable solution?

5. To find the cos 135°, you need to know its reference angle. How do you determine the reference angle?

6. What is the exact value of cos 45°?

Performance Task Preview

You can use the concepts and skills in the chapter to perform error analysis for a teacher. Understanding trigonometric identities and equations will help you finish the Performance Task at the end of the chapter.

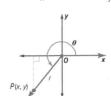

In this Performance Task you will:
- make sense of problems and persevere in solving them
- reason abstractly and quantitatively
- construct viable arguments and critique the reasoning of others
- attend to precision
- look for and make use of structure

New Vocabulary

English		Español
trigonometric identity	p. 655	identidad trigonométrica
trigonometric equation	p. 683	ecuación trigonométrica

Review Vocabulary

formula *fórmula* a mathematical sentence that expresses the relationship between certain quantities

identity *identidad* an equality that remains true regardless of the values of any variables that are in it

trigonometric functions *funciones rigonométricas* For any angle, with measure θ, a point $P(x, y)$ on its terminal side, $r = \sqrt{x^2 + y^2}$, the trigonometric functions of θ are as follows.

$$\sin \theta = \frac{y}{r} \qquad \cos \theta = \frac{x}{r} \qquad \tan \theta = \frac{y}{x}$$

$$\csc \theta = \frac{r}{y} \qquad \sec \theta = \frac{r}{x} \qquad \cot \theta = \frac{x}{y}$$

Key Vocabulary ELL

Introduce the key vocabulary in the chapter using the routine below.

Define A trigonometric identity is an equation involving a trigonometric function that is true for all values of the variable.

Example The identity $\sin(-\theta) = -\sin \theta$ is a Negative Angle Identity.

Ask What other identities have you learned about this year? Answers may vary; students may mention identity matrices or the identity function.

Trigonometric Identities

SUGGESTED PACING (DAYS)

| 90 min. | 0.5 |
| 45 min. | 1.0 |

Instruction

Track Your Progress

Objectives

1 Use trigonometric identities to find trigonometric values.

2 Use trigonometric identities to simplify expressions.

Mathematical Background

Students must have good recall of the quotient, reciprocal, and Pythagorean identities. While it is helpful to remember other identities, the other identities can be derived from the basic identities.

THEN	NOW	NEXT
F.TF.2 Explain how the unit circle in the coordinate plane enables the extension of trigonometric functions to all real numbers, interpreted as radian measures of angles traversed counterclockwise around the unit circle.	**F.TF.8** Prove the Pythagorean identity $\sin^2(\theta) + \cos^2(\theta) = 1$ and use it to find $\sin(\theta)$, $\cos(\theta)$, or $\tan(\theta)$ given $\sin(\theta)$, $\cos(\theta)$, or $\tan(\theta)$ and the quadrant of the angle.	**F.TF.9** Prove the addition and subtraction formulas for sine, cosine, and tangent and use them to solve problems.

Go Online! All of these resources and more are available at connectED.mcgraw-hill.com

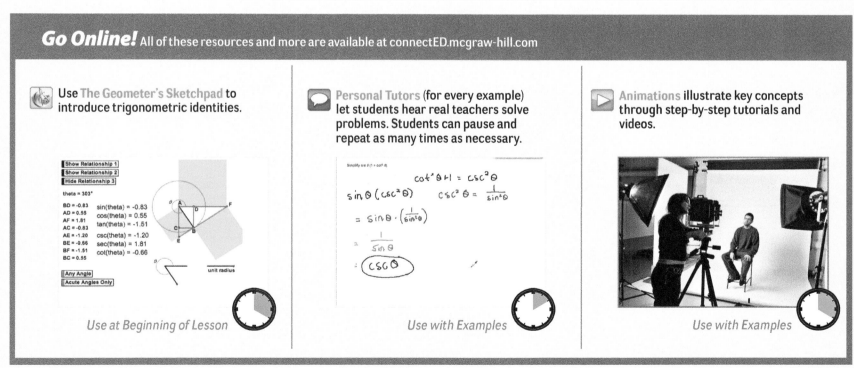

Use **The Geometer's Sketchpad** to introduce trigonometric identities.

Use at Beginning of Lesson

Personal Tutors (for every example) let students hear real teachers solve problems. Students can pause and repeat as many times as necessary.

Use with Examples

Animations illustrate key concepts through step-by-step tutorials and videos.

Use with Examples

OER Using Open Educational Resources

Tools Have students work individually to create flashcards and notes using **Study Blue** to help them remember the trigonometric identities in this chapter. Students can add to their notes as they learn new identities and skills. *Use as homework or review*

Go Online!
connectED.mcgraw-hill.com

Worksheets

Differentiate Your Resources

Extra Practice Additional practice or homework; Skills Practice is best for approaching-level students and Practice is best for on-level and beyond-level students

Skills Practice

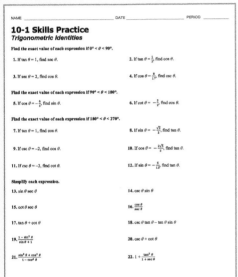

Practice

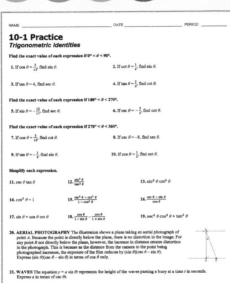

Word Problem Practice

Intervention Reteaching and vocabulary activities that can be used with struggling or absent students and as ELL support

Extension Activities that can be used to extend lesson concepts

Study Guide and Intervention

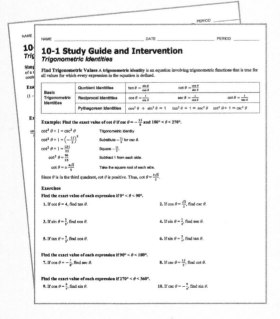

Study Notebook

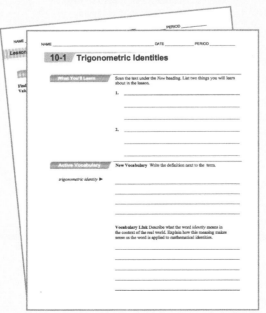

Enrichment

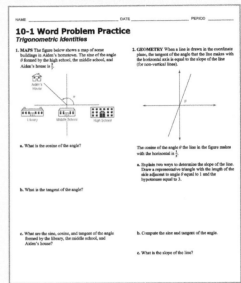

LESSON 1

Trigonometric Identities

:: Then	:: Now	:: Why?
• You evaluated trigonometric functions.	**1** Use trigonometric identities to find trigonometric values. **2** Use trigonometric identities to simplify expressions.	• The amount of light that a source provides to a surface is called the *illuminance*. The illuminance E in foot candles on a surface is related to the distance R in feet from the light source. The formula $\sec \theta = \frac{I}{ER^2}$, where I is the intensity of the light source measured in candles and θ is the angle between the light beam and a line perpendicular to the surface, can be used in situations in which lighting is important, as in photography.

 New Vocabulary
trigonometric identity

 Mathematical Practices
2 Reason abstractly and quantitatively.
7 Look for and make use of structure.

Content Standards
F.TF.8 Prove the Pythagorean identity $\sin^2(\theta) + \cos^2(\theta) = 1$ and use it to find $\sin(\theta)$, $\cos(\theta)$, or $\tan(\theta)$ given $\sin(\theta)$, $\cos(\theta)$, or $\tan(\theta)$ and the quadrant of the angle.

1 Find Trigonometric Values The equation above can also be written as $E = \frac{I \cos \theta}{R^2}$. This is an example of a trigonometric identity. A **trigonometric identity** is an equation involving trigonometric functions that is true for all values for which every expression in the equation is defined.

If you can show that a specific value of the variable in an equation makes the equation false, then you have produced a *counterexample*. It only takes one counterexample to prove that an equation is not an identity.

⬦ Key Concept Basic Trigonometric Identities

Quotient Identities	
$\tan \theta = \frac{\sin \theta}{\cos \theta}$, $\cos \theta \neq 0$	$\cot \theta = \frac{\cos \theta}{\sin \theta}$, $\sin \theta \neq 0$

Reciprocal Identities	
$\sin \theta = \frac{1}{\csc \theta}$, $\csc \theta \neq 0$	$\csc \theta = \frac{1}{\sin \theta}$, $\sin \theta \neq 0$
$\cos \theta = \frac{1}{\sec \theta}$, $\sec \theta \neq 0$	$\sec \theta = \frac{1}{\cos \theta}$, $\cos \theta \neq 0$
$\tan \theta = \frac{1}{\cot \theta}$, $\cot \theta \neq 0$	$\cot \theta = \frac{1}{\tan \theta}$, $\tan \theta \neq 0$

Pythagorean Identities		
$\cos^2 \theta + \sin^2 \theta = 1$	$\tan^2 \theta + 1 = \sec^2 \theta$	$\cot^2 \theta + 1 = \csc^2 \theta$

Cofunction Identities		
$\sin\left(\frac{\pi}{2} - \theta\right) = \cos \theta$	$\cos\left(\frac{\pi}{2} - \theta\right) = \sin \theta$	$\tan\left(\frac{\pi}{2} - \theta\right) = \cot \theta$

Negative Angle Identities		
$\sin(-\theta) = -\sin \theta$	$\cos(-\theta) = \cos \theta$	$\tan(-\theta) = -\tan \theta$

The negative angle identities are sometimes called *odd-even* identities.

The identity $\tan \theta = \frac{\sin \theta}{\cos \theta}$ is true except for angle measures such as 90°, 270°, ... , $90° + k180°$, where k is an integer. The cosine of each of these angle measures is 0, so $\tan \theta$ is not defined when $\cos \theta = 0$. An identity similar to this is $\cot \theta = \frac{\cos \theta}{\sin \theta}$.

Mathematical Practices Strategies

Reason abstractly and quantitatively. Help students understand the relationship of the trigonometric identities to the right triangle and the unit circle. For example, ask:

• How are the Pythagorean identities derived using the unit circle? Because $\cos \theta$ and $\sin \theta$ are the legs of a right triangle in the unit circle, the Pythagorean Theorem states that the sum of the squares of the legs of a right triangle equals the square of the hypotenuse. In the unit circle, the hypotenuse has radius 1. The other identities can be found by solving the original identity for each leg.

• How can you explain the cofunction identities using a right triangle? In a right triangle, the two acute angles are complementary. If one measure is known, then the other can be found as a difference between $\frac{\pi}{2} - \theta$. The trigonometric ratios are equal for the cofunction of the complement.

Launch

Have students read the Why? section of the lesson. Ask:

• On the right-hand side of the illuminance formula, which variables appear in the numerator? In the denominator? intensity; illuminance and distance

• In a right triangle, what is the ratio for $\sec \theta$?
$\sec \theta = \frac{\text{hypotenuse}}{\text{adjacent side}}$

• What is the reciprocal identity of $\sec \theta$? $\frac{1}{\cos \theta}$, $\cos \theta \neq 0$

Go Online!

Interactive Whiteboard
Use the *eLesson* or *Lesson Presentation* to present this lesson.

Teach

Ask the scaffolded questions for each example to build conceptual understanding for students at all levels.

1 Find Trigonometric Values

Example 1 Use Trigonometric Identities

AL Why is the denominator of the answer equal to 4? When we take the square root of the right hand side we take the square root of the numerator and the denominator and the square root of 16 is 4.

OL How could you use your solution to find θ? Take $\cos^{-1}$ of both sides.

BL Show how you can derive the identity $\tan^2 \theta + 1 = \sec^2 \theta$ from $\sin^2 \theta + \cos^2 \theta = 1$. Divide all terms by $\cos^2 \theta$: $\dfrac{\sin^2 \theta}{\cos^2 \theta} + \dfrac{\cos^2 \theta}{\cos^2 \theta} = \dfrac{1}{\cos^2 \theta}$. $\dfrac{\sin^2 \theta}{\cos^2 \theta} = \tan^2 \theta$, $\dfrac{\cos^2 \theta}{\cos^2 \theta} = 1$, and $\dfrac{1}{\cos^2 \theta} = \sec^2 \theta$, so we have $\tan^2 \theta + 1 = \sec^2 \theta$.

Need Another Example?

a. Find the exact value of $\tan \theta$ if $\sec \theta = -2$ and $180° < \theta < 270°$. $\tan \theta = \sqrt{3}$

b. Find the exact value of $\sin \theta$ if $\cos \theta = -\dfrac{1}{2}$ and $90° < \theta < 180°$. $\sin \theta = \dfrac{\sqrt{3}}{2}$

2 Simplify Expressions

Example 2 Simplify an Expression

AL How could you check your solution? Graph both expressions in your calculator.

OL Why is $\sin \theta \csc \theta = 1$? They are reciprocals of each other.

BL Does it matter which terms you simplify first? No, you could have simplified $\dfrac{1}{\cot \theta}$ to $\tan \theta$ first and then simplified $\sin \theta \csc \theta$ to 1.

Need Another Example?
Simplify $\sin \theta (\csc \theta - \sin \theta)$. $\cos^2 \theta$

You can use trigonometric identities to find exact values of trigonometric functions. You can find approximate values by using a graphing calculator.

F.TF.8

Example 1 Use Trigonometric Identities

a. Find the exact value of $\cos \theta$ if $\sin \theta = \frac{1}{4}$ and $90° < \theta < 180°$.

$$\cos^2 \theta + \sin^2 \theta = 1 \qquad \text{Pythagorean identity}$$
$$\cos^2 \theta = 1 - \sin^2 \theta \qquad \text{Subtract } \sin^2 \theta \text{ from each side.}$$
$$\cos^2 \theta = 1 - \left(\frac{1}{4}\right)^2 \qquad \text{Substitute } \frac{1}{4} \text{ for } \sin \theta$$
$$\cos^2 \theta = 1 - \frac{1}{16} \qquad \text{Square } \frac{1}{4}.$$
$$\cos^2 \theta = \frac{15}{16} \qquad \text{Subtract: } \frac{16}{16} - \frac{1}{16} = \frac{15}{16}.$$
$$\cos \theta = \pm \frac{\sqrt{15}}{4} \qquad \text{Take the square root of each side.}$$

Because θ is in the second quadrant, $\cos \theta$ is negative. Thus, $\cos \theta = -\dfrac{\sqrt{15}}{4}$.

CHECK Use a calculator to find an approximate answer.

Step 1 Find Arcsin $\frac{1}{4}$.

$$\sin^{-1} \frac{1}{4} \approx 14.48° \qquad \text{Use a calculator.}$$

Because $90° < \theta < 180°$, $\theta \approx 180° - 14.48°$ or about $165.52°$.

Step 2 Find $\cos \theta$.

Replace θ with $165.52°$.

$$\cos 165.52° \approx -0.97$$

Step 3 Compare with the exact value.

$$-\frac{\sqrt{15}}{4} \overset{?}{\approx} 0.97$$
$$-0.968 \approx 0.97 \checkmark$$

b. Find the exact value of $\csc \theta$ if $\cot \theta = -\frac{3}{5}$ and $270° < \theta < 360°$.

$$\cot^2 \theta + 1 = \csc^2 \theta \qquad \text{Pythagorean identity}$$
$$\left(-\frac{3}{5}\right)^2 + 1 = \csc^2 \theta \qquad \text{Substitute } -\frac{3}{5} \text{ for } \cot \theta$$
$$\frac{9}{25} + 1 = \csc^2 \theta \qquad \text{Square } -\frac{3}{5}$$
$$\frac{34}{25} = \csc^2 \theta \qquad \text{Add: } \frac{9}{25} + \frac{25}{25} = \frac{34}{25}$$
$$\pm \frac{\sqrt{34}}{5} = \csc \theta \qquad \text{Take the square root of each side.}$$

Because θ is in the fourth quadrant, $\csc \theta$ is negative. Thus, $\csc \theta = -\dfrac{\sqrt{34}}{5}$.

Guided Practice

1A. Find $\sin \theta$ if $\cos \theta = \frac{1}{3}$ and $270° < \theta < 360°$. $-\dfrac{2\sqrt{2}}{3}$

1B. Find $\sec \theta$ if $\sin \theta = -\frac{2}{7}$ and $180° < \theta < 270°$. $-\dfrac{7\sqrt{5}}{15}$

2 Simplify Expressions Simplifying an expression that contains trigonometric functions means that the expression is written as a numerical value or in terms of a single trigonometric function, if possible.

Study Tip

Quadrants Here is a table to help you remember which ratios are positive and which are negative in each quadrant.

Function	+	−
$\sin \theta$	1, 2	3, 4
$\cos \theta$	1, 4	2, 3
$\tan \theta$	1, 3	2, 4
$\csc \theta$	1, 2	3, 4
$\sec \theta$	1, 4	2, 3
$\cot \theta$	1, 3	2, 4

Go Online!

You will want to reference the Basic Trigonometric Identities as you study this chapter. Log into your eStudent Edition to bookmark this lesson.

Differentiated Instruction **AL** **OL**

IF students have difficulty with trigonometric identities,

THEN have students work in groups of three. Ask each group to choose one of the identities in the Key Concept box on p. 655 and work together to demonstrate that it is true. Students should verify their results using the definitions of sine, cosine, and tangent in terms of the sides of a right triangle.

Example 2 Simplify an Expression

F.TF.8

Simplify $\dfrac{\sin \theta \csc \theta}{\cot \theta}$.

$$\dfrac{\sin \theta \csc \theta}{\cot \theta} = \dfrac{\sin \theta \left(\dfrac{1}{\sin \theta}\right)}{\dfrac{1}{\tan \theta}} \qquad \csc \theta = \dfrac{1}{\sin \theta} \text{ and } \cot \theta = \dfrac{1}{\tan \theta}$$

$$= \dfrac{1}{\dfrac{1}{\tan \theta}} \qquad \dfrac{\sin \theta}{\sin \theta} = 1$$

$$= \dfrac{1}{1} \cdot \dfrac{\tan \theta}{1} \text{ or } \tan \theta \qquad \dfrac{a}{b} \div \dfrac{c}{d} = \dfrac{a}{b} \cdot \dfrac{d}{c}$$

▶ **Guided Practice**

Simplify each expression.

2A. $\dfrac{\tan^2 \theta \csc^2 \theta - 1}{\sec^2 \theta}$ $\sin^2 \theta$ **2B.** $\dfrac{\sec \theta}{\sin \theta}(1 - \cos^2 \theta)$ $\tan \theta$

Simplifying trigonometric expressions can be helpful when solving real-world problems.

F.TF.8

Real-World Example 3 Simplify and Use an Expression

LIGHTING Refer to the beginning of the lesson.

a. Solve the formula in terms of E.

$$\sec \theta = \dfrac{I}{ER^2} \qquad \text{Original equation}$$

$$ER^2 \sec \theta = I \qquad \text{Multiply each side by } ER^2$$

$$ER^2 \dfrac{1}{\cos \theta} = I \qquad \dfrac{1}{\cos \theta} = \sec \theta$$

$$\dfrac{E}{\cos \theta} = \dfrac{I}{R^2} \qquad \text{Divide each side by } R^2$$

$$E = \dfrac{I \cos \theta}{R^2} \qquad \text{Multiply each side by } \cos \theta$$

b. Is the equation in part **a** equivalent to $R^2 = \dfrac{I \tan \theta \cos \theta}{E}$? Explain.

$$R^2 = \dfrac{I \tan \theta \cos \theta}{E} \qquad \text{Original equation}$$

$$ER^2 = I \tan \theta \cos \theta \qquad \text{Multiply each side by } E$$

$$E = \dfrac{I \tan \theta \cos \theta}{R^2} \qquad \text{Divide each side by } R^2$$

$$E = \dfrac{I \left(\dfrac{\sin \theta}{\cos \theta}\right) \cos \theta}{R^2} \qquad \tan \theta = \dfrac{\sin \theta}{\cos \theta}$$

$$E = \dfrac{I \sin \theta}{R^2} \qquad \text{Simplify}$$

No; the equations are not equivalent. $R^2 = \dfrac{I \tan \theta \cos \theta}{E}$ simplifies to $E = \dfrac{I \sin \theta}{R^2}$.

▶ **Guided Practice**

3. Rewrite $\cot^2 \theta - \tan^2 \theta$ in terms of $\sin \theta$. $\dfrac{1 - 2 \sin^2 \theta}{\sin^2 \theta - \sin^4 \theta}$

Example 3 Simplify and Use an Expression

AL What do E and R represent? E is the illuminance and R is the distance in feet from the light source.

OL Explain how you can determine if two expressions are equivalent. See if they simplify to the same expression.

BL Give an upper bound for E in terms of I and R. $E \le \dfrac{I}{R^2}$

Need Another Example?

Lighting Refer to the beginning of the lesson.
a. Solve the illuminance formula in terms of R.

$$R = \sqrt{\dfrac{I \cos \theta}{E}}$$

b. Is the equation in part **a** equivalent to $\dfrac{1}{R^2} = \dfrac{E}{I \sec \theta}$? no

e Essential Question

How do you simplify a trigonometric expression? Convert the expression into terms of sine and cosine. Use trigonometric identities to eliminate terms in the expression.

Differentiated Instruction **BL**

Extension Beyond-level students are often interested in learning how certain mathematical ideas came into existence. Provide an extension to the math history link about the Indian mathematician, Aryabhatta. Students can investigate the prior mathematical discoveries of Aryabhatta that led to the need for inverse trigonometric functions in the context of his work.

Go Online!

The most up-to-date resources available for your program can be found at connectED.mcgraw-hill.com.

Practice

Formative Assessment Use Exercises 1–8 to assess students' understanding of the concepts in this lesson.

The Practice and Problem Solving exercises assess the content taught in the lesson. The Preparing for Assessment page is meant to be used as preparation for assessment.

Teaching the Mathematical Practices

Perseverance Mathematically proficient students start by explaining to themselves the meaning of a problem and looking for entry points to its solution. They analyze givens, constraints, relationships, and goals and make conjectures about the form and meaning of the solution and plan a solution pathway rather than simply jumping into a solution attempt.

Extra Practice

See page R10 for extra exercises for students who are approaching level or for on-level students who need additional reinforcement.

Teaching Tip

Trigonometric Ratios You can use the familiar definitions of sine, cosine, and tangent as the ratios of the opposite side, adjacent side, and hypotenuse of a right triangle to show why $\frac{\sin\theta}{\cos\theta} = \tan\theta$.

Check Your Understanding ○ = Step-by-Step Solutions begin on page R11.

Go Online! for a Self-Check Quiz

Example 1
F.TF.8

Find the exact value of each expression if $0° < \theta < 90°$.

1. If $\cot\theta = 2$, find $\tan\theta$. $\frac{1}{2}$
2. If $\sin\theta = \frac{4}{5}$, find $\cos\theta$. $\frac{3}{5}$
3. If $\cos\theta = \frac{2}{3}$, find $\sin\theta$. $\frac{\sqrt{5}}{3}$
4. If $\cos\theta = \frac{2}{3}$, find $\csc\theta$. $\frac{3\sqrt{5}}{5}$

Example 2
F.TF.8

Simplify each expression.

5. $\tan\theta\cos^2\theta$ $\sin\theta\cos\theta$
6. $\csc^2\theta - \cot^2\theta$ 1
7. $\frac{\cos\theta\csc\theta}{\tan\theta}$ $\cot^2\theta$

Example 3
F.TF.8

8. **PERSEVERANCE** When unpolarized light passes through polarized sunglass lenses, the intensity of the light is cut in half. If the light then passes through another polarized lens with its axis at an angle of θ to the first, the intensity of the light is again diminished. The intensity of the emerging light can be found by using the formula $I = I_0 - \frac{I_0}{\csc^2\theta}$, where I_0 is the intensity of the light incoming to the second polarized lens, I is the intensity of the emerging light, and θ is the angle between the axes of polarization.

a. Simplify the formula in terms of $\cos\theta$. $I = I_0\cos^2\theta$

b. Use the simplified formula to determine the intensity of light that passes through a second polarizing lens with axis at 30° to the original. $I = \frac{3}{4}I_0$; The light has three fourths the intensity it had before passing through the second polarizing lens.

Axis 2
Axis 1
Unpolarized light

Practice and Problem Solving Extra Practice is on page R10.

Example 1
F.TF.8

Find the exact value of each expression if $0° < \theta < 90°$.

9. If $\cos\theta = \frac{3}{5}$, find $\csc\theta$. $\frac{5}{4}$
10. If $\sin\theta = \frac{1}{2}$, find $\tan\theta$. $\frac{\sqrt{3}}{3}$
11. If $\sin\theta = \frac{3}{5}$, find $\cos\theta$. $\frac{4}{5}$
12. If $\tan\theta = 2$, find $\sec\theta$. $\sqrt{5}$

Find the exact value of each expression if $180° < \theta < 270°$.

13. If $\cos\theta = -\frac{3}{5}$, find $\csc\theta$. $-\frac{5}{4}$
14. If $\sec\theta = -3$, find $\tan\theta$. $2\sqrt{2}$
15. If $\cot\theta = \frac{1}{4}$, find $\csc\theta$. $\frac{-\sqrt{17}}{4}$
16. If $\sin\theta = -\frac{1}{2}$, find $\cos\theta$. $-\frac{\sqrt{3}}{2}$

Find the exact value of each expression if $270° < \theta < 360°$.

17. If $\cos\theta = \frac{5}{13}$, find $\sin\theta$. $-\frac{12}{13}$
18. If $\tan\theta = -1$, find $\sec\theta$. $\sqrt{2}$
19. If $\sec\theta = \frac{5}{3}$, find $\cos\theta$. $\frac{3}{5}$
20. If $\csc\theta = -\frac{5}{3}$, find $\cos\theta$. $\frac{4}{5}$

Example 2
F.TF.8

Simplify each expression.

21. $\sec\theta\tan^2\theta + \sec\theta$ $\sec^3\theta$
22. $\cos\left(\frac{\pi}{2} - \theta\right)\cot\theta$ $\cos\theta$
23. $\cot\theta\sec\theta$ $\csc\theta$
24. $\sin\theta(1 + \cot^2\theta)$ $\csc\theta$
25. $\sin\left(\frac{\pi}{2} - \theta\right)\sec\theta$ 1
26. $\frac{\cos(-\theta)}{\sin(-\theta)}$ $-\cot\theta$

Differentiated Homework Options

Levels	AL Basic	OL Core	BL Advanced
Exercises	9–27, 42–59	9–33 odd, 34–37, 39, 41–59	28–59
2-Day Option	9–27 odd, 51–59	9–27, 51–59	
	10–26 even, 42–50	28–50	

 You can use ALEKS to provide additional remediation support with personalized instruction and practice.

Example 3
F.TF.8

(27) **ELECTRONICS** When there is a current in a wire in a magnetic field, such as in a hairdryer, a force acts on the wire. The strength of the magnetic field can be determined using the formula $B = \frac{F \csc \theta}{I\ell}$, where F is the force on the wire, I is the current in the wire, ℓ is the length of the wire, and θ is the angle the wire makes with the magnetic field. Rewrite the equation in terms of $\sin \theta$. (*Hint: Solve for F.*) $F = I\ell B \sin \theta$

B Simplify each expression.

28. $\frac{1 - \sin^2 \theta}{\sin^2 \theta} \cot^2 \theta$ **29.** $\tan \theta \csc \theta \sec \theta$ **30.** $\frac{1}{\sin^2 \theta} - \frac{\cos^2 \theta}{\sin^2 \theta}$ 1

31. $2(\csc^2 \theta - \cot^2 \theta)$ 2   **32.** $(1 + \sin \theta)(1 - \sin \theta)$ **33.** $2 - 2\sin^2 \theta$ $2\cos^2 \theta$
                                                        $\cos^2 \theta$

34. **SUN** The ability of an object to absorb energy is related to a factor called the emissivity e of the object. The emissivity can be calculated by using the formula $e = \frac{W \sec \theta}{AS}$, where W is the rate at which a person's skin absorbs energy from the Sun, S is the energy from the Sun in watts per square meter, A is the surface area exposed to the Sun, and θ is the angle between the Sun's rays and a line perpendicular to the body.

   a. Solve the equation for W. Write your answer using only $\sin \theta$ or $\cos \theta$. $W = eAS\cos\theta$

   b. Find W if $e = 0.80$, $\theta = 40°$, $A = 0.75$ m^2, and $S = 1000$ W/m^2. Round to the nearest hundredth. 459.63 W

35. (MP) **MODELING** The map shows some of the buildings in Maria's neighborhood that she visits on a regular basis. The sine of the angle θ formed by the roads connecting the dance studio, the school, and Maria's house is $\frac{4}{9}$.

   a. What is the cosine of the angle? $\frac{\sqrt{65}}{9}$

   b. What is the tangent of the angle? $\frac{4\sqrt{65}}{65}$

35c. $\frac{4}{9}$, $\frac{\sqrt{65}}{9}$,
   $\frac{4\sqrt{65}}{65}$

   c. What are the sine, cosine, and tangent of the angle formed by the roads connecting the piano teacher's house, the school, and Maria's house?

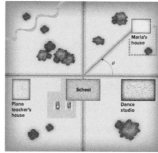

36. **MULTIPLE REPRESENTATIONS** In this problem, you will use a graphing calculator to determine whether an equation may be a trigonometric identity. Consider the trigonometric identity $\tan^2 \theta - \sin^2 \theta = \tan^2 \theta \sin^2 \theta$.

   a. **Tabular** Copy and complete the table below.

θ	0°	30°	45°	60°
$\tan^2 \theta - \sin^2 \theta$	0	$\frac{1}{12}$	$\frac{1}{2}$	$\frac{9}{4}$
$\tan^2 \theta \sin^2 \theta$	0	$\frac{1}{12}$	$\frac{1}{2}$	$\frac{9}{4}$

   b. **Graphical** Use a graphing calculator to graph $\tan^2 \theta - \sin^2 \theta = \tan^2 \theta \sin^2 \theta$ as two separate functions. Sketch the graph. See margin.

   c. **Analytical** If the graphs of the two functions do not match, then the equation is not an identity. Do the graphs coincide? yes

   d. **Analytical** Use a graphing calculator to determine whether the equation $\sec^2 x - 1 = \sin^2 x \sec^2 x$ may be an identity. (Be sure your calculator is in degree mode.) yes

---

MP **Teaching the Mathematical Practices**

Modeling Mathematically proficient students can apply the mathematics they know to solve problems arising in everyday life, analyze relationships mathematically to draw conclusions, and interpret their mathematical results in the context of a situation.

Levels of Complexity Chart

The levels of the exercises progress from 1 to 3, with Level 1 indicating the lowest level of complexity.

Exercises	9–27	28–36, 51–59	37–50
▶ Level 3			●
▶ Level 2		●	
Level 1	●		

Additional Answer

36b.

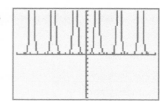

$[-540, 540]$ scl: 90 by $[-10, 10]$ scl: 1

Differentiated Instruction (AL) (OL) (BL) (ELL)

Kinesthetic Learners Put students into pairs. Have each pair of students make up a set of Basic Trigonometric Identities cards. Pairs should make up two cards for every type of identity within each category (quotient, reciprocal, Pythagorean, cofunction, and negative angle). For example, for the quotient identities, four cards will be made: $\tan \theta$, $\frac{\sin \theta}{\cos \theta}$, $\cot \theta$, and $\frac{\cos \theta}{\sin \theta}$. A total of 28 should be made. Then, pairs will play a game of "Memory." The first partner will turn over two cards. The two cards are removed if they represent equal basic trigonometric values. If not, both are turned back over. The other partner then takes a turn.

$\cot \theta$	$\frac{\cos \theta}{\sin \theta}$					

Go Online! (e)

eSolutions Manual

Create worksheets, answer keys, and solutions handouts for your assignments.

Assess

Crystal Ball Tell students to look ahead to Lesson 10-2. Ask them to write how they think what they learned today will help them with Lesson 10-2.

Watch Out!

Error Analysis For Exercise 42, students should see that Rosalina is correct. Explain to students that inductive reasoning (generalizing from several examples) can never prove that an identity is true, but that any specific counterexample is enough to show that an equation is not an identity. For Exercise 50, students should see that Ebony is correct (because $\cos^2 \theta + \sin^2 \theta = 1$) and that Jordan is not correct $\left(\text{because } \dfrac{a}{b+c} \neq \dfrac{a}{b} + \dfrac{a}{c}\right)$. Explain to students that when they simplify a trigonometric expression, they use the same properties as when they simplify any rational expression.

Standards for Mathematical Practice

Emphasis On	Exercises
1 Make sense of problems and persevere in solving them.	8
3 Construct viable arguments and critique the reasoning of others.	42
5 Model with mathematics.	27, 34, 35, 37
8 Look for and express regularity in repeated reasoning.	44, 48

Additional Answers

44. $\sec \theta = \dfrac{I}{ER^2}$

$\dfrac{1}{\cos \theta} = \dfrac{I}{ER^2}$ $\qquad \sec \theta = \dfrac{1}{\cos \theta}$

$ER^2 = I \cos \theta$ $\qquad$ Cross multiply.

$\cos \theta = \dfrac{ER^2}{I}$ $\qquad$ Divide each side by I.

46. $\tan(-A) = \dfrac{\sin(-A)}{\cos(-A)}$

$\qquad = \dfrac{-\sin A}{\cos A}$

$\qquad = -\dfrac{\sin A}{\cos A}$

$\qquad = -\tan A$

---

37. **SKIING** A skier of mass m descends a θ-degree hill at a constant speed. When Newton's laws are applied to the situation, the following system of equations is produced: $F_n - mg \cos \theta = 0$ and $mg \sin \theta - \mu_k F_n = 0$, where g is the acceleration due to gravity, F_n is the normal force exerted on the skier, and μ_k is the coefficient of friction. Use the system to define μ_k as a function of θ. $\mu_k = \tan \theta$

Simplify each expression.

38. $\dfrac{\tan\left(\frac{\pi}{2} - \theta\right)\sec \theta}{1 - \csc^2 \theta}$ $-\tan \theta \sec \theta$

39. $\dfrac{\cos\left(\frac{\pi}{2} - \theta\right) - 1}{1 + \sin(-\theta)}$ -1

40. $\dfrac{\sec \theta \sin \theta + \cos\left(\frac{\pi}{2} - \theta\right)}{1 + \sec \theta}$ $\sin \theta$

41. $\dfrac{\cot \theta \cos \theta}{\tan(-\theta) \sin\left(\frac{\pi}{2} - \theta\right)}$ $-\cot^2 \theta$

H.O.T. Problems   Use Higher-Order Thinking Skills

42. **CRITIQUE ARGUMENTS** Clyde and Rosalina are debating whether an equation from their homework assignment is an identity. Clyde says that since he has tried ten specific values for the variable and all of them worked, it must be an identity. Rosalina argues that specific values could only be used as counterexamples to prove that an equation is not an identity. Is either of them correct? Explain your reasoning.
Rosalina; there may be other values for which the equation is not true.

43. **CHALLENGE** Find a counterexample to show that $1 - \sin x = \cos x$ is *not* an identity. Sample answer: $x = 45°$

44. **REASONING** Demonstrate how the formula about illuminance from the beginning of the lesson can be rewritten to show that $\cos \theta = \dfrac{ER^2}{I}$. See margin.

45. **WRITING IN MATH** Pythagoras is most famous for the Pythagorean Theorem. The identity $\cos^2 \theta + \sin^2 \theta = 1$ is an example of a Pythagorean identity. Why do you think that this identity is classified in this way?

46. **PROOF** Prove that $\tan(-a) = -\tan a$ by using the quotient and negative angle identities. See margin.

47. **OPEN-ENDED** Write two expressions that are equivalent to $\tan \theta \sin \theta$. Sample answer: $\dfrac{\sin \theta}{\cos \theta} \cdot \sin \theta$ and $\dfrac{\sin^2 \theta}{\cos \theta}$

48. **REASONING** Explain how you can use division to rewrite $\sin^2 \theta + \cos^2 \theta = 1$ as $1 + \cot^2 \theta = \csc^2 \theta$. Divide all of the terms by $\sin^2 \theta$.

49. **CHALLENGE** Find $\cot \theta$ if $\sin \theta = \frac{3}{5}$ and $90° \leq \theta < 180°$. $-\frac{4}{3}$

50. **ERROR ANALYSIS** Jordan and Ebony are simplifying $\dfrac{\sin^2 \theta}{\cos^2 \theta + \sin^2 \theta}$. Is either of them correct? Explain your reasoning.

Jordan	Ebony
$\dfrac{\sin^2 \theta}{\cos^2 \theta + \sin^2 \theta} = \dfrac{\sin^2 \theta}{\cos^2 \theta} + \dfrac{\sin^2 \theta}{\sin^2 \theta}$ $= \tan^2 \theta + 1$ $= \sec^2 \theta$	$\dfrac{\sin^2 \theta}{\cos^2 \theta + \sin^2 \theta} = \dfrac{\sin^2 \theta}{1}$ $= \sin^2 \theta$

45. Sample answer: The functions $\cos \theta$ and $\sin \theta$ can be thought of as the lengths of the legs of a right triangle, and the number 1 can be thought of as the measure of the corresponding hypotenuse.

50. Ebony; Jordan did not use the identity that $\sin^2 \theta + \cos^2 \theta = 1$ and made an error adding rational expressions.

Differentiated Instruction   OL   BL

Extension Ask students if it is always possible to simplify a trigonometric expression by writing it in terms of one trigonometric function. Ask them to provide an example if it is not possible or explain why it is possible.

55b. $\dfrac{\sin(90° - \theta)}{\cos(\theta - 90°)} = \dfrac{\sin(90° - \theta)}{\cos(90° - \theta)}$ Negative angle identity, $\cos(\theta - 90) = \cos(90 - \theta)$

$\qquad = \dfrac{\cos \theta}{\sin \theta}$ Substitute the cofunction identities.

$\qquad = \cot \theta$ Substitute the cotangent quotient identity.

Preparing for Assessment

51. Of the following expressions, which is/are equivalent to $\cos \theta$? ⓂP 2, 7 F.TF.8 **E**

$$\textbf{I} \quad \sin (90° - \theta)$$
$$\textbf{II} \quad \frac{1}{\sec \theta}$$
$$\textbf{III} \quad \cos (-\theta)$$

- ○ **A** I only
- ○ **B** III only
- ○ **C** I and II only
- ○ **D** II and III only
- ○ **E** I, II, and III

52. Which expression, when defined, can be used in place of $\cos \left(\frac{\pi}{2} - x\right) \sec x$? ⓂP 2, 7 F.TF.8 **D**

- ○ **A** −1
- ○ **B** −tan x
- ○ **C** 1
- ○ **D** tan x
- ○ **E** sin² x

53. Suppose $90° < \theta < 180°$ and $\cos \theta = -\frac{4}{5}$. What is $\sin^2\theta$? ⓂP 2, 7 F.TF.8 **D**

- ○ **A** $-\frac{5}{4}$
- ○ **B** $\frac{3}{5}$
- ○ **C** $-\frac{9}{25}$
- ○ **D** $\frac{9}{25}$
- ○ **E** $\frac{16}{25}$

54. For $0° < \theta < 90°$, which expression is **NOT** valid? ⓂP 2, 7 F.TF.8 **A**

- ○ **A** $\cos (-\theta) = -\cos \theta$
- ○ **B** $\sin (-\theta) = -\sin \theta$
- ○ **C** $\tan (-\theta) = -\tan \theta$
- ○ **D** $\cos (-\theta) = \cos \theta$
- ○ **E** $\tan (-\theta) = -\frac{\sin \theta}{\cos \theta}$

55. MULTI-STEP

a. Which of the following trigonometric expressions is equivalent to $\frac{\sin (90° - \theta)}{\cos (\theta - 90°)}$? ⓂP 2, 7 F.TF.8 **D**

- ○ **A** sin θ
- ○ **B** cos θ
- ○ **C** tan θ
- ○ **D** cot θ
- ○ **E** sec θ

b. Use the trigonometric identities to simplify $\frac{\sin (90° - \theta)}{\cos (\theta - 90°)}$. See margin

56. To what is $\frac{\tan \theta}{\sec \theta}$ equivalent? ⓂP 2, 7 F.TF.8

> $\sin \theta$

57. To what is $(1 + \cos x)(1 - \cos x)$ equivalent? ⓂP 2, 7 F.TF.8

> $\sin^2 \theta$

58. If $\sec \theta = \frac{3}{5}$, what is $\cos \theta$ for $0° < \theta < 90°$? ⓂP 7 F.TF.8 **B**

- ○ **A** $\frac{3}{4}$
- ○ **B** $\frac{5}{3}$
- ○ **C** $\frac{4}{3}$
- ○ **D** $\frac{5}{4}$
- ○ **E** $\frac{4}{5}$

59. Which is **NOT** a Pythagorean identity? Choose all that apply. ⓂP 2 F.TF.8 **C, D**

- ☐ **A** $\cos^2 \theta = 1 - \sin^2 \theta$
- ☐ **B** $\tan^2 \theta + 1 = \sec^2 \theta$
- ☐ **C** $\sin^2 \theta + 1 = \cos^2 \theta$
- ☐ **D** $\tan^2 \theta + \cot^2 \theta = 1$
- ☐ **E** $\cot^2 \theta = \csc^2 \theta - 1$

54.

A	CORRECT
B	Chose a correct negative angle identity for sine
C	Chose a correct negative angle identity for tangent
D	Chose a correct negative angle identity for cosine
E	Chose a correct negative angle identity for tangent since $-\frac{\sin \theta}{\cos \theta} = -\tan \theta$

55.

A	See part **b** work.
B	See part **b** work.
C	See part **b** work.
D	CORRECT

Preparing for Assessment

Dual Coding		
Items	Content Standards	ⓂP Mathematical Practices
51	F.TF.8	2, 7
52	F.TF.8	2, 7
53	F.TF.8	2, 7
54	F.TF.8	2, 7
55-57	F.TF.8	2, 7
58	F.TF.8	7
59	F.TF.8	2

Diagnose Student Errors

Survey student responses for each item. Class trends may indicate common errors and misconceptions.

51.

A	Did not recognize the reciprocal identity and the negative angle identity as correct
B	Did not recognize the cofunction identity and reciprocal identity as correct
C	Did not recognize the negative angle identity as correct
D	Did not recognize the cofunction identity as correct
E	CORRECT

52.

A	Interpreted the cofunction identity as −cos x
B	Interpreted the cofunction identity as −sin x
C	Interpreted the cofunction identity as cos x
D	CORRECT
E	Interpreted sec x as sin x

53.

A	Found the reciprocal of cos θ
B	Computed sin θ
C	Assumed the sine function in Quadrant II was negative
D	CORRECT
E	Computed $\cos^2\theta$

Go Online!

Self-Check Quiz

Students can use *Self-Check Quizzes* to check their understanding of this lesson.

Verifying Trigonometric Identities

Track Your Progress

Objectives

1 Verify trigonometric identities by transforming one side of an equation into the form of the other side.

2 Verify trigonometric identities by transforming each side of the equation into the same form.

Mathematical Background

An identity can be verified by transforming one or both sides of an equation at the same time. However, the transformation of both sides must be performed independently.

THEN	NOW	NEXT
A.CED.2 Create equations in two or more variables to represent relationships between quantities; graph equations on coordinate axes with labels and scales.	**F.TF.8** Prove the Pythagorean identity $\sin^2(\theta) + \cos^2(\theta) = 1$ and use it to find $\sin(\theta)$, $\cos(\theta)$, or $\tan(\theta)$ given $\sin(\theta)$, $\cos(\theta)$, or $\tan(\theta)$ and the quadrant of the angle.	**F.TF.9** Prove the addition and subtraction formulas for sine, cosine, and tangent and use them to solve problems.

Go Online! All of these resources and more are available at connectED.mcgraw-hill.com

Personal Tutors (for every example) let students hear real teachers solve problems. Students can pause and repeat as many times as necessary.

Use **Self-Check Quiz** to assess students' understanding of the concepts in this lesson.

Chapter Project allows students to create and customize a project as a nontraditional method of assessment.

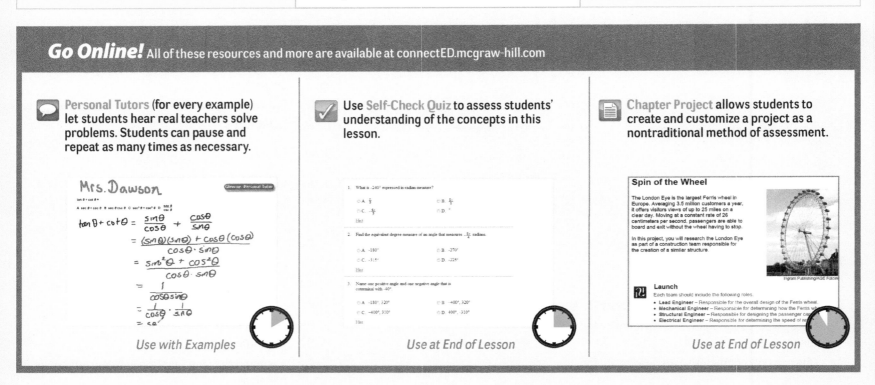

Use with Examples — *Use at End of Lesson* — *Use at End of Lesson*

OER Using Open Educational Resources

Lesson Sharing You can access premade slide presentations on **slideshare.net** about trigonometric identities. You can use these slides as ideas to add to your lesson or present them via your interactive white board. The comment section is a great place to learn from other educators who have already used these presentations in their own classrooms. *Use as planning*

Differentiate Your Resources

Extra Practice Additional practice or homework; Skills Practice is best for approaching-level students and Practice is best for on-level and beyond-level students

Skills Practice

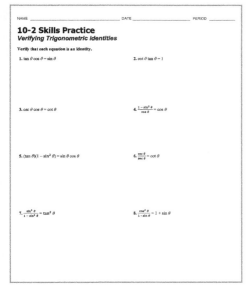

Practice

Word Problem Practice

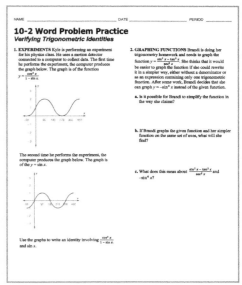

Intervention Reteaching and vocabulary activities that can be used with struggling or absent students and as ELL support

Study Guide and Intervention

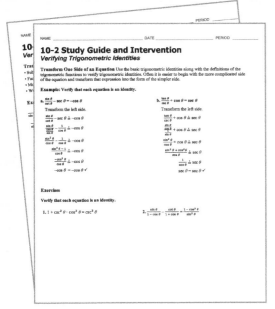

Study Notebook

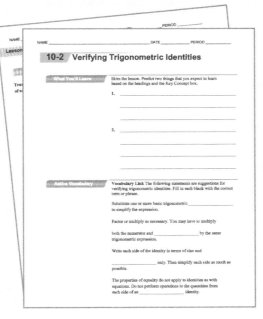

Extension Activities that can be used to extend lesson concepts

Enrichment

Launch

Have students read the Why? section of the lesson. Ask:

- On the right-hand side of the *angle of incline* equation, which variables appear in the numerator? In the denominator? *v; g* and *R*

- How can you express tan θ in terms of sin θ and cos θ? $\tan\theta = \dfrac{\sin\theta}{\cos\theta}$

- Does $\dfrac{\sin\theta}{\cos\theta}$ equal $\dfrac{v^2}{gR}$ or $\dfrac{gR}{v^2}$? $\dfrac{v^2}{gR}$

Teach

Ask the scaffolded questions for each example to build conceptual understanding for students at all levels.

1 Transform One Side of an Equation

Example 1 Transform One Side of an Equation

AL Why is $\sin^2\theta = 1 - \cos^2\theta$? Because if we subtract $\cos^2\theta$ from both sides of the Pythagorean identity $\sin^2\theta + \cos^2\theta = 1$ we get $\sin^2\theta = 1 - \cos^2\theta$.

OL Why don't we have to multiply both sides of the equation by $\dfrac{1+\cos\theta}{1+\cos\theta}$? $\dfrac{1+\cos\theta}{1+\cos\theta} = 1$, so when we multiply the left hand side of the equation by $\dfrac{1+\cos\theta}{1+\cos\theta}$ it does not change the expression.

BL Explain how you could verify the identity by first substituting $\sin^2\theta = 1 - \cos^2\theta$ in the numerator. $\dfrac{\sin^2\theta}{1-\cos\theta} = \dfrac{1-\cos^2\theta}{1-\cos\theta}$
$= \dfrac{(1-\cos\theta)(1+\cos\theta)}{1-\cos\theta} = 1 + \cos\theta$

(continued on the next page)

:: Then	:: Now	:: Why?
You used identities to find trigonometric values and simplify expressions.	**1** Verify trigonometric identities by transforming one side of an equation into the form of the other side. **2** Verify trigonometric identities by transforming each side of the equation into the same form.	While running on a circular track, Lamont notices that his body is not perpendicular to the ground. Instead, it leans away from a vertical position. The nonnegative acute angle θ that Lamont's body makes with the vertical is called the *angle of incline* and is described by the equation $\tan\theta = \dfrac{v^2}{gR}$. This is not the only equation that describes the angle of incline in terms of trigonometric functions. Another such equation is $\sin\theta = \cos\dfrac{v^2}{gR}\theta$, where $0 \le \theta \le 90°$. Are these two equations completely independent of one another or are they merely different versions of the same relationship?

MP **Mathematical Practices**

1 Make sense of problems and persevere in solving them.

7 Look for and express regularity in repeated reasoning.

Content Standards
F.TF.8 Prove the Pythagorean identity $\sin^2(\theta) + \cos^2(\theta) = 1$ and use it to find $\sin(\theta)$, $\cos(\theta)$, or $\tan(\theta)$ given $\sin(\theta)$, $\cos(\theta)$, or $\tan(\theta)$ and the quadrant of the angle.

1 Transform One Side of an Equation You can use the basic trigonometric identities along with the definitions of the trigonometric functions to verify identities. If you wish to show an identity, you need to show that it is true for all values of θ.

> **Key Concept** Verifying Identities by Transforming One Side
>
> **Step 1** Simplify one side of an equation until the two sides of the equation are the same. It is often easier to work with the more complicated side of the equation.
>
> **Step 2** Transform that expression into the form of the simpler side.

F.TF.8

Example 1 Transform One Side of an Equation

Verify that $\dfrac{\sin^2\theta}{1-\cos\theta} = 1 + \cos\theta$ is an identity.

$\dfrac{\sin^2\theta}{1-\cos\theta} \overset{?}{=} 1 + \cos\theta$ — Original equation

$\dfrac{1+\cos\theta}{1+\cos\theta} \cdot \dfrac{\sin^2\theta}{1-\cos\theta} \overset{?}{=} 1 + \cos\theta$ — Multiply the numerator and denominator by $1 + \cos\theta$

$\dfrac{\sin^2\theta(1+\cos\theta)}{1-\cos^2\theta} \overset{?}{=} 1 + \cos\theta$ — $(1+\cos\theta)(1-\cos\theta) = 1 - \cos^2\theta$

$\dfrac{\sin^2\theta(1+\cos\theta)}{\sin^2\theta} \overset{?}{=} 1 + \cos\theta$ — $\sin^2\theta = 1 - \cos^2\theta$

$1 + \cos\theta = 1 + \cos\theta$ ✓ — Divide the numerator and denominator by $\sin^2\theta$

> **Guided Practice**
>
> 1. Verify that $\cot^2\theta - \cos^2\theta = \cot^2\theta \cos^2\theta$ is an identity. See margin.

MP **Mathematical Practices Strategies**

Make sense of problems and persevere in solving them. Help students understand that they can verify trigonometric identities using deduction. For example, ask:

- How would you check to see if $x = 5$ is a solution to $5x - 4 = 20$? Is $x = 5$ a solution? Why or why not? Substitute the value $x = 5$ into the expression on the left. Since the substitution does not result in 20, the value on the right, it is not a solution.

- To verify identities, what could you do? Substitute identities to simplify one side in order to see if it is equal to the more simplified side.

- If the sides are not equal, what do you know? The proposed identity is false.

When verifying trigonometric identities, it is often helpful to rewrite an expression in terms of sine and cosine before simplifying.

F.TF.8

Example 2 Rewrite Expressions

Verify that $\dfrac{1 - \cos \theta}{1 + \cos \theta} = (\csc \theta - \cot \theta)^2$.

Rewrite the expression on the right side of the equation in terms of $\sin \theta$ and $\cos \theta$. Then simplify.

$\dfrac{1 - \cos \theta}{1 + \cos \theta} \overset{?}{=} (\csc \theta - \cot \theta)^2$ Original equation

$\dfrac{1 - \cos \theta}{1 + \cos \theta} \overset{?}{=} \csc^2\theta - 2 \cot \theta \csc \theta + \cot^2 \theta$ Multiply

$\dfrac{1 - \cos \theta}{1 + \cos \theta} \overset{?}{=} \dfrac{1}{\sin^2 \theta} - 2 \cdot \dfrac{\cos \theta}{\sin \theta} \cdot \dfrac{1}{\sin \theta} + \dfrac{\cos^2 \theta}{\sin^2 \theta}$ $\csc \theta = \frac{1}{\sin \theta}, \cot \theta = \frac{\cos \theta}{\sin \theta}$

$\dfrac{1 - \cos \theta}{1 + \cos \theta} \overset{?}{=} \dfrac{1}{\sin^2\theta} - \dfrac{2 \cos \theta}{\sin^2 \theta} + \dfrac{\cos^2 \theta}{\sin^2 \theta}$ Simplify

$\dfrac{1 - \cos \theta}{1 + \cos \theta} \overset{?}{=} \dfrac{1 - 2\cos\theta + \cos^2 \theta}{\sin^2\theta}$ Add and subtract.

$\dfrac{1 - \cos \theta}{1 + \cos \theta} \overset{?}{=} \dfrac{(1 - \cos \theta)(1 - \cos \theta)}{1 - \cos^2 \theta}$ Factor the numerator; $\sin^2 \theta = 1 - \cos^2 \theta$.

$\dfrac{1 - \cos \theta}{1 + \cos \theta} \overset{?}{=} \dfrac{(1 - \cos \theta)(1 - \cos \theta)}{(1 - \cos \theta)(1 + \cos \theta)}$ Factor the denominator.

$\dfrac{1 - \cos \theta}{1 + \cos \theta} = \dfrac{1 - \cos \theta}{1 + \cos \theta}$ ✓ Simplify.

> **Watch Out!**
> **Verify Identities** Verifying an identity is like checking the solution of an equation. You must simplify one or both sides separately until they are the same.

▶ **Guided Practice**

Verify that each equation is an identity. 2A–B. See margin.

2A. $\tan^2 \theta (\cot^2 \theta - \cos^2 \theta) = \cos^2 \theta$ **2B.** $\cos^2 \theta \sec \theta \csc \theta = \cot \theta$

2 Transform Each Side of an Equation Sometimes it is easier to transform each side of an equation separately into a common form. The following suggestions may be helpful as you verify trigonometric identities.

Key Concept Suggestions for Verifying Identities

- Substitute one or more basic trigonometric identities to simplify the expression.

- Factor or multiply as necessary. You may have to multiply both the numerator and denominator by the same trigonometric expression.

- Write each side of the identity in terms of sine and cosine only. Then simplify each side as much as possible.

- The properties of equality do not apply to identities as with equations. Do not perform operations to the quantities on each side of an unverified identity.

Differentiated Instruction AL OL BL ELL

Interpersonal Learners Have groups or pairs of students work together to verify some of the identities in Exercises 10–19. Have students record the techniques they found helpful. Ask students to compare their list of techniques to the list of suggestions given at the bottom of p. 663. Also have them discuss failed strategies. What did and did not work, and why?

Need Another Example?
Verify that $\csc \theta \cos \theta \tan \theta = 1$ is an identity.

$\csc \theta \cos \theta \tan \theta \overset{?}{=} 1$

$\dfrac{1}{\sin \theta} \cdot \cos \theta \cdot \dfrac{\sin \theta}{\cos \theta} \overset{?}{=} 1$

$1 = 1$ ✓

Example 2 Rewrite Expressions

AL How do we expand $(\csc \theta - \cot \theta)^2$? FOIL

OL What type of expression is $1 - \cos^2 \theta$? difference of squares

BL Why is it best to start with the right-hand side of this equation? It is the more complicated side, so it is better to simplify it to the left-hand side.

Need Another Example?
Verify that $\dfrac{\csc \theta}{\cos \theta} - \tan \theta = \cot \theta$

$\dfrac{\csc \theta}{\cos \theta} - \tan \theta \overset{?}{=} \cot \theta$

$\dfrac{1}{\cos \theta \sin \theta} - \dfrac{\sin\theta}{\cos \theta} \overset{?}{=} \cot \theta$

$\dfrac{1 - \sin^2 \theta}{\cos\theta \sin\theta} \overset{?}{=} \cot \theta$

$\dfrac{\cos^2 \theta}{\cos\theta \sin\theta} \overset{?}{=} \cot \theta$

$\dfrac{\cos\theta}{\sin\theta} \overset{?}{=} \cot \theta$

$\cot \theta = \cot \theta$ ✓

Additional Answers (Guided Practice)

1. $\cot^2 \theta - \cos^2 \theta \overset{?}{=} \cot^2 \theta \cos^2 \theta$

$\dfrac{\cos^2 \theta}{\sin^2 \theta} - \cos^2 \theta \overset{?}{=} \cot^2 \theta \cos^2 \theta$

$\cos^2 \theta \left(\dfrac{1}{\sin^2 \theta} - 1 \right) \overset{?}{=} \cot^2 \theta \cos^2 \theta$

$\cos^2 \theta (\csc^2 \theta - 1) \overset{?}{=} \cot^2 \theta \cos^2 \theta$

$\cot^2 \theta \cos^2 \theta = \cot^2 \theta \cos^2 \theta$ ✓

2A. $\tan^2 \theta - \cos^2 \theta \overset{?}{=} \cos^2 \theta$

$\dfrac{\sin^2 \theta}{\cos^2 \theta} \left(\dfrac{\cos^2 \theta}{\sin^2 \theta} - \cos^2 \theta \right) \overset{?}{=} \cos^2 \theta$

$\dfrac{\sin^2 \theta \cos^2 \theta}{\sin^2 \theta \cos^2 \theta} - \dfrac{\sin^2 \theta \cos^2\theta}{\cos^2\theta} \overset{?}{=} \cos^2 \theta$

$1 - \sin^2 \theta \overset{?}{=} \cos^2 \theta$

$\cos^2 \theta = \cos^2 \theta$ ✓

2B. $\cos^2 \theta \sec \theta \csc \theta \overset{?}{=} \cot \theta$

$\cos^2 \theta \dfrac{1}{\cos \theta} \dfrac{1}{\sin \theta} \overset{?}{=} \cot \theta$

$\cos \theta \left(\dfrac{1}{\sin \theta} \right) \overset{?}{=} \cot \theta$

$\dfrac{\cos \theta}{\sin \theta} \overset{?}{=} \cot \theta$

$\cot \theta = \cot \theta$ ✓

2 Transform Each Side of an Equation

Example 3 Verify by Transforming Each Side

AL When have you verified the identity? when each side of the equation is the same

OL Why is $\sec^2 \theta (2 - \sec^2 \theta) = (2 - \sec^2 \theta) \sec^2 \theta$? because multiplication is commutative

BL Why is $\sec^2 \theta (2 - \sec^2 \theta)$ always less than 1? $\tan^4 \theta$ is always positive, so $1 - \tan^4 \theta$ is always less than 1.

Need Another Example?
Verify that $\csc \theta + \sec \theta = \dfrac{1 + \cot \theta}{\cos \theta}$ is an identity.

$$\csc \theta + \sec \theta \overset{?}{=} \frac{1 + \cot \theta}{\cos \theta}$$

$$\frac{1}{\sin \theta} + \frac{1}{\cos \theta} \overset{?}{=} \frac{1 + \frac{\cos \theta}{\sin \theta}}{\cos \theta}$$

$$\frac{\cos \theta + \sin \theta}{\sin \theta \cos \theta} \overset{?}{=} \frac{\sin \theta \left(1 + \frac{\cos \theta}{\sin \theta}\right)}{\sin \theta \cos \theta}$$

$$\frac{\cos \theta + \sin \theta}{\sin \theta \cos \theta} = \frac{\sin \theta + \cos \theta}{\sin \theta \cos \theta} \checkmark$$

Teaching the Mathematical Practices

Precision Mathematically proficient students try to use clear definitions in their reasoning, calculate accurately and efficiently, and make explicit use of definitions.

Additional Answer (Guided Practice)

3. $\csc^2 \theta - \cot^2 \theta \overset{?}{=} \cot \theta \tan \theta$

$$\frac{1}{\sin^2 \theta} - \frac{\cos^2 \theta}{\sin^2 \theta} \overset{?}{=} \frac{\cos \theta}{\sin \theta} \cdot \frac{\sin \theta}{\cos \theta}$$

$$\frac{1 - \cos^2 \theta}{\sin^2 \theta} \overset{?}{=} 1$$

$$\frac{\sin^2 \theta}{\sin^2 \theta} \overset{?}{=} 1$$

$$1 = 1 \checkmark$$

Go Online! eBook

Interactive Student Guide
Use the *Interactive Student Guide* to deepen conceptual understanding.
· Verifying the Pythagorean Identity

F.TF.8

Go Online!
Watch *Personal Tutor* videos to hear descriptions of how to verify trigonometric identities. Try describing how to verify an identity for a partner. **ELL**

Example 3   Verify by Transforming Each Side

Verify that $1 - \tan^4 \theta = 2 \sec^2 \theta - \sec^4 \theta$ is an identity.

$1 - \tan^4 \theta \overset{?}{=} 2 \sec^2 \theta - \sec^4 \theta$	Original equation
$(1 - \tan^2 \theta)(1 + \tan^2 \theta) \overset{?}{=} \sec^2 \theta (2 - \sec^2 \theta)$	Factor each side
$[1 - (\sec^2 \theta - 1)] \sec^2 \theta \overset{?}{=} (2 - \sec^2 \theta) \sec^2 \theta$	$1 + \tan^2 \theta = \sec^2 \theta$
$(2 - \sec^2 \theta) \sec^2 \theta = (2 - \sec^2 \theta) \sec^2 \theta \checkmark$	Simplify

▶ **Guided Practice**

3. Verify that $\csc^2 \theta - \cot^2 \theta = \cot \theta \tan \theta$ is an identity. **See margin.**

Go Online! for a Self-Check Quiz

Check Your Understanding   ◯ = Step-by-Step Solutions begin on page R11.

Examples 1–3   **MP** **PRECISION** Verify that each equation is an identity. 1–9. See Ch. 10 Answer Appendix.
F.TF.8

1. $\cot \theta + \tan \theta = \dfrac{\sec^2 \theta}{\tan \theta}$

2. $\cos^2 \theta = (1 + \sin \theta)(1 - \sin \theta)$

3. $\sin \theta = \dfrac{\sec \theta}{\tan \theta + \cot \theta}$

4. $\tan^2 \theta = \dfrac{1 - \cos^2 \theta}{\cos^2 \theta}$

5. $\tan^2 \theta \csc^2 \theta = 1 + \tan^2 \theta$

6. $\tan^2 \theta = (\sec \theta + 1)(\sec \theta - 1)$

7. $\dfrac{\tan^2 \theta + 1}{\tan^2 \theta} = \csc^2 \theta$

8. $\cot \theta + \sec \theta = \dfrac{\cos^2 \theta + \sin \theta}{\sin \theta \cos \theta}$

9. $\sin^2 \theta + \tan^2 \theta = (1 - \cos^2 \theta) + \dfrac{\sec^2 \theta}{\csc^2 \theta}$

Practice and Problem Solving   Extra Practice is on page R10.

Example 1   Verify that each equation is an identity. 10–19. See Ch. 10 Answer Appendix.
F.TF.8

10. $\cos^2 \theta + \tan^2 \theta \cos^2 \theta = 1$

11. $\cot \theta (\cot \theta + \tan \theta) = \csc^2 \theta$

12. $1 + \sec^2 \theta \sin^2 \theta = \sec^2 \theta$

13. $\sin \theta \sec \theta \cot \theta = 1$

14. $\dfrac{1 - \cos \theta}{1 + \cos \theta} = (\csc \theta - \cot \theta)^2$

15. $\dfrac{1 - 2 \cos^2 \theta}{\sin \theta \cos \theta} = \tan \theta - \cot \theta$

16. $\tan \theta = \dfrac{\sec \theta}{\csc \theta}$

17. $\cos \theta = \sin \theta \cot \theta$

18. $(\sin \theta - 1)(\tan \theta + \sec \theta) = -\cos \theta$

19. $\cos \theta \cos (-\theta) - \sin \theta \sin (-\theta) = 1$

Example 2   **20. LADDER** Some students derived an expression for the length of a ladder that, when carried flat, could fit around a corner from a 5-foot-wide hallway into a 7-foot-wide hallway, as shown. They determined that the maximum length ℓ of a ladder that would fit was given by $\ell(\theta) = \dfrac{7 \sin \theta + 5 \cos \theta}{\sin \theta \cos \theta}$. When their teacher worked the problem, she concluded that $\ell(\theta) = 7 \sec \theta + 5 \csc \theta$. Are the two expressions equivalent? **yes**
F.TF.8

7 ft

5 ft

Differentiated Homework Options

Levels	**AL** Basic	**OL** Core	**BL** Advanced
Exercises	10–34, 54–69	11–33 odd, 35, 36, 37–51 odd, 52–69	35–69
2-Day Option	11–33 odd, 62–69	10–34, 62–69	
	10–34 even, 54–61	35–61	

 You can use ALEKS to provide additional remediation support with personalized instruction and practice.

Example 3
F.TF.8

Verify that each equation is an identity. 21–34. See Ch. 10 Answer Appendix.

21. $\sec \theta - \tan \theta = \dfrac{1 - \sin \theta}{\cos \theta}$

22. $\dfrac{1 + \tan \theta}{\sin \theta + \cos \theta} = \sec \theta$

23. $\sec \theta \csc \theta = \tan \theta + \cot \theta$

24. $\sin \theta + \cos \theta = \dfrac{2 \sin^2 \theta - 1}{\sin \theta - \cos \theta}$

25. $(\sin \theta + \cos \theta)^2 = \dfrac{2 + \sec \theta \csc \theta}{\sec \theta \csc \theta}$

26. $\dfrac{\cos \theta}{1 - \sin \theta} = \dfrac{1 + \sin \theta}{\cos \theta}$

27. $\csc \theta - 1 = \dfrac{\cot^2 \theta}{\csc \theta + 1}$

28. $\cos \theta \cot \theta = \csc \theta - \sin \theta$

29. $\sin \theta \cos \theta \tan \theta + \cos^2 \theta = 1$

30. $(\csc \theta - \cot \theta)^2 = \dfrac{1 - \cos \theta}{1 + \cos \theta}$

31. $\csc^2 \theta = \cot^2 \theta + \sin \theta \csc \theta$

32. $\dfrac{\sec \theta - \csc \theta}{\csc \theta \sec \theta} = \sin \theta - \cos \theta$

33. $\sin^2 \theta + \cos^2 \theta = \sec^2 \theta - \tan^2 \theta$

34. $\sec \theta - \cos \theta = \tan \theta \sin \theta$

 35. **SENSE-MAKING** The diagram at the right represents a game of tetherball. As the ball rotates around the pole, a conical surface is swept out by the line segment $\overline{SP}$. A formula for the relationship between the length L of the string and the angle θ that the string makes with the pole is given by the equation $L = \dfrac{g \sec \theta}{\omega^2}$. Is $L = \dfrac{g \tan \theta}{\omega^2 \sin \theta}$ also an equation for the relationship between L and θ? yes

36. RUNNING A portion of a racetrack has the shape of a circular arc with a radius of 16.7 meters. As a runner races along the arc, the sine of her angle of incline θ is found to be $\frac{1}{4}$. Find the speed of the runner. Use the Angle of Incline Formula given at the beginning of the lesson, $\tan \theta = \dfrac{v^2}{gR}$, where $g = 9.8$ and R is the radius. (*Hint*: Find $\cos \theta$ first.) 6.5 m/s

When simplified, would the expression be equal to 1 or -1?

37. $\cot (-\theta) \tan(-\theta)$ 1
38. $\sin \theta \csc (-\theta)$ -1
39. $\sin^2 (-\theta) + \cos^2 (-\theta)$ 1

40. $\sec (-\theta) \cos (-\theta)$ 1
41. $\sec^2 (-\theta) - \tan^2 (-\theta)$ 1
42. $\cot (-\theta) \cot \left(\frac{\pi}{2} - \theta\right)$ -1

Simplify the expression to either a constant or a basic trigonometric function.

43. $\dfrac{\tan \left(\frac{\pi}{2} - \theta\right) \csc \theta}{\csc^2 \theta}$ $\cos \theta$

44. $\dfrac{1 + \tan \theta}{1 + \cot \theta}$ $\tan \theta$

45. $(\sec^2 \theta + \csc^2 \theta) - (\tan^2 \theta + \cot^2 \theta)$ 2

46. $\dfrac{\sec^2 \theta - \tan^2 \theta}{\cos^2 x + \sin^2 x}$ 1

47. $\tan \theta \cos \theta$ $\sin \theta$

48. $\cot \theta \tan \theta$ 1

49. $\sec \theta \sin \left(\frac{\pi}{2} - \theta\right)$ 1

50. $\dfrac{1 + \tan^2 \theta}{\csc^2 \theta}$ $\tan^2 \theta$

51. $y = -\dfrac{gx^2}{2v_0^2}(1 + \tan^2 \theta) + x \tan \theta$

51. PHYSICS When a firework is fired from the ground, its height y and horizontal displacement x are related by the equation $y = \dfrac{-gx^2}{2v_0^2 \cos^2 \theta} + \dfrac{x \sin \theta}{\cos \theta}$, where v_0 is the initial velocity of the projectile, θ is the angle at which it was fired, and g is the acceleration due to gravity. Rewrite this equation so that $\tan \theta$ is the only trigonometric function that appears in the equation.

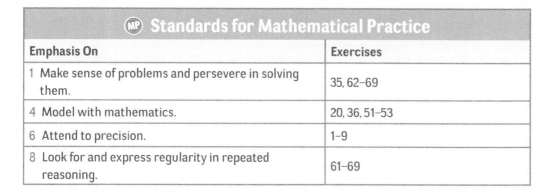

21–34. See Ch. 10 Answer Appendix.

Practice

Formative Assessment Use Exercises 1–9 to assess students' understanding of the concepts in this lesson.

The Practice and Problem Solving exercises assess the content taught in the lesson. The Preparing for Assessment page is meant to be used as preparation for assessment.

Extra Practice

See page R10 for extra exercises for students who are approaching level or for on-level students who need additional reinforcement.

MP **Teaching the Mathematical Practices**

Sense-Making Mathematically proficient students start by explaining to themselves the meaning of a problem and looking for entry points to its solution. They analyze givens, constraints, relationships, and goals. They check their answers to problems using a different method, and they continually ask themselves, "Does this make sense?"

Follow-Up

Students have explored and verified trigonometric identities.
Ask:
- **Why are trigonometric identities useful?** Sample answer: Trigonometric identities provide a way to simplify complex trigonometric functions by rewriting the expressions found within the functions in equivalent, but more convenient forms.

MP **Standards for Mathematical Practice**	
Emphasis On	**Exercises**
1 Make sense of problems and persevere in solving them.	35, 62–69
4 Model with mathematics.	20, 36, 51–53
6 Attend to precision.	1–9
8 Look for and express regularity in repeated reasoning.	61–69

Levels of Complexity Chart

The levels of the exercises progress from 1 to 3, with Level 1 indicating the lowest level of complexity.

Exercises	10–34	35–51, 62–69	52–61
Level 3			●
Level 2		●	
Level 1	●		

Assess

Yesterday's News Have students write how learning basic trigonometric identities in Lesson 10-1 has helped them with verifying the more complex identities in today's lesson.

Additional Answers

59. Sample answer: Sine and cosine are the trigonometric functions with which most people are familiar, and all trigonometric expressions can be written in terms of sine and cosine. Also, by rewriting complex trigonometric expressions in terms of sine and cosine it may be easier to perform operations and to apply trigonometric properties.

65b. $\dfrac{\sin^2 \theta + \cos^2 \theta}{\tan \theta \cos \theta} = \dfrac{1}{\tan \theta \cos \theta}$ Use the Pythagorean Identity.

$= \dfrac{1}{\left(\dfrac{\sin \theta}{\cos \theta}\right) \cos \theta}$ Substitute the quotient identity for $\tan \theta$.

$= \dfrac{1}{\sin \theta}$ Simplify.

66. $\dfrac{\tan \theta}{\sin \theta} = \tan \theta \left(\dfrac{1}{\sin \theta}\right)$

$= \dfrac{\sin \theta}{\cos \theta}\left(\dfrac{1}{\sin \theta}\right)$

$= \dfrac{1}{\cos \theta}$

$= \sec \theta$

67. $\sin x \cos x \tan x = \sin x \cos x \left(\dfrac{\sin x}{\cos x}\right)$

$= \sin^2 x$

$= 1 - \cos^2 x$

---

52. ELECTRONICS When an alternating current of frequency f and peak current I_0 passes through a resistance R, the power delivered to the resistance at time t seconds is $P = I_0{}^2 R \sin^2 2\pi ft$.

    a. Write an expression for the power in terms of $\cos^2 2\pi ft$. $P = I_0{}^2 R(1 - \cos^2 2\pi ft)$

    b. Write an expression for the power in terms of $\csc^2 2\pi ft$. $P = \dfrac{I_0{}^2 R}{\csc^2 2\pi ft}$

53 THROWING A BALL In this problem, you will investigate the path of a ball represented by the equation $h = \dfrac{v_0{}^2 \sin^2 \theta}{2g}$, where θ is the measure of the angle between the ground and the path of the ball, v_0 is its initial velocity in meters per second, and g is the acceleration due to gravity. The value of g is 9.8 m/s^2.

53a–c. See Ch. 10 Answer Appendix.

    a. If the initial velocity of the ball is 47 meters per second, find the height of the ball at $30°$, $45°$, $60°$, and $90°$. Round to the nearest tenth.

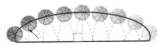

    b. Graph the equation on a graphing calculator.

    c. Show that the formula $h = \dfrac{v_0{}^2 \tan^2 \theta}{2g \sec^2 \theta}$ is equivalent to the one given above.

F.TF.8

H.O.T. Problems   Use Higher-Order Thinking Skills

54. WHICH ONE DOESN'T BELONG? Identify the equation that does not belong with the other three. Explain your reasoning.

54. $\sin^2 \theta - \cos^2 \theta = 2\sin^2 \theta$; the other three are Pythagorean identities, but this is not.

$\sin^2 \theta + \cos^2 \theta = 1$	$1 + \cot^2 \theta = \csc^2 \theta$
$\sin^2 \theta - \cos^2 \theta = 2 \sin^2 \theta$	$\tan^2 \theta + 1 = \sec^2 \theta$

55. $\tan^2 \theta = \dfrac{\sin^2 \theta}{\cos^2 \theta}$

$= \dfrac{1 - \cos^2 \theta}{\cos^2 \theta}$

$= \dfrac{1}{\cos^2 \theta} - \dfrac{\cos^2 \theta}{\cos^2 \theta}$

$= \sec^2 \theta - 1$

55. CHALLENGE Transform the right side of $\tan^2 \theta = \dfrac{\sin^2 \theta}{\cos^2 \theta}$ to show that $\tan^2 \theta = \sec^2 \theta - 1$.

56. WRITING IN MATH Explain why you cannot square each side of an equation when verifying a trigonometric identity. The properties of equality do not apply to identities as they do with equations. Do not perform operations to the quantities from each side of an unverified identity.

57. **REASONING** Explain why $\sin^2 \theta + \cos^2 \theta = 1$ is an identity, but $\sin \theta = \sqrt{1 - \cos \theta}$ is not.   Sample answer: counterexample $45°$, $30°$

58. WRITE A QUESTION A classmate is having trouble trying to verify a trigonometric identity involving multiple trigonometric functions to multiple degrees. Write a question to help her work through the problem. Sample answer: Have you tried using the most common identity, $\sin^2 \sigma + \cos^2 \sigma = 1$, to simplify?

59. ⓔ **WRITING IN MATH** Why do you think expressions in trigonometric identities are often rewritten in terms of sine and cosine?   See margin.

60. CHALLENGE Let $x = \frac{1}{2} \tan \theta$, where $-\frac{\pi}{2} < \theta < \frac{\pi}{2}$. Write $f(x) = \dfrac{x}{\sqrt{1 + 4x^2}}$ in terms of a single trigonometric function of θ. $f(\theta) = \frac{1}{2}\sin \theta$

61. **REASONING** Justify the three basic Pythagorean identities.   See Ch. 10 Answer Appendix.

Differentiated Instruction   BL

Extension $\sin x + \cos x = 1$ is not an identity, meaning that it is not true for all real values of x. Find the values of x for which the equation is true. $0° + k \cdot 360°$ or $90° + k \cdot 360°$, where k is any integer

Preparing for Assessment

62. If $\tan(-\theta) \cos(-\theta) = Y$, then which of the following expressions could be Y? ⊕ 1, 8 F.TF.8 **E**

- ○ **A** $\dfrac{\cos^2 \theta}{\sin \theta}$
- ○ **B** 1
- ○ **C** -1
- ○ **D** $-\csc \theta$
- ○ **E** $-\sin \theta$

63. Which trigonometric function when defined is equivalent to $\dfrac{\cos^2 \alpha}{\sin \alpha \cot \alpha}$? ⊕ 1, 8 F.TF.8 **A**

- ○ **A** $\cos \alpha$
- ○ **B** $\dfrac{\cos \alpha}{\sin^2 \alpha}$
- ○ **C** $\dfrac{1}{\cos \alpha}$
- ○ **D** $\cot^2 \alpha$
- ○ **E** $\dfrac{\cos^3 \alpha}{\sin^2 \alpha}$

64. For which value(s) for h is $h \sin \theta = \sin(h\theta)$ an identity? ⊕ 1, 8 F.TF.8 **E**

    I $h = -1$
    II $h = 0$
    III $h = 1$

- ○ **A** I and II only
- ○ **B** II only
- ○ **C** III only
- ○ **D** II and III only
- ○ **E** I, II, and III

65. MULTI-STEP

a. For all $\theta \neq \dfrac{n\pi}{2}$, the identity shown below is complete for which of the following expressions? ⊕ 1, 8 F.TF.8 **C**

$$\frac{\sin^2 \theta + \cos^2 \theta}{\tan \theta \cos \theta} = \underline{\hspace{2cm}}$$

- ○ **A** 1　　○ **D** $\sin \theta + \cos \theta$
- ○ **B** $\dfrac{1}{\cos \theta}$　　○ **E** $\sin \theta + \dfrac{\cos^2 \theta}{\sin \theta}$
- ○ **C** $\dfrac{1}{\sin \theta}$

b. Use trigonometric identities to simplify the expression in part **a**. See margin

66. Prove that $\dfrac{\tan \theta}{\sin \theta} = \sec \theta$. ⊕ 1, 8 F.TF.8 See margin

67. Prove that $\sin x \cos x \tan x = 1 - \cos^2 x$. ⊕ 1, 8 F.TF.8 See margin

68. When simplified, which expressions would be equal to -1? Check all that apply. ⊕ 1, 8 F.TF.8 **A, D**

- ☐ **A** $\csc \theta \sin(-\theta)$
- ☐ **B** $\cos \theta \sec(-\theta)$
- ☐ **C** $\sin^2 \theta + \cos^2 \theta$
- ☐ **D** $\cot(-\theta) \tan \theta$
- ☐ **E** $\sec^2(-\theta) - \tan^2(-\theta)$

69. What is the simplified form of $\cot \theta \sec \theta$? ⊕ 1, 8 **D** F.TF.8

- ○ **A** 1
- ○ **B** -1
- ○ **C** $\sin \theta$
- ○ **D** $\csc \theta$
- ○ **E** $\tan \theta$

Preparing for Assessment

Dual Coding		
Items	Content Standards	⊕ Mathematical Practices
62	F.TF.8	1, 8
63	F.TF.8	1, 8
64	F.TF.8	1, 8
65	F.TF.8	1, 8
66-67	F.TF.8	1, 8
68-69	F.TF.8	1, 8

Diagnose Student Errors

Survey student responses for each item. Class trends may indicate common errors and misconceptions.

62.

A	Substituted cotangent ratio for tangent
B	Substituted $-\dfrac{1}{\tan \theta}$ for $\cos(-\theta)$
C	Substituted $\dfrac{1}{\tan \theta}$ for $\cos(-\theta)$
D	Attempted to use reciprocal ratio for sine
E	CORRECT

63.

A	CORRECT
B	Canceled cosine terms instead of sine terms
C	Used reciprocal of answer
D	Incorrectly substituted cotangent ratio
E	Substituted tangent ratio for cotangent

64.

A	Did not find that $h = 1$ resulted in an equivalent statement
B	Did not appropriately apply the negative angle identity and did not find $h = 1$ to be true
C	Did not appropriately apply the negative angle identity and did not know $\sin 0 = 0$
D	Did not appropriately apply the negative angle identity
E	CORRECT

Go Online!

Self-Check Quiz

Students can use *Self-Check Quizzes* to check their understanding of this lesson. You can also give the *Chapter Quiz*, which covers the content in Lessons 10-1 and 10-2.

Sum and Difference Identities

Track Your Progress

Objectives

1 Find values of sine and cosine by using sum and difference identities.

2 Verify trigonometric identities by using sum and difference identities.

Mathematical Background

The identities for the sum and difference of two angles, sin $(a \pm b)$, cos $(a \pm b)$, and tan $(a \pm b)$, can be used to find values of sine and cosine. They can also be used to help verify identities such as sin $(180° + \theta) = -\sin \theta$.

THEN	NOW	NEXT
F.TF.3 Use special triangles to determine geometrically the values of sine, cosine, tangent for $\frac{\pi}{3}$, $\frac{\pi}{4}$ and $\frac{\pi}{6}$, and use the unit circle to express the values of sine, cosine, and tangent for $\pi - x$, $\pi + x$, and $2\pi - x$ in terms of their values for x, where x is any real number.	**F.TF.8** Prove the Pythagorean identity $\sin^2(\theta) + \cos^2(\theta) = 1$ and use it to find $\sin(\theta)$, $\cos(\theta)$, or $\tan(\theta)$ given $\sin(\theta)$, $\cos(\theta)$, or $\tan(\theta)$ and the quadrant of the angle.	**F.TF.9** Prove the addition and subtraction formulas for sine, cosine, and tangent and use them to solve problems.

Go Online! All of these resources and more are available at connectED.mcgraw-hill.com

Personal Tutors (for every example) let students hear real teachers solve problems. Students can pause and repeat as many times as necessary.

Animations illustrate key concepts through step-by-step tutorials and videos.

Use **Self-Check Quiz** to assess students' understanding of the concepts in this lesson.

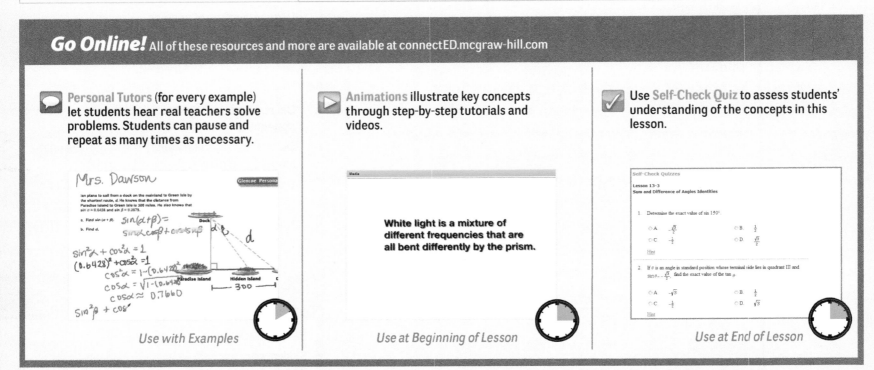

Use with Examples

White light is a mixture of different frequencies that are all bent differently by the prism.

Use at Beginning of Lesson

Use at End of Lesson

OER **Using Open Educational Resources**

Practice Have students play the Trigonometric Identities games on **ixl.com** to practice working with the trigonometric identities they have learned throughout the chapter. The site tracks students' scores, and the questions automatically increase in difficulty as they progress through the game. *Use as homework or review*

Differentiate Your Resources

Extra Practice Additional practice or homework; Skills Practice is best for approaching-level students and Practice is best for on-level and beyond-level students

Skills Practice

NAME _____ DATE _____ PERIOD _____

10-3 Skills Practice
Sum and Difference Identities

Find the exact value of each expression.

1. sin 330°   2. cos (−165°)   3. sin (−225°)

4. cos 135°   5. sin (−45°)   6. cos 210°

7. cos (−135°)   8. sin 75°   9. sin (−195°)

Verify that each equation is an identity.

10. sin (90° + θ) = cos θ

11. sin (180° + θ) = −sin θ

12. cos (270° − θ) = −sin θ

13. cos (θ − 90°) = sin θ

14. sin (θ − π/2) = −cos θ

15. cos (π + θ) = −cos θ

Practice

NAME _____ DATE _____ PERIOD _____

10-3 Practice
Sum and Difference Identities

Find the exact value of each expression.

1. cos 75°   2. cos 375°   3. sin (−165°)

4. sin (−105°)   5. sin 150°   6. cos 240°

7. sin 225°   8. sin (−75°)   9. sin 195°

Verify that each equation is an identity.

10. cos (180° − θ) = −cos θ

11. sin (360° + θ) = sin θ

12. sin (45° + θ) − sin (45° − θ) = √2 sin θ

13. cos (x − π/6) + sin (x − π/3) = sin x

14. **SOLAR ENERGY** On March 21, the maximum amount of solar energy that falls on a square foot of ground at a certain location is given by E sin (90° − φ), where φ is the latitude of the location and E is a constant. Use the difference of angles formula to find the amount of solar energy, in terms of cos φ, for a location that has a latitude of φ.

15. **ELECTRICITY** In a certain circuit carrying alternating current, the formula c = 2 sin (120t) can be used to find the current c in amperes after t seconds.

a. Rewrite the formula using the sum of two angles.

b. Use the sum of angles formula to find the exact current at t = 1 second.

Word Problem Practice

NAME _____ DATE _____ PERIOD _____

10-3 Word Problem Practice
Sum and Difference Identities

1. **ART** As part of a mosaic that an artist is making, she places two right triangular tiles together to make a new triangular piece. One tile has lengths of 3 inches, 4 inches, and 5 inches. The other tile has lengths 4 inches, 4√3 inches, and 8 inches. The pieces are placed with the sides of 4 inches against each other, as shown in the figure below.

a. What is the exact value of the sine of angle A?

b. What is the exact value of the cosine of angle A?

c. What is the measure of angle A?

d. Is the new triangle formed from the two triangles also a right triangle?

2. **MANUFACTURING** A robotic arm performs two operations in the course of manufacturing each car door that comes down the assembly line. As the door arrives in front of the robotic arm, the arm is parallel to the assembly line. Once the door is in place, the arm rotates counterclockwise 72° to perform the first operation. After performing the first operation, the arm rotates another 33° counterclockwise to perform the second operation. After the second operation is completed, the robotic arm rotates clockwise back to its starting position.

a. After completing the second operation, through what angle does the robotic arm rotate to return to its starting position?

b. What is the exact value of the sine of the angle through which the arm rotates to return to its starting position?

c. What is the exact value of the cosine of the angle through which the arm rotates to return to its starting position?

Intervention Reteaching and vocabulary activities that can be used with struggling or absent students and as ELL support

Study Guide and Intervention

NAME _____ DATE _____ PERIOD _____

10-3 Study Guide and Intervention
Sum and Difference Identities

Sum and Difference Identities The following formulas are useful for evaluating an expression like sin 15° from the known values of sine and cosine of 60° and 45°.

Sum and Difference of Angles	The following identities hold true for all values of α and β. cos (α ± β) = cos α · cos β ∓ sin α · sin β sin (α ± β) = sin α · cos β ± cos α · sin β

Example: Find the exact value of each expression.

a. cos 345°
cos 345° = cos (300° + 45°)
= cos 300° · cos 45° − sin 300° · sin 45°
= ½ · √2/2 − (−√3/2) · √2/2
= (√2 + √6)/4

b. sin (−105°)
sin (−105°) = sin (45° − 150°)
= sin 45° · cos 150° − cos 45° · sin 150°
= √2/2 · (−√3/2) − √2/2 · ½
= −(√2 + √6)/4

Exercises

Find the exact value of each expression.

1. sin 105°   2. cos 285°   3. cos (−75°)

4. cos (−165°)   5. sin 195°   6. cos 420°

7. sin (−75°)   8. cos 135°   9. cos (−15°)

10. sin 345°   11. cos (−105°)   12. sin 495°

Study Notebook

NAME _____ DATE _____ PERIOD _____

10-3   Sum and Difference Identities

What You'll Learn Scan the text in this lesson. Write two facts you learned about sum and difference of angles identities as you scanned the text.

1. _____

2. _____

Active Vocabulary

Review Vocabulary Fill in each blank with the correct term or phrase. (Lesson 13-1)

trigonometric identity ▶ A trigonometric identity is an equation involving trigonometric _____ that is true for all values for which every expression in the equation is _____.

Vocabulary Link Fill in the blanks to complete each identity.

sin(A + B) = _____ A _____ B + _____ A _____ B

cos(A − B) = _____ A _____ B + _____ A _____ B

tan(A + B) = (tan _____ + tan _____)/(1 − tan _____ tan _____)

Extension Activities that can be used to extend lesson concepts

Enrichment

NAME _____ DATE _____ PERIOD _____

10-3 Enrichment
Identities for the Products of Sines and Cosines

By adding the identities for the sines of the sum and difference of the measures of two angles, a new identity is obtained.

sin (α + β) = sin α cos β + cos α sin β
sin (α − β) = sin α cos β − cos α sin β
(i)   sin (α + β) + sin (α − β) = 2 sin α cos β

This new identity is useful for expressing certain products as sums.

Example: Write sin 3θ cos θ as a sum.
In the identity let α = 3θ and β = θ so that
2 sin 3θ cos θ = sin (3θ + θ) + sin (3θ − θ).
Thus, sin 3θ cos θ = ½ sin 4θ + ½ sin 2θ.

By subtracting the identities for sin (α + β) and sin (α − β), a similar identity for expressing a product as a difference is obtained.

(ii)   sin (α + β) − sin (α − β) = 2 cos α sin β

1. Use the identities for sin (α + β) and cos (α − β) to find identities for expressing the products 2 cos α cos β and 2 sin α sin β as a sum or difference.

2. Find the value of sin 105° cos 75° without using tables.

3. Express cos θ sin θ/2 as a difference.

Launch

Have students read the Why? section of the lesson. Ask:

- Where else have you heard the term *interference*? Sample answer: television and radio

- At its peak, how does the amplitude of the combined wave compare to the amplitude of the initial two waves? The amplitude of the combined wave is the sum of the amplitudes of the two initial waves.

- Why does the combined wave cross the *x*-axis at a point where neither of the two initial waves are crossing the axis? The combined wave is the sum of the other two waves. It crosses the *x*-axis at points where one of the initial waves is above the *x*-axis and the other wave is an equal distance below the *x*-axis.

Go Online!

Use the *eLesson* or *Lesson Presentation* to present this lesson.

Interactive Whiteboard

LESSON 3

Sum and Difference Identities

:Then	:Now	:Why?
You found values of trigonometric functions for general angles.	**1** Find values of sine and cosine by using sum and difference identities. **2** Verify trigonometric identities by using sum and difference identities.	Have you ever been using a wireless Internet provider and temporarily lost the signal? Waves that pass through the same place at the same time cause interference. Interference occurs when two waves combine to have a greater, or smaller, amplitude than either of the component waves.

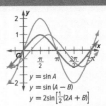

$y = \sin A$
$y = \sin (A - B)$
$y = 2\sin\left[\frac{1}{2}(2A + B)\right]$

MP Mathematical Practices

3 Construct viable arguments and critique the reasoning of others.

6 Attend to precision.

Content Standards

F.TF.8 Prove the Pythagorean identity $\sin^2(\theta) + \cos^2(\theta) = 1$ and use it to find $\sin(\theta)$, $\cos(\theta)$, or $\tan(\theta)$ given $\sin(\theta)$, $\cos(\theta)$, or $\tan(\theta)$ and the quadrant of the angle.

1 Sum and Difference Identities Notice that the third equation shown above involves the sum of A and B. It is often helpful to use formulas for the trigonometric values of the difference or sum of two angles. For example, you could find the exact value of sin 15° by evaluating sin (60° − 45°). Formulas exist that can be used to evaluate expressions like sin $(A - B)$ or cos $(A + B)$.

Key Concept Sum and Difference Identities

Sum Identities	Difference Identities
• $\sin (A + B) = \sin A \cos B + \cos A \sin B$	• $\sin (A - B) = \sin A \cos B - \cos A \sin B$
• $\cos (A + B) = \cos A \cos B - \sin A \sin B$	• $\cos (A - B) = \cos A \cos B + \sin A \sin B$
• $\tan (A + B) = \frac{\tan A + \tan B}{1 - \tan A \tan B}$	• $\tan (A - B) = \frac{\tan A - \tan B}{1 + \tan A \tan B}$

F.TF.8

Example 1　Find Trigonometric Values

Find the exact value of each expression.

a. sin 105°

Use the identity sin $(A + B) = \sin A \cos B + \cos A \sin B$.

$\sin 105° = \sin (60° + 45°)$ 　　　　$A = 60°$ and $B = 45°$

$= \sin 60° \cos 45° + \cos 60° \sin 45°$ 　　Sum identity

$= \left(\frac{\sqrt{3}}{2} \cdot \frac{\sqrt{2}}{2}\right) + \left(\frac{1}{2} \cdot \frac{\sqrt{2}}{2}\right)$ 　　Evaluate each expression.

$= \frac{\sqrt{6}}{4} + \frac{\sqrt{2}}{4}$ or $\frac{\sqrt{6} + \sqrt{2}}{4}$ 　　Multiply.

b. cos (−120°)

Use the identity cos $(A - B) = \cos A \cos B + \sin A \sin B$.

$\cos (-120) = \cos (60° - 180°)$ 　　$A = 60°$ and $B = 180°$

$= \cos 60° \cos 180° + \sin 60° \sin 180°$ 　　Difference identity

$= \frac{1}{2} \cdot (-1) + \frac{\sqrt{3}}{2} \cdot 0$ 　　Evaluate each expression.

$= -\frac{1}{2}$ 　　Multiply.

Guided Practice

1A. sin 15° $\frac{\sqrt{6} - \sqrt{2}}{4}$ 　　　　**1B.** cos (−15°) $\frac{\sqrt{6} + \sqrt{2}}{4}$

MP Mathematical Practices Strategies

Attend to precision. Help students understand that they can find the trigonometric values exactly. For example, ask:

- Suppose you were trying to find the sine of 75°. How does knowing that 45 + 30 = 75 help? 45 and 30 are special angles for which the trigonometric values are specifically defined.

- What happens when you use a calculator to find the value? You get an answer that is irrational with a decimal approximation.

- What happens when you use sin $(A + B) = \sin A \cos B + \cos A \sin B$? You get an exact value, $\frac{1 + \sqrt{3}}{2\sqrt{2}}$.

Real-World Example 2   Sum and Difference of Angles Identities

F.TF.8

SURVEYING A surveyor measures the angle between one side of a rectangular lot and the line from her position to the opposite corner of the lot as 30°. She then measures the angle between that line and the west bank of a creek as 45°. She is standing 100 yards from the opposite corner of the property. How long is the west bank of the creek within the lot?

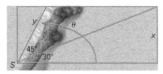

Understand The question asks for the length of the creek within the lot, or y.

Plan Draw a picture that labels all the things that you know from the information given.

Solve Solve for x.

$\sin 30° = \dfrac{x}{100}$ Definition of sine

$x = 100 \sin 30°$

$x = 50$ Since the lot is rectangular, opposite sides are equal.

Now look at the triangle on the far left and solve for y.

$\cos 15° = \dfrac{50}{y}$ Definition of cosine

$\cos (45° - 30°) = \dfrac{50}{y}$ $15 = 45 - 30$

$\cos 45° \cos 30° + \sin 45° \sin 30° = \dfrac{50}{y}$ Difference identity

$\dfrac{\sqrt{2}}{2} \cdot \dfrac{\sqrt{3}}{2} + \dfrac{\sqrt{2}}{2} \cdot \dfrac{1}{2} = \dfrac{50}{y}$ Evaluate

$\dfrac{\sqrt{6} + \sqrt{2}}{4} = \dfrac{50}{y}$ Simplify

$(\sqrt{6} + \sqrt{2})y = 200$ Cross products

$y = \dfrac{200}{(\sqrt{6} + \sqrt{2})} \cdot \dfrac{(\sqrt{6} - \sqrt{2})}{(\sqrt{6} - \sqrt{2})}$

$y = 50(\sqrt{6} - \sqrt{2})$

$y = 50\sqrt{6} - 50\sqrt{2}$ or about 51.8

The length of the creek within the lot is about 51.8 yards.

Check Use a calculator to find $\text{Arccos} \dfrac{50}{51.8} \approx 15°$. ✓

Guided Practice

2. The harmonic motion of an object can be described by $x = 4 \cos \left(2\pi t - \dfrac{\pi}{4}\right)$, where x is the distance from the equilibrium point in inches and t is time in minutes. Find the exact distance from the equilibrium point at 45 seconds. $2\sqrt{2}$ inches below

---

Differentiated Instruction OL BL ELL

Verbal/Linguistic Learners Ask students to identify patterns, similarities, and differences in the sum and difference identities. Then ask the students to write short sentences describing what they have identified.

Teach

Ask the scaffolded questions for each example to build conceptual understanding for students at all levels.

1 Sum and Difference Identities

Example 1   Find Trigonometric Values

AL What special angles do you want to use to rewrite other angles in terms of? any angles that have a reference angle of 30, 45, 60 or any of the quadrantal angles

OL How could you rewrite $-120°$ as a sum? $-60° + -60°$

BL Explain why the identity for $\sin (A - B)$ is equivalent to $\sin (A + (-B))$. $\sin (A + (-B)) = \sin A \cos (-B) + \cos A \sin (-B) = \sin A \cos B - \cos A \sin B = \sin (A - B)$

Need Another Example?
Find the exact value of each expression.

a. $\sin 75°$ $\dfrac{\sqrt{2} + \sqrt{6}}{4}$

b. $\cos (-75°)$ $\dfrac{\sqrt{6} - \sqrt{2}}{4}$

Example 2   Sum and Difference of Angles Identities

AL Why do we use sine to solve for x? Because x is the opposite side of the 30° angle and the other length we have is the hypotenuse

OL Why is the adjacent side of the 15° angle also x? Because opposite sides of a rectangle are equal

BL Give another identity you could have used to solve for y. $\cos (60° - 45°)$

Need Another Example?

Distance Refer to Example 2. If z represents the distance between the upper left-hand corner of the property and the point where the creek crosses the top boundary, then $\tan 15° = \dfrac{z}{50}$ or $\tan (45° - 30°) = \dfrac{z}{50}$. Use the identity for $\tan (A - B)$ to find an exact value for z. $z = \dfrac{50\left(1 - \dfrac{1}{\sqrt{3}}\right)}{1 + \dfrac{1}{\sqrt{3}}}$, which can be simplified to $z = 100 - 50\sqrt{3}$.

2 Verify Trigonometric Identities

Example 3 Verify Trigonometric Identities

AL What type of angles are θ and $90° - \theta$?
complementary

OL To what is $\sin(90° - \theta)$ equal? $\cos \theta$

BL To what is $\tan(90° - \theta)$ equal? $\dfrac{\cos \theta}{\sin \theta}$

Need Another Example?

Verify that each equation is an identity.

a. $\cos(360° - \theta) = \cos \theta$

$\cos(360° - \theta) \stackrel{?}{=} \cos \theta$

$\cos 360° \cos \theta +$
$\sin 360° \sin \theta \stackrel{?}{=} \cos \theta$

$1 \cdot \cos \theta + 0 \cdot \sin \theta \stackrel{?}{=} \cos \theta$

$\cos \theta = \cos \theta \checkmark$

b. $\cos(\pi - \theta) = -\cos \theta$

$\cos(\pi - \theta) \stackrel{?}{=} -\cos \theta$

$\cos \pi \cos \theta + \sin \pi$
$\sin \theta \stackrel{?}{=} -\cos \theta$

$-1 \cdot \cos \theta + 0 \cdot \sin \theta \stackrel{?}{=} -\cos \theta$

$-\cos \theta = -\cos \theta \checkmark$

MP Teaching the Mathematical Practices

Sense-Making Mathematically proficient students analyze givens, constraints, relationships, and goals. Encourage students to become familiar with angle measures for which identities can be easily applied.

Modeling Mathematically proficient students can apply the mathematics they know to solve problems arising in everyday life, analyze relationships mathematically to draw conclusions, and interpret their mathematical results in the context of a situation.

Go Online!

eBook

Interactive Student Guide

Use the *Interactive Student Guide* to deepen conceptual understanding.
• Angle Sum and Difference Formulas

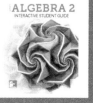

ALGEBRA 2
INTERACTIVE STUDENT GUIDE

Study Tip

MP Sense-Making Make a list of the trigonometric values for the angles between 0° and 360° for which the sum and difference identities can be easily used. Use your list as a reference.

3A. $\sin(90° - \theta) \stackrel{?}{=} \cos \theta$;
$\sin 90° \cos \theta -$
$\cos 90° \sin \theta \stackrel{?}{=} \cos \theta$;
$1 \cos \theta - 0 \sin \theta \stackrel{?}{=} \cos \theta$;
$\cos \theta = \cos \theta$
3B. $\cos(90° + \theta) \stackrel{?}{=} -\sin \theta$;
$\cos 90° \cos \theta -$
$\sin 90° \sin \theta \stackrel{?}{=} -\sin \theta$;
$0 \cos \theta - 1 \sin \theta \stackrel{?}{=} -\sin \theta$;
$-\sin \theta = -\sin \theta$

2 Verify Trigonometric Identities You can also use the sum and difference identities to verify identities.

F.TF.8

Example 3 Verify Trigonometric Identities

Verify that each equation is an identity.

a. $\cos(90° - \theta) = \sin \theta$

$\cos(90° - \theta) \stackrel{?}{=} \sin \theta$	Original equation
$\cos 90° \cos \theta + \sin 90° \sin \theta \stackrel{?}{=} \sin \theta$	Sum identity
$0 \cdot \cos \theta + 1 \cdot \sin \theta \stackrel{?}{=} \sin \theta$	Evaluate each expression.
$\sin \theta = \sin \theta \checkmark$	Simplify.

b. $\sin\left(\theta + \dfrac{\pi}{2}\right) = \cos \theta$

$\sin\left(\theta + \dfrac{\pi}{2}\right) \stackrel{?}{=} \cos \theta$	Original equation
$\sin \theta \cos \dfrac{\pi}{2} + \cos \theta \sin \dfrac{\pi}{2} \stackrel{?}{=} \cos \theta$	Sum identity
$\sin \theta \cdot 0 + \cos \theta \cdot 1 \stackrel{?}{=} \cos \theta$	Evaluate each expression.
$\cos \theta = \cos \theta \checkmark$	Simplify.

▶ **Guided Practice**

3A. $\sin(90° - \theta) = \cos \theta$ **3B.** $\cos(90° + \theta) = -\sin \theta$

Check Your Understanding ◯ = Step-by-Step Solutions begin on page R11.

✓ **Go Online!** for a Self-Check Quiz

Example 1
F.TF.8

Find the exact value of each expression.

1 $\cos 165°$ $-\dfrac{\sqrt{2}+\sqrt{6}}{4}$ **2.** $\cos 105°$ $\dfrac{\sqrt{2}-\sqrt{6}}{4}$ **3.** $\cos 75°$ $\dfrac{\sqrt{6}-\sqrt{2}}{4}$

4. $\sin(-30°)$ $-\dfrac{1}{2}$ **5.** $\sin 135°$ $\dfrac{\sqrt{2}}{2}$ **6.** $\sin(-210°)$ $\dfrac{1}{2}$

Example 2
F.TF.8

7. **MP** MODELING Refer to the beginning of the lesson. *Constructive interference* occurs when two waves combine to have a greater amplitude than either of the component waves. *Destructive interference* occurs when the component waves combine to have a smaller amplitude. The first signal can be modeled by the equation $y = 20 \sin(3\theta + 45°)$. The second signal can be modeled by the equation $y = 20 \sin(3\theta + 225°)$.

a. Find the sum of the two functions. 0

b. What type of interference results when signals modeled by the two equations are combined? The interference is destructive. The signals cancel each other completely.

Example 3
F.TF.8

Verify that each equation is an identity. 8–11. See margin.

8. $\sin(90° + \theta) = \cos \theta$ **9.** $\cos\left(\dfrac{3\pi}{2} - \theta\right) = -\sin \theta$

10. $\tan\left(\theta + \dfrac{\pi}{2}\right) = -\cot \theta$ **11.** $\sin(\theta + \pi) = -\sin \theta$

Additional Answers

8.

$\sin(90° + \theta) \stackrel{?}{=} \cos \theta$

$\sin 90° \cos \theta + \cos 90° \sin \theta \stackrel{?}{=} \cos \theta$

$1 \cdot \cos \theta + 0 \cdot \sin \theta \stackrel{?}{=} \cos \theta$

$\cos \theta = \cos \theta \checkmark$

9.

$\cos\left(\dfrac{3\pi}{2} - \theta\right) \stackrel{?}{=} -\sin \theta$

$\cos \dfrac{3\pi}{2} \cos \theta + \sin \dfrac{3\pi}{2} \sin \theta \stackrel{?}{=} -\sin \theta$

$0 \cdot \cos \theta - 1 \cdot \sin \theta \stackrel{?}{=} -\sin \theta$

$-\sin \theta = -\sin \theta \checkmark$

Practice

Formative Assessment Use Exercises 1–11 to assess students' understanding of the concepts in this lesson.

The Practice and Problem Solving exercises assess the content taught in the lesson. The Preparing for Assessment page is meant to be used as preparation for assessment.

Practice and Problem Solving

Extra Practice is on page R10.

Example 1
F.TF.8

Find the exact value of each expression.

12. $\sin 165°$ $\dfrac{\sqrt{6}-\sqrt{2}}{4}$

13. $\cos 135°$ $-\dfrac{\sqrt{2}}{2}$

14. $\cos \dfrac{7\pi}{12}$ $\dfrac{\sqrt{2}-\sqrt{6}}{4}$

15. $\sin \dfrac{\pi}{12}$ $\dfrac{\sqrt{6}-\sqrt{2}}{4}$

16. $\tan 195°$ $2-\sqrt{3}$

17. $\cos\left(-\dfrac{\pi}{12}\right)$ $\dfrac{\sqrt{2}+\sqrt{6}}{4}$

Example 2
F.TF.8

18. ELECTRONICS In a certain circuit carrying alternating current, the formula $c = 2 \sin (120t)$ can be used to find the current c in amperes after t seconds.

  a. Rewrite the formula using the sum of two angles.   Sample answer: $c = 2 \sin (90t + 30t)$

  b. Use the sum of angles formula to find the exact current at $t = 1$ second. $\sqrt{3}$ amperes

Example 3
F.TF.8

Verify that each equation is an identity.   **19–22. See margin.**

19. $\cos\left(\dfrac{\pi}{2} + \theta\right) = -\sin\theta$

20. $\cos(60° + \theta) = \sin(30° - \theta)$

21. $\cos(180° + \theta) = -\cos\theta$

22. $\tan(\theta + 45°) = \dfrac{1 + \tan\theta}{1 - \tan\theta}$

23. **REASONING** The monthly high temperatures for Minneapolis, Minnesota, can be modeled by the equation $y = 31.65 \sin\left(\dfrac{\pi}{6}x - 2.09\right) + 52.35$, where the months x are represented by January = 1, February = 2, and so on. The monthly low temperatures for Minneapolis can be modeled by the equation $y = 30.15 \sin\left(\dfrac{\pi}{6}x - 2.09\right) + 32.95$.

  a. Write a new function by adding the expressions on the right side of each equation and dividing the result by 2. $y = 30.9 \sin\left(\dfrac{\pi}{6}x - 2.09\right) + 42.65$

  b. What is the meaning of the function you wrote in part **a**?
The new function represents the average of the high and low temperatures for each month.

Find the exact value of each expression.

24. $\tan 165°$ $-2+\sqrt{3}$

25. $\sec 1275°$ $\sqrt{2}-\sqrt{6}$

26. $\sin 735°$ $\dfrac{\sqrt{6}-\sqrt{2}}{4}$

27. $\tan\dfrac{23\pi}{12}$ $-2+\sqrt{3}$

28. $\csc\dfrac{5\pi}{12}$ $\sqrt{6}-\sqrt{2}$

29. $\cot\dfrac{113\pi}{12}$ $2-\sqrt{3}$

30. FORCE In the figure at the right, the effort F necessary to hold a safe in position on a ramp is given by $F = \dfrac{W(\sin A + \mu \cos A)}{\cos A - \mu \sin A}$, where W is the weight of the safe and $\mu = \tan\theta$. Show that $F = W \tan(A + \theta)$.
See Ch. 10 Answer Appendix.

31 QUILTING As part of a quilt that is being made, the quilter places two right triangular swatches together to make a new triangular piece. One swatch has sides 6 inches, 8 inches, and 10 inches long. The other swatch has sides 8 inches, $8\sqrt{3}$ inches, and 16 inches long. The pieces are placed with the sides of eight inches against each other, as shown in the figure, to form triangle ABC.

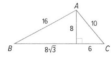

  a. What is the exact value of the sine of angle BAC? $\dfrac{3 + 4\sqrt{3}}{10}$

  b. What is the exact value of the cosine of angle BAC? $\dfrac{4 - 3\sqrt{3}}{10}$

  c. What is the measure of angle BAC? $\approx 96.9°$

  d. Is the new triangle formed from the two triangles also a right triangle?   no

Differentiated Homework Options

Levels	AL Basic	OL Core	BL Advanced
Exercises	12–22, 38–50	13–29 odd, 30–33, 35, 37–50	35–50
2-Day Option	13–21 odd, 43–50	12–22, 43–50	
	12–22 even, 38–42	23–42	

 You can use ALEKS to provide additional remediation support with personalized instruction and practice.

Additional Answers

10.
$$\tan\left(\theta + \dfrac{\pi}{2}\right) \overset{?}{=} -\cot\theta$$
$$\dfrac{\sin\left(\theta + \dfrac{\pi}{2}\right)}{\cos\left(\theta + \dfrac{\pi}{2}\right)} \overset{?}{=} -\cot\theta$$
$$\dfrac{\sin\theta\cos\dfrac{\pi}{2} + \cos\theta\sin\dfrac{\pi}{2}}{\cos\theta\cos\dfrac{\pi}{2} - \sin\theta\sin\dfrac{\pi}{2}} \overset{?}{=} -\cot\theta$$
$$\dfrac{(\sin\theta)\cdot 0 + (\cos\theta)\cdot 1}{(\cos\theta)\cdot 0 - (\sin\theta)\cdot 1} \overset{?}{=} -\cot\theta$$
$$-\dfrac{\cos\theta}{\sin\theta} \overset{?}{=} -\cot\theta$$
$$-\cot\theta = -\cot\theta \checkmark$$

11.
$$\sin(\theta + \pi) \overset{?}{=} -\sin\theta$$
$$\sin\theta\cos\pi + \cos\theta\sin\pi \overset{?}{=} -\sin\theta$$
$$(\sin\theta)(-1) + (\cos\theta)(0) \overset{?}{=} -\sin\theta$$
$$-\sin\theta = -\sin\theta \checkmark$$

19.
$$\cos\left(\dfrac{\pi}{2} + \theta\right) \overset{?}{=} -\sin\theta$$
$$\cos\dfrac{\pi}{2}\cos\theta - \sin\dfrac{\pi}{2}\sin\theta \overset{?}{=} -\sin\theta$$
$$(0)(\cos\theta) - (1)(\sin\theta) \overset{?}{=} -\sin\theta$$
$$-\sin\theta = -\sin\theta \checkmark$$

20.
$$\cos(60° + \theta) \overset{?}{=} \sin(30° - \theta)$$
$$\cos 60°\cos\theta - \sin 60°\sin\theta \overset{?}{=}$$
$$\sin 30°\cos\theta - \cos 30°\sin\theta$$
$$\dfrac{1}{2}\cos\theta - \dfrac{\sqrt{3}}{2}\sin\theta \overset{?}{=}$$
$$\dfrac{1}{2}\cos\theta - \dfrac{\sqrt{3}}{2}\sin\theta \checkmark$$

21.
$$\cos(180° + \theta) \overset{?}{=} -\cos\theta$$
$$\cos 180°\cos\theta - \sin 180°$$
$$\sin\theta \overset{?}{=} -\cos\theta$$
$$-1\cdot\cos\theta - 0\cdot\sin\theta \overset{?}{=} -\cos\theta$$
$$-\cos\theta = -\cos\theta \checkmark$$

22.
$$\tan(\theta + 45°) \overset{?}{=} \dfrac{1 + \tan\theta}{1 - \tan\theta}$$
$$\dfrac{\tan\theta + \tan 45°}{1 - \tan\theta\tan 45°} \overset{?}{=} \dfrac{1 + \tan\theta}{1 - \tan\theta}$$
$$\dfrac{\tan\theta + 1}{1 - (\tan\theta)(1)} \overset{?}{=} \dfrac{1 + \tan\theta}{1 - \tan\theta}$$
$$\dfrac{1 + \tan\theta}{1 - \tan\theta} = \dfrac{1 + \tan\theta}{1 - \tan\theta} \checkmark$$

Go Online!

eSolutions Manual

Create worksheets, answer keys, and solutions handouts for your assignments.

Extra Practice

See page R10 for extra exercises for students who are approaching level or for on-level students who need additional reinforcement.

Levels of Complexity Chart

The levels of the exercises progress from 1 to 3, with Level 1 indicating the lowest level of complexity.

Exercises	12–22	23–31, 43–50	32–42
Level 3			●
B Level 2		●	
Level 1	●		

Assess

Ticket Out the Door Have students make a list of angles between 0° and 360° for which the sum and difference formulas can easily be used. Then have them tell what the angles have in common. **ELL**

Additional Answer

32a.
$$\frac{\sin\left[\frac{1}{2}(a+b)\right]}{\sin\frac{b}{2}}$$

$$=\frac{\sin\left[\frac{1}{2}(a+60°)\right]}{\sin\frac{60°}{2}}$$

$$=\frac{\sin\left(\frac{a}{2}+30°\right)}{\sin 30°}$$

$$=\frac{\sin\frac{a}{2}\cos 30°+\cos\frac{a}{2}\sin 30°}{\sin 30°}$$

$$=\frac{\left(\sin\frac{a}{2}\right)\left(\frac{\sqrt{3}}{2}\right)+\left(\cos\frac{a}{2}\right)\left(\frac{1}{2}\right)}{\frac{1}{2}}$$

$$=\sqrt{3}\sin\frac{a}{2}+\cos\frac{a}{2}$$

Go Online!

Self-Check Quiz

Students can use *Self-Check Quizzes* to check their understanding of this lesson. You can also give the *Chapter Quiz*, which covers the content in Lesson 10-3.

---

39. Sample answer: To determine wireless Internet interference, you need to determine the sine or cosine of the sum or difference of two angles. Interference occurs when waves pass through the same space at the same time. When the combined waves have a greater amplitude, constructive interference results. When the combined waves have a smaller amplitude, destructive interference results.

C **32. OPTICS** When light passes symmetrically through a prism, the index of refraction n of the glass with respect to air is

$$n = \frac{\sin\left[\frac{1}{2}(a+b)\right]}{\sin\frac{b}{2}},$$ where a is the measure of the deviation angle and b is the measure of the prism apex angle.

prism apex

a. Show that for the prism shown, $n = \sqrt{3}\sin\frac{a}{2}+\cos\frac{a}{2}$. See margin.

b. Find n for the prism shown. $\sqrt{3}$

33. MULTIPLE REPRESENTATIONS In this problem, you will disprove the hypothesis that sin $(A + B) = \sin A + \sin B$. **b, c. See Ch. 10 Answer Appendix.**

a. Tabular Copy and complete the table.

b. Graphical Assume that B is always 15° less than A. Use a graphing calculator to graph $y = \sin(x + x - 15)$ and $y = \sin x + \sin(x - 15)$ on the same screen.

c. Analytical Determine whether $\cos(A + B) = \cos A + \cos B$ is an identity. Explain your reasoning.

A	B	sin A	sin B	sin (A + B)	sin A + sin B
30°	90°	$\frac{1}{2}$	1	$\frac{\sqrt{3}}{2}$	$\frac{3}{2}$
45°	60°	$\frac{\sqrt{2}}{2}$	$\frac{\sqrt{3}}{2}$	$\frac{\sqrt{2}+\sqrt{6}}{4}$	$\frac{\sqrt{2}+\sqrt{3}}{2}$
60°	45°	$\frac{\sqrt{3}}{2}$	$\frac{\sqrt{2}}{2}$	$\frac{\sqrt{2}+\sqrt{6}}{4}$	$\frac{\sqrt{2}+\sqrt{3}}{2}$
90°	30°	1	$\frac{1}{2}$	$\frac{\sqrt{3}}{2}$	$\frac{3}{2}$

Verify that each equation is an identity. 34–37. See Ch. 10 Answer Appendix.

34. $\sin(A + B) = \frac{\tan A + \tan B}{\sec A \sec B}$

35 $\cos(A + B) = \frac{1 - \tan A \tan B}{\sec A \sec B}$

36. $\sec(A - B) = \frac{\sec A \sec B}{1 + \tan A \tan B}$

37. $\sin(A + B)\sin(A - B) = \sin^2 A - \sin^2 B$

42. Sample answer: $A = 35°$, $B = 60°$, $C = 85°$; $0.7002 + 1.7321 + 11.4301 \stackrel{?}{=} (0.7002)(1.7321)(11.4301)$; $13.86 = 13.86$ ✓

F.TF.8

H.O.T. Problems   Use Higher-Order Thinking Skills

38. **MP** **REASONING** Simplify the following expression without expanding any of the sums or differences. $\sin(-2\theta)$

$$\sin\left(\frac{\pi}{3} - \theta\right)\cos\left(\frac{\pi}{3} + \theta\right) - \cos\left(\frac{\pi}{3} - \theta\right)\sin\left(\frac{\pi}{3} + \theta\right)$$

39. **WRITING IN MATH** Use the information at the beginning of the lesson and in Exercise 7 to explain how the sum and difference identities are used to describe wireless Internet interference. Include an explanation of the difference between constructive and destructive interference.

40. **CHALLENGE** Derive an identity for cot $(A + B)$ in terms of cot A and cot B. See Ch. 10 Answer Appendix.

41. **MP** **REASONING** The figure shows two angles A and B in standard position on the unit circle. Use the Distance Formula to find d, where $(x_1, y_1) = (\cos B, \sin B)$ and $(x_2, y_2) = (\cos A, \sin A)$.
See Ch. 10 Answer Appendix.

42. **OPEN-ENDED** Consider the following theorem. *If A, B, and C are the angles of an oblique triangle, then* $\tan A + \tan B + \tan C = \tan A \tan B \tan C$. Choose values for A, B, and C. Verify that the conclusion is true for your specific values.

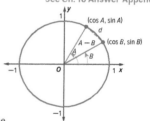

Differentiated Instruction **OL** **BL**

Extension Tell students that sin 20° is approximately −0.3420. Tell them to use this information to find sin 65° and cos 65°. 0.9063, 0.4226

Preparing for Assessment

43. The expression $\sin 45° \cos 30° + \cos 45° \sin 30°$ can be used to evaluate which of the following measures? ⓐ 1 F.TF.8 **D**

- ○ **A** $\sin 15°$
- ○ **B** $\cos 15°$
- ○ **C** $\tan 75°$
- ○ **D** $\sin 75°$
- ○ **E** $\cos 75°$

44. What is the exact measure of $\tan 15°$? ⓜ 6 F.TF.8 **E**

- ○ **A** $\dfrac{1 - \sqrt{3}}{1 + \sqrt{3}}$
- ○ **B** $1 - \dfrac{\sqrt{3}}{3}$
- ○ **C** 1
- ○ **D** $2 + \sqrt{3}$
- ○ **E** $2 - \sqrt{3}$

45. If $\cos 60° = \dfrac{1}{2}$, which of the following expressions is equal to 1? ⓜ 6 F.TF.8 **B**

- ○ **A** $\sin^2 30° - \cos^2 30°$
- ○ **B** $\sin^2 30° + \cos^2 30°$
- ○ **C** $\sin^2 60° + \cos^2 0°$
- ○ **D** $-\sin 60° + \cos 0°$
- ○ **E** $-\sin 30° + \cos 30°$

46. What is the exact value of $\tan \dfrac{5\pi}{12}$? ⓜ 6 F.TF.8

$\boxed{2 + \sqrt{3}}$

47. Which expression can be used to find $\cos 10°$? ⓐ 1 F.TF.8 **D**

- ○ **A** $\sin 60° \cos 50° - \cos 60° \sin 50°$
- ○ **B** $\cos 60° \cos 50° - \sin 60° \sin 50°$
- ○ **C** $\sin 60° \cos 50° + \cos 60° \sin 50°$
- ○ **D** $\cos 60° \cos 50° + \sin 60° \sin 50°$
- ○ **E** $\cos 60° \cos 50° - \cos 60° \cos 50°$

48. MULTI-STEP

a. Which of the following expressions has the same value as $\cos(\pi + \theta)$? ⓜ 2 F.TF.8 **A**

- ○ **A** $-\cos \theta$
- ○ **B** $-\sin \theta$
- ○ **C** $\dfrac{1}{\sin \theta}$
- ○ **D** $\dfrac{1}{\cos \theta}$
- ○ **E** $\sin \theta - \cos \theta$

b. Use the sum identity to determine the value of $\cos(\pi + \theta)$. **See margin**

49. Which expressions have a value of $\dfrac{1}{\sqrt{2}}$? Check all that apply. ⓜ 6, 8 F.TF.8 **B, C, D**

- ☐ **A** $\sin(-45°)$
- ☐ **B** $\cos(-45°)$
- ☐ **C** $\sin 45°$
- ☐ **D** $\sin 135°$
- ☐ **E** $\cos(-135°)$

50. Which expression can be evaluated using the Angle Difference Identity? ⓜ 2 F.TF.8 **D**

- ○ **A** $\dfrac{\tan 30° + \tan 45°}{1 - \tan 30° \tan 45°}$
- ○ **B** $\cos 30° \cos 45° - \sin 30° \sin 45°$
- ○ **C** $\sin 30° \cos 45° + \cos 30° \sin 45°$
- ○ **D** $\cos 30° \cos 45° + \sin 30° \sin 45°$

ⓜ Standards for Mathematical Practice

Emphasis On	Exercises
4 Model with mathematics.	7, 18, 30–32
8 Look for and express regularity in repeated reasoning.	23, 38, 41

Additional Answer

48b. $\cos(\pi + \theta) = \cos \pi \cos \theta + \sin \pi \sin \theta$?   Use the Cosine Sum identity.

$\qquad = (-1) \cos \theta + (0) \sin \theta$ Evaluate each expression.

$\qquad = -\cos \theta + 0$ Simplify.

$\qquad = -\cos \theta$

Preparing for Assessment

	Dual Coding	
Items	Content Standards	ⓜ Mathematical Practices
43	F.TF.8	1
44	F.TF.8	6
45	F.TF.8	6
46	F.TF.8	6
47	F.TF.8	1
48	F.TF.8	2
49	F.TF.8	6, 8
50	F.TF.8	2

Diagnose Student Errors

Survey student responses for each item. Class trends may indicate common errors and misconceptions.

43.

A	Misinterpreted difference identity for sine
B	Used difference identity for cosine
C	Associated sum identity with tangent
D	CORRECT
E	Associated sum identity with cosine

44.

A	Swapped values for tan 45° and tan 30° in difference identity for tangent
B	Found tan 45° − tan 30°
C	Added values in numerator of difference identity for tangent
D	Used the tangent sum identity
E	CORRECT

45.

A	Subtracted measures instead of adding
B	CORRECT
C	Attempted to use sum identity
D	Attempted to use sum/difference identity
E	Misinterpreted sum/difference identity

47.

A	Chose an identity for sin 10°
B	Chose an identity for cos 110°
C	Chose an identity for cos 110°
D	CORRECT
E	Chose an expression that is not an identity

RtI Response to Intervention

Use the Intervention Planner to help you determine your Response to Intervention

Intervention Planner

TIER 1 On Level OL

IF students miss 25% of the exercises or less,

THEN choose a resource:

SE Lessons 10-1 through 10-3

Go Online!
 Skills Practice
 Chapter Project
 Self-Check Quizzes

TIER 2 Strategic Intervention AL
Approaching grade level

IF students miss 50% of the exercises,

THEN **Go Online!**
 Study Guide and Intervention
 Extra Examples
 Personal Tutors
 Homework Help

TIER 3 Intensive Intervention
2 or more grades below level

IF students miss 75% of the exercises,

THEN choose a resource:

Use *Math Triumphs, Alg. 2*

Go Online!
 Extra Examples
 Personal Tutors
 Homework Help
 Review Vocabulary

Go Online!

ᵉAssessment

You can use the premade Mid-Chapter Test to assess students' progress in the first half of the chapter. Customize and create multiple versions of your Mid-Chapter Quiz and answer keys that align to your standards can be delivered on paper or online.

CHAPTER 10
Mid-Chapter Quiz
Lessons 10-1 through 10-3

Simplify each expression. (Lesson 10-1)

1. $\cot \theta \sec \theta$ $\csc \theta$

2. $\dfrac{1 - \cos^2 \theta}{\sin^2 \theta}$ 1

3. $\dfrac{1}{\cos \theta} - \dfrac{\sin^2 \theta}{\cos \theta}$ $\csc \theta$

4. $\cos \left(\dfrac{\pi}{2} - \theta \right) \csc \theta$ 1

5. **HISTORY** In 1861, the United States 34-star flag was adopted. For this flag, $\tan \theta = \dfrac{31.5}{51}$. Find $\sin \theta$. See margin.

Find the value of each expression. (Lesson 10-1)

6. $\sin \theta$, if $\cos \theta = \dfrac{3}{5}$; $0° < \theta < 90°$ $\dfrac{4}{5}$

7. $\csc \theta$, if $\cot \theta = \dfrac{1}{2}$; $270° < \theta < 360°$ $-\dfrac{\sqrt{5}}{2}$

8. $\tan \theta$, if $\sec \theta = \dfrac{4}{3}$; $0° < \theta < 90°$ $\dfrac{\sqrt{7}}{3}$

9. **MULTIPLE CHOICE** Which of the following is equivalent to $\dfrac{\cos \theta}{1 - \sin^2 \theta}$? (Lesson 10-1) **D**

A $\cos \theta$　　C $\tan \theta$

B $\csc \theta$　　D $\sec \theta$

10. **AMUSEMENT PARKS** Suppose a child on a merry-go-round is seated on an outside horse. The diameter of the merry-go-round is 16 meters. The angle of inclination is represented by the equation $\tan \theta = \dfrac{v^2}{gR}$, where R is the radius of the circular path, v is the speed in meters per second, and g is 9.8 meters per second squared. (Lesson 10-1)

a. If the sine of the angle of inclination of the child is $\dfrac{1}{5}$, what is the angle of inclination made by the child? about 11.5°

b. What is the velocity of the merry-go-round? about 4 m/s

c. What mathematical practice did you use to solve this problem? See students' work.

Verify that each of the following is an identity. (Lesson 10-2)

11. $\cot^2 \theta + 1 = \dfrac{\cot \theta}{\cos \theta \cdot \sin \theta}$

12. $\dfrac{\cos \theta \csc \theta}{\cot \theta} = 1$

13. $\dfrac{\sin \theta \tan \theta}{1 - \cos \theta} = (1 + \cos \theta) \sec \theta$

14. $\tan \theta (1 - \sin \theta) = \dfrac{\cos \theta \sin \theta}{1 + \sin \theta}$

11–14. See Ch. 10 Answer Appendix.

15. **COMPUTER** The front of a computer monitor is usually measured along the diagonal of the screen as shown below. (Lesson 10-2)

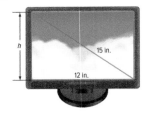

a. Find h. 9

b. Using the diagram shown, show that $\cot \theta = \dfrac{\cos \theta}{\sin \theta}$. See Ch. 10 Answer Appendix.

Verify that each of the following is an identity. (Lesson 10-2)

16. $\tan^2 \theta + 1 = \dfrac{\tan \theta}{\cos \theta \cdot \sin \theta}$

17. $\dfrac{\sin \theta \cdot \sec \theta}{\sec \theta - 1} = (\sec \theta + 1) \cot \theta$

18. $\sin^2 \theta \cdot \tan^2 \theta = \tan^2 \theta - \sin^2 \theta$

19. $\cot \theta (1 - \cos \theta) = \dfrac{\cos \theta \cdot \sin \theta}{1 + \cos \theta}$

16–19. See Ch. 10 Answer Appendix.

Find the exact value of each expression. (Lesson 10-3)

20. $\cos 105°$ $\dfrac{\sqrt{2} - \sqrt{6}}{4}$

21. $\sin (-135°)$ $\dfrac{-\sqrt{2}}{2}$

22. $\tan 15°$ $2 - \sqrt{3}$

23. $\cot 75°$ $2 - \sqrt{3}$

24. **MULTIPLE CHOICE** What is the exact value of $\cos \dfrac{5\pi}{12}$? (Lesson 10-3) **C**

A $\sqrt{2}$　　C $\dfrac{\sqrt{6} - \sqrt{2}}{4}$

B $\dfrac{\sqrt{6} + \sqrt{2}}{2}$　　D $\dfrac{\sqrt{6} + \sqrt{2}}{4}$

25. See Ch. 10 Answer Appendix.

25. Verify that $\cos 30° \cos \theta + \sin 30° \sin \theta = \sin 60° \cos \theta + \cos 60° \sin \theta$ is an identity. (Lesson 10-3)

Foldables Study Organizer

Dinah Zike's **FOLDABLES**

Before students complete the Mid-Chapter Quiz, encourage them to review the information for Lessons 10-1 through 10-3 in their Foldables. Ask students to share the items they have added to their Foldables that have been helpful as they study trigonometry.

ALEKS can be used as a formative assessment tool to target learning gaps for those who are struggling, while providing enhanced learning for those who have mastered the concepts.

Additional Answer

5. $\dfrac{31.5\sqrt{3593.25}}{3593.25}$

LESSON 10-4

Double-Angle and Half-Angle Identities

SUGGESTED PACING (DAYS)

90 min. **0.5**
45 min. **1.0**

Instruction

Track Your Progress

Objectives

1 Find values of sine and cosine by using double-angle identities.

2 Find values of sine and cosine by using half-angle identities.

Mathematical Background

The Double-Angle Formulas are special cases of the Sum of Angles Formulas studied in the previous lesson. The Double-Angle Formulas are obtained by setting A and B equal to each other in the Sum of Angles Formulas.

THEN	NOW	NEXT
F.TF.1 Understand radian measure of an angle as the length of the arc on the unit circle subtended by the angle.	**F.TF.8** Prove the Pythagorean identity $\sin^2(\theta) + \cos^2(\theta) = 1$ and use it to find $\sin(\theta)$, $\cos(\theta)$, or $\tan(\theta)$ given $\sin(\theta)$, $\cos(\theta)$, or $\tan(\theta)$ and the quadrant of the angle.	**F.TF.9** Prove the addition and subtraction formulas for sine, cosine, and tangent and use them to solve problems.

Go Online! All of these resources and more are available at connectED.mcgraw-hill.com

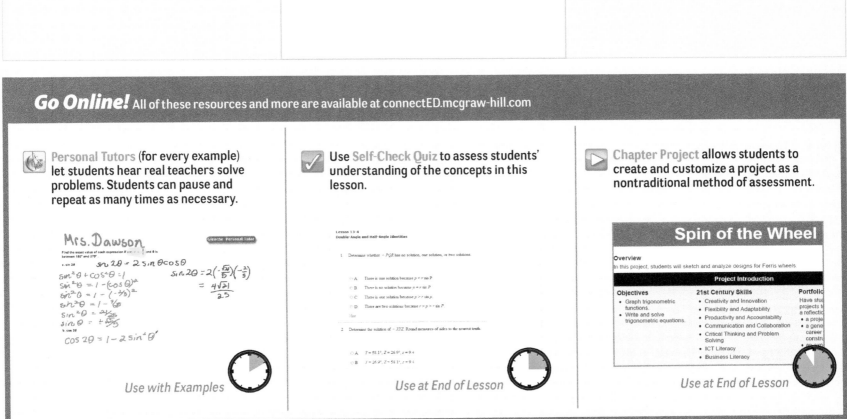

Personal Tutors (for every example) let students hear real teachers solve problems. Students can pause and repeat as many times as necessary.

Use with Examples

Use **Self-Check Quiz** to assess students' understanding of the concepts in this lesson.

Use at End of Lesson

Chapter Project allows students to create and customize a project as a nontraditional method of assessment.

Use at End of Lesson

OER Using Open Educational Resources

Publishing Have students work in groups to create brochures on **ReadWriteThink Printing Press** about trigonometric identities to check their understanding of the concepts. *Use as homework*

Go Online!
connectED.mcgraw-hill.com
Worksheets

Differentiate Your Resources

Extra Practice Additional practice or homework; Skills Practice is best for approaching-level students and Practice is best for on-level and beyond-level students

Skills Practice

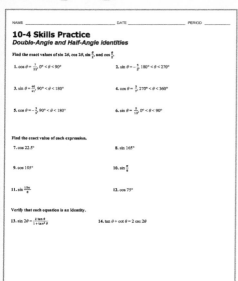

Practice

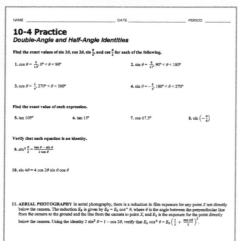

Word Problem Practice

Intervention Reteaching and vocabulary activities that can be used with struggling or absent students and as ELL support

Study Guide and Intervention

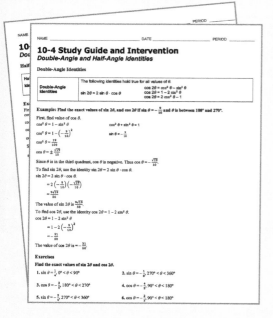

Study Notebook

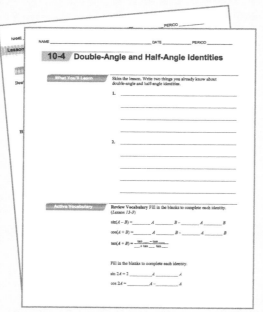

Extension Activities that can be used to extend lesson concepts

Enrichment

LESSON 4

Double-Angle and Half-Angle Identities

:: Then	:: Now	:: Why?
You found values of sine and cosine by using sum and difference identities.	**1** Find values of sine and cosine by using double-angle identities. **2** Find values of sine and cosine by using half-angle identities.	Chicago's Buckingham Fountain contains jets placed at specific angles to create arcs. When a stream of water shoots into the air with velocity v at an angle of θ with the horizontal, the model predicts that the water will travel a horizontal distance of $D = \frac{v^2}{g} \sin 2\theta$ and reach a maximum height of $H = \frac{v^2}{2g} \sin^2 \theta$. The ratio of H to D helps determine the total height and width of the fountain. Express $\frac{H}{D}$ as a function of θ.

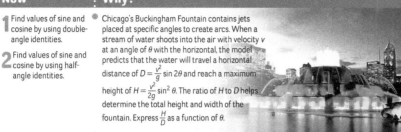

MP Mathematical Practices

3 Construct viable arguments and critique the reasoning of others.

6 Attend to precision.

Content Standards
F.TF.8 Prove the Pythagorean identity $\sin^2(\theta) + \cos^2(\theta) = 1$ and use it to find $\sin(\theta)$, $\cos(\theta)$, or $\tan(\theta)$ given $\sin(\theta)$, $\cos(\theta)$, or $\tan(\theta)$ and the quadrant of the angle.

1 Double-Angle Identities It is sometimes useful to have identities to find the value of a function of twice an angle or half an angle.

Key Concept Double-Angle Identities

The following identities hold true for all values of θ.

$$\cos 2\theta = \cos^2 \theta - \sin^2 \theta$$
$$\sin 2\theta = 2 \sin \theta \cos \theta \qquad \cos 2\theta = 2 \cos^2 \theta - 1 \qquad \tan 2\theta = \frac{2 \tan \theta}{1 - \tan^2 \theta}$$
$$\cos 2\theta = 1 - 2 \sin^2 \theta$$

F.TF.8

Example 1 Double-Angle Identities

Find the exact value of $\sin 2\theta$ if $\sin \theta = \frac{2}{3}$ and θ is between 0° and 90°.

Step 1 Use the identity $\sin 2\theta = 2 \sin \theta \cos \theta$ to find the value of $\cos \theta$.

$$\cos^2 \theta = 1 - \sin^2 \theta \qquad \cos^2 \theta + \sin^2 \theta = 1$$
$$\cos^2 \theta = 1 - \left(\frac{2}{3}\right)^2 \qquad \sin \theta = \frac{2}{3}$$
$$\cos^2 \theta = \frac{5}{9} \qquad \text{Subtract.}$$
$$\cos \theta = \pm \frac{\sqrt{5}}{3} \qquad \text{Take the square root of each side.}$$

Because θ is in the first quadrant, cosine is positive. Thus, $\cos \theta = \frac{\sqrt{5}}{3}$.

Step 2 Find $\sin 2\theta$.

$$\sin 2\theta = 2 \sin \theta \cos \theta \qquad \text{Double-angle identity}$$
$$= 2 \left(\frac{2}{3}\right)\left(\frac{\sqrt{5}}{3}\right) \qquad \sin \theta = \frac{2}{3} \text{ and } \cos \theta = \frac{\sqrt{5}}{3}$$
$$= \frac{4\sqrt{5}}{9} \qquad \text{Multiply.}$$

Guided Practice

1. Find the exact value of $\sin 2\theta$ if $\cos \theta = -\frac{1}{3}$ and $90° < \theta < 180°$. $\quad -\frac{4\sqrt{2}}{9}$

MP Mathematical Practices Strategies

Reason abstractly and quantitatively. Help students understand how to use trigonometric identities like tools in a game to solve problems. Ask them to assess their own learning and comprehension by asking leading questions such as the ones below.

• The Double-Angle Formulas are special cases of which formulas that you have already studied? Sum of Angles Formulas and the Pythagorean Identity

• How can you use these formulas to derive the Double-Angle Formulas? In the Sum of Angles Formulas, set A and B equal to one-another to derive the sine and cosine double-angle identities. In the Pythagorean Identity, subtract or to derive the other cosine double-angle identities.

Launch

Have students read the Why? section of the lesson.

Ask:

• What is the difference between $\sin 2\theta$ and $\sin^2 \theta$? Explain. $\sin 2\theta$ represents the sine of the angle that is two times θ; $\sin^2 \theta$ represents the square of the value of $\sin \theta$.

• Will expressing $\frac{H}{D}$ as a function of θ include the variable v? No, when simplifying, $\frac{v^2}{v^2} = 1$.

• Will it include g? Explain. No, $\frac{1}{2g} \div \frac{1}{g}$ is $\frac{1}{2}$, so it will not include the variable g.

Teach

Ask the scaffolded questions for each example to build conceptual understanding for students at all levels.

1 Double-Angle Identities

Example 1 Double-Angle Identities

AL What is $\csc \theta$? $\frac{3}{2}$

OL What interval is 2θ in if θ is between 0° and 90°? 2θ is between 0° and 180°.

BL Show that $\cos^2 \theta - \sin^2 \theta$ is equivalent to $2 \cos^2 \theta - 1$. $\cos^2 \theta - \sin^2 \theta = \cos^2 \theta - (1 - \cos^2 \theta) = \cos^2 \theta - 1 + \cos^2 \theta = 2 \cos^2 \theta - 1$

Need Another Example?
Find the exact value of $\cos 2\theta$ if $\sin \theta = \frac{3}{4}$ and θ is between 0° and 90°. -0.125

Go Online!

Interactive Whiteboard

Use the *eLesson* or *Lesson Presentation* to present this lesson.

Example 2 Double-Angle Identities

AL What is sec 2θ? 9

OL What would the identity to find cot 2θ be?
$$\frac{1 - \tan^2 \theta}{2 \tan \theta}$$

BL Explain how you could find tan 2θ without using the double-angle identity. $\tan 2\theta = \dfrac{\sin 2\theta}{\cos 2\theta}$
$$= \frac{\frac{4\sqrt{5}}{9}}{\frac{1}{9}} = 4\sqrt{5}$$

Need Another Example?

Find the exact value of each expression if $\cos \theta = \dfrac{4}{5}$ and θ is between 0° and 90°.

a. $\tan 2\theta$ $\dfrac{24}{7}$

b. $\sin 2\theta$ $\dfrac{24}{25}$

Teaching Tips

Sense-Making Remind students of the identity $\sin^2 \theta + \cos^2 \theta = 1$ and its two variations that occur by subtracting either $\sin^2 \theta$ or $\cos^2 \theta$ from both sides. Explain that there are three versions of the formula for $\cos 2\theta$ because of the three variations of the identity $\sin^2 \theta + \cos^2 \theta = 1$.

Study Tip

Deriving Formulas
You can use the identity for $\sin (A + B)$ to find the sine of twice an angle θ, $\sin 2\theta$, and the identity for $\cos (A + B)$ to find the cosine of twice an angle θ, $\cos 2\theta$.

Real-World Career

Electrician An electrician specializes in the wiring of electrical components. Electricians serve an apprenticeship lasting 3–5 years. Schooling in electrical theory and building codes is required. Certification requires work experience and a passing score on a written test.

F.TF.8

Example 2 Double-Angle Identities

Find the exact value of each expression if $\sin \theta = \dfrac{2}{3}$ and θ is between 0° and 90°.

a. $\cos 2\theta$

Because we know the values of $\cos \theta$ and $\sin \theta$, we can use any of the double-angle identities for cosine. We will use the identity $\cos 2\theta = 1 - 2 \sin^2 \theta$.

$\cos 2\theta = 1 - 2 \sin^2 \theta$ Double-angle identity

$\quad = 1 - 2\left(\dfrac{2}{3}\right)^2$ or $\dfrac{1}{9}$ $\sin \theta = \dfrac{2}{3}$

b. $\tan 2\theta$

Step 1 Find $\tan \theta$ to use the double-angle identity for $\tan 2\theta$.

$\tan \theta = \dfrac{\sin \theta}{\cos \theta}$ Definition of tangent

$\quad = \dfrac{\frac{2}{3}}{\frac{\sqrt{5}}{3}}$ $\sin \theta = \dfrac{2}{3}$ and $\cos \theta = \dfrac{\sqrt{5}}{3}$

$\quad = \dfrac{2}{\sqrt{5}}$ or $\dfrac{2\sqrt{5}}{5}$ Rationalize the denominator.

Step 2 Find $\tan 2\theta$.

$\tan 2\theta = \dfrac{2 \tan \theta}{1 - \tan^2 \theta}$ Double-angle identity

$\quad = \dfrac{2\left(\frac{2\sqrt{5}}{5}\right)}{1 - \left(\frac{2\sqrt{5}}{5}\right)^2}$ $\tan \theta = \dfrac{2\sqrt{5}}{5}$

$\quad = \dfrac{2\left(\frac{2\sqrt{5}}{5}\right)}{\frac{25}{25} - \frac{20}{25}}$ Square the denominator.

$\quad = \dfrac{\frac{4\sqrt{5}}{5}}{\frac{1}{5}}$ Simplify.

$\quad = \dfrac{4\sqrt{5}}{5} \cdot \dfrac{5}{1}$ or $4\sqrt{5}$ $\dfrac{a}{b} \div \dfrac{c}{d} = \dfrac{a}{b} \cdot \dfrac{d}{c}$

▸ **Guided Practice**

Find the exact value of each expression if $\cos \theta = -\dfrac{1}{3}$ and $90° < \theta < 180°$.

2A. $\cos 2\theta$ $-\dfrac{7}{9}$ **2B.** $\tan 2\theta$ $\dfrac{4\sqrt{2}}{7}$

2 Half-Angle Identities It is sometimes useful to have identities to find the value of a function of half an angle.

Key Concept Half-Angle Identities

The following identities hold true for all values of θ.

$\sin \dfrac{\theta}{2} = \pm\sqrt{\dfrac{1 - \cos \theta}{2}}$ $\cos \dfrac{\theta}{2} = \pm\sqrt{\dfrac{1 + \cos \theta}{2}}$ $\tan \dfrac{\theta}{2} = \pm\sqrt{\dfrac{1 - \cos \theta}{1 + \cos \theta}}, \cos \theta \neq -1$

Differentiated Instruction

Auditory/Musical Learners If possible, ask a music teacher at your school to talk to students about harmonics. Students playing stringed instruments may also be willing to share what they have learned about harmonics and waves. If a music teacher is not available, a physics teacher may also be able to demonstrate harmonics or bring a device that creates standing waves in class.

Study Tip

Choosing the Sign In the first step of the solution, you may want to determine the quadrant in which the terminal side of $\frac{\theta}{2}$ will lie. Then you can use the correct sign from that point on.

Reading Math

MP Precision The first sign of the half-angle identity is read *plus or minus*. Unlike with the double-angle identities, you must determine the sign.

Example 3 Half-Angle Identities

a. Find the exact value of $\cos \frac{\theta}{2}$ if $\sin \theta = -\frac{4}{5}$ and θ is in the third quadrant.

$$\cos^2 \theta = 1 - \sin^2 \theta \qquad \text{Use a Pythagorean identity to find } \cos \theta.$$

$$\cos^2 \theta = 1 - \left(-\frac{4}{5}\right)^2 \qquad \sin \theta = -\frac{4}{5}$$

$$\cos^2 \theta = 1 - \frac{16}{25} \qquad \text{Evaluate exponent.}$$

$$\cos^2 \theta = \frac{9}{25} \qquad \text{Subtract.}$$

$$\cos \theta = \pm\frac{3}{5} \qquad \text{Take the square root of each side.}$$

Because θ is in the third quadrant, $\cos \theta = -\frac{3}{5}$.

$$\cos \frac{\theta}{2} = \pm\sqrt{\frac{1 + \cos \theta}{2}} \qquad \text{Half-angle identity}$$

$$= \pm\sqrt{\frac{1 - \frac{3}{5}}{2}} \qquad \cos \theta = -\frac{3}{5}$$

$$= \pm\sqrt{\frac{1}{5}} \qquad \text{Simplify.}$$

$$= \pm\frac{1}{\sqrt{5}} \cdot \frac{\sqrt{5}}{\sqrt{5}} \text{ or } \pm\frac{\sqrt{5}}{5} \qquad \text{Rationalize the denominator.}$$

If θ is between $180°$ and $270°$, $\frac{\theta}{2}$ is between $90°$ and $135°$. So, $\cos \frac{\theta}{2}$ is $-\frac{\sqrt{5}}{5}$.

b. Find the exact value of $\cos 67.5°$.

$$\cos 67.5° = \cos \frac{135°}{2} \qquad 67.5° = \frac{135°}{2}$$

$$= \sqrt{\frac{1 + \cos 135°}{2}} \qquad \cos \frac{\theta}{2} = \pm\sqrt{\frac{1 + \cos \theta}{2}}$$

$$= \sqrt{\frac{1 - \frac{\sqrt{2}}{2}}{2}} \qquad 67.5° \text{ is in Quadrant I; the value is positive.}$$

$$= \sqrt{\frac{\frac{2}{2} - \frac{\sqrt{2}}{2}}{2}} \qquad 1 = \frac{2}{2}$$

$$= \sqrt{\frac{\frac{2 - \sqrt{2}}{2}}{2}} \qquad \text{Subtract fractions.}$$

$$= \sqrt{\frac{2 - \sqrt{2}}{2} \cdot \frac{1}{2}} \qquad \frac{a}{b} \div \frac{c}{d} = \frac{a}{b} \cdot \frac{d}{c}$$

$$= \sqrt{\frac{2 - \sqrt{2}}{4}} \qquad \text{Multiply}$$

$$= \frac{\sqrt{2 - \sqrt{2}}}{\sqrt{4}} \qquad \sqrt{\frac{a}{b}} = \frac{\sqrt{a}}{\sqrt{b}}$$

$$= \frac{\sqrt{2 - \sqrt{2}}}{2} \qquad \text{Simplify}$$

Guided Practice

3. Find the exact value of $\sin \frac{\theta}{2}$ if $\sin \theta = \frac{2}{3}$ and θ is in the second quadrant. $\dfrac{\sqrt{18 + 6\sqrt{5}}}{6}$

2 Half-Angle Identities

Example 3 Half-Angle Identities

AL How do you determine which sign to use in the half-angle identities? which quadrant θ is in

OL If θ is in the fourth quadrant, in which quadrant is $\frac{\theta}{2}$? second

BL Using the half angle identity, for what value(s) of θ is tangent undefined? Explain. $\frac{\pi}{2} + \pi n$; That is when the cosine in the denominator of the half-angle identity is -1, which causes the expression in the denominator of the identity to equal 0.

Need Another Example?

a. Find the exact value of $\cos \frac{\theta}{2}$ if $\sin \theta = \frac{3}{5}$ and θ is in the second quadrant. $\sqrt{\dfrac{1}{10}}$ or $\dfrac{\sqrt{10}}{10}$

b. Find the exact value of $\sin 165°$. $\dfrac{\sqrt{2 - \sqrt{3}}}{2}$

Watch Out!

Preventing Errors Stress the Study Tip provided in the margin next to Example 3. Determining the proper sign for the answer at the beginning of the computation will help some students avoid forgetting this step at the end of their computations.

Go Online!

The most up-to-date resources available for your program can be found at connectED.mcgraw-hill.com.

Example 4 Simplify Using Double-Angle Identities

AL What do H and D represent? H is the maximum height and D is the horizontal distance.

OL Find H if $\theta = 45°$ and $D = 2$ feet. $H = \dfrac{1}{2}$ foot

BL Solve the equation for D. $D = 4H \cot \theta$

Need Another Example?

Fountain Refer to the beginning of the lesson. Find $\dfrac{D}{H}$. $4 \cot \theta$ or $\dfrac{4}{\tan \theta}$

Example 5 Verify Identities

AL What other expressions would have been equal to $\cos 2\theta$? $2 \cos^2 \theta - 1, 1 - 2 \sin^2 \theta$

OL What types of identities are used in this problem? quotient identities, double-angle... identities

BL Show how the left-hand side could simplify to the right-hand side. $\dfrac{\cos 2\theta}{1 + \sin 2\theta}$

$$= \frac{\cos^2 \theta - \sin^2 \theta}{1 + 2 \cos \theta \sin \theta} = \frac{(\cos \theta + \sin \theta)(\cos \theta - \sin \theta)}{\cos^2 \theta + \sin^2 \theta + 2 \cos \theta \sin \theta}$$

$$= \frac{(\cos \theta + \sin \theta)(\cos \theta - \sin \theta)}{(\cos \theta + \sin \theta)^2} = \frac{\cos \theta - \sin \theta}{\cos \theta + \sin \theta}$$

$$= \frac{\dfrac{\cos \theta}{\sin \theta} - 1}{\dfrac{\cos \theta}{\sin \theta} + 1} = \frac{\cot \theta - 1}{\cot \theta + 1}$$

Need Another Example?

Verify that $\sin \theta \, (\cos^2 \theta - \cos 2\theta) = \sin^3 \theta$ is an identity.

$$\sin \theta \, (\cos^2 \theta - \cos 2\theta) \overset{?}{=} \sin^3 \theta$$

$$\sin \theta \, [\cos^2 \theta - (\cos^2 \theta - \sin^2 \theta)] \overset{?}{=} \sin^3 \theta$$

$$\sin \theta \, (\cos^2 \theta - \cos^2 \theta + \sin^2 \theta) \overset{?}{=} \sin^3 \theta$$

$$\sin \theta \, (\sin^2 \theta) \overset{?}{=} \sin^3 \theta$$

$$\sin^3 \theta = \sin^3 \theta \checkmark$$

Additional Answer (Guided Practice)

5.
$$4 \cos^2 x - \sin^2 2x \overset{?}{=} 4 \cos^4 x$$

$$4 \cos^2 x - 4 \sin^2 x \cos^2 x \overset{?}{=} 4 \cos^4 x$$

$$4 \cos^2 x \, (1 - \sin^2 x) \overset{?}{=} 4 \cos^4 x$$

$$4 \cos^2 x \cos^2 x \overset{?}{=} 4 \cos^4 x$$

$$4 \cos^4 x = 4 \cos^4 x \checkmark$$

F.TF.8

Real-World Example 4 Simplify Using Double-Angle Identities

FOUNTAIN Refer to the beginning of the lesson. Find $\dfrac{H}{D}$.

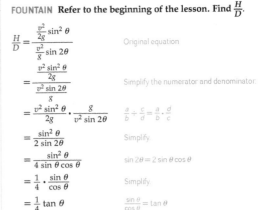

Real-World Link
The City Hall Park Fountain in New York City is located in the heart of Manhattan in front of City Hall.
Source: Fodor's

$$\frac{H}{D} = \frac{\dfrac{v^2}{2g} \sin^2 \theta}{\dfrac{v^2}{g} \sin 2\theta} \qquad \text{Original equation}$$

$$= \frac{\dfrac{v^2 \sin^2 \theta}{2g}}{\dfrac{v^2 \sin 2\theta}{g}} \qquad \text{Simplify the numerator and denominator.}$$

$$= \frac{v^2 \sin^2 \theta}{2g} \cdot \frac{g}{v^2 \sin 2\theta} \qquad \frac{a}{b} \div \frac{c}{d} = \frac{a}{b} \cdot \frac{d}{c}$$

$$= \frac{\sin^2 \theta}{2 \sin 2\theta} \qquad \text{Simplify.}$$

$$= \frac{\sin^2 \theta}{4 \sin \theta \cos \theta} \qquad \sin 2\theta = 2 \sin \theta \cos \theta$$

$$= \frac{1}{4} \cdot \frac{\sin \theta}{\cos \theta} \qquad \text{Simplify.}$$

$$= \frac{1}{4} \tan \theta \qquad \frac{\sin \theta}{\cos \theta} = \tan \theta$$

▸ **Guided Practice**

Find each value.

4A. $\sin 135°$ $\dfrac{\sqrt{2}}{2}$

4B. $\cos \dfrac{7\pi}{8}$ $\dfrac{\sqrt{2 + \sqrt{2}}}{2}$

Recall that you can use the sum and difference identities to verify identities. Double- and half-angle identities can also be used to verify identities.

F.TF.8

Example 5 Verify Identities

Verify that $\dfrac{\cos 2\theta}{1 + \sin 2\theta} = \dfrac{\cot \theta - 1}{\cot \theta + 1}$ is an identity.

$$\frac{\cos 2\theta}{1 + \sin 2\theta} \overset{?}{=} \frac{\cot \theta - 1}{\cot \theta + 1} \qquad \text{Original equation}$$

$$\frac{\cos 2\theta}{1 + \sin 2\theta} \overset{?}{=} \frac{\dfrac{\cos \theta}{\sin \theta} - 1}{\dfrac{\cos \theta}{\sin \theta} + 1} \qquad \cot \theta = \frac{\cos \theta}{\sin \theta}$$

$$\frac{\cos 2\theta}{1 + \sin 2\theta} \overset{?}{=} \frac{\cos \theta - \sin \theta}{\cos \theta + \sin \theta} \qquad \text{Multiply numerator and denominator by } \sin \theta$$

$$\frac{\cos 2\theta}{1 + \sin 2\theta} \overset{?}{=} \frac{\cos \theta - \sin \theta}{\cos \theta + \sin \theta} \cdot \frac{\cos \theta + \sin \theta}{\cos \theta + \sin \theta} \qquad \text{Multiply the right side by 1.}$$

$$\frac{\cos 2\theta}{1 + \sin 2\theta} \overset{?}{=} \frac{\cos^2 \theta - \sin^2 \theta}{\cos^2 \theta + 2 \cos \theta \sin \theta + \sin^2 \theta} \qquad \text{Multiply.}$$

$$\frac{\cos 2\theta}{1 + \sin 2\theta} \overset{?}{=} \frac{\cos^2 \theta - \sin^2 \theta}{1 + 2 \cos \theta \sin \theta} \qquad \text{Simplify.}$$

$$\frac{\cos 2\theta}{1 + \sin 2\theta} = \frac{\cos 2\theta}{1 + \sin 2\theta} \checkmark \qquad \cos^2 \theta - \sin^2 \theta = \cos 2\theta; \, 2 \cos \theta \sin \theta = \sin 2\theta$$

▸ **Guided Practice**

5. Verify that $4 \cos^2 x - \sin^2 2x = 4 \cos^4 x$. See margin.

Go Online!

Check your mastery of double-angle and half-angle identities by using the Self-Check Quiz in ConnectED.

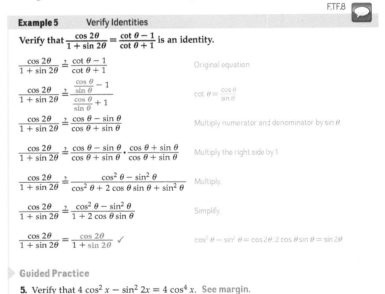

Follow-Up

Students have explored trigonometric identities and equations.
Ask:

- How do you decide what techniques to use when verifying a trigonometric identity?
  Sample answer: If possible, simplify the most complex side of the identity by substituting basic trigonometric identities. When dealing with a more complex identity, work each side separately to obtain a common expression.

Thinkstock/Stockbyte/Getty Images

Check Your Understanding

 = Step-by-Step Solutions begin on page R11.

Examples 1–3
F.TF.8

PRECISION Find the exact values of $\sin 2\theta$, $\cos 2\theta$, $\sin \frac{\theta}{2}$, and $\cos \frac{\theta}{2}$.

1. $\frac{\sqrt{15}}{8}, \frac{7}{8}$,

$\frac{\sqrt{8-2\sqrt{15}}}{4}$,

$\frac{\sqrt{8+2\sqrt{15}}}{4}$

1. $\sin \theta = \frac{1}{4}$; $0° < \theta < 90°$

2. $\sin \theta = \frac{4}{5}$; $90° < \theta < 180°$ 2. $\frac{24}{25}, -\frac{7}{25}, \frac{\sqrt{5}}{5}, \frac{2\sqrt{5}}{5}$

3. $\cos \theta = -\frac{5}{13}$; $\frac{\pi}{2} < \theta < \pi$

4. $\cos \theta = \frac{3}{5}$; $270° < \theta < 360°$ 3. $-\frac{120}{169}, -\frac{119}{169}, \frac{3\sqrt{13}}{13}, \frac{2\sqrt{13}}{13}$

5. $\tan \theta = -\frac{8}{15}$; $90° < \theta < 180°$

6. $\tan \theta = \frac{5}{12}$; $\pi < \theta < \frac{3\pi}{2}$ 4. $-\frac{24}{25}, \frac{7}{25}, \frac{\sqrt{5}}{5}, \frac{2\sqrt{5}}{5}$

5. $-\frac{240}{289}, \frac{161}{289}, \frac{4\sqrt{17}}{17}, \frac{\sqrt{17}}{17}$

Find the exact value of each expression.

7. $\sin \frac{\pi}{8}$ $\frac{\sqrt{2-\sqrt{2}}}{2}$

8. $\cos 15°$ $\frac{\sqrt{2+\sqrt{3}}}{2}$

6. $\frac{120}{169}, \frac{119}{169}, \frac{5\sqrt{26}}{26}, \frac{\sqrt{26}}{26}$

Example 4
F.TF.8

9. **SOCCER** A soccer player kicks a ball at an angle of 37° with the ground with an initial velocity of 52 feet per second. The distance d that the ball will go in the air if it is not blocked is given by $d = \frac{2v^2 \sin \theta \cos \theta}{g}$. In this formula, g is the acceleration due to gravity and is equal to 32 feet per second squared, and v is the initial velocity.

37°

a. Simplify this formula by using a double-angle identity. $d = \frac{v^2 \sin 2\theta}{g}$

b. Using the simplified formula, how far will this ball go? $\approx 81\,\text{ft}$

Example 5
F.TF.8

Verify that each equation is an identity. **10, 11. See margin.**

10. $\tan \theta = \frac{1 - \cos 2\theta}{\sin 2\theta}$

11. $(\sin \theta + \cos \theta)^2 = 1 + 2 \sin \theta \cos \theta$

12. $-\frac{4\sqrt{5}}{9}, \frac{1}{9}, \frac{\sqrt{6} \cdot \sqrt{3+\sqrt{5}}}{6}, \frac{\sqrt{6} \cdot \sqrt{3-\sqrt{5}}}{6}$

13. $\frac{240}{289}, \frac{161}{289}, \frac{5\sqrt{34}}{34}, \frac{3\sqrt{34}}{34}$

Practice and Problem Solving

Extra Practice is on page R10.

Examples 1–3
F.TF.8

Find the exact values of $\sin 2\theta$, $\cos 2\theta$, $\sin \frac{\theta}{2}$, and $\cos \frac{\theta}{2}$. 14. $-\frac{24}{25}, \frac{7}{25}, \frac{\sqrt{5}}{5}, \frac{2\sqrt{5}}{5}$

12. $\sin \theta = \frac{2}{3}$; $90° < \theta < 180°$

13. $\sin \theta = -\frac{15}{17}$; $\pi < \theta < \frac{3\pi}{2}$

14. $\cos \theta = \frac{3}{5}$; $\frac{3\pi}{2} < \theta < 2\pi$

15. $\cos \theta = \frac{1}{5}$; $270° < \theta < 360°$ 15. $-\frac{4\sqrt{6}}{25}, -\frac{23}{25}, \frac{\sqrt{10}}{5}, \frac{\sqrt{15}}{5}$

16. $\tan \theta = \frac{4}{3}$; $180° < \theta < 270°$

17. $\tan \theta = -2$; $\frac{\pi}{2} < \theta < \pi$

16. $\frac{24}{25}, -\frac{7}{25}, \frac{2\sqrt{5}}{5}, -\frac{\sqrt{5}}{5}$

17. $-\frac{4}{5}, -\frac{3}{5}, \sqrt{\frac{\sqrt{5}+1}{2\sqrt{5}}}, \sqrt{\frac{\sqrt{5}-1}{2\sqrt{5}}}$

Find the exact value of each expression.

18. $\sin 75°$ $\frac{\sqrt{2+\sqrt{3}}}{2}$

19. $\sin \frac{3\pi}{8}$ $\frac{\sqrt{2+\sqrt{2}}}{2}$

20. $\cos \frac{7\pi}{12}$ $-\frac{\sqrt{2-\sqrt{3}}}{2}$

21. $\tan 165°$ $-\sqrt{7-4\sqrt{3}}$

22. $\tan \frac{5\pi}{12}$ $\sqrt{7+4\sqrt{3}}$

23. $\tan 22.5°$ $\sqrt{3-2\sqrt{2}}$

24. **GEOGRAPHY** The Mercator projection of the globe is a projection on which the distance between the lines of latitude increases with their distance from the equator. The calculation of the location of a point on this projection involves the expression $\tan \left(45° + \frac{L}{2}\right)$, where L is the latitude of the point.

a. Write this expression in terms of a trigonometric function of L. **See margin.**

b. The latitude of Tallahassee, Florida, is 30° north. Find the value of the expression if $L = 30°$. $\sqrt{3}$

Differentiated Homework Options

Levels	AL Basic	OL Core	BL Advanced
Exercises	12–29, 37–47	13–29 odd, 30, 31–35 odd, 37–47	30–47
2-Day Option	13–29 odd, 43, 44	12–29, 43, 44	
	12–28 even, 37–47	30–47	

 You can use ALEKS to provide additional remediation support with personalized instruction and practice.

Practice

Formative Assessment Use Exercises 1–11 to assess students' understanding of the concepts in this lesson.

The Practice and Problem Solving exercises assess the content taught in the lesson. The Preparing for Assessment page is meant to be used as preparation for assessment.

MP Teaching the Mathematical Practices

Precision Mathematically proficient students try to use clear definitions in their reasoning, calculate accurately and efficiently, and make explicit use of definitions.

Extra Practice

See page R10 for extra exercises for students who are approaching level or for on-level students who need additional reinforcement.

Additional Answers

10. $\tan \theta \stackrel{?}{=} \frac{1 - \cos 2\theta}{\sin 2\theta}$

$\stackrel{?}{=} \frac{1 - (1 - 2\sin^2 \theta)}{2 \sin \theta \cos \theta}$

$\stackrel{?}{=} \frac{2 \sin^2 \theta}{2 \sin \theta \cos \theta}$

$\stackrel{?}{=} \frac{\sin \theta}{\cos \theta}$

$= \tan \theta \checkmark$

11. $(\sin \theta + \cos \theta)^2 \stackrel{?}{=} 1 + 2 \sin \theta \cos \theta$

$(\sin \theta + \cos \theta)(\sin \theta + \cos \theta) \stackrel{?}{=} 1 + 2 \sin \theta \cos \theta$

$\sin^2 \theta + 2 \sin \theta \cos \theta + \cos^2 \theta \stackrel{?}{=} 1 + 2 \sin \theta \cos \theta$

$1 + 2 \sin \theta \cos \theta = 1 + 2 \sin \theta \cos \theta \checkmark$

24a. $\dfrac{1 \pm \sqrt{\dfrac{1 - \cos L}{1 + \cos L}}}{1 \mp \sqrt{\dfrac{1 - \cos L}{1 + \cos L}}}$

WatchOut!

Error Analysis For Exercise 37, students should see that Teresa incorrectly replaced $\frac{\sqrt{6}}{4} - \frac{\sqrt{2}}{4}$ with $\frac{\sqrt{4}}{4}$, while Nathan incorrectly replaced $\cos 30°$ with $\frac{1}{2}$ rather than with $\frac{\sqrt{3}}{2}$. Explain to students that both first steps, $\sin 15° = \sin(45° - 30°)$ and $\sin 15° = \sin \frac{30°}{2}$, are valid first steps. However, Teresa's fourth step should be $\frac{\sqrt{6} - \sqrt{2}}{4}$ and Nathan's steps, after the first line, should be

$$\sin \frac{30°}{2} = \sqrt{\frac{1 - \frac{\sqrt{3}}{2}}{2}} = \sqrt{\frac{2 - \sqrt{3}}{4}} = \frac{\sqrt{2 - \sqrt{3}}}{2}.$$

Levels of Complexity Chart

The levels of the exercises progress from 1 to 3, with Level 1 indicating the lowest level of complexity.

Exercises	12–29	30–36, 43–47	37–42
C Level 3			●
B Level 2		●	
Level 1	●		

Teaching the Mathematical Practices

Critique Arguments Mathematically proficient students are also able to compare the effectiveness of two plausible arguments, distinguish correct logic or reasoning for that which is flawed, and—if there is a flaw in an argument—explain what it is.

Assess

Name the Math Have students tell how they determine whether a problem involves a double-angle or half-angle identity and how they use each type of equation. **ELL**

Go Online!

eSolutions Manual

Create worksheets, answer keys, and solutions handouts for your assignments.

Example 4
F.TF.8

25 ELECTRONICS Consider an AC circuit consisting of a power supply and a resistor. If the current I_0 in the circuit at time t is $I_0 \sin t\theta$, then the power delivered to the resistor is $P = I_0^2 R \sin^2 t\theta$, where R is the resistance. Express the power in terms of $\cos 2t\theta$. $P = \frac{1}{2} I_0^2 R - \frac{1}{2} I_0^2 R \cos 2t\theta$

Example 5
F.TF.8

Verify that each equation is an identity. 26–29. See Ch. 10 Answer Appendix.

26. $\tan 2\theta = \dfrac{2}{\cot \theta - \tan \theta}$ **27.** $1 + \dfrac{1}{2} \sin 2\theta = \dfrac{\sec \theta + \sin \theta}{\sec \theta}$

28. $\sin \dfrac{\theta}{2} \cos \dfrac{\theta}{2} = \dfrac{\sin \theta}{2}$ **29.** $\tan \dfrac{\theta}{2} = \dfrac{\sin \theta}{1 + \cos \theta}$

B **30. FOOTBALL** Suppose a place kicker consistently kicks a football with an initial velocity of 95 feet per second. Prove that the horizontal distance the ball travels in the air will be the same for $\theta = 45° + A$ as for $\theta = 45° - A$. Use the formula given in Exercise 9. See Ch. 10 Answer Appendix.

Find the exact values of $\sin 2\theta$, $\cos 2\theta$, and $\tan 2\theta$.

31. $\cos \theta = \dfrac{4}{5}; 0° < \theta < 90°$ $\dfrac{24}{25}, \dfrac{7}{25}, \dfrac{24}{7}$ **32.** $\sin \theta = \dfrac{1}{3}; 0 < \theta < \dfrac{\pi}{2}$ $\dfrac{4\sqrt{2}}{9}, \dfrac{7}{9}, \dfrac{4\sqrt{2}}{7}$

33. $\tan \theta = -3; 90° < \theta < 180°$ $-\dfrac{3}{5}, -\dfrac{4}{5}, \dfrac{3}{4}$ **34.** $\sec \theta = -\dfrac{4}{3}; 90° < \theta < 180°$ $\dfrac{3\sqrt{7}}{8}, \dfrac{1}{8}, -3\sqrt{7}$

35. $\csc \theta = -\dfrac{5}{2}; \dfrac{3\pi}{2} < \theta < 2\pi$ $-\dfrac{4\sqrt{21}}{25}, \dfrac{17}{25}, -\dfrac{4\sqrt{21}}{17}$ **36.** $\cot \theta = \dfrac{3}{2}; 180° < \theta < 270°$ $\dfrac{12}{13}, \dfrac{5}{13}, \dfrac{12}{5}$

F.TF.8

H.O.T. Problems Use Higher-Order Thinking Skills

C **37. ERROR ANALYSIS** Teresa and Nathan are calculating the exact value of $\sin 15°$. Is either of them correct? Explain your reasoning. See Ch. 10 Answer Appendix.

38. $\angle PBD$ is an inscribed angle that subtends the same arc as the central angle $\angle POD$, so $m \angle PBD = \frac{1}{2}\theta$. By right triangle trigonometry, $\tan \frac{1}{2}\theta = \frac{PA}{BA} =$

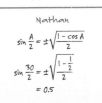

Teresa	Nathan
$\sin(A - B) = \sin A \cos B - \cos A \sin B$ $\sin(45 - 30) = \sin 45 \cos 30 - \cos 45 \sin 30$ $= \dfrac{\sqrt{2}}{2} \cdot \dfrac{\sqrt{3}}{2} - \dfrac{\sqrt{2}}{2} \cdot \dfrac{1}{2}$ $= \dfrac{\sqrt{4}}{4}$	$\sin \dfrac{A}{2} = \pm\sqrt{\dfrac{1 - \cos A}{2}}$ $\sin \dfrac{30}{2} = \pm\sqrt{\dfrac{1 - \frac{1}{2}}{2}}$ $= 0.5$

38. $\dfrac{PA}{1 + OA} = \dfrac{\sin \theta}{1 + \cos \theta}.$

MP CONSTRUCT ARGUMENTS Circle O is a unit circle. Use the figure to prove that $\tan \frac{1}{2}\theta = \dfrac{\sin \theta}{1 + \cos \theta}.$

39. WRITING IN MATH Write a short paragraph about the conditions under which you would use each of the three identities for $\cos 2\theta$. See Ch. 10 Answer Appendix.

42. Sample answer: Because $d = \frac{v^2 \sin 2\theta}{g}$, d is at a maximum when $\sin 2\theta = 1$, that is, when $2\theta = 90°$ or $\theta = 45°$.

40. PROOF Use the formula for $\sin(A + B)$ to derive the formula for $\sin 2\theta$, and use the formula for $\cos(A + B)$ to derive the formula for $\cos 2\theta$. See Ch. 10 Answer Appendix.

41. MP REASONING Derive the half-angle identities from the double-angle identities.
See Ch. 10 Answer Appendix.

42. OPEN-ENDED Suppose a golfer consistently hits the ball so that it leaves the tee with an initial velocity of 115 feet per second and $d = \frac{2v^2 \sin \theta \cos \theta}{g}$. Explain why the maximum distance is attained when $\theta = 45°$.

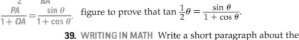

Differentiated Instruction **BL**

Extension Ask students to write an expression for $\sin 4\theta$ in terms of $\sin \theta$ and $\cos \theta$. Sample answer: $\sin 4\theta = 4 \sin \theta \cos \theta - 8 \sin^3 \theta \cos \theta$

Preparing for Assessment

43. Suppose θ is an angle in Quadrant IV of the coordinate plane such that $\sin\theta = -\frac{12}{13}$ and $\cos\theta = \frac{5}{13}$. What is $\sin 2\theta$? (MP) 2, 6 F.TF.8 **B**

- ○ **A** $-\frac{24}{26}$
- ○ **B** $-\frac{120}{169}$
- ○ **C** $-\frac{7}{13}$
- ○ **D** $\frac{25}{169}$
- ○ **E** $\frac{144}{169}$

44. An angle θ exists such that $270° < \theta < 360°$ and $\cos\theta = \frac{3}{5}$. What is the exact value of $\cos\frac{\theta}{2}$? (MP) 2, 6 F.TF.8 **A**

- ○ **A** $-\frac{2\sqrt{5}}{5}$
- ○ **B** $-\frac{3}{10}$
- ○ **C** $\frac{9}{25}$
- ○ **D** $\frac{\sqrt{5}}{5}$
- ○ **E** $\frac{6}{5}$

45. MULTI-STEP When the tree in the image casts a 12-foot shadow, the angle of elevation of the sun is θ. When the sun rises higher in the sky and the angle of elevation is 2θ, the length of shadow from the same tree is reduced to 4 feet. (MP) 1, 2, 3, 6 F.TF.8

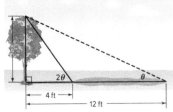

4 ft   12 ft

a. Which trigonometric identity can be used to solve this problem? **C**

- ○ **A** double-angle cosine identity
- ○ **B** half-angle tangent identity
- ○ **C** double-angle tangent identity
- ○ **D** half-angle sine identity
- ○ **E** half-angle cosine identity

b. Carey wants to calculate the height of the tree and performs the following algebra. Which lines contain errors? Choose all that apply. **A, B**

- ☐ **A** $\tan 2\theta = \frac{h}{12}$
- ☐ **B** $\tan\theta = \frac{h}{4}$
- ☐ **C** $\tan 2\theta = \frac{2\tan\theta}{1-\tan^2\theta}$
- ☐ **D** $\frac{h}{4} = \frac{\frac{2h}{12}}{1-\frac{h^2}{12^2}}$
- ☐ **E** $\frac{h}{4} = \frac{\frac{h}{6}}{\frac{144-h^2}{144}}$
- ☐ **F** $\frac{h}{4}\left(\frac{144-h^2}{144}\right) = \frac{h}{6}$
- ☐ **G** $144 - h^2 = 96$

c. To the nearest whole foot, what is the height h of the tree? **B**

- ○ **A** 5 ft
- ○ **B** 7 ft
- ○ **C** 8 ft
- ○ **D** 10 ft
- ○ **E** 16 ft

46. What is the exact value of $-\cos\frac{5\pi}{8}$? (MP) 6 F.TF.8

$$\frac{\sqrt{2-\sqrt{2}}}{2}$$

47. For $\cos\theta = -\frac{4}{5}$, $90 < \theta < 180$, find the exact values. (MP) 6 F.TF.8

- **a.** $\sin 2\theta$ $-\frac{24}{25}$
- **b.** $\cos 2\theta$ $\frac{7}{25}$
- **c.** $\tan 2\theta$ $-\frac{24}{7}$
- **d.** $\sin\frac{\theta}{2}$ $\frac{3\sqrt{10}}{10}$
- **e.** $\cos\frac{\theta}{2}$ $\frac{\sqrt{10}}{10}$
- **f.** $\tan\frac{\theta}{2}$ -3

Standards for Mathematical Practice

Emphasis On	Exercises
1 Make sense of problems and persevere in solving them.	9, 24, 25, 30
3 Construct viable arguments and critique the reasoning of others.	10, 11, 26–29, 37–42
6 Attend to precision.	1–8, 12–25, 31–36

45c.

A	Subtracted the length of the tree from the length of the shadow
B	CORRECT
C	Calculated the hypotenuse of the small triangle
D	Doubled the height of the tree
E	Added the length of the shadows

Preparing for Assessment

	Dual Coding		
Items	Content Standards	(MP) Mathematical Practices
43	F.TF.8	2, 6
44	F.TF.8	2, 6
45	F.TF.8	1, 2, 3, 6
46-47	F.TF.8	6

Diagnose Student Errors

Survey student responses for each item. Class trends may indicate common errors and misconceptions.

43.

A	Multiplied $\sin\theta$ by $\frac{2}{2}$
B	CORRECT
C	Computed $\sin\theta + \cos\theta$
D	Computed $\cos^2\theta$
E	Computed $\sin^2\theta$

44.

A	CORRECT
B	Divided $-\cos\theta$ by 2
C	Computed $\cos^2\theta$
D	Found $\sin\frac{\theta}{2}$
E	Multiplied numerator of $\cos\theta$ by 2

45a.

A	Identified wrong identity
B	Identified wrong identity
C	CORRECT
D	Identified wrong identity
E	Identified wrong identity

Go Online!

Self-Check Quiz

Students can use *Self-Check Quizzes* to check their understanding of this lesson. You can also give the *Chapter Quiz*, which covers the content in Lesson 10-4.

LESSON 10-5
Solving Trigonometric Equations

SUGGESTED PACING (DAYS)		
90 min.	.25	0.75
45 min.	0.5	1.5
	Explore Lab	Instruction

Track Your Progress

Objectives

1 Solve trigonometric equations.

2 Find extraneous solutions from trigonometric equations.

Mathematical Background

Many trigonometric equations have an infinite number of solutions. If no interval is specified to restrict the number of solutions, the infinite number of solutions must be specified by using an expression rather than by simply using a number. There is a solution that occurs for each full rotation about the origin, a solution has the form $a° + k \cdot 360°$, where k is any integer.

THEN	NOW	NEXT
F.TF.1 Understand radian measure of an angle as the length of the arc on the unit circle subtended by the angle.	**F.TF.8** Prove the Pythagorean identity $\sin^2(\theta) + \cos^2(\theta) = 1$ and use it to find $\sin(\theta)$, $\cos(\theta)$, or $\tan(\theta)$ given $\sin(\theta)$, $\cos(\theta)$, or $\tan(\theta)$ and the quadrant of the angle.	**F.TF.9** Prove the addition and subtraction formulas for sine, cosine, and tangent and use them to solve problems.

Go Online! All of these resources and more are available at connectED.mcgraw-hill.com

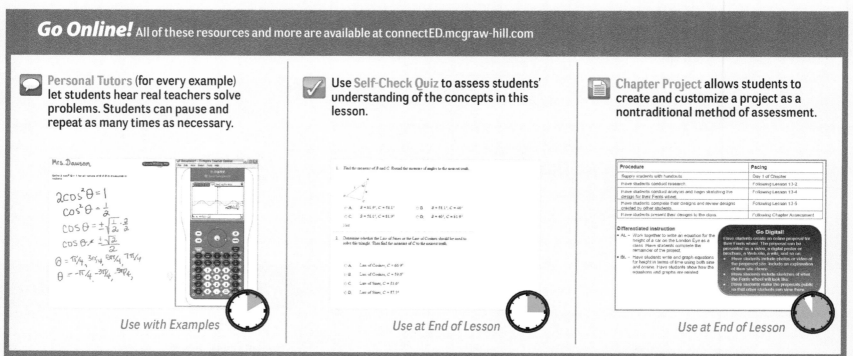

Personal Tutors (for every example) let students hear real teachers solve problems. Students can pause and repeat as many times as necessary.

Use with Examples

Use **Self-Check Quiz** to assess students' understanding of the concepts in this lesson.

Use at End of Lesson

Chapter Project allows students to create and customize a project as a nontraditional method of assessment.

Use at End of Lesson

OER **Using Open Educational Resources**

Assessment Have students work in groups to create quizzes, stories, or poems on **Quizilla** to help them remember key concepts about trigonometric identities and solving trigonometric equations. Students can then share their work with the class. *Use as homework*

Go Online!
connectED.mcgraw-hill.com

Worksheets

Differentiate Your Resources

Extra Practice Additional practice or homework; Skills Practice is best for approaching-level students and Practice is best for on-level and beyond-level students

Skills Practice

10-5 Skills Practice
Solving Trigonometric Equations

Practice

10-5 Practice
Solving Trigonometric Equations

Word Problem Practice

10-5 Word Problem Practice
Solving Trigonometric Equations

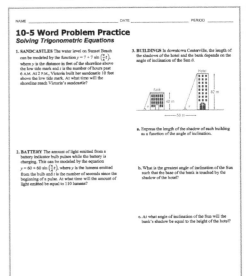

Intervention Reteaching and vocabulary activities that can be used with struggling or absent students and as ELL support

Extension Activities that can be used to extend lesson concepts

Study Guide and Intervention

10-5 Study Guide and Intervention
Solving Trigonometric Equations

Study Notebook

10-5 Solving Trigonometric Equations

Enrichment

10-5 Enrichment
Mechanical Engineering

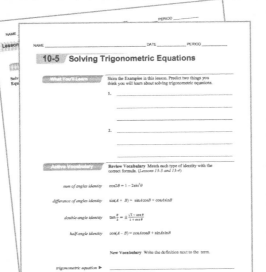

Launch

Objective Use a graphing calculator to solve trigonometric equations.

Materials for Each Student

- TI-83/84 Plus or other graphing calculator

Teaching Tip

In Activity 1, approximate solutions can also be found by using the **TRACE** feature. In most situations, however, the intersect feature will give more accurate solutions.

Remind students to use appropriate windows for their graphs.

Teach

Working in Cooperative Groups Have students work in pairs so they can help each other correct keystroke errors. Have them complete Activities 1 and 2 and Exercise 1. **ELL**

Ask:

- How are the solutions of the equations related to the points where the graphs intersect? The solutions are the *x*-values of the points of intersection.

- How can you tell if an equation has no solution? The graphs of Y1 and Y2 do not intersect.

Practice Have students complete Exercises 2–6.

Assess

Formative Assessment

Use Exercise 6 to assess whether students understand how the specified interval affects the solutions.

Go Online!

Graphic Calculators

Students can use the Graphing Calculator Personal Tutors to review the use of the graphing calculator to represent functions. They can also use the Other Calculator Keystrokes, which cover lab content for students with calculators other than the TI-84 Plus.

Graphing Technology Lab
Solving Trigonometric Equations

The graph of a trigonometric function is made up of points that represent all values that satisfy the function. To solve a trigonometric equation, you need to find all values of the variable that satisfy the equation. You can use a TI-83/84 Plus graphing calculator to solve trigonometric equations by graphing each side of the equation as a function and then locating the points of intersection.

Mathematical Practices
5 Use appropriate tools strategically.

Activity 1   Real Solutions

Work cooperatively. Use a graphing calculator to solve $\sin x = 0.4$ **if** $0° \leq x < 360°$.

Step 1 Enter and graph related equations. Rewrite the equation as two equations, $Y1 = \sin x$ and $Y2 = 0.4$. Then graph the two equations. Because the interval is in degrees, set your calculator to degree mode.

KEYSTROKES: MODE ▼ ▼ ► ENTER
Y= SIN X,T,θ,n)
ENTER 0.4 ENTER GRAPH

[0, 360] scl: 90 by [−1, 1] scl: 0.1

Step 2 Approximate the solutions. Based on the graph, you can see that there are two points of intersection in the interval $0° \leq x < 360°$. Use the **CALC** feature to determine the *x*-values at which the two graphs intersect.

The solutions are $x \approx 23.57°$ and $x \approx 156.4°$.

Activity 2   No Real Solutions

Work cooperatively. Use a graphing calculator to solve $\tan^2 x \cos x + 3 \cos x = 0$ **if** $0° \leq x < 360°$.

Step 1 Enter and graph related equations. The related equations to be graphed are $Y1 = \tan^2 x \cos x + 3 \cos x$ and $Y2 = 0$.

KEYSTROKES: Y= TAN X,T,θ,n) x² COS
X,T,θ,n) + 3 COS X,T,θ,n
) ENTER 0 ENTER

[0, 360] scl: 90 by [−15, 15] scl: 1

Step 2 These two functions do not intersect.

Therefore, the equation $\tan^2 x \cos x + 3 \cos x = 0$ has no real solutions.

Exercises

Work cooperatively. Use a graphing calculator to solve each equation for the values of *x* indicated.   6. −360°, −180°, 0°, 180°

1. $\sin x = 0.7$; $0° \leq x < 360°$ 44.4°, 135.6°

2. $\tan x = \cos x$; $0° \leq x < 360°$ 38.17°, 141.8°

3. $3 \cos x + 4 = 0.5$; $0° \leq x < 360°$ no real solution

4. $0.25 \cos x = 3.4$; $-720° \leq x < 720°$ no real solution

5. $\sin 2x = \sin x$; $0° \leq x < 360°$ 0°, 60°, 180°, 300°

6. $\sin 2x - 3 \sin x = 0$ if $-360° \leq x < 360°$

From Concrete to Abstract

Have students consider Exercises 1–6 with a different interval or no interval and explain how this affects the solutions.

LESSON 5

Solving Trigonometric Equations

::Then	::Now	::Why?
• You verified trigonometric identities.	**1** Solve trigonometric equations. **2** Find extraneous solutions from trigonometric equations.	• When you ride a Ferris wheel that has a diameter of 40 meters and turns at a rate of 1.5 revolutions per minute, the height above the ground, in meters, of your seat after t minutes can be modeled by the equation $h = 21 - 20 \cos 3\pi t$. After the ride begins, how long is it before your seat is 31 meters above the ground for the first time?

New Vocabulary
trigonometric equations

MP Mathematical Practices
4 Model with mathematics.
6 Attend to precision.
7 Look for and make use of structure.

Content Standards
F.TF.8 Prove the Pythagorean identity $\sin^2(\theta) + \cos^2(\theta) = 1$ and use it to find $\sin(\theta)$, $\cos(\theta)$, or $\tan(\theta)$ given $\sin(\theta)$, $\cos(\theta)$, or $\tan(\theta)$ and the quadrant of the angle.

1 Solve Trigonometric Equations So far in this chapter, you have studied a special type of trigonometric equation called an identity. Trigonometric identities are equations that are true for all values of the variable for which both sides are defined. In this lesson, you will examine **trigonometric equations** that are true only for certain values of the variable. Solving these equations resembles solving algebraic equations.

F.TF.8

Example 1   Solve Equations for a Given Interval

Solve $\sin \theta \cos \theta - \frac{1}{2} \cos \theta = 0$ if $0 \le \theta \le 180°$.

$\sin \theta \cos \theta - \frac{1}{2} \cos \theta = 0$ Original equation

$\cos \theta \left(\sin \theta - \frac{1}{2} \right) = 0$ Factor

$\cos \theta = 0$ or $\sin \theta - \frac{1}{2} = 0$ Zero Product Property

$\theta = 90°$ $\sin \theta = \frac{1}{2}$

$\theta = 30°$ or $150°$

The solutions are 30°, 90°, and 150°.

CHECK You can check the answer by graphing $y = \sin \theta \cos \theta$ and $y = \frac{1}{2} \cos \theta$ in the same coordinate plane on a graphing calculator. Then find the points where the graphs intersect. You can see that there are infinitely many such points, but we are only interested in the points between 0° and 180°.

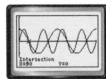

[0, 720] scl: 90 by [−1, 1] scl: 0.5

Guided Practice

1. Find all solutions of $\sin 2\theta = \cos \theta$ if $0 \le \theta \le 2\pi$. $\frac{\pi}{6}, \frac{\pi}{2}, \frac{5\pi}{6}, \frac{3\pi}{2}$

Trigonometric equations are usually solved for values of the variable between 0° and 360° or between 0 radians and 2π radians. There are solutions outside that interval. These other solutions differ by integral multiples of the period of the function.

MP Mathematical Practices Strategies

Reason abstractly and quantitatively.
Help students understand the meaning of trigonometric equations and their solutions. For example, ask:

• When can a trigonometric equation have infinitely many solutions? When the function is periodic and the domain is not limited, it will have infinitely many solutions.

• How can you limit the solutions to an equation that has infinitely many solutions? Limit the domain.

• How many solutions would the equation $1 = \sin 2\theta$ have on the domain $0 < \theta < 2\pi$? 2

• How many solutions would the equation $1 = \sin 2\theta$ have on the domain $-2\pi < \theta < 2\pi$? 4

• How many solutions would the equation $1 = \sin 2\theta$ have on the domain $\frac{3\pi}{2} < \theta < 2\pi$? 0

• Answer the above questions for $1 = \sin \theta$. What do you notice? The answers are the same.

• Would the pattern you noticed always be true for θ compared to 2θ? No. If the double angle is in a different quadrant, the sign of the value would change.

Launch

Have students read the Why? section of the lesson. Ask:

• Take any point on the Ferris wheel. How far does that point travel in 1 revolution? 40π or about 125.7 meters

• How far does a position on the Ferris wheel travel in 1 minute? 60π or about 188.5 meters

• When $t = 0$, what is the value of $20 \cos 3\pi t$? 20

Teach

Ask the scaffolded questions for each example to build conceptual understanding for students at all levels.

1 Solve Trigonometric Equations

Example 1 Solve equations for a Given Interval

AL Which quadrants could θ be in if $0 \le \theta \le 180°$? One or two

OL Why isn't 270° a solution? 270° is not between 0 and 180°.

BL Write an expression for all solutions to $\sin \theta \cos \theta - \frac{1}{2}\cos \theta = 0$. $90° + 180°k, 30° + 360°k, 150° + 360°k$

Need Another Example?
Solve $2 \cos^2 \theta - 1 = \sin \theta$ if $0° \le \theta \le 180°$. 30°, 150°

Go Online!

Interactive Whiteboard
Use the *eLesson* or *Lesson Presentation* to present this lesson.

Example 2 Infinitely Many Solutions

AL What would the solutions be in degrees?
$180° + 360°k$

OL What transformation is performed on cos θ to graph cos θ + 1? It is shifted up 1 unit.

BL Explain why there are infinitely many solutions. The graph of cos θ + 1 is periodic, so it intersects the x-axis infinitely many times.

Need Another Example?
a. Solve $\cos θ + \frac{1}{4} = \sin^2 θ$ for all values of θ if θ is measured in degrees. $60° + k \cdot 360°, 300° + k \cdot 360°$, where k is any integer
b. Solve $2 \cos θ = -1$ for all values of θ if θ is measured in radians. $\frac{2π}{3} + 2kπ$ and $\frac{4π}{3} + 2kπ$, where k is any integer

Example 3 Solve Trigonometric Equations

AL What do h and t represent? h is the height of your seat off the ground, t is the time in minutes.

OL What would the height of your seat be after 10 seconds? 21 meters

BL Why is this situation modeled with a periodic function? The Ferris wheel is circular, so as it revolves the height of the seat makes a circle.

Need Another Example?
Amusement Parks Refer to the beginning of the lesson. How long after the Ferris wheel starts will your seat first be $(10\sqrt{2} + 21)$ meters above the ground?

$10\sqrt{2} + 21 = 21 - 20 \cos 3πt;$
$10\sqrt{2} = -20 \cos 3πt;$
$-\frac{\sqrt{2}}{2} = \cos 3πt;$
$\text{Cos}^{-1}\left(-\frac{\sqrt{2}}{2}\right) = 3πt;$
$\frac{3}{4}π = 3πt$
$\frac{1}{4} = t$

$\frac{1}{4}$ min or 15 sec

Teaching Tip

Reasoning In Example 2, show students how to look for patterns in the solutions. Students should look for pairs of solutions that differ by exactly π or 2π.

F.TF.8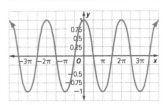

Example 2   Infinitely Many Solutions

Solve cos θ + 1 = 0 for all values of θ if θ is measured in radians.

$\cos θ + 1 = 0$
  $\cos θ = -1$

Look at the graph of $y = \cos θ$ to find solutions of $\cos θ = -1$.

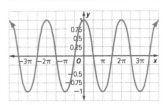

The solutions are π, 3π, 5π, and so on, and −π, −3π, −5π, and so on. The only solution in the interval 0 radians to 2π radians is π. The period of the cosine function is 2π radians. So the solutions can be written as $π + 2kπ$, where k is any integer.

Study Tip
Expressing Solutions as Multiples The expression $π + 2kπ$ includes all multiples of π in the solution set, so it is not necessary to list them separately.

2A. $90° + k \cdot 180°$, $120° + k \cdot 360°$, and $240° + k \cdot 360°$
2B. $\frac{7π}{6} + 2kπ$ and $\frac{11π}{6} + 2kπ$

▷ **Guided Practice**

2A. Solve $\cos 2θ + \cos θ + 1 = 0$ for all values of θ if θ is measured in degrees.
2B. Solve $2 \sin θ = -1$ for all values of θ if θ is measured in radians.

Trigonometric equations are often used to solve real-world problems.

F.TF.8

Real-World Example 3   Solve Trigonometric Equations

AMUSEMENT PARKS Refer to the beginning of the lesson. How long after the Ferris wheel starts will your seat first be 31 meters above the ground?

$h = 21 - 20 \cos 3πt$	Original equation
$31 = 21 - 20 \cos 3πt$	Replace h with 31.
$10 = -20 \cos 3πt$	Subtract 21 from each side.
$-\frac{1}{2} = \cos 3πt$	Divide each side by −20.
$\cos^{-1}\left(-\frac{1}{2}\right) = 3πt$	Take the Arccosine.
$\frac{2π}{3} = 3πt$	The Arccosine of $-\frac{1}{2}$ is $\frac{2π}{3}$.
$\frac{2π}{3} + 2πk = 3πt$	k is any integer.
$\frac{2}{9} + \frac{2}{3}k = t$	Divide each term by 3π.

The least positive value for t is obtained by letting $k = 0$ in the first expression. Therefore, $t = \frac{2}{9}$ of a minute or about 13 seconds.

▷ **Guided Practice**

3. How long after the Ferris wheel starts will your seat first be 41 meters above the ground? about 20 seconds

---

Differentiated Instruction **AL** **OL**

Interpersonal Learners As students work through this lesson, have them create a class list on the board that identifies common errors they made. Encourage students to add suggestions for how to avoid their errors. For example, one common error is having one's calculator set to degrees when it needs to be set to radians for a problem, and vice versa.

2 Extraneous Solutions

2 **Extraneous Solutions** Some trigonometric equations have no solution. For example, the equation $\cos \theta = 4$ has no solution because all values of $\cos \theta$ are between -1 and 1, inclusive. Thus, the solution set for $\cos \theta = 4$ is empty. F.TF.8

Example 4 Determine Whether a Solution Exists

Solve each equation.

a. $2 \sin^2 \theta - 3 \sin \theta - 2 = 0$ if $0 \leq \theta \leq 2\pi$

$2 \sin^2 \theta - 3 \sin \theta - 2 = 0$ Original equation

$(\sin \theta - 2)(2 \sin \theta + 1) = 0$ Factor

$\sin \theta - 2 = 0$ or $2 \sin \theta + 1 = 0$ Zero Product Property

$\sin \theta = 2$ $2 \sin \theta = -1$

This is not a solution $\sin \theta = -\frac{1}{2}$

because all values of

$\sin \theta$ are between -1 $\theta = \frac{7\pi}{6}$ or $\frac{11\pi}{6}$

and 1, inclusive.

The solutions are $\frac{7\pi}{6}$ or $\frac{11\pi}{6}$.

CHECK $2 \sin \theta - 3 \sin \theta - 2 = 0$ $2 \sin^2 \theta - 3 \sin \theta - 2 = 0$

$2 \sin^2 \left(\frac{7\pi}{6}\right) - 3 \sin \left(\frac{7\pi}{6}\right) - 2 \stackrel{?}{=} 0$ $2 \sin^2 \left(\frac{11\pi}{6}\right) - 3 \sin \left(\frac{11\pi}{6}\right) - 2 \stackrel{?}{=} 0$

$2\left(\frac{1}{4}\right) - 3\left(-\frac{1}{2}\right) - 2 \stackrel{?}{=} 0$ $2\left(\frac{1}{4}\right) - 3\left(-\frac{1}{2}\right) - 2 \stackrel{?}{=} 0$

$\frac{1}{2} + \frac{3}{2} - 2 \stackrel{?}{=} 0$ $\frac{1}{2} + \frac{3}{2} - 2 \stackrel{?}{=} 0$

$0 = 0 ✓$ $0 = 0 ✓$

b. $\sin \theta = 1 + \cos \theta$ if $0° \leq \theta < 360°$

$\sin \theta = 1 + \cos \theta$ Original equation

$\sin^2 \theta = (1 + \cos \theta)^2$ Square each side.

$1 - \cos^2 \theta = 1 + 2 \cos \theta + \cos^2 \theta$ $\sin^2 \theta = 1 - \cos^2 \theta$

$0 = 2 \cos \theta + 2 \cos^2 \theta$ Set the left side equal to 0.

$0 = 2 \cos \theta (1 + \cos \theta)$ Factor.

$1 + \cos \theta = 0$ or $2 \cos \theta = 0$ Zero Product Property

$\cos \theta = -1$ $\cos \theta = 0$

$\theta = 180$ $\theta = 90°$ or $270°$

CHECK $\sin \theta = 1 + \cos \theta$ $\sin \theta = 1 + \cos \theta$

$\sin 90° \stackrel{?}{=} 1 + \cos 90°$ $\sin 180° \stackrel{?}{=} 1 + \cos 180°$

$1 \stackrel{?}{=} 1 + 0$ $0 \stackrel{?}{=} 1 + (-1)$

$1 = 1 ✓$ $0 = 0 ✓$

$\sin \theta = 1 + \cos \theta$

$\sin 270° \stackrel{?}{=} 1 + \cos 270°$

$-1 \stackrel{?}{=} 1 + 0$

$-1 \neq 1 ✗$

The solutions are $90°$ and $180°$.

Guided Practice

4A. $\sin^2 \theta + 2 \cos^2 \theta = 4$ no solution   **4B.** $\cos^2 \theta + 3 = 4 - \sin^2 \theta$

4B. identity; therefore, infinitely many solutions

If an equation cannot be solved easily by factoring, try rewriting the expression using trigonometric identities. However, using identities and some algebraic operations, such as squaring, may result in extraneous solutions. So, it is necessary to check your solutions using the original equation.

Problem-Solving Tip

MP **Regularity** Look for patterns in your solutions. Look for pairs of solutions that differ by exactly π or 2π and write your solutions with the simplest possible pattern.

---

2 Extraneous Solutions

Example 4 Determine Whether a Solution Exists

AL How do you determine if a solution is extraneous? It does not satisfy the original equation.

OL What would the answers to part **a** be in degrees? $210°$ and $330°$

BL Give two inequalities that describe when $\sin \theta$ would not have any solutions. $\sin \theta < -1$ and $\sin \theta > 1$

Need Another Example?

Solve each equation.

a. $\sin \theta \cos \theta = \cos^2 \theta$ if $0 \leq \theta \leq 2\pi$ $\frac{\pi}{4}, \frac{\pi}{2}, \frac{5\pi}{4}, \frac{3\pi}{2}$

b. $\cos \theta = 1 - \sin \theta$ if $0° \leq \theta \leq 360°$ $0°, 90°, 360°$

Regularity Mathematically proficient students notice if calculations are repeated, and look both for general methods and for shortcuts. They continually evaluate the reasonableness of their intermediate results. Encourage students to examine their solutions closely, and to write them in the simplest possible form.

Differentiated Instruction ELL

Beginning/Intermediate Have students work in small groups. This strategy allows every student to have an opportunity to speak several times. Ask a question or give a prompt about trigonometric equations such as "Name one way that solving trigonometric equations is similar to, or different from, solving quadratic equations." Then pass a stick or other object to the student. The student speaks, everyone listens, and then passes the object to the next person. The next student speaks, everyone listens then the student passes the object on until everyone has had one or two turns.

Go Online!

The most up-to-date resources available for your program can be found at connectED.mcgraw-hill.com.

Example 5 Solve Trigonometric Equations by Using Identities

AL What type of identity is used here? a Pythagorean identity

OL Why can't $\tan^2 \theta$ ever be negative? Any number squared is positive or zero.

BL Give four angles that would be solutions to this equation. $60°, -60°, 240°, -120°$

Need Another Example?

Solve $\tan^4 \theta - 4 \sec^2 \theta = -7$ for all values of θ if θ is measured in degrees. $\theta = 60° + 180°k, 120° + 180°k,$ or $45° + 90° \cdot k$, where k is any integer.

Practice

Formative Assessment Use Exercises 1–29 to assess students' understanding of the concepts in this lesson.

The Practice and Problem Solving exercises assess the content taught in the lesson. The Preparing for Assessment page is meant to be used as preparation for assessment.

Teaching the Mathematical Practices

MP

Regularity Mathematically proficient students notice if calculations are repeated, and look both for general methods and for shortcuts. As they work to solve a problem, mathematically proficient students maintain oversight of the process, while attending to the details and continually evaluate the reasonableness of their intermediate results.

Teaching Tip

Notation In all answers in this lesson involving k, k is any integer.

Additional Answer

21b. Every day from February 19 to October 20; sample explanation: Since the longest day of the year occurs around June 22, the days between February 19 and October 20 must increase in length until June 22 and then decrease in length until October 20.

F.TF.8

Example 5 Solve Trigonometric Equations by Using Identities

Solve $2 \sec^2 \theta - \tan^4 \theta = -1$ for all values of θ if θ is measured in degrees.

$$2 \sec^2 \theta - \tan^4 \theta = -1 \qquad \text{Original equation}$$
$$2(1 + \tan^2 \theta) - \tan^4 \theta = -1 \qquad \sec^2 \theta = 1 + \tan^2 \theta$$
$$2 + 2 \tan^2 \theta - \tan^4 \theta = -1 \qquad \text{Distributive Property}$$
$$\tan^4 \theta - 2 \tan^2 \theta - 3 = 0 \qquad \text{Set one side of the equation equal to 0.}$$
$$(\tan^2 \theta - 3)(\tan^2 \theta + 1) = 0 \qquad \text{Factor.}$$
$$\tan^2 \theta - 3 = 0 \qquad \text{or} \qquad \tan^2 \theta + 1 = 0 \qquad \text{Zero Product Property}$$
$$\tan^2 \theta = 3 \qquad\qquad \tan^2 \theta = -1$$
$$\tan \theta = \pm\sqrt{3} \qquad \text{This part gives no solutions since } \tan^2 \theta \text{ is never negative.}$$

$\theta = 60° + 180°k$ and $\theta = -60° + 180°k$, where k is any integer. The solutions are $60° + 180°k$ and $-60° + 180°k$.

Study Tip
Solving Trigonometric Equations Remember that *solving a trigonometric equation* means solving for all values of the variable.

Guided Practice

Solve each equation.

5A. $\sin \theta \cot \theta - \cos^2 \theta = 0$

5B. $\frac{\cos \theta}{\cot \theta} + 2 \sin^2 \theta = 0$

5A. $90° + k \cdot 180°$
5B. $\frac{7\pi}{6} + 2k \cdot \pi$, and $\frac{11\pi}{6} + 2k \cdot \pi$

Check Your Understanding ◯ = Step-by-Step Solutions begin on page R11.

✓ **Go Online!** for a Self-Check Quiz

Example 1
F.TF.8

MP REGULARITY Solve each equation if $0° \leq \theta \leq 360°$.

1. $2 \sin \theta + 1 = 0$ $210°, 330°$

2. $\cos^2 \theta + 2 \cos \theta + 1 = 0$ $180°$

3. $\cos 2\theta + \cos \theta = 0$ $60°, 180°, 300°$

4. $2 \cos \theta = 1$ $60°, 300°$

5. $\cos \theta = -\frac{\sqrt{3}}{2}$ $150°, 210°$

6. $\sin 2\theta = -\frac{\sqrt{3}}{2}$ $120°, 150°, 300°, 330°$

7 $\cos 2\theta = 8 - 15 \sin \theta$ $30°, 150°$

8. $\sin \theta + \cos \theta = 1$ $0°, 90°, 360°$

Example 2
F.TF.8

Solve each equation for all values of θ if θ is measured in radians.

9. $4 \sin^2 \theta - 1 = 0$

10. $2 \cos^2 \theta = 1$ $\frac{\pi}{4} + \frac{k}{2}\pi$

11. $\cos 2\theta \sin \theta = 1$ $\frac{3\pi}{2} + 2k\pi$

12. $\sin \frac{\theta}{2} + \cos \frac{\theta}{2} = \sqrt{2}$ $\frac{\pi}{2} + 4\pi k$

13. $\cos 2\theta + 4 \cos \theta = -3$ $\pi + 2k\pi$

14. $\sin \frac{\theta}{2} + \cos \theta = 1$

9. $\pm\frac{\pi}{6} + 2k\pi$ or $\pm\frac{5\pi}{6} + 2k\pi$

14. $0 + 2k\pi; \frac{\pi}{3} + 4k\pi; \frac{5\pi}{3} + 4k\pi$

Solve each equation for all values of θ if θ is measured in degrees.

15. $\cos 2\theta - \sin^2 \theta + 2 = 0$

16. $\sin^2 \theta - \sin \theta = 0$

17. $2 \sin^2 \theta - 1 = 0$ $45° + k \cdot 90°$

18. $\cos \theta - 2 \cos \theta \sin \theta = 0$

19. $\cos 2\theta \sin \theta = 1$ $270° + k \cdot 360°$

20. $\sin \theta \tan \theta - \tan \theta = 0$ $0° + k \cdot 180°, 90° + k \cdot 360°$

15. $90° + k \cdot 180°$
16. $0° + k \cdot 180°,$ $90° + k \cdot 360°$
18. $30° + k \cdot 360°,$ $150° + k \cdot 360°,$ $90° + k \cdot 180°$

Example 3
F.TF.8

21. LIGHT The number of hours of daylight d in Hartford, Connecticut, may be approximated by the equation $d = 3 \sin \frac{2\pi}{365}t + 12$, where t is the number of days after March 21.

a. On what days will Hartford have exactly $10\frac{1}{2}$ hours of daylight?

b. Using the results in part **a**, tell what days of the year have at least $10\frac{1}{2}$ hours of daylight. Explain how you know. See margin.

21a. There will be $10\frac{1}{2}$ hours of daylight 213 and 335 days after March 21; that is, on October 20 and February 19.

Differentiated Homework Options

Levels	**AL** Basic	**OL** Core	**BL** Advanced
Exercises	30–48, 59–68	31–55 odd, 56–68	49–68
2-Day Option	31–47 odd, 64–68	30–48, 64–68	
	30–48 even, 59–63	49–62	

 You can use ALEKS to provide additional remediation support with personalized instruction and practice.

Examples 4-5 Solve each equation.

25. $\frac{\pi}{2} + \pi k, \frac{\pi}{6} + 2\pi k, \frac{5\pi}{6} + 2\pi k$

F.TF.8 **22.** $\sin^2 2\theta + \cos^2 \theta = 0$ $\frac{\pi}{2} + \pi k$

23. $\tan^2 \theta + 2\tan\theta + 1 = 0$ $\frac{3\pi}{4} + \pi k$

24. $\cos^2 \theta + 3\cos\theta = -2$ $\pi + 2\pi k$

25. $\sin 2\theta - \cos\theta = 0$

26. $\tan\theta = 1$ $45° + k \cdot 180°$ or $\frac{\pi}{4} + k \cdot \pi$

27. $\cos 8\theta = 1$ $0° + k \cdot 45°$ or $0 + k \cdot \frac{\pi}{4}$

28. $\sin\theta + 1 = \cos 2\theta$ $0° + k \cdot 180°, 210° + k \cdot 360°, 330° + k \cdot 360°$

29. $2\cos^2 \theta = \cos\theta$ $\frac{\pi}{3} + 2k\pi, \frac{\pi}{2} + k\pi, \frac{5\pi}{3} + 2k\pi$

Practice and Problem Solving

Extra Practice is on page R10.

Example 1 Solve each equation for the given interval. **30.** $60°, 120°, 240°, 300°$ **32.** $\frac{\pi}{6}, \frac{\pi}{2}, \frac{5\pi}{6}, \frac{3\pi}{2}$

F.TF.8 **30.** $\cos^2 \theta = \frac{1}{4}; 0° \le \theta \le 360°$

31. $2\sin^2 \theta = 1; 90° < \theta < 270°$ $135°, 225°$

32. $\sin 2\theta - \cos\theta = 0; 0 \le \theta \le 2\pi$

33. $3\sin^2 \theta = \cos^2 \theta; 0 \le \theta \le \frac{\pi}{2}$ $\frac{\pi}{6}$

34. $2\sin\theta + \sqrt{3} = 0; 180° < \theta < 360°$ $240°, 300°$

35. $4\sin^2 \theta - 1 = 0; 180° < \theta < 360°$ $210°, 330°$

Example 2 Solve each equation for all values of θ if θ is measured in radians.

36. $\frac{\pi}{3} + 2k\pi, \frac{5\pi}{3} + 2k\pi$

F.TF.8 **36.** $\cos 2\theta + 3\cos\theta = 1$

37 $2\sin^2 \theta = \cos\theta + 1$

37. $\pi + 2k\pi, \frac{\pi}{3} + 2k\pi, \frac{5\pi}{3} + 2k\pi$

45. $\frac{7\pi}{6} + 2k\pi, \frac{11\pi}{6} + 2k\pi$ or $210° + k \cdot 360°, 330° + k \cdot 360°$

38. $\cos^2 \theta - \frac{3}{2} = \frac{5}{2}\cos\theta$

39. $3\cos\theta - \cos\theta = 2$ $0 + 2k\pi$

38. $\frac{2\pi}{3} + 2k\pi, \frac{4\pi}{3} + 2k\pi$

Solve each equation for all values of θ if θ is measured in degrees.

40. $\sin\theta - \cos\theta = 0$ $45° + k \cdot 180°$

41. $\tan\theta - \sin\theta = 0$ $0° + k \cdot 180°$

42. $\sin^2 \theta = 2\sin\theta + 3$ $270° + k \cdot 360°$

43. $4\sin^2 \theta = 4\sin\theta - 1$ $30° + k \cdot 360°, 150° + k \cdot 360°$

Example 3
F.TF.8 **44. ELECTRONICS** The tallest structure in Tessa's hometown is a television transmitting tower with a height of 1626 feet. What is the measure of θ if the length of the shadow is 1 mile? about 17°

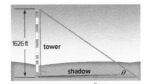
1626 ft
tower
shadow θ

Examples 4-5 Solve each equation.

F.TF.8 **45.** $2\sin^2 \theta = 3\sin\theta + 2$

46. $2\cos^2 \theta + 3\sin\theta = 3$

46. $\frac{\pi}{6} + 2k\pi, \frac{5\pi}{6} + 2k\pi, \frac{\pi}{2} + 2k\pi$ or $30° + k \cdot 360°, 150° + k \cdot 360°, 90° + k \cdot 360°$

47. $\sin^2 \theta + \cos 2\theta = \cos\theta$ $0 + 2k\pi, \frac{\pi}{2} + k\pi$ or $0° + k \cdot 360°, 90° + k \cdot 180°$

48. $2\cos^2 \theta = -\cos\theta$

48. $\frac{\pi}{2} + k\pi, \frac{2\pi}{3} + 2k\pi, \frac{4\pi}{3} + 2k\pi$ or $90° + k \cdot 180°, 120° + k \cdot 360°, 240° + k \cdot 360°$

 49. **SENSE-MAKING** Due to ocean tides, the depth y in meters of the River Thames in London varies as a sine function of x, the hour of the day. On a certain day that function was $y = 3\sin\left[\frac{\pi}{6}(x - 4)\right] + 8$, where $x = 0, 1, 2, ..., 24$ corresponds to 12:00 midnight, 1:00 A.M., 2:00 A.M., ..., 12:00 midnight the next night.

a. What is the maximum depth of the River Thames on that day? 11 m

b. At what times does the maximum depth occur? 7:00 A.M. and 7:00 P.M.

Solve each equation if θ is measured in radians.

50. $(\cos\theta)(\sin 2\theta) - 2\sin\theta + 2 = 0$ $\frac{\pi}{2} + 2\pi k$

51. $2\sin^2 \theta + (\sqrt{2} - 1)\sin\theta = \frac{\sqrt{2}}{2}$

51. $\frac{\pi}{6} + 2\pi k, \frac{5\pi}{6} + 2\pi k, \frac{5\pi}{4} + 2\pi k, \frac{7\pi}{4} + 2\pi k$

52. $30° + 360°k, 150° + 360°k, 330° + 360°k$

Solve each equation if θ is measured in degrees.

52. $\sin 2\theta + \frac{\sqrt{3}}{2} = \sqrt{3}\sin\theta + \cos\theta$

53. $1 - \sin^2 \theta - \cos\theta = \frac{3}{4}$ $120° + 360°k, 240° + 360°k$

Levels of Complexity Chart

The levels of the exercises progress from 1 to 3, with Level 1 indicating the lowest level of complexity.

Exercises	30–48	49–56, 64–68	57–63
Level 3			●
Level 2		○	
Level 1	●		

Standards for Mathematical Practice

Emphasis On	Exercises
1 Make sense of problems and persevere in solving them.	21, 44–56
3 Construct viable arguments and critique the reasoning of others.	59–63
6 Attend to precision.	1–8, 22–43
8 Look for and express regularity in repeated reasoning.	9–20

Go Online!

eSolutions Manual

Create worksheets, answer keys, and solutions handouts for your assignments.

Assess

Ticket Out the Door Have students write an equation involving $\sin^2 \theta$ that has exactly one solution for the interval $90° < \theta < 270°$.

Teaching the Mathematical Practices

Perseverance Mathematically proficient students start by explaining to themselves the meaning of a problem and looking for entry points to its solution. They analyze givens, constraints, relationships, and goals and make conjectures about the form and meaning of the solution and plan a solution pathway rather than simply jumping into a solution attemp.

Additional Answers

60. Each type of equation may require adding, subtracting, multiplying, or dividing each side by the same number. Quadratic and trigonometric equations can often be solved by factoring. Linear and quadratic equations do not require identities. All linear and quadratic equations can be solved algebraically, whereas some trigonometric equations may be solved more easily by using a graphing calculator. A linear equation has at most one solution. A quadratic equation has at most two solutions. A trigonometric equation usually has infinitely many solutions, unless the values of the variable are restricted.

61. Sample answer: All trigonometric functions are periodic. Therefore, once one or more solutions are found for a certain interval, there will be additional solutions that can be found by adding integral multiples of the period of the function to those solutions.

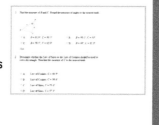

Solve each equation.

54. $2 \sin \theta = \sin 2\theta$ $\quad \pi k$

55. $\cos \theta \tan \theta - 2\cos^2 \theta = -1$ $\quad \frac{\pi}{6} + 2\pi k, \frac{5\pi}{6} + 2\pi k$

56. DIAMONDS According to Snell's Law, $n_1 \sin i = n_2 \sin r$, where n_1 is the index of refraction of the medium the light is exiting, n_2 is the index of refraction of the medium the light is entering, i is the degree measure of the angle of incidence, and r is the degree measure of the angle of refraction.

   a. The index of refraction of a diamond is 2.42, and the index of refraction of air is 1.00. If a beam of light strikes a diamond at an angle of 35°, what is the angle of refraction? **13.71°**

   b. Explain how a gemologist might use Snell's Law to determine whether a diamond is genuine. **Measure the angles of incidence and refraction to determine the index of refraction. If the index is 2.42, the diamond is genuine.**

57. **PERSEVERANCE** A wave traveling in a guitar string can be modeled by the equation $D = 0.5 \sin (6.5x)° \sin (2500t)°$, where D is the displacement in millimeters at the position x millimeters from the left end of the string at time t seconds. Find the first positive time when the point 0.5 meter from the left end has a displacement of 0.01 millimeter. **0.0026 second**

58. MULTIPLE REPRESENTATIONS Consider the trigonometric inequality $\sin \theta \geq \frac{1}{2}$.

   a. **Tabular** Construct a table of values for $0° \leq \theta \leq 360°$. For what values of θ is $\sin \theta \geq \frac{1}{2}$? **a–d. See Ch. 10 Answer Appendix.**

   b. **Graphical** Graph $y = \sin \theta$ and $y = \frac{1}{2}$ on the same graph for $0° \leq \theta \leq 360°$. For what values of θ is the graph of $y = \sin \theta$ above the graph of $y = \frac{1}{2}$?

   c. **Analytic** Based on your answers for parts **a** and **b**, solve $\sin \theta \geq \frac{1}{2}$ for all values of θ.

   d. **Algebraic** Solve each inequality if $0 \leq \theta \leq 360°$. Then solve each for all values of θ.

     i. $\cos \theta \geq \frac{\sqrt{2}}{2}$ **ii.** $2 \sin \theta \leq \sqrt{3}$

     iii. $-\sin \theta \geq 0$ **iv.** $\cos \theta - 1 < -\frac{1}{2}$

F.TF.8

H.O.T. Problems   Use Higher-Order Thinking Skills

59. CHALLENGE Solve $\sin 2x < \sin x$ for $0 \leq x \leq 2\pi$ without a calculator. $\quad \frac{\pi}{3} < x < \pi$ or $\frac{5\pi}{3} < x < 2\pi$

60. **REASONING** Compare and contrast solving trigonometric equations with solving linear and quadratic equations. What techniques are the same? What techniques are different? How many solutions do you expect? **See margin.**

61. **WRITING IN MATH** Why do trigonometric equations often have infinitely many solutions? **See margin.**

62. OPEN-ENDED Write an example of a trigonometric equation that has exactly two solutions if $0° \leq \theta \leq 360°$. **Sample answer: $2\cos \theta = 0$, 90° and 270°**

63. CHALLENGE How many solutions in the interval $0° \leq \theta \leq 360°$ should you expect for $a \sin (b\theta + c) = d$, if $a \neq 0$ and b is a positive integer? **0, b, or $2b$**

Differentiated Instruction BL

Extension Ask students to explore the solutions of the equation $\sin x = \frac{x}{k}$, where k is a positive integer. Ask them to determine how the number of solutions to the equation changes as k changes. What value of x is a solution of the equation for every value of k? **0**

Preparing for Assessment

64. The solution(s) for $\sec^2 \theta - 5 = -1$, in the interval $180° \le \theta \le 360°$, is given by which of the following? ⓜ 2,6 F.TF.8 **D**

    I 120°

    II 210°

    III 240°

    IV 300°

  ○ **A** I only

  ○ **B** II only

  ○ **C** I and II

  ○ **D** III and IV

  ○ **E** I, III, and IV

65. How many solutions does $1 + \sin \theta = \cos \theta$ have in the interval $0° \le \theta < 360°$? ⓜ 2,6 F.TF.8 **C**

  ○ **A** 0

  ○ **B** 1

  ○ **C** 2

  ○ **D** 4

  ○ **E** infinitely many

66. Which of the following is a solution to $\tan \theta \sin \theta = \sin \theta$ in the interval $0° < \theta < 180°$? ⓜ 2,6 F.TF.8 **B**

  ○ **A** 30°

  ○ **B** 45°

  ○ **C** 60°

  ○ **D** 90°

  ○ **E** 135°

67. What is the solution to $4 \sin \theta - \sin \theta = 3$ on the interval $0 \le \theta \le 2\pi$? ⓜ 2,6 F.TF.8 **B**

  ○ **A** 0

  ○ **B** $\frac{\pi}{2}$

  ○ **C** π

  ○ **D** $\frac{3\pi}{2}$

  ○ **E** 2π

68. a. MULTI-STEP The graph of $y = \sin \theta$ is shown. Which of the following shows all of the solutions for $\sin \theta = -1$ in the interval $-\pi < 0 < \pi$? ⓜ 2,6 F.TF.8 **B**

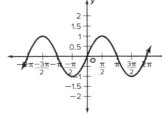

  ○ **A** $-\pi$ and π

  ○ **B** $-\frac{\pi}{2}$

  ○ **C** 0

  ○ **D** $-\frac{\pi}{2}$ and $\frac{3\pi}{2}$

  ○ **E** $-\pi, -\frac{\pi}{2}, \frac{3\pi}{2}$ and π

  b. How many solutions are there to $\sin \theta = 1$ for the domain $-2\pi < \theta < 2\pi$? **C**

  ○ **A** 0

  ○ **B** 1

  ○ **C** 2

  ○ **D** 4

68a.

A	Found a measure where $\sin \theta = 0$
B	CORRECT
C	Used the origin
D	Found a solution outside of the interval
E	Found incorrect solutions

Preparing for Assessment

Dual Coding

Exercise	Content Objective	ⓜ Mathematical Practices
64	F.TF.8	2, 6
65	F.TF.8	2, 6
66	F.TF.8	2, 6
67	F.TF.8	2, 6
68	F.TF.8	2, 6

Diagnose Student Errors

Survey student responses for each item. Class trends may indicate common errors and misconceptions.

64.

A	Selected a solution not in the given interval
B	Confused reciprocal ratio for secant with sine
C	Selected solutions only where secant is negative
D	CORRECT
E	Selected a solution not in the given interval

65.

A	Did not consider 0° or 270° as solutions
B	Overlooked 0° as being in the interval
C	CORRECT
D	Assumed solutions were located on all four axes
E	Did not consider the restrictions on θ

66.

A	Made an error in simplifying $\frac{\sin \theta}{\cos \theta}$
B	CORRECT
C	Made an error in simplifying $\frac{\sin \theta}{\cos \theta}$
D	Selected a measure that is undefined for tangent
E	Simplified $\frac{\sin \theta}{\cos \theta}$ as -1 instead of 1

67.

A	Found a measure where $\sin \theta = 0$
B	CORRECT
C	Found a measure where $\sin \theta = 0$
D	Found a measure where $\sin \theta = -1$
E	Found a measure where $\sin \theta = 0$

68b.

A	Used the wrong interval
B	Used the wrong interval
C	CORRECT
D	Used the wrong interval

FOLDABLES® Study Organizer

A completed Foldable for this chapter should include the Key Concepts related to trigonometric identities and equations.

Key Vocabulary ⟨ELL⟩

The page reference after each word denotes where that term was first introduced. If students have difficulty answering questions 1–7, remind them that they can use these page references to refresh their memories about the vocabulary terms.

Have students work in pairs to discuss each term in the Key Vocabulary list as they complete the Vocabulary Check. Encourage students to review using their notes and the text.

You can use the detailed reports in ALEKS to automatically monitor students' progress and pinpoint remediation needs prior to the chapter test.

Additional Answers

8. Simplify one side of the equation until the two sides of the equation are the same.

9. Trigonometric identities are equations that are true for all values of the variable for which both sides are defined. Trigonometric equations are true only for certain values of the variable.

Study Guide and Review

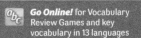 *Go Online!* for Vocabulary Review Games and key vocabulary in 13 languages

Study Guide

Key Concepts

Trigonometric Identities (Lessons 10-1, 10-2, and 10-5)

- Trigonometric identities describe the relationships between trigonometric functions.

- Trigonometric identities can be used to simplify, verify, and solve trigonometric equations and expressions.

Sum and Difference Identities (Lesson 10-3)

- For all values of A and B:

$$\cos (A \pm B) = \cos A \cos B \mp \sin A \sin B$$
$$\sin (A \pm B) = \sin A \cos B \pm \cos A \sin B$$
$$\tan (A \pm B) = \frac{\tan A \pm \tan B}{1 \mp \tan A \tan B}$$

Double-Angle and Half-Angle Identities (Lesson 10-4)

- Double-angle identities:

$$\sin 2\theta = 2 \sin \theta \cos \theta$$
$$\cos 2\theta = \cos^2 \theta - \sin^2 \theta$$
$$\cos 2\theta = 1 - 2 \sin^2 \theta$$
$$\cos 2\theta = 2 \cos^2 \theta - 1$$
$$\tan 2\theta = \frac{2 \tan \theta}{1 - \tan^2 \theta}$$

- Half-angle identities:

$$\sin \frac{\theta}{2} = \pm\sqrt{\frac{1 - \cos \theta}{2}}$$
$$\cos \frac{\theta}{2} = \pm\sqrt{\frac{1 + \cos \theta}{2}}$$
$$\tan \frac{\theta}{2} = \pm\sqrt{\frac{1 - \cos \theta}{1 + \cos \theta}}, \cos \theta \neq -1$$

FOLDABLES® Study Organizer

Use your Foldable to review the chapter. Working with a partner can be helpful. Ask for clarification of concepts as needed.

Key Vocabulary

cofunction identity (p. 655)

negative angle identity (p. 655)

Pythagorean identity (p. 655)

quotient identity (p. 655)

reciprocal identity (p. 655)

trigonometric equation (p. 683)

trigonometric identity (p. 655)

Vocabulary Check

Choose the correct term to complete each sentence.

1. The _____ can be used to find the sine or cosine of 75° if the sine and cosine of 90° and 15° are known.
   difference of angles identity

2. The identities $\tan \theta = \frac{\sin \theta}{\cos \theta}$ and $\cot \theta = \frac{\cos \theta}{\sin \theta}$ are examples of _____. **quotient identities**

3. The _____ can be used to find sin 60° using 30° as a reference. **double-angle identity**

4. The _____ identity can be used to find $\cos 22\frac{1}{2}°$. **half-angle**

5. The identities $\csc \theta = \frac{1}{\sin \theta}$ and $\sec \theta = \frac{1}{\cos \theta}$ are examples of _____. **reciprocal identities**

6. The _____ can be used to find the sine or cosine of 120° if the sine and cosine of 90° and 30° are known. **sum of angles identity**

7. $\cos^2 \theta + \sin^2 \theta = 1$ is an example of a(n) _____.
   7. Pythagorean identity

Concept Check

8. Explain how to verify a trigonometric identity. **See margin.**

9. Explain the difference between trigonometric identities and trigonometric equations. **See margin.**

Answering the Essential Question

Before answering the Essential Question, have students review their answers to the *Building on the Essential Question* exercises found throughout the chapter.

- How can representing the same mathematical concept in different ways be helpful? (p. 652)

- How do you simplify a trigonometric expression? (p. 657)

- Why are trigonometric identities useful? (p. 665)

- How do you decide what techniques to use when verifying a trigonometric identity? (p. 678)

Go Online!

Vocabulary Review

Students can use the *Vocabulary Review Games* to check their understanding of the vocabulary terms in this chapter. Students should refer to the *Student-Built Glossary* they have created as they went through the chapter to review important terms. You can also give a *Vocabulary Test* over the content of this chapter.

Lesson-by-Lesson Review

10-1 Trigonometric Identities

F.TF.8

Find the value of each expression.

10. $\sin \theta$, if $\cos \theta = \frac{\sqrt{2}}{2}$ and $270° < \theta < 360°$ $\quad -\frac{\sqrt{2}}{2}$

11. $\sec \theta$, if $\cot \theta = \frac{\sqrt{2}}{2}$ and $90° < \theta < 180°$ $\quad -\sqrt{3}$

12. $\tan \theta$, if $\cot \theta = 2$ and $0° < \theta < 90°$ $\quad \frac{1}{2}$

13. $\cos \theta$, if $\sin \theta = -\frac{3}{5}$ and $180° < \theta < 270°$ $\quad -\frac{4}{5}$

14. $\csc \theta$, if $\cot \theta = -\frac{4}{5}$ and $270° < \theta < 360°$ $\quad \frac{\sqrt{41}}{5}$

15. **SOCCER** For international matches, the maximum dimensions of a soccer field are 110 meters by 75 meters. Find $\sin \theta$. **See margin.**

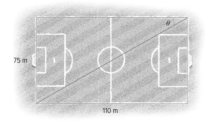

75 m

110 m

Simplify each expression.   16. $\cos^2 \theta$ 18. $\csc \theta$

16. $1 - \tan \theta \sin \theta \cos \theta$

17. $\tan \theta \csc \theta$ $\quad \sec \theta$

18. $\sin \theta + \cos \theta \cot \theta$

19. $\cos \theta (1 + \tan^2 \theta)$ $\quad \sec \theta$

Example 1

Find $\sin \theta$ if $\cos \theta = \frac{3}{4}$ and $0° < \theta < 90°$.

$\cos^2 \theta + \sin^2 \theta = 1$ — Trigonometric identity

$\sin^2 \theta = 1 - \cos^2 \theta$ — Subtract $\cos^2 \theta$ from each side.

$\sin^2 \theta = 1 - \left(\frac{3}{4}\right)^2$ — Substitute $\frac{3}{4}$ for $\cos \theta$

$\sin^2 \theta = 1 - \frac{9}{16}$ — Square $\frac{3}{4}$

$\sin^2 \theta = \frac{7}{16}$ — Subtract.

$\sin \theta = \pm\frac{\sqrt{7}}{4}$ — Take the square root of each side.

Because θ is in the first quadrant, $\sin \theta$ is positive.

Thus, $\sin \theta = \frac{\sqrt{7}}{4}$.

Example 2

Simplify $\cos \theta \sec \theta \cot \theta$.

$\cos \theta \sec \theta \cot \theta = \cos \theta \left(\frac{1}{\cos \theta}\right)\left(\frac{\cos \theta}{\sin \theta}\right)$

$= \cot \theta$

10-2 Verifying Trigonometric Identities

F.TF.8

Verify that each of the following is an identity.

20. $\tan \theta \cos \theta + \cot \theta \sin \theta = \sin \theta + \cos \theta$

21. $\frac{\cos \theta}{\cot \theta} + \frac{\sin \theta}{\tan \theta} = \sin \theta + \cos \theta$

22. $\sec^2 \theta - 1 = \frac{\sin^2 \theta}{1 - \sin^2 \theta}$ **20–23. See margin.**

23. **GEOMETRY** The right triangle shown at the right is used in a special quilt. Use the measures of the sides of the triangle to show that $\tan^2 \theta + 1 = \sec^2 \theta$.

3   4 $\sqrt{7}$

Example 3

Verify that $\frac{\cos \theta + 1}{\sin \theta} = \cot \theta + \csc \theta$ is an identity.

$\frac{\cos \theta + 1}{\sin \theta} \stackrel{?}{=} \cot \theta + \csc \theta$ — Original equation

$\frac{\cos \theta}{\sin \theta} + \frac{1}{\sin \theta} \stackrel{?}{=} \cot \theta + \csc \theta$ — Simplify.

$\cot \theta + \csc \theta = \cot \theta + \csc \theta \checkmark$ — Simplify.

---

22. $\sec^2 \theta - 1 \stackrel{?}{=} \frac{\sin^2 \theta}{1 - \sin^2 \theta}$

$\sec^2 \theta - 1 \stackrel{?}{=} \frac{\sin^2 \theta}{\cos^2 \theta}$

$\sec^2 \theta - 1 \stackrel{?}{=} \tan^2 \theta$

$\sec^2 \theta - 1 = \sec^2 \theta - 1 \checkmark$

23. $\tan^2 \theta + 1 = \left(\frac{\sqrt{7}}{3}\right)^2 + 1 = \frac{7}{9} + 1$

$= \frac{7}{9} + \frac{9}{9} = \frac{16}{9};$

$\sec^2 \theta = \left(\frac{4}{3}\right)^2 = \frac{16}{9}$

Lesson-by-Lesson Review

Intervention If the given examples are not sufficient to review the topics covered by the questions, remind students that the lesson references tell them where to review that topic in their textbook.

Two-Day Option Have students complete the Lesson-by-Lesson Review. Then you can use McGraw-Hill eAssessment to customize another review worksheet that practices all the objectives of this chapter or only the objectives on which your students need more help.

Additional Answers

15. $\frac{15\sqrt{709}}{709}$

20. $\tan \theta \cos \theta + \cot \theta \sin \theta \stackrel{?}{=} \sin \theta + \cos \theta$

$\frac{\sin \theta}{\cos \theta} \cdot \cos \theta + \frac{\cos \theta}{\sin \theta} \cdot \sin \theta \stackrel{?}{=} \sin \theta + \cos \theta$

$\sin \theta + \cos \theta = \sin \theta + \cos \theta \checkmark$

21. $\frac{\cos \theta}{\cot \theta} + \frac{\sin \theta}{\tan \theta} \stackrel{?}{=} \sin \theta + \cos \theta$

$\cos \theta \div \frac{\cos \theta}{\sin \theta} + \sin \theta \div \frac{\sin \theta}{\cos \theta} \stackrel{?}{=} \sin \theta + \cos \theta$

$\cos \theta \cdot \frac{\sin \theta}{\cos \theta} + \sin \theta \cdot \frac{\cos \theta}{\sin \theta} \stackrel{?}{=} \sin \theta + \cos \theta$

$\sin \theta + \cos \theta = \sin \theta + \cos \theta \checkmark$

Go Online!

Anticipation Guide

Students should complete the *Chapter 10 Anticipation Guide*, and discuss how their responses have changed now that they have completed Chapter 10.

Additional Answers (Practice Test)

2. $\cos(30° - \theta) \overset{?}{=} \sin(60° + \theta)$

$\cos 30° \cos \theta + \sin 30° \sin \theta \overset{?}{=}$
$\sin 60° \cos \theta + \cos 60° \sin \theta$

$\dfrac{\sqrt{3}}{2} \cos \theta + \dfrac{1}{2} \sin \theta =$
$\dfrac{\sqrt{3}}{2} \cos \theta + \dfrac{1}{2} \sin \theta \checkmark$

3. $\cos(\theta - \pi) \overset{?}{=} -\cos \theta$

$\cos \theta \cos \pi + \sin \theta \sin \pi \overset{?}{=} -\cos \theta$

$(\cos \theta)(-1) + (\sin \theta)(0) \overset{?}{=} -\cos \theta$

$-\cos \theta = -\cos \theta \checkmark$

10. $\sin \theta (\cot \theta + \tan \theta) \overset{?}{=} \sec \theta$

$\sin \theta \left(\dfrac{\cos \theta}{\sin \theta} + \dfrac{\sin \theta}{\cos \theta} \right) \overset{?}{=} \sec \theta$

$\cos \theta + \dfrac{\sin^2 \theta}{\cos \theta} \overset{?}{=} \sec \theta$

$\dfrac{\cos^2 \theta + \sin^2 \theta}{\cos \theta} \overset{?}{=} \sec \theta$

$\dfrac{1}{\cos \theta} \overset{?}{=} \sec \theta$

$\sec \theta = \sec \theta \checkmark$

11. $\dfrac{\cos^2 \theta}{1 - \sin \theta} \overset{?}{=} \dfrac{\cos \theta}{\sec \theta - \tan \theta}$

$\dfrac{\cos^2 \theta}{1 - \sin \theta} \overset{?}{=} \dfrac{\cos \theta}{\dfrac{1}{\cos \theta} - \dfrac{\sin \theta}{\cos \theta}}$

$\dfrac{\cos^2 \theta}{1 - \sin \theta} \overset{?}{=} \dfrac{\cos \theta}{\dfrac{1 - \sin \theta}{\cos \theta}}$

$\dfrac{\cos^2 \theta}{1 - \sin \theta} \overset{?}{=} \cos \theta \cdot \dfrac{\cos \theta}{1 - \sin \theta}$

$\dfrac{\cos^2 \theta}{1 - \sin \theta} = \dfrac{\cos^2 \theta}{1 - \sin \theta} \checkmark$

12. $(\tan \theta + \cot \theta)^2 \overset{?}{=} \csc^2 \theta \sec^2 \theta$

$\left(\dfrac{\sin \theta}{\cos \theta} + \dfrac{\cos \theta}{\sin \theta} \right)^2 \overset{?}{=} \csc^2 \theta \sec^2 \theta$

$\left(\dfrac{\sin^2 \theta + \cos^2 \theta}{\cos \theta \sin \theta} \right)^2 \overset{?}{=} \csc^2 \theta \sec^2 \theta$

$\left(\dfrac{1}{\cos \theta \sin \theta} \right)^2 \overset{?}{=} \csc^2 \theta \sec^2 \theta$

$\dfrac{1}{\cos^2 \theta} \cdot \dfrac{1}{\sin^2 \theta} \overset{?}{=} \csc^2 \theta \sec^2 \theta$

$\sec^2 \theta \csc^2 \theta = \csc^2 \theta \sec^2 \theta \checkmark$

13. $\dfrac{1 + \sec \theta}{\sec \theta} \overset{?}{=} \dfrac{\sin^2 \theta}{1 - \cos \theta}$

$\dfrac{1}{\sec \theta} + \dfrac{\sec \theta}{\sec \theta} \overset{?}{=} \dfrac{\sin^2 \theta}{1 - \cos \theta} \cdot \dfrac{1 + \cos \theta}{1 + \cos \theta}$

$\cos \theta + 1 \overset{?}{=} \dfrac{\sin^2 \theta (1 + \cos \theta)}{1 - \cos^2 \theta}$

$\cos \theta + 1 \overset{?}{=} \dfrac{\sin^2 \theta (1 + \cos \theta)}{\sin^2 \theta}$

$\cos \theta + 1 = 1 + \cos \theta \checkmark$

10-3 Sum and Difference Identities

F.TF.8

Find the exact value of each expression.

24. $\cos(-135°)$ $-\dfrac{\sqrt{2}}{2}$

25. $\cos 15°$ $\dfrac{\sqrt{6} + \sqrt{2}}{4}$

26. $\sin 210°$ $-\dfrac{1}{2}$

27. $\sin 105°$ $\dfrac{\sqrt{6} + \sqrt{2}}{4}$

28. $\tan 75°$ $\sqrt{3} + 2$

29. $\cos 105°$ $\dfrac{-\sqrt{6} + \sqrt{2}}{4}$

Verify that each of the following is an identity.

30. $\sin(\theta + 90) = \cos \theta$

31. $\sin\left(\dfrac{3\pi}{2} - \theta \right) = -\cos \theta$

32. $\tan(\theta - \pi) = \tan \theta$

30–32. See Ch. 10 Answer Appendix.

Example 4

Find the exact value of sin 75°.

Use $\sin(A + B) = \sin A \cos B + \cos A \sin B$.

$\sin 75° = \sin(30° + 45°)$
$= \sin 30° \cos 45° + \cos 30° \sin 45°$
$= \left(\dfrac{1}{2} \right) \left(\dfrac{\sqrt{2}}{2} \right) + \left(\dfrac{\sqrt{3}}{2} \right) \left(\dfrac{\sqrt{2}}{2} \right)$
$= \dfrac{\sqrt{2}}{4} + \dfrac{\sqrt{6}}{4}$ or $\dfrac{\sqrt{2} + \sqrt{6}}{4}$

10-4 Double-Angle and Half-Angle Identities

F.TF.8

Find the exact values of $\sin 2\theta$, $\cos 2\theta$, $\sin \dfrac{\theta}{2}$, and $\cos \dfrac{\theta}{2}$ for each of the following.

33. $\cos \theta = \dfrac{4}{5}$; $0° < \theta < 90°$

34. $\sin \theta = -\dfrac{1}{4}$; $180° < \theta < 270°$

35. $\cos \theta = -\dfrac{2}{3}$; $\dfrac{\pi}{2} < \theta < \pi$

33–36. See Ch. 10 Answer Appendix.

36. BASEBALL The infield of a baseball diamond is a square with side length 90 feet.

  a. Find the length of the diagonal.

  b. Write the ratio for sin 45° using the lengths of the baseball diamond.

  c. Use the formula $\sin \dfrac{\theta}{2} = \pm\sqrt{\dfrac{1 - \cos \theta}{2}}$ to verify the ratio you wrote in part **b.**

Example 5

Find the exact value of $\sin \dfrac{\theta}{2}$ if $\cos \theta = -\dfrac{3}{5}$ and θ is in the second quadrant.

$\sin \dfrac{\theta}{2} = \pm\sqrt{\dfrac{1 - \cos \theta}{2}}$ Half-angle identity

$= \pm\sqrt{\dfrac{1 - \left(-\dfrac{3}{5} \right)}{2}}$ $\cos \theta = -\dfrac{3}{5}$

$= \pm\sqrt{\dfrac{\dfrac{8}{5}}{2}}$ Subtract.

$= \pm\sqrt{\dfrac{4}{5}}$ Divide.

$= \pm\dfrac{2\sqrt{5}}{5}$ Simplify.

Because θ is in the second quadrant, $\sin \dfrac{\theta}{2} = \dfrac{2\sqrt{5}}{5}$.

10-5 Solving Trigonometric Equations

F.TF.8

Find all solutions of each equation for the given interval.

37. $2 \cos \theta - 1 = 0$; $0° \le \theta < 360°$ $60°, 300°$

38. $4 \cos^2 \theta - 1 = 0$; $0 \le \theta < 2\pi$ $\dfrac{\pi}{3}, \dfrac{2\pi}{3}, \dfrac{4\pi}{3}, \dfrac{5\pi}{3}$

39. $\sin 2\theta + \cos \theta = 0$; $0° \le \theta < 360°$ $90°, 210°, 270°, 330°$

40. $\sin^2 \theta = 2 \sin \theta + 3$; $0° \le \theta < 360°$ $270°$

41. $4 \cos^2 \theta - 4 \cos \theta + 1 = 0$; $0 \le \theta < 2\pi$ $\dfrac{\pi}{3}, \dfrac{5\pi}{3}$

Example 6

Find all solutions of $\sin 2\theta - \cos \theta = 0$ if $0 \le \theta < 2\pi$.

$\sin 2\theta - \cos \theta = 0$ Original equation

$2 \sin \theta \cos \theta - \cos \theta = 0$ Double-angle identity

$\cos \theta (2 \sin \theta - 1) = 0$ Factor.

$\cos \theta = 0$ or $2 \sin \theta - 1 = 0$

$\theta = \dfrac{\pi}{2}, \dfrac{3\pi}{2}$ $\sin \theta = \dfrac{1}{2}; \theta = \dfrac{\pi}{6}, \dfrac{5\pi}{6}$

Go Online!

ᵉAssessment

Customize and create multiple versions of chapter tests and answer keys that align to your standards. Tests can be delivered on paper or online.

CHAPTER 10
Practice Test

Go Online! for another Chapter Test

1. **MULTIPLE CHOICE** Which expression is equivalent to $\sin \theta + \cos \theta \cot \theta$? **D**

   A $\cot \theta$ C $\sec \theta$

   B $\tan \theta$ D $\csc \theta$

2. Verify that $\cos(30° - \theta) = \sin(60° + \theta)$ is an identity. **See margin.**

3. Verify that $\cos(\theta - \pi) = -\cos \theta$ is an identity. **See margin.**

4. **MULTIPLE CHOICE** What is the exact value of $\sin \theta$, if $\cos \theta = -\frac{3}{5}$ and $90° < \theta < 180°$? **D**

   A $\frac{5}{3}$ C $-\frac{4}{5}$

   B $\frac{\sqrt{34}}{8}$ D $\frac{4}{5}$

Find the value of each expression.

5. $\cot \theta$, if $\sec \theta = \frac{4}{3}$; $270° < \theta < 360°$ $-\frac{3\sqrt{7}}{7}$

6. $\tan \theta$, if $\cos \theta = -\frac{1}{2}$; $90° < \theta < 180°$ $-\sqrt{3}$

7. $\sec \theta$, if $\csc \theta = -2$; $180° < \theta < 270°$ $-\frac{2\sqrt{3}}{3}$

8. $\cot \theta$, if $\csc \theta = -\frac{5}{3}$; $270° < \theta < 360°$ $-\frac{4}{3}$

9. $\sec \theta$, if $\sin \theta = \frac{1}{2}$; $0° \le \theta < 90°$ $\frac{2\sqrt{3}}{3}$

Verify that each of the following is an identity.

10–13. See margin.

10. $\sin \theta (\cot \theta + \tan \theta) = \sec \theta$

11. $\frac{\cos^2 \theta}{1 - \sin \theta} = \frac{\cos \theta}{\sec \theta - \tan \theta}$

12. $(\tan \theta + \cot \theta)^2 = \csc^2 \theta \sec^2 \theta$

13. $\frac{1 + \sec \theta}{\sec \theta} = \frac{\sin^2 \theta}{1 - \cos \theta}$

14. $\frac{\sin \theta}{1 - \cos \theta} = \csc \theta + \cot \theta$

See Ch. 10 Answer Appendix.

15. **MULTIPLE CHOICE** What is the exact value of $\tan \frac{\pi}{8}$? **B**

   A $\frac{\sqrt{2 - \sqrt{3}}}{2}$

   B $\sqrt{2} - 1$

   C $1 - \sqrt{2}$

   D $-\frac{\sqrt{2 - \sqrt{3}}}{2}$

16. **HISTORY** Some researchers believe that the builders of ancient pyramids, such as the Great Pyramid of Khufu, may have tried to build the faces as equilateral triangles. Later they had to change to other types of triangles. Suppose a pyramid is built such that a face is an equilateral triangle of side length 18 feet. **a, b. See Ch. 10 Answer Appendix.**

18 ft

a. Find the height of the equilateral triangle.

b. Use the formula $\sin 2\theta = 2 \sin \theta \cos \theta$ and the measures of the equilateral triangle and its height to show that $\sin 2(30°) = \sin 60°$. Find the exact values.

Find the exact value of each expression.

17. $\cos(-225°)$ $-\frac{\sqrt{2}}{2}$ 18. $\sin 480°$ $\frac{\sqrt{3}}{2}$

19. $\cos 75°$ $\frac{\sqrt{6} - \sqrt{2}}{4}$ 20. $\sin 165°$ $\frac{\sqrt{6} - \sqrt{2}}{4}$

21. **ROCKETS** A model rocket is launched with an initial velocity of 20 meters per second. The range of a projectile is given by the formula $R = \frac{v^2}{g} \sin 2\theta$, where R is the range, v is the initial velocity, g is acceleration due to gravity or 9.8 meters per second squared, and θ is the launch angle. What angle is needed in order for the rocket to reach a range of 25 meters? **See Ch. 10 Answer Appendix.**

Solve each equation for all values of θ if θ is measured in radians.

22. $2 \cos^2 \theta - 3 \cos \theta - 2 = 0$ $\frac{2\pi}{3} + 2\pi k, \frac{4\pi}{3} + 2\pi k$

23. $2 \sin 3\theta - 1 = 0$ $\frac{\pi}{18} + \frac{2k\pi}{3}, \frac{5\pi}{18} + \frac{2k\pi}{3}$

Solve each equation for $0° \le \theta \le 360°$ if θ is measured in degrees.

24. $\cos 2\theta + \cos \theta = 2$ $0°, 360°$

25. $\sin \theta \cos \theta - \frac{1}{2} \sin \theta = 0$ $0°, 60°, 180°, 300°, 360°$

Go Online!

Chapter Tests

You can use premade leveled *Chapter Tests* to differentiate assessment for your students. Students can also take self-checking *Chapter Tests* to plan and prepare for chapter assessments.

MC = multiple-choice questions

FR = free-response questions

Form	Type	Level
1	MC	AL
2A	MC	OL
2B	FR	OL
2C	FR	OL
3	FR	BL
Vocabulary Test		
Extended-Response Test		

Before the Test

Have students complete the Study Notebook Tie it Together activity to review topics and skills presented in the chapter.

RtI Response to Intervention

Use the Intervention Planner to help you determine your Response to Intervention.

Intervention Planner

TIER 1 **On Level** OL

IF students miss 25% of the exercises or less,

THEN choose a resource:

   SE   Lessons 10-1 through 10-5

   Go Online!

     Skills Practice

     Chapter Project

     Self-Check Quizzes

TIER 2 **Strategic Intervention** AL
Approaching grade level

IF students miss 50% of the exercises,

THEN *Go Online!*

     Study Guide and Intervention

     Extra Examples

     Personal Tutors

     Homework Help

TIER 3 **Intensive Intervention**
2 or more grades below level

IF students miss 75% of the exercises,

THEN choose a resource:

   Use Math Triumphs, Alg. 2

   Go Online!

     Extra Examples

     Personal Tutors

     Homework Help

     Review Vocabulary

Launch

Objective Apply concepts and skills from this chapter in a real-world setting.

Teach

Ask:

- **What is the significance of θ being between 90° and 180°?** It means θ terminates in Quadrant II, so the answer will be negative because cosine is negative in Quadrant II.

- **How can you get from tangent and cotangent to secant and cosecant?** Rewrite the tangent and cotangent in terms of sine and cosine, and then rewrite the sine and cosine in terms of secant and cosecant.

- **Which identity will be useful in Part C?** the cosine sum identity

- **What is the first step in solving a trigonometric equation?** Sample answer: Check to see if the equation can be solved using a trigonometric identity.

The Performance Task focuses on the following content standards and standards for mathematical practice.

Dual Coding

Parts	Content Standards	(MP) Mathematical Practices
A	F.TF.8	2, 3, 6
B	F.TF.8	1, 6, 7
C	F.TF.8	2, 3, 6
D	F.TF.8	2, 3, 6

Levels of Complexity Chart

Parts	Level 1	Level 2	Level 3
A		●	
B	●		
C		●	
D			●

Go Online!   eBook

Interactive student guide
Refer to *Interactive Student Guide* for an additional Performance Task.

ALGEBRA 2
INTERACTIVE STUDENT GUIDE

Performance Task

Provide a clear solution to each part of the task. Be sure to show all of your work, include all relevant drawings, and justify your answers.

ERROR ANALYSIS A math teacher assigns her students a project that consists of grading and providing feedback on a hypothetical student's test.

For each item, explain whether the student's answer is correct. If the answer is incorrect, or if any steps are missing in their work, explain the mistake and provide a correction.

Part A
Find the exact value of $\cos \theta$ if $\sin = \frac{1}{2}$ and $90° < \theta < 180°$.

$$\cos^2\theta + \sin^2\theta = 1$$
$$\cos^2\theta = 1 - \sin^2\theta$$
$$\cos^2\theta = 1 - \frac{1}{2}$$
$$\cos^2\theta = \frac{1}{2}$$
$$\cos\theta = \frac{1}{\sqrt{2}}$$
$$\cos\theta = \frac{\sqrt{2}}{2}$$

Part B
Prove the identity $\cot(x) + \tan(x) = \sec(x)\csc(x)$.

$$\cot(x) + \tan(x) = \frac{\cos(x)}{\sin(x)} + \frac{\sin(x)}{\cos(x)}$$
$$= \frac{\cos^2(x)}{\sin(x)\cos(x)} + \frac{\sin^2(x)}{\sin(x)\cos(x)}$$
$$= \frac{\cos^2(x) + \sin^2(x)}{\sin(x)\cos(x)}$$
$$= \frac{1}{\sin(x)\cos(x)}$$
$$= \left(\frac{1}{\sin(x)}\right)\left(\frac{1}{\cos x}\right)$$
$$= \sec(x)\csc(x)$$

Part C
Find the exact value of $\cos(135°)$.

$$\cos(135°) = \cos(90°) + \cos(45°)$$
$$= \sin(90°)\cos(45°) + \cos(90°)\sin(45°)$$
$$= (1)\left(\frac{\sqrt{2}}{2}\right) + (0)\left(\frac{\sqrt{2}}{2}\right)$$
$$= \left(\frac{\sqrt{2}}{2}\right) + 0$$
$$= \frac{\sqrt{2}}{2}$$

Part D
Solve $\cos^2\theta + \cos\theta = \sin^2\theta$ for all values of θ if $0° \le \theta < 360°$.

$$\cos^2\theta + \cos\theta = \sin^2\theta$$
$$\cos^2\theta + \cos\theta = 1 - \cos^2\theta$$
$$2\cos^2\theta + \cos\theta - 1 = 0$$
$$(2\cos\theta - 1)(\cos\theta + 1) = 0$$
$$\cos\theta = \frac{1}{2} \quad \text{or} \quad \cos\theta = -1$$
$$\theta = 60° \text{ and } 180°$$

Test-Taking Strategy

Example

Solve the problem below.

Simplify the trigonometric expression shown below by writing it in terms of sin θ.

$$\frac{\cos \theta}{\sec \theta + \tan \theta}$$

Step 1 Are there any mathematical operations you can apply? Are there are any laws or identities you can apply?
I can use trigonometric identities.

Step 2 How will you apply them?
I will rewrite sec θ and tan θ in terms of sine and cosine using the definitions of secant and tangent. Then I'll simplify the result and apply the Pythagorean identity to get cosine in terms of sine.

Step 3 What is the correct answer?
The answer is $1 - \sin \theta$.

> **Test-Taking Tip**
> **Strategies for Simplifying Expressions** Some problems, especially those involving trigonometric identities, require you to use the properties of algebra to simplify expressions. Follow the steps below to help prepare to solve these kinds of problems.

Step 1 Study the expression you are being asked to simplify. Determine whether there are any mathematical operations, laws, or identities that would be applicable.

Step 2 Solve the problem using the order of operations, by combining like terms, and applying applicable laws and identities.

Step 3 If time permits, check your answer.

Need Another Example?

Simplify $(\cot \theta + 1)^2 - 2 \cot \theta$ by writing it in terms of csc θ.

a. Are there any mathematical operations you can apply? Are there are any laws or identities you can apply? I can use trigonometric identities.

b. How will you apply them? I'll rewrite $(\cot \theta + 1)^2 - 2 \cot \theta$ as $\cot^2 \theta + 1$ and then rewrite in terms of cosecant. Then I'll simplify the result.

c. What is the correct answer? $\csc^2 \theta$

Apply the Strategy

Solve the problem.

Simplify $\dfrac{\sec \theta}{\cot \theta + \tan \theta}$ by writing it in terms of sin θ.

Answer the questions below.

a. Are there any mathematical operations you can apply? Are there are any laws or identities you can apply? I can use trigonometric identities.

b. How will you apply them? I'll rewrite cot θ and tan θ in terms of sine and cosine and sec θ in terms of cosine. Then I'll simplify the result.

c. What is the correct answer? $\sin \theta$

Go Online!

The most up-to-date resources available for your program can be found at connectED.mcgraw-hill.com.

Diagnose Student Errors

Survey student responses for each item. Class trends may indicate common errors and misconceptions.

1. A | CORRECT
B | Computed $-\cos^2\theta$
C | Computed $\cos^2\theta$
D | Found the reciprocal of $\tan\theta$
E | Computed $\sec^2\theta$

2. A | CORRECT
B | Failed to use Pythagorean Identity in I
C | Misinterpreted negative angle identity for sine
D | Misinterpreted negative angle identity for sine
E | Misinterpreted negative angle identity for sine

4. A | CORRECT
B | CORRECT
C | Assumed the secant ratio in Quadrant IV was negative
D | Confused tangent with cosine
E | CORRECT

5. A | Assumed time varied directly with number of workers and inversely with number of rows
B | Assumed time varied directly with both the number of workers and the number of rows
C | Assumed time varied inversely with both the number of workers and the number of rows
D | CORRECT

7. A | Confused addition and multiplication
B | Confused $\dfrac{\log 2}{\log b}$ with $\dfrac{\log b}{\log 2}$
C | Confused $\dfrac{\log a}{\log b}$ with $\log\dfrac{a}{b}$
D | CORRECT
E | Calculated $\log 2 = 1$

8. A | CORRECT
B | Swapped values for $\sin 270°$ and $\cos 270°$
C | Found $\sin 270° = 1$ instead of -1
D | Found $\cos 270°$ and $\sin 270°$ to be the same value, -1
E | Failed to realize $\cos 270° = 0$

Go Online!

Standardized Test Practice

Students can take self-checking tests in standardized format to plan and prepare for assessments.

Preparing for Assessment
Cumulative Review

Read each question. Then fill in the correct answer on the answer document provided by your teacher or on a sheet of paper.

1. Suppose $180° < \theta < 270°$ and $\tan\theta = \frac{4}{3}$. What is $\cos\theta$? **A**

- A $-\frac{3}{5}$
- B $-\frac{9}{25}$
- C $\frac{9}{25}$
- D $\frac{3}{4}$
- E $\frac{25}{9}$

2. When $P = 1$, which of the following trigonometric expressions, when defined, is considered an identity? **A**

$$\text{I} \quad -\sin^2\theta - \cos^2\theta = -P$$
$$\text{II} \quad \tan^2\theta - \sec^2\theta = -P$$
$$\text{III} \quad \frac{\sin\theta}{\sin(-\theta)} = P$$

- A I and II only
- B II only
- C III only
- D II and III only
- E I, II, and III

3. What is the remainder when $x^2 + 3x + 5$ is divided by $x + 1$?

> 3

Test-Taking Tip
Question 2 Because III is listed in three of the answer choices, first determine if III is an identity. If it is not, you can quickly eliminate choices C, D, and E.

4. A line segment extends from the origin of the coordinate plane $O(0, 0)$ to point $B(a, -b)$ in Quadrant IV, as shown in the figure below. If a, b, and c are positive real numbers, select all trigonometric ratios that represent the measure of the acute angle θ formed by $\overline{OB}$ and the x-axis. **A, B, E**

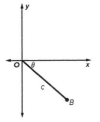

- ☐ A $\sin\theta = -\frac{b}{c}$
- ☐ B $\cos\theta = \frac{a}{c}$
- ☐ C $\sec\theta = -\frac{c}{a}$
- ☐ D $\tan\theta = \frac{a}{c}$
- ☐ E $\csc\theta = -\frac{c}{b}$

5. The time required to harvest the eggplant crop varies inversely with the size of the picking crew and directly with the number of rows planted. Last year a crew of 12 workers harvested 70 rows of eggplant in 2 days. This year the field was expanded to 100 rows. How many days will it take a crew of 8 workers to harvest this year's crop? **D**

- A $\frac{14}{15}$
- B $1\frac{19}{21}$
- C $2\frac{1}{10}$
- D $4\frac{2}{7}$

6. What is the value of $\cos\theta$ when $\sin\theta = -\frac{\sqrt{3}}{2}$ and $180° < \theta < 270°$?

> $-\frac{1}{2}$

9. A | Used the sine difference identity
B | Used the cosine difference identity
C | CORRECT
D | Confused terms in the sine difference identity
E | Selected a sum that is not an identity

10. A | Computed $\cot 2\theta$
B | Found sum of $\sin 2\theta + \cos 2\theta$
C | Divided numerator of $\sin 2\theta$ by 2
D | CORRECT
E | Computed $\tan^2 2\theta$

Go Online! for Standardized Test Practice

7. Which shows a property of logarithms? **D**

- ○ **A** $\log b + \log c = \log (b + c)$
- ○ **B** $\log_2 b = \dfrac{\log 2}{\log b}$
- ○ **C** $\dfrac{\log a}{\log b} = \log a - \log b$
- ○ **D** $\log b + \log c = \log (bc)$
- ○ **E** $\log 2^b = b$

8. Which of the following trigonometric expressions has the same value as $\sin (270° + \theta)$? **A**

- ○ **A** $-\cos \theta$
- ○ **B** $-\sin \theta$
- ○ **C** $\cos \theta$
- ○ **D** $-\cos \theta - \sin \theta$
- ○ **E** $-\cos \theta + \sin \theta$

9. Which equation can be used to determine $\sin \angle DAF$, shown in the diagram below? **C**

- ○ **A** $\sin \angle DAF = \sin \alpha \cos \beta - \cos \alpha \sin \beta$
- ○ **B** $\sin \angle DAF = \sin \alpha \sin \beta + \cos \alpha \cos \beta$
- ○ **C** $\sin \angle DAF = \sin \alpha \cos \beta + \cos \alpha \sin \beta$
- ○ **D** $\sin \angle DAF = \sin \alpha \sin \beta - \cos \alpha \cos \beta$
- ○ **E** $\sin \angle DAF = \sin \alpha \sin \beta + \sin \alpha \sin \beta$

10. For a given angle θ such that $0° < \theta < 90°$, $\sin 2\theta = \frac{24}{25}$, and $\cos 2\theta = \frac{7}{25}$, the measure of $\tan 2\theta$ is defined by what numerical value? **D**

- ○ **A** $\frac{7}{24}$
- ○ **B** $\frac{31}{25}$
- ○ **C** $\frac{12}{7}$
- ○ **D** $\frac{24}{7}$
- ○ **E** $\frac{576}{49}$

11. When a stop sign casts an 8-foot shadow, the angle of elevation of the Sun is θ. When the angle of elevation is 2θ, the length of the shadow is 2 feet.

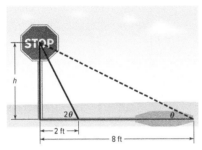

- **a.** What is the height h of the stop sign? Round to the nearest tenth of a foot.

  | 5.7 | ft

- **b.** ⊙ What mathematical practice did you use to solve this problem? **See students' work.**

12. For what value of θ, such that $0° < \theta < 360°$, is the equation $\sin^2 \theta - \sin \theta = 2$ true?

  | 270 | °

Need Extra Help?											
If you missed Question...	1	2	3	4	5	6	7	8	9	10	11	12
Go to Lesson...	10-1	10-2	4-3	10-1	7-5	10-1	6-7	10-3	10-3	10-4	10-4	10-5

Formative Assessment

You can use these pages to benchmark student progress.

📄 Standardized Test Practice

Test Item Formats

In the Cumulative Review, students will encounter different formats for assessment questions to prepare them for standardized tests.

Question Type	Exercises
Multiple-Choice	1–2, 5, 7–10
Multiple Correct Answers	4
Type Entry: Short Response	3, 6, 11, 12
Type Entry: Extended Response	11

Answer Sheet Practice

Have students simulate taking a standardized test by recording their answers on a practice recording sheet.

LS LEARNSMART®

Use LearnSmart as part of your test-preparation plan to measure student topic retention. You can create a student assignment in LearnSmart for additional practice on these topics.

· Trigonometric Identities

Go Online!

eAssessment

Customize and create multiple versions of chapter tests and answer keys that align to your standards. Tests can be delivered on paper or online.

1. $\cot\theta + \tan\theta \overset{?}{=} \dfrac{\sec^2\theta}{\tan\theta}$

$\cot\theta + \tan\theta \overset{?}{=} \dfrac{\tan^2\theta + 1}{\tan\theta}$

$\cot\theta + \tan\theta \overset{?}{=} \dfrac{\tan^2\theta}{\tan\theta} + \dfrac{1}{\tan\theta}$

$\cot\theta + \tan\theta = \tan\theta + \cot\theta$ ✓

2. $\cos^2\theta \overset{?}{=} (1 + \sin\theta)(1 - \sin\theta)$

$\cos^2\theta \overset{?}{=} 1 - \sin^2\theta$

$\cos^2\theta = \cos^2\theta$ ✓

3. $\sin\theta \overset{?}{=} \dfrac{\sec\theta}{\tan\theta + \cot\theta}$

$\sin\theta \overset{?}{=} \dfrac{\frac{1}{\cos\theta}}{\frac{\sin\theta}{\cos\theta} + \frac{\cos\theta}{\sin\theta}}$

$\sin\theta \overset{?}{=} \dfrac{\frac{1}{\cos\theta}}{\frac{\sin^2\theta + \cos^2\theta}{\cos\theta\sin\theta}}$

$\sin\theta \overset{?}{=} \dfrac{\frac{1}{\cos\theta}}{\frac{1}{\cos\theta\sin\theta}}$

$\sin\theta \overset{?}{=} \dfrac{1}{\cos\theta} \cdot \dfrac{\cos\theta\sin\theta}{1}$

$\sin\theta = \sin\theta$ ✓

4. $\tan^2\theta \overset{?}{=} \dfrac{1 - \cos^2\theta}{\cos^2\theta}$

$\tan^2\theta \overset{?}{=} \dfrac{\sin^2\theta}{\cos^2\theta}$

$\tan^2\theta = \tan^2\theta$ ✓

5. $\tan^2\theta\csc^2\theta \overset{?}{=} 1 + \tan^2\theta$

$\dfrac{\sin^2\theta}{\cos^2\theta} \cdot \dfrac{1}{\sin^2\theta} \overset{?}{=} \sec^2\theta$

$\dfrac{1}{\cos^2\theta} \overset{?}{=} \sec^2\theta$

$\sec^2\theta = \sec^2\theta$ ✓

6. $\tan^2\theta \overset{?}{=} (\sec\theta + 1)(\sec\theta - 1)$

$\tan^2\theta \overset{?}{=} \sin^2\theta - 1$

$\tan^2\theta = \tan^2\theta$ ✓

7. $\dfrac{\tan^2\theta + 1}{\tan^2\theta} = \csc^2\theta$

$\dfrac{\sec^2\theta}{\tan^2\theta} = \csc^2\theta$

$\dfrac{\frac{1}{\cos^2\theta}}{\frac{\sin^2\theta}{\cos^2\theta}} = \csc^2\theta$

$\dfrac{1}{\cos^2\theta} \cdot \dfrac{\cos^2\theta}{\sin^2\theta} = \csc^2\theta$

$\dfrac{1}{\sin^2\theta} = \csc^2\theta$

$\csc^2\theta = \csc^2\theta$

8. $\cot\theta + \sec\theta \overset{?}{=} \dfrac{\cos^2\theta + \sin\theta}{\sin\theta\cos\theta}$

$\cot\theta + \sec\theta \overset{?}{=} \dfrac{\cos^2\theta}{\sin\theta\cos\theta} + \dfrac{\sin\theta}{\sin\theta\cos\theta}$

$\cot\theta + \sec\theta \overset{?}{=} \dfrac{\cos\theta}{\sin\theta} + \dfrac{1}{\cos\theta}$

$\cot\theta + \sec\theta = \cot\theta + \sec\theta$

9. $\sin^2\theta + \tan^2\theta \overset{?}{=} (1 - \cos^2\theta) + \dfrac{\sec^2\theta}{\csc^2\theta}$

$\sin^2\theta + \tan^2\theta \overset{?}{=} \sin^2\theta + \dfrac{\sec^2\theta}{\csc^2\theta}$

$\sin^2\theta + \tan^2\theta \overset{?}{=} \sin^2\theta + \dfrac{1}{\cos^2\theta} \div \dfrac{1}{\sin^2\theta}$

$\sin^2\theta + \tan^2\theta \overset{?}{=} \sin^2\theta + \dfrac{\sin^2\theta}{\cos^2\theta}$

$\sin^2\theta + \tan^2\theta = \sin^2\theta + \tan^2\theta$

10. $\cos^2\theta + \tan^2\theta\cos^2\theta \overset{?}{=} 1$

$\cos^2\theta + \dfrac{\sin^2\theta}{\cos^2\theta} \cdot \cos^2\theta \overset{?}{=} 1$

$\cos^2\theta + \sin^2\theta \overset{?}{=} 1$

$1 = 1$ ✓

11. $\cot\theta(\cot\theta + \tan\theta) \overset{?}{=} \csc^2\theta$

$\cot^2\theta + \cot\theta\tan\theta \overset{?}{=} \csc^2\theta$

$\cot^2\theta + \dfrac{\sin\theta}{\cos\theta} \cdot \dfrac{\cos\theta}{\sin\theta} \overset{?}{=} \csc^2\theta$

$\cot^2\theta + 1 \overset{?}{=} \csc^2\theta$

$\csc^2\theta = \csc^2\theta$ ✓

12. $1 + \sec^2\theta\sin^2\theta \overset{?}{=} \sec^2\theta$

$1 + \dfrac{1}{\cos^2\theta} \cdot \sin^2\theta \overset{?}{=} \sec^2\theta$

$1 + \tan^2\theta \overset{?}{=} \sec^2\theta$

$\sec^2\theta = \sec^2\theta$ ✓

13. $\sin\theta\sec\theta\cot\theta \overset{?}{=} 1$

$\sin\theta \cdot \dfrac{1}{\cos\theta} \cdot \dfrac{\cos\theta}{\sin\theta} \overset{?}{=} 1$

$1 = 1$ ✓

14. $\dfrac{1 - \cos\theta}{1 + \cos\theta} \overset{?}{=} (\csc\theta - \cot\theta)^2$

$\dfrac{1 - \cos\theta}{1 + \cos\theta} \overset{?}{=} \csc^2\theta - 2\cot\theta\csc\theta + \cot^2\theta$

$\dfrac{1 - \cos\theta}{1 + \cos\theta} \overset{?}{=} \dfrac{1}{\sin^2\theta} - 2 \cdot \dfrac{\cos\theta}{\sin\theta} \cdot \dfrac{1}{\sin\theta} + \dfrac{\cos^2\theta}{\sin^2\theta}$

$\dfrac{1 - \cos\theta}{1 + \cos\theta} \overset{?}{=} \dfrac{1}{\sin^2\theta} - \dfrac{2\cos\theta}{\sin^2\theta} + \dfrac{\cos^2\theta}{\sin^2\theta}$

$\dfrac{1 - \cos\theta}{1 + \cos\theta} \overset{?}{=} \dfrac{1 - 2\cos\theta + \cos^2\theta}{\sin^2\theta}$

$\dfrac{1 - \cos\theta}{1 + \cos\theta} \overset{?}{=} \dfrac{(1 - \cos\theta)(1 - \cos\theta)}{1 - \cos^2\theta}$

$\dfrac{1 - \cos\theta}{1 + \cos\theta} \overset{?}{=} \dfrac{(1 - \cos\theta)(1 - \cos\theta)}{(1 - \cos\theta)(1 + \cos\theta)}$

$\dfrac{1 - \cos\theta}{1 + \cos\theta} = \dfrac{1 - \cos\theta}{1 + \cos\theta}$ ✓

15. $\dfrac{1 - 2\cos^2\theta}{\sin\theta\cos\theta} \overset{?}{=} \tan\theta - \cot\theta$

$\dfrac{(1 - \cos^2\theta) - \cos^2\theta}{\sin\theta\cos\theta} \overset{?}{=} \tan\theta - \cot\theta$

$\dfrac{\sin^2\theta - \cos^2\theta}{\sin\theta\cos\theta} \overset{?}{=} \tan\theta - \cot\theta$

$\dfrac{\sin^2\theta}{\sin\theta\cos\theta} - \dfrac{\cos^2\theta}{\sin\theta\cos\theta} \overset{?}{=} \tan\theta - \cot\theta$

$\dfrac{\sin\theta}{\cos\theta} - \dfrac{\cos\theta}{\sin\theta} \overset{?}{=} \tan\theta - \cot\theta$

$\tan\theta - \cot\theta = \tan\theta - \cot\theta$ ✓

16. $\tan \theta \overset{?}{=} \dfrac{\sec \theta}{\csc \theta}$

$\tan \theta \overset{?}{=} \dfrac{\frac{1}{\cos \theta}}{\frac{1}{\sin \theta}}$

$\tan \theta \overset{?}{=} \dfrac{\sin \theta}{\cos \theta}$

$\tan \theta = \tan \theta \checkmark$

17. $\cos \theta \overset{?}{=} \sin \theta \cot \theta$

$\cos \theta \overset{?}{=} \sin \theta \left(\dfrac{\cos \theta}{\sin \theta}\right)$

$\cos \theta = \cos \theta \checkmark$

18.
$(\sin \theta - 1)(\tan \theta + \sec \theta) \overset{?}{=} -\cos \theta$

$\sin \theta \tan \theta + \sin \theta \sec \theta - \tan \theta - \sec \theta \overset{?}{=} -\cos \theta$

$\dfrac{\sin^2 \theta}{\cos \theta} + \dfrac{\sin \theta}{\cos \theta} - \dfrac{\sin \theta}{\cos \theta} - \dfrac{1}{\cos \theta} \overset{?}{=} -\cos \theta$

$\dfrac{\sin^2 \theta}{\cos \theta} - \dfrac{1}{\cos \theta} \overset{?}{=} -\cos \theta$

$\dfrac{\sin^2 \theta - 1}{\cos \theta} \overset{?}{=} -\cos \theta$

$\dfrac{-\cos^2 \theta}{\cos \theta} \overset{?}{=} -\cos \theta$

$-\cos \theta = -\cos \theta \checkmark$

19. $\cos \theta \cos(-\theta) - \sin \theta \sin(-\theta) \overset{?}{=} 1$

$\cos \theta \cos \theta - \sin \theta (-\sin \theta) \overset{?}{=} 1$

$\cos^2 \theta + \sin^2 \theta \overset{?}{=} 1$

$1 = 1 \checkmark$

21. $\sec \theta - \tan \theta \overset{?}{=} \dfrac{1 - \sin \theta}{\cos \theta}$

$\dfrac{1}{\cos \theta} - \dfrac{\sin \theta}{\cos \theta} \overset{?}{=} \dfrac{1 - \sin \theta}{\cos \theta}$

$\dfrac{1 - \sin \theta}{\cos \theta} = \dfrac{1 - \sin \theta}{\cos \theta} \checkmark$

22.
$\dfrac{1 + \tan \theta}{\sin \theta + \cos \theta} \overset{?}{=} \sec \theta$

$\dfrac{1 + \frac{\sin \theta}{\cos \theta}}{\sin \theta + \cos \theta} \overset{?}{=} \sec \theta$

$\dfrac{\frac{\cos \theta + \sin \theta}{\cos \theta}}{\sin \theta + \cos \theta} \overset{?}{=} \sec \theta$

$\dfrac{\cos \theta + \sin \theta}{\cos \theta} \cdot \dfrac{1}{\sin \theta + \cos \theta} \overset{?}{=} \sec \theta$

$\dfrac{1}{\cos \theta} \overset{?}{=} \sec \theta$

$\sec \theta = \sec \theta \checkmark$

23. $\sec \theta \csc \theta \overset{?}{=} \tan \theta + \cot \theta$

$\dfrac{1}{\cos \theta} \cdot \dfrac{1}{\sin \theta} \overset{?}{=} \dfrac{\sin \theta}{\cos \theta} + \dfrac{\cos \theta}{\sin \theta}$

$\dfrac{1}{\cos \theta \sin \theta} \overset{?}{=} \dfrac{\sin^2 \theta}{\sin \theta \cos \theta} + \dfrac{\cos^2 \theta}{\sin \theta \cos \theta}$

$\dfrac{1}{\cos \theta \sin \theta} \overset{?}{=} \dfrac{\sin^2 \theta + \cos^2 \theta}{\sin \theta \cos \theta}$

$\dfrac{1}{\cos \theta \sin \theta} = \dfrac{1}{\cos \theta \sin \theta} \checkmark$

24. $\sin \theta + \cos \theta \overset{?}{=} \dfrac{2 \sin^2 \theta - 1}{\sin \theta - \cos \theta}$

$\sin \theta + \cos \theta \overset{?}{=} \dfrac{2 \sin^2 \theta - (\sin^2 \theta + \cos^2 \theta)}{\sin \theta - \cos \theta}$

$\sin \theta + \cos \theta \overset{?}{=} \dfrac{\sin^2 \theta - \cos^2 \theta}{\sin \theta - \cos \theta}$

$\sin \theta + \cos \theta \overset{?}{=} \dfrac{(\sin \theta - \cos \theta)(\sin \theta - \cos \theta)}{\sin \theta - \cos \theta}$

$\sin \theta + \cos \theta = \sin \theta + \cos \theta \checkmark$

25. $(\sin \theta + \cos \theta)^2 \overset{?}{=} \dfrac{2 + \sec \theta \csc \theta}{\sec \theta \csc \theta}$

$(\sin \theta + \cos \theta)^2 \overset{?}{=} \dfrac{2 + \frac{1}{\cos \theta} \cdot \frac{1}{\sin \theta}}{\frac{1}{\cos \theta} \cdot \frac{1}{\sin \theta}}$

$(\sin \theta + \cos \theta)^2 \overset{?}{=} \left(2 + \dfrac{1}{\cos \theta \sin \theta}\right) \cdot \dfrac{\cos \theta \sin \theta}{1}$

$(\sin \theta + \cos \theta)^2 \overset{?}{=} 2 \cos \theta \sin \theta + 1$

$(\sin \theta + \cos \theta)^2 \overset{?}{=} 2 \cos \theta \sin \theta + \cos^2 \theta + \sin^2 \theta$

$(\sin \theta + \cos \theta)^2 = (\sin \theta + \cos \theta)^2 \checkmark$

26. $\dfrac{\cos \theta}{1 - \sin \theta} \overset{?}{=} \dfrac{1 + \sin \theta}{\cos \theta}$

$\dfrac{\cos \theta}{1 - \sin \theta} \overset{?}{=} \dfrac{1 + \sin \theta}{\cos \theta} \cdot \dfrac{1 - \sin \theta}{1 - \sin \theta}$

$\dfrac{\cos \theta}{1 - \sin \theta} \overset{?}{=} \dfrac{1 - \sin^2 \theta}{\cos \theta (1 - \sin \theta)}$

$\dfrac{\cos \theta}{1 - \sin \theta} \overset{?}{=} \dfrac{\cos^2 \theta}{\cos \theta (1 - \sin \theta)}$

$\dfrac{\cos \theta}{1 - \sin \theta} = \dfrac{\cos \theta}{1 - \sin \theta} \checkmark$

27. $\csc \theta - 1 \overset{?}{=} \dfrac{\cot^2 \theta}{\csc \theta + 1}$

$\csc \theta - 1 \overset{?}{=} \dfrac{\csc^2 \theta - 1}{\csc \theta + 1}$

$\csc \theta - 1 \overset{?}{=} \dfrac{(\csc \theta - 1)(\csc \theta + 1)}{\csc \theta + 1}$

$\csc \theta - 1 = \csc \theta - 1 \checkmark$

28. $\cos \theta \cot \theta \overset{?}{=} \csc \theta - \sin \theta$

$(\cos \theta) \dfrac{\cos \theta}{\sin \theta} \overset{?}{=} \dfrac{1}{\sin \theta} - \sin \theta$

$\dfrac{\cos^2 \theta}{\sin \theta} \overset{?}{=} \dfrac{1}{\sin \theta} - \dfrac{\sin \theta}{1} \cdot \dfrac{\sin \theta}{\sin \theta}$

$\dfrac{\cos^2 \theta}{\sin \theta} \overset{?}{=} \dfrac{1}{\sin \theta} - \dfrac{\sin^2 \theta}{\sin \theta}$

$\dfrac{\cos^2 \theta}{\sin \theta} \overset{?}{=} \dfrac{1 - \sin^2 \theta}{\sin \theta}$

$\dfrac{\cos^2 \theta}{\sin \theta} = \dfrac{\cos^2 \theta}{\sin \theta} \checkmark$

29. $\sin \theta \cos \theta \tan \theta + \cos^2 \theta \overset{?}{=} 1$

$\sin \theta \cos \theta \cdot \dfrac{\sin \theta}{\cos \theta} + \cos^2 \theta \overset{?}{=} 1$

$\sin^2 \theta + \cos^2 \theta \overset{?}{=} 1$

$1 = 1 \checkmark$

30. $(\csc\theta - \cot\theta)^2 \stackrel{?}{=} \dfrac{1-\cos\theta}{1+\cos\theta}$

$\left(\dfrac{1}{\sin\theta} - \dfrac{\cos\theta}{\sin\theta}\right)^2 \stackrel{?}{=} \dfrac{1-\cos\theta}{1+\cos\theta}$

$\left(\dfrac{1-\cos\theta}{\sin\theta}\right)^2 \stackrel{?}{=} \dfrac{1-\cos\theta}{1+\cos\theta}$

$\dfrac{(1-\cos\theta)^2}{\sin^2\theta} \stackrel{?}{=} \dfrac{1-\cos\theta}{1+\cos\theta}$

$\dfrac{(1-\cos\theta)(1-\cos\theta)}{\sin^2\theta} \stackrel{?}{=} \dfrac{1-\cos\theta}{1+\cos\theta}$

$\dfrac{(1-\cos\theta)(1-\cos\theta)}{1-\cos^2\theta} \stackrel{?}{=} \dfrac{1-\cos\theta}{1+\cos\theta}$

$\dfrac{(1-\cos\theta)(1-\cos\theta)}{(1-\cos\theta)(1+\cos\theta)} \stackrel{?}{=} \dfrac{1-\cos\theta}{1+\cos\theta}$

$\dfrac{1-\cos\theta}{1+\cos\theta} = \dfrac{1-\cos\theta}{1+\cos\theta}$ ✓

31. $\csc^2\theta \stackrel{?}{=} \cot^2\theta + \sin\theta\csc\theta$

$\csc^2\theta \stackrel{?}{=} \cot^2\theta + \sin\theta \cdot \dfrac{1}{\sin\theta}$

$\csc^2\theta \stackrel{?}{=} \cot^2\theta + 1$

$\csc^2\theta = \csc^2\theta$ ✓

32. $\dfrac{\sec\theta - \csc\theta}{\csc\theta\sec\theta} \stackrel{?}{=} \sin\theta - \cos\theta$

$\dfrac{\sec\theta}{\csc\theta\sec\theta} - \dfrac{\csc\theta}{\csc\theta\sec\theta} \stackrel{?}{=} \sin\theta - \cos\theta$

$\dfrac{1}{\csc\theta} - \dfrac{1}{\sec\theta} \stackrel{?}{=} \sin\theta - \cos\theta$

$\sin\theta - \cos\theta = \sin\theta - \cos\theta$ ✓

33. $\sin^2\theta + \cos^2\theta \stackrel{?}{=} \sec^2\theta - \tan^2\theta$

$1 \stackrel{?}{=} \tan^2\theta + 1 - \tan^2\theta$

$1 = 1$ ✓

34. $\sec\theta - \cos\theta \stackrel{?}{=} \tan\theta\sin\theta$

$\dfrac{1}{\cos\theta} - \cos\theta \stackrel{?}{=} \tan\theta\sin\theta$

$\dfrac{1}{\cos\theta} - \dfrac{\cos^2\theta}{\cos\theta} \stackrel{?}{=} \tan\theta\sin\theta$

$\dfrac{1-\cos^2\theta}{\cos\theta} \stackrel{?}{=} \tan\theta\sin\theta$

$\dfrac{\sin^2\theta}{\cos\theta} \stackrel{?}{=} \tan\theta\sin\theta$

$\left(\dfrac{\sin\theta}{\cos\theta}\right)\sin\theta \stackrel{?}{=} \tan\theta\sin\theta$

$\tan\theta\sin\theta = \tan\theta\sin\theta$ ✓

53a.

Angle measure	Height
30°	28.2 m
45°	56.4 m
60°	84.5 m
90°	112.7 m

53b.

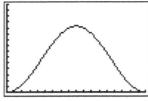

[0, 180] scl: 10 by [0, 150] scl: 10

53c. $\dfrac{v_0^2\tan^2\theta}{2g\sec^2\theta} \stackrel{?}{=} \dfrac{v_0^2\sin^2\theta}{2g}$

$\dfrac{v_0^2\left(\dfrac{\sin^2\theta}{\cos^2\theta}\right)}{2g\left(\dfrac{1}{\cos^2\theta}\right)} \stackrel{?}{=} \dfrac{v_0^2\sin^2\theta}{2g}$

$\dfrac{v_0^2\sin^2\theta}{2g} = \dfrac{v_0^2\sin^2\theta}{2g}$ ✓

61. Using the unit circle and the Pythagorean Theorem, we can justify $\cos^2\theta + \sin^2\theta = 1$.

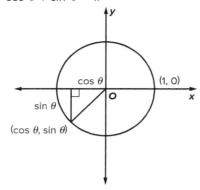

If we divide each term of the identity $\cos^2\theta + \sin^2\theta = 1$ by $\cos^2\theta$, we can justify $1 + \tan^2\theta = \sec^2\theta$.

$\dfrac{\cos^2\theta}{\cos^2\theta} + \dfrac{\sin^2\theta}{\cos^2\theta} = \dfrac{1}{\cos^2\theta}$

$1 + \tan^2\theta = \sec^2\theta$

If we divide each term of the identity $\cos^2\theta + \sin^2\theta = 1$ by $\sin^2\theta$, we can justify $\cot^2\theta + 1 = \csc^2\theta$.

$\dfrac{\cos^2\theta}{\sin^2\theta} + \dfrac{\sin^2\theta}{\sin^2\theta} = \dfrac{1}{\sin^2\theta}$

$\cot^2\theta + 1 = \csc^2\theta$

Lesson 10-3

30. $F = \dfrac{W(\sin A + \mu\cos A)}{\cos A - \mu\sin A}$

$= \dfrac{W(\sin A + \tan\theta\cos A)}{\cos A - \tan\theta\sin A}$

$= \dfrac{W\left(\dfrac{\sin A}{\cos A} + \dfrac{\tan\theta\cos A}{\cos A}\right)}{\dfrac{\cos A}{\cos A} - \dfrac{\tan\theta\sin A}{\cos A}}$

$= \dfrac{W(\tan A + \tan\theta)}{1 - \tan A\tan\theta}$

$= W\tan(A + \theta)$

33b.

33c. No; a counterexample is: $\cos(30° + 45°) = \cos 30° + \cos 45°$, which equals $\dfrac{\sqrt{3}}{2} + \dfrac{\sqrt{2}}{2}$ or about 1.5731. Since a cosine value cannot be greater than 1, this statement must be false.

34. $\sin(A + B) \stackrel{?}{=} \dfrac{\tan A + \tan B}{\sec A \sec B}$

$\sin(A + B) \stackrel{?}{=} \dfrac{\dfrac{\sin A}{\cos A} + \dfrac{\sin B}{\cos B}}{\dfrac{1}{\cos A} \cdot \dfrac{1}{\cos B}}$

$\sin(A + B) \stackrel{?}{=} \dfrac{\dfrac{\sin A}{\cos A} + \dfrac{\sin B}{\cos B}}{\dfrac{1}{\cos A} \cdot \dfrac{1}{\cos B}} \cdot \dfrac{\cos A \cos B}{\cos A \cos B}$

$\sin(A + B) \stackrel{?}{=} \dfrac{\sin A \cos B + \cos A \sin B}{1}$

$\sin(A + B) = \sin(A + B)$ ✓

35. $\cos(A + B) \stackrel{?}{=} \dfrac{1 - \tan A \tan B}{\sec A \sec B}$

$\cos(A + B) \stackrel{?}{=} \dfrac{1 - \dfrac{\sin A}{\cos A} \cdot \dfrac{\sin B}{\cos B}}{\dfrac{1}{\cos A} \cdot \dfrac{1}{\cos B}}$

$\cos(A + B) \stackrel{?}{=} \dfrac{1 - \dfrac{\sin A}{\cos A} \cdot \dfrac{\sin B}{\cos B}}{\dfrac{1}{\cos A} \cdot \dfrac{1}{\cos B}} \cdot \dfrac{\cos A \cos B}{\cos A \cos B}$

$\cos(A + B) \stackrel{?}{=} \dfrac{\cos A \cos B - \sin A \sin B}{1}$

$\cos(A + B) = \cos(A + B)$ ✓

36. $\sec(A - B) \stackrel{?}{=} \dfrac{\sec A \sec B}{1 + \tan A \tan B}$

$\sec(A - B) \stackrel{?}{=} \dfrac{\dfrac{1}{\cos A} \cdot \dfrac{1}{\cos B}}{1 + \dfrac{\sin A}{\cos A} \cdot \dfrac{\sin B}{\cos B}}$

$\sec(A - B) \stackrel{?}{=} \dfrac{\dfrac{1}{\cos A} \cdot \dfrac{1}{\cos B}}{1 + \dfrac{\sin A}{\cos A} \cdot \dfrac{\sin B}{\cos B}} \cdot \dfrac{\cos A \cos B}{\cos A \cos B}$

$\sec(A - B) \stackrel{?}{=} \dfrac{1}{\cos A \cos B + \sin A \sin B}$

$\sec(A - B) \stackrel{?}{=} \dfrac{1}{\cos(A - B)}$

$\sec(A - B) = \sec(A - B)$ ✓

37.

$\sin(A + B) \sin(A - B) \stackrel{?}{=} \sin^2 A - \sin^2 B$

$(\sin A \cos B + \cos A \sin B)$
$(\sin A \cos B - \cos A \sin B) \stackrel{?}{=} \sin^2 A - \sin^2 B$

$(\sin A \cos B)^2 - (\cos A \sin B)^2 \stackrel{?}{=} \sin^2 A - \sin^2 B$

$\sin^2 B \cos^2 B - \cos^2 A \sin^2 B \stackrel{?}{=} \sin^2 A - \sin^2 B$

$\sin^2 A \cos^2 B + \sin^2 A \sin^2 B -$
$\sin^2 A \sin^2 B - \cos^2 A \sin^2 B \stackrel{?}{=} \sin^2 A - \sin^2 B$

$\sin^2 A(\cos^2 B + \sin^2 B) - \sin^2 B(\sin^2 A + \cos^2 A) \stackrel{?}{=} \sin^2 A - \sin^2 B$

$(\sin^2 A)(1) - (\sin^2 B)(1) \stackrel{?}{=} \sin^2 A - \sin^2 B$

$\sin^2 A - \sin^2 B = \sin^2 A - \sin^2 B$ ✓

40. $\cot(A + B) = \dfrac{1}{\tan(A + B)}$

$= \dfrac{1}{\dfrac{\tan A + \tan B}{1 - \tan A \tan B}}$

$= \dfrac{1 - \tan A \tan B}{\tan A + \tan B}$

$= \dfrac{1 - \dfrac{1}{\cot A} \cdot \dfrac{1}{\cot B}}{\dfrac{1}{\cot A} + \dfrac{1}{\cot B}} \cdot \dfrac{\cot A \cot B}{\cot A \cot B}$

$= \dfrac{\cot A \cot B - 1}{\cot A + \cot B}$

41. $d = \sqrt{(\cos A - \cos B)^2 + (\sin A - \sin B)^2}$

$d^2 = (\cos A - \cos B)^2 + (\sin A - \sin B)^2$

$d^2 = (\cos^2 A - 2\cos A \cos B + \cos^2 B) + (\sin^2 A - 2\sin A \sin B + \sin^2 B)$

$d^2 = \cos^2 A + \sin^2 A + \cos^2 B + \sin^2 B - 2\cos A \cos B - 2\sin A \sin B$

$d^2 = 1 + 1 - 2\cos A \cos B - 2\sin A \sin B$

$d^2 = 2 - 2\cos A \cos B - 2\sin A \sin B$

Mid-Chapter Quiz

11. $\sec^2 \theta + 1 \stackrel{?}{=} \dfrac{\cot \theta}{\cos \theta \cdot \sin \theta}$

$\csc^2 \theta \stackrel{?}{=} \dfrac{\cos \theta}{\cos \theta \cdot \sin \theta \cdot \sin \theta}$

$\csc^2 \theta \stackrel{?}{=} \dfrac{1}{\sin^2 \theta}$

$\csc^2 \theta = \csc^2 \theta$ ✓

12. $\dfrac{\cos \theta \csc \theta}{\cot \theta} \stackrel{?}{=} 1$

$\dfrac{\cos \theta}{\sin \theta} \cdot \tan \theta \stackrel{?}{=} 1$

$\dfrac{\cos \theta}{\sin \theta} \cdot \dfrac{\sin \theta}{\cos \theta} \stackrel{?}{=} 1$

$1 = 1$ ✓

13. $\dfrac{\sin\theta\tan\theta}{1-\cos\theta} \overset{?}{=} (1+\cos\theta)\sec\theta$

$\dfrac{\sin\theta\left(\dfrac{\sin\theta}{\cos\theta}\right)}{1-\cos\theta} \overset{?}{=} (1+\cos\theta)\dfrac{1}{\cos\theta}$

$\dfrac{\dfrac{\sin^2\theta}{\cos\theta}}{1-\cos\theta} \overset{?}{=} \dfrac{1}{\cos\theta}+1$

$\dfrac{\sin^2\theta}{\cos\theta}\cdot\dfrac{1}{1-\cos\theta} \overset{?}{=} \dfrac{1}{\cos\theta}+1$

$\dfrac{\sin^2\theta}{\cos\theta\,(1-\cos\theta)} \overset{?}{=} \dfrac{1}{\cos\theta}+1$

$\dfrac{1-\cos^2\theta}{\cos\theta\,(1-\cos\theta)} \overset{?}{=} \dfrac{1}{\cos\theta}+1$

$\dfrac{(1-\cos\theta)(1+\cos\theta)}{\cos\theta(1-\cos\theta)} \overset{?}{=} \dfrac{1}{\cos\theta}+1$

$\dfrac{1+\cos\theta}{\cos\theta} \overset{?}{=} \dfrac{1}{\cos\theta}+1$

$\dfrac{1}{\cos\theta}+1 = \dfrac{1}{\cos\theta}+1 \checkmark$

14. $\tan\theta\,(1-\sin\theta) \overset{?}{=} \dfrac{\cos\theta\sin\theta}{1+\sin\theta}$

$\tan\theta\,(1-\sin\theta) \overset{?}{=} \dfrac{\cos\theta\sin\theta}{1+\sin\theta}\cdot\dfrac{1-\sin\theta}{1-\sin\theta}$

$\tan\theta\,(1-\sin\theta) \overset{?}{=} \dfrac{\cos\theta\sin\theta(1-\sin\theta)}{1-\sin^2\theta}$

$\tan\theta\,(1-\sin\theta) \overset{?}{=} \dfrac{\cos\theta\sin\theta(1-\sin\theta)}{\cos^2\theta}$

$\tan\theta\,(1-\sin\theta) \overset{?}{=} \dfrac{\sin\theta(1-\sin\theta)}{\cos\theta}$

$\tan\theta\,(1-\sin\theta) \overset{?}{=} \dfrac{\sin\theta}{\cos\theta}\cdot(1-\sin\theta)$

$\tan\theta\,(1-\sin\theta) = \tan\theta\,(1-\sin\theta) \checkmark$

15b. $\cot\theta = \dfrac{12}{9}$, $\dfrac{\cos\theta}{\sin\theta} = \dfrac{\frac{12}{15}}{\frac{9}{15}} = \dfrac{12}{9}$, so $\dfrac{12}{9} = \dfrac{12}{9}$

16. $\tan^2\theta + 1 \overset{?}{=} \dfrac{\tan\theta\frac{15}{}}{\cos\theta\cdot\sin\theta}$

$\sec^2\theta \overset{?}{=} \dfrac{\tan\theta}{\cos\theta\cdot\sin\theta}$

$\sec^2\theta \overset{?}{=} \dfrac{\frac{\sin\theta}{\cos\theta}}{\cos\theta\cdot\sin\theta}$

$\sec^2\theta \overset{?}{=} \dfrac{\sin\theta}{\cos^2\theta\cdot\sin\theta}$

$\sec^2\theta \overset{?}{=} \dfrac{1}{\cos^2\theta}$

$\sec^2\theta = \sec^2\theta \checkmark$

17. $\dfrac{\sin\theta\cdot\sec\theta}{\sec\theta-1} \overset{?}{=} (\sec\theta+1)\cot\theta$

$\dfrac{\sec\theta+1}{\sec\theta+1}\cdot\dfrac{\sin\theta\cdot\sec\theta}{\sec\theta-1} \overset{?}{=} (\sec\theta+1)\cot\theta$

$\dfrac{\sin\theta\cdot\sec\theta(\sec\theta+1)}{\sec^2\theta-1} \overset{?}{=} (\sec\theta+1)\cot\theta$

$\dfrac{\sin\theta\cdot\dfrac{1}{\cos\theta}(\sec\theta+1)}{\tan^2\theta} \overset{?}{=} (\sec\theta+1)\cot\theta$

$\dfrac{\dfrac{\sin\theta}{\cos\theta}(\sec\theta+1)}{\tan^2\theta} \overset{?}{=} (\sec\theta+1)\cot\theta$

$\dfrac{\tan\theta(\sec\theta+1)}{\tan^2\theta} \overset{?}{=} (\sec\theta+1)\cot\theta$

$\dfrac{\sec\theta+1}{\tan\theta} \overset{?}{=} (\sec\theta+1)\cot\theta$

$\dfrac{\sec\theta+1}{1}\cdot\dfrac{1}{\tan\theta} \overset{?}{=} (\sec\theta+1)\cot\theta$

$(\sec\theta+1)\cot\theta = (\sec\theta+1)\cot\theta \checkmark$

18. $\sin^2\theta\cdot\tan^2\theta \overset{?}{=} \tan^2\theta - \sin^2\theta$

$\sin^2\theta\cdot\tan^2\theta \overset{?}{=} \dfrac{\sin^2\theta}{\cos^2\theta} - \sin^2\theta$

$\sin^2\theta\cdot\tan^2\theta \overset{?}{=} \dfrac{\sin^2\theta - \sin^2\theta\cos^2\theta}{\cos^2\theta}$

$\sin^2\theta\cdot\tan^2\theta \overset{?}{=} \dfrac{\sin^2\theta\,(1-\cos^2\theta)}{\cos^2\theta}$

$\sin^2\theta\cdot\tan^2\theta \overset{?}{=} \dfrac{\sin^2\theta\,(\sin^2\theta)}{\cos^2\theta}$

$\sin^2\theta\cdot\tan^2\theta \overset{?}{=} \sin^2\theta\cdot\dfrac{\sin^2\theta}{\cos^2\theta}$

$\sin^2\theta\cdot\tan^2\theta = \sin^2\theta\cdot\tan^2\theta \checkmark$

19. $\cot\theta(1-\cos\theta) \overset{?}{=} \dfrac{\cos\theta\cdot\sin\theta}{1+\cos\theta}$

$\cot\theta(1-\cos\theta) \overset{?}{=} \dfrac{\cos\theta\cdot\sin\theta}{1+\cos\theta}\cdot\dfrac{1-\cos\theta}{1-\cos\theta}$

$\cot\theta(1-\cos\theta) \overset{?}{=} \dfrac{\cos\theta\cdot\sin\theta\,(1-\cos\theta)}{1-\cos^2\theta}$

$\cot\theta(1-\cos\theta) \overset{?}{=} \dfrac{\cos\theta\cdot\sin\theta\,(1-\cos\theta)}{\sin^2\theta}$

$\cot\theta(1-\cos\theta) \overset{?}{=} \dfrac{\cos\theta\,(1-\cos\theta)}{\sin\theta}$

$\cot\theta(1-\cos\theta) \overset{?}{=} \dfrac{\cos\theta}{\sin\theta}\,(1-\cos\theta)$

$\cot\theta(1-\cos\theta) = \cot\theta\,(1-\cos\theta) \checkmark$

25. $\cos 30°\cos\theta + \sin 30°\sin\theta \overset{?}{=} \sin 60°\cos\theta + \cos 60°\sin\theta$

$\dfrac{\sqrt{3}}{2}\cos\theta + \dfrac{1}{2}\sin\theta = \dfrac{\sqrt{3}}{2}\cos\theta + \dfrac{1}{2}\sin\theta \checkmark$

26. $\tan 2\theta \overset{?}{=} \dfrac{2}{\cot \theta - \tan \theta}$

$\tan 2\theta \overset{?}{=} \dfrac{2}{\cot \theta - \tan \theta} \cdot \dfrac{\tan \theta}{\tan \theta}$

$\tan 2\theta \overset{?}{=} \dfrac{2 \tan \theta}{\cot \theta \tan \theta - \tan^2 \theta}$

$\tan 2\theta \overset{?}{=} \dfrac{2 \tan \theta}{1 - \tan^2 \theta}$

$\tan 2\theta = \tan 2\theta$ ✓

27. $1 + \dfrac{1}{2} \sin 2\theta \overset{?}{=} \dfrac{\sec \theta + \sin \theta}{\sec \theta}$

$\overset{?}{=} \dfrac{\dfrac{1}{\cos \theta} + \sin \theta}{\dfrac{1}{\cos \theta}}$

$\overset{?}{=} \dfrac{\dfrac{1}{\cos \theta} + \sin \theta}{\dfrac{1}{\cos \theta}} \cdot \dfrac{\cos \theta}{\cos \theta}$

$\overset{?}{=} 1 + \dfrac{1}{2} \cdot 2 \sin \theta \cos \theta$

$= 1 + \dfrac{1}{2} \sin 2\theta$ ✓

28. $\sin \dfrac{\theta}{2} \cos \dfrac{\theta}{2} \overset{?}{=} \dfrac{\sin \theta}{2}$

$\dfrac{2 \sin \dfrac{\theta}{2} \cos \dfrac{\theta}{2}}{2} \overset{?}{=} \dfrac{\sin \theta}{2}$

$\dfrac{\sin 2\left(\dfrac{\theta}{2}\right)}{2} \overset{?}{=} \dfrac{\sin \theta}{2}$

$\dfrac{\sin \theta}{2} = \dfrac{\sin \theta}{2}$ ✓

29. $\tan \dfrac{\theta}{2} \overset{?}{=} \dfrac{\sin \theta}{1 + \cos \theta}$

$\tan \dfrac{\theta}{2} \overset{?}{=} \dfrac{\sin 2\left(\dfrac{\theta}{2}\right)}{1 + \cos 2\left(\dfrac{\theta}{2}\right)}$

$\tan \dfrac{\theta}{2} \overset{?}{=} \dfrac{2 \sin \dfrac{\theta}{2} \cos \dfrac{\theta}{2}}{1 + 2 \cos^2 \dfrac{\theta}{2} - 1}$

$\tan \dfrac{\theta}{2} \overset{?}{=} \dfrac{2 \sin \dfrac{\theta}{2} \cos \dfrac{\theta}{2}}{2 \cos^2 \dfrac{\theta}{2}}$

$\tan \dfrac{\theta}{2} \overset{?}{=} \dfrac{\sin \dfrac{\theta}{2}}{\cos \dfrac{\theta}{2}}$

$\tan \dfrac{\theta}{2} = \tan \dfrac{\theta}{2}$ ✓

30. For $\theta = 45° + \alpha$,

$d = \dfrac{v^2 \sin 2(45° + \alpha)}{g}$

$= \dfrac{v^2 \sin (90° + 2\alpha)}{g}$

$= \dfrac{v^2(\sin 90° \cos 2\alpha + \cos 90° \sin 2\alpha)}{g}$

$= \dfrac{v^2(1 \cdot \cos 2\alpha + 0 \cdot \sin 2\alpha)}{g}$

$= \dfrac{v^2 \cos 2\alpha}{g}$

For $\theta = 45° - \alpha$,

$d = \dfrac{v^2 \sin 2(45° - \alpha)}{g}$

$= \dfrac{v^2 \sin (90° - 2\alpha)}{g}$

$= \dfrac{v^2(\sin 90° \cos 2\alpha - \cos 90° \sin 2\alpha)}{g}$

$= \dfrac{v^2(1 \cdot \cos 2\alpha - 0 \cdot \sin 2\alpha)}{g}$

$= \dfrac{v^2 \cos 2\alpha}{g}$

37. No; Teresa incorrectly added the square roots, and Nathan used the half-angle identity incorrectly. He used sin 30° in the formula instead of first finding the cosine.

39. If you are only given the value of cos θ, then $\cos 2\theta = 2 \cos^2 \theta - 1$ is the best identity to use. If you are only given the value of sin θ, then $\cos 2\theta = 1 - 2 \sin^2 \theta$ is the best identity to use. If you are given the values of both cos θ and sin θ, then $\cos 2\theta = \cos^2 \theta - \sin^2 \theta$ works just as well as the other two.

40. $\sin 2\theta = \sin(\theta + \theta)$

$= \sin \theta \cos \theta + \cos \theta \sin \theta$

$= 2 \sin \theta \cos \theta$

$\cos 2\theta = \cos(\theta + \theta)$

$= \cos \theta \cos \theta - \sin \theta \sin \theta$

$= \cos^2 \theta - \sin^2 \theta$

You can find alternate forms for cos 2θ by making substitutions into the expression $\cos^2 \theta - \sin^2 \theta$.

$\cos^2 \theta - \sin^2 \theta = (1 - \sin^2 \theta) - \sin^2 \theta$

$= 1 - 2 \sin^2 \theta$

Substitute $1 - \sin^2 \theta$ for $\cos^2 \theta$.
Simplify.

$\cos^2 \theta - \sin^2 \theta = \cos^2 \theta - (1 - \cos^2 \theta)$

$= 2 \cos^2 \theta - 1$

Substitute $1 - \cos^2 \theta$ for $\sin^2 \theta$.
Simplify.

Chapter 10 Answer Appendix

41. Find $\sin \frac{A}{2}$.

$1 - 2\sin^2 \theta = \cos 2\theta$	Double-angle identity
$1 - 2\sin^2 \frac{A}{2} = \cos A$	Substitute $\frac{A}{2}$ for θ and A for 2θ.
$\sin^2 \frac{A}{2} = \frac{1 - \cos A}{2}$	Solve for $\sin^2 \frac{A}{2}$.
$\sin \frac{A}{2} = \pm\sqrt{\frac{1 - \cos A}{2}}$	Take the square root of each side.

Find $\cos \frac{A}{2}$.

$2\cos^2 \theta - 1 = \cos 2\theta$	Double-angle identity
$2\cos^2 \frac{A}{2} - 1 = \cos A$	Substitute $\frac{A}{2}$ for θ and A for 2θ.
$\cos^2 \frac{A}{2} = \frac{1 + \cos A}{2}$	Solve for $\cos^2 \frac{A}{2}$.
$\cos \frac{A}{2} = \pm\sqrt{\frac{1 + \cos A}{2}}$	Take the square root of each side.

Find $\tan \frac{A}{2}$.

$\tan \frac{A}{2} = \dfrac{\sin \frac{A}{2}}{\cos \frac{A}{2}}$	Quotient Identity
$\tan \frac{A}{2} = \dfrac{\pm\sqrt{\frac{1 - \cos A}{2}}}{\pm\sqrt{\frac{1 + \cos A}{2}}}$	Half-Angle Identities
$\tan \frac{A}{2} = \pm\sqrt{\dfrac{\frac{1 - \cos A}{2}}{\frac{1 + \cos A}{2}}}$	Quotient Property of Radicals
$\tan \frac{A}{2} = \pm\sqrt{\dfrac{1 - \cos A}{1 + \cos A}}$	Simplify.

Lesson 10-5

58a.

θ	$\sin \theta$	θ	$\sin \theta$
0°	1	210°	$-\frac{1}{2}$
30°	$\frac{1}{2}$	225°	$-\frac{\sqrt{2}}{2}$
45°	$\frac{\sqrt{2}}{2}$	240°	$-\frac{\sqrt{3}}{2}$
60°	$\frac{\sqrt{3}}{2}$	270°	-1
90°	1	300°	$-\frac{\sqrt{3}}{2}$
120°	$\frac{\sqrt{3}}{2}$	315°	$-\frac{\sqrt{2}}{2}$
135°	$\frac{\sqrt{2}}{2}$	330°	$-\frac{1}{2}$
150°	$\frac{1}{2}$	360°	0
180°	0		

58b. The graph of $y = \sin \theta$ is above the graph of $y = \frac{1}{2}$ for $30° < \theta < 150°$.

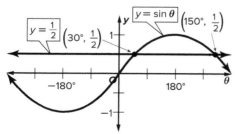

58c. The graph of $y = \sin \theta$ is above the graph of $y = \frac{1}{2}$ for $30° < \theta < 150°$ and the period of $\sin \theta$ is $[-360°, 360°]$, so the solutions of $\sin \theta > \frac{1}{2}$ are $30° + k \cdot 360° < \theta < 150° + k \cdot 360°$.

58d. i. $0° \le \theta \le 45°$ and $315° \le \theta \le 360°$; $0° + k \cdot 360° \le \theta \le 45° + k \cdot 360°$ and $315° + k \cdot 360° \le \theta \le 360° + k \cdot 360°$; ii. $0° \le \theta \le 60°$ and $120° \le \theta \le 360°$; $0° + k \cdot 360° \le \theta \le 60° + k \cdot 360°$ and $120° + k \cdot 360° \le \theta \le 360° + k \cdot 360°$; iii. $180° \le \theta \le 360°$; $180° + k \cdot 360° \le \theta \le 360° + k \cdot 360°$; iv. $60° \le \theta \le 300°$; $60° + k \cdot 360° \le \theta \le 300° + k \cdot 360°$

Study Guide and Review

30.
$$\sin(\theta + 90°) \overset{?}{=} \cos \theta$$
$$\sin \theta \cos 90° + \cos \theta \sin 90° \overset{?}{=} \cos \theta$$
$$(\sin \theta)(0) + (\cos \theta)(1) \overset{?}{=} \cos \theta$$
$$\cos \theta = \cos \theta \checkmark$$

31.
$$\sin\left(\frac{3\pi}{2} - \theta\right) \overset{?}{=} -\cos \theta$$
$$\sin \frac{3\pi}{2} \cos \theta - \cos \frac{3\pi}{2} \sin \theta \overset{?}{=} -\cos \theta$$
$$(-1) \cos \theta - (0) \sin \theta \overset{?}{=} -\cos \theta$$
$$-\cos \theta = -\cos \theta \checkmark$$

32.
$$\tan(\theta - \pi) \overset{?}{=} \tan \theta$$
$$\frac{\tan \theta - \tan \pi}{1 + \tan \theta \tan \pi} \overset{?}{=} \tan \theta$$
$$\frac{\tan \theta - 0}{1 + (\tan \theta)(0)} \overset{?}{=} \tan \theta$$
$$\frac{\tan \theta}{1} \overset{?}{=} \tan \theta$$
$$\tan \theta = \tan \theta \checkmark$$

33. $\sin 2\theta = \frac{24}{25}$, $\cos 2\theta = \frac{7}{25}$, $\sin \frac{\theta}{2} = \frac{\sqrt{10}}{10}$, and $\cos \frac{\theta}{2} = \frac{3\sqrt{10}}{10}$

34. $\sin 2\theta = \frac{\sqrt{15}}{8}$, $\cos 2\theta = \frac{7}{8}$, $\sin \frac{\theta}{2} = \frac{\sqrt{2}\sqrt{4 + \sqrt{15}}}{4}$, and
$$\cos \frac{\theta}{2} = -\frac{\sqrt{2}\sqrt{4 - \sqrt{15}}}{4}$$

35. $\sin 2\theta = -\frac{4\sqrt{5}}{9}$, $\cos 2\theta = -\frac{1}{9}$, $\sin \frac{\theta}{2} = \frac{\sqrt{30}}{6}$, and $\cos \frac{\theta}{2} = \frac{\sqrt{6}}{6}$

36a. $c^2 = 90^2 + 90^2$; $c^2 = 8100 + 8100$; $c^2 = 16{,}200$; $c = 90\sqrt{2}$

36b. $\sin 45° = \frac{90}{90\sqrt{2}} = \frac{1}{\sqrt{2}} = \frac{\sqrt{2}}{2}$

36c. $\sin \frac{\theta}{2} = \pm\sqrt{\frac{1 - \cos \theta}{2}}$; $\sin \frac{90}{2} = \pm\sqrt{\frac{1 - \cos 90}{2}}$; $\sin \frac{90}{2} = \pm\sqrt{\frac{1 - 0}{2}} = \frac{1}{\sqrt{2}} = \frac{\sqrt{2}}{2}$

Practice Test

14. $\dfrac{\sin \theta}{1 - \cos \theta} \stackrel{?}{=} \cos \theta + \cot \theta$

$\dfrac{\sin \theta}{1 - \cos \theta} \stackrel{?}{=} \dfrac{1}{\sin \theta} + \dfrac{\cos \theta}{\sin \theta}$

$\dfrac{\sin \theta}{1 - \cos \theta} \stackrel{?}{=} \dfrac{1 + \cos \theta}{\sin \theta}$

$\dfrac{\sin \theta}{1 - \cos \theta} \stackrel{?}{=} \dfrac{1 + \cos \theta}{\sin \theta} \cdot \dfrac{1 - \cos \theta}{1 - \cos \theta}$

$\dfrac{\sin \theta}{1 - \cos \theta} \stackrel{?}{=} \dfrac{1 - \cos^2 \theta}{\sin \theta(1 - \cos \theta)}$

$\dfrac{\sin \theta}{1 - \cos \theta} \stackrel{?}{=} \dfrac{\sin^2 \theta}{\sin \theta(1 - \cos \theta)}$

$\dfrac{\sin \theta}{1 - \cos \theta} \stackrel{?}{=} \dfrac{\sin \theta}{1 - \cos \theta}$ ✓

16a. $18^2 = 9^2 + a^2; 324 = 81 + a^2; 243 = a^2; a = 9\sqrt{3}$

16b. $\sin 2\theta = 2 \sin \theta \cos \theta;$

$\sin 2(30°) = 2 \sin 30° \cos 30°;$

$\sin 60° = 2\left(\dfrac{9}{18}\right)\left(\dfrac{9\sqrt{3}}{18}\right) = \dfrac{162\sqrt{3}}{324} = \dfrac{\sqrt{3}}{2};$

$\sin 60° = 9\dfrac{\sqrt{3}}{18} = \dfrac{\sqrt{3}}{2}$

21. $R = \dfrac{v^2}{g} \sin 2\theta; 25 = \dfrac{20^2}{9.8} \sin 2\theta; 0.6125 = \sin 2\theta; \theta \approx 18.9°$

Student Handbook

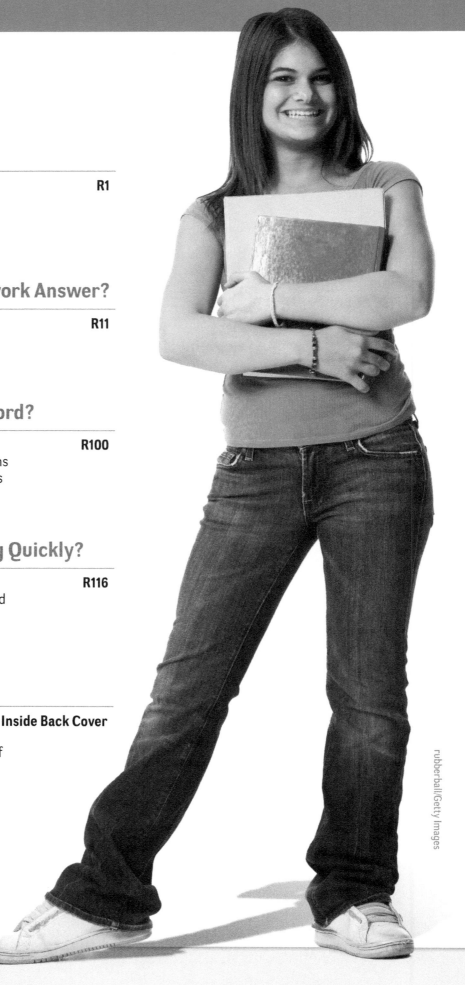

This **Student Handbook** can help you answer these questions.

What if I Need More Practice?

Extra Practice   R1

The **Extra Practice** section provides additional problems for each lesson so you have ample opportunity to practice new skills.
See Teacher Edition Volume 1 for Chapters 1–5;
see Volume 2 for Chapters 6–10.

What if I Need to Check a Homework Answer?

Selected Answers and Solutions   R11

The answers to odd-numbered problems are included in **Selected Answers and Solutions**.
See Teacher Edition Volume 1 for Chapters 1–5;
see Volume 2 for Chapters 6–10.

What if I Forget a Vocabulary Word?

Glossary/Glosario   R100

The **English-Spanish Glossary** provides definitions and page numbers of important or difficult words used throughout the textbook.

What if I Need to Find Something Quickly?

Index   R116

The **Index** alphabetically lists the subjects covered throughout the entire textbook and the pages on which each subject can be found.

What if I Forget a Formula?

Formulas, Symbols,   **Inside Back Cover**
and Parent Functions

Inside the back cover of your math book is a list of **Formulas and Symbols** that are used in the book.

CHAPTER 6 — Exponential and Logarithmic Functions

Graph each function. State the domain and range. (Lesson 6-1) 1-2. See Extra Practice Answer Appendix for graphs.

1. $f(x) = 3(4)^x$ $D = \{$all real numbers$\}$;
$R = \{f(x) \mid f(x) > 0\}$

2. $f(x) = 2^{3x} - 3$ $D = \{$all real numbers$\}$;
$R = \{f(x) \mid f(x) > -3\}$

Solve each equation. (Lesson 6-2)

3. $2^{x-1} = 8^{x+3}$ -5

4. $5^{2x+12} = 25^{10x-12}$ 2

Write an equation for the nth term of each geometric sequence. (Lesson 6-3)

5. $0.75, 3, 12, \ldots$ $a_n = 0.75(4)^{n-1}$

6. $0.4, -2, 10, \ldots$ $a_n = 0.4(-5)^{n-1}$

Evaluate each expression. (Lesson 6-4)

7. $\log_8 64$ 2

8. $\log_7 1$ 0

9. $\log_5 \frac{1}{25}$ -2

10. DOLPHINS The number of dolphins living in an ocean region after t months can be approximated by $n(t) = 920 \log_{10}(t - 1)$. (Lesson 6-4)

  a. How many dolphins are in the ocean region after 4 months?   ≈439

  b. How many dolphins are in the ocean region after 2 years?   ≈1253

11. TAXI FARES Diego collected data on the distance in miles and the fare in dollars for several taxi rides in his town. The ordered pairs are (2, 9.25), (5, 19.75), (8, 29.75), and (12, 44.50). (Lesson 6-5)

  a. What type of regression model works best with the data?   linear

  b. If Diego rides a taxi for 10 miles, how much should he expect to pay? $37.20

Solve each equation. Check your solution. (Lesson 6-6)

12. $\log_4 3 + \log_4 x = \log_4 12$ 4

13. $\log_7 6 - \log_7 2 = \log_7 x$ 3

14. $4 \log_2 x = \log_2 81$ 3

15. $\log_6 x + \log_6 (x + 5) = 2$ 4

16. POPULATION The wolf population per year at a national park is listed in the table below. (Lesson 6-6)

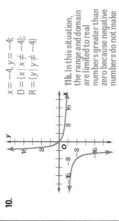

Year	Population
2005	420
2006	399
2007	379
2008	360
2009	342
2020	?

  a. Determine the annual percent of decrease of the population.   5%

  b. Write a logarithmic function for the time in years based upon population from 2005.

  c. How many wolves will there be in the national park in 2020?   ≈195 wolves

Solve each equation. Round to the nearest ten thousandth. (Lesson 6-7)

17. $5^x = 60$ 2.5440 **18.** $4.4^{x^2} = 21$ 1.4819

18b. $T = \log_{0.95} \left(\frac{P}{420} \right)$

19. $3^{x-2} = 4^x$ -7.6377

Solve each equation. Round to the nearest ten thousandth. (Lesson 6-8)

20. $4e^x - 6 = 11$ 1.4469 **21.** $3e^{-x} + 4 = 17$ -1.4663

22. $2x^{3x} - 8 = 21$ 0.8914

Solve each equation. (Lesson 6-9)

23. $\log_{25} x = \frac{3}{2}$ 125

24. $\log_9 81 = x$ 2

25. $\log_6 (5x - 3) = \log_6 (x + 9)$ 3

26. $\log_8 (x^2 + 6) = \log_8 (5x)$ 2 or 3

27. INTEREST Stefanie deposited $2000 into a savings account paying 2.5% interest compounded continuously. (Lesson 6-10)

  a. Determine how long it will take Stefanie to have a balance of $2400.   7.3 yr

  b. How long will it take Stefanie to double her investment?   27.7 yr

  c. If Stefanie wanted to have $5000 after 10 years, how much should she have invested? $3894

CHAPTER 7 — Rational Functions

Simplify each expression. (Lesson 7-1)

1. $\frac{(x-2)(x+5)}{x^2 + 2x - 8} \cdot \frac{x+5}{x+4}$

2. $\frac{(x+3)(x-4)}{x^2 + x - 20} \cdot \frac{x+3}{x+5}$

3. $\frac{r^2 - 16}{r^2 - 8x + 12} \cdot \frac{r^2 + 3x - 18}{r^2 - 7x + 12}$ $\frac{(x+4)(x+6)}{(x-6)(x-2)}$

4. $\frac{x^2 + 10x + 21}{x^2 - 5x + 6} \div \frac{r^2 + 3x - 28}{r^2 - 8x + 15}$ $\frac{(x+3)(x-5)}{(x-2)(x-4)}$

5. BUSINESS The average hourly income I of a college student x years after graduation can be modeled by the function $I(x) = \frac{11x^2 + 33x + 22}{x + 2}$. Simplify the function. (Lesson 7-2) $11(x+1)$

Simplify each expression. (Lesson 7-2)

6. $\frac{2}{r^2 - 7x + 12} + \frac{6}{r^2 + x - 20}$ $\frac{8(x-1)}{(x-3)(x-4)(x+5)}$

7. $\frac{4}{r^2 - 11x + 30} - \frac{3}{r^2 - 13x + 40}$ $\frac{x - 14}{(x-5)(x-6)(x-8)}$

Identify the asymptotes, domain, and range of each function. (Lesson 7-3)

8.

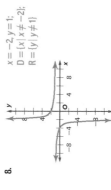

$x = -2, y = 1;$
$D = \{x \mid x \neq -2\};$
$R = \{y \mid y \neq 1\}$

9.

$x = 0, y = 2;$
$D = \{x \mid x \neq 0\};$
$R = \{y \mid y \neq 2\}$

Identify the asymptotes, domain, and range of each function. (Lesson 7-3)

10.

$x = -4, y = -4;$
$D = \{x \mid x \neq -4\};$
$R = \{y \mid y \neq -4\}$

11b. In this situation, the range and domain are limited to real numbers greater than zero because negative numbers do not make sense.

11. FIELD TRIP The American history class is planning a field trip to Washington, D.C. It costs $300 to rent the bus, and each ticket to the museum costs $12. (Lesson 7-3)

  a. If C represents the cost for each student and n represents the number of students, write an equation to represent the cost to each student as a function of how many students go on the field trip. $C(n) = \frac{12n + 300}{n}$

  b. Are there any limitations to the range or domain in this situation? Explain.

Graph each function. (Lesson 7-4)

12. $f(x) = \frac{2x}{x - 3}$ **13.** $f(x) = \frac{x - 4}{x + 2}$

12-13. See Extra Practice Answer Appendix.

If a varies jointly as b and c, find a when $b = 2$ and $c = -3$, given each of the following. (Lesson 7-5)

14. $a = -12$ when $b = 4$ and $c = 6$ 3

15. $a = -36$ when $b = 12$ and $c = -4$ -4.5

Solve each equation. Check your solution. (Lesson 7-6)

16. $\frac{3}{x+4} + \frac{1}{x-1} = \frac{17}{24}$ $4, -\frac{23}{17}$

17. $\frac{12}{x+4} - \frac{3}{x-3} = \frac{24}{r^2 + x - 12}$ 8

18. $\frac{14}{x-5} - \frac{7}{x+5} = \frac{-7}{r^2 - 25}$ -16

19. EXERCISE Natalie can bike 10 kilometers per hour faster than Aaron. By the time Natalie travels 60 kilometers, Aaron has gone 40 kilometers. (Lesson 7-6)

  a. Write an expression for Natalie's time. $r + 10$

  b. Write an expression for Aaron's time. $t = \frac{40}{r}$

  c. Write and solve the rational equation to determine the speed of each biker.

19c. $\frac{60}{r+10} = \frac{40}{r}$; Natalie: 30 km/h, Aaron 20 km/h $t = \frac{60}{r+10}$

CHAPTER 8   Statistics and Probability

Determine whether each situation describes a survey, an experiment, or an observational study. Then identify the sample, and suggest a population from which it may have been selected. (Lesson 8-1)

1. Every fifth person coming out of a football stadium is asked how often they go to a football game.   survey; sample: every fifth person; population: everyone

2. A company posts a controversial advertisement in their store and records the reactions of the customers.   observational study; sample: the participants in the study; population: potential customers

3. A randomly selected group of people are chosen to test the effects of a new drug on energy levels.   See Extra Practice Answer Appendix.

4. A randomly selected group of people are asked their opinions on political topics.   survey; sample: the participants in the survey; population: eligible voters

Design and conduct a simulation using a probability model. Then report the results. (Lesson 8-2)

5. Natasha plays on the basketball team. This season she has made 80% of her free throws.   See Extra Practice Answer Appendix.

6. Pilar is playing a board game with eight different categories, each with the same number of questions that must be answered correctly in order to win. See Extra Practice Answer Appendix.

For a sample survey of 120 soccer players, find the following population proportions. (Lesson 8-3)

7. 60 prefer weekend games. 0.5

8. 48 prefer to practice on Wednesdays. 0.4

9. 100 prefer to play outside. 0.83

10. 15 prefer to be goalie. 0.125

Compute the margin of error for each sample size. (Lesson 8-3)

11. 80 ±11.2%

12. 250 ±6.3%

Find the sample size for each margin of error. (Lesson 8-3)

13. ±1% 10,000

14. ±5% 400

15. **COLLEGE** The SAT scores of Mr. Williams's calculus class are given below. (Lesson 8-4)

SAT Scores						
2020	1250	1500	1700	1450	1600
1830	1230	1330	1450	1430	1250
1560	1450	1350	2200	2050	1640
2210	1550	1900	2140	1280	1750
2400					1800

a. Use a graphing calculator to create a histogram. Then describe the shape of the distribution.   See Extra Practice Answer Appendix; positively skewed.

b. Describe the center and spread of the data using either the mean and standard deviation or the five-number summary. Justify your choice.   See Extra Practice Answer Appendix.

16. Ron wants to create a misleading histogram using the data in exercise 15. Identify intervals along the horizontal axis that will make the data appear evenly distributed. (Lesson 8-5)   Sample answer: 1200–1340, 1350–1510, 1520–1720, 1730–2030, 2040–2400

17. **ASSEMBLIES** The number of assemblies at West High School each year is normally distributed with a mean of 12.4 and a standard deviation of 1.6. (Lesson 8-6)

a. What is the probability that there will be more than 10 assemblies in a given year? 93.3%

b. If the school has existed for 30 years, in how many of those years were there between 11 and 13 assemblies? about 14 yr

18. **STUDENT COUNCIL** The number of students that run for student council each year is normally distributed with a mean of 16.8 students and a standard deviation of 3.7. (Lesson 8-6)

a. What is the probability that fewer than 10 students run in a given year? 3.3%

b. If the school has kept records for 20 years, in how many of those years were there between 15 and 20 students who ran for student council? about 10 yr

19. **FOOD PROCESSING** It is known that 0.1% of the cans produced at a food processing plant are an incorrect weight. An employee has a scale that is 98% accurate. The employee weighs some cans and discards any that the scale shows to be an incorrect weight. Do you think this is a good decision? Explain. (Lesson 8-7)   See Extra Practice Answer Appendix.

CHAPTER 9   Trigonometric Functions

Use a trigonometric function to find the value of x. Round to the nearest tenth. (Lesson 9-1)

1. 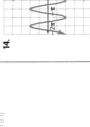 6.6

2. 6.1

3. **FLAGPOLE** Tina is standing away from a flagpole. The angle of elevation from the ground to the top of the flagpole is 15°. The flagpole is 50 feet high. Determine how far Tina is standing from the flagpole. (Lesson 9-1) 187 ft

4. **SAILS** One leg of a triangular sail of a sailboat is 25 feet long. The angle opposite this leg is 40°. What is the perimeter of the sail to the nearest tenth? (Lesson 9-1) 93.7 ft

Rewrite each degree measure in radians and each radian measure in degrees. (Lesson 9-2)

5. 300° $\frac{5\pi}{3}$

6. $-80°$ $-\frac{4\pi}{9}$

7. $\frac{2\pi}{3}$ $-120°$

8. $-\frac{\pi}{6}$ $-30°$

Find the exact value of each trigonometric function. (Lesson 9-3)

9. sin 120° $\frac{\sqrt{3}}{2}$

10. csc $\frac{3\pi}{4}$ $\sqrt{2}$

The terminal side of angle θ in standard position intersects the unit circle at each point P. Find cos θ and sin θ. (Lesson 9-4)

11. $P\left(\frac{35}{37},\frac{12}{37}\right)$ $\cos\theta=\frac{35}{37}$ and $\sin\theta=\frac{12}{37}$

12. $P\left(-\frac{\sqrt{3}}{2},\frac{1}{2}\right)$ $\cos\theta=-\frac{\sqrt{3}}{2}$ and $\sin\theta=\frac{1}{2}$

13. $P\left(\frac{\sqrt{2}}{2},-\frac{\sqrt{2}}{2}\right)$ $\cos\theta=\frac{\sqrt{2}}{2}$ and $\sin\theta=-\frac{\sqrt{2}}{2}$

Determine the period of each function. (Lesson 9-4)

14.

15.

Find the exact value of each expression. (Lesson 9-4)

16. $\cos(-60)°$ $\frac{1}{2}$

17. $\sin\left(-\frac{3\pi}{4}\right)$ $-\frac{\sqrt{2}}{2}$

18. $\cos\left(\frac{13\pi}{4}\right)$ $-\frac{\sqrt{2}}{2}$

19. sin $(-990)°$ 1

20–21. See Extra Practice Answer Appendix for graphs. **Find the amplitude and period of each function. Then graph the function.** (Lesson 9-5)

20. $y = 3\cos\theta$ amplitude 3; period 360°

21. $y = \frac{1}{2}\sin 3\theta$ amplitude $\frac{1}{2}$; period 120°

22–23. See Extra Practice Answer Appendix for graphs. **State the amplitude, period, phase shift, and vertical shift for each function. Then graph the function.** (Lesson 9-6)

22. $y = 2\tan(\theta + 30) + 3$ 2; 360°; $h = -30$°; $k = 3$

23. $y = 4\cos\left(\theta - \frac{\pi}{2}\right) - 2$ 4; 360°; $h = \frac{\pi}{2}$; $k = -2$

CHAPTER 10   Trigonometric Identities and Equations

Find the exact value of each expression if $0° < \theta < 90°$. (Lesson 10-1)

1. If $\cos \theta = \dfrac{3}{5}$, find $\sin \theta$. $\dfrac{4}{5}$

2. If $\tan \theta = 2$, find $\cot \theta$. $\dfrac{1}{2}$

3. If $\sin \theta = \dfrac{\sqrt{5}}{3}$, find $\cos \theta$. $\dfrac{2}{3}$

4. If $\csc \theta = \dfrac{3\sqrt{5}}{5}$, find $\tan \theta$. $\dfrac{\sqrt{5}}{2}$

5–8. See Extra Practice Answer Appendix.
Verify that each equation is an identity. (Lesson 10-2)

5. $\dfrac{\sin^2 \theta}{\cos^2 \theta} \cdot \csc^2 \theta = 1 + \tan^2 \theta$

6. $\sec \theta \cot \theta = \csc \theta$

7. $\sin \theta \cot \theta = \cos \theta$

8. $\dfrac{\sec^2 \theta}{\csc^2 \theta} = \tan^2 \theta$

9. CONSTRUCTION A window has the dimensions shown below. Use the measures of the sides of the triangle to show that $\sin^2 \theta + \cos^2 \theta = 1$. (Lesson 10-2)

$$
\begin{aligned}
\sin^2 \theta + \cos^2 \theta &= \left[\frac{3\sqrt{5}}{9}\right]^2 + \left[\frac{6}{9}\right]^2 \\
&= \frac{5}{9} + \frac{36}{81} \\
&= \frac{5}{9} + \frac{4}{9} \\
&= 1
\end{aligned}
$$

(triangle with hypotenuse 9, sides $3\sqrt{5}$ and 6, angle θ)

Find the exact value of each expression. (Lesson 10-3)

10. $\cos(-30°)$ $\dfrac{\sqrt{3}}{2}$

11. $\sin(-120°)$ $-\dfrac{\sqrt{3}}{2}$

12. $\sin 135°$ $\dfrac{\sqrt{2}}{2}$

13. $\cos 225°$ $-\dfrac{\sqrt{2}}{2}$

Find the exact value of each expression. (Lesson 10-4)

14. $\sin 15°$ $\dfrac{\sqrt{2} - \sqrt{3}}{2}$

15. $\cos \dfrac{\pi}{12}$ $\dfrac{\sqrt{2} + \sqrt{3}}{2}$

16. $\tan 67.5°$ $\sqrt{3 + 2\sqrt{2}}$

17. $\sin 165°$ $\dfrac{\sqrt{2} - \sqrt{3}}{2}$

Solve each equation if $0° \le \theta \le 360°$. (Lesson 10-5)

18. $2 \sin \theta = 1$ $30°, 150°$

19. $2 \cos \theta + 1 = 0$ $120°, 240°$

20. $4 \cos^2 \theta - 1 = 0$ $60°, 120°, 240°, 300°$

21. TOYS A toy's height h in inches can be modeled by the equation $h = 4 \sin \dfrac{11\pi}{12} t$, where t represents time in seconds. (Lesson 10-5)

  a. At what point will the toy be 4 inches above its resting position for the first time? $\dfrac{6}{11} \approx 0.55$ second

  b. At what point will the toy be 2 inches above its resting position for the first time? $\dfrac{2}{11} \approx 0.18$ second

22. BUOYS As waves move past a buoy, the height in feet of the buoy oscillates according to the equation $y = 6 \cos 30t$, where t is the time in seconds. If the buoy is at the top of the wave at 0 seconds, during what time intervals will the buoy be 3 feet below the midline? (Lesson 10-5)

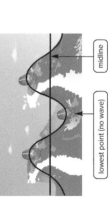

midline

lowest point (no wave)

$4 + 12n$ and $6 + 12n$ seconds where n is an integer

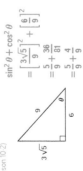

Chapter 6

1.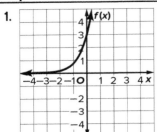

2.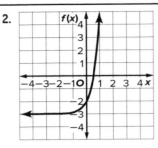

Chapter 7

12.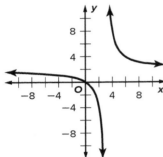

13.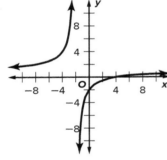

Chapter 8

3. experiment; sample: the participants in the experiment; population: the population of people that could take the drug

5. Sample answer: Use a spinner that is divided into two sectors, one containing 80% or 288° and the other containing 20% or 72°. Do 20 trials and record the results in a frequency table.

Outcome	Frequency
Makes Free Throw	15
Misses Free Throw	5
Total	20

The experimental probability of making a free throw is 75%, which is less than the theoretical probability.

6. Sample answer: Use a spinner that is divided into 8 equal sectors, each 45°. Do 50 trails and record the results in a frequency table.

Outcome	Frequency
Category 1	3
Category 2	3
Category 3	6
Category 4	13
Category 5	4
Category 6	9
Category 7	7
Category 8	5
Total	50

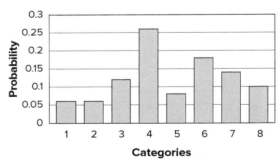

The probability of landing on Categories 1 and 2 is 0.06, Category 3 is 0.12, Category 4 is 0.26, Category 5 is 0.08, Category 6 is 0.18, Category 7 is 0.14, and Category 8 is 0.1.

15a.

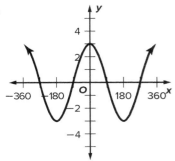

[1200, 2600] scl: 200 by [−2, 8] scl: 1

15b. Sample answer: The distribution is skewed, so use the five-number summary.

The range is 1230 to 2400. The median is 1600, and half of the scores are between 1390 and 1960.

19. No; the probability that a can that is identified as an incorrect weight is actually an incorrect weight is about 4.9%, so it may be wasteful to discard every can that the scale identifies as an incorrect weight.

Chapter 9

20.

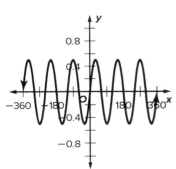

21.

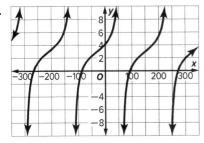

22.

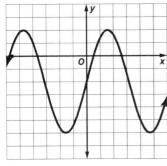

23.

Chapter 10

5. $\dfrac{\sin^2\theta}{\cos^2\theta} \cdot \csc^2\theta \overset{?}{=} 1 + \tan^2\theta$

$\dfrac{\sin^2\theta}{\cos^2\theta} \cdot \dfrac{1}{\sin^2\theta} \overset{?}{=} \sec^2\theta$

$\dfrac{1}{\cos^2\theta} \overset{?}{=} \sec^2\theta$

$\sec^2\theta = \sec^2\theta$ ✔

6. $\sec\theta \cot\theta \overset{?}{=} \csc\theta$

$\dfrac{1}{\cos\theta} \cdot \dfrac{\cos\theta}{\sin\theta} \overset{?}{=} \csc\theta$

$\dfrac{1}{\sin\theta} \overset{?}{=} \csc\theta$

$\csc\theta = \csc\theta$ ✔

7. $\sin\theta \cot\theta \overset{?}{=} \cos\theta$

$\sin\theta \cdot \dfrac{\cos\theta}{\sin\theta} \overset{?}{=} \cos\theta$

$\cos\theta = \cos\theta$ ✔

8. $\dfrac{\sec^2\theta}{\csc^2\theta} \overset{?}{=} \tan^2\theta$

$\dfrac{\dfrac{1}{\cos^2\theta}}{\dfrac{1}{\sin^2\theta}} \overset{?}{=} \tan^2\theta$

$\dfrac{\sin^2\theta}{\cos^2\theta} \overset{?}{=} \tan^2\theta$

$\tan^2\theta = \tan^2\theta$ ✔

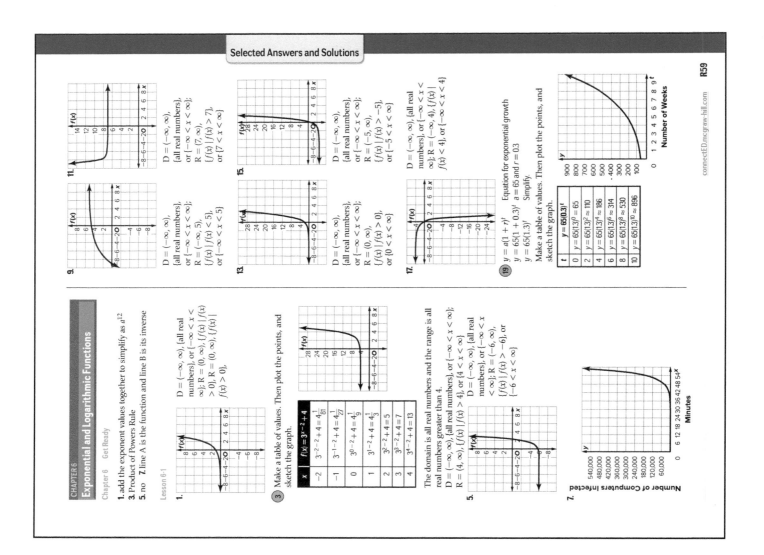

11. graph; $D = (-\infty, \infty)$, [all real numbers], or $\{-\infty < x < \infty\}$; $R = (7, \infty)$, $\{f(x) \mid f(x) > 7\}$, or $\{7 < x < \infty\}$

9. graph; $D = (-\infty, \infty)$, [all real numbers], or $\{-\infty < x < \infty\}$; $R = (-\infty, 5)$, $\{f(x) \mid f(x) < 5\}$, or $\{-\infty < x < 5\}$

15. graph; $D = (-\infty, \infty)$, [all real numbers], or $\{-\infty < x < \infty\}$; $R = (-5, \infty)$, $\{f(x) \mid f(x) > -5\}$, or $\{-5 < x < \infty\}$

13. graph; $D = (-\infty, \infty)$, [all real numbers], or $\{-\infty < x < \infty\}$; $R = (0, \infty)$, $\{f(x) \mid f(x) > 0\}$, or $\{0 < x < \infty\}$

17. graph; $D = (-\infty, \infty)$, [all real numbers], or $\{-\infty < x < \infty\}$; $R = (-\infty, 4)$, $\{f(x) \mid f(x) < 4\}$, or $\{-\infty < x < 4\}$

19. $y = a(1 + r)^t$ Equation for exponential growth
$y = 65(1 + 0.3)^t$ $a = 65$ and $r = 0.3$
$y = 65(1.3)^t$ Simplify.
Make a table of values. Then plot the points, and sketch the graph.

t	$y = 65(1.3)^t$
0	$y = 65(1.3)^0 = 65$
2	$y = 65(1.3)^2 \approx 110$
4	$y = 65(1.3)^4 \approx 186$
6	$y = 65(1.3)^6 \approx 314$
8	$y = 65(1.3)^8 \approx 530$
10	$y = 65(1.3)^{10} \approx 896$

connectED.mcgraw-hill.com   R59

CHAPTER 6

Exponential and Logarithmic Functions

Chapter 6   Get Ready

1. add the exponent values together to simplify as a^{12}
3. Product of Powers Rule
5. no   **7.** line A is the function and line B is its inverse

Lesson 6-1

1. graph; $D = (-\infty, \infty)$, [all real numbers], or $\{-\infty < x < \infty\}$; $R = (0, \infty)$, $\{f(x) \mid f(x) > 0\}$, $R = (0, \infty)$, $\{f(x) \mid f(x) > 0\}$

3. Make a table of values. Then plot the points, and sketch the graph.

x	$f(x) = 3^{x-2} + 4$
-2	$3^{-2-2} + 4 = 4\frac{1}{81}$
-1	$3^{-1-2} + 4 = 4\frac{1}{27}$
0	$3^{0-2} + 4 = 4\frac{1}{9}$
1	$3^{1-2} + 4 = 4\frac{1}{3}$
2	$3^{2-2} + 4 = 5$
3	$3^{3-2} + 4 = 7$
4	$3^{4-2} + 4 = 13$

The domain is all real numbers and the range is all real numbers greater than 4.
$D = (-\infty, \infty)$, [all real numbers], or $\{-\infty < x < \infty\}$;
$R = (4, \infty)$, $\{f(x) \mid f(x) > 4\}$, or $\{4 < x < \infty\}$

5. graph; $D = (-\infty, \infty)$, [all real numbers], or $\{-\infty < x < \infty\}$; $R = (-6, \infty)$, $\{f(x) \mid f(x) > -6\}$, or $\{-6 < x < \infty\}$

7. graph; Number of Computers Infected vs Minutes

Selected Answers and Solutions

23. $55,085.44 **25.** $\left\{b \mid b > \frac{1}{5}\right\}$ **27.** $\{d \mid d \geq -1\}$

29. $\left\{w \mid w < \frac{2}{5}\right\}$

31. **a.**
$$\frac{a}{w^{1.31}} = \frac{170}{45^{1.31}} \quad \text{Write a proportion.}$$
$$a \cdot 45^{1.31} = w^{1.31} \cdot 170 \quad \text{Cross Products Property}$$
$$a = \frac{w^{1.31} \cdot 170}{45^{1.31}} \quad \text{Divide each side by } 45^{1.31}.$$
$$a = 1.16 w^{1.31} \quad \text{Use a calculator.}$$

b. $a = 1.16 w^{1.31}$
$= 1.16(430)^{1.31}$ $w = 430$
$\approx 3268 \text{ yd}^2$ Use a calculator.

33. $\frac{1}{7}$ **35.** $-\frac{3}{13}$ **37.** 1   **39a.** $d = 1.30 h^2$ **39b.** about
1001 cm   **41a.** 2, 4, 8, 16

41b.

Cuts	Pieces
1	2
2	4
3	8
4	16

41c. $y = 2^x$ **41d.** $y = 0.003(2)^x$
41e. about 3,221,225.47 in.

43. Sample answer: Beth; Liz added the exponents instead of multiplying them when taking the power of a power.   **45.** Reducing the term will be more beneficial. The multiplier is 1.3756 for the 4-year and 1.3828 for the 6.5%.   **47.** Sample answer: $4^x < 4^2$
49. Sample answer: Divide the final amount by the initial amount. If n is the number of time intervals that pass, take the nth root of the answer.   **51.** C
53. E   **55a.** $-\frac{3}{7}$ **55b.** -1 **57a.** $x = 2$ **57b.** $x > 3$
57c. $x \geq 3$

Lesson 6-3

1. 1024   **3.** $a_n = 18 \cdot \left(\frac{1}{3}\right)^{n-1}$

5. $a_n = a_1 r^{n-1}$ nth term of a geometric sequence
$4 = a_1(3^{2-1})$ $a_n = 4, r = 3$ and $n = 2$
$4 = a_1(3)$ Evaluate the power.
$\frac{4}{3} = a_1$ Divide each side by 3.
$a_n = a_1 r^{n-1}$ nth term of a geometric sequence
$a_n = \frac{4}{3}(3)^{n-1}$ $a_1 = \frac{4}{3}, r = 3$

7. $a_n = 12(-8)^{n-1}$ **9.** 1, 5, 25 or -1, 5, -25
11. 4095   **13.** $\frac{13}{16}$ **15.** 512   **17.** 93 in.   **19.** 25
21. 512   **23.** $a_n = (-3)(-2)^{n-1}$ **25.** $a_n = (-1)(-1)^{n-1}$
27. $a_n = 8 \cdot \left(\frac{1}{4}\right)^{n-1}$ **29.** $a_n = 7(2)^{n-1}$
31. $a_n = \frac{1}{15,552}(6)^{n-1}$ **33.** $a_n = 648\left(\frac{1}{3}\right)^{n-1}$
35. 270, 90, 30 or -270, 90, -30
37. $\frac{7}{3}, \frac{14}{9}, \frac{28}{27}$ or $-\frac{7}{3}, \frac{14}{9}, -\frac{28}{27}$ **39.** 15 and 75
41. 99.19%   **43.** 31.9375   **45.** 9707.82   **47.** 2188
49. $-87,381$

51. $S_n = \frac{a_1 - a_1 r^n}{1 - r}$ Sum formula
$-2912 = \frac{a_1 - a_1(3^6)}{1 - 3}$ $S_n = -2912, r = 3$ and $n = 6$

$-2912 = \frac{a_1(1 - 3^6)}{1 - 3}$ Distributive Property
$-2912 = \frac{-728 a_1}{-2}$ Subtract.
$-2912 = 364 a_1$ Simplify.
$-8 = a_1$ Divide each side by 364.

53. 64   **55.** 0.25

57. $S_n = \frac{a_1 - a_1 r^n}{1 - r}$ Alternate sum formula
$= \frac{100 - 100(0.5)^5}{1 - (0.5)}$ Substitution
$= 193.75 \text{ ft}$ Use a calculator.

59. 524,288   **61.** about 471 cm   **63a.** $53.24, $64.42, $94.32   **63b.** $855.37   **63c.** Each payment made is rounded to the nearest cent, so the sum of the payments may be off by several cents.

65. $S_n = \frac{a_1 - a_1 r}{1 - r}$ Alternate sum formula
$= a_1 \cdot \frac{a_1 r \cdot r^{n-1}}{r^{n-1}}$ Formula for nth term
$= \frac{a_n - \frac{a_n r}{r^{n-1}}}{1 - r}$ Divide both sides by r^{n-1}.
$S_n = \frac{\frac{a_n}{r^{n-1}} - \frac{a_n r \cdot r^{n-1}}{r^{n-1}}}{1 - r}$ Substitution.
$= \frac{\frac{a_n}{r^{n-1}} - a_n r}{1 - r}$ Multiply by $\frac{1}{r^{n-1}}$.
$= \frac{\frac{a_n(1 - r^n)}{r^{n-1}}}{1 - r}$ Simplify.
$= \frac{a_n(1 - r^n)}{r^{n-1}(1 - r)}$ Divide by $(1 - r)$.
$= \frac{a_n(1 - r^n)}{r^{n-1}}$ Simplify.

67. Sample answer: $n - 1$ needs to change to n, and the 10 needs to change to a 9. When this happens, the terms for both series will be identical (a_1 in the first series will equal a_1 in the second series, and so on), and the series will be equal to each other.   **69.** 234
71. Sample answer: $4 + 8 + 16 + 32 + 64 + 128$
73. 78,732   **75.** 0.9375   **77.** C   **79a.** 1000   **79b.** 16,000

Lesson 6-4

1. $8^3 = 512$ **3.** $\log_{11} 1331 = 3$ **5.** 2   **7.** 0

9.

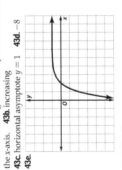

11.

13. $2^4 = 16$ **15.** $9^{-2} = \frac{1}{81}$ **17.** $12^2 = 144$
19. $\log_9 \frac{1}{9} = -1$ **21.** $\log_2 256 = 8$ **23.** $\log_{27} 9 = \frac{2}{3}$
25. -2 **27.** 3   **29.** $\frac{2}{3}$ **31.** $\frac{1}{2}$

33. $\log_5 3125 = y$ Let the logarithm equal y.
$3125 = 5^y$ Definition of logarithm

21.
$D = (-\infty, \infty)$, {all real numbers}, or $\{-\infty < x < \infty\}$;
$R = (-6, \infty)$, $\{f(x) \mid f(x) > -6\}$, or $\{-6 < x < \infty\}$

23.
$D = (-\infty, \infty)$, {all real numbers}, or $\{-\infty < x < \infty\}$;
$R = (-2, \infty)$, $\{f(x) \mid f(x) > -2\}$, or $\{-2 < x < \infty\}$

25.
$D = (-\infty, \infty)$, {all real numbers}, or $\{-\infty < x < \infty\}$; $R = (-\infty, 2)$, $\{f(x) \mid f(x) < 2\}$, or $\{-\infty < x < 2\}$

27a. decay; 0.9
27b. The $P(x)$-intercept represents the number of inline skaters in 2004. The asymptote is the x-axis. The number of inline skaters can approach 0, but will never equal 0. This makes sense as there will probably always be some who continue to skate.   **29a.** $f(x) = 18(1.25)^{x-1}$
29b. growth; 1.25   **29c.** 134
31. $g(x) = 4(2)^{x-3}$ or $g(x) = \frac{1}{2}(2^x)$

33. a.

b. Sample answer: The graph of $f(x)$ appears to be the graph of b^x reflected across the x-axis. As the values of x increase, the output values decrease.
c. Sample answer: The graphs of $g(x)$ and $h(x)$ appear to be translated to the left.
d. Sample answer: $f(x)$ and $g(x)$ are growth and $h(x)$ is decay; the absolute value of the output is increasing for the growth functions and decreasing for the decay function.
35. Vince; the graphs of the function would be the same.   **37.** Sample answer: 10   **39.** A   **41.** B
43a. translated right 2 units, up 1 unit, and reflected in the x-axis.   **43b.** increasing
43c. horizontal asymptote $y = 1$ **43d.** -8
43e.

45. $1636.65

Lesson 6-2

1. 12   **3.** -10 **5a.** $c = 2^{\frac{1}{15}}$ **5b.** 16 cells   **7.** $x \geq 4.5$ **9.** 0

11. $81^{a+2} = 3^{3a+1}$ Original equation
$(3^4)^{a+2} = 3^{3a+1}$ Rewrite 81 as 3^4.
$3^{4(a+2)} = 3^{3a+1}$ Power of a Power
$3^{4a+8} = 3^{3a+1}$ Distributive Property
$4a + 8 = 3a + 1$ Property of Equality for Exponential Functions
$a + 8 = 1$ Subtract 3a from each side.
$a = -7$ Subtract 8 from each side.

13. $\frac{5}{3}$ **15a.** $y = 10,000(1.045)^x$ **15b.** about $26,336.52
17. $y = 256(0.75)^x$ **19.** $y = 144(3.5)^x$ **21.** $16,755.63

Selected Answers and Solutions

$5^5 = 5^{-1}y$ $\frac{1}{5} = 5^{-1}$ Property of Equality for Exponential Functions

$5 = -1y$ Divide each side by -1.

$-5 = y$

So, $\log_{\frac{1}{5}} 3125 = -5$.

35. 4

37.

39.

41.

43.

45.

47.

49a. 2 **49b.**

49c. $\frac{1}{8}$; less light

51. This represents a transformation of the graph of $f(x) = \log_2 x$. $|a| = 4$: The graph expands vertically. $h = 4$: The graph is translated 4 units to the right. $k = 6$: The graph is translated 6 units up.

53.

55.

57a. $S(3) \approx 30$, $S(15) \approx 50$, $S(63) \approx 70$
57b. If $3000 is spent on advertising, $30,000 is returned in sales. If $15,000 is spent on advertising, $50,000 is returned in sales. If $63,000 is spent on advertising, $70,000 is returned in sales.

57c.

Sales versus Money Spent on Advertising

57d. Sample answer: Because eventually the graph plateaus, and no matter how much money you spend you are still returning about the same in sales.

59. a. $\log_{\left(1 + \frac{0.24}{12}\right)} \frac{A}{2000} = 12t$ Original formula

$\log_{1.02} \frac{A}{2000} = 12t$ Simplify.

$\frac{A}{2000} = 1.02^{12t}$ Definition of logarithm

$A = 2000 \cdot 1.02^{12t}$ Multiply each side by 2000.

Make a table of values. Then plot the points, and sketch the graph.

t	$A = 2000 \cdot 1.02^{12t}$
0	$A = 2000 \cdot 1.02^{12(0)} = 2000$
2	$A = 2000 \cdot 1.02^{12(2)} \approx 3217$
4	$A = 2000 \cdot 1.02^{12(4)} \approx 5174$
6	$A = 2000 \cdot 1.02^{12(6)} \approx 8322$
8	$A = 2000 \cdot 1.02^{12(8)} \approx 13,386$
10	$A = 2000 \cdot 1.02^{12(10)} \approx 21,530$

b. From the graph, $A = 4000$ at about $t = 3$. So, it will take approximately 3 years for the debt to double.

c. From the graph, $A = 6000$ at about $t = 4.5$. So, it will take approximately 4.5 years for the debt to triple.

61. Never; if zero were in the domain, the equation would be $y = \log_b 0$. Then $b^y = 0$. However, for any real number b, there is no real power that would let $b^y = 0$.
63. $\log_7 51$; Sample answer: $\log_7 51$ equals a little more than 2. $\log_6 61$ equals a little less than 2. $\log_7 71$ equals a little less than 2. Therefore, $\log_7 51$ is the greatest.
65. No; Elisa was closer. She should have $-y = 2$ or $y = -2$ instead of $y = 2$. Matthew used the definition of logarithms incorrectly. **67.** C **69.** B **71.** B, C, F
73. $a = 2$, $b = 3$

Lesson 6-5

1. nonlinear, positive, weak
3. linear, negative, strong
5. The x variable is sometimes referred to as the predictor variable because when you solve for x you are finding the solution to a prediction of unknown events.
7. linear; $y = \frac{1}{2}x + 3$
9. exponential
11. $x = 0$
13. $x = 2$
15. $x = 4$
17. 8 hours
19. nonlinear, no correlation
21. linear; $y = 1.5x + 4$
23. $x = 4$
25. $x = 2.68$
27. $x = 0$
29. power **31.** linear
33. linear; $y = 0.062756x + 0.32753$
35. weak **37.** perfect
39. Sample answer: If the right side of the graph is higher than the left side, the direction is positive. If the right side of the graph is lower than the left side, the direction is negative.

41. a. To find the values for the table, start with with $x = 0$ and a population of 502,000. Multiply the population by 1.032 (an increase of 3.2%) to find the population for each successive value of x.

x	Population
0	502,000
1	518,064
2	534,642
3	551,751
4	569,407

41b.

41c. Exponential growth is modeled by the equation $y = a(1 + r)^t$, where a is the initial amount and r is the growth rate. In this case, $a = 502,000$ and $r = 0.032$, so $y = 502,000(1 + 0.032)^t$. **41d.** exponential
41e. 2018 is 24 years after 1994, so let $t = 24$ in the equation $y = 502,000(1 + 0.032)^t$. $y \approx 1,069,095$
41f. Use technology to make a table of values and look for the first year in which the population is at least 1.5 million; find the corresponding value of x; 35 years.
43. Sample answer: It helps you eliminate possibilities and check your work. The linear regression model is very different then the quadratic regression model and by studying a scatterplot, you can sometimes see those differences.

45. a.

b. Sample answer: $0 \leq x \leq 10$ by 1 and $10 \leq y \leq 50$ by 5
c. Enter the data in a graphing calculator and try different regression models to find trhe one with the greatest value of r. A power regression is the best regression model.
d. Yes; the correlation coefficient is close to 1.
e. The power regression equation is $y = ax^b$, where $a \approx 11.0163$ and $b \approx 0.6595$. Solve $30.57 = 11.0163x^{0.6595}$. $x \approx 4.7$ cm.

47. Sample answer: We need to have the line of best fit to solve for x. If we choose the wrong model, the regression model will not have the highest correlation rate and the prediction will be innacurate.

49a. Sketching the graph could be more accurate if the data has a few outliers that are skewing the data.

49b–d. The sketched line of best fit would be linear, but with the graphing calculator the regression model would be a curve. So the graphing calculator would be more accurate in these cases.

51. B **53.** A **55a.** 0, 1, 2, 3, 4, 5, 6, 7, 8, 9
55b. 512, 256, 128, 64, 32, 16, 8, 4, 2, 1
55c. nonlinear, negative direction, strong correlation
55d. Exponential, because it has the strongest correlation. **55e.** 16 **55f.** 2 **55g.** 9

Lesson 6-6

1. 2.085 **3.** 0.3685 **5.** Guadalupe Peak: 67,287.7 Pa; Emory Peak: 70,166.37 Pa; Anthony's Nose: 73,081.33 Pa; Panther Peak: 75,006.13 Pa; Buck Mountain: 76,662.09 Pa **7.** 2.4182 **9.** 2 **11.** 13.4403 **13.** 2.1610

 15. $\log_4 \frac{4}{3}$ **17.** 1.5 **19.** 2.1606 **21.** 3.4818 **23.** 8 **25.** 2

 27a.
$$P = \log_{10}\left(1 + \frac{1}{d}\right)$$ Original equation

$10^P = 1 + \frac{1}{d}$ Definition of logarithm

$10^P - 1 = \frac{1}{d}$ Subtract 1 from each side.

$d(10^P - 1) = 1$ Multiply each side by d.

$d = \frac{1}{10^P - 1}$ Divide each side by $10^P - 1$.

b. $d = \frac{1}{10^P - 1}$

$= \frac{1}{10^{0.097} - 1}$ $P = 0.097$

≈ 4 Use a calculator.

c. $P = \log_{10}\left(1 + \frac{1}{d}\right)$ Original equation

$= \log_{10}\left(1 + \frac{1}{1}\right)$ $d = 1$

$= \log_{10} 2$ Simplify.

≈ 0.30103 Replace $\log_{10} 2$ with 0.30103

The probability is about 30.1%.

29. 2.1133 **31.** 0.1788 **33.** 1.7228 **35.** 2.0478 **37.** 3
39. 5 **41.** $85\frac{1}{3}$ **43.** $\left(\frac{x-2}{256}\right)^{\frac{1}{6}}$ **45.** $\sqrt{6}$, $-\sqrt{6}$ **47.** 5
49. 12 **51.** false **53.** false **55.** true **57.** false
59a. Sample answer: about 6 yr **59b.** Sample answer: She will earn $1050 × 12 or $12,600 from Social Security. She needs to earn $50,000 − $12,600 or $37,400 from interest. The principal in her CD account will need to be $37,400 ÷ 0.05 or $748,000. Set up the exponential growth equation and solve for t.

$A = P(1 + r)^t$ Exponential Growth
$748,000 = 320,000(1.15)^t$ $A = 748,000$. $P = 320,000$. and $r = .015$
$2.3375 = 1.15^t$ Divide each side by 320000.
$2.3375 = 1.15^t$ Simplify.
$\log 2.3375 = \log 1.15^t$ Property of Equality of Logarithms
$\log 2.3375 = t \log 1.15$ Power Property of Logarithms
$\frac{\log 2.3375}{\log 1.15} = t$ Divide each side by log 1.15.
$6.07 \approx t$ Simplify.

59c. Sample answers: The annual increase in her account remains at 15%. She needs to pay taxes on Social Security. Each CD will earn 5% interest.

61a. Sample answer: $\log_{b_0} \frac{xz}{5} = \log_{b_0} x + \log_{b_0} z - \log_{b_0} 5$

61b. Sample answer: $\log_{b_0} m^4 p^6 = 4 \log_{b_0} m + 6 \log_{b_0} p$

61c. Sample answer: $\log_{b_0} \frac{j^4 k}{h^5} = 8 \log_{b_0} j + \log_{b_0} k - 5 \log_{b_0} h$

63a. $\log_{b_0} 1 = 0$, because $b^0 = 1$. **63b.** $\log_{b_0} b = 1$, because $b^1 = b$. **63c.** $\log_{b_0} b^x = x$, because $b^x = b^x$.

65. $\log_{b_0} 24 \neq \log_{b_0} 20 + \log_{b_0} 4$; all other choices are equal to $\log_{b_0} 24$.

67. $x^3 \log_{b_0} 2 - \log_{b_0} 5 = x^{\log_{b_0} 2^5 - \log_{b_0} 5}$
$= x^{\log_{b_0} \frac{8}{5}}$
$= x^{\log_{b_0} \frac{8}{5}}$
$= \frac{8}{5}$

69. C **71.** C **73.** 4.902 **75a.** all numbers greater than or equal to 0 **75b.** all real numbers greater than or equal to 0 **75c.** 4.3 **75d.** 5.3 **75e.** 6.3 **75f.** Sample answer: The magnitude of each earthquake increases by 1 unit when the intensity increases by a factor of 10.

Lesson 6-7

1. 0.6990 **3.** −0.3979 **5.** 3.55 × 10²⁴ ergs **7.** 0.8442

9. $11^{b-3} = 5^b$ Original equation
$\log 11^{b-3} = \log 5^b$ Property of Equality for Logarithmic Functions
$(b-3) \log 11 = b \log 5$ Power Property of Logarithms
$b \log 11 - 3 \log 11 = b \log 5$ Distributive Property
$-3 \log 11 = b \log 5 - b \log 11$ Subtract $b \log 11$ from each side.
$-3 \log 11 = b(\log 5 - \log 11)$ Distributive Property
$\frac{-3 \log 11}{\log 5 - \log 11} = b$ Divide each side by log 5 − log 11.
$9.1237 \approx b$ Use a calculator.

11. $\{p \mid p \le 4.4190\}$ **13.** $\frac{\log 23}{\log 4} \approx 2.2618$
15. $\frac{\log 5}{\log 2} \approx 2.3219$ **17.** 1.0414 **19.** 0.9138 **21.** −1.3979
23. 1.7740 **25.** 5.9647 **27.** ±1.1691 **29.** $\{n \mid n > 0.6667\}$
31. $\{y \mid y \ge -3.8188\}$ **33.** $\frac{\log 18}{\log 7} \approx 1.4854$
35. $\frac{\log 16}{\log 2} = 4$ **37.** $\frac{\log 11}{\log 3} \approx 2.1827$

---

the solution of 1.85. They all should produce the same result because you are starting with the same equation. If they do not, then an error was made.

73. $\log_{\sqrt{a}} 3 = \log_a x$ Original equation
$\frac{\log_a 3}{\log_a \sqrt{a}} = \log_a x$ Change of Base Formula
$\frac{\log_a 3}{\frac{1}{2}} = \log_a x$ Multiply numerator and denominator by 2.
$2 \log_a 3 = \log_a x$ Power Property of Logarithms
$\log_a 3^2 = \log_a x$ Property of Equality for Logarithmic Functions
$3^2 = x$
$9 = x$

75. $\log_9 27 = 3$ and $\log_{27} 3 = \frac{1}{3}$; Conjecture:

$\log_b \frac{1}{p} = \frac{1}{\log_p a}$

Proof: $\log_a b \ge \frac{1}{\log_b a}$
$\frac{\log_{b_0} b}{\log_{b_0} a} \ge \frac{1}{\frac{\log_{b_0} a}{\log_{b_0} a}}$ Change of Base Formula
$\frac{\log_{b_0} b}{\log_{b_0} a} \ge \frac{\log_{b_0} a}{\log_{b_0} a}$ Inverse Property of Exponents and Logarithms

77. B **79.** B **81.** B **83a.** 4491 **83b.** 8360
83c. 2079. Yes; Sample answer: The year seems reasonable based on the model of the equation.
83d. The domain is all real numbers greater than −5. These numbers allow for the value inside the parentheses to be positive.
83e. The range is all real numbers greater than or equal to 0. These numbers allow for a positive or 0 population.
83f. 1934 **85a.** $\log_3 30$ **85b.** $x < \log_2 30$
85c. $x > 2 \log_4 7$ **85d.** $x \ge \log_{12} 10$ **85e.** $x \le \log_{12} 0.05$
85f. $x \le \log_{12} 0.05$ **85g.** $\log_{2.5}\left(\frac{4}{3}\right)$ **85h.** $x \ge \log_{2.5}\left(\frac{4}{3}\right)$

Lesson 6-8

1. ln 30 = x **3.** ln x = 3 **5.** 7 ln 2 **7.** ln 17,496
9. 0.1352 **11.** 993.6527 **13.** $\{x \mid -25.0855 < x < 15.0855, x \neq -5\}$ **15.** $\{x \mid x > 3.3673\}$ **17a.** about 243,092
17b. about 58 min **17c.** about 7 **19.** ln 0.1 = −5x
21. 5.4 = e^x **23.** e^{36} = x + 4 **25.** e^7 = e^x **27.** 7 ln 10
29. $7 \ln \frac{1}{2} + 5 \ln 2 = 7 \ln 2^{-1} + 5 \ln 2$ Rewrite $\frac{1}{2}$ as 2^{-1}.
$= -\ln 2^{-7} + \ln 2^5$ Power Property of Logarithms
$= \ln (2^{-7})(2^5)$ Product Property of Logarithms
$= \ln 2^{-2}$ Simplify.
$= -2 \ln 2$ Power Property of Logarithms

31. ln 81^{-6} **33.** 3.7955 **35.** 0.6931 **37.** −0.5596
39. $\{x \mid x \le 2.1633\}$ **41.** $\{x \mid x > 8.0105\}$
43. $\{x \mid x < -239.8802 \text{ or } x > 239.8802\}$

45. The graph of g(x) is a dilation of the graph of f(x), which is compressed vertically by a factor of 0.5.

$[-2, 8]$ scl: 1 by $[-5, 5]$ scl: 1

39a. $n = 35[\log_4 (t + 2)]$ Original equation
$= 35[\log_4 (10 + 2)]$ $t = 10$
$= 35[\log_4 (12)]$ Simplify.
$= 35 \cdot \frac{\log_{10} 12}{\log_{10} 4}$ Change of Base Formula
≈ 62.737 Use a calculator.

In 2010, there are about 62.737 thousand, or 62,737 pet owners.

b. $n = 35[\log_4 (t + 2)]$ Original equation
$80 = 35[\log_4 (t + 2)]$ $n = 80$
$2.2857 = \log_4 (t + 2)$ Divide each side by 35.
$4^{2.2857} = t + 2$ Definition of logarithm
$22 = t$ Simplify.

In 22 years after 2000, or in 2022, there will be 80,000 pet owners. An increase of about 20,000 pet owners since 2010 seems reasonable.

41. 3.3578 **43.** −0.0710 **45.** 4.7393 **47.** $\{x \mid x \ge 2.3223\}$
49. $\{x \mid x \le 0.9732\}$ **51.** $\{p \mid p \ge 2.9437\}$
53. $\frac{\log 12}{\log 4} \approx 1.7925$ **55.** $\frac{\log 7.29}{\log 5} \approx 1.2343$
57. 113.03 cents

59. The graph of g(x) is a dilation of the graph of f(x), which is stretched vertically. This graph is also reflected across the x-axis.

$[-2, 8]$ scl: 1 by $[-5, 5]$ scl: 1

61. The graph of g(x) is a translation of the graph of f(x), which is translated down 5 units.

$[-2, 8]$ scl: 2 by $[-5, 5]$ scl: 2

63. The graph of g(x) is a translation of the graph of f(x), which is translated right 6 units.

$[-2, 18]$ scl: 2 by $[-5, 5]$ scl: 1

65 $4^{x^2-3} = 16$ Original equation
$4^{x^2-3} = 4^2$ Rewrite 16 as 4².
$x^2 - 3 = 2$ Property of Equality for Exponential Functions
$x^2 = 5$ Add 3 to each side.
$x = \pm\sqrt{5}$ Take the square root of each side.
$x \approx \pm 2.2361$ Use a calculator.

67. 3.5 **69.** −3.8188 **71a.** The solution is between 1.8 and 1.9. **71b.** (1.85, 13) **71c.** Yes; all methods produce

c.

$$35e^{0.0978t} > 80e^{0.071t}$$ Original equation

$$\ln 35e^{0.0978t} > \ln 80e^{0.071t}$$ Property of Inequality for Logarithms

$$\ln 35 + \ln e^{0.0978t} > \ln 80 + \ln e^{0.071t}$$ Product Property of Logarithms

$$\ln 35 + 0.0978t > \ln 80 + 0.071t$$ $\ln e^x = x$

$$0.0268t > \ln 80 - \ln 35$$ Subtract (0.071t + ln 35) from each side.

$$t > \frac{\ln 80 - \ln 35}{0.0268}$$ Divide each side by 0.0268.

$$t > 30.85$$ Use a calculator.

The number of cells of this bacteria exceed the number of cells in the other bacteria in about 30.85 minutes.

7

$$y = ae^{-0.00012t}$$ Equation for the decay of Carbon-14

$$0.85a = ae^{-0.00012t}$$ y = 0.85a

$$0.85 = e^{-0.00012t}$$ Divide each side by a.

$$\ln 0.85 = \ln e^{-0.00012t}$$ Property of Equality for Logarithmic Functions

$$\ln 0.85 = -0.00012t$$ $\ln e^x = x$

$$\frac{\ln 0.85}{-0.00012} = t$$ Divide each side by −0.00012.

$$1354 \approx t$$ Use a calculator.

The bone is about 1354 years old.

9. about 14.85 billion yr   **11.** about 20.1 yr

13a.

Population (millions) vs Time (yr)

13b. The graphs intersect at $t = 20.79$. Sample answer: This intersection indicates the point at which both functions determine the same population at the same time.

13c. Sample answer: The logistic function $g(t)$ is a more accurate estimate of the country's population because $f(t)$ will continue to grow exponentially and $g(t)$ considers limitations on population growth such as food supply. **15.** $t = 113.45$ **17.** Sample answer: The spread of the flu throughout a small town. The growth of this is limited to the population of the town itself. **19.** 2.30 **21.** A **23.** C **25.** C **27a.** 100°C

27b. about −0.1099

27c. 24.6 min

---

b. $\beta = 10 \log_{10}\left(\dfrac{I}{10^{-12}}\right)$ Original equation

$$= 10 \log_{10}\left(\frac{10^{-2}}{10^{-12}}\right)$$ $I = 10^{-2}$

$$= 10 \log_{10} 10^{10}$$ Quotient of Powers Property

$$= 10(10) \text{ or } 100$$ Definition of logarithm

c. Sample answer: The power of the logarithm only changes by 2. The power is the answer to the logarithm. That 2 is multiplied by the 10 before the logarithm. So we expect the decibels to change by 20.

39. $6\frac{17}{20}$ **41.** The logarithmic function of the form $y = \log_b x$ is the inverse of the exponential function of the form $y = b^x$. The domain of one of the two inverse functions is the range of the other. The range of one of the two inverse functions is the domain of the other. **43a.** less than **43b.** less than **43c.** no **43d.** infinitely many **45a.** 19,952,623.2 **45b.** 665,087.4 **45c.** 5.8 **47.** D **49.** B **51.** C **53.** B

Lesson 6-10

1a. 5.545×10^{-10} **1b.** 1,578,843,530 yr
1c. about 30.48 mg   **1d.** 3,750, 120,003 yr

3a.

Population vs Time (yr)

3b. $P(t) = 16,500$ **3c.** 16,500   **3d.** about 102 years

5 a.

$$y = 80e^{kt}$$ Original formula

$$675 = 80e^{k(30)}$$ $y = 675, t = 30$

$$8.4375 = e^{30k}$$ Divide each side by 80.

$$\ln 8.4375 = \ln e^{30k}$$ Property of Equality for Logarithmic Functions

$$\ln 8.4375 = 30k$$ $\ln e^x = x$

$$\frac{\ln 8.4375}{30} = k$$ Divide each side by 30.

$$0.071 \approx k$$ Use a calculator.

b. $y = 80e^{kt}$ Original formula

$$6000 = 80e^{0.071t}$$ $y = 6000, k = 0.071$

$$75 = e^{0.071t}$$ Divide each side by 80.

$$\ln 75 = \ln e^{0.071t}$$ Property of Equality for Logarithmic Functions

$$\ln 75 = 0.071t$$ $\ln e^x = x$

$$\frac{\ln 75}{0.071} = t$$ Divide each side by 0.071.

$$60.8 \approx t$$ Use a calculator.

The bacteria will reach a population of 6000 cells in about 60.8 minutes.

Selected Answers and Solutions

47. The graph of $g(x)$ is a translation of the graph of $f(x)$, which is translated up 8 units.

[−2, 8] scl 1 by [−10, 10] scl 2

49. The graph of $g(x)$ is a translation of the graph of $f(x)$, which is translated left 5 units.

[−10, 10] scl 2 by [−10, 10] scl 2

51 a. $A = Pe^{rt}$ Exponential Formula

$$800e^{(0.045)(5)} = 800 \quad r = 0.045, t = 5$$

$$P = 800e^{0.225}$$ Simplify.

$$\approx 1001.86$$ Use a calculator.

About \$1001.86 will be in the account.

b. $\ln \dfrac{A}{P} = rt$ Logarithmic Formula

$$\ln 2 = 0.045t \quad A = 2P, r = 0.045$$

$$\frac{\ln 2}{0.045} = t$$ Divide each side by 0.045.

$$15.4 \approx t$$ Use a calculator.

It would take about 15.4 years to double your money.

c. $\ln \dfrac{A}{P} = rt$ Logarithmic Formula

$$\ln 2 = 9r \quad A = 2P, t = 9$$

$$\frac{\ln 2}{9} = r$$ Divide each side by 9.

$$0.077 \approx r$$ Use a calculator.

You would need a rate of about 7.7%.

d. $A = Pe^{rt}$ Exponential Formula

$$10,000 = Pe^{(0.0475)(12)} \quad A = 10,000, r = 0.0475, t = 12$$

$$10,000 = Pe^{0.57}$$ Simplify.

$$\frac{10,000}{e^{0.57}} = P$$ Divide each side by $e^{0.57}$.

$$5655.25 \approx P$$ Use a calculator.

You would need to deposit about \$5655.25.

53. $4 \ln 2 - 3 \ln 5$ **55.** $\ln x + 4 \ln y - 3 \ln z$
57. −0.8340   **59.** 1.1301

61a.

61b. y-axis; $a(x) = -e^x$

61c.

61d. Sample answer: No; these functions are reflections over $y = -x$, which indicates that they are not inverses.

63. Let $p = \ln a$ and $q = \ln b$. That means that $e^p = a$ and $e^q = b$.

$$ab = e^p \times e^q$$

$$ab = e^{p+q}$$

$$\ln(ab) = (p + q)$$

$$\ln(ab) = \ln a + \ln b$$

65 Sample answer: $e^{\ln 3}$ **67a.** $P(6) = 8$ billion
67b. $P(-16) = 6.01$ billion   **67c.** 53.319 years
67d. 1968   **69.** $x > e^{\frac{7}{9}}$ **71.** B   **73.** 0.403   **75.** B

Lesson 6-9

1. 16   **3.** C   **5.** $\left\{x \mid 0 < x \le \frac{1}{64}\right\}$ **7.** $\left\{x \mid \frac{4}{3} < x < 2\right\}$

9. 3125   **11.** −2   **13.** 9

15 $\log_{12}(x^2 - 7) = \log_{12}(x + 5)$ Original equation

$$x^2 - 7 = x + 5$$ Property of Equality for Exponential Functions

$$x^2 - x - 7 = 5$$ Subtract x from each side.

$$x^2 - x - 12 = 0$$ Subtract 5 from each side.

$$(x - 4)(x + 3) = 0$$ Factor.

$$x - 4 = 0 \text{ or } x + 3 = 0$$ Zero Product Property

$$x = 4 \qquad x = -3$$ Solve each equation.

17. 5   **19.** −3   **21.** 318 mph; The wind speed of 318 mph seems high because most tornadoes have wind speeds of about 110 mph.   **23.** $\{x \mid x \ge 256\}$

25 $\log_2 x \le -2$ Original inequality

$$0 < x \le 2^{-2}$$ Property of Inequality for Exponential Functions

$$0 < x \le \frac{1}{4}$$ Simplify.

The solution is $\left\{x \mid 0 < x \le \frac{1}{4}\right\}$ or $\left(0, \frac{1}{4}\right]$.

27. $\left\{x \mid 0 < x < \frac{1}{7}\right\}$ **29.** $\left\{x \mid \frac{1}{2} < x \le 1\right\}$

31. $\left\{x \mid -\frac{5}{12} < x \le 1\right\}$ **33.** $\{x \mid x \ge 8\}$

35a. 37   **35b.** 61

37 a. $\beta = 10 \log_{10}\left(\dfrac{I}{10^{-12}}\right)$ Original equation

$$= 10 \log_{10}\left(\frac{1}{10^{-12}}\right) \quad I = 10$$

$$= 10 \log_{10} 10^{12}$$ Write $\frac{1}{10^{-12}}$ as 10^{12}.

$$= 10(12) \text{ or } 120$$ Definition of logarithm

Chapter 6 Study Guide and Review

1. logarithm **3.** change of base formula
5. logarithmic function **7.** Common logarithms have
base 10, while natural logarithms have base e.

9.

$D = (-\infty, \infty)$ or {all real numbers};
$R = (0, \infty)$ or $\{f(x) \mid f(x) > 0\}$

11.

$D = (-\infty, \infty)$ or {all real numbers};
$R = (-6, \infty)$ or $\{f(x) \mid f(x) > -6\}$

13.

$D = (-\infty, \infty)$ or {all real
numbers}; $R = (-1, \infty)$ or $\{f(x) \mid f(x) > -1\}$

15a. $f(x) = 120,000(0.97)^x$
15b. about 88,491
17. −7 **19.** $\frac{9}{41}$
21. −2 **23.** $x \leq -\frac{2}{5}$
25. $x \geq -1$ **27.** 320 **29.** 8 **31.** 18, 54 **33.** 12, −36
35. 225 **37.** 215 **39.** 129 **41.** $2^{-4} = \frac{1}{16}$ **43.** 4

45.

47. logarithmic; $y = -10.9503 + 26.2680\ln x$
49. exponential; $y = 0.8383(1.6997)^x$
51. 2.5841 **53.** −1.2921 **55.** 30 **57.** $2\sqrt{3}$ **59.** $\sqrt{\frac{5}{2}}$
61. 1.16 times **63.** $x \approx \pm 1.3637$ **65.** $r \approx 4.6102$
67. $\{x \mid x \leq -6.3013\}$ **69.** $\{w \mid w \geq 3.3750\}$
71. 1.9769 **73.** 2.8424 **75.** 2.4935 **77.** 1.9459
79. 201.7144 **81.** $\{x \mid -3 < x < -0.2817\}$ **87.** −6
83. $\{x \mid x < 1.9459\}$ **85.** about 17.2 years
89. $\left\{ x \mid 0 < x \leq \frac{1}{125} \right\}$
91. −6 **93.** 1000 **95.** ≈ 5.8 days

CHAPTER 7
Rational Functions

Chapter 7 Get Ready

1. Sample answer: multiply each side by 8 **3.** The GCF
of 72 and 77 is 1. **5.** $a = -4$, $b = 16$, $c = -1$
7. Associative Property **9.** Write the equation
$5(11) = 8p$ and then solve for p.

Lesson 7-1

1. $\frac{x+3}{x+8}$; $x \neq -8$ **3.** D
5. $\frac{a^2x - b^2x}{by - ay} = \frac{x(a^2 - b^2)}{by - ay}$ Factor.
$= \frac{x(b-a)(a+b)}{-y(b-a)}$ Factor.
$= \frac{-x(b-a)(a+b)}{y(b-a)}$ $a - b = -1(b-a)$
$= \frac{-x(a+b)}{y}$ Eliminate common factors.
$= \frac{-x(a+b)}{y}$

7. $\frac{2by}{3ax}$ **9.** $\frac{(a-1)}{12(a-1)}$ **11.** 4 **13.** $\frac{x(x+6)}{x+4}$, $x \neq -4$ Simplify.
15. $\frac{4}{x+3}$, $x \neq -5$ **17.** $\frac{x(x+2)}{6(x+5)^2}$, $x \neq -5$ **19.** J **21.** $\frac{x^2}{x+6}$
23. $\frac{c+4}{c+5}$ **25.** $\frac{1}{4ab^2c^2}$ **27.** $\frac{32b}{3ac^2}$ **29.** $\frac{5a^4c}{3b}$
31. $\frac{y^2 + 8y + 15}{y^2 - 9y + 18} \div \frac{y^2 - 9}{y^2 - 9}$
$= \frac{(y+3)(y+5)}{(y-3)(y-6)}$
$= \frac{(y+3)(y+5)}{(y-3)(y-6)} \cdot \frac{(y-3)(y-6)}{(y-3)(y-6)}$
$= \frac{(y+5)}{(y-6)}$
$= y + 5$
33. $\frac{2(x-5)}{33}$ **35.** $\frac{(x-3)(x+1)}{6(x+7)}$ **37.** $\frac{-a^2(a+b)}{b^4}$ Simplify.
39a. $\frac{33+m}{121+a}$ **39b.** $\frac{33+m}{121+a}$ **41a.** T(x) **41b.** about 3.9 mm thick **43.** $1\frac{1}{4}$
45. $\frac{x+3}{(x+2)(x-1)}$ **47.** $\frac{20\pi^2 f_1 c^{-2}}{3a^2 c^2} \cdot \left(\frac{16x^3y^3}{9acz} \right)^{-1} = \frac{9acz}{16x^3y^3}$
$= \frac{20\pi^2 f_1 c^{-2}}{3a^2 c^2} \cdot \frac{9acz}{16x^3y^3}$
$= \frac{20\pi^2 f_1 c^{-2}}{3a^2 c^2} \cdot \frac{9acz}{16x^3y^3}$ $r^{-2} = \frac{1}{r^2}$
$= \frac{2 \cdot 2 \cdot 5 \cdot x \cdot y \cdot y \cdot y \cdot y \cdot 3 \cdot 3 \cdot a \cdot c \cdot z}{3 \cdot a \cdot a \cdot c \cdot c \cdot 2 \cdot 2 \cdot 2 \cdot 2 \cdot x \cdot x \cdot x \cdot y \cdot y \cdot y}$ Factor.
$= \frac{\overset{1}{2} \cdot \overset{1}{2} \cdot 5 \cdot \overset{1}{x} \cdot y \cdot y \cdot y \cdot \overset{1}{3} \cdot \overset{1}{3} \cdot \overset{1}{a} \cdot \overset{1}{c} \cdot z}{\underset{1}{3} \cdot \underset{1}{a} \cdot a \cdot \underset{1}{c} \cdot c \cdot 2 \cdot 2 \cdot 2 \cdot 2 \cdot \underset{1}{x} \cdot x \cdot \underset{1}{y} \cdot \underset{1}{y} \cdot \underset{1}{y}}$ Eliminate common factors.
$= \frac{5 \cdot y \cdot y \cdot y \cdot 3}{a \cdot a \cdot c \cdot z \cdot 2 \cdot 2 \cdot x}$ Simplify.
$= \frac{15y^3}{4a^2czx}$ Simplify.

Lesson 7-2

1. $80x^3y^3$
7. $\frac{21b-4}{36a}$ **9.** $\frac{9x+15}{(x+3)(x+6)}$ **11.** $\frac{x-11}{3(x+2)(x-2)}$
13. $\frac{14x-10}{(x+1)(x-2)}$ **15.** $\frac{3y+2}{y+3}$ **17.** $\frac{2a+5b}{3b-8a}$
19. $180x^2y^4z^2$ **21.** $6(x+4)(2x-1)(2x+3)$
23. $\frac{28by^2 - 9bx}{105x^3y^4z}$ **25.** $\frac{20x^2y + 120y + 6x^2}{15x^3y}$
27. $\frac{15b^3 + 100ab^2 - 216a}{240ab^3}$
29. $\frac{6}{y^2 - 2y - 35} + \frac{4}{y^2 + 9y + 20}$ Factor denominators.
$= \frac{6}{(y-7)(y+5)} + \frac{4}{(y+4)(y+5)}$
$= \frac{6(y+4)}{(y-7)(y+5)(y+4)} + \frac{4(y-7)}{(y+4)(y+5)(y-7)}$ Multiply by missing factors.
$= \frac{6y + 24 + 4y - 28}{(y-7)(y+5)(y+4)}$ Add the numerators.
$= \frac{10y - 4}{(y-7)(y+5)(y+4)}$ Simplify.
31. $\frac{10x-10}{(2x-1)(x+6)(x-3)}$ **33.** $\frac{2x^2 + 32x}{3(x-2)(x+3)(2x+5)}$
35. $\frac{1000x + 800y}{x(x+2y)}$

---

49. $\frac{2(4x+1)(2x+1)}{5(2x-1)(x+2)}$ **51.** $\frac{2x+1}{-9(x+2)}$
53. $\frac{x(x+8)}{2(2x-1)(3x+1)}$ **55.** $\frac{-2(x-8)(x+4)(x-2)(x+1)}{(2x+1)^2(x^2+2x-6)}$

57. **a.** 5 tracks $\cdot \frac{2 \text{ miles}}{1 \text{ track}} \cdot \frac{5280 \text{ feet}}{1 \text{ mile}} \cdot \frac{1 \text{ car}}{75 \text{ feet}}$
$= 5 \text{ tracks} \cdot \frac{2 \text{ miles}}{1 \text{ track}} \cdot \frac{15 \cdot 352 \text{ feet}}{1 \text{ mile}} \cdot \frac{1 \text{ car}}{5 \cdot 15 \text{ feet}}$
$= \frac{1 \cdot 2 \cdot 352 \cdot 1 \text{ car}}{1 \cdot 1 \cdot 1}$
$= 704 \text{ cars}$
c. 704 cars $\cdot \frac{8 \text{ attendants}}{1 \text{ car}} \cdot \frac{45 \text{ s}}{1 \text{ attendant}} \cdot \frac{1 \text{ min}}{60 \text{ s}} \cdot \frac{1 \text{ h}}{60 \text{ min}}$
$= 704 \text{ cars} \cdot \frac{8 \text{ attendants}}{1 \text{ car}} \cdot \frac{45 \text{ s}}{1 \text{ attendant}} \cdot \frac{1 \text{ min}}{60 \text{ s}} \cdot \frac{1 \text{ h}}{60 \text{ min}}$
$= \frac{704 \cdot 8 \cdot 45 \cdot 1 \cdot 1 \text{ h}}{1 \cdot 1 \cdot 60 \cdot 60}$
$= 70.4 \text{ hours}$
b. 5 tracks $\cdot \frac{2 \text{ miles}}{1 \text{ track}} \cdot \frac{5280 \text{ feet}}{1 \text{ mile}} \cdot \frac{1 \text{ car}}{75 \text{ feet}}$

59. Sample answer: The two expressions are equivalent
except that the rational expression is undefined at
$x = 3$. **61.** $x^2 + x - 6$ **63.** Sample answer: Sometimes;
with a denominator like $x^2 + 2$, in which the denominator
cannot equal 0, the rational expression can be defined for
all values of x. **65.** Sample answer: When the original
expression was simplified, a factor of x was taken out of
the denominator. If x were to equal 0, then this expression
would be undefined. So, the simplified expression is also
undefined for x. **67.** D **69.** D

29. a. $rt = d$ rate • time = distance
$rt = 60.5$ $d = 60.5$
$r = \dfrac{60.5}{t}$ Divide each side by t.

b. This represents a transformation of the graph of $f(x) = \dfrac{1}{x}$. There are asymptotes at $t = 0$ and $r = 0$. Because $a = 60.5$, the graph is stretched vertically.

$r = \dfrac{60.5}{t}$

Write the equation.

c. $r = \dfrac{60.5}{t}$
$= \dfrac{60.5}{0.48}$ $t = 0.48$
≈ 126 ft/s   Use a calculator.

31. D $= \{x \mid x \neq -2\}$; R $= \{f(x) \mid f(x) \neq -5\}$; $x = -2$, $f(x) = -5$

$f(x) = \dfrac{-4}{x+2} - 5$

33. D $= \{x \mid x \neq 4\}$; R $= \{f(x) \mid f(x) \neq 3\}$; $x = 4$, $f(x) = 3$

$f(x) = \dfrac{2}{x-4} + 3$

35. D $= \{x \mid x \neq 7\}$; R $= \{f(x) \mid f(x) \neq -8\}$; $x = 7$, $f(x) = -8$

$f(x) = \dfrac{2}{x-7} - 8$

17.
$f(x) = \dfrac{-2}{x-5}$

D $= \{x \mid x \neq 5\}$;
R $= \{f(x) \mid f(x) \neq 0\}$

19.
$f(x) = \dfrac{9}{x+3} + 6$

D $= \{x \mid x \neq -3\}$;
R $= \{f(x) \mid f(x) \neq 6\}$

21.
$f(x) = \dfrac{-6}{x+4} - 2$

D $= \{x \mid x \neq -4\}$;
R $= \{f(x) \mid f(x) \neq -2\}$

23a. $m = \dfrac{5000}{d}$

23b.
$m = \dfrac{5000}{d}$

23c. 13.7 mi

25.
$f(x) = \dfrac{3}{2x-4} - 2$

D $= \{x \mid x \neq 2\}$;
R $= \{f(x) \mid f(x) \neq 0\}$

27.
$f(x) = \dfrac{2}{4x+1}$

D $= \left\{ x \mid x \neq -\dfrac{1}{4}\right\}$;
R $= \{f(x) \mid f(x) \neq 0\}$

3.
$f(x) = \dfrac{5}{x}$

1 or D $= \{x \mid x \neq 1\}$. The range is all real numbers not equal to 0 or R $= \{f(x) \mid f(x) \neq 0\}$.

5.
$f(x) = \dfrac{-1}{x-2} + 4$

D $= \{x \mid x \neq 1\}$;
R $= \{f(x) \mid f(x) \neq 0\}$

7. $x = -4$, $f(x) = 0$; D $= \{x \mid x \neq -4\}$; R $= \{f(x) \mid f(x) \neq 0\}$ **9.** $x = -6$, $f(x) = -2$; D $= \{x \mid x \neq -6\}$; R $= \{f(x) \mid f(x) \neq -2\}$

11.
$f(x) = \dfrac{3}{x}$

D $= \{x \mid x \neq 0\}$;
R $= \{f(x) \mid f(x) \neq 0\}$

13.
$f(x) = \dfrac{2}{x-6}$

D $= \{x \mid x \neq 6\}$;
R $= \{f(x) \mid f(x) \neq 0\}$

15. This represents a transformation of the graph of $f(x) = \dfrac{1}{x}$. $a = 2$: The graph is stretched vertically. $k = 3$: The graph is translated 3 units up. There is a horizontal asymptote at $f(x) = 3$. Domain: D $= \{x \mid x \neq 0\}$. Range: R $= \{f(x) \mid f(x) \neq 3\}$.

$f(x) = \dfrac{2}{x} + 3$

---

37. $\dfrac{4}{x+5} + \dfrac{9}{x-6} = \dfrac{4(x-6)}{(x+5)(x-6)} + \dfrac{9(x+5)}{(x+5)(x-6)}$

$\dfrac{5}{x-6} - \dfrac{8}{x+5} = \dfrac{5(x+5)}{(x+5)(x-6)} - \dfrac{8(x-6)}{(x+5)(x-6)}$ Simplify the numerator and denominator.

$= \dfrac{4x - 24 + 9x + 45}{(x+5)(x-6)}$

$= \dfrac{5x + 25 - 8x + 48}{(x+5)(x-6)}$ Combine like terms.

$= \dfrac{13x + 21}{(x+5)(x-6)}$
$= \dfrac{-3x + 73}{(x+5)(x-6)}$

$= \dfrac{13x + 21}{(x+5)(x-6)} \div \dfrac{-3x + 73}{(x+5)(x-6)}$ Write as a division expression.

$= \dfrac{13x + 21}{(x+5)(x-6)} \cdot \dfrac{(x+5)(x-6)}{-3x + 73}$ Multiply by the reciprocal of the divisor.

$= \dfrac{13x + 21}{-3x + 73}$ Simplify.

39. $\dfrac{-x^2 + 33x + 16}{12x^2 + 11x - 27}$ **41.** $420x^5y^4z^3$

43. $(x+4)(x-4)(2x+1)(x-7)$ **45.** $\dfrac{360a^2 + 5a - 36}{60a^2}$

47. $\dfrac{42x + 41}{6(3x-1)(x+8)(2x+3)}$ **49.** 0   **51.** $\dfrac{5a - 11}{6}$

53. $(x-4)(x+5)$ **55.** $-\dfrac{3}{2}$ **57.** -1

59a. $y = \dfrac{70x}{x - 70}$ **59b.** Sample answer: When the object is 70 mm away, y needs to be 0, which is impossible.

61. a. $P_0\left(\dfrac{s_0}{s_0 - y}\right) - P_0\left(\dfrac{s_0}{s_0 - y}\right) = \dfrac{P_0 s_0}{s_0 - x} - \dfrac{P_0 s_0}{s_0 - y}$

$= \dfrac{P_0 s_0(s_0 - y)}{(s_0 - x)(s_0 - y)} - \dfrac{P_0 s_0(s_0 - x)}{(s_0 - x)(s_0 - y)}$

$= \dfrac{P_0 s_0(s_0 - y) - P_0 s_0(s_0 - x)}{(s_0 - x)(s_0 - y)}$

$= \dfrac{P_0 s_0 s_0 - P_0 s_0 y - P_0 s_0 s_0 - P_0 s_0 x}{(s_0 - x)(s_0 - y)}$

$= \dfrac{P_0 s_0 x - P_0 s_0 y}{(s_0 - x)(s_0 - y)}$

b. $\dfrac{(s_0 - x)(s_0 - y)}{P_0 s_0 x - P_0 s_0 y} = \dfrac{(500)(332)(70) - (500)(332)(45)}{(332 - 70)(332 - 45)}$

$P_0 = 500$, $s_0 = 332$, $x = 70$, $y = 45$

$= \dfrac{4,150,000}{75,194}$ Simplify.

≈ 55.2 Hz   Simplify.

63. $\dfrac{-3x^3 - 2x^2 + 16x - 5}{4x^3 + 18x^2 - 6x}$ **65.** Sample answer: $20a^4b^2c$,

$15ab^6$, $9abc$ **67.** C   **69.** B   **71a.** C   **71b.** C   **71c.** B

Lesson 7-3

1. $x - 1 = 0$
$x = 1$

$f(x)$ is not defined when $x = 1$. So, there is a vertical asymptote at $x = 1$.

From $x = 1$, as x-values decrease, $f(x)$ values approach 0, and as x-values increase, $f(x)$ values approach 0. So there is a horizontal asymptote at $f(x) = 0$. The domain is all real numbers not equal to

17.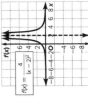
$f(x) = \dfrac{1}{(x+4)^2}$

19.
$f(x) = \dfrac{(x-4)^2}{x+2}$

21.
$f(x) = \dfrac{x^3+1}{x^2-4}$

23.
$f(x) = \dfrac{3x^2+8}{2x-1}$

25.
$f(x) = \dfrac{x^4-2x^2+1}{x^3+2}$

27a. $f(x) = \dfrac{x^2}{x+1}$ **27b.** Sample answer: I took the information I was given and made a visual. Then I set the denominator equal to 0 when x was −1, giving me x + 1. Because this is a rational function with no horizontal asymptotes I made the numerator x^2, meeting all the requirements I was given. **27c.** Sample answer: I made the assumption that there were no oblique asymptotes and the graph had no point discontinuity.

29.
$f(x) = \dfrac{x^2+4x-12}{x-2}$

31.
$f(x) = \dfrac{x^2-64}{x-8}$

$$x+2 \enclose{longdiv}{x^2+8x+20}$$
$$\dfrac{x+6}{\quad}$$
$$(-)\ x^2 + 2x$$
$$\dfrac{6x+10}{\quad}$$
$$(-)\ 6x+12$$
$$-2$$

The oblique asymptote is $y = x + 6$.

7.
$f(x) = \dfrac{x^2+x-12}{x+4}$

9.
$g(x) = \dfrac{x^3}{8x-4}$

11.
$f(x) = \dfrac{x^2+8x+20}{x+2}$

13.
$f(x) = \dfrac{x^2+64}{16x-24}$

15. Because $a(x) = 4$, there are no zeros. The function is undefined for $x = 2$, so there is a vertical asymptote at $x = 2$. Because the degree of the numerator is less than the degree of the denominator, there is a horizontal asymptote at $f(x) = 0$. The difference between the degree of the numerator and the degree of the denominator is 2, so there is no oblique asymptote.

$f(x) = \dfrac{4}{(x-2)^2}$

---

graph is translated 5 units to the left and has a vertical asymptote at $x = -5$ and a horizontal asymptote at $y = 0$. **39d.**
$y - 7 = 4\left(\dfrac{1}{x} + 5\right)$

41. Sample answer: $f(x) = \dfrac{2}{x-3} + 4$ and
$g(x) = \dfrac{5}{x-3} + 4$
$g(x) = \left(\dfrac{-5}{x}, -3\right) + 4$

43. D **45a.** B, C, F **45b.** B

Lesson 7-4

1.
$f(x) = \dfrac{x^4-2}{x^2-1}$

3a.
$P(x) = \dfrac{7+x}{11+x}$

3b. The part in the first quadrant. **3c.** It represents his original field goal percentage of 63.6%. **3d.** $y = 1$; this represents 100% which he cannot achieve because he has already missed 4 field goals.

5. $x^2 + 8x + 20 = 0$ Set $a(x) = 0$.

Because $b^2 - 4ac = 8^2 - 4(1)(20)$ or −16, there are no real roots. So, there are no zeros.

$x + 2 = 0$ Set $b(x) = 0$.
$x = -2$ Subtract 2 from each side.

There is a vertical asymptote at $x = -2$. The degree of the numerator is greater than the degree of the denominator, so there is no horizontal asymptote. The difference between the degree of the numerator and the degree of the denominator is 1, so there is an oblique asymptote.

37a.

x	$a(x)=x^2$	$b(x)=x^{-2}$	$c(x)=x^3$	$d(x)=x^{-3}$
−4	16	$\frac{1}{16}$	−64	$-\frac{1}{64}$
−3	9	$\frac{1}{9}$	−27	$-\frac{1}{27}$
−2	4	$\frac{1}{4}$	−8	$-\frac{1}{8}$
−1	1	1	−1	−1
0	0	undefined	0	undefined
1	1	1	1	1
2	4	$\frac{1}{4}$	8	$\frac{1}{8}$
3	9	$\frac{1}{9}$	27	$\frac{1}{27}$
4	16	$\frac{1}{16}$	64	$\frac{1}{64}$

37b.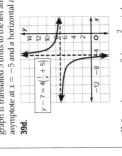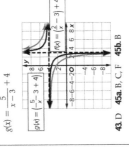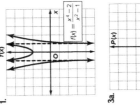
$a(x)=x^2$
$b(x)=x^{-2}$

37c. $a(x)$: D = {all real numbers}, R = {$a(x) \mid a(x) \geq 0$}; as $x \to -\infty$, $a(x) \to \infty$, as $x \to \infty$, $a(x) \to \infty$; At $x = 0$, $a(x) = 0$, so there is a zero at $x = 0$. $b(x)$: D = {$x \mid x \neq 0$}, R = {$b(x) \mid b(x) > 0$}; as $x \to -\infty$, $b(x) \to 0$, as $x \to \infty$, $b(x) \to 0$; At $x = 0$, $b(x)$ is undefined, so there is an asymptote at $x = 0$.

37d.
$c(x)=x^3$
$d(x)=x^{-3}$

37e. $c(x)$: D = {all real numbers}, R = {all real numbers}; as $x \to -\infty$, $c(x) \to -\infty$, as $x \to \infty$, $c(x) \to \infty$; At $x = 0$, $c(x) = 0$, so there is a zero at $x = 0$. $d(x)$: D = {$x \mid x \neq 0$}, R = {$d(x) \mid d(x) \neq 0$}; as $x \to -\infty$, $d(x) \to 0$, as $x \to \infty$, $d(x) \to 0$; At $x = 0$, $d(x)$ is undefined, so there is an asymptote at $x = 0$.

37f. For two power functions $f(x) = ax^n$ and $g(x) = ax^{-n}$, for every x, $f(x)$ and $g(x)$ are reciprocals. The domains are similar except that for $g(x)$, $x \neq 0$. Additionally, wherever $f(x)$ has a zero, $g(x)$ is undefined.

39a. The first graph has a vertical asymptote at $x = 0$ and a horizontal asymptote at $y = 0$. The second graph is translated 7 units up and has a vertical asymptote at $x = 0$ and a horizontal asymptote at $y = 7$. **39b.** Both graphs have a vertical asymptote at $x = 0$ and a horizontal asymptote at $y = 0$. The second graph is stretched by a factor of 4. **39c.** The second graph has a vertical asymptote at $x = 0$ and a horizontal asymptote at $y = 0$. The second

7.14

9a. between 20 and 30 pounds

	Pounds	Price per Pound	Total Price
dried fruit	10	$6.25	6.25(10)
mixed nuts	m	$4.50	4.50m
trail mix	$10 + m$	$5.00	5(10 + m)

9b. $62.5 + 4.5m = 50 + 5m$ **9c.** 25; yes, the answer falls within the estimate **11a.** Kendal: $\frac{1}{60}$ Chandi: $\frac{1}{80}$ **11b.** Kendal: $\frac{x}{60}$; Chandi: $\frac{x}{80}$ **11c.** $\frac{x}{60} + \frac{x}{80} = 1$; between 30 and 40 mins **11d.** about 34.3 min; yes, the answer falls within the estimate **13.** $c < 0$, or $\frac{13}{18} < c$ **15.** $b < 0$, or $\frac{35}{12} < b$ **17.** 2 **19.** 1 **21.** ∅

23. cost of 3 pounds of bananas for $0.90/pound = 0.9(3)
cost of x pounds of apples for $1.25/pound = 1.25$x$
total weight = 3 + x

$$\text{average monthly cost} = \frac{\text{total cost}}{\text{total weight}}$$

$$\frac{0.9(3) + 1.25x}{3 + x} = 1 \quad \text{Write an equation.}$$

$$\frac{2.7 + 1.25x}{3 + x} = 1 \quad \text{Simplify the numerator.}$$

$$(3 + x)\frac{(2.7 + 1.25x)}{3 + x} = (3 + x)(1) \quad \text{LCD is } (3+x). \text{ Multiply by the LCD.}$$

$$\frac{1}{(3+x)}\frac{(2.7 + 1.25x)}{3 + x} = (3 + x)(1) \quad \text{Divide out common factors.}$$

$$2.7 + 1.25x = 3 + x \quad \text{Simplify.}$$
$$2.7 + 0.25x = 3 \quad \text{Subtract } x \text{ from each side.}$$
$$0.25x = 0.3 \quad \text{Subtract 2.7 from each side.}$$
$$x = 1.2 \quad \text{Divide each side by 0.25.}$$

She must purchase 1.2 pounds of apples.

25. $x < 0$ or $x > 1.75$ **27.** $x < -2$, or $2 < x < 14$ **29.** $x < -5$ or $4 < x < \frac{17}{3}$ **31.** 55.56 mph **33a.** 1; yes; 3

33b.

33c. 1; no

33d. Graph both sides of the equation. Where the graphs intersect, there is a solution. Where they do not, then the possible solution is extraneous.

35. ∅ **37.** all real numbers except 5, −5, 0 **39.** Sample answer: Multiplying each sides of a rational inequality can produce extraneous solutions. Therefore, you should check all solutions to make sure that they satisfy the original equation or inequality. **41.** D **43.** D

45b. $44.\overline{4}$ mph

47. **a.** $F = G\frac{m_1 m_2}{d^2}$ Law of Universal Gravitation
$$= (6.67 \times 10^{-11})\frac{(7.36 \times 10^{22})(5.97 \times 10^{24})}{(3.84 \times 10^8)^2}$$
$$\approx 2 \times 10^{20} \text{ newtons}$$

b. $F = G\frac{m_1 m_2}{d^2}$ Law of Universal Gravitation
$$= (6.67 \times 10^{-11})\frac{(1.99 \times 10^{30})(5.97 \times 10^{24})}{(1.5 \times 10^{11})^2}$$
$$\approx 3.5 \times 10^{22} \text{ newtons}$$

c. $F = G\frac{m_1 m_2}{d^2}$ Law of Universal Gravitation
$$= (6.67 \times 10^{-11})\frac{(1000)(1000)}{(0.1)^2}$$
$$= 6.67 \times 10^{-3} \text{ newtons}$$

49. a and g are directly related. **51.** Sample answer: The force of an object varies jointly as its mass and acceleration. **53a.** A **53b.** D

53c.

Number of Painters	Time to Paint the House (h)
10	3.75
15	2.5
20	1.86
25	1.5

53d.

55. D **57.** C

Lesson 7-6

1.11 **3.** 7

5. The LCD for the terms is $(x − 5)(x − 4)$.

$$\frac{8}{x - 5} - \frac{9}{x - 4} = \frac{x^2 - 9x + 20}{5} \quad \text{Original equation}$$

$$(x - 5)(x - 4)\frac{8}{x - 5} - (x - 5)(x - 4)\frac{9}{x - 4} = (x - 5)(x - 4)\frac{5}{x^2 - 9x + 20} \quad \text{Multiply by the LCD.}$$

$$\underbrace{(x-4)}(8) - \underbrace{(x-5)}(9) = \frac{(x-5)(x-4)(5)}{x^2 - 9x + 20} \quad \text{Divide common factors.}$$

$$(x - 4)(8) - (x - 5)(9) = 5 \quad \text{Simplify.}$$
$$8x - 32 - 9x + 45 = 5 \quad \text{Distribute.}$$
$$-x + 13 = 5 \quad \text{Simplify.}$$
$$-x = -8 \quad \text{Subtract 13 from each side}$$
$$x = 8 \quad \text{Divide each side by −1.}$$

41.

43. Similarities: Both have vertical asymptotes at $x = 0$. Both approach 0 as x approaches $-\infty$ and approach 0 as x approaches ∞. Differences: $f(x)$ has holes at $x = 1$ and $x = -1$, while $g(x)$ has vertical asymptotes at $x = \sqrt{2}$ and $x = -\sqrt{2}$. $f(x)$ has no zeros, but $g(x)$ has zeros at at $x = 1$ and $x = -1$.

45. $f(x) = \frac{x}{a - b} + \frac{(a - b)}{b}$
$$= \frac{x + ca - cb}{a - b}$$

47. D **49.** C **51a.** C **51b.** A **51c.** B **51d.** D

Lesson 7-5

1. 21 **3.** −32 **5.** −48 **7.** 1.5 **9.** −56 **11.** $m = \frac{1}{6}w$

13. $\frac{a_1}{b_1 c_1} = \frac{a_2}{b_2 c_2}$ Joint variation

$$\frac{-60}{-5(4)} = \frac{a_2}{4(-3)}$$
$$-60(4)(-3) = -5(4)(a_2) \quad \text{Cross multiply.}$$
$$720 = -20a_2 \quad \text{Simplify.}$$
$$-36 = a_2 \quad \text{Divide each side by −20.}$$

15. −3 **17.** −22.5 **19.** 38 **21a.** $s = \frac{48}{t}$

21b. 3.2 hours **23.** −10 **25.** direct **27.** neither

29. $\frac{x_1}{y_2} = \frac{x_2}{y_1}$ Inverse variation

$$\frac{16}{20} = \frac{x_2}{5} \quad x_1 = 16, y_1 = 5, y_2 = 20$$
$$16(5) = 20(x_2) \quad \text{Cross multiply.}$$
$$80 = 20x_2 \quad \text{Simplify.}$$
$$4 = x_2 \quad \text{Divide each side by 20.}$$

31. −12 **33.** inverse; −2 **35.** combined; 10 **37.** direct; 4 **39.** direct; −2 **41.** inverse; 7 **43.** joint; 20

45a. $800 = rt$

33.

35.

37. **a.** total cost = phone cost + monthly usage charge
$$= 150 + 40x$$

$$\text{average monthly cost} = \frac{\text{total cost}}{\text{number of months}}$$

$$f(x) = \frac{150 + 40x}{x}$$

b. The vertical asymptote is $x = 0$. Because the degree of the numerator equals the degree of the denominator, the horizontal asymptote is at $f(x) = \frac{40}{1} = 40$.

c. Sample answer: The number of months and the average cost cannot have negative values.

d. $f(x) = \frac{150 + 40x}{x}$ Write the equation.
$$45 = \frac{150 + 40x}{x} \quad f(x) = 45$$
$$45x = 150 + 40x \quad \text{Multiply each side by } x.$$
$$5x = 150 \quad \text{Subtract } 40x \text{ from each side.}$$
$$x = 30 \quad \text{Divide each side by 5.}$$

After 30 months, the average monthly charge will be $45.

39.

Selected Answers and Solutions

45a.

	Distance (mi)	Avg Speed (mph)	Time (h)
With Current	4	$2 + r$	$\frac{4}{2 + r}$
Against Current	4	$2 - r$	$\frac{4}{2 - r}$

45b. $\frac{4}{2} + r + \frac{4}{2} - r = 6$

45c. $(2 + r)(2 - r)$

45d. about 1.15 miles per hour

Chapter 7 Study Guide and Review

1. complex fraction **3.** rational equations **5.** joint variation **7.** point discontinuity **9.** If the degree of $a(x)$ is greater than the degree of $b(x)$, there is no horizontal asymptote. If the degree of $a(x)$ is less than the degree of $b(x)$, the horizontal asymptote is $y = 0$. If the degree of $a(x)$ is equal to the degree of $b(x)$, the horizontal asymptote is $y = \dfrac{\text{leading coefficient of } a(x)}{\text{leading coefficient of } b(x)}$.

11. $\dfrac{10yz^2}{9x}$

13. $\dfrac{x - 1}{x - 2}$ **15.** $\dfrac{x - 3}{x + 1}$ **17.** $\dfrac{27b + 10a^2}{12ab^2}$ **19.** $\dfrac{3xy^3 + 8y^3 - 5x}{6x^2y^2}$

21. $\dfrac{12x^2 - 10x + 6}{2(x + 2)(3x - 4)(x + 1)}$ **23.** $\dfrac{10x + 20}{(x + 6)(x + 1)}$

25.

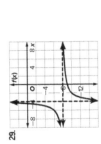

$D = \{x \mid x \neq 0\}, R = \{f(x) \mid f(x) \neq 2\}$

27.

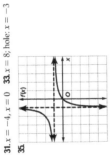

$D = \{x \mid x \neq 9\}, R = \{f(x) \mid f(x) \neq 0\}$

29.

(graph)

$D = \{x \mid x \neq -4\}, R = \{f(x) \mid f(x) \neq -8\}$

31. $x = -4, x = 0$ **33.** $x = 8$; hole: $x = -3$

35.

(graph)

37.

(graph)

39. $a = 15$ **41.** $y = -\dfrac{1}{3}$ **43.** $y = \dfrac{48}{5}$ **45.** $x = \dfrac{46}{17}$

47. $x = -7$ **49.** $x = 8$ **51.** $-\dfrac{9}{10} < x < 0$

CHAPTER 8
Statistics and Probability

Chapter 8 Get Ready

1. the average of the data: Add all the values together and divide by the total number of values.

3. the value that occurs most often in the set

5. Multiply the individual probabilities.

7. An outlier in a set of data will not affect the shape of the distribution plot.

Lesson 8-1

1. survey; sample: the students in the study; population: the student body

3. Observation study; sample answer: The scores of the participants are observed and compared without them being affected by the study. **5.** 909

7. The students' grades are being observed, so this is an observational study. The sample consists of the physics students selected. The population is the set all college students that take a physics course.

9. Survey; sample answer: The data will be obtained from opinions given by members of the sample population. **11.** Observational study; sample answer: The eating habits of the participants will be observed and compared without them being affected by the study. **13.** 199 **15.** 448 **17.** 1674 **19.** 2721 **21.** There is no bias.

23. a. The sample group is the 8- to 18-year-olds who were actually surveyed. The population is represented by the sample, so the population is all 8- to 18-year-olds in the United States.

b. average time

c. Interpret the bar graph for each group. The red bar represents talking and the blue bar represents texting. Sample answer: The 8- to 10-year-old group talked for about 10 minutes a day and did not text at all. The 11- to 14-year-old group talked for about 30 minutes a day and texted for about 70 minutes a day. The 15- to 18-year-old group talked for about 40 minutes a day and texted for about 110 minutes a day.

d. Sample answer: A cell phone company might use a report like this to determine which age group to target in their ads.

25a. See students' work.

25b. Sample answer for Product A:

Product A	
Number	Frequency
0–6	卌 卌 卌 IIII
7–9	卌 卌 I

Sample answer for Product B: $\approx 76.7\%$

Sample answer for Product B:

Product B	
Number	Frequency
0–7	卌 卌 卌 卌 III
8–9	卌 II

25c. Sample answer: Yes; the probability that Product B is effective is 13.4% higher than that of Product A.

25d. Sample answer: It depends on what the product is and how it is being used. For example, if the product is a pencil sharpener, then the lower price may be more important than the effectiveness, and therefore, might not justify the price difference. However, if the product is a life-saving medicine, the effectiveness may be more important than the price, and therefore, might justify the price difference. **27.** true

29. An invalid sampling method and type of sample can produce bias. For example, if a sample is not random, the person conducting the study can influence the results by selecting a specific sample of people. Also, if an experiment is used when an observational study is the more logical type of study to be used, the study can be unreliable. For example, if someone wants to analyze the speeds of vehicles on a specific stretch of highway and decides to place an empty police car on the side of the road, the data will be affected by the police car. The results of this study will show lower speeds than are normally driven on the highway. Biased survey questions and incorrect procedures can affect the reliability of a study as well. A survey question that is poorly written may result in a response that does not accurately reflect the opinion of the participant.

31. D **33.** A, D **35.** D

Lesson 8-2

1. Sample answer: Use a spinner that is divided into two sectors, one containing 80% or 288° and the other containing 20% or 72°. Do 20 trials and record the results in a frequency table.

Outcome	Frequency
A	17
Below an A	3
Total	20

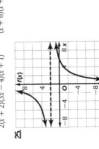
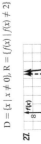
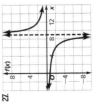

The probability of Clara getting an A on her next quiz is 0.85. The probability of earning any other grade is $1 - 0.85$ or 0.15.

3. Sample answer: Use a random number generator to generate integers 1 through 20 where 1–9 represents tae kwon do, 10–15 represents yoga, 16–18 represents swimming, and 19–20 represents kickboxing. Do 20 trials and record the results in a frequency table.

Outcome	Frequency
tae kwon do	9
yoga	7
swimming	1
kickboxing	3
Total	20

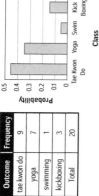

Class

The probability of a customer taking the tae kwon do class is 0.45, taking yoga is 0.35, swimming is 0.05, and taking kickboxing is 0.15.

5 Sample answer: Use a spinner that is divided into two sectors. One sector should be 95% of the circle or $(0.95)360° = 342°$. The other should be 5% of the circle or $(0.05)360° = 18°$. Do 50 trials and record the results in a frequency table.

Outcome	Frequency
Sale	46
No Sale	4
Total	50

The probability of Ian selling a game is 0.92. The probability of not selling a game is $1 - 0.92$ or 0.08.

7. Sample answer: Use a random number generator to generate integers 1 through 20, where 1–8 represents drama, 9–14 represents mystery, 15–19 represents comedy, and 20 represents action. Do 20 trials and record the results in a frequency table.

Outcome	Frequency
drama	11
mystery	3
comedy	6
action	0
Total	20

Movie Genres

The probability of a customer choosing a drama is 0.55, choosing a mystery is 0.15, choosing a comedy is 0.3, and choosing an action film is 0. **9.** Sample answer: Use a random number generator to generate integers 1 through 20 where 1–9 represents Europe, 10–14 represents Asia, 15–17 represents South America, 18–19 represents Africa, and 20 represents Australia. Do 20 trials and record the results in a frequency table.

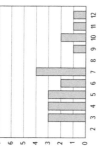

Outcome	Frequency
Europe	7
Asia	6
South America	5
Africa	2
Australia	0
Total	20

Travel Destinations

The probability of a customer traveling to Europe is 0.35, to Asia is 0.3, to South America is 0.25, to Africa is 0.1, and to Australia is 0.

11a. Sample answer: The theoretical probability that the player gets a hit is 30%, and the theoretical probability that he does not get a hit is 70%. Use a random number generator to generate integers 1 through 10. The integers 1–3 will represent a hit, and the integers 4–10 will represent the player not getting a hit. The simulation will consist of 50 trials.

11b. Sample answer: $P(\text{hit}) = 28\%$, $P(\text{not a hit}) = 72\%$

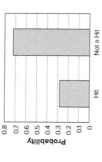

13a. Sample answer: The theoretical probability that a spin results in each prize is 20%. Use a random number generator to generate integers 1 through 5. The integer 1 will represent a hot pretzel, the integer 2 will represent a burger, the integer 3 will represent a large drink, the integer 4 will represent nachos, and the integer 5 will represent a small popcorn. The simulation will consist of 50 trials.

13b. Sample answer: $P(\text{hot pretzel}) = 20\%$, $P(\text{burger}) = 20\%$, $P(\text{large drink}) = 28\%$, $P(\text{nachos}) = 14\%$, $P(\text{small popcorn}) = 18\%$

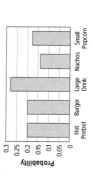

15a. Sample answer: 9, 10, 6, 6, 7, 9, 5, 9, 5, 7, 6, 5, 7, 3, 9, 7, 6, 7, 8, 7 **15b.** Sample answer: 4, 10, 5, 10, 6, 7, 12, 3, 7, 4, 7, 9, 3, 6, 4, 11, 5, 7, 5, 3

15c. Sample answer:

Trial	Sum of Roll	Sum of Output from Random Number Generator
1	9	4
2	10	10
3	6	5
4	6	10
5	7	6
6	9	7
7	5	12
8	9	3
9	5	7
10	7	4
11	6	7
12	5	9
13	7	3
14	3	6
15	9	4
16	7	11
17	6	5
18	7	7
19	8	5
20	7	3

15d. Sample answer:

Dice – 5 Rolls

Dice – 10 Rolls

Dice – 20 Rolls

15e. Sample answer: The bar graph has more data points at the middle sums as more trials are added.

15f. Sample answer:

Random Number Generator

15g. Sample answer: They both have the most data points at the middle sums. **17 Yes**; sample answer: If the spinner were going to be divided equally into three outcomes, each sector would measure 120. Because you only want to know the probability of outcome C, you can record spins that end in the red area as a success, or the occurrence of outcome C, and spins that end in the blue area as a failure, or an outcome of A or B.

19. Sample answer: You should consider the design of the simulation, how many trials were used, and whether the theoretical and experimental probabilities are reasonably close. **21a.** Sample answer: Using a spinner with 8 equal regions **21b.** Sample answer: Cat 1 – 0.14, Cat 2 – 0.15, Cat 3 – 0.12, Cat 4 – 0.10, Cat 5 – 0.13, Cat 6 – 0.16, Cat 7 – 0.09, Cat 8 – 0.11 **26.** A **23.** A, D, G

Lesson 8-3

1. 15 **3.** ±10% **5.** $\mu = 52.5$ **7.** $p = 0.54$ **9.** $p = 0.60$

11. margin of error = ±5.77%

13. margin of error = ±3.64%

15. margin of error = ±2%

17 $\pm 3\% = \pm \dfrac{1}{\sqrt{n}}(100)$

$0.03\sqrt{n} = 1$

$\sqrt{n} = 33.333$

$n = 1111.11$

Round to the nearest individual; $n = 1111$.

19. $n = 204$ **21.** Fitness center A: 316.6, Fitness center B: 310.1. The claim is true.

23 Margin of error = $\pm \dfrac{1}{\sqrt{n}}(100)$ where n is the sample size, so $n = 250$.

margin of error = $\pm \dfrac{1}{\sqrt{250}}(100) \approx \pm 6.3 \approx \pm 6.3\%$

25. With a margin of error of ±0.7% the range is 85.3% to 86.7%. **27.** Add the sum of all ratings and divide that by 100, the number of candy bars.

29. $p = \frac{250}{300} = 0.83$

31. $p = 0.75$ **33.** B   **35.** A   **37.** A, C

39a. ± 8.2%   **39b.** $n = 473$ **39c.** For better results, Cassy's best option is to survey 473 homeowners with a margin of error of ± 4.6%.

Lesson 8-4

1. a. First, press STAT ENTER and enter each data value. Then, press 2nd [STAT PLOT] ENTER ENTER and choose ⊞. Finally, adjust the window to dimensions appropriate for the data.

[4, 32] scl: 4 by [0, 8] scl: 1

Because the majority of the data is on the right and there is a tail on the left, the distribution is negatively skewed.

b. Sample answer: The distribution is skewed, so use the five-number summary. Press STAT ▶ ENTER and scroll down to display the statistics for the data set.

```
1-Var Stats
↑n=24
minX=7
Q1=15.5
Med=22.5
Q3=26
maxX=30
```

The range is 7 to 30 minutes. The median is 22.5 minutes, and half of the data are between 15.5 and 26.

3a.

Mrs. Johnson's Class   Mr. Edmunds' Class

[5, 40] scl: 5 by [0, 8] scl: 1   [5, 40] scl: 5 by [0, 8] scl: 1

Mrs. Johnson's class, positively skewed; Mr. Edmunds' class, negatively skewed

3b. Sample answer: The distributions are skewed, so use the five-number summaries. The range for both classes is the same. However, the median for Mrs. Johnson's class is 17 and the median for Mr. Edmunds' class is 28. The lower quartile for Mr. Edmunds' class is 20. Because this is greater than the median for Mrs. Johnson's class, this means that 75% of the data from Mr. Edmunds' class is greater than 50% of the data from Mrs. Johnson's class. Therefore, we can conclude that the students in Mr. Edmunds' class had slightly higher sales overall than the students in Mrs. Johnson's class.

5a.

[50, 200] scl: 25 by [0, 5] scl: 1   [50, 200] scl: 25 by [0, 5] scl: 1

negatively skewed

5b. Sample answer: The distribution is skewed, so use the five-number summary. The range is 53 to 179 points. The median is 138.5 points, and half of the data are between 106.5 and 157 points.

7a. Sophomore Year

[1200, 1900] scl: 100 by [0, 8] scl: 1

Junior Year

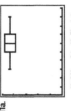

[1300, 2200] scl: 100 by [0, 8] scl: 1

both symmetric

7b. Sample answer: The distributions are symmetric, so use the means and standard deviations. The mean score for sophomore year is about 1552.9 with standard deviation of about 147.2. The mean score for junior year is about 1753.8 with standard deviation of about 159.1. We can conclude that the scores and the variation of the scores from the mean both increased from sophomore year to junior year.

9. a. Enter the tuitions for the public colleges as **L1**. Graph these data as **Plot1** by pressing 2nd [STAT PLOT] ENTER ENTER and choosing ⊡⊞. Enter the tuitions for the private colleges as **L2**. Graph these data as **Plot2** by pressing 2nd [STAT PLOT] ▶ ENTER ENTER and choosing ⊡⊞. For **Xlist**, enter **L2**. Adjust the window to dimensions appropriate for the data.

[0, 35,000] scl: 5000 by [−10, 100] scl: 1

For both sets of data, the whiskers are approximately equal, and the median is in the middle of the data. The distributions appear to be symmetric.

b. Sample answer: The distributions appear to be symmetric, so use the means and standard deviations. Press STAT ▶ ENTER ENTER to display the statistics for the public colleges.

```
1-Var Stats
x̄=7367.375
Σx=176817
Σx²=1334978605
Sx=11185.0786
σx=1169.125782
↓n=24
```

Press STAT ▶ ENTER 2nd [L1] ENTER to display the statistics for the private colleges.

```
1-Var Stats
x̄=15762.70833
Σx=378305
Σx²=7255117107
Sx=7494.942442
σx=7337.136507
↓n=24
```

The mean for the public colleges is $7367.38 with standard deviation of about $1160.13. The mean for private colleges is about $15,762.71 with standard deviation of about $7337.14. We can conclude that not only is the average cost of private schools far greater than the average cost of public schools, but the variation of the costs from the mean is also much greater.

11a.

[0, 30] scl: 3 by [0, 5] scl: 1

Sample answer: The distribution is symmetric, so use the mean and standard deviation. The mean of the data is 18 with standard deviation of about 5.2 points.

11b.

mean: 14.6; median: 17

[0, 30] scl: 3 by [0, 5] scl: 1

11c. Sample answer: Adding the scores from the first four games causes the shape of the distribution to go from being symmetric to being negatively skewed. Therefore, the center and spread should be described using the five-number summary.

13. a. Sample answer: Because the distribution is positively skewed, the median will be to the left of the mean closer to the majority of the data. An estimate for the median is 10. The mean will be more affected by the tail, and will be to the right of the majority of the data. An estimate for the mean is 14.   **b.** Sample answer: Because the distribution is negatively skewed, the median will be to the right of the mean closer to the majority of the data. An estimate for the median is 24. The mean will be more affected by the tail, and will be to the left of the majority of the data. An estimate for the mean is 20.   **c.** Sample answer: Because the distribution is symmetric, the mean and median will be approximately equal near the middle of the data. An estimate for the mean and median is 17.

15. Sample answer: The heights of the players on the Pittsburgh Steelers roster appear to represent a normal distribution.

Heights of the Players on the 2009 Pittsburgh Steelers Roster (inches)

75	74	74	75	75	77	77	72
77	72	70	75	78	71	75	
77	71	71	73	76	74	72	73
75	70	74	73	76	75		
71	69	77	77	80	75	77	
67	74	69	71	77	73	73	

The mean of the data is about 73.61 in. or 6 ft 1.61 in. The standard deviation is about 2.97 in.

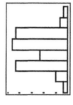

[66, 82] scl: 2 by [0, 15] scl: 3

The birth months of the players do not display central tendency.

Birth Months of the Players on the 2009 Pittsburgh Steelers Roster

1	12	10	3	11	10	5
4	8	9	11	1	11	5
8	6	11	4	3	4	8	2
3	7	2	1	11	4	5
1	1	6	1	6	8	9
3	3	1	6	9	1	9
6	5	10	11	12	9	

[0, 12] scl: 2 by [0, 15] scl: 3

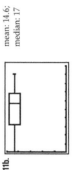

17.C 19.E 21a.9A: max 70, min 20; 9B: max 80, min 45 21b.75% 21c.25% 21d.50%

Lesson 8-5

1. Sample answer: The conclusion may not be valid because the report is based on an observational study, so a cause-and-effect conclusion cannot be made. There may be other reasons why students who read a newspaper had higher test scores. For example, they may be more likely to read books or may spend more time studying.

3. Sample answer: The scale on the y-axis makes it appear as if the number of subscribers was almost constant, although the number of subscribers actually decreased by 25% from 2012 to 2016.

5. The conclusion is valid because the study was a controlled experiment with a large sample and individuals randomly assigned to a control group and a treatment group.

7. To analyze the graph, check the values on each axis. The least value on the y-axis is 30. This means the y-axis does not begin at 0.
The y-axis does not begin at a zero, which makes it appear as if sales revenue is increasing rapidly.

9. The conclusion is not valid because 6 out of 5235 people surveyed is approximately 0.1%; the population of the United States is about 320,000,000, and 0.1% of this population is about 320,000. The conclusion that millions of Americans know viral dance moves is not valid.

11. The formula for the margin of error is $\pm \frac{1}{\sqrt{n}}(100)$ where n is the sample size.
The sample size in this situation is 2050, because 2050 registered voters were surveyed.
So, the margin of error is $\pm \frac{1}{\sqrt{2050}}(100) \approx 2.2$.
The margin of error is about 2.2%, which means that it is unlikely that Rodriguez would receive less than 53% − 2.2% = 50.8% of the vote. Therefore, according to this survey, Rodrigues is likely to receive more than 50% of the vote, the election is not a toss-up, and Rodriguez is likely to win.

13. Disagree; it is possible to draw a cause-and-effect conclusion from a study if the study is a controlled experiment in which individuals are randomly assigned to a control group and to a treatment group.

15. Sample answer:

Ages of Museum Visitors (Frequency vs Age; bars at 1–20, 21–30, 31–70)

17.B 19.B, E 21.D

Lesson 8-6

1. 251 < X < 581 3a. about 386 teens 3b. 84%
5. 2.05; 2.05 standard deviations greater than the mean 7. 3.08; 2.40 standard deviations less than the mean 9. X < 15.9 or X > 42.7 11a. about 509
11b. 0.15% 13 −1.33; 1.33 standard deviations less than the mean 15. 177.7; 1.73 standard deviations greater than the mean

17. a. Find the z-values associated with 90,000 and 110,000. The mean μ is 113,627 and the standard deviation σ is 14,266.
$z = \frac{X - \mu}{\sigma}$
$= \frac{90,000 - 113,627}{14,266}$
≈ -1.656
$z = \frac{X - \mu}{\sigma}$
$= \frac{110,000 - 113,627}{14,266}$
≈ -0.254
Use a graphing calculator to find the area between the z-values.

```
normalcdf(-1.656
,-.254)
         .3508869415
```

The value 0.35 is the percentage of batteries that will last between 90,000 and 110,000 miles. The total number of batteries in this group is 0.35 × 20,000 or 7000.

b. Find the z-value associated with 125,000.
$z = \frac{X - \mu}{\sigma}$
$= \frac{125,000 - 113,627}{14,266}$
≈ 0.797
We are looking for values greater than 125,000, so we can use a graphing calculator to find the area between z = 0.797 and z = 4.

```
normalcdf(.797,4
)
         .2126937641
```

The value 0.21 is the percentage of batteries that will last between more than 125,000 miles. The total number of batteries in this group is 0.21 × 20,000 or 4200.

c. Find the z-value associated with 100,000.
$z = \frac{X - \mu}{\sigma}$
$= \frac{100,000 - 113,627}{14,266}$
≈ -0.955
We are looking for values less than 100,000, so we can use a graphing calculator to find the area between z = −4 and z = −0.955.

```
normalcdf(-4,-.9
55)
         .1697571502
```

The probability that if you buy a car battery at random it will last less than 100,000 miles is about 17.0%.

19. a. The middle 68% represents all data values within one standard deviation of the mean. Add ±$115 to $829. The range of rates is $714 to $944.
b. Find the z-value associated with 1000.
$z = \frac{X - \mu}{\sigma}$
$= \frac{1000 - 829}{115}$
≈ 1.487
We are looking for values more than 1000, so we can use a graphing calculator to find the area between z = 1.487 and z = 4.

```
normalcdf(1.487,
4)
         .0684757505
```

The probability that a customer selected at random will pay more than $1000 is about 6.8%. Out of 900 people, about 0.068 × 900 or 62 people will pay more than $1000.

c. Sample answer: I would expect people with several traffic citations to lie to the far right of the distribution where insurance costs are highest, because I think insurance companies would charge them more.

d. Sample answer: I think auto insurance companies would charge younger people more than older people because they have not been driving as long. I think they would charge more for expensive cars and sports cars and less for cars that have good safety ratings. I think they would charge a person less if they have a good driving record and more if they have had tickets and accidents.

21. Sample answer: Hiroko; Monica's solution would work with a uniform distribution. 23. Sample answer: True; according to the Empirical Rule, 99% of the data lie within 3 standard deviations of the mean. Therefore, only 1% will fall outside of three sigma. An infinitely small amount will fall outside of six-sigma. 25. Sample answer: The scores per team in each game of the first round of the 2010 NBA playoffs. The mean is 96.56 and the standard deviation is 11.06. The middle 68% of the distribution is 85.50 < X < 107.62. The middle 95% is 74.44 < X < 118.68. The middle 99.7% is 63.38 < X < 129.74. 27.C 29.D 31.C 33a. 0.9332 33b. 0.7734 33c. 0.2956 33d. 0.9199

Lesson 8-7

1a. $P(\text{Alexa}) = \frac{5}{18}$, $P(\text{Gayle}) = \frac{4}{9}$, $P(\text{Mei}) = \frac{5}{18}$; because these probabilities are not equal, the method is not fair. 1b. Sample answer: If the sum is 2 through 5, Alexa is the Captain; if the sum is 6 or 8, Gayle is the captain; if the sum is 9 through 12, Mei is the captain; if the sum is 7, roll again.

3 The expected number of tokens from each roll is 20(0.02) + 10(0.1) + 5(0.18) + 2(0.3) + 1(0.4) = 3.3. By playing 10 times, Latrell can expect to win 10(3.3) = 33 tokens.
It takes 15 tickets to get a prize, so 33 tickets can be redeemed for 2 prizes, with 3 tickets leftover.
The value of the 2 prizes is 2($4.95) = $9.90, which is more than the cost of the 10 rolls, because 10($0.25) = $2.50.
The value of the prizes is worth more than the cost of playing the games, so this is a good decision.

5. Yes; $P(\text{Micah}) = P(\text{Carrie}) = \frac{1}{2}$
7. No; $P(\text{Micah}) = \frac{1}{3}$, $P(\text{Carrie}) = \frac{2}{3}$
9. No; the expected number on a box top is 2.16, so Ricardo can expect to have a sum of 17.28 with 8 boxes. This results in one prize worth $18.95, which is less than the cost of the 8 boxes ($29.20).
11. Sample answer: Latoya presents if the sum is 6 or less; David presents if the sum is 8 or more; roll again if the sum is 7. 13. Sample answer: Latoya presents if the number is 5 through 13; David presents if the number is 14 through 22; if the number is 23, choose a new number.
15 a. The probability that a pop is a berry pop is $\frac{1}{4}$ or 0.25.
The probability that a pop is not a berry pop is 1 − P(berry) = 1 − 0.25 = 0.75.
The probability that none of the 6 pops is a berry pop is $(0.75)^6 \approx 17.8\%$.
This means that approximately 1 out of 5 boxes will not contain any berry pops, so it is not a very unlikely occurrence. For this reason, it makes sense to disagree with Brian's decision.

b. From part a, the probability that a box contains no berry pops is about 0.178.
A case contains 5 boxes. So, the probability of no berry pops in a case is $(0.178)^5 \approx 0.02\%$. Because this

R84

probability is close to 0, this is very unlikely to happen by chance. For this reason, it makes sense to agree with Brian's decision in this case.

17. Disagree; Sample answer: Assign one option to the outcomes 1–3, assign the other option to the outcomes 4–6, and spin again if the outcome is 7.

19. $P(\text{red}) = 0.5$, but $P(\text{spade}) = 0.25$, so the probabilities of the options are not all equal. Instead, he could assign the first option to the hearts. If the chosen card is a diamond, he should choose a new card until one of the options is chosen. **21.** B, C, E, F **23.** B **25.** It's a bad decision because the cost of cans is greater than the expected value of the prizes.

Chapter 8 Study Guide and Review

1. bias **3.** Empirical Rule **5.** A negatively skewed distribution has most of the data to the right of the mean and a positively skewed distribution has most of the data to the left of the mean. **7.** survey; sample: every tenth shopper; population: all potential shoppers **9.** survey; sample: every fifth person; population: student body

11a. Sample answer: The theoretical probability that the bus is late is 60%, and the theoretical probability that the bus is not late is 40%. Use a random number generator to generate integers 1 through 5. The integers 1–3 will represent the bus being late, and the integers 4–5 will represent the bus not being late. The simulation will consist of 50 trials.

11b. Sample answer: $P(\text{late}) = 58\%$

13. There is a population mean of 1.5 pets.

15a.

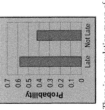

[8, 18] scl: 1 by [0, 10] scl: 1
positively skewed

15b. Sample answer: The distribution is skewed, so use the five-number summary. The values range from 8.7 to 18.1 days. The median is 11.55 days, and half of the data are between 9.55 and 13.3 days.

17. The conclusion may not be valid. The sample may not have been chosen using a simple random sample and may have been too small. This is an observational study, not an experiment. The cause-and-effect relationship cannot be determined.

19a. 68.5%

19b. about 2 games

21a. The president has double the chance of being chosen as everybody else. Everyone should have an equally likely chance of being chosen.

21b. The advisor should assign each member 1 number and if the other number is rolled, ignore it and roll again.

R85

CHAPTER 9
Trigonometric Functions

Chapter 9 Get Ready

1. $a^2 + b^2 = c^2$ **3.** Yes; they both measure 45°. **5.** $x = 12$ **7.** Each leg length will be $\dfrac{x\sqrt{2}}{2}$.

Lesson 9-1

1. $\sin \theta = \dfrac{4}{5}$, $\cos \theta = \dfrac{3}{5}$, $\tan \theta = \dfrac{4}{3}$, $\csc \theta = \dfrac{5}{4}$, $\sec \theta = \dfrac{5}{3}$, $\cot \theta = \dfrac{3}{4}$ **3.** $\sin A = \dfrac{\sqrt{33}}{7}$, $\csc A = \dfrac{7\sqrt{33}}{33}$, $\cot A = \dfrac{4\sqrt{33}}{33}$, $\sec A = \dfrac{7}{4}$, $\tan A = \dfrac{\sqrt{33}}{4}$

5. The side opposite the 60° angle is given. The missing measure is the hypotenuse. Use the sine function to find x.

$$\sin \theta = \dfrac{\text{opp}}{\text{hyp}} \quad\quad \text{Sine function}$$
$$\sin 60° = \dfrac{22}{x} \quad\quad \theta = 60° \text{ and opp} = 22$$
$$\dfrac{\sqrt{3}}{2} = \dfrac{22}{x} \quad\quad \sin 60° = \dfrac{\sqrt{3}}{2}$$
$$\sqrt{3}x = 44 \quad\quad \text{Cross multiply.}$$
$$x = \dfrac{44}{\sqrt{3}} \quad\quad \text{Divide each side by } \sqrt{3}.$$
$$x \approx 25.4 \quad\quad \text{Use a calculator.}$$

7. 8.3 **9.** 25.4 **11.** about 274.7 ft **13.** $\sin \theta = \dfrac{12}{13}$, $\cos \theta = \dfrac{5}{13}$, $\tan \theta = \dfrac{12}{5}$, $\csc \theta = \dfrac{13}{12}$, $\sec \theta = \dfrac{13}{5}$, $\cot \theta = \dfrac{5}{12}$ **15.** $\sin \theta = \dfrac{\sqrt{51}}{10}$, $\cos \theta = \dfrac{7}{10}$, $\tan \theta = \dfrac{\sqrt{51}}{7}$, $\csc \theta = \dfrac{10\sqrt{51}}{51}$, $\sec \theta = \dfrac{10}{7}$, $\cot \theta = \dfrac{7\sqrt{51}}{51}$

17. $\sin A = \dfrac{8}{17}$, $\cos A = \dfrac{15}{17}$, $\csc A = \dfrac{17}{8}$, $\sec A = \dfrac{17}{15}$, $\cot A = \dfrac{15}{8}$ **19.** $\sin B = \dfrac{3\sqrt{10}}{10}$, $\cos B = \dfrac{\sqrt{10}}{10}$, $\csc B = \dfrac{\sqrt{10}}{3}$, $\sec B = \sqrt{10}$, $\cot B = \dfrac{1}{3}$ **21.** 12.7

23.
$$\tan \theta = \dfrac{\text{opp}}{\text{adj}} \quad\quad \text{Tangent function}$$
$$\tan 30° = \dfrac{x}{18} \quad\quad \text{Replace } \theta \text{ with 30°, opp with } x, \text{ and adj with 18.}$$
$$\dfrac{\sqrt{3}}{3} = \dfrac{x}{18} \quad\quad \tan 30° = \dfrac{\sqrt{3}}{3}$$
$$\dfrac{18\sqrt{3}}{3} = x \quad\quad \text{Multiply each side by 18.}$$
$$10.4 \approx x \quad\quad \text{Use a calculator.}$$

25. 8.7 **27.** 132.5 ft **29.** 30 **31.** 36.9 **33.** 32.5 **35.** 25.3 ft higher **37.** $x = 21.9$, $y = 20.8$ **39.** $x = 19.3$, $y = 70.7$ **41.** 54.9 **43.** 20.5 **45.** 11.5

47.

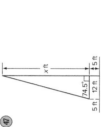

$$\tan \theta = \dfrac{\text{opp}}{\text{adj}} \quad\quad \text{Tangent function}$$
$$\tan 74.5° = \dfrac{x}{12} \quad\quad \text{Replace } \theta \text{ with 74.5°, opp with } x \text{ and adj with 12.}$$
$$12 \cdot \tan 74.5° = x \quad\quad \text{Multiply each side by 12.}$$
$$43 \approx x \quad\quad \text{Use a calculator.}$$

So, the height of the bird's nest is 43 + 5 or 48 feet.

49a. about 647.2 ft **49b.** about 239.4 ft **51.** $m\angle A = 59°$, $a = 31.6$, $c = 36.9$ **53.** $m\angle A = 38.7$, $m\angle B = 51.3°$, $b = 7.5$, $c = 9.6$ **55.** True; $\sin \theta = \dfrac{\text{opp}}{\text{hyp}}$ and the values of the opposite side and the hypotenuse of an acute triangle are positive, so the value of the sine function is positive. **57.** Sample answer: The slope describes the ratio of the vertical rise to the horizontal run of the roof. The vertical rise is opposite the angle that the roof makes with the horizontal. The horizontal run is the adjacent side. So, the tangent of the angle of elevation equals the ratio of the rise to the run, or the slope of the roof; $\theta = 33.7$. **59.** C **61a.** D **61b.** D **61c.** D **61d.** C **61e.** $28.3^2 \approx 15^2 + 24^2$

Lesson 9-2

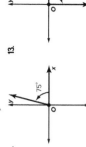

1. (140°)

3. (390°)

5. Sample answer: positive angle: $175° + 360° = 535°$; negative angle: $175° - 360° = -185°$

7. 45° **9.** $-\dfrac{2\pi}{9}$

11. (75°)

13. (−90°)

15. (295°)

17. (240°)

R86 — Selected Answers and Solutions

19. Sample answer: 410°, -310° **21.** Sample answer: 565°, -155° **23.** Sample answer: 280°, -440°

25. $330° = 330 \cdot \dfrac{\pi \text{ radians}}{180°}$
$= \dfrac{330\pi}{180}$ or $\dfrac{11\pi}{6}$ radians

27. -60° **29.** $\dfrac{19\pi}{18}$ **31.** Sample answer: 260°, -100°
35. 1 h 15 min **33.** 6.7 cm **37.** about 12.6 ft
39. Sample answer: $\dfrac{5\pi}{4}, -\dfrac{11\pi}{4}$

41. a.
[figure 165°]

b. $165° = 165 \cdot \dfrac{\pi \text{ radians}}{180°}$
$= \dfrac{165\pi}{180}$ or $\dfrac{11\pi}{12}$ radians

c. $s = r\theta$ Formula for arc length
$= 6.5 \cdot \dfrac{11\pi}{12}$ $r = 6.5$ and $\theta = \dfrac{11\pi}{12}$
≈ 18.7 ft Use a calculator.

d. The arc length would double. Because $s = r\theta$, if r is doubled and θ remains unchanged, then the value of s is also doubled.

43. 472.5° **45.** $\dfrac{10\pi}{9}$ **47a.** 3 **47b.** 0.1 ft **49.** $x = 2$
51. Sample answer: 440° and -280°

[figure 80°]

53. $\dfrac{\theta}{2\pi} = \dfrac{s}{2\pi r}$ Substitute.
$2\pi r\theta = 2\pi r s$ Find the cross products.
$r\theta = s$ Divide each side by 2π.
One degree represents an angle measure that equals $\dfrac{1}{360}$ rotation around a circle. One radian represents the measure of an arc of length r that intercepts an arc of length r. To change from degrees to radians, multiply the number of degrees by $\dfrac{\pi \text{ radians}}{180°}$. To change from radians to degrees, multiply the number of radians by $\dfrac{180°}{\pi \text{ radians}}$.

55. C **57a.** D **57b.** A **57c.** 142.5° **57d.** 9.9 ft
59a. 3 radians **59b.** $\dfrac{540}{\pi}$ degrees

Lesson 9-3

1. $\sin \theta = \dfrac{2\sqrt{5}}{5}$, $\cos \theta = \dfrac{\sqrt{5}}{5}$, $\tan \theta = 2$, $\csc \theta = \dfrac{\sqrt{5}}{2}$,
$\sec \theta = \sqrt{5}$, $\cot \theta = \dfrac{1}{2}$ **3.** $\sin \theta = -1$, $\cos \theta = 0$,
$\tan \theta =$ undefined, $\csc \theta = -1$, $\sec \theta =$ undefined,
$\cot \theta = 0$

5. 65°
[figure $\theta = 115°$]

7. $\dfrac{\sqrt{2}}{2}$ **9.** -2

11a. [figure 125°]
11b. 55°; $\cos 55° = \dfrac{d}{5\frac{1}{2}}$
11c. 3.2 in.

13. [figure (-6, 8)]
$r = \sqrt{x^2 + y^2}$
$= \sqrt{(-6)^2 + 8^2}$
$= \sqrt{100}$ or 10
Use $x = -6$, $y = 8$, and $r = 10$.
$\sin \theta = \dfrac{y}{r} = \dfrac{8}{10}$ or $\dfrac{4}{5}$
$\cos \theta = \dfrac{x}{r} = \dfrac{-6}{10}$ or $-\dfrac{3}{5}$
$\tan \theta = \dfrac{y}{x} = \dfrac{8}{-6}$ or $-\dfrac{4}{3}$
$\csc \theta = \dfrac{r}{y} = \dfrac{10}{8}$ or $\dfrac{5}{4}$
$\sec \theta = \dfrac{r}{x} = \dfrac{10}{-6}$ or $-\dfrac{5}{3}$
$\cot \theta = \dfrac{x}{y} = \dfrac{-6}{8}$ or $-\dfrac{3}{4}$

15. $\sin \theta = -1$, $\cos \theta = 0$, $\tan \theta =$ undefined, $\csc \theta = -1$, $\sec \theta =$ undefined, $\cot \theta = 0$
17. $\sin \theta = \dfrac{\sqrt{10}}{10}$, $\cos \theta = \dfrac{3\sqrt{10}}{10}$, $\tan \theta = \dfrac{1}{3}$,
$\csc \theta = \sqrt{10}$, $\sec \theta = -\sqrt{10}$, $\cot \theta = 3$
19. 75°
[figure $\theta = 285°$]
21. $\dfrac{\pi}{4}$
[figure $\theta = \dfrac{7\pi}{4}$]

R87 — Selected Answers and Solutions

23. 40°
[figure 400°]

25. -1 **27.** $-\sqrt{2}$
29. $\dfrac{1}{2}$ **31.** $\dfrac{2\sqrt{3}}{3}$

33. [figure 145°, 10 ft, 35°, d]
$\cos 35° = \dfrac{d}{10}$
$10 \cdot \cos 35° = d$
$8.2 \approx d$
The water reaches about 8.2 feet to the left of the sprinkler.

35. about 10.1 m **37.** $\cos \theta = -\dfrac{3}{5}$, $\tan \theta = -\dfrac{4}{3}$,
$\csc \theta = \dfrac{5}{4}$, $\sec \theta = -\dfrac{5}{3}$, $\cot \theta = -\dfrac{3}{4}$ **39.** $\sin \theta = \dfrac{15}{17}$,
$\tan \theta = \dfrac{15}{8}$, $\csc \theta = \dfrac{17}{15}$, $\sec \theta = -\dfrac{17}{8}$, $\cot \theta = \dfrac{8}{15}$
41. 0 **43.** $-\dfrac{1}{2}$ **45.** $\dfrac{\sqrt{3}}{2}$ **47.** No; for $\sin \theta = \dfrac{\sqrt{2}}{2}$ and $\tan \theta = -1$, the reference angle is 45°. However, for $\sin \theta$ to be positive and $\tan \theta$ to be negative, the reference angle must be in the second quadrant. So, the value of θ must be 135° or an angle coterminal with 135°. **49.** Sample answer: We know that $\cot \theta = \dfrac{x}{y}$, $\sin \theta = \dfrac{y}{r}$, and $\cos \theta = \dfrac{x}{r}$. Because $\sin 180 = 0$, it must be true that $y = 0$. Thus, $\cot \theta = \dfrac{x}{0}$ which is undefined. **51.** Sample answer: First, sketch the angle and determine in which quadrant it is located. Then use the appropriate rule for finding its reference angle θ'. A reference angle is the acute angle formed by the terminal side of θ and the x-axis. Next, find the value of the trigonometric function for θ'. Finally, use the quadrant location to determine the sign of the trigonometric function value of θ. **53.** E
55a. A, D, E, G **55b.** E **55c.** C, G

Lesson 9-4

1. $\cos \theta = \dfrac{15}{17}$, $\sin \theta = \dfrac{8}{17}$ **3.** 2 **5a.** 4 seconds
5b. Sample answer:
[graph — Height of Swing (ft) vs Time (s)]

7. $\dfrac{\sqrt{3}}{2}$ **9.** $\cos \theta = \dfrac{3}{5}$, $\sin \theta = \dfrac{4}{5}$; $P(\cos \theta, \sin \theta)$
11. $P\left(\dfrac{\sqrt{3}}{2}, \dfrac{1}{2}\right) = P(\cos \theta, \sin \theta)$

13. 3 **15.** 12 **17.** 180°
$\cos \theta = \dfrac{\sqrt{3}}{2}$, $\sin \theta = \dfrac{1}{2}$

19a.
[graph — Temperature (°F) vs Month]

19b. 12 mo or 1 yr **21.** $\dfrac{1}{2}$ **23.** $\dfrac{\sqrt{2}}{2}$ **25.** $\dfrac{\sqrt{3}}{2}$
27. a. The period is the time it takes to complete one rotation. So, the period is 60 seconds ÷ 2.5 or 24 seconds.
b. Because the siren is 1 mile from Ms. Miller's house and the beam of sound has a radius of 1 mile, then the minimum distance of the sound beam from the house is 0 miles and the maximum distance is 2 miles. Draw a sine curve with the pattern repeating every 24 seconds. Sample answer:
[graph — d vs t]

29a.
[figures: circles with 60°, 210°, 30°, 150°, 315°]

period $= \dfrac{2\pi}{|b|}$ Write the relationship between the period and b.

$2 = \dfrac{2\pi}{|b|}$ Substitution

$b = \pi$ Solve for b.

$y = a \sin b\theta$ General equation for the sine function
$h = 1 \sin \pi t$ Replace y with 1, b with π, and θ with t.
$h = \sin \pi t$ Simplify.

$h = \sin (180 \cdot 20.5)$
$= 1$ m

b. $h = \sin \pi t$

31a. $y = \cos 260\pi t$
The amplitude remains the same. The period decreases because it is the reciprocal of the frequency.
31b. The amplitude remains the same. The period decreases because it is the reciprocal of the frequency.

33. amplitude: $\dfrac{1}{2}$; period: 480° $\quad y = \dfrac{1}{2}\cos\dfrac{3}{4}\theta$

35. amplitude: does not exist; period: 450° $\quad y = 2\sec\dfrac{4}{5}\theta$

37. amplitude: does not exist; period: 30° $\quad y = 2\cot 6\theta$

19. amplitude: 3; period: 180° $\quad y = 3\cos 2\theta$

21a. $h = 4 \sin \dfrac{2}{3}\pi t$
21b. $h = 4\sin\dfrac{2}{3}\pi t$

23. period: 360° $\quad y = \tan\dfrac{1}{2}\theta$

25. period: 180° $\quad y = 2\cot\theta$

27. period: 180° $\quad y = 2\tan\theta$

29 **a.** Because the frequency is 0.5, the period is $\dfrac{1}{0.5}$ or 2.

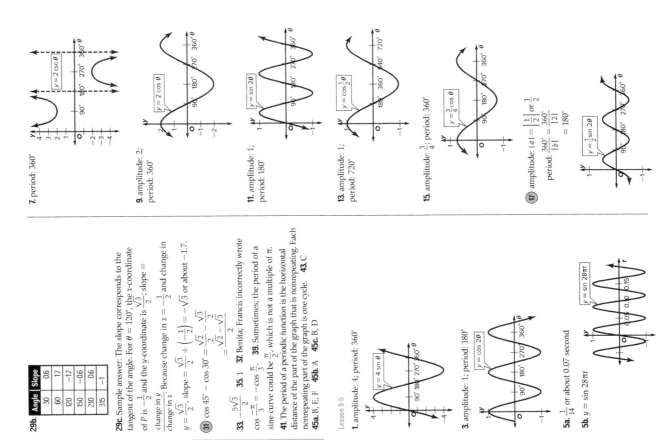

7. period: 360° $\quad y = 2\csc\theta$

9. amplitude: 2; period: 360° $\quad y = 2\cos\theta$

11. amplitude: 1; period: 180° $\quad y = \sin 2\theta$

13. amplitude: 1; period: 720° $\quad y = \cos\dfrac{1}{2}\theta$

15. amplitude: $\dfrac{3}{4}$; period: 360° $\quad y = \dfrac{3}{4}\cos\theta$

17 amplitude: $|a| = \left|\dfrac{1}{2}\right|$ or $\dfrac{1}{2}$
period: $\dfrac{360°}{|b|} = \dfrac{360°}{|2|} = 180°$ $\quad y = \dfrac{1}{2}\sin 2\theta$

29b.

Angle	Slope
30	0.6
60	1.7
120	−1.7
150	−0.6
210	0.6
315	−1

29c. Sample answer: The slope corresponds to the tangent of the angle. For $\theta = 120°$, the x-coordinate of P is $-\dfrac{1}{2}$ and the y-coordinate is $\dfrac{\sqrt{3}}{2}$, slope $= \dfrac{\text{change in } y}{\text{change in } x}$. Because change in $x = -\dfrac{1}{2}$ and change in $y = \dfrac{\sqrt{3}}{2}$, slope $= \dfrac{\sqrt{3}}{2} \div \left(-\dfrac{1}{2}\right) = -\sqrt{3}$ or about -1.7.

31 $y = \dfrac{\sqrt{3}}{2}$, slope $= \dfrac{\sqrt{3}}{2}$

33. $\dfrac{5\sqrt{3}}{2}$ **35.** 1 **37.** Benita; Francis incorrectly wrote $\cos -\dfrac{\pi}{6} = -\cos\dfrac{\pi}{3}$. **39.** Sometimes; the period of a sine curve could be $\dfrac{\pi}{2}$, which is not a multiple of π.

41. The period of a periodic function is the horizontal distance of the part of the graph that is nonrepeating. Each nonrepeating part of the graph is one cycle. **43.** C

$\cos 45° - \cos 30° = \dfrac{\sqrt{2}}{2} - \dfrac{\sqrt{3}}{2} = \dfrac{\sqrt{2}-\sqrt{3}}{2}$

45a. B, E, F **45b.** A **45c.** B, D

Lesson 9-5

1. amplitude: 4; period: 360° $\quad y = 4\sin\theta$

3. amplitude: 1; period: 180° $\quad y = \cos 2\theta$

5a. $\dfrac{1}{14}$ or about 0.07 second
5b. $y = \sin 28\pi t$

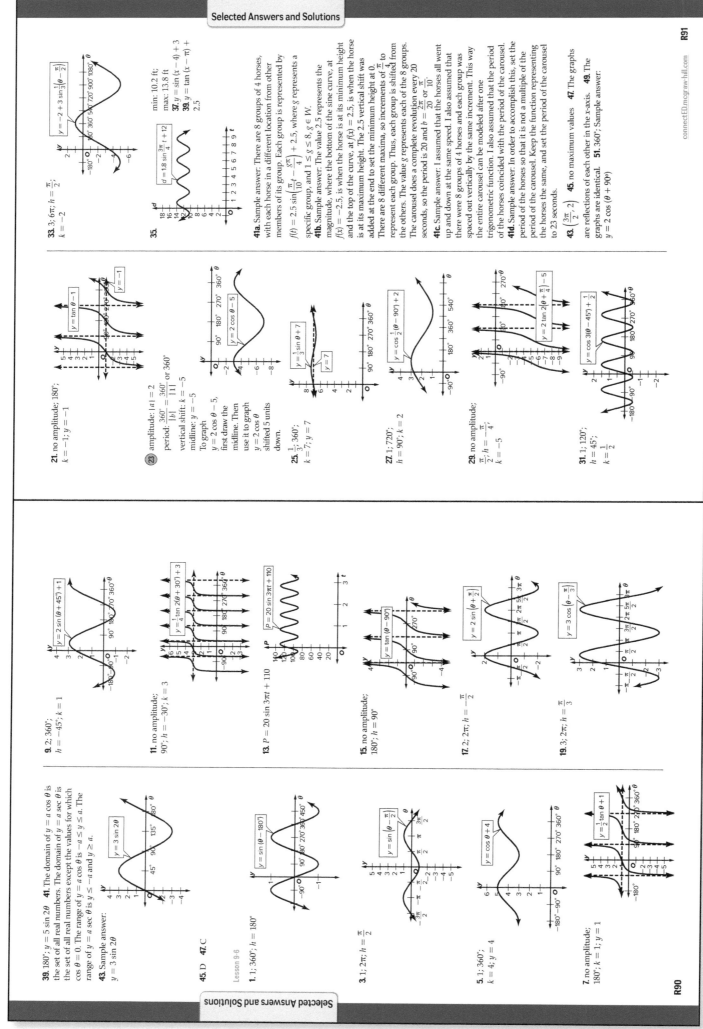

Selected Answers and Solutions

33. 3; 6π; $h = \frac{\pi}{2}$; $k = -2$

$y = -2 + 3 \sin \frac{1}{3}\left(\theta - \frac{\pi}{2}\right)$

35.

$d = 1.8 \sin \frac{3\pi}{4}t + 12$

min: 10.2 ft;
max: 13.8 ft

37. $y = \sin(x - 4) + 3$

39. $y = \tan(x - \pi) + 2.5$

41a. Sample answer: There are 8 groups of 4 horses, with each horse in a different location from other members of its group. Each group is represented by $f(t) = 2.5 \sin\left(\frac{\pi}{10}t - \frac{g\pi}{4}\right) + 2.5$, where g represents a specific group, and $1 \le g \le 8$, $g \in W$.

41b. Sample answer: The value 2.5 represents the magnitude, where the bottom of the sine curve, at $f(x) = -2.5$, is when the horse is at its minimum height and the top of the curve, at $f(x) = 2.5$, is when the horse is at its maximum height. The 2.5 vertical shift was added at the end to set the minimum height at 0. There are 8 different maxima, so increments of $\frac{\pi}{4}$ to represent each group. Thus, each group is shifted from the others. The value g represents each of the 8 groups. The carousel does a complete revolution every 20 seconds, so the period is 20 and $b = \frac{2\pi}{20}$ or $\frac{\pi}{10}$.

41c. Sample answer: I assumed that the horses all went up and down at the same speed. I also assumed that there were 8 groups of 4 horses and each group was spaced out vertically by the same increment. This way the entire carousel can be modeled after one trigonometric function. I also assumed that the period of the horses coincided with the period of the carousel.

41d. Sample answer: In order to accomplish this, set the period of the horses so that it is not a multiple of the period of the carousel. Keep the function representing the horses the same, and set the period of the carousel to 23 seconds.

43. $\left(\frac{3\pi}{2}, 2\right)$ **45.** no maximum values **47.** The graphs are reflections of each other in the x-axis. **49.** The graphs are identical. **51.** 360°; Sample answer: $y = 2 \cos(\theta + 90°)$

21. no amplitude; 180°; $k = -1$; $y = -1$

$y = \tan \theta - 1$
$y = -1$

23 amplitude: $|a| = 2$
period: $\frac{360°}{|b|} = \frac{360°}{|1|}$ or 360°
vertical shift: $k = -5$
midline: $y = -5$
To graph $y = 2\cos\theta - 5$, first draw the midline. Then use it to graph $y = 2\cos\theta$ shifted 5 units down.

$y = 2\cos\theta - 5$

25. $\frac{1}{3}$; 360°; $k = 7$; $y = 7$

$y = \frac{1}{3}\sin\theta + 7$
$y = 7$

27. 1; 720°; $h = 90°$; $k = 2$

$y = \cos\frac{1}{2}(\theta - 90°) + 2$

29. no amplitude; $\frac{\pi}{2}$; $h = -\frac{\pi}{4}$; $k = -5$

$y = \tan 2\left(\theta + \frac{\pi}{4}\right) - 5$

31. 1; 120°; $h = 45°$; $k = \frac{1}{2}$

$y = \cos 3(\theta - 45°) + \frac{1}{2}$

9. 2; 360°; $h = -45°$; $k = 1$

$y = 2\sin(\theta + 45°) + 1$

11. no amplitude; 90°; $h = -30°$; $k = 3$

$y = \frac{1}{4}\tan 2(\theta + 30°) + 3$

13. $P = 20 \sin 3\pi t + 110$

$P = 20 \sin 3\pi t + 110$

15. no amplitude; 180°; $h = 90°$

$y = \tan(\theta - 90°)$

17. 2; 2π; $h = -\frac{\pi}{2}$

$y = 2\sin\left(\theta + \frac{\pi}{2}\right)$

19. 3; 2π; $h = \frac{\pi}{3}$

$y = 3\cos\left(\theta - \frac{\pi}{3}\right)$

39. 180°; $y = 5\sin 2\theta$ **41.** The domain of $y = a\cos\theta$ is the set of all real numbers. The domain of $y = a\sec\theta$ is the set of all real numbers except the values for which $\cos\theta = 0$. The range of $y = a\cos\theta$ is $-a \le y \le a$. The range of $y = a\sec\theta$ is $y \le -a$ and $y \ge a$.

43. Sample answer: $y = 3\sin 2\theta$

$y = 3\sin 2\theta$

45. D   **47.** C

Lesson 9-6

1. 1; 360°; $h = 180°$

$y = \sin(\theta - 180°)$

3. 1; 2π; $h = \frac{\pi}{2}$

$y = \sin\left(\theta - \frac{\pi}{2}\right)$

5. 1; 360°; $k = 4$; $y = 4$

$y = \cos\theta + 4$

7. no amplitude; 180°; $k = 1$; $y = 1$

$y = \frac{1}{2}\tan\theta + 1$

53 The midline lies halfway between the maximum and minimum values. So $y = \frac{4+2}{2}$ or 3. Because the midline is $y = 3$, the vertical shift is $k = 3$.

The amplitude is the difference between the midline value and the maximum value. So $|a| = |4 - 3|$ or 1.

Because the cycle repeats every 180°, the period is 180°.

$period = \frac{360°}{|b|}$ Write the relationship between the period and b.

$180° = \frac{360°}{|b|}$ Substitution

$b = 2$ Solve for b.

The graph is the sine curve shifted 45° to the right. So the phase shift is $h = 45°$.

$y = a \sin b(\theta - h) + k$ General equation for the sine function

$y = 1 \sin 2(\theta - 45°) + 3$ Replace a with 1, b with 2, h with 45° and k with 3

$y = \sin 2(\theta - 45°) + 3$ Simplify.

55. 180°; no phase shift; $k = 6$

57. 120°; $h = 45°$; $k = 1$

59. π; $h = -\frac{\pi}{2}$, $k = -3$

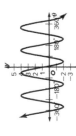
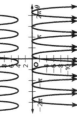
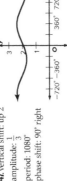

$y = \cot \theta + 6$

$y = \frac{1}{2} \sec 3(\theta - 45°) + 1$

$y = 4 \sec 2\left(\theta + \frac{\pi}{2}\right) - 3$

61. The graph of $y = 3 \sin 2\theta + 1$ has an amplitude of 3 rather than an amplitude of 1. It is shifted up 1 unit from the parent graph and is compressed so that it has a period of 180°.

63. Sample answer: $y = 2 \sin \theta - 3$

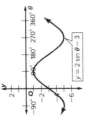

$y = 2 \sin \theta - 3$

65. A   **67.** C   **69a.** B, D   **69b.** B   **69c.** C   **69d.** A, C
69e. E

Chapter 9   Study Guide and Review

1. true   **3.** true   **5.** Use the given trigonometric ratio to label two sides of a right triangle, and then use the Pythagorean Theorem to find the third side. Apply definitions to determine the other ratios.   **7.** $a = 10.9$; $A = 65°$; $B = 25°$ **9.** $A = 15°$; $a = 4.0$; $c = 15.5$
11. $B = 55°$; $a = 12.6$; $b = 18.0$
13. about 8.8 feet   **15.** 450°   **17.** $\frac{7\pi}{4}$ **19.** 295°, −425°
21. $\frac{16}{\pi}$ **23.** $\frac{\sqrt{3}}{3}$ **25.** 0   **27.** $\sin \theta = \frac{12}{13}$, $\cos \theta = \frac{5}{13}$,
$\tan \theta = \frac{12}{5}$, $\csc \theta = \frac{13}{12}$, $\sec \theta = \frac{13}{5}$, $\cot \theta = \frac{5}{12}$
29. about 17.1 meters   **31.** $\frac{-\sqrt{6}}{4}$ **33.** 0   **35.** 15 seconds
37. amplitude: 1, period: 720°

$y = \cos \frac{1}{2}\theta$

39. amplitude: not defined, period: 360°

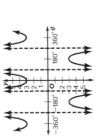

41. amplitude: not defined, period: 720°

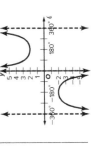

43. vertical shift: up 1; amplitude: 3; period 180°; phase shift: 90° right

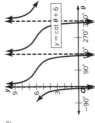

45. vertical shift: up 2
amplitude: not defined
period: $\frac{2\pi}{3}$
phase shift: right

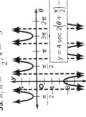

47. vertical shift: up 2
amplitude: $\frac{1}{3}$
period: 1080°
phase shift: 90° right

Selected Answers and Solutions

CHAPTER 10

Trigonometric Identities and Equations

Chapter 10 Get Ready

1. factorable; $-4a(4a-1)$ **3.** $x^2 + x - 42 = 0$
5. $180°, -135°$

Lesson 10-1

1. $\frac{1}{2}$ **3.** $\frac{\sqrt{5}}{3}$ **5.** $\sin\theta\cos\theta$ **7.** $\cot^2\theta$ **9.** $\frac{5}{4}$ **11.** $\frac{4}{5}$

13. $-\frac{5}{4}$

15. $\cot^2\theta + 1 = \csc^2\theta$ Pythagorean Identity
$\left(\frac{1}{4}\right)^2 + 1 = \csc^2\theta$ Substitute $\frac{1}{4}$ for $\cot\theta$.
$\frac{1}{16} + 1 = \csc^2\theta$ Square $\frac{1}{4}$.
$\frac{17}{16} = \csc^2\theta$ Add.
$\pm\frac{\sqrt{17}}{4} = \csc\theta$ Take the square root of each side.

Because θ is in the third quadrant, $\csc\theta$ is negative.

So, $\csc\theta = -\frac{\sqrt{17}}{4}$.

17. $\frac{12}{13}$ **19.** $\frac{3}{5}$ **21.** $\sec^3\theta$ **23.** $\csc\frac{1}{2}\theta$ **25.** 1

27. $\ell B = \frac{F\csc\theta}{\ell\ell}$
$\ell \cdot B = F\csc\theta$ Original equation
$\ell\ell \cdot B = F\csc\theta$ Multiply each side by $\ell\ell$.
$\ell\ell B = F \cdot \frac{1}{\sin\theta}$ $\csc\theta = \frac{1}{\sin\theta}$
$\ell\ell B\sin\theta = F$ Multiply each side by $\sin\theta$.

The equation can be written as $F = \ell\ell B\sin\theta$.

29. $\sec\theta$ **31.** 2 **33.** $2\cos^2\theta$ **35a.** $\frac{\sqrt{65}}{9}$ **35b.** $\frac{4\sqrt{65}}{65}$

35c. $\frac{4}{9}$, $\frac{\sqrt{65}}{9}$, $\frac{4\sqrt{65}}{65}$ **37.** $\mu_k = \tan\theta$

39. $\frac{\cos\left(\frac{\pi}{2}-\theta\right)}{1+\sin(-\theta)} - 1 = \frac{\sin\theta - 1}{1+\sin(-\theta)} - 1$ $\cos\left(\frac{\pi}{2}-\theta\right) = \sin\theta$
$= \frac{\sin\theta - 1}{1 - \sin\theta} - 1$ $\sin(-\theta) = -\sin\theta$
$= \frac{\sin\theta - 1}{-1(\sin\theta - 1)} - 1$ $1 - \sin\theta = -1(\sin\theta - 1)$
$= \frac{1}{-1}$ or -1 Simplify.

41. $-\cot^2\theta$ **43.** Sample answer: $x = 45°$ **45.** The functions $\cos\theta$ and $\sin\theta$ can be thought of as the lengths of the legs of a right triangle, and the number 1 can be thought of as the measure of the corresponding hypotenuse. **47.** Sample answer: $\frac{\sin\theta}{\cos\theta} \cdot \sin\theta$ and $\frac{\sin^2\theta}{\cos\theta}$ **49.** $\frac{4}{3}$ **51.** E **53.** D **55a.** D

55b.
$\frac{\sin(90° - \theta)}{\cos(\theta - 90°)} = \frac{\sin(90° - \theta)}{\cos(90° - \theta)}$ Negative angle identity, $\cos(\theta - 90) = \cos(90 - \theta)$
$= \frac{\cos\theta}{\sin\theta}$ Substitute the cofunction identities.
$= \cot\theta$ Substitute the cotangent quotient identity.

57. $\sin^2\theta$ **59.** C, D

Lesson 10-2

1. $\cot\theta + \tan\theta \stackrel{?}{=} \frac{\sec^2\theta}{\tan\theta}$
$\cot\theta + \tan\theta \stackrel{?}{=} \frac{\tan^2\theta + 1}{\tan\theta}$
$\cot\theta + \tan\theta \stackrel{?}{=} \frac{\tan^2\theta}{\tan\theta} + \frac{1}{\tan\theta}$
$\cot\theta + \tan\theta = \tan\theta + \cot\theta$ ✓

3. $\sin^2\theta \stackrel{?}{=} \frac{\sec\theta}{\tan\theta + \cot\theta}$
$\sin^2\theta \stackrel{?}{=} \frac{\frac{1}{\cos\theta}}{\frac{\sin\theta}{\cos\theta} + \frac{\cos\theta}{\sin\theta}}$
$\sin^2\theta \stackrel{?}{=} \frac{\frac{1}{\cos\theta}}{\frac{\sin^2\theta + \cos^2\theta}{\cos\theta\sin\theta}}$
$\sin^2\theta \stackrel{?}{=} \frac{1}{\cos\theta} \cdot \frac{\cos\theta\sin\theta}{1}$
$\sin^2\theta = \sin\theta$ ✓

5. $\tan^2\theta\csc^2\theta \stackrel{?}{=} 1 + \tan^2\theta$
$\frac{\sin^2\theta}{\cos^2\theta} \cdot \frac{1}{\sin^2\theta} \stackrel{?}{=} \sec^2\theta$
$\frac{1}{\cos^2\theta} \stackrel{?}{=} \sec^2\theta$
$\sec^2\theta = \sec^2\theta$ ✓

7. $\frac{\tan^2\theta + 1}{\tan^2\theta} \stackrel{?}{=} \csc^2\theta$
$\frac{\sec^2\theta}{\tan^2\theta} \stackrel{?}{=} \csc^2\theta$
$\frac{\frac{1}{\cos^2\theta}}{\frac{\sin^2\theta}{\cos^2\theta}} \stackrel{?}{=} \csc^2\theta$
$\frac{1}{\cos^2\theta} \cdot \frac{\cos^2\theta}{\sin^2\theta} \stackrel{?}{=} \csc^2\theta$
$\frac{1}{\sin^2\theta} \stackrel{?}{=} \csc^2\theta$
$\csc^2\theta = \csc^2\theta$ ✓

9. $\sin^2\theta + \tan^2\theta \stackrel{?}{=} (1 - \cos^2\theta) + \frac{\sec^2\theta}{\csc^2\theta}$
$\sin^2\theta + \tan^2\theta \stackrel{?}{=} \sin^2\theta + \frac{\sec^2\theta}{\csc^2\theta}$
$\sin^2\theta + \tan^2\theta \stackrel{?}{=} \sin^2\theta + \frac{\frac{1}{\cos^2\theta}}{\frac{1}{\sin^2\theta}}$
$\sin^2\theta + \tan^2\theta \stackrel{?}{=} \sin^2\theta + \frac{1}{\cos^2\theta} \cdot \frac{\sin^2\theta}{\cos^2\theta} \div \frac{1}{\sin^2\theta}$
$\sin^2\theta + \tan^2\theta \stackrel{?}{=} \sin^2\theta + \tan^2\theta$ ✓

11. $\cot\theta(\cot\theta + \tan\theta) \stackrel{?}{=} \csc^2\theta$
$\cot^2\theta + \cot\theta\tan\theta \stackrel{?}{=} \csc^2\theta$
$\cot^2\theta + \frac{\cos\theta}{\sin\theta} \cdot \frac{\sin\theta}{\cos\theta} \stackrel{?}{=} \csc^2\theta$
$\cot^2\theta + 1 \stackrel{?}{=} \csc^2\theta$
$\csc^2\theta = \csc^2\theta$ ✓

13. $\sin\theta\sec\theta\cot\theta \stackrel{?}{=} 1$
$\sin\theta \cdot \frac{1}{\cos\theta} \cdot \frac{\cos\theta}{\sin\theta} \stackrel{?}{=} 1$
$1 = 1$ ✓

15. $\frac{1 - 2\cos^2\theta}{\sin\theta\cos\theta} \stackrel{?}{=} \tan\theta - \cot\theta$
$\frac{(1 - \cos^2\theta) - \cos^2\theta}{\sin\theta\cos\theta} \stackrel{?}{=} \tan\theta - \cot\theta$

$\frac{\sin^2\theta - \cos^2\theta}{\sin\theta\cos\theta} \stackrel{?}{=} \tan\theta - \cot\theta$
$\frac{\sin^2\theta}{\sin\theta\cos\theta} - \frac{\cos^2\theta}{\sin\theta\cos\theta} \stackrel{?}{=} \tan\theta - \cot\theta$
$\frac{\sin\theta}{\cos\theta} - \frac{\cos\theta}{\sin\theta} \stackrel{?}{=} \tan\theta - \cot\theta$
$\tan\theta - \cot\theta = \tan\theta - \cot\theta$ ✓

17. $\cos\theta \stackrel{?}{=} \sin\theta\cot\theta$
$\cos\theta \stackrel{?}{=} \sin\theta \frac{\cos\theta}{\sin\theta}$
$\cos\theta = \cos\theta$ ✓

19. $\cos\theta\cos(-\theta) - \sin\theta\sin(-\theta) \stackrel{?}{=} 1$
$\cos\theta\cos\theta - \sin\theta(-\sin\theta) \stackrel{?}{=} 1$
$\cos^2\theta + \sin^2\theta \stackrel{?}{=} 1$
$1 = 1$ ✓

21. $\sec\theta - \tan\theta \stackrel{?}{=} \frac{1 - \sin\theta}{\cos\theta}$
$\frac{1}{\cos\theta} - \frac{\sin\theta}{\cos\theta} \stackrel{?}{=} \frac{1 - \sin\theta}{\cos\theta}$
$\frac{1 - \sin\theta}{\cos\theta} = \frac{1 - \sin\theta}{\cos\theta}$ ✓

23. $\sec\theta\csc\theta \stackrel{?}{=} \tan\theta + \cot\theta$
$\frac{1}{\cos\theta} \cdot \frac{1}{\sin\theta} \stackrel{?}{=} \frac{\sin\theta}{\cos\theta} + \frac{\cos\theta}{\sin\theta}$
$\frac{1}{\sin\theta\cos\theta} \stackrel{?}{=} \frac{\sin^2\theta + \cos^2\theta}{\sin\theta\cos\theta}$
$\frac{1}{\sin\theta\cos\theta} = \frac{1}{\sin\theta\cos\theta}$ ✓

25. $(\sin\theta + \cos\theta)^2 \stackrel{?}{=} 2 + \sec\theta\csc\theta$
$(\sin\theta + \cos\theta)^2 \stackrel{?}{=} \frac{2 + \sec\theta\csc\theta}{\sec\theta\csc\theta}$
$(\sin\theta + \cos\theta)^2 \stackrel{?}{=} \frac{2 + \frac{1}{\cos\theta} \cdot \frac{1}{\sin\theta}}{\frac{1}{\cos\theta\sin\theta}} \cdot \frac{\cos\theta\sin\theta}{1}$
$(\sin\theta + \cos\theta)^2 \stackrel{?}{=} 2\cos\theta\sin\theta + 1$
$(\sin\theta + \cos\theta)^2 \stackrel{?}{=} 2\cos\theta\sin\theta + \cos^2\theta + \sin^2\theta$
$(\sin\theta + \cos\theta)^2 = (\sin\theta + \cos\theta)^2$ ✓

27. $\csc\theta - 1 \stackrel{?}{=} \frac{\cot^2\theta}{\csc\theta + 1}$
$\csc\theta - 1 \stackrel{?}{=} \frac{\csc^2\theta - 1}{\csc\theta + 1}$
$\csc\theta - 1 \stackrel{?}{=} \frac{(\csc\theta - 1)(\csc\theta + 1)}{\csc\theta + 1}$
$\csc\theta - 1 = \csc\theta - 1$ ✓

29. $\sin\theta\cos\theta\tan\theta + \cos^2\theta \stackrel{?}{=} 1$
$\sin\theta\cos\theta \cdot \frac{\sin\theta}{\cos\theta} + \cos^2\theta \stackrel{?}{=} 1$
$\sin^2\theta + \cos^2\theta \stackrel{?}{=} 1$
$1 = 1$ ✓

31. $\csc^2\theta \stackrel{?}{=} \cot^2\theta + \sin\theta\csc\theta$
$\csc^2\theta \stackrel{?}{=} \cot^2\theta + \sin\theta \cdot \frac{1}{\sin\theta}$
$\csc^2\theta \stackrel{?}{=} \cot^2\theta + 1$
$\csc^2\theta = \csc^2\theta$ ✓

33. $\sin^2\theta + \cos^2\theta \stackrel{?}{=} \sec^2\theta - \tan^2\theta$
$1 \stackrel{?}{=} \tan^2\theta + 1 - \tan^2\theta$
$1 = 1$ ✓

35. yes

37. $\cot(-\theta)\tan(-\theta) = \frac{1}{\tan(-\theta)} \cdot \tan(-\theta)$
$= 1$
$= \frac{\tan(-\theta)}{\cot(-\theta)}$ Simplify.

39. 1 **41.** 1 **43.** $\cos\theta$ **45.** 2 **47.** $\sin\theta$ **49.** 1

51. $y = \frac{gt^2}{2v_0^2}(1 + \tan^2\theta) + x\tan\theta$

53a. $h = \frac{v_0^2\sin^2\theta}{2g} = \frac{47^2\sin^2\theta}{2(9.8)}$ Replace v_0 with 47 and g with 9.8
$= \frac{2209\sin^2\theta}{19.6}$ Simplify.

$\frac{2209\sin^2 30°}{19.6} \approx 28.2$ m $\theta = 30°$
$\frac{2209\sin^2 45°}{19.6} \approx 56.4$ m $\theta = 45°$
$\frac{2209\sin^2 60°}{19.6} \approx 84.5$ m $\theta = 60°$
$\frac{2209\sin^2 90°}{19.6} \approx 112.7$ m $\theta = 90°$

b. Sample answer: Enter the equation $y = \frac{2209(\sin\theta)^2}{19.6}$.
Use the window Xmin = 0, Xmax = 180, Xscl = 10, Ymin = 0, Ymax = 150, Yscl = 10, Xres = 1.

[0, 180] scl: 10 by [0, 150] scl: 10

c. $\frac{v_0^2\tan^2\theta}{2g\sec^2\theta} \stackrel{?}{=} \frac{v_0^2\sin^2\theta}{2g}$
$\frac{v_0^2\left(\frac{\sin^2\theta}{\cos^2\theta}\right)}{2g\left(\frac{1}{\cos^2\theta}\right)} \stackrel{?}{=} \frac{v_0^2\sin^2\theta}{2g}$
$\frac{v_0^2\sin^2\theta}{2g} = \frac{v_0^2\sin^2\theta}{2g}$ ✓

55. $\tan^2\theta = \frac{\sin^2\theta}{\cos^2\theta} = \frac{1 - \cos^2\theta}{\cos^2\theta} = \frac{1}{\cos^2\theta} - \frac{\cos^2\theta}{\cos^2\theta}$
$= \sec^2\theta - 1$

57. Sample answer: counterexample $45°, 30°$

59. Sample answer: Sine and cosine are the trigonometric functions with which most people are familiar, and all trigonometric expressions can be written in terms of sine and cosine. Also, by rewriting complex trigonometric expressions in terms of sine and cosine it may be easier to perform operations and to apply trigonometric properties.

R96 (left page)

61. Using the unit circle and the Pythagorean Theorem, we can justify $\cos^2\theta + \sin^2\theta = 1$.

If we divide each term of the identity
$\cos^2\theta + \sin^2\theta = 1$
by $\cos^2\theta$, we can justify $1 + \tan^2\theta = \sec^2\theta$.

$$\frac{\cos^2\theta}{\cos^2\theta} + \frac{\sin^2\theta}{\cos^2\theta} = \frac{1}{\cos^2\theta}$$
$$1 + \tan^2\theta = \sec^2\theta$$

If we divide each term of the identity $\cos^2\theta + \sin^2\theta = 1$ by $\sin^2\theta$, we can justify $\cot^2\theta + 1 = \csc^2\theta$.

$$\frac{\cos^2\theta}{\sin^2\theta} + \frac{\sin^2\theta}{\sin^2\theta} = \frac{1}{\sin^2\theta}$$
$$\cot^2\theta + 1 = \csc^2\theta$$

63. A **65a.** C

65b.
$$\frac{\sin^2\theta + \cos^2\theta}{\tan\theta\cos\theta} = \frac{1}{\tan\theta\cos\theta} \quad \text{Use the Pythagorean Identity.}$$
$$= \frac{1}{\left(\frac{\sin\theta}{\cos\theta}\right)\cos\theta} \quad \text{Substitute the quotient identity for } \tan\theta.$$
$$= \frac{1}{\sin\theta} \quad \text{Simplify.}$$

67. $\sin x \cos x \tan x = \sin x \cos x \left(\frac{\sin x}{\cos x}\right)$
$$= \sin^2 x$$
$$= 1 - \cos^2 x$$

69. D

Lesson 10-3

1 $\cos 165° = \cos(120° + 45°)$

3. $\dfrac{\sqrt{6}-\sqrt{2}}{4}$ **5.** $\dfrac{\sqrt{2}}{2}$ **7a.** 0 **7b.** The interference is destructive. The signals cancel each other completely.

9.
$$\cos\frac{3\pi}{2}\cos\theta + \sin\frac{3\pi}{2}\sin\theta \stackrel{?}{=} -\sin\theta$$
$$0 \cdot \cos\theta - 1 \cdot \sin\theta \stackrel{?}{=} -\sin\theta$$
$$-\sin\theta = -\sin\theta\ \checkmark$$

11.
$$\sin(\theta + \pi) \stackrel{?}{=} -\sin\theta$$
$$\sin\theta\cos\pi + \cos\theta\sin\pi \stackrel{?}{=} -\sin\theta$$
$$(\sin\theta)(-1) + (\cos\theta)(0) \stackrel{?}{=} -\sin\theta$$
$$-\sin\theta = -\sin\theta\ \checkmark$$

13. $-\dfrac{\sqrt{2}}{2}$ **15.** $\dfrac{\sqrt{6}-\sqrt{2}}{4}$ **17.** $\dfrac{\sqrt{2}+\sqrt{6}}{4}$

19.
$$\cos\left(\frac{\pi}{2}+\theta\right) \stackrel{?}{=} -\sin\theta$$
$$\cos\frac{\pi}{2}\cos\theta - \sin\frac{\pi}{2}\sin\theta \stackrel{?}{=} -\sin\theta$$
$$(0)(\cos\theta) - (1)(\sin\theta) \stackrel{?}{=} -\sin\theta$$
$$-\sin\theta = -\sin\theta\ \checkmark$$

21.
$$\cos(180° + \theta) \stackrel{?}{=} -\cos\theta$$
$$\cos 180°\cos\theta - \sin 180°\sin\theta \stackrel{?}{=} -\cos\theta$$
$$-1 \cdot \cos\theta - 0 \cdot \sin\theta \stackrel{?}{=} -\cos\theta$$
$$-\cos\theta = -\cos\theta\ \checkmark$$

23a. $y = 19.76\sin\left(\frac{5\pi}{32}x - 1.79\right) + 64.27$ **23b.** The new function represents the middle average of the high and low temperatures for each month. **25.** $\sqrt{2} - \sqrt{6}$

27. $-2 + \sqrt{3}$ **29.** $2 - \sqrt{3}$

31 a. Let X be the endpoint of the segment that is 8 inches long.
$$\sin(m\angle BAC)$$
$$= \sin(m\angle BAX + m\angle XAC)$$
$$= \sin(m\angle BAX)\cos(m\angle XAC) + \cos(m\angle BAX)\sin(m\angle XAC)$$
$$= \frac{8\sqrt{3}}{16}\cdot\frac{8}{10} + \frac{8}{16}\cdot\frac{6}{10} \quad \sin = \frac{opp}{hyp} \text{ and } \cos = \frac{adj}{hyp}$$
$$= \frac{8\sqrt{3}}{10} + \frac{3}{10} \quad \text{Multiply.}$$
$$= \frac{4\sqrt{3}}{10} + \frac{3}{10} \quad \text{Add.}$$
$$= \frac{3 + 4\sqrt{3}}{10}$$

b. Let X be the endpoint of the segment that is 8 inches long.
$$\cos(m\angle BAC)$$
$$= \cos(m\angle BAX + m\angle XAC)$$
$$= \cos(m\angle BAX)\cos(m\angle XAC) - \sin(m\angle BAX)\sin(m\angle XAC)$$
$$= \frac{8}{16}\cdot\frac{8}{10} - \frac{8\sqrt{3}}{16}\cdot\frac{6}{10} \quad \sin = \frac{opp}{hyp} \text{ and } \cos = \frac{adj}{hyp}$$
$$= \frac{4}{10} - \frac{3\sqrt{3}}{10} \quad \text{Multiply.}$$
$$= \frac{4 - 3\sqrt{3}}{10} \quad \text{Add.}$$

c. $\cos(m\angle BAC) = \dfrac{4 - 3\sqrt{3}}{10}$
$$m\angle BAC = \cos^{-1}\left(\frac{4 - 3\sqrt{3}}{10}\right)$$
$$\approx 96.9°$$

d. Because $m\angle BAC \neq 90$, the triangle formed is not a right triangle.

33a.

A	B	sin A	sin B	sin (A + B)	sin (A + B)	sin A + sin B
30°	90°	$\frac{1}{2}$	1	$\frac{\sqrt{3}}{2}$	$\frac{\sqrt{3}}{2}$	$\frac{3}{2}$
45°	60°	$\frac{\sqrt{2}}{2}$	$\frac{\sqrt{3}}{2}$	$\frac{\sqrt{2}+\sqrt{6}}{4}$	$\frac{\sqrt{2}+\sqrt{6}}{4}$	$\frac{\sqrt{2}+\sqrt{3}}{2}$
60°	45°	$\frac{\sqrt{3}}{2}$	$\frac{\sqrt{2}}{2}$	$\frac{\sqrt{2}+\sqrt{6}}{4}$	$\frac{\sqrt{2}-\sqrt{6}}{4}$	$\frac{\sqrt{2}+\sqrt{3}}{2}$
90°	30°	1	$\frac{1}{2}$	$\frac{\sqrt{3}}{2}$	$\frac{\sqrt{3}}{2}$	$\frac{3}{2}$

R97 (right page)

33b.

33c. No; a counterexample is: $\cos(30° + 45°) = \cos 30° + \cos 45°$, which equals $\frac{\sqrt{3}}{2} + \frac{\sqrt{2}}{2}$ or about 1.5731. Because a cosine value cannot be greater than 1, this statement must be false.

35 $\cos(A + B) \stackrel{?}{=} \dfrac{1 - \tan A\tan B}{\sec A\sec B}$
$$\cos(A + B) \stackrel{?}{=} \frac{1 - \frac{\sin A}{\cos A}\cdot\frac{\sin B}{\cos B}}{\frac{1}{\cos A}\cdot\frac{1}{\cos B}}$$
$$\cos(A + B) \stackrel{?}{=} \frac{1 - \frac{\sin A\sin B}{\cos A\cos B}}{\frac{1}{\cos A\cos B}}$$
$$\cos(A + B) \stackrel{?}{=} \frac{\cos A\cos B - \sin A\sin B}{\cos A\cos B}\cdot\frac{\cos A\cos B}{1} \quad \text{Simplify.}$$
$$\cos(A + B) = \cos(A + B)\ \checkmark \quad \text{Difference Identity}$$

37. $\sin(A + B)\sin(A - B) \stackrel{?}{=} \sin^2 A - \sin^2 B$
$(\sin A\cos B + \cos A\sin B)(\sin A\cos B - \cos A\sin B) \stackrel{?}{=} \sin^2 A - \sin^2 B$
$(\sin A\cos B)^2 - (\cos A\sin B)^2 \stackrel{?}{=} \sin^2 A - \sin^2 B$
$\sin^2 B\cos^2 B - \cos^2 A\sin^2 B \stackrel{?}{=} \sin^2 A - \sin^2 B$
$\cos^2 A\cos^2 B \stackrel{?}{=} \sin^2 A - \sin^2 A\sin^2 B - \sin^2 A\cos^2 B + \sin^2 A(\cos^2 B + \sin^2 B) - \sin^2 B(\sin^2 A + \cos^2 A) \stackrel{?}{=} \sin^2 A - \sin^2 B$
$(\sin^2 A)(1) - (\sin^2 B)(1) \stackrel{?}{=} \sin^2 A - \sin^2 B$
$\sin^2 A - \sin^2 B = \sin^2 A - \sin^2 B\ \checkmark$

39. Sample answer: To determine wireless Internet interference, you need to determine the sine or cosine of the sum or difference of two angles. Interference occurs when waves pass through the same space at the same time. When the combined waves have a greater amplitude, constructive interference results. When the combined waves have a smaller amplitude, destructive interference results.

41. $d = \sqrt{(\cos A - \cos B)^2 + (\sin A - \sin B)^2}$
$d^2 = (\cos A - \cos B)^2 + (\sin A - \sin B)^2$
$d^2 = (\cos^2 A - 2\cos A\cos B + \cos^2 B) + (\sin^2 A - 2\sin A\sin B + \sin^2 B)$
$d^2 = \cos^2 A + \cos^2 B + \sin^2 A + \sin^2 B - 2\cos A\cos B - 2\sin A\sin B$
$d^2 = 1 + 1 - 2\cos A\cos B - 2\sin A\sin B \quad \begin{array}{l}\sin^2 A + \cos^2 A \\ = 1 \text{ and } \sin^2 B \\ + \cos^2 B = 1\end{array}$
$d^2 = 2 - 2\cos A\cos B - 2\sin A\sin B$

Now find the value of d^2 when the angle having measure $A - B$ is in standard position on the unit circle, as shown in the figure below.

[cos $(A - B)$, sin $(A - B)$]

$d = \sqrt{[\cos(A - B) - 1]^2 + [\sin(A - B) - 0]^2}$
$d^2 = [\cos(A - B) - 1]^2 + [\sin(A - B) - 0]^2$
$= \cos^2(A - B) - 2\cos(A - B) + 1 + \sin^2(A - B)$
$= 1 - 2\cos(A - B) + 1$
$= 2 - 2\cos(A - B)$

43. D **45.** B **47.** D **49.** B, C, D

Lesson 10-4

1. $\dfrac{\sqrt{15}}{8}, \dfrac{7}{8}, \dfrac{\sqrt{15}}{7}$ **3.** $\dfrac{\sqrt{8 + 2\sqrt{15}}}{4}, \dfrac{\sqrt{8 + 2\sqrt{15}}}{4}$

$\dfrac{3\sqrt{13}}{13}, \dfrac{2\sqrt{13}}{13}$ **5.** $\dfrac{240}{289}, \dfrac{161}{289}, \dfrac{240}{161}$ **7.** $\dfrac{\sqrt{2 - \sqrt{2}}}{2}$

9a. $d = \dfrac{v^2\sin 2\theta}{g}$ **9b.** ≈ 81 ft

11.
$(\sin\theta + \cos\theta)^2 \stackrel{?}{=} 1 + 2\sin\theta\cos\theta$
$(\sin\theta + \cos\theta)(\sin\theta + \cos\theta) \stackrel{?}{=} 1 + 2\sin\theta\cos\theta$
$\sin^2\theta + 2\sin\theta\cos\theta + \cos^2\theta \stackrel{?}{=} 1 + 2\sin\theta\cos\theta$
$1 + 2\sin\theta\cos\theta = 1 + 2\sin\theta\cos\theta\ \checkmark$

13. $\dfrac{240}{289}, \dfrac{161}{289}, \dfrac{5\sqrt{34}}{34}, \dfrac{3\sqrt{34}}{34}$

15
$\sin^2\theta = 1 - \cos^2\theta$
$\sin^2\theta = 1 - \left(\frac{1}{5}\right)^2 \quad \cos\theta = \frac{1}{5}$
$\sin^2\theta = 1 - \frac{1}{25} \quad \text{Subtract.}$
$\sin^2\theta = \frac{24}{25}$
$\sin\theta = \pm\frac{2\sqrt{6}}{5} \quad \text{Take the square root of each side.}$

Because θ is in the fourth quadrant, sine is negative.
So, $\sin\theta = -\frac{2\sqrt{6}}{5}$.

$\sin 2\theta = 2\sin\theta\cos\theta \quad \text{Double-angle identity}$
$= 2\left(-\frac{2\sqrt{6}}{5}\right)\left(\frac{1}{5}\right) \quad \sin\theta = -\frac{2\sqrt{6}}{5} \text{ and } \cos\theta = \frac{1}{5}$
$= -\frac{4\sqrt{6}}{25} \quad \text{Simplify.}$

$\cos 2\theta = 1 - 2\sin^2\theta \quad \text{Double-angle identity}$
$= 1 - 2\left(\frac{24}{25}\right) \quad \sin^2\theta = \frac{24}{25}$
$= -\frac{23}{25} \quad \text{Simplify.}$

$\sin\frac{\theta}{2} = \pm\sqrt{\frac{1 - \cos\theta}{2}} \quad \text{Half-angle identity}$

connectED.mcgraw-hill.com

61. Sample answer: All trigonometric functions are periodic. Therefore, once one or more solutions are found for a certain interval, there will be additional solutions that can be found by adding integral multiples of the period of the function to those solutions.

63. 0, b, 2b **65.** C **67.** B

Chapter 10 Study Guide and Review

1. difference of angles identity **3.** double-angle identity **5.** half-angle identity **7.** Pythagorean identity **9.** Trigonometric identities are equations that are true for all values of the variable for which both sides are defined. Trigonometric equations are true only for certain values of the variable.

11. $-\sqrt{3}$ **13.** $\frac{4}{5}$ **15.** $\frac{15\sqrt{709}}{709}$

17. sec θ **19.** sec θ

21.
$$\frac{\cos\theta + \sin\theta}{\cot\theta} \cdot \frac{\frac{1}{\tan\theta}}{\frac{1}{\cos\theta}} = -\cos\theta$$

$$\cos\theta \div \cos\theta + \sin\theta \div \frac{1}{\sin\theta} = \sin\theta + \cos\theta$$

$$\cos\theta \cdot \frac{\sin\theta}{\cos\theta} + \sin\theta \cdot \frac{\cos\theta}{\sin\theta} = \sin\theta + \cos\theta$$

$$\sin\theta + \cos\theta = \sin\theta + \cos\theta \checkmark$$

23. $\tan^2\theta + 1 = \left(\frac{\sqrt{7}}{3}\right)^2 + 1 = \frac{7}{9} + 1 = \frac{7}{9} + \frac{9}{9} = \frac{16}{9}$,

$\sec^2\theta = \left(\frac{4}{3}\right)^2 = \frac{16}{9}$

25. $\frac{\sqrt{6}+\sqrt{2}}{4}$ **27.** $\frac{\sqrt{6}+\sqrt{2}}{4}$ **29.** $\frac{-\sqrt{6}+\sqrt{2}}{4}$

31.
$$\sin\left(\frac{3\pi}{2} - \theta\right) = \cos\frac{3\pi}{2}\sin\frac{z}{2} - \cos\theta$$

$$\sin\frac{3\pi}{2}\cos\theta - \cos\frac{3\pi}{2}\sin\frac{z}{2} = -\cos\theta$$

$$(-1)\cos\theta - (0)\sin\theta = -\cos\theta$$

$$-\cos\theta = -\cos\theta \checkmark$$

33. $\sin\frac{\theta}{2} = \frac{24}{25}$, $\cos 2\theta = \frac{7}{25}$, $\sin\frac{\theta}{2} = \frac{\sqrt{10}}{10}$, and

$\cos\frac{\theta}{2} = \frac{3\sqrt{10}}{10}$

35. $\sin 2\theta = -\frac{4\sqrt{5}}{9}$, $\cos 2\theta = -\frac{1}{9}$,

$\sin\frac{\theta}{2} = \frac{\sqrt{30}}{6}$, and $\cos\frac{\theta}{2} = \frac{\sqrt{6}}{6}$

37. 60°, 300°

39. 90°, 210°, 270°, 330°

41. $\frac{\pi}{3}, \frac{5\pi}{3}$

$$-2\sin\theta + 1 = 0 \text{ or } \sin\theta - 7 = 0 \quad \text{Zero Product Property}$$

$$\sin\theta = \frac{1}{2} \qquad \sin\theta = 7$$

$$\theta = 30° \text{ or } 150° \qquad \text{no solution because } 0 \le \sin\theta \le 1$$

9. $\frac{\pi}{6} + 2k\pi$ or $\pm\frac{5\pi}{6} + 2k\pi$ **11.** $\frac{3\pi}{2} + 2k\pi$ **13.** $\pi + 2k\pi$

15. 90° + $k \cdot$ 180° **17.** 45° + $k \cdot$ 90° **19.** 270° + $k \cdot$ 360°

21a. There will be $11\frac{1}{2}$ hours of daylight 205 and 342 days after March 21; that is, on October 13 and February 26. **21b.** Every day from March 4 to October 13; sample explanation: Because the longest day of the year occurs around June 22, the days between February 26 and October 13 must increase in length until June 22 and then decrease in length until October 13.

23. $\frac{3\pi}{4} + \pi k$ **25.** $\frac{\pi}{6} + \pi k, \frac{\pi}{6} + 2\pi k, \frac{5\pi}{6} + 2\pi k$

27. 0° + $k \cdot$ 45° or 0 + $k \cdot \frac{\pi}{4}$

29. $\frac{\pi}{3} + 2k\pi, \frac{\pi}{3} + k\pi, \frac{5\pi}{4} + 2k\pi$

31. 135°, 225° **33.** $\frac{\pi}{6}$ **35.** 210°, 330°

37.
$$2\sin^2\theta = \cos\theta + 1 \qquad \text{Original equation}$$

$$2(1 - \cos^2\theta) = \cos\theta + 1$$

$$2 - 2\cos^2\theta = \cos\theta + 1 \qquad \text{Simplify.}$$

$$-2\cos^2\theta - \cos\theta + 1 = 0 \qquad \text{Subtract } \cos\theta + 1 \text{ from each side and simplify.}$$

$$(-2\cos\theta + 1)(\cos\theta + 1) = 0 \qquad \text{Factor.}$$

$$-2\cos\theta + 1 = 0 \text{ or } \cos\theta + 1 = 0 \qquad \text{Zero Product Property}$$

$$\cos\theta = \frac{1}{2} \qquad \cos\theta = -1$$

The solutions of $\cos\theta = \frac{1}{2}$ are $\frac{\pi}{3} + 2\pi k$ and $\frac{5\pi}{3} + 2\pi k$.

The solution of $\cos\theta = -1$ is $\pi + 2k\pi$.

39. 0 + 2$k\pi$ **41.** 0° + $k \cdot$ 180° **43.** 30° + $k \cdot$ 360°, 150° + $k \cdot$ 360° **45.** $\frac{7\pi}{6} + 2\pi k, \frac{11\pi}{6} + 2\pi k$ **47.** 0 + 2$k\pi, \frac{\pi}{2} + k\pi$ or 0° + $k \cdot$ 360°, 90° + $k \cdot$ 180° **49a.** 11 m

49b. 7:00 A.M. and 7:00 P.M.

51. $\frac{\pi}{6} + 2\pi k, \frac{5\pi}{6} + 2\pi k, \frac{5\pi}{4} + 2\pi k, \frac{7\pi}{4} + 2\pi k$

53. 120° + 360°k, 240° + 360°k

55. $\frac{5\pi}{6} + 2\pi k, \frac{5\pi}{6} + 2\pi k$

57.
$$D = 0.5\sin(6.5x)\sin(2500t) \qquad \text{Original equation}$$
$$0.01 = 0.5\sin(6.5 + 500)\sin(2500t)$$
$$D = 0.01 \text{ mm and } x = 0.5 \cdot 1000 \text{ or } 500 \text{ mm}$$
$$0.01 = 0.5\sin(3250)\sin(2500t) \qquad \text{Simplify.}$$
$$0.1152 \approx \sin(2500t) \qquad \text{Divide each side by } 0.5\sin(3250).$$
$$\text{Sin}^{-1}(0.1152) \approx 2500t \qquad \text{Use the } \sin^{-1} \text{ function.}$$
$$6.6152 \approx 2500t \qquad \text{Use a calculator.}$$
$$0.0026 \approx t \qquad \text{Divide each side by } 2500.$$

The time is about 0.0026 second.

59. $\frac{\pi}{3} < x < \pi$ or $\frac{5\pi}{3} < x < 2\pi$

$$= \pm\sqrt{\frac{1 - \frac{1}{5}}{2}} \qquad \cos\theta = \frac{1}{5}$$

$$= \pm\sqrt{\frac{4}{5}}{2} \qquad \text{Simplify.}$$

$$= \pm\frac{\sqrt{2}}{\sqrt{5}} = \frac{\sqrt{5}}{\sqrt{5}} \text{ or } \pm\frac{\sqrt{10}}{5} \qquad \text{Rationalize the denominator.}$$

If θ is between 270° and 360°, $\frac{\theta}{2}$ is between 135° and 180°. So, $\sin\frac{\theta}{2}$ is $\frac{\sqrt{10}}{5}$.

$$\cos\frac{\theta}{2} = \pm\sqrt{\frac{1 + \cos\theta}{2}} \qquad \text{Half-angle identity}$$

$$= \pm\sqrt{\frac{1 + \frac{1}{5}}{2}} \qquad \cos\theta = \frac{1}{5}$$

$$= \pm\sqrt{\frac{3}{5}}{2} \qquad \text{Simplify.}$$

$$= \pm\frac{\sqrt{3}}{\sqrt{5}} \cdot \frac{\sqrt{5}}{\sqrt{5}} \text{ or } \pm\frac{\sqrt{15}}{5} \qquad \text{Rationalize the denominator.}$$

If θ is between 270° and 360°, $\frac{\theta}{2}$ is between 135° and 180°. So, $\cos\frac{\theta}{2}$ is $-\frac{\sqrt{15}}{5}$.

17. $\frac{4}{5}, -\frac{3}{5}, \sqrt{\frac{\sqrt{5}+1}{2\sqrt{5}}}, \sqrt{\frac{\sqrt{5}-1}{2\sqrt{5}}}$ **19.** $\frac{\sqrt{2+\sqrt{2}}}{2}$

21. $-\sqrt{7-4\sqrt{3}}$ **23.** $\sqrt{3}-2\sqrt{2}$

25. $P = I_0^2 R\sin^2\theta t$

$$P = I_0^2 R(\cos^2\theta t - \cos 2\theta t)$$

$$P = I_0^2 R\left(\frac{1}{2}\cos 2\theta t + \frac{1}{2} - \cos 2\theta t\right)$$

$$P = I_0^2 R\left(\frac{1}{2} - \frac{1}{2}\cos 2\theta t\right)$$

$$P = \frac{1}{2}I_0^2 R - \frac{1}{2}I_0^2 R\cos 2\theta t$$

27. $1 + \frac{1}{2}\sin 2\theta = \frac{\sec\theta + \sin\theta}{\sec\theta}$

$$= \frac{\sec\theta + \sin\theta}{\frac{1}{\cos\theta}} \qquad \text{Original equation}$$

$$= \frac{\frac{1}{\cos\theta} + \sin\theta}{\frac{1}{\cos\theta}} \qquad \sin^2\theta t = \cos^2\theta t - \cos 2\theta t$$

$$= \frac{\frac{1}{\cos\theta} + \sin\theta}{\frac{1}{\cos\theta}} \cdot \frac{\cos\theta}{\cos\theta} \qquad \frac{1}{2}\cos^2\theta t = \frac{1}{2}\cos 2\theta t + \frac{1}{2}$$

$$= 1 + \frac{1}{2} \cdot 2\sin\theta\cos\theta \qquad \text{Simplify.}$$

$$= 1 + \frac{1}{2}\sin 2\theta \checkmark \qquad \text{Distributive Property}$$

29. $\tan\frac{\theta}{2} = \frac{\sin\theta}{1 + \cos\theta}$

$$\tan\frac{\theta}{2} = \frac{\sin 2\left(\frac{\theta}{2}\right)}{1 + \cos 2\left(\frac{\theta}{2}\right)}$$

$$\tan\frac{\theta}{2} = \frac{2\sin\frac{\theta}{2}\cos\frac{\theta}{2}}{1 + 2\cos^2\frac{\theta}{2} - 1}$$

$$\tan\frac{\theta}{2} = \frac{2\sin\frac{\theta}{2}\cos\frac{\theta}{2}}{2\cos^2\frac{\theta}{2}}$$

$$\tan\frac{\theta}{2} = \frac{\sin\frac{\theta}{2}}{\cos\frac{\theta}{2}}$$

$$\tan\frac{\theta}{2} = \tan\frac{\theta}{2} \checkmark$$

31. $\frac{24}{25}, \frac{7}{25}, \frac{24}{7}$ **33.** $-\frac{3}{5}, \frac{4}{5}, -\frac{3}{4}$ **35.** $\frac{4\sqrt{21}}{25}, \frac{17}{25}, \frac{4\sqrt{21}}{17}$

37. No; Teresa incorrectly added the square roots, and Nathan used the half-angle identity incorrectly. He used sin 30° in the formula instead of first finding the cosine. **39.** If you are only given the value of $\cos\theta$, then $\cos 2\theta = 2\cos^2\theta - 1$ is the best identity to use. If you are only given the value of $\sin\theta$, then $\cos 2\theta = 1 - 2\sin^2\theta$ is the best identity to use. If you are given the values of both $\cos\theta$ and $\sin\theta$, then $\cos 2\theta = \cos^2\theta - \sin^2\theta$ works just as well as the other two.

41. Find $\sin\frac{A}{2}$.

$$1 - 2\sin^2\theta = \cos 2\theta \qquad \text{Double-angle identity}$$

$$1 - 2\sin^2\frac{A}{2} = \cos A \qquad \text{Substitute } \frac{A}{2} \text{ for } \theta \text{ and } A \text{ for } 2\theta.$$

$$\sin^2\frac{A}{2} = \frac{1 - \cos A}{2} \qquad \text{Solve for } \sin^2\frac{A}{2}.$$

$$\sin\frac{A}{2} = \pm\sqrt{\frac{1 - \cos A}{2}} \qquad \text{Take the square root of each side.}$$

Find $\cos\frac{A}{2}$.

$$2\cos^2\theta - 1 = \cos 2\theta \qquad \text{Double-angle identity}$$

$$2\cos^2\frac{A}{2} - 1 = \cos A \qquad \text{Substitute } \frac{A}{2} \text{ for } \theta \text{ and } A \text{ for } 2\theta.$$

$$\cos^2\frac{A}{2} = \frac{1 + \cos A}{2} \qquad \text{Solve for } \cos^2\frac{A}{2}.$$

$$\cos\frac{A}{2} = \pm\sqrt{\frac{1 + \cos A}{2}} \qquad \text{Take the square root of each side.}$$

Find $\tan\frac{A}{2}$.

$$\tan\frac{A}{2} = \frac{\sin\frac{A}{2}}{\cos\frac{A}{2}} \qquad \text{Quotient Identity}$$

$$\tan\frac{A}{2} = \frac{\pm\sqrt{\frac{1 - \cos A}{2}}}{\pm\sqrt{\frac{1 + \cos A}{2}}} \qquad \text{Half-Angle Identities}$$

$$\tan\frac{A}{2} = \pm\sqrt{\frac{\frac{1 - \cos A}{2}}{\frac{1 + \cos A}{2}}} \qquad \text{Quotient Property of Radicals}$$

$$\tan\frac{A}{2} = \pm\sqrt{\frac{1 - \cos A}{1 + \cos A}} \qquad \text{Simplify.}$$

43. B **45a.** A, B **45b.** A **45c.** A **47a.** $\frac{24}{25}$ **47b.** $\frac{7}{25}$

47c. $\frac{24}{7}$ **47d.** $\frac{3\sqrt{10}}{10}$ **47e.** $\frac{\sqrt{10}}{10}$ **47f.** -3

Lesson 10-5

1. 210°, 330° **3.** 60°, 180°, or 300° **5.** 150°, 210°

$$\cos 2\theta = 8 - 15\sin\theta$$

$$\cos 2\theta = 8 - 15\sin\theta \qquad \text{Original equation}$$

$$1 - 2\sin^2\theta + 15\sin\theta - 8 = 0 \qquad \text{Add } 15\sin\theta - 8 \text{ to each side.}$$

$$-2\sin^2\theta + 15\sin\theta - 7 = 0 \qquad \text{Double-angle identity}$$

$$2\sin^2\theta + 15\sin\theta - 7 = 0 \qquad \text{Simplify.}$$

$$(-2\sin\theta + 1)(\sin\theta - 7) = 0 \qquad \text{Factor.}$$

Glossary/Glosario

Multilingual eGlossary

Go to connectED.mcgraw-hill.com for a glossary of terms in these additional languages:

Arabic	Chinese	Hmong
Bengali	English	Korean
Brazilian Portugese	Haitian Creole	Russian

Spanish	Vietnamese
Tagalog	
Urdu	

English

A

absolute value function (p. 120) A function written as $f(x) = |x|$, where $f(x) = \begin{cases} x \text{ if } x > 0 \\ 0 \text{ if } x = 0. \\ -x \text{ if } x < 0. \end{cases}$ values of x

amplitude (p. 627) For functions in the form $y = a \sin b\theta$ or $y = a \cos b\theta$, the amplitude is $|a|$.

angle of depression (p. 598) The angle between a horizontal line and the line of sight from the observer to an object at a lower level.

angle of elevation (p. 598) The angle between a horizontal line and the line of sight from the observer to an object at a higher level.

asymptote (p. 373) A line that a graph approaches.

axis of symmetry (p. 152) A line about which a figure is symmetric.

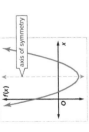

Español

A

función del valor absoluto Una función que se escribe $f(x) = |x|$, donde $f(x) = \begin{cases} x \text{ si } x > 0 \\ 0 \text{ si } x = 0. \\ -x \text{ si } x < 0. \end{cases}$

amplitud Para funciones de la forma $y = a \sin b\theta$ o $y = a \cos b\theta$, la amplitud es $|a|$.

ángulo de depresión Ángulo entre una recta horizontal y la línea visual de un observador a una figura en un nivel inferior.

ángulo de elevación Ángulo entre una recta horizontal y la línea visual de un observador a una figura en un nivel superior.

asíntota Recta a la que se aproxima una gráfica.

eje de simetría Recta respecto a la cual una figura es simétrica.

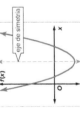

B

bias (p. 531) An error that results in a mispresentation of members of a population.

Binomial Theorem (p. 237) If n is a nonnegative integer, then $(a + b)^n = 1a^nb^0 + \frac{n}{1} a^{n-1}b^1 + \frac{n(n+1)}{1 \cdot 2} a^{n-2}b^2 + \cdots + 1a^0b^n$.

boundary (p. 35) A line or curve that separates the coordinate plane into two regions.

bounded (p. 60) A region is bounded when the graph of a system of constraints is a polygonal region.

C

central angle (p. 607) An angle with a vertex at the center of the circle and sides that are radii.

Change of Base Formula (p. 425) For all positive numbers a, b, and n, where $a \neq 1$ and $b \neq 1$, $\log_a n = \frac{\log_b n}{\log_b a}$.

circular function (p. 620) A function defined using a unit circle.

combination (p. 911) An arrangement of objects in which order is not important.

combined variation (p. 503) When one quantity varies directly and/or inversely as two or more other quantities.

common logarithms (p. 423) Logarithms that use 10 as the base.

completing the square (p. 192) A process used to make a quadratic expression into a perfect square trinomial.

complex conjugates (p. 175) Two complex numbers of the form $a + bi$ and $a - bi$.

complex fraction (p. 470) A rational expression whose numerator and/or denominator contains a rational expression.

complex number (p. 173) Any number that can be written in the form $a + bi$, where a and b are real numbers and i is the imaginary unit.

composition of functions (p. 322) A function is performed, and then a second function is performed on the result of the first function. The composition of f and g is denoted by $f \circ g$, and $[f \circ g](x) = f[g(x)]$.

sesgo Error que resulta en la representación errónea de los miembros de una población.

teorema del binomio Si n es un entero no negativo, entonces $(a + b)^n = 1a^nb^0 + \frac{n}{1} a^{n-1}b^1 + \frac{n(n+1)}{1 \cdot 2} a^{n-2}b^2 + \cdots + 1a^0b^n$.

frontera Recta o curva que divide un plano de coordenadas en dos regiones.

acotada Una región está acotada cuando la gráfica de un sistema de restricciones es una región poligonal.

ángulo central Ángulo cuyo vértice es el centro del círculo y cuyos lados son radios.

fórmula del cambio de base Para todo número positivo a, b y n, donde $a \neq 1$ y $b \neq 1$, $\log_a n = \frac{\log_b n}{\log_b a}$.

funciones circulares Funciones definidas en un círculo unitario.

combinación Arreglo de elementos en que el orden no es importante.

variación combinada Cuando una cantidad varia directamente e inverso como dos o más otras cantidades.

logaritmos comunes El logaritmo de base 10.

completar el cuadrado Proceso mediante el cual una expresión cuadrática se transforma en un trinomio cuadrado perfecto.

conjugados complejos Dos números complejos de la forma $a + bi$ y $a - bi$.

fracción compleja Expresión racional cuyo numerador o denominador contiene una expresión racional.

número complejo Cualquier número que puede escribirse de la forma $a + bi$, donde a y b son números reales e i es la unidad imaginaria.

composición de funciones Se evalúa una función y luego se evalúa una segunda función en el resultado de la primera función. La composición de f y g se define con $f \circ g$, y $[f \circ g](x) = f[g(x)]$.

R102 (English)

compound event **(p. 14)** Two or more simple events.

compound interest **(p. 384)** Interest paid on the principal of an investment and any previously earned interest.

conditional probability **(p. P16)** The probability of an event occurring given that another event has already occurred.

confidence level **(p. 546)** The probability that survey results from several samples of the same population will have a confidence interval that includes the actual value of the population parameter.

consistent **(p. 43)** A system of equations that has at least one ordered pair that satisfies both equations.

constant of variation **(p. 500)** The constant k used with direct or inverse variation.

constant term **(p. 151)** In $f(x) = ax^2 + bx + c$, c is the constant term.

constraints **(p. 36)** Conditions given to variables, often expressed as linear inequalities.

continuous relation **(p. 88)** A relation that can be graphed with a line or smooth curve.

correlation coefficient **(p. 409)** A measure that shows how well data are modeled by a linear equation.

cosecant **(p. 594)** For any angle, with measure α, a point $P(x, y)$ on its terminal side, $r = \sqrt{x^2 + y^2}$. $\csc \alpha = \frac{r}{y}$

cosine **(p. 594)** For any angle, with measure α, a point $P(x, y)$ on its terminal side, $r = \sqrt{x^2 + y^2}$. $\cos \alpha = \frac{x}{r}$

cotangent **(p. 594)** For any angle, with measure α, a point $P(x, y)$ on its terminal side, $r = \sqrt{x^2 + y^2}$. $\cot \alpha = \frac{x}{y}$

coterminal angles **(p. 605)** Two angles in standard position that have the same terminal side.

cube root function **(p. 345)** A function described by the equation $f(x) = a\sqrt[3]{x - h} + k$

cycle **(p. 621)** One complete pattern of a periodic function.

D

decay factor **(p. 376)** In exponential decay, the base of the exponential expression, $1 - r$.

R102 (Español)

evento compuesto Dos o más eventos simples.

interés compuesto Interés obtenido tanto sobre la inversión inicial como sobre el interés conseguido.

probabilidad condicional Probabilidad de un evento dado que otro evento ya ha ocurrido.

nivel de confianza La probabilidad de que los resultados de un estudio basado en varias muestras de la misma población tengan un intervalo de confianza que incluye el valor real del parámetro de la población.

consistente Sistema de ecuaciones para el cual existe al menos un par ordenado que satisface ambas ecuaciones.

constante de variación La constante k que se usa en variación directa o inversa.

término constante En $f(x) = ax^2 + bx + c$, c es el término constante.

restricciones Condiciones a que están sujetas las variables, a menudo escritas como desigualdades lineales.

relación continua Relación cuya gráfica puede ser una recta o una curva suave.

coeficiente de correlación Una medida que demuestra cómo los datos bien son modelados por una ecuación lineal.

cosecante Para cualquier ángulo de medida α, un punto $P(x, y)$ en su lado terminal. $r = \sqrt{x^2 + y^2}$. $\csc \alpha = \frac{r}{y}$

coseno Para cualquier ángulo de medida α, un punto $P(x, y)$ en su lado terminal. $r = \sqrt{x^2 + y^2}$. $\cos \alpha = \frac{x}{r}$

cotangente Para cualquier ángulo de medida α, un punto $P(x, y)$ en su lado terminal. $r = \sqrt{x^2 + y^2}$. $\cot \alpha = \frac{x}{y}$

ángulos coterminales Dos ángulos en posición estándar que tienen el mismo lado terminal.

función de raíz cúbica Función que se describe con la ecuación $f(x) = a\sqrt[3]{x - h} + k$

ciclo Un patrón completo de una función periódica.

D

factor de decaimiento En decaimiento exponencial, la base de la expresión exponencial, $1 - r$.

R103 (English)

degree of a polynomial **(p. 231)** The greatest degree of any term in the polynomial.

dependent **(p. 43)** A system of equations that has an infinite number of solutions.

dependent events **(p. P16)** Events in which the outcome of one event affects the outcome of another event.

dependent variable **(p. 89)** The other variable in a function, usually y, whose values depend on x.

depressed polynomial **(p. 289)** The quotient when a polynomial is divided by one of its binomial factors.

descriptive statistics **(p. P24)** The branch of statistics that focuses on collecting, summarizing, and displaying data.

dilation **(p. 126)** A transformation in which a geometric figure is enlarged or reduced.

dimensional analysis **(p. 236)** Performing operations with units.

direct variation **(p. 500)** y varies directly as x if there is some nonzero constant k such that $y = kx$. k is called the constant of variation.

discrete relation **(p. 88)** A relation in which the domain is a set of individual points.

discriminant **(p. 202)** In the Quadratic Formula, the expression $b^2 - 4ac$.

domain **(p. P4)** The set of all x-coordinates of the ordered pairs of a relation.

E

e **(p. 430)** The irrational number 2.71828... is the base of the natural logarithms.

elimination method **(p. 45)** Eliminate one of the variables in a system of equations by adding or subtracting the equations.

end behavior **(pp. 103, 255)** The behavior of the graph as x approaches positive infinity $(+\infty)$ or negative infinity $(-\infty)$.

equation **(p. 5)** A mathematical sentence stating that two mathematical expressions are equal.

experiment **(p. 531)** Something that is intentionally done to people, animals, or objects, and then the response is observed.

experimental probability **(pp. P13, 538)** What is estimated from observed simulations or experiments.

R103 (Español)

grado de un polinomio Grado máximo de cualquier término del polinomio.

dependiente Sistema de ecuaciones que posee un número infinito de soluciones.

evento dependiente Eventos en que el resultado de uno de los eventos afecta el resultado de otro de los eventos.

variable dependiente La otra variable en una función, por lo general y, cuyo valor depende de x.

polinomio reducido El cociente cuando se divide un polinomio entre uno de sus factores binomiales.

estadística descriptiva Rama de la estadística cuyo enfoque es la recopilación, resumen y demostración de los datos.

homotecia Transformación en que se amplía o se reduce un figura geométrica.

análisis dimensional Realizar operaciones con unidades.

variación directa y varía directamente con x si hay una constante no nula k tal que $y = kx$. k se llama la constante de variación.

relación discreta Relación en la cual el dominio es un conjunto de puntos individuales.

discriminante En la fórmula cuadrática, la expresión $b^2 - 4ac$.

dominio El conjunto de todas las coordenadas x de los pares ordenados de una relación.

E

e El número irracional 2.71828... es la base de los logaritmos naturales.

método de eliminación Eliminar una de las variables de un sistema de ecuaciones sumando o restando las ecuaciones.

comportamiento final El comportamiento de una gráfica a medida que x tiende a más infinito $(+\infty)$ o menos infinito $(-\infty)$.

ecuación Enunciado matemático que afirma la igualdad de dos expresiones matemáticas.

experimento Algo que se hace intencionalmente poblar, los animales, o los objetos, y entonces la respuesta se observa.

probabilidad experimental Qué se estima de simulaciones o de experimentos observados.

R104 (Glossary / Glosario)

exponential decay (p. 375) Exponential decay occurs when a quantity decreases exponentially over time.

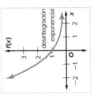

exponential equation (p. 383) An equation in which the variables occur as exponents.

exponential function (p. 373) A function of the form $y = ab^x$, where $a \neq 0$, $b > 0$, and $b \neq 1$.

exponential growth (p. 373) Exponential growth occurs when a quantity increases exponentially over time.

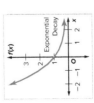

exponential inequality (p. 385) An inequality involving exponential functions.

extraneous solution (p. 352) A number that does not satisfy the original equation.

extrema (pp. 105, 263) The maximum and minimum values of a function.

F

factored form (p. 179) The form of a polynomial showing all of its factors. $y = a(x - p)(x - q)$ is the factored form of a quadratic equation.

Factor Theorem (p. 289) The binomial $x - r$ is a factor of the polynomial $P(x)$ if and only if $P(r) = 0$.

fair (p. 573) An experiment or decision is fair if all possible outcomes or choices have the same probability of being chosen.

feasible region (p. 60) The intersection of the graphs in a system of constraints.

five-number summary (p. P26) The three quartiles and the maximum and minimum values in a set of data.

---

desintegración exponencial Ocurre cuando una cantidad disminuye exponencialmente con el tiempo.

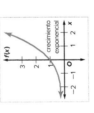

ecuación exponencial Ecuación en que las variables aparecen en los exponentes.

función exponencial Una función de la forma $y = ab^x$, donde $a \neq 0$, $b > 0$ y $b \neq 1$.

crecimiento exponencial El que ocurre cuando una cantidad aumenta exponencialmente con el tiempo.

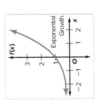

desigualdad exponencial Desigualdad que contiene funciones exponenciales.

solución extraña Número que no satisface la ecuación original.

extrema Son los valores máximos y mínimos de una función.

F

forma reducida La forma de un polinomio que demuestra todos sus factores. $y = a(x - p)(x - q)$ es la forma descompuesta en factores de una ecuación cuadrática.

teorema factor El binomio $x - r$ es un factor del polinomio $P(x)$ si $P(r) = 0$.

imparcial Un experimento, o una decisión, es imparcial si todos los resultados u opciones posibles tienen la misma probabilidad de ser elegidos.

región viable Intersección de las gráficas de un sistema de restricciones.

resumen de cinco números Los tres cuartiles y los valores máximo y mínimo de un conjunto de datos.

R105

FOIL method (p. 179) The product of two binomials is the sum of the products of **F** the first terms, **O** the outer terms, **I** the inner terms, and **L** the last terms.

frequency (p. 628) The number of cycles in a given unit of time.

function (p. P4) A relation in which each element of the domain is paired with exactly one element in the range.

function notation (p. 89) An equation of y in terms of x can be rewritten so that $y = f(x)$. For example, $y = 2x + 1$ can be written as $f(x) = 2x + 1$.

G

geometric means (p. 391) The terms between any two nonsuccessive terms of a geometric sequence.

geometric series (p. 392) The sum of the terms of a geometric sequence.

greatest integer function (p. 119) A step function, written as $f(x) = [\![x]\!]$, where $f(x)$ is the greatest integer less than or equal to x.

growth factor (p. 375) In exponential growth, the base of the exponential expression, $1 + r$.

H

horizontal asymptote (p. 491) A horizontal line which a graph approaches.

hyperbola (p. 483) The set of all points in the plane such that the absolute value of the difference of the distances from two given points in the plane, called foci, is constant.

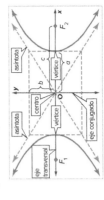

I

imaginary unit (p. 172) i or the principal square root of -1.

inconsistent (p. 43) A system of equations with no ordered pair that satisfy both equations.

---

método FOIL El producto de dos binomios es la suma de los productos de los primeros (First) términos, los términos exteriores (Outer), los términos interiores (Inner) y los últimos (Last) términos.

frecuencia El número de ciclos en una unidad del tiempo dada.

función Relación en que a cada elemento del dominio le corresponde un solo elemento del rango.

notación funcional Una ecuación de y en términos de x puede escribirse en la forma $y = f(x)$. Por ejemplo, $y = 2x + 1$ puede escribirse como $f(x) = 2x + 1$.

G

media geométrica Cualquier término entre dos términos no consecutivos de una sucesión geométrica.

serie geométrica La suma de los términos de una sucesión geométrica.

función del máximo entero Una función etapa que se escribe $f(x) = [\![x]\!]$, donde $f(x)$ es el máximo entero que es menor que o igual a x.

factor del crecimiento En el crecimiento exponencial, la base de la expresión exponencial, $1 + r$.

H

asíntota horizontal Una línea horizontal a que un gráfico acerca.

hipérbola Conjunto de todos los puntos de un plano en los que el valor absoluto de la diferencia de sus distancias a dos puntos dados del plano, llamados focos, es constante.

I

unidad imaginaria i o la raíz cuadrada principal de -1.

inconsistente Un sistema de ecuaciones para el cual no existe par ordenado alguno que satisfaga ambas ecuaciones.

independent (p. 43) A system of equations with exactly one solution.

independent events (p. P16) Events in which the outcome of one event does not affect the outcome of another event.

independent variable (p. 89) In a function, the variable, usually x, whose values make up the domain.

inflection point (p. 345) A point of a curve at which a change in the direction of curvature happens.

infinity (p. 20) Without bound, or continues without end.

initial side (p. 604) The fixed ray of an angle.

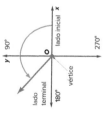

interquartile range (IQR) (p. P27) The range of the middle half of a set of data. It is the difference between the upper quartile and the lower quartile.

interval notation (p. 20) A way to describe the solution set of an inequality.

inverse function (p. 329) Two functions f and g are inverse functions if and only if both of their compositions are the identity function.

inverse relation (p. 329) Two relations are inverse relations if and only if whenever one relation contains the element (a, b) the other relation contains the element (b, a).

inverse variation (p. 502) y varies inversely as x if there is some nonzero constant k such that $xy = k$ or $y = \frac{k}{x}$ where $x \neq 0$ and $y \neq 0$.

J

joint variation (p. 501) y varies jointly as x and z if there is some nonzero constant k such that $y = kxz$.

---

independiente Un sistema de ecuaciones que posee una única solución.

eventos independientes Eventos en que el resultado de uno de los eventos no afecta el resultado de otro evento.

variable independiente En una función, la variable, por lo general x, cuyos valores forman el dominio.

punto de inflexión Punto de una curva en el que ocurre un cambio de dirección en su curvatura.

infinito Sin límite, o continúa sin extremo.

lado inicial de un ángulo El rayo fijo de un ángulo.

amplitud intercuartílica Amplitud de la mitad central de un conjunto de datos. Es la diferencia entre el cuartil superior y el inferior.

notación del intervalo Una manera de describir el sistema de la solución de una desigualdad.

función inversa Dos funciones f y g son inversas mutuas si y sólo si las composiciones de ambas son la función identidad.

relaciones inversas Dos relaciones son relaciones inversas mutuas si y sólo si cada vez que una de las relaciones contiene el elemento (a, b), la otra contiene el elemento (b, a).

variación inversa y varía inversamente con x si hay una constante no nula k tal que $xy = k$ o $y = \frac{k}{x}$ donde $x \neq 0$ y $y \neq 0$.

variación conjunta y varía conjuntamente con x y z si hay una constante no nula k tal que $y = kxz$.

L

leading coefficient (p. 253) The coefficient of the term with the highest degree.

linear equation (p. 95) An equation that has no operations other than addition, subtraction, and multiplication of a variable by a constant.

linear function (p. 95) A function whose ordered pairs satisfy a linear equation.

linear inequality (p. 35) An inequality that describes a half-plane with a boundary that is a straight line.

linear programming (p. 60) The process of finding the maximum or minimum values of a function for a region defined by inequalities.

line of reflection (p. 126) The line over which a reflection flips a figure.

line of symmetry (p. 97) A line that divides a figure into two halves that are reflections of each other.

line symmetry (p. 97) Figures that match exactly when folded in half have line symmetry.

Location Principle (p. 262) Suppose $y = f(x)$ represents a polynomial function and a and b are two numbers such that $f(a) < 0$ and $f(b) > 0$. Then the function has at least one real zero between a and b.

linear term (p. 151) In the equation $f(x) = ax^2 + bx + c$, bx is the linear term.

logarithm (p. 397) In the function $x = b^y$, y is called the logarithm, base b, of x. Usually written as $y = \log_b x$ and is read "y equals log base b of x."

logarithmic equation (p. 437) An equation that contains one or more logarithms.

logarithmic function (p. 398) The function $y = \log_b x$, where $b > 0$ and $b \neq 1$, which is the inverse of the exponential function $y = b^x$.

logarithmic inequality (p. 438) An inequality that contains one or more logarithms.

logistic growth model (p. 448) A growth model that represents growth that has a limiting factor. Logistic models are the most accurate models for representing population growth.

lower quartile (p. 26) The median of the lower half of a set of data, indicated by LQ.

---

coeficiente líder Coeficiente del término de mayor grado.

ecuación lineal Ecuación sin otras operaciones que las de adición, sustracción y multiplicación de una variable por una constante.

función lineal Función cuyos pares ordenados satisfacen una ecuación lineal.

desigualdad lineal Desigualdad lineal es una desigualdad que describe un semiplano con un límite que es una línea recta.

programación lineal Proceso de hallar los valores máximo o mínimo de una función lineal en una región definida por las desigualdades.

línea de la reflexión La línea excedente que una reflexión mueve de un tirón una figura.

eje de simetría Línea que divide una figura en dos mitades que son reflejos mutuos.

simetría axial Las figuras que coinciden exactamente cuando se doblan por la mitad tienen simetría axial.

principio de ubicación Sea $y = f(x)$ una función polinómica con a y b dos números tales que $f(a) < 0$ y $f(b) > 0$. Entonces la función tiene por lo menos un resultado real entre a y b.

término lineal En la ecuación $f(x) = ax^2 + bx + c$, el término lineal es bx.

logaritmo En la función $x = b^y$, y es el logaritmo en base b, de x. Generalmente escrito como $y = \log_b x$ y se lee "y es igual al logaritmo en base b de x".

ecuación logarítmica Ecuación que contiene uno o más logaritmos.

función logarítmica La función $y = \log_b x$ donde $b > 0$ y $b \neq 1$, inversa de la función exponencial $y = b^x$.

desigualdad logarítmica Desigualdad que contiene uno o más logaritmos.

modelo logístico del crecimiento Un modelo del crecimiento que representa el crecimiento que tiene un factor limitador. Los modelos logísticos son los modelos más exactos para representar crecimiento de la población.

cuartil inferior Mediana de la mitad inferior de un conjunto de datos, se denota con LQ.

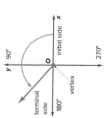

M

mapping (p. P4) How each member of the domain is paired with each member of the range.

transformaciones La correspondencia entre cada miembro del dominio con cada miembro del rango.

margin of error (p. 546) The limit on the difference between how a sample responds and how the total population would respond.

margen de error muestral Límite en la diferencia entre las respuestas obtenidas con una muestra y cómo pudiera responder la población entera.

maximum value (p. 154) The y-coordinate of the vertex of the quadratic function $f(x) = ax^2 + bx + c$, where $a < 0$.

valor máximo La coordenada y del vértice de la función cuadrática $f(x) = ax^2 + bx + c$, where $a < 0$.

mean (p. P24) The sum of the values in a set of data divided by the total number of values in the set.

media Suma de los valores en un conjunto de datos dividida entre el número total de valores en el conjunto.

median (p. P24) The middle value or the mean of the two middle values in a set of data when the data are arranged in numerical order.

mediana Valor del medio o la media de los valores del medio en un conjunto de datos, cuando los datos están ordenados numéricamente.

midline (p. 636) A horizontal axis used as the reference line about which the graph of a periodic function oscillates.

línea media Eje horizontal que se usa como recta de referencia alrededor de la cual oscila la gráfica de una función periódica.

minimum value (p. 154) The y-coordinate of the vertex of the quadratic function $f(x) = ax^2 + bx + c$, where $a > 0$.

valor mínimo La coordenada y del vértice de la función cuadrática $f(x) = ax^2 + bx + c$, donde $a > 0$.

mode (p. P24) The value or values that appear most often in a set of data.

moda Valor o valores que aparecen con más frecuencia en un conjunto de datos.

mutually exclusive (p. P14) Two events that cannot occur at the same time.

mutuamente exclusivos Dos eventos que no pueden ocurrir simultáneamente.

N

natural base, e (p. 430) An irrational number approximately equal to 2.71828. . .

base natural, e Número irracional aproximadamente igual a 2.71828. . .

natural base exponential function (p. 430) An exponential function with base e, $y = e^x$.

función exponencial natural La función exponencial de base e, $y = e^x$.

natural logarithm (p. 430) Logarithms with base e, written ln x.

logaritmo natural Logaritmo de base e, el que se escribe ln x.

nonlinear function (p. 95) A function that is not modeled by a linear equation.

función no lineal Función que no se representa con una ecuación lineal.

normal distribution (p. 566) A continuous, symmetric, bell-shaped distribution of a random variable.

distribución normal Distribución con forma de campana, simétrica y continua de una variable aleatoria.

Normal Distribution

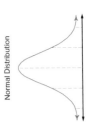

Distribución normal

O

oblique asymptote (p. 493) An asymptote that is neither horizontal nor vertical and is sometimes called a slant asymptote.

asíntota oblicuo Una asíntota que es ni horizontal ni la vertical y a veces se llama una asíntota inclinada.

observational study (p. 531) Individuals are observed and no attempt is made to influence the results.

estudio de observación Observan a los individuos y no se hace ninguna tentativa de influenciar los resultados.

odds (p. P15) A ratio that compares the number of ways an event can occur to the number of ways that it cannot occur.

posibilidad Razón que compara el número de maneras en que puede ocurrir un evento al número de maneras en que no puede ocurrir dicho evento.

one-to-one function (p. 87) 1. A function where each element of the range is paired with exactly one element of the domain. 2. A function whose inverse is a function.

función biunívoca 1. Función en la que a cada elemento del rango le corresponde solo un elemento del dominio. 2. Función cuya inversa es una función.

onto function (p. 87) Each element of the range corresponds to an element of the domain.

sobre la función Cada elemento de la gama corresponde a un elemento del dominio.

open sentence (p. 5) A mathematical sentence containing one or more variables.

enunciado abierto Enunciado matemático que contiene una o más variables.

optimize (p. 62) To seek the optimal price or amount that is desired to minimize costs or maximize profits.

optimice Buscar el precio óptimo o ascender que se desea para reducir al mínimo costes o para maximizar de los beneficios.

ordered triple (p. 67) 1. The coordinates of a point in space 2. The solution of a system of equations in three variables x, y, and z

triple ordenado 1. Las coordenadas de un punto en el espacio 2. Solución de un sistema de ecuaciones en tres variables x, y y z

outcome (p. P9) The results of a probability experiment or an event.

resultados Lo que produce un experimento o evento probabilístico.

outlier (p. P27) A data point that does not appear to belong to the rest of the set.

valor atípico Dato que no parece pertenecer al resto el conjunto.

P

parabola (p. 151) The graph of a quadratic function. The set of all points in a plane that are the same distance from a given point, called the focus, and a given line, called the directrix.

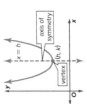

parábola La gráfica de una función cuadrática. Conjunto de todos los puntos de un plano que están a la misma distancia de un punto dado, llamado foco, y de una recta dada, llamada directriz.

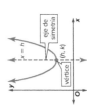

parallel lines (p. 30) Nonvertical coplanar lines with the same slope.

rectas paralelas Rectas coplanares no verticales con la misma pendiente.

parameter (p. 531) A measure that describes a characteristic of a population.

parámetro Una medida que describe una característica de una población.

Pascal's triangle (p. 237) A triangular array of numbers such that the $(n + 1)$th row is the coefficient of the terms of the expansion $(x + y)^n$ for $n = 0, 1, 2 \ldots$

period (p. 621) The least possible value of a for which $f(x) = f(x + a)$.

periodic function (p. 621) 1. A function with y-values that repeat at regular intervals. 2. A function is called periodic if there is a number a such that $f(x) = f(x + a)$ for all x in the domain of the function.

permutation (p. P9) An arrangement of objects in which order is important.

perpendicular lines (p. 30) In a plane, any two oblique lines, the product of whose slopes is -1.

phase shift (p. 635) A horizontal translation of a trigonometric function.

piecewise-defined function (p. 118) A function that is written using two or more expressions.

piecewise-linear function (p. 119) A function in which the equation for each interval is linear.

point discontinuity (p. 494) If the original function is undefined for $x = a$ but the related rational expression of the function in simplest form is defined for $x = a$, then there is a hole in the graph at $x = a$.

point of symmetry (p. 98) The common point of reflection for all points of a figure.

point-slope form (p. 29) An equation in the form $y - y_1 = m(x - x_1)$ where (x_1, y_1) are the coordinates of a point on the line and m is the slope of the line.

point symmetry (p. 98) Two distinct points P and P' are symmetric with respect to point M if and only if M is the midpoint of $\overline{PP'}$. Point M is symmetric with respect to itself.

polynomial function (p. 254) A function that is represented by a polynomial equation.

polynomial identity (p. 282) A polynomial equation that is true for any values that are substituted for the variables.

polynomial in one variable (p. 253)
$a_n x^n + a_{n-1} x^{n-1} + \cdots + a_2 x^2 + a_1 x + a_0$ where the coefficients $a_n, a_{n-1}, a_{n-2}, \ldots, a_0$ represent real numbers, and a_n is not zero and n is a nonnegative integer.

population (p. 24) An entire group of living things or objects.

population mean (p. 545) The mean, or average, of a characteristic for the entire population. The population mean can be represented by μ.

population proportion (p. 545) The ratio of the number of members in the population that have a particular characteristic to the number of members in the population.

positive correlation (p. 92) When the values in a scatter plot are closely linked in a positive manner.

power function (pp. 251, 254) An equation in the form $f(x) = ax^b$ where a and b are nonzero real numbers.

prime polynomial (p. 274) A polynomial that cannot be factored.

probability (p. P13) A measure of the chance that a given event will occur.

probability model (p. P13, 538) A mathematical model used to represent the outcomes of an experiment.

pure imaginary numbers (p. 172) The square roots of negative real numbers. For any positive real number b,
$$\sqrt{-b^2} = \sqrt{b^2} \cdot \sqrt{-1}, \text{ or } bi$$

Q

quadrantal angle (p. 613) An angle in standard position whose terminal side coincides with one of the axes.

quadrants (p. P4) The four areas of a Cartesian coordinate plane.

quadratic equation (p. 163) A quadratic function set equal to a value, in the form $ax^2 + bx + c = 0$, where $a \neq 0$.

quadratic form (p. 277) For any numbers a, b, and c, except for $a = 0$, an equation that can be written in the form $u^2 + u + c = 0$, where u is some expression in x.

Quadratic Formula (p. 199) The solutions of a quadratic equation of the form $ax^2 + bx + c = 0$, where $a \neq 0$, are given by the Quadratic Formula, which is
$$x = \frac{-b \pm \sqrt{b^2 - 4ac}}{2a}$$

quadratic function (p. 151) A function described by the equation $f(x) = ax^2 + bx + c$, where $a \neq 0$.

quadratic inequality (p. 209) An inequality of the form $y > ax^2 + bx + c$, $y \geq ax^2 + bx + c$, $y < ax^2 + bx + c$, or $y \leq ax^2 + bx + c$.

quadratic term (p. 151) In the equation $f(x) = ax^2 + bx + c$, ax^2 is the quadratic term.

---

triángulo de Pascal Arreglo triangular de números tal que la fila de $(n + 1)$th está compuesta de proporciona los coeficientes de los términos de la expansión de $(x + y)^n$ para $n = 0, 1, 2 \ldots$

período El menor valor positivo posible para a, para el cual $f(x) = f(x + a)$.

función periódica 1. Una función con y-valores aquella repetición con regularidad. 2. Función para la cual hay un número a tal que $f(x) = f(x + a)$ para todo x en el dominio de la función.

permutación Arreglo de elementos en que el orden es importante.

rectas perpendiculares En un plano, dos rectas oblicuas cualesquiera cuyas pendientes tienen un producto igual a -1.

desplazamiento de fase Traslación horizontal de una función trigonométrica.

función por trozos-definida Una función se escribe que usando dos o más expresiones.

función lineal por partes Una función en que la ecuación para cada intervalo es lineal.

discontinuidad evitable Si la función original no está definida en $x = a$ pero la expresión racional reducida correspondiente de la función está definida en $x = a$, entonces la gráfica tiene una ruptura o corte en $x = a$.

punto de simetría Punto común de reflexión para todos los puntos de una figura.

forma punto-pendiente Ecuación de la forma $y - y_1 = m(x - x_1)$ donde (x_1, y_1) es un punto en la recta y m es la pendiente de la recta.

simetría central Dos puntos distintos P y P' son simétricos respecto de un punto M, si y sólo si M es el punto medio de $\overline{PP'}$. El punto M es simétrico respecto de sí mismo.

función polinomial Función representada por una ecuación polinomial.

identidad polinómica Ecuación polinómica que es verdadera para cualquier valor por el que se reemplazan las variables.

polinomio de una variable
$a_n x^n + a_{n-1} x^{n-1} + \cdots + a_2 x^2 + a_1 x + a_0$ donde los coeficientes $a_n, a_{n-1}, a_{n-2}, \ldots, a_0$ son números reales, a_n no es nulo y n es un entero no negativo.

población Un grupo entero de cosas o de objetos vivos.

media de la población La media, o el promedio, de una característica de toda la población. La media de la población se puede representar con el símbolo μ.

proporción de la población La razón de la cantidad de miembros de la población que tienen una característica particular a la cantidad de miembros que hay en la población.

correlación positiva Cuando los valores en un diagrama de la dispersión se ligan de cerca de una manera positiva.

función potencia Ecuación de la forma $f(x) = ax^b$, donde a y b son números reales.

polinomio primero Un polinomio que no puede ser descompuesto en factores.

probabilidad Medida de la posibilidad de que ocurra un evento dado.

modelo de probabilidad Modelo matemático que se usa para representar los resultados de un experimento.

número imaginario puro Raíz cuadrada de un número real negativo. Para cualquier número real positivo b,
$$\sqrt{-b^2} = \sqrt{b^2} \cdot \sqrt{-1}, \text{ o } bi$$

Q

ángulo de cuadrante Ángulo en posición estándar cuyo lado terminal coincide con uno de los ejes.

cuadrantes Las cuatro regiones de un plano de coordenadas Cartesiano.

ecuación cuadrática Función cuadrática igual a un valor, de la forma $ax^2 + bx + c = 0$, donde $a \neq 0$.

forma de ecuación cuadrática Para cualquier número a, b y c excepto $a = 0$, una ecuación que puede escribirse de la forma $u^2 + u + c = 0$, donde u es una expresión en x.

fórmula cuadrática Las soluciones de una ecuación cuadrática de la forma $ax^2 + bx + c = 0$, donde $a \neq 0$, se dan por la fórmula cuadrática, que es
$$x = \frac{-b \pm \sqrt{b^2 - 4ac}}{2a}$$

función cuadrática Función descrita por la ecuación $f(x) = ax^2 + bx + c$, donde $a \neq 0$.

desigualdad cuadrática Desigualdad cuadrática de la forma $y > ax^2 + bx + c$, $y \geq ax^2 + bx + c$, $y < ax^2 + bx + c$, o $y \leq ax^2 + bx + c$.

término cuadrático En la ecuación $f(x) = ax^2 + bx + c$, ax^2 el término cuadrático es ax^2.

quartic function (p.255) A fourth-degree function.

función quartic Una función del cuarto-grado.

quartiles (p.P26) The values that divide a set of data into four equal parts.

cuartiles Valores que dividen un conjunto de datos en cuatro partes iguales.

quintic function (p.255) A fifth-degree function.

función quintic Una función del quinto-grado.

R

radian (p.606) The measure of an angle θ in standard position whose rays intercept an arc of length 1 unit on the unit circle.

radián Medida de un ángulo θ en posición normal cuyos rayos intersecan un arco de 1 unidad de longitud en el círculo unitario.

radical equation (p.352) An equation with radicals that have variables in the radicands.

ecuación radical Ecuación con radicales que tienen variables en el radicando.

radical function (p.338) A function that contains the root of a variable.

función radical Una función que contiene la raíz de una variable.

radical inequality (p.354) An inequality that has a variable in the radicand.

desigualdad radical Desigualdad que tiene una variable en el radicando.

random sample (p.531) A sample in which every member of the population has an equal chance of being selected.

muestra aleatoria Muestra en la cual cada miembro de la población tiene igual posibilidad de ser elegido.

range (pp.P4, P25) 1. The set of all y-coordinates of a relation. 2. The difference between the greatest and least values in a set of data.

rango 1. Conjunto de todas las coordenadas y de una relación. 2. Diferencia entre el valor mayor y el menor en un conjunto de datos.

rate of change (p.21) How much a quantity changes on average, relative to the change in another quantity, over time.

tasa de cambio Lo que cambia una cantidad en promedio, respecto al cambio en otra cantidad, por lo general el tiempo.

rate of continuous decay (p.445) The rate at which something decays continuously. Represented by a constant k in the exponential decay function $f(x) = ae^{-kt}$, where a is the initial value, and t is time in years.

índice de desintegración continua Ritmo al cual algo se desintegra continuamente. Representado por la constante k en la función de desintegración exponencial $f(x) = ae^{-kt}$, donde a es el valor inicial y t es el tiempo en años.

rate of continuous growth (p.445) The rate at which something grows continuously. The value of k in the exponential growth function, $f(x) = ae^{kt}$.

el índice de crecimiento continuo Es la tasa en la cual algo crece continuamente. El valor de k en la función exponencial del crecimiento, $f(x) = ae^{kt}$.

rational equation (p.508) Any equation that contains one or more rational expressions.

ecuación racional Cualquier ecuación que contiene una o más expresiones racionales.

rational expression (p.467) A ratio of two polynomial expressions.

expresión racional Razón de dos expresiones polinomiales.

rational function (p.491) An equation of the form $f(x) = \dfrac{p(x)}{q(x)}$, where $p(x)$ and $q(x)$ are polynomial functions, and $q(x) \neq 0$.

función racional Ecuación de la forma $f(x) = \dfrac{p(x)}{q(x)}$, donde $p(x)$ y $q(x)$ son funciones polinomiales y $q(x) \neq 0$.

rational inequality (p.513) Any inequality that contains one or more rational expressions.

desigualdad racional Cualquier desigualdad que contiene una o más expresiones racionales.

reciprocal function (pp.483, 595) 1. A function of the form $f(x) = \dfrac{1}{a(x)}$, where $a(x)$ is a linear function and $a(x) \neq 0$. 2. Trigonometric functions that are reciprocals of each other.

funciones recíprocas 1. Una función de la forma $f(x) = \dfrac{1}{a(x)}$, donde $a(x)$ es una función linear y $a(x) \neq 0$. 2. Funciones trigonométricas de la función que son recíprocals de una a.

reference angle (p.613) The acute angle formed by the terminal side of an angle in standard position and the x-axis.

ángulo de referencia El ángulo agudo formado por el lado terminal de un ángulo en posición estándar y el eje x.

reflection (p.126) A transformation in which every point of a figure is mapped to a corresponding image across a line of symmetry.

reflexión Transformación en que cada punto de una figura se aplica a través de una recta de simetría a su imagen correspondiente.

regression curve (p.408) A curve of best fit for a data set that is not modeled by a linear equation.

curva de regresión Curva de mejor ajuste para un conjunto de datos que no está representado por una ecuación lineal.

regression line (p.408) A line of best fit.

recta de regresión Una recta de óptimo ajuste.

relation (p.P4) A set of ordered pairs.

relación Conjunto de pares ordenados.

relative frequency (p.538) The ratio of the number of observations in a category to the total number of observations.

frecuencia relativa Razón del número de observaciones en una categoría al número total de observaciones.

relative maximum (pp.105, 263) A point on the graph of a function where no other nearby points have a greater y-coordinate.

máximo relativo Punto en la gráfica de una función en donde ningún otro punto cercano tiene una coordenada y mayor.

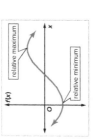

relative minimum (pp.105, 263) A point on the graph of a function where no other nearby points have a lesser y-coordinate.

mínimo relativo Punto en la gráfica de una función en donde ningún otro punto cercano tiene una coordenada y menor.

roots (p.163) The solutions of a quadratic equation.

raíz Las soluciones de una ecuación cuadrática.

S

sample (p.P24) A part of a population.

muestra Parte de una población.

sample space (p.P9) The set of all possible outcomes of an experiment.

espacio muestral Conjunto de todos los resultados posibles de un experimento probabilístico.

secant (p.594) For any angle, with measure α, a point $P(x, y)$ on its terminal side, $r = \sqrt{x^2 + y^2}$. $\sec \alpha = \dfrac{r}{x}$.

secante Para cualquier ángulo de medida α, un punto $P(x, y)$ en su lado terminal, $r = \sqrt{x^2 + y^2}$. $\sec \alpha = \dfrac{r}{x}$.

set-builder notation (p.P5) The expression of the solution set of an inequality, for example $\{x \mid x > 9\}$.

notación de construcción de conjuntos Escritura del conjunto solución de una desigualdad, por ejemplo, $\{x \mid x > 9\}$.

simple event (p.P14) One event.

evento simple Un solo evento.

simplify (p.229) To rewrite an expression without parentheses or negative exponents.

reducir Escribir una expresión sin paréntesis o exponentes negativos.

Glossary/Glosario

simulation (p. 538) The use of a probability experiment to mimic a real-life situation.

sine (p. 594) For any angle, with measure α, a point $P(x, y)$ on its terminal side, $r = \sqrt{x^2 + y^2}$.
$$\sin \alpha = \frac{y}{r}.$$

slope (p. 22) The ratio of the change in y-coordinates to the change in x-coordinates.

slope-intercept form (p. 28) The equation of a line in the form $y = mx + b$, where m is the slope and b is the y-intercept.

solution (p. 5) A replacement for the variable in an open sentence that results in a true sentence.

square root function (p. 338) A function that contains a square root of a variable.

Square Root Property (p. 173) For any real number n, if $x^2 = n$, then $x = \pm\sqrt{n}$.

standard deviation (p. P25) The square root of the variance.

standard form (p. 163) 1. A linear equation written in the form $Ax + By = C$, where A, B, and C are integers whose greatest common factor is 1, $A \geq 0$, and A and B are not both zero. 2. A quadratic equation written in the form $ax^2 + bx + c = 0$, where a, b, and c are integers, and $a \neq 0$.

standard normal distribution (p. 568) A normal distribution with a mean of 0 and a standard deviation of 1.

standard position (p. 604) An angle positioned so that its vertex is at the origin and its initial side is along the positive x-axis.

statistic (pp. P24, 531) A measure that describes a characteristic of a sample.

step function (p. 119) A function whose graph is a series of line segments.

substitution method (p. 44) A method of solving a system of equations in which one equation is solved for one variable in terms of the other.

survey (p. 531) Used to collect information about a population.

synthetic division (p. 244) A method used to divide a polynomial by a binomial.

synthetic substitution (p. 287) The use of synthetic division to evaluate a function.

---

simulación Uso de un experimento probabilístico para imitar una situación de la vida real.

seno Para cualquier ángulo de medida α, un punto $P(x, y)$ en su lado terminal, $r = \sqrt{x^2 + y^2}$.
$$\sin \alpha = \frac{y}{r}.$$

pendiente La razón del cambio en coordenadas y al cambio en coordenadas x.

forma pendiente-intersección Ecuación de una recta de la forma $y = mx + b$, donde m es la pendiente y b la intersección.

solución Sustitución de la variable de un enunciado abierto que resulta en un enunciado verdadero.

función radical Función que contiene la raíz cuadrada de una variable.

Propiedad de la raíz cuadrada Para cualquier número real n, si $x^2 = n$, entonces $x = \pm\sqrt{n}$.

desviación estándar La raíz cuadrada de la varianza.

forma estándar 1. Ecuación lineal escrita de la forma $Ax + By = C$, donde A, B, y C son enteros cuyo máximo común divisores 1, $A \geq 0$ y A y B no son cero simultáneamente. 2. Una ecuación cuadrática escrita en la forma $ax^2 + bx + c = 0$, donde a, b, y c son números enteros, y $a \neq 0$.

distribución normal estándar Distribución normal con una media de 0 y una desviación estándar de 1.

posición estándar Ángulo en posición tal que su vértice está en el origen y su lado inicial está a lo largo del eje x positivo.

estadística Una medida que describe una característica de una muestra.

función etapa Función cuya gráfica es una serie de segmentos de recta.

método de sustitución Método para resolver un sistema de ecuaciones en que una de las ecuaciones se resuelve en una de las variables en términos de la otra.

encuesta Reúnia información acerca de una población.

división sintética Método que se usa para dividir un polinomio entre un binomio.

sustitución sintética Uso de la división sintética para evaluar una función polinomial.

---

system of equations (p. 42) A set of equations with the same variables.

system of inequalities (p. 52) A set of inequalities with the same variables.

T

tangent (p. 594) 1. A line that intersects a circle at exactly one point. 2. For any angle, with measure α, a point $P(x, y)$ on its terminal side, $r = \sqrt{x^2 + y^2}$.
$$\tan \alpha = \frac{y}{x}.$$

terminal side (p. 604) A ray of an angle that rotates about the center.

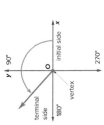

theoretical probability (pp. P13, 538) What should occur in a probability experiment.

translation (p. 125) A figure is moved from one location to another on the coordinate plane without changing its size, shape, or orientation.

tree diagram (p. P9) A diagram that shows all possible outcomes of an event.

trigonometric equations (p. 683) Equations containing at least one trigonometric function that is true for some but not all values of the variable.

trigonometric functions (p. 594) For any angle, with measure α, a point $P(x, y)$ on its terminal side, $r = \sqrt{x^2 + y^2}$, the trigonometric functions of α are as follows:

$$\sin \alpha = \frac{y}{r} \qquad \cos \alpha = \frac{x}{r} \qquad \tan \alpha = \frac{y}{x}$$
$$\csc \alpha = \frac{r}{y} \qquad \sec \alpha = \frac{r}{x} \qquad \cot \alpha = \frac{x}{y}$$

trigonometric identity (p. 655) An equation involving a trigonometric function that is true for all values of the variable for which the function is defined.

trigonometric ratio (p. 594) Compares the side lengths of a right triangle.

---

sistema de ecuaciones Conjunto de ecuaciones con las mismas variables.

sistema de desigualdades Conjunto de desigualdades con las mismas variables.

T

tangente 1. Recta que interseca un círculo en un solo punto. 2. Para cualquier ángulo, de medida α, un punto $P(x, y)$ en su lado terminal, $r = \sqrt{x^2 + y^2}$.
$$\tan \alpha = \frac{y}{x}.$$

lado terminal Rayo de un ángulo que gira alrededor de un centro.

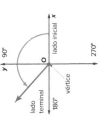

probabilidad teórica Lo que debería ocurrir en un experimento probabilístico.

traslación Se mueve una figura de un lugar a otro en un plano de coordenadas sin cambiar su tamaño, forma u orientación.

diagrama de árbol Diagrama que muestra todos los posibles resultados de un evento.

ecuación trigonométrica Ecuación que contiene por lo menos una función trigonométrica y que sólo se cumple para algunos valores de la variable.

funciones trigonométricas Para cualquier ángulo, de medida α, un punto $P(x, y)$ en su lado terminal, $r = \sqrt{x^2 + y^2}$, las funciones trigonométricas de α son las siguientes:

$$\sin \alpha = \frac{y}{r} \qquad \cos \alpha = \frac{x}{r} \qquad \tan \alpha = \frac{y}{x}$$
$$\csc \alpha = \frac{r}{y} \qquad \sec \alpha = \frac{r}{x} \qquad \cot \alpha = \frac{x}{y}$$

identidad trigonométrica Ecuación que involucra una o más funciones trigonométricas y que se cumple para todos los valores de la variable en que el función es definido.

razón trigonométrica Compara las longitudes laterales de un triángulo derecho.

trigonometry (p. 594) The study of the relationships between the angles and sides of a right triangle.

turning point (pp. 105, 263) Point at which a graph turns. The location of relative maxima or minima.

unbounded (p. 60) A system of inequalities that forms a region that is open.

unit analysis (p. 236) The process of including unit measurement when computing.

unit circle (p. 620) A circle of radius 1 unit whose center is at the origin of a coordinate system.

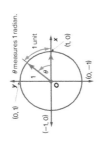

univariate data (p. 24) Data with one variable.

variable (p. P24) 1. A characteristic of a population that can assume different values called data. 2. A symbol, usually a letter, used to represent an unknown quantity.

variance (p. 25) The mean of the squares of the deviations from the arithmetic mean.

vertex (p. 152) 1. Any of the points of intersection of the graphs of the constraints that determine a feasible region. 2. The point at which the axis of symmetry intersects a parabola. 3. The point on each branch nearest the center of a hyperbola.

vertical asymptote (p. 491) If the related rational expression of a function is written in simplest form and is undefined for $x = a$, then $x = a$ is a vertical asymptote.

vertical line test (p. 88) If no vertical line intersects a graph in more than one point, then the graph represents a function.

vertical shift (p. 636) When graphs of trigonometric functions are translated vertically.

---

trigonometría Estudio de las relaciones entre los lados y ángulos de un triángulo rectángulo.

momento crucial Un punto en el cual un gráfico da vuelta. La localización de máximos o de mínimos relativos.

U

no acotado Sistema de desigualdades que forma una región abierta.

análisis de la unidad Proceso de incluir unidades de medida al computar.

círculo unitario Círculo de radio 1 cuyo centro es el origen de un sistema de coordenadas.

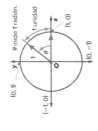

datos univariados Datos con una variable.

V

variable 1. Característica de una población que puede tomar diferentes valores llamados datos. 2. Símbolo, generalmente una letra, que se usa para representar una cantidad desconocida.

varianza Media de los cuadrados de las desviaciones de la media aritmética.

vértice 1. Cualquiera de los puntos de intersección de las gráficas que los contienen y que determinan una región viable. 2. Punto en el que el eje de simetría interseca una parábola. 3. El punto en cada rama más cercano al centro de una hipérbola.

asíntota vertical Si la expresión racional que corresponde a una función racional se reduce y está no definida en $x = a$, entonces $x = a$ es una asíntota vertical.

prueba de la recta vertical Si ninguna recta vertical interseca una gráfica en más de un punto, entonces la gráfica representa una función.

cambio vertical Cuando los gráficos de funciones trigonométricas son translado verticales.

W

weighted average (p. 510) A method for finding the mean of a set of numbers in which some elements of the set carry more importance, or weight, than others.

X

x-intercept (p. 133) The x-coordinate of the point at which a graph crosses the x-axis.

Y

y-intercept (p. 133) The y-coordinate of the point at which a graph crosses the y-axis.

Z

zeros (pp. 134, 163) The x-intercepts of the graph of a function; the points for which $f(x) = 0$.

z-value (p. 568) The number of standard deviations that a given data value is from the mean.

---

promedio ponderado Un método para encontrar el medio de un sistema de los números en los cuales algunos elementos del sistema llevan más importancia, o peso, que otros.

X

intersección x La coordenada x del punto o puntos en que una gráfica interseca o cruza el eje x.

Y

intersección y La coordenada y del punto o puntos en que una gráfica interseca o cruza el eje y.

Z

ceros Las intersecciones x de la gráfica de una función; los puntos x para los que $f(x) = 0$.

valor Z Número de variaciones estándar que separa un valor dado de la media.

Index

Index

E

Index

Index

Labs. *See* Algebra Labs; Geometry Labs; Graphing Technology Labs; Spreadsheet Lab

L

Learning Styles

Index

M

N

Naturalist Learners. *See* Learning Styles

O

P

Q

Quizzes. *See* Go Online!

R

S

T

Teaching the Mathematical Practices. *See* Mathematical Practices